RNA Methodologies

A Laboratory Guide for Isolation and Characterization

Sixth Edition

Robert E. Farrell, Jr.
Pennsylvania State University, York, PA, United States

ELSEVIER

ACADEMIC PRESS
An imprint of Elsevier

Academic Press is an imprint of Elsevier
125 London Wall, London EC2Y 5AS, United Kingdom
525 B Street, Suite 1650, San Diego, CA 92101, United States
50 Hampshire Street, 5th Floor, Cambridge, MA 02139, United States
The Boulevard, Langford Lane, Kidlington, Oxford OX5 1GB, United Kingdom

ISBN: 978-0-323-90221-2

For Information on all Academic Press publications
visit our website at https://www.elsevier.com/books-and-journals

Publisher: Andre G. Wolff
Acquisitions Editor: Michelle Fisher
Editorial Project Manager: Pat Gonzalez
Production Project Manager: Sreejith Viswanathan
Cover Designer: Matthew Limbert

Typeset by MPS Limited, Chennai, India

Dedication

For Catherine Ann,
Sean, Megan, and Kieran
Emma and Jon
Liam and Jenny
Patrick

Contents

Preface

RNA Rules!

More than 30 years have passed since work began on the first edition of RNA Methodologies, which appeared in 1993. At that time, RNA was viewed as an "interesting" molecule and many molecular biologists were happy if they could do a decent Northern blot. Thirty years later, we have just begun to appreciate the diversity of RNA both in form and function. This volume has been expanded to present contemporary and traditional laboratory techniques and protocols to tell the RNA story from multiple perspectives. By unifying these themes into a coherent start-to-finish format, a holistic understanding of gene expression is presented. It is an exciting time to be an RNA scientist studying this fascinating if not enigmatic molecule, which one should approach with fearless confidence.

RNA Methodologies is organized in a skills-type progression, beginning with basic biological and chemical information about RNA and the role of the various types of transcripts in the society of the cell. From there, it is about strategies and protocols for isolating RNA, and the importance of using good RNase-free technique at all times; some things never change, and the meticulous processing of RNA and proper storage procedures are among them. It has been especially rewarding over the years to see this book often cited in the user manuals for various RNA-associated products as a recommended resource for learning RNA skills. An important focal point in this regard is the chapter on RNA quality control because the success of every downstream application following RNA isolation from its biological source depends on sample purity and molecular integrity. As a frequent consultant, it is amazing to me to see how often these important measures are overlooked. Updated standard methods and protocols are then presented, including some traditional techniques which are enjoying a resurgence in popularity as a means of validating results produced using other methods. The direction of this book then moves into newer applications, some of which require very specialized equipment, such as RNA-seq and microarray analysis. The last chapters speak to some RNA success stories and the book concludes with ancillary techniques and supporting information in the appendices, including a section on the importance of good recordkeeping.

My enthusiasm for all things RNA has not changed; if anything, it has become considerably more robust. Nor have the goals of the sixth edition of

RNA Methodologies changed; rather, they have matured. The first goal is to provide detailed hands-on protocols for procuring RNA of the best possible quality from cells and tissue to support any of a number of applications. The second goal is to provide troubleshooting advice for those (frequent) times when things do not quite work out in the lab. The third goal is to present various methods for assessing transcriptional activity in a way that they support or complement each other, rather than being viewed as standalone procedures. There is a fair amount of cross-referencing throughout this book because many of the strategies and methods used to characterize RNA are interdependent upon each other. The fourth goal is to make procedural and application-type connections wherever possible. The fifth goal of this book is to help the reader develop greater confidence in the laboratory. To that end, this text is written for the principal investigator, bench scientist, physician, veterinarian, lab technician, graduate student, undergraduate research assistant, and anyone else capable of performing basic research techniques—there is something in it for everyone. What has changed are many of the tools through which these goals are accomplished. Although there is now a kit for almost everything, this book contains protocols for many hands-on procedures should an investigator find a kit insufficient for his or her needs. Sophisticated instrumentation for performing advanced techniques has also been developed as a result of various technological achievements. These improvements are for the better, allowing the contemporary investigator to be much more productive than was possible even as recently as five years ago. The biotechnology revolution has certainly had a profound and lasting impact in all areas of the life sciences.

The most significant advancement in the RNA world since the publication of the fifth edition of this book has been the successful production of mRNA vaccines to combat a worldwide pandemic. While the notion of harnessing the natural gene expression pathway to evoke an immune response is not a terribly new idea, the rise of the novel SARS-CoV-2 virus was the driving force responsible for their development in record time. The speed and entrepreneurial manner in which these important pharmaceuticals became available has demonstrated clearly and without question that this is the advent of the RNA vaccine. The success, effectiveness, and the economics behind the production of these vaccines portend the development of many more types of RNA vaccines in the near future. Had the pandemic not occurred, RNA vaccines would have inevitably become part of the repertoire of modern medicine, though just not as quickly. Although the perishable nature of mRNA in vaccines remains a concern, one of the better qualities of this type of vaccine is the ease with which the mRNA sequence can be changed to match viral mutations when they arise.

Noteworthy also are the RNA-related technological refinements and lower per-reaction cost of next-generation sequencing, thereby providing broader access to this important tool, even if the sequencing is outsourced. The wealth of available transcriptome sequencing data has provided considerable insight into the realm of noncoding RNA. These data have begun to reveal the astonishing biochemical circuitry and magnitude of the influence of miRNA, circRNA, and other noncoding species on gene expression. It is safe to say that we are in the midst of the early years in terms of our understanding of the several faces of RNA. Without a doubt, there are many other functions associated with RNA that are awaiting discovery. Back in the 1970s and early 1980s, who would have ever thought that there was much more than mRNA, tRNA, and rRNA? Where RNA is involved, one quickly learns to expect the unexpected.

The sixth edition of RNA methodologies is an evolving collection of tried, tested, and optimized laboratory protocols for the isolation and characterization of eukaryotic RNA. All of the background information and the updates included herein emphasize the "what," the "why," and the "how" from the onset of the study so as to avoid unsound tactics for RNA handling and subsequent characterization. It is intended to provide a rationale to assist in the decision-making process for individuals at all levels of experience by presenting realistic alternatives for achieving the same experimental goals and demonstrating how various techniques contribute to the understanding of gene expression and regulation thereof. Many of the incorporated notations and hints are based upon personal experience and pave the way for the expedient recovery and study of RNA and the most judicious use of resources. The liberal incorporation of representative data, flow charts, and tables facilitate learning and will assist in the planning and implementation phases of a project. Inclusion of this material is timely and appropriate since the RNA novice lacks the historical perspective and frame of reference that more experienced investigators often enjoy. If the reader takes away an "Oh, so that's why it's done this way" moment, then this title will continue making valuable service contributions to our colleagues who are engaged in any of a variety of research domains.

I wish every reader great success whether you are working with RNA for the first time or if you are an experienced molecular biologist looking to hone your skill set. Some very good advice is to always think two steps ahead in a research plan and reflect upon how the ensuing protocols might affect or even bias the interpretation of data. Perhaps most importantly, do not ever hesitate to ask for help. You have a lot of colleagues in the RNA community. And remember, you are limited only by your own ingenuity.

I wish to acknowledge with sincere thanks and appreciation the intellectual encouragement of the many colleagues and friends who, in some way, supported the preparation of the sixth edition of this book. The support and patience of my family are also gratefully acknowledged and are very much appreciated. Lastly, a word of special thanks to Elsevier Academic Press for their longstanding support of RNA Methodologies, beginning with the inception of this project in 1991. It has been and continues to be an amazing adventure.

Initium sapientiae timor Domini

Chapter 1

RNA and the cellular biochemistry revisited

Why study ribonucleic acid?

One could rightly argue that all cell and tissue functions are ultimately governed by gene expression. Consequently, the reasons for electing to study the modulation of RNA levels as at least one parameter of cellular biochemistry may be as diverse as the intracellular RNA population itself. Generally speaking, the characterization of RNA is almost always related to transcription, that is, gene expression questions being asked in the context of a particular scientific inquiry, and most often revolves around measuring the dynamic abundance level of one or several transcripts.

The goals in any experimental design involving RNA generally revolve around one or more fundamental themes including but not limited to:

1. Measurement of the steady-state abundance of cellular transcripts. Steady-state RNA refers to the net accumulation of transcription products in the cell, or in a subcellular compartment such as the nucleus or the cytoplasm. It is the combined result of RNA synthesis, stability, and degradation. This is the most common reason RNA is isolated from cells and tissues. Analysis may focus on one transcript, a few transcripts, or all transcripts simultaneously, the latter approach is commonly referred to as global analysis of gene expression or whole transcriptome profiling. Given the ease with which RNA can be purified from biological sources, highly accurate quantitative and qualitative profiles of RNA populations are attainable using a variety of contemporary laboratory techniques.
2. Synthesis of complementary DNA (cDNA). Unstable, single-stranded messenger RNA (mRNA) is the template for the in vitro synthesis of very stable single- or double-stranded cDNA molecules. This is the first step for subsequent amplification by the polymerase chain reaction (PCR), often for quantification purposes, for direct ligation into a vector for sequencing or for expression of the encoded protein, for the physical separation of two or more cDNA species, or for the older strategy of synthesizing and archiving an entire cDNA library for long-term storage and analysis. In any event, the construction of cDNA is the creation of a

RNA Methodologies. DOI: https://doi.org/10.1016/B978-0-323-90221-2.00044-8

permanent biochemical record of the cell at the moment of cellular disruption. Historically, the synthesis of highly representative cDNA is one of the most important methodologies in the molecular biology laboratory and, in some hands, remains something of a challenge.

3. Detection of viruses which harbor an RNA genome. This proceeds via the synthesis of cDNA, as described above, followed by PCR or another cDNA amplification method.
4. Mapping of the transcription start site (TSS). This is performed through some variant of 5′- or 3′ RACE (rapid amplification of cDNA ends; see Chapter 10: RT-PCR: A Science and an Art Form). As it is well known that a single genetic locus often has the potential to produce multiple RNAs, each with a different TSS and often in a tissue-specific manner, TSS mapping is an invaluable technique.
5. Measurement of the *rate* of transcription of gene sequences or the pathways of RNA processing. This may be deduced, at least in part, by the nuclear run-on assay in which radiolabeled ribonucleotide precursors are incorporated into nascent transcripts in direct proportion to the rate at which each species of RNA is being transcribed (see Chapter 19: Analysis of Nuclear RNA). When used in conjunction with other methods that examine steady-state RNA levels, the regulation of genes can often be assigned as transcriptional or due to posttranscriptional events.
6. In vitro transcription of specific RNAs. Any transcript can be mass-produced by simply performing transcription reactions featuring a unique DNA sequence (cloned or artificially synthesized) acting as a template and accompanied by all of the raw materials needed to support transcription. The ability to produce large quantities of a unique RNA species in vitro is particularly important in the new world of RNA vaccine production.
7. In vitro translation of purified mRNA. The resulting polypeptide may be further characterized by immunoprecipitation or Western analysis. In applications such as rational drug design, in vitro translation is helpful because understanding the three-dimensional architecture of a protein, and its wild-type or mutated function(s), may suggest novel applications in the area of functional proteomics. Moreover, the protein can be produced in whatever quantity is desired.

What is ribonucleic acid?

RNA is a long, unbranched polymer of ribonucleoside monophosphate moieties joined together by phosphodiester linkages. Both eukaryotic and prokaryotic RNAs are single-stranded molecules. The unlinked monomer building blocks of both RNA and DNA are known generically as nucleotides. Each nucleotide consists of three key components: a pentose (five-carbon sugar), at least one phosphate group (nucleotides may contain as many as three phosphate groups), and a nitrogenous base (Fig. 1.1). A nitrogenous base joined to a pentose sugar is

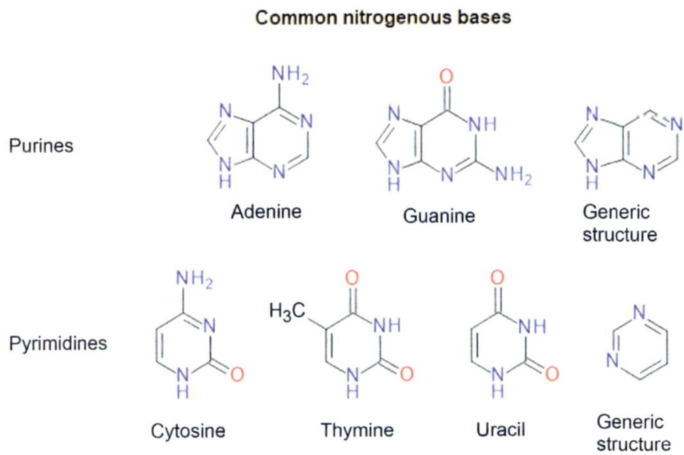

FIGURE 1.1 The identity of a nucleotide is defined by the base that is attached to the 1′ carbon. In practice, the nucleotides that make up an RNA or DNA molecule are represented by the standard one-letter abbreviation for the base each contains: adenine (A), cytosine (C), guanine (G), thymine (T), or uracil (U).

TABLE 1.1 Comparative nucleotide structure.

	RNA nucleotides	DNA nucleotides
Key nucleotide components	Five-carbon ribose	Five-carbon deoxyribose
	Phosphate group(s)	Phosphate group(s)
	Nitrogenous base	Nitrogenous base
Common nitrogenous bases		
Purines	Adenine	Adenine
	Guanine	Guanine
Pyrimidines	Cytosine	Cytosine
	Uracil	Thymine

known as a nucleoside. When a phosphate group is added, the composite, a phosphate ester of the nucleoside, is known as a nucleotide.

$$\text{base} + \text{sugar} = \text{nucleoside}$$

$$\text{nucleoside} + \text{phosphate} = \text{nucleoside}$$

The components of RNA and DNA nucleosides and nucleotides are compared in Table 1.1.

The key chemical difference between RNA and DNA resides in the structure at the 2′ position of the five-carbon sugar ribose. In the case of RNA, a hydroxyl group (−OH) is joined to the 2′ carbon, while the 2′ −H group in DNA is the underlying basis of the name of the sugar "deoxyribose," meaning the absence of an oxygen atom. In addition, the base uracil is found in RNA, substituted in DNA by the closely pyrimidine thymine (chemically, thymine is 5-methyluracil), though it is possible to find deoxynucleotides containing uracil in certain situations. More precisely, RNA is assembled from ribonucleotide precursors and DNA is assembled from deoxyribonucleotide precursors. Hence, RNA is so-named because of the ribose sugar it contains, just as DNA is named from its constituent 2′-deoxyribose sugar. Essential base, nucleoside, and nucleotide nomenclature are summarized in Table 1.2.

Nitrogenous bases and the pentose sugar components of nucleosides are both cyclic. By convention, the numbering system for the carbon and nitrogen atoms that make up the bases is 1, 2, 3, and so forth, while the numbering system for the constituent carbon atoms of the sugar (ribose or deoxyribose) is 1′, 2′, 3′, 4′, and 5′. The purpose of this nomenclature is to avoid confusion when referring to the constituent atoms of the sugar vs those found in the base of a particular nucleotide or nucleoside.

The ribonucleoside triphosphates are collectively referred to as NTP; in various molecular biology protocols, the symbol NTP refers to an equimolar cocktail of ATP, CTP, GTP, and UTP. Similarly, the deoxy form of a nucleotide is denoted by the placement of a lower case "d" preceding the nucleotide triphosphate, as in dATP, dCTP, dGTP, and dTTP, and the symbol

TABLE 1.2 Base, nucleoside, and nucleotide nomenclature.

	Base	Nucleoside	Nucleotide triphosphate precursors
RNA	Adenine (A)	Adenosine	Adenosine-5′-triphosphate (ATP)
	Cytosine (C)	Cytidine	Cytidine-5′-triphosphate (CTP)
	Guanine (G)	Guanosine	Guanosine-5′-triphosphate (GTP)
	Uracil (U)	Uridine	Uridine-5′-triphosphate (UTP)
DNA	Adenine (A)	2′-deoxyadenosine	2′-deoxyadenosine-5′-triphosphate (dATP)
	Cytosine (C)	2′-deoxycytidine	2′-deoxycytidine-5′-triphosphate (dCTP)
	Guanine (G)	2′-deoxyguanosine	2′-deoxyguanosine-5′-triphosphate (dGTP)
	Thymine (T)	2′-deoxythymidine	2′-deoxythymidine-5′-triphosphate (dTTP)

Nucleosides consist of a base and a sugar only. A nucleoside is promoted to a nucleotide upon addition of at least one phosphate group.

dNTP (also, dXTP) refers to an equimolar cocktail of the four deoxynucleoside triphosphates in protocols capable of supporting the synthesis of cDNA or PCR products. It is the triphosphate form of a nucleotide that is utilized as a precursor during nucleic acid synthesis. The phosphate nearest to the sugar is known as the α phosphate, followed by the β phosphate, followed by the γ phosphate, which is furthest from the nucleoside moiety (Fig. 1.2). During nucleic acid polymerization, the β and γ phosphates (PPi; inorganic phosphate) are cleaved (released) from the nucleotide, and the resulting single-phosphate nucleotide, a nucleoside monophosphate, is then incorporated into the nascent polynucleotide chain.

Polynucleotide synthesis

Any enzyme with an associated polymerase activity is capable of synthesizing nucleic acid molecules from nucleotide precursors. The synthesis of RNA is mediated by the activity of enzymes known as RNA polymerases while DNA is synthesized, not unexpectedly, by DNA polymerases. A nucleic acid molecule is the result of linking nucleotides together via phosphodiester bonds. The formation of these bonds involves the hydrophilic attack by the 3′ hydroxyl group of the last nucleotide added to the nascent polynucleotide on the 5′ phosphate group of the incoming nucleotide (Fig. 1.3). For this reason, nucleic acid synthesis is said to proceed 5′→3′, and there are no known exceptions to this process.

In order for the synthesis of nucleic acids to occur in vivo or in vitro, there are two fundamental requirements that must be fulfilled and maintained to initiate and support continued nucleic acid polymerization:

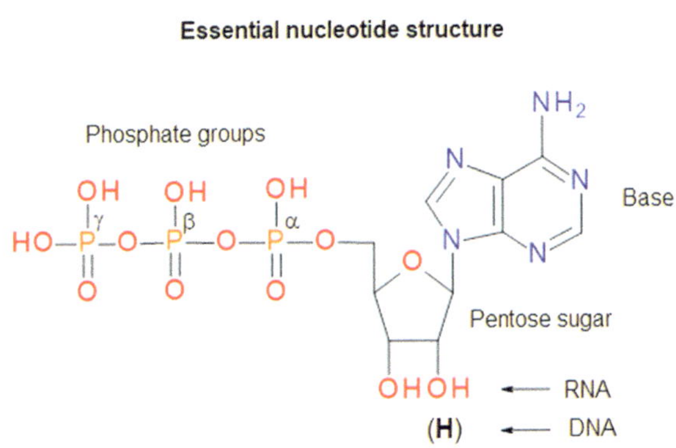

FIGURE 1.2 Adenosine-5′-triphosphate. The three constituent phosphate groups are designated α, β, and γ based on the proximity of each group to the nucleoside (base + sugar) component of the molecule. The replacement of the 2′-OH with H would convert this molecule to a deoxynucleotide.

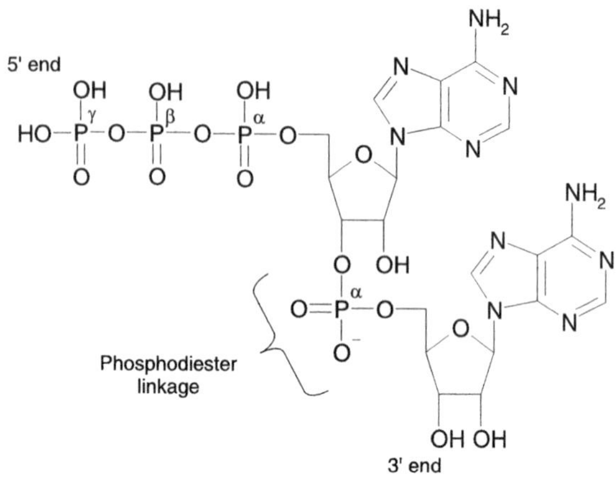

FIGURE 1.3 The dinucleotide that results from the formation of the first phosphodiester linkage has structurally different ends, namely a phosphate group at the 5′ end and a hydroxyl group at the 3′ end. The structural differences at the 5′- and 3′-ends are maintained regardless of the number of nucleotides that are added to the elongating RNA or DNA molecule.

1. There must be a template (a strand or an oligomer) to direct the polymerase-mediated insertion of the correct (complementary) nucleotide into the nascent chain. This occurs predictably, according to the conventions set down in Chargaff's Rule (Zamenhof et al., 1952), which succinctly states that adenine ordinarily base-pairs with thymine or uracil by the formation of two hydrogen bonds (A::T, A::U) and that guanine ordinarily base-pairs to cytosine with three hydrogen bonds (G:::C). Polymerases capable of adding nucleotides without template information are said to exhibit rare terminal transferase activity, as in the case the unusual enzyme terminal deoxynucleotidyl transferase. This enzyme has broad applications in the area of cDNA synthesis as well as certain forms of 5′-RACE. These special cases are discussed in detail in Chapter 9, cDNA: A Permanent Biochemical Record of the Cell, and Chapter 10, RT-PCR: A Science and an Art Form.
2. For initiation and elongation to occur, there must be a free 3′-OH to which the next nucleotide in the chain can be joined via a phosphodiester linkage. Thus, the entire process of transcription requires that each newly added nucleotide must possess a requisite 3′-OH with which to form the next phosphodiester bond. This applies equally to RNA and DNA synthesis, both in vivo and in vitro. The lack or inaccessibility of a 3′-OH will result in immediate cessation of strand elongation.

This results in a polynucleotide with a consistent pattern of 5′→3′ linkages between adjacent nucleotides; elongation is frequently referred to as the 5′→3′ polymerase activity associated with the enzyme. Upon completion, nucleic acid molecules are assembled in such a way that:

TABLE 1.3 Stereochemical preferences of purines and pyrimidines.

	Base	Preferred	Rare
Purines	Adenine	Amino form	Imino form
	Guanine	Keto form	Enol form
Pyrimidines	Cytosine	Amino form	Imino form
	Thymine	Keto form	Enol form
	Uracil	Keto form	Enol form

Tautomeric forms of complementary nitrogenous bases favor canonical Watson-Crick base pairing resulting in antiparallel strand orientation whenever double-stranded molecules form.

1. The ends of the molecule are structurally different from one another. The first nucleotide of the molecule has a free 5′ (tri)phosphate, constituting the so-called 5′ end of the molecule. The last nucleotide that was added exhibits a free 3′ hydroxyl group, and this is known as the 3′ end of the molecule.
2. The backbone of the molecule consists of an alternating pattern of sugar and phosphate groups. Known as the phosphodiester backbone, or simply the backbone, of the molecule, it imparts a net negative charge to the molecule by virtue of its constituent phosphate groups.
3. The base associated with each nucleotide protrudes away from the backbone of the molecule. This stereochemistry makes the bases very accessible for hydrogen bonding (base pairing) to a complementary polynucleotide sequence. This proclivity is at the very heart of molecular hybridization in the laboratory.

The nitrogenous bases found in nucleotides are categorized as either purines (adenine and guanine) or pyrimidines (cytosine, thymine, and uracil), both of which are flat aromatic molecules. The specificity of base pairing (purine with pyrimidine) is maintained by the stereochemical preferences of the bases listed here. In other words, what is commonly known as "Watson–Crick" base-pairing is predicated on the bases involved being in their preferred tautomeric forms, as shown in Table 1.3.

Hydrogen bonds, which are highly directional, form between complementary bases when an electropositive hydrogen atom is attracted to an electronegative atom such as oxygen or nitrogen. Because of the manner in which bases protrude from their respective phosphodiester backbones, antiparallel base pairing or hybridization of complementary strands is strongly favored. Thus, the 5′ end of one strand is opposite the 3′ end of the complementary strand to which it is base-paired and often represented as shown here.

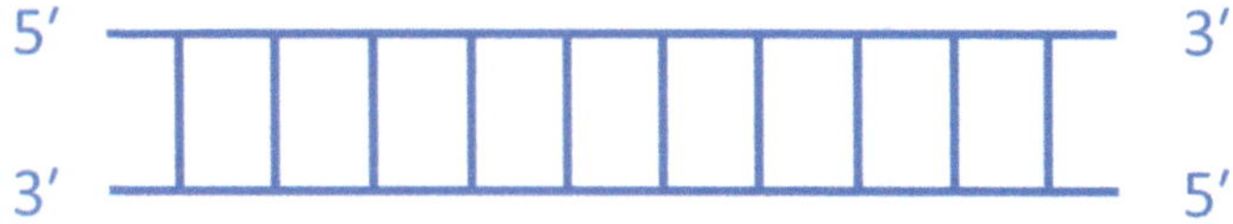

This is true for all double-stranded molecules: dsDNA, dsRNA, and DNA:RNA hybrids. The ability to promote, or prevent, base-pairing in this manner is a central act in the molecular biology laboratory.

The obvious structural differences at the 5′ and 3′ ends of a molecule support a convention by which one may unambiguously refer to the position of any feature of a nucleic acid molecule in relation to any other feature:

Upstream means that a structure or feature is closer to or in the direction of the 5′ end of the molecule, relative to some other point of reference; it can also mean in the opposite direction of gene expression.

Downstream means that a structure or feature is closer to or in the direction of the 3′ end of the molecule, relative to some other point of reference; it can also mean in the direction of gene expression.

For the sake of simplicity, upstream and downstream are most often used to mean "in the opposite direction of expression" and "in the direction of expression," respectively. This nomenclature may be especially useful when describing features or regions of a double-stranded nucleic acid molecule, in discussions pertaining to either the structure or the expression of a gene and, in particular, for the purpose of primer design to support PCR (Chapter 10: RT-PCR: A Science and an Art Form).

The actual base sequence, that is, the linear order of ribonucleotides, is known as the primary (1°) structure of an RNA molecule, and this order is dictated by the order of nucleotides on the DNA template strand. There is a tremendous proclivity for a single RNA molecule to exhibit extensive intramolecular base pairing, resulting in what is known as a secondary (2°) structure. The variety of possible interactions is impressive and often described using such colorful nomenclature as RNA hairpins, stems, interior loops, bulge loops, multibranched loops, kissing loops, cruciform structures, and pseudoknots (Fig. 1.4). Higher-order three-dimensional folding, the so-called tertiary (3°) structure which RNA molecules exhibit, is best described as the collection of 2° structural elements arranged in such a way that an RNA molecule is able to perform its biological function. Much has been suggested, for example, about the role of folding by careful study of transfer RNAs, the classical example of intramolecular base-pairing *par excellence*. It is important to note that some of the 2° and 3° structure of tRNA is attributed to the formation of noncanonical base pairs. The canonical base pairs are G · C, A · T, and A · U. The noncanonical base pairs include G · U, A · C, A · G, C · U, U · U, G · G, A · Ψ (Ψ = pseudouridine), G · Ψ, A · A · U trimers, and others. Studies have also demonstrated that mRNA assumes varying degrees

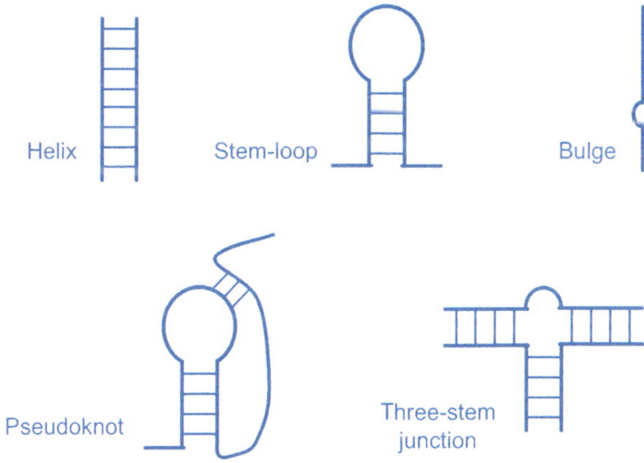

FIGURE 1.4 Examples of secondary structure commonly observed in single-strand ribonucleic acid molecules. Note how a single molecule is able to exhibit intramolecular base-pairing by the antiparallel juxtaposition of complementary regions. Double-stranded regions may be perfectly or imperfectly base-paired. To a large extent, the variety and locations of stems, hairpins, and loops will influence the ensuing tertiary structure and function.

of transient 2° and 3° structure which, in no small measure, influence its function in the cytoplasm. For most laboratory applications, higher-order folding must be disrupted, as described below, before an assay with a quantitative component can be performed using an RNA sample. Failure to do so generally has a severe negative impact on accurate profiling of the sample. A contemporary method for resolving transcriptome intra- and intermolecular interactions is psoralen analysis of RNA interactions and structures (PARIS), information from which is considered to be part of the "RNA structurome" (Lu et al., 2018; Zhang et al., 2021).

Types of ribonucleic acid

Transcription is the process by which single-stranded RNA molecules are synthesized and the resultant products are generically referred to as *transcripts*. In the past, cellular transcripts were broadly classified as rRNA, tRNA, hnRNA, or mRNA, as well as a collection of small RNAs of previously unknown function. Now, however, one must include the very diverse population of noncoding RNA (ncRNA), all of which are of immense interest in the study of the regulation of gene expression (Table 1.4). Each category of RNA, which in eukaryotic cells is synthesized by a different type of RNA polymerase, performs a different function in the cell. In contrast, all transcripts in bacteria are produced by a single type of RNA polymerase. The various types of RNA are not represented in equal amounts in the cell. The term "abundance" is used to refer to the prevalence of each transcript species, which is directly related to the physiology of the cell.

TABLE 1.4 Ribonucleic acid types and functions.

RNA type	Name	Symbol	Primary function	Prokaryotic	Eukaryotic
Coding	Messenger RNA	mRNA	Template for the synthesis of proteins	Yes	Yes
	Heterogeneous nuclear RNA	hnRNA	Large unspliced precursor of mRNA (pre-mRNA)	No	Yes
Long noncoding (lnc)	Ribosomal RNA	rRNA	Forms scaffolding of the ribosomal subunits	Yes	Yes
	Transfer RNA	tRNA	Transports amino acids to the ribosome to support translation	Yes	Yes
	Long intergenic (or intervening) noncoding RNA	lincRNA	Product of transcription of intergenic regions	No	Yes
	XIST	Xist	Sex-linked lncRNA involved in X-chromosome inactivation and Barr body formation in female mammals	No	Yes
Small noncoding (snc)	Small nuclear RNA	snRNA	Facilitates splicing of hnRNA into mature mRNA as well as rRNA processing. snRNA molecules exist as an RNA-protein complex, referred to as a snRNP, or snurp	No	Yes
	Small nucleolar RNA	snoRNA	Processing of immature rRNA transcripts in the nucleolus; some have a role in gene silencing	No	Yes
	Small cytoplasmic RNA	scRNA	Facilitates protein trafficking, secretion, and possibly mRNA degradation. scRNA molecules exist as an RNA-protein complex, referred to as a scRNP, or scyrp	Yes	Yes
	microRNA	miRNA	Short antisense RNAs that regulate gene expression by blocking mRNA and inhibiting translation via the transient formation of a dsRNA region with various target transcripts	No	Yes

	Small interfering RNA	siRNA	Short antisense RNAs that mediate the destruction of a specific mRNA through the formation of a short dsRNA region with the target transcript	No	Yes
	Piwi-interacting RNA	piRNA	Involved in epigenetics and posttranscriptional silencing via Argonaute protein interactions. Largest group of sncRNA in animals; usually 23–30 nts	No	Yes
Catalytic RNA	Ribozyme	–	An RNA molecule with a catalytic function	Yes	Yes
	Telomerase RNA	–	RNA portion of the enzyme/RNA complex that repairs chromosome telomeres (TERC: telomerase RNA component)	No	Yes

rRNA is the most abundant RNA component in the cell. In prokaryotic cells, the major rRNA species are the 23S rRNA, 16S rRNA, and 5S rRNA. The eukaryotic counterparts are identified as the 28S rRNA, 18S rRNA, and 5S rRNA, as well as a fourth ribosomal transcript, the 5.8S rRNA. These molecules form the scaffolding of ribosomes, which become translationally competent when decorated with myriad ribosomal proteins. At present, there are 55 known prokaryotic ribosomal proteins and 82 known eukaryotic (mammalian) ribosomal proteins. Not all ribosomes are functional at any given time, and the existence of a pool of transiently inactive ribosomes is itself a regulator of gene expression. The super abundance of rRNA in a purified RNA sample is often used as both an RNA mass loading control (Chapter 11: Quantitative PCR Techniques) as well as internal electrophoresis molecular weight markers (Chapter 14: Electrophoresis of RNA).

tRNA is responsible for the transportation of amino acids to the ribosome to support protein synthesis. Amino acid molecules are small, ordinarily ranging from 74 to 95 nts. When shuttling an amino acid covalently linked to its 3′ end, a tRNA is said to be "charged." Placement of the correct amino acid into the nascent polypeptide depends on the recognition of the mRNA codon (a group of three nucleotides) within the coding region of mRNA by a complementary trinucleotide motif carried on one arm of the tRNA known as the anticodon. The tRNA anticodon base-pairs to the mRNA codon within the ribosome, thereby supporting protein elongation (for review, see Suzuki, 2021). While neither as large nor as abundant as rRNA, the smaller tRNA species play a central role in translation.

mRNA is the most diverse of all the transcripts. Ironically, even though mRNA is by far the least abundant of all transcript types, it is the mRNA that drives the phenotype of the cell. mRNA alone directs the synthesis of proteins through the use of the cellular translation apparatus. There is wide variation in the number and abundance of RNA species in the cell; the abundance of a specific type of RNA is subject to dramatic change as the demands on the cell change. Some mRNAs are present in hundreds of copies per cell while others are present in only a few copies per cell; this aspect of the RNA profile of the cell can be problematic because very low abundance transcripts are sometimes difficult to detect even with sensitive contemporary techniques.

The variegated and widespread noncoding RNAs (ncRNAs) arise from areas of the genome long thought to be transcriptionally silent. These molecules are exceptionally diverse both in form and function, though a common theme is their direct or indirect involvement in the regulation of gene expression, among other possible roles. Noncoding RNAs are presented in greater detail in Chapter 12, miRNA and Other Noncoding RNAs.

Transcription and the central dogma

According to the central dogma of molecular biology, a term coined by Francis Crick in his 1957 paper "On Protein Synthesis," the expression of hereditary information flows from genomic sequences (DNA), through an mRNA intermediate, to ultimate phenotypic manifestation in the form of a functional polypeptide (Fig. 1.5). Whereas this design mirrors what occurs naturally in both prokaryotic and eukaryotic cells, certain "violations" have been observed in nature, (1) accompanying the discovery of the retroviral enzyme reverse transcriptase, that is, RNA-dependent DNA polymerase (Baltimore, 1970; Temin and Mizutani, 1970), by which RNA may serve as the template for the synthesis of DNA, and (2) the discovery of RNA editing (Benne, et al., 1986; reviewed by Eisenberg and Levanon, 2018), in which a transcribed sequence is subject to single base changes.

The first of several steps in what is commonly referred to as RNA biogenesis is the synthesis of the transcript at a specific chromosome locus. Transcription occurs in the nucleus (and mitochondria and chloroplasts) of eukaryotic cells, and in the common cellular compartment in prokaryotic cells. All phases of transcription are subject to variation and are potential control points in the regulation of gene expression. A transcriptional unit is best thought of as a DNA sequence that manifests appropriate signals for the initiation and termination of transcription and is capable of supporting the synthesis of a primary RNA transcript. The process of *transcription* is so named because the transfer of information from DNA to RNA is in the same language, namely the language of nucleic acids. In contrast, the process known as *translation* is so named because nucleic acid instructions in the form of mRNA are used to direct the assembly of a primary polypeptide from amino acid precursors: the nucleic acid instructions are executed in (translated to) the language of proteins. The ribosome is the organelle of polypeptide synthesis in all cells, and each ribosome independently directs the sequential linkage of amino acids as the associated mRNA is interpreted. Upon completion and release

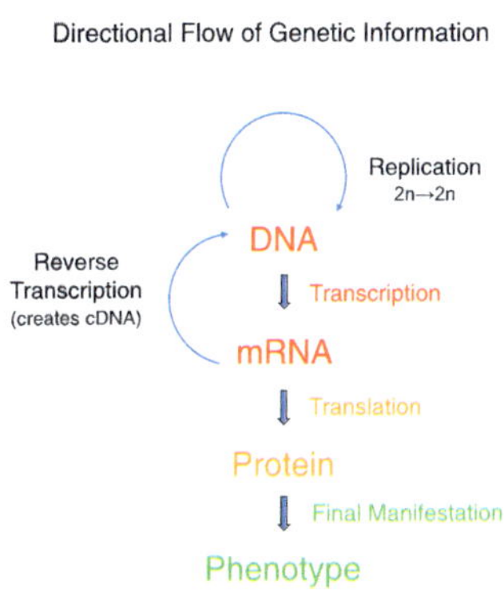

FIGURE 1.5 The central dogma of molecular biology. The process of transcription produces messenger ribonucleic acid while the process of translation produces protein. Replication, the process by which DNA is duplicated, occurs during S phase in the eukaryotic cell cycle. cDNA, in contrast, is not found in the cell but is synthesized in vitro and is commonly used to measure transcriptional activity or to assay for the presence of an RNA virus.

from the ribosomes, eukaryotic proteins are typically modified, sorted, packaged, and directed to their proper subcellular location as they move through the endomembrane system, of which the endoplasmic reticulum and the Golgi apparatus are key components; prokaryotic and other eukaryotic proteins are often under the influence of various small cytoplasmic RNA (scRNA) species that guide them to their proper destination. As with RNA, and to a lesser extent DNA, proteins exhibit a marked capacity for higher-order folding (Table 1.5). As with RNA, the functionality of a protein molecule is associated with its shape. Unlike RNA, however, in which the shape of the molecule is naturally dynamic, the distortion of the tertiary (3°) or quaternary (4°) structure of protein is associated with altered or rapid loss of function.

Promoters, transcription factors, and regulatory elements

Transcription is mediated by enzymes known as RNA polymerases. These enzymes, in conjunction with myriad proteins known as transcription factors, recognize very specific and highly conserved *promoter*, or initiation,

TABLE 1.5 Higher-order folding of nucleic acids and protein.

	RNA	Protein	DNA
1° structure	Nucleotide order	Amino acid order	Nucleotide order
2° structure	Stem-loop structures and hairpins, which may include mismatches	α-helices and β- pleated sheets	Antiparallel base-pairing between two complementary DNA strands
3° structure	Three-dimensional folding	Three-dimensional folding	Three-dimensional folding for form A-DNA, B-DNA, or Z-DNA double-helices.
4° structure	Interaction of two or more folded RNA molecules, often by association with RNA binding proteins.	Aggregation of two or more protein subunits	Interaction of double helical DNA with proteins, such as histone proteins, as in chromatin formation
Catalytic variants	Ribozymes	Enzymes	DNAzymes

The primary structure of nucleic acids and proteins is the order of monomers. The secondary structure of a molecule is the first level of folding that occurs as a consequence of its primary structure. The tertiary structure is the three-dimensional arrangement of atoms within the molecule. The quaternary structure of a molecule, when it forms, is higher-order folding the results from interaction of the molecule with one more identical or nonidentical molecules.
For DNAzyme review, see Morrison, D., Rothenbroker, M., Li, Y. (2018). DNAzymes: selected for applications. Small Methods 2(3), 1700319 https://doi.org/10.1002/smtd.201700319.

sequences within the enormous complexity of genomic DNA. Promoters are spatially associated with the structural portion (body) of a gene (Fig. 1.6) and consist of several recognizable upstream nucleotide sequence motifs. These sequences are known as consensus sequences, a term used to describe the most commonly observed pattern of nucleotides at a particular location. For example, the symbol $T_{80}A_{95}T_{45}A_{60}A_{50}T_{96}$ indicates that thymine is the first base associated with this consensus motif 80% of the time, A appears 95% of the time in the second position, and so forth. The exact sequence and precise geometry of these regulatory elements can either promote or prevent the onset of transcription, and do so with varying degrees of efficiency.

Any promoter component that is located 5′, or upstream, from the transcription start site (TSS) is indicated with a "minus" sign in front of the actual nucleotide distance from the TSS. By convention, the first transcribed nucleotide is designated as +1, and any other nucleotides or features located 3′, or downstream, from the TSS are likewise designated with a "plus" sign placed in front of the actual nucleotide distance. Knowledge of promoter consensus sequence function is due largely to experiments involving standard DNA cloning techniques, site-directed mutagenesis, DNA sequencing, and in silico analysis.

In prokaryotic systems, the essential elements of the promoter region include the so-called −10 sequence, formerly known as the Pribnow box or the Pribnow–Schaller box, consisting of the consensus sequence $T_{80}A_{95}T_{45}A_{60}A_{50}T_{96}$, and another conserved region located further upstream is known as the −35 sequence ($T_{82}T_{84}G_{78}A_{65}C_{54}A_{45}$). In some organisms, an AT-rich domain (the UP element) is also observed further upstream. The

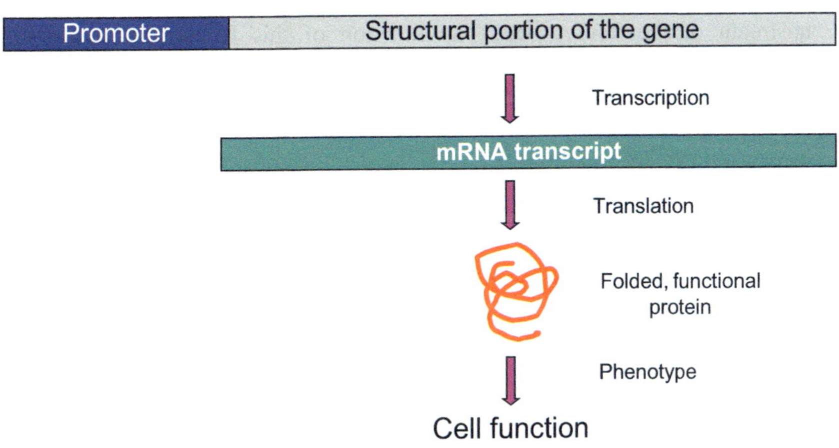

FIGURE 1.6 Genes, some of which encode messenger ribonucleic acid which, in turn, encode proteins, are under the direct influence of a regulatory element known as a promoter.

spacing between the −10 sequence and the −35 sequence is tightly regulated, with seventeen base pairs being optimal, and variations in the length of the region between these two elements can reduce the efficiency of the promoter.

In eukaryotic cells, promoters associated with nuclear genes are variable in the structure; these variations are due to the presence of multiple nuclear RNA polymerases as well as the requisite transcription factor initiation complex that must form. Transcription factors are small proteins that are continually binding to and altering the shape of the chromatin. The remodeling of chromatin in the promoter locale is characterized by changes in the association between genomic DNA and the histone proteins which decorate it. Best thought of as a type of histone displacement, the objective is to facilitate access to the gene promoter by altering the local architecture of the chromatin. This is an ATP-dependent process. Transient covalent modifications to histone proteins include acetylation, methylation, and phosphorylation. Generally speaking, histone acetylation is associated with the activation of transcription, while methylation commonly correlates with gene silencing. The net result is the activation, or silencing, of various subsets of genes in a temporal or environmentally-induced manner.

Interestingly, promoters recognized by RNA polymerase II, the enzyme responsible for the synthesis of mRNA (discussed below), often display similar sequence homology with prokaryotic gene promoters (Fig. 1.7). The eukaryotic promoter counterpart is known as the "TATA box," formerly known as the Hogness box or the Goldberg-Hogness box, and so named because of the prevalence of the highly conserved TATAA motif. Point mutations involving any of these five bases strongly downregulate the function of that promoter. Another promoter component, the transcription initiation factor TFIIB recognition element (BRE), is directly adjacent to and upstream of the TATA box. The function of this heptanucleotide motif (often, GGGCGCC) is to attract TFIIB, a key element in the assembly of the transcription apparatus associated with RNA polymerase II. While at one time it was thought that all eukaryotic promoters manifest a TATA box, this is now known to be untrue. Instead, these rather prevalent TATA-less promoters are typically characterized by an initiator region (INR) and a downstream promoter element (DPE), which is observed approximately thirty base

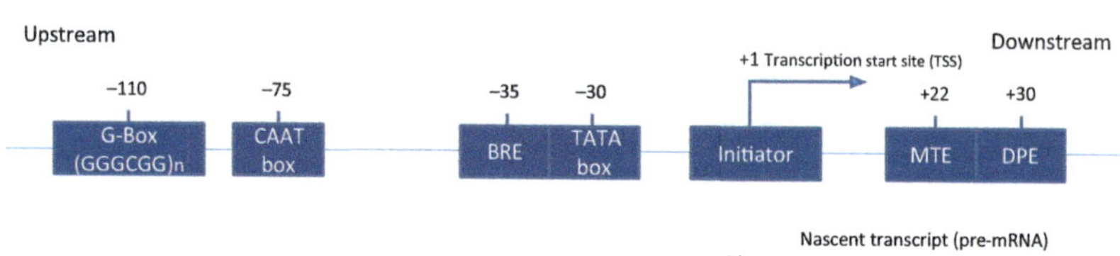

FIGURE 1.7 Generalized structure of a eukaryotic gene promoter. See text for details.

pairs downstream (+30) from the TSS. The motif ten elements (MTEs) exclusively maps to +18 through +27 and is located downstream from INR and immediately upstream from the DPE. At least one function associated with the MTE is its ability to act in place of an absent TATA box. In addition to the TATA, DPE, and MTE promoter structural components, another promoter motif is the "CAAT box," found in several but not all promoters, and so named because of the conservation of its sequence. When present in eukaryotic promoters, the TATA box is usually centered at −30 and the CAAT box appears around −75, though the CAAT box has been shown to function quite effectively much further upstream, and even in reverse orientation. These elements appear to control the initial binding of the RNA polymerase and promoter efficiency, respectively. Another frequently observed promoter element is the sequence $(GGGCGG)_n$, known as the G-box element or simply as the GC box. Present in one or more copies, this GC-rich region is generally observed between −90 and −120 within the promoter region. Interestingly, it appears that there is no one component or organization that is shared by all promoters, though the particular permutation of promoter elements and distances between them is recognizable as a transcription initiation regulator. Succinctly, by comparison with transcription in prokaryotic cells, the elaborate initiation of eukaryotic transcription requires the presence of numerous transcription factors, coactivators, and transcription activator proteins that bind to these *cis*-acting components which, collectively, make up a promoter. The widely accepted role of early transcription factor binding to gene promoters is to recruit RNA polymerase to that site so as to ultimately initiate transcription. Rather than being thought of as merely an on–off switch associated with a particular gene, a promoter functions more like a thermostat that increases (upregulates) and decreases (downregulates) the expression of a gene in response to the prevailing local conditions acting upon a cell.

Eukaryotic promoters do not always function alone. Transcription in eukaryotic cells can be influenced profoundly by the presence of a regulatory element known as an *enhancer*, the function of which appears to be the stimulation of transcription. First discovered in the early 1980s, the precise location and orientation of an enhancer relative to the gene promoter varies from one gene to the next. It is not uncommon for the chromatin to actually fold back over on itself in order to bring the enhancer and the promoter into proximity. This transient remodeling explains how enhancers located thousands of bases upstream from the promoter are able to exert their influence. Some genes, including those which encode immunoglobulins, carry enhancers within the structural portion of the gene itself. Removal of enhancer sequences can reduce the transcriptional efficiency at a locus normally under the influence of that enhancer sequence, as can the binding of repressor proteins to functionally disparate DNA sequences known as silencers.

In vitro transcription of genes that are not naturally associated with an enhancer element can be increased significantly if an enhancer is ligated to the DNA construct, usually in no particular orientation, and often at a considerable distance (hundreds, if not thousands of base pairs) from the TSS. In vivo, a translocation event that brings a promoter and a gene into proximity can result in inappropriate expression of the gene, often with potentially catastrophic consequences, as in the case of Burkitt's lymphoma (Taub et al., 1982). The transcriptional influence of upstream and downstream enhancer sequences, and antagonistic silencer sequences, on gene promoters is well documented.

Many enhancers, but not all, are transcriptionally active (Djebali et al., 2012; Andersson et al., 2014), producing enhancer RNAs, or simply eRNAs. At present, the number of known eRNAs in human cells is in the tens of thousands and their transcription points to enhancer functionality in terms of promoting expression of the cognate gene (reviewed by Field and Adelman, 2020). These noncoding transcripts are produced from promoters located upstream and in the opposite (antisense) orientation by comparison with the mRNA promoter. The eRNAs are believed to recruit components of the transcription initiation complex and the RNA polymerase II may track to a transcription promoter by first identifying the enhancer. In the intergenic area between the enhancer and the promoter, histone acetylation and ensuing chromatin remodeling facilitate transcription initiation at the enhancer-promoter locus. The sequential binding of transcription factors and ancillary components in the immediate vicinity of the gene locus ultimately results in the formation of a loop and concomitant spatial juxtaposition of the components of the template DNA needed to support the initiation of transcription. Succinctly, enhancers perform their function by increasing the concentration of transcription activator proteins in the vicinity of the associated mRNA promoter.

During transcription, both strands of the gene being transcribed have different names and different roles. The strand that actually serves as the template upon which RNA is polymerized is properly referred to as the *template strand*. The other strand, which does not act in a template capacity, is called the *coding strand*. The coding strand is also known in some circles as the sense strand, while the template strand may be referred to as the antisense strand (Fig. 1.8). The choice of nomenclature is purely a matter of personal preference. When publishing a gene sequence, the convention is to report the sequence of only the coding strand, written 5′ to 3′, from left to right. The implication is that the template strand is base-paired to the coding strand and lying antiparallel to it and therefore does not need to be reported. The DNA template strand is so named because the precise sequence of nucleotides inserted into the nascent RNA transcript is determined by, and complementary to, the template strand nucleotide sequence. It is important to realize that the coding strand and template strand may switch roles depending upon the placement of transcriptional promoter sequences (Fig. 1.9). One powerful example of this phenomenon in vitro is the cloning of a double-stranded

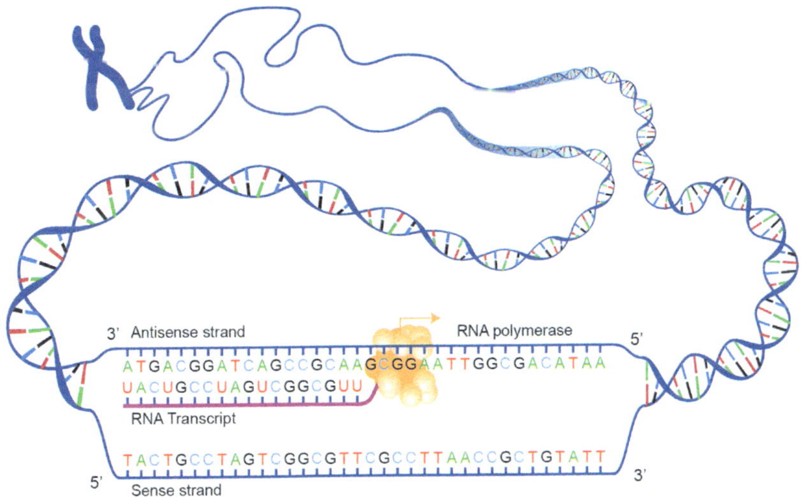

FIGURE 1.8 Ribonucleic acid is synthesized by transcription. In this process, one of the two DNA strands acts as the template (antisense strand) upon which the transcript is assembled $5' \rightarrow 3'$. The other DNA strand, known as the coding (sense) strand, exhibits the same base sequence as the RNA except for the substitution of uracil (U) in place of thymine (T). *Courtesy the National Human Genome Research Institute (NHGRI; http://www.genome.gov).*

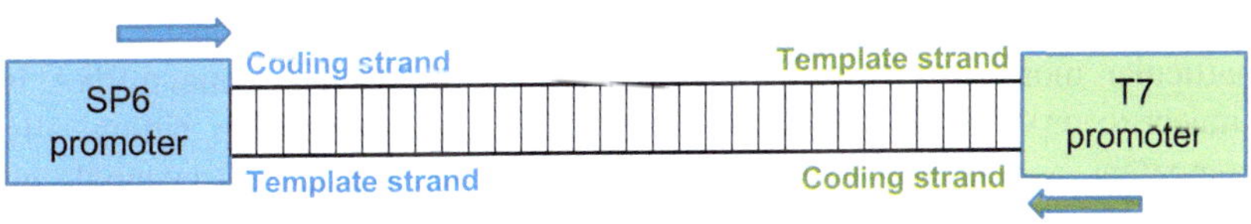

FIGURE 1.9 Promoters positioned in opposite orientation relative to a DNA sequence allow the template and coding strands to switch roles during transcription. This arrangement permits the synthesis of + RNA and −RNA from the same DNA construct.

DNA between two different transcription promoters in opposite orientations; often the bacteriophage polymerase promoters SP6, T3, or T7 are selected because of their high efficiency. Constructions such as these are frequently employed to accommodate in vitro transcription of large amounts of sense and/or antisense RNA for use as nucleic acid probes (Chapter 17: Nucleic Acid Probe Technology), for RNAi (Chapter 13: RNA Interference and Gene Editing), or for other applications.

Gene and genome organization affect transcription

In order to understand the significance of the products of transcription, it is first essential to understand the organization of the genes themselves. The typical prokaryotic genome exhibits little extraneous baggage. Frequently, genes that encode proteins associated with a common metabolic pathway are clustered together, as suggested by the operon model (Jacob and Monod, 1961). The *lac* operon, the gene products of which facilitate the metabolism of lactose as a carbon source in bacteria, is but one very well-characterized example. The RNA molecule that results from the transcription of an operon is usually polycistronic, meaning that more than one polypeptide is encoded in a single RNA transcript.

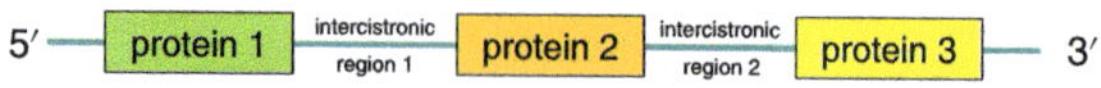

The coding information within a polycistronic mRNA for each polypeptide is contiguous: there are no interruptions in the coding sequences by noncoding information. This design favors maximum efficiency of energy resource utilization in unicellular organisms. In fact, the kinetics of prokaryotic gene expression are so rapid that bacterial mRNA is usually being transcribed, undergoing translation, and degraded simultaneously. The rapid turnover of RNA in this manner has, in the past, frustrated valiant attempts to clone or otherwise characterize prokaryotic mRNA. While significant improvements favoring the isolation of high-quality RNA from both Gram-negative and Gram-positive bacteria have been made, and many of these innovations are available in kit form, the isolation of intact prokaryotic RNA remains something of a challenge in many laboratories.

In contrast to prokaryotes, nearly all eukaryotic mRNAs are monocistronic. Although a single polypeptide species results from the translation of a particular monocistronic eukaryotic mRNA molecule, that same mRNA is subject to repeated translation as long as the transcript remains biologically competent and chemically stable. To maximize translation potential, an mRNA transcript is often engaged by several ribosomes that are all involved in simultaneous, orderly translation of that transcript. Such a cluster of ribosomes attached to a single mRNA molecule is known as a polysome (or polyribosome) which, to many, resembles a translation assembly line. Polysomes are observed both in prokaryotic and eukaryotic cells, though eukaryotic polysomes, with 7–8 ribosomes per polysome complex, tend to be smaller than their prokaryotic counterparts. Succinctly, a large number of polypeptide molecules can be manufactured from a single RNA molecule. Polysomes are observed free-floating in the cytoplasm, they can be membrane-bound, and sometimes they are attached to the cytoskeleton. The entirety of mRNAs so engaged in a cell at any given moment is known as the polysome fraction, which can be used to assess the translational competence of a cell under a defined set of experimental conditions.

Close examination of eukaryotic genes reveals that for a vast majority of genes there are considerably more nucleotides within a particular locus than are necessary to direct the synthesis of the corresponding polypeptide, that is, the DNA sequence and the amino acid sequence are not colinear over the span of the locus. This size differential can also be observed at the level of the mature mRNA in the cytoplasm, which is usually quite a bit shorter than the gene locus from whence it was transcribed. Upon further scrutiny, this discrepancy can be resolved at the level of the organization of the structural portion of the gene itself, the sequences within which fall into one of two categories:

1. *Exons* are regions of DNA that are represented in the corresponding mature mRNA. Exons may or may not have a peptide coding function.
2. *Introns* are regions of DNA that are transcribed but generally are not represented in the corresponding mRNA. Introns are usually spliced out of the primary RNA transcript (the immediate product of transcription), accompanied by the joining of adjacent exon sequences. The majority of introns do not direct polypeptide synthesis, though there are several noteworthy exceptions (see Farrell and Bassett, 2007).

The number and length of exons and introns associated with a gene are quite variable depending upon locus and this variability even pertains to loci that are strongly conserved across evolutionary time. By comparison with introns which can be several thousand base pairs in length, exons tend to be rather short, each encoding fewer than 100 amino acids in most organisms. In some cases, the high sequence conservation in one or more exons of a gene has been directly responsible for the isolation of a related gene (an ortholog) from a different organism. In less common cases, some genes lack introns altogether, of which human β-interferon and thrombomodulin are but two examples.

The base sequence of the primary RNA transcript correlates precisely with the DNA from which it is derived, meaning that it contains both exon and intron sequences. These primary transcription products are only a precursor to functional mRNA, and are confined to the eukaryotic nucleus where, appropriately, it is collectively known as *heterogeneous nuclear RNA* (hnRNA) or simply pre-mRNA. hnRNA and specific nuclear proteins that bind to it form rather abundant heterogeneous nuclear ribonucleoprotein complexes (hnRNPs). Following intron removal and export through nuclear membrane channels, mRNAs exist as messenger ribonucleoprotein (mRNP) complexes in the cytoplasm.

Introns vary dramatically in number, length, and base sequence and often exhibit multiple translation termination (stop) codons in all reading frames. This is not entirely unexpected because the noncoding nature of introns favors the accumulation of mutations that might otherwise be detrimental if they were to occur within an exon or other critical area. Examination of the splice junctions of introns, however, reveals strict conservation of two dinucleotide consensus sequences contained entirely within the intron. Proceeding from the 5′ end of the RNA, introns are found to begin with a

GU dinucleotide (known as the left or donor site) and end with an AG dinucleotide (known as the right or acceptor site). This is the so-called GU-AG rule, describing exon-intron splice sites, and refers to the RNA sequence. The corresponding DNA coding strand dinucleotide at the 5′ end (beginning) of an intron is GT; while once believed to occur 100% of the time in higher eukaryotes, some exceptions have been noted (Szafranski et al., 2007) and the consensus phenomenon does not apply to yeast mitochondrial or tRNA genes, nor to chloroplast loci.

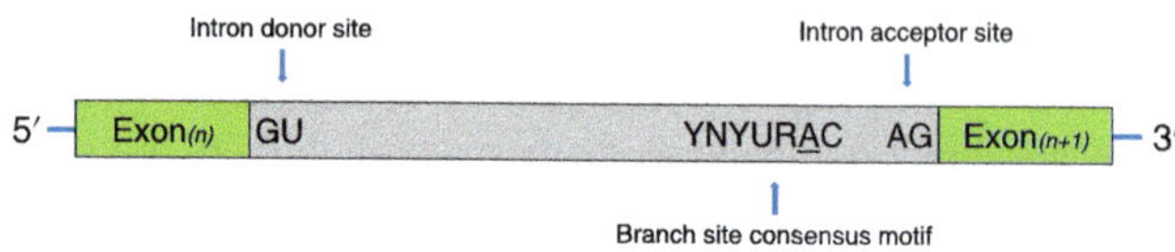

The nucleotides immediately adjacent to both sides of the GU and AG intron boundaries are also conserved to an extent, typically 60%–80%. Also intimately involved in mediating the intron removal process is the highly conserved branch site consensus sequence (YNYURAC, in vertebrates); R is the symbol for a purine, Y is the symbol for a pyrimidine, and N is the symbol for any base. A point mutation at a splice site or within the consensus generally results in the inactivation of that site. In some cases, splice-site mutations can result in the production of an aberrant mRNA through the use of an alternative splice site, often located within the intron, as in certain β-thalassemic individuals. Knowledge of the high conservation of splice sites is the basis of a method for exon identification known as exon trapping (Péterfy et al., 2000) in which a putative exon-containing sequence is cloned into a specialized vector that consists of an intron flanked by two known exons (exon–intron–exon); if an exon is present in the experimental DNA, it will be trapped by ligation into the vector intron and will result in a longer transcript that can be can be detected electrophoretically or by melting curve analysis. The exon trapping method has fallen out of favor due to the low cost and ready availability of next-generation sequencing (NGS) and an extensive repertoire of tools for in silico analysis.

The mechanics of intron removal and exon ligation, which occur cotranscriptionally, that is, while the RNA polymerase is still active, are mediated in part by a highly conserved family of small nuclear RNAs (snRNA; 100–300 bases). These molecules exist as the RNA-protein complexes known as U1, U2, U4, U5, and U6 (major species) as well as U11, U12, U4atac, and U6atac (minor species with specialized functions), and are confined to the nucleus where they are referred to as small nuclear ribonucleoproteins (snRNPs, or snurps). The snRNPs, along with many other proteins as splicing factors, form large complexes known as spliceosomes, which are known to mediate pre-mRNA splicing and which are dependent upon various epigenetic modifications (reviewed by Morais et al., 2021).

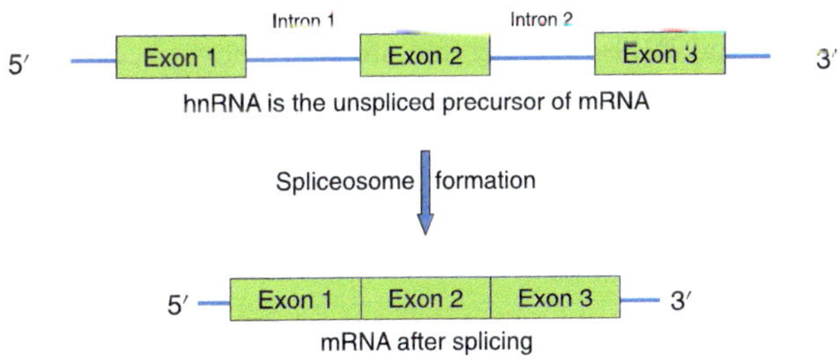

Closely associated with the capacity for transcript splicing are Cajal bodies (CBs), small organelle-like, punctulate structures in the nucleoplasm that were first observed more than a century ago (Cajal, 1903). Unlike organelles, these structures are nonmembrane bound; they are spherical in appearance, with a typical diameter of 0.4–1.0 μm. CBs are characterized by a high concentration of the protein coilin and are packed with RNA. They are dynamic in that they are visible at certain times and not others, which may be related to differentiation, development, and even progression through the cell cycle. Found in higher eukaryotic plant and animal cells, CBs represent transcriptionally active regions, including histone loci, and in the nucleolus where they are linked to ribosome biogenesis and telomere upkeep. However, one of the best-known functions CBs is their factory-like role in the assembly of snRNPs associated with mRNA splicing.

Intron removal has also been demonstrated to have a role in the nuclear export of the spliced and matured mRNA. The proteins involved in the splicing mechanism and exon concatenation recruit additional proteins that are specifically required for nuclear egress. Among the proteins in the resulting exon junction complex (EJC) is the ALY/REF export adapter, which binds directly to the RNA, and the TAP-p15 export receptor complex, each of which has a direct role in nuclear pore engagement (reviewed by Grünwald et al., 2011). Once on the cytoplasmic side of a nuclear pore, the mRNA sheds the array of proteins which facilitated its nuclear egress. This ensures unidirectional movement. Improperly spliced or otherwise compromised mRNAs fail to associate with the correct combination of proteins required for nucleocytoplasmic transport, thereby promoting their retention in the nucleus and rapid degradation. Although intron removal and the splicing together of exons in and of itself is not required for transport from the nucleus, since intronless transcripts move efficiently into the cytoplasm, splicing clearly enhances transport. The export process used for rRNA and other RNAs is less clear. Splicing of pre-mRNA molecules also produces some unexpected results, namely, the production of enigmatic circular RNAs (circRNA), which are introduced in Chapter 12, miRNA and Other Noncoding RNAs.

Finally, the excision of certain introns in RNA can also occur as the result of RNA self-cleavage. Catalytic RNAs, generically referred to as ribozymes were first described by Cech et al. (1981). In particular, the group I intron ribozyme and RNase P ribozyme are well-known because they are the first two ribozymes to be discovered (Kruger et al., 1982; Guerrier-Takada and Altman, 1984). These discoveries led to the awarding of the Nobel Prize in Chemistry in 1989 to Thomas Cech and Sidney Altman. Extensive information about this fascinating aspect of RNA biology can be found elsewhere (for review, see Park et al., 2019).

Ribonucleic acid polymerases and the products of transcription

Genes are transcribed by enzymes known as RNA polymerases which produce the different types of RNA in the cell, including rRNA, tRNA, and mRNA, as well as a variety of noncoding (large and small) RNA species. In the nucleus, eukaryotic genes are transcribed by one of three major RNA polymerases; these enzymes are among the largest and most complex proteins in the cell and consist of more subunits than their prokaryotic counterpart. The eukaryotic enzymes are properly known as RNA polymerase I (RNAP I), RNA polymerase II (RNAP II), and RNA polymerase III (RNAP III), each of which is responsible for transcribing a different class of genes (Table 1.6). RNA polymerases are active only in the presence of DNA, and require nucleotide precursors (ATP, CTP, GTP, and UTP), myriad transcription factors, and function in a Mg^{2+}-dependent manner. In addition, eukaryotic RNA polymerases differ in the sensitivity each exhibits to the bicyclic octapeptide toxin α-amanitin from the poisonous mushroom *Amanita phalloides* (Roeder, 1976), applications for which are described in Chapter 19, Analysis of Nuclear RNA. Prokaryotes, in contrast, exhibit only one type of RNA polymerase, which transcribes all classes of RNA.

In plants, RNA polymerase IV (RNAP IV) and RNA polymerase V (RNAP V) have also been identified. These enzymes are homologs of the well-characterized, mRNA-producing enzyme RNA polymerase II. RNAP IV and RNAP V have collaborative roles in the synthesis and activation of miRNA/siRNA species involved in silencing pathways via RNA-directed DNA methylation (RdDM) (Onodera et al., 2005; Herr et al., 2005; Kanno et al., 2005; Pontier et al., 2005; Zhang et al., 2007; reviewed by Rymen et al., 2020). As such, the products of RNAP IV and RNAP V have a direct influence on numerous biochemical processes, including growth and development and defense mechanisms against viruses.

As with all nucleic acid molecules, RNA transcripts are assembled only in the $5' \rightarrow 3'$ direction. Transcription involves three distinct phases, namely, initiation, elongation, and termination, the details of which are beyond the scope of this volume. Briefly, initiation involves the attachment of RNA

TABLE 1.6 Eukaryotic ribonucleic acid polymerases.

Eukaryotic enzyme	Site of action	Products	Sensitivity to α-amanitin
RNA Polymerase I; RNAP I	Nucleolus	rRNA (47 S → 28 S, 8 S, and 5.8 S rRNA)[a]	−
RNA Polymerase II; RNAP II	Nucleoplasm	hnRNA → mRNA; lincRNA, miRNA[b], snRNA, scRNA[c]	+ + + +
RNA Polymerase III; RNAP III	Nucleoplasm	tRNA, 5S rRNA, snRNA, snoRNA, scRNA, miRNA	+
RNA Polymerase IV; RNAP IV (plants)	Nucleoplasm	miRNA/siRNA involved in heterochromatin methylation and transcriptional gene silencing	−
RNA Polymerase V; Pol V or RNAP V (plants)	Nucleoplasm	miRNA involved in heterochromatin methylation and transcriptional gene silencing	−
RNA-dependent RNA polymerases, RdRPs	Cellular	Amplification of miRNA for gene silencing; viral replication	−
Chloroplast RNA polymerases • PEP; plastid-encoded RNA polymerase • RPOTp and RPOTmp; nuclear-encoded RNA polymerases	Chloroplasts	Chloroplast gene transcripts	−
Mitochondrial RNA polymerase; mtRNAP	Mitochondria, but encoded in the nucleus	Mitochondrial gene transcripts	−

[a]*RNAP I is considered the most specialized of the three canonical nuclear polymerases because the members of this highly repetitive gene family that it transcribes are virtually identical, as are the resulting transcripts.*
[b]*Most known miRNAs are transcribed by RNAP II, though some of these important regulatory transcripts are transcribed by RNAP III.*
[c]*Certain scRNAs and snRNAs are transcribed by RNA polymerase II while others are transcribed by RNA polymerase III.*

polymerase to a DNA gene promoter by association with transcription-, activation-, and initiation factors, followed by the acquisition of what will be the first ribonucleotide in the RNA molecule. Elongation involves the sequential addition of ribonucleotides to the nascent chain, a process also involving accessory elongation factors. Termination is the completion of RNA synthesis and the disengagement of both the newly synthesized RNA and the RNA polymerase which manufactured it from the DNA template. Transcription termination, as with initiation and elongation, is sequence-dependent and, at least in prokaryotes, is influenced by the presence of small proteins (termination factors) and often involves the transient formation of RNA 2° (hairpin) structures; different transcription termination strategies are used by each of the eukaryotic RNA polymerases. Transcription errors notwithstanding, the nucleotide sequence of the resulting RNA molecule is identical to the coding strand of the DNA from which it is derived, the only difference being the substitution of the base uracil for thymine.

In eukaryotic cells, transcription of the genes encoding ribosomal RNA is mediated by RNA polymerase I. The primary product of RNA polymerase I transcription in higher eukaryotes is the very large unspliced 47S precursor rRNA which, at approximately 14,000 bases in humans, is the largest known precursor RNA in mammals. The 47S rRNA eventually yields the smaller 28S, 18S, and 5.8S rRNAs after processing (Fig. 1.10; reviewed by Henras et al., 2015). Between 80% and 85% of cellular RNA is found in ribosomes in the form of the 28S, 18S, 5.8S, and 5S rRNAs; these transcripts form

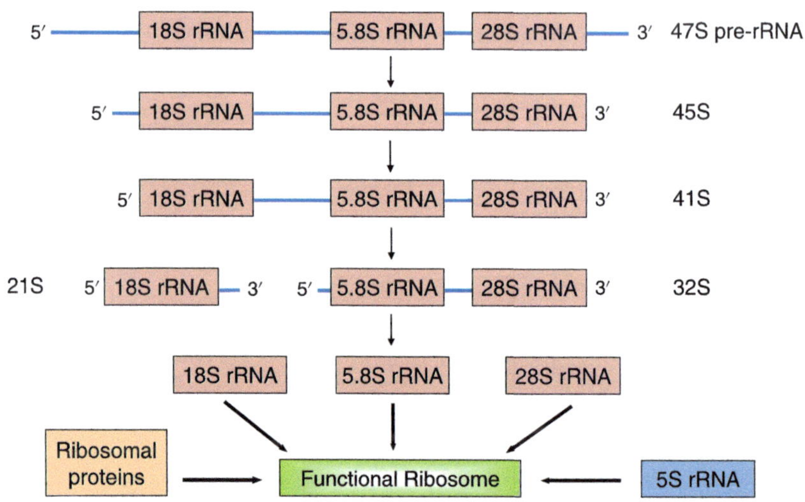

FIGURE 1.10 One pathway of rRNA biogenesis in human cells. The 28S, 18S, and 5.8S rRNA are liberated from a common 47S transcript that is the product of RNA polymerase I. The other essential transcript, 5S mRNA, is produced independently by RNA polymerase III. *Adopted, in part, from Lewin (1997), by permission of Oxford University Press.*

complexes with myriad ribosome-specific proteins to form the 60S and 40S eukaryotic ribosomal subunits, respectively. As described in Chapter 14, Electrophoresis of RNA, the abundant 28S and 18S rRNAs are useful as excellent natural molecular weight size markers when either total cellular RNA or total cytoplasmic RNA is electrophoresed. RNA polymerase III is responsible for transcribing the genes that encode tRNA molecules, the 5S rRNA, certain repetitive elements, snRNA, snoRNA, scRNA, and some miRNA transcripts described above. Approximately 10%−15% of the total cellular RNA mass is tRNA. Thus, the vast majority of RNA in the cell represents the transcriptional products of RNA polymerase I and RNA polymerase III; rRNA and tRNA are rather undiversified and generally not subject to significant alteration of their expression profile as a function of the cell state.

In sharp contrast to RNA polymerase I and RNA polymerase III, the transcription products of RNA polymerase II are as diverse as the cellular biochemistry itself. This is not at all unexpected because the mRNA in great measure drives the phenotype of the cell. Of the 1 to 5×10^{-5} μg RNA harbored in a typical mammalian cell, the relative mass contribution of RNA polymerase II transcription products is generally between 20% and 40%, though only 1%−4% of which is mature mRNA because a great deal of all transcribed hnRNA is degraded in the nucleus, strongly suggesting a rationale for studying posttranscriptional regulation of gene expression. Interestingly, rRNA was first believed to have a template role in the synthesis of proteins. The first indications that a new, separate class of RNA, mRNA, acts as the template molecule emerged from studies involving T4 phage-infected *Escherichia coli* cells (Volkin and Astrachan, 1956; Brenner et al., 1961; Gros et al., 1961; Hall and Spiegelman, 1961). Later, multiple 5′ ends were observed among late SV40 mRNAs (Ghosh et al., 1978) and late polyoma mRNAs (Flavell et al., 1979), the first indications of the heterogeneous nature of RNA polymerase II transcription initiation. The amount of mRNA may vary depending on a number of factors, including the degree of differentiation.

Genes transcribed by RNA polymerase II which, after processing, renders functional mRNA are localized mostly within the nonrepetitive regions of the genome, as demonstrated by R_0t kinetics studies. A typical cell is transcribing a subset of several thousand different genes at any given moment depending on cell type and cell state. Given the complexity of the cellular biochemistry, this observed heterogeneity within the mRNA population is necessary to satisfy even basal-level requirements for viability, though it is remarkable to note that as much as 70% of all transcribed hnRNA is degraded in the nucleus. RNA polymerase II also has major responsibility for the transcription of genes encoding miRNA.

Another category of RNA polymerase, RNA-dependent RNA polymerase (RdRP), also known as RNA replicase, is a versatile enzyme that can synthesize RNA from an RNA template, rather than the requisite DNA template associated with the other RNA polymerases. RdRP is well known for its role in the replication of poliovirus and other RNA viruses. It has come under much closer scrutiny because of the documented role of this enzyme in the RNA interference pathways, especially in higher plants by mediating the synthesis of double-stranded RNA (dsRNA) molecules which, upon processing, give rise to mRNA-hampering miRNAs/siRNAs.

Hallmarks of a typical messenger ribonucleic acid

Many genes are transcribed constitutively by RNA polymerase II, and this enzyme has a role in the integration of associated nuclear events such as splicing and polyadenylation. Of these numerous transcripts, large quantities of heterogeneous nuclear RNA (hnRNA) are turned over in the nucleus. This may partially be due to errors in transcription or posttranscriptional processing which must be managed by the gene expression machinery in the cell; in the RNA world, it's all about quality control. It is also possible that not an insignificant amount of hnRNA is retained in the nucleus to facilitate some yet-to-be-discovered aspect of gene regulation. In any event, mRNAs in eukaryotic cells emerge from precursor hnRNA through a series of end-modifying reactions such as the formation of a unique 5′ cap and 3′-end processing and polyadenylation, as well as internal modifications such as splicing and methylation. Transcripts are produced at different rates from different loci; therefore, each mRNA species is classified based on its cytoplasmic prevalence or, more properly, its abundance. There are three official such categories, high abundance, medium abundance, and low abundance mRNAs and, in the mind of this Author, the unofficial *very low* abundance category.

Highly abundant transcripts are present in hundreds of copies per cell. These are most often observed when a cell is producing an enormous quantity of a particular protein or is highly specialized or *differentiated* to perform a unique function. Medium abundance transcripts are best thought of as being present in dozens of copies per cell, and many of which have housekeeping functions. A gene is said to have a housekeeping function if the encoded gene product plays a maintenance role or if the basal expression is needed to maintain the physiology of the cell or even viability. Since the expression of these genes is generally constant, and not expected to change as a function of cell state, housekeeping genes are assayed as internal controls in various quantitative assays (Chapter 11: Quantitative PCR Techniques). Low abundance mRNAs are generally present in ten or fewer copies per cell and often are difficult to assay by many of the older classical techniques, such as Northern analysis (Chapter 16: Northern Analysis). *Very low* abundance mRNAs are those present, on average, in fewer than one copy per cell, a designation which is generally associated with heterogeneous tissue samples or, very commonly, in cases where cancer cells manifest a high degree of aneuploidy. In the past, low and very low abundance mRNAs were said to represent the "hard to clone genes," though newer methods for the assay of gene expression have revealed a plethora of previously unknown transcripts of various persuasions. Most importantly, the prevalence or abundance of an mRNA species in a cell is subject to change of monumental proportions. Such changes may occur in response to natural changes in the cellular milieu or due to experimental manipulation.

A typical human fibroblast cell contains approximately 10 picograms (pg) of RNA, which is equivalent to about 10^6 molecules transcribed from a particular subset of the 20,000 or so genes estimated to make up the human genome. While this mRNA heterogeneity reflects the diversity of proteins that these mRNAs encode, a typical eukaryotic mRNA molecule shares several structural features with nearly all other mRNA molecules (Fig. 1.11). As will become evident from the descriptions which follow, producing a functional mRNA molecule is amazingly complex.

5′ Cap

Mature eukaryotic mRNA molecules are characteristically monocistronic polyribonucleotides produced by RNA polymerase II. The immediate product of transcription, a precursor hnRNA molecule, displays the following structure at the 5′ end of the molecule

$$5'\,\text{pppRNNN.....}3'$$

meaning that the first *transcribed* nucleotide contains a purine base (symbol for a purine = R), either adenine or guanine. Not unexpectedly, the 5′ triphosphate exhibited by the nucleotide remains intact and phosphodiester bonds join ribonucleotides sequentially in a 5′→3′ orientation. A structure known as the 5′ cap is assembled in a step-wise manner just after the initiation of transcription. Capping occurs as soon as the 5′ end of the transcript emerges from the RNA polymerase complex, when nascent polynucleotides are generally between 20 and 30 bases long. The net result of the ensuing reactions that are required for capping is the formation of an unusual 5′→5′ triphosphate linkage between the first transcribed nucleotide (the original RNA) and a

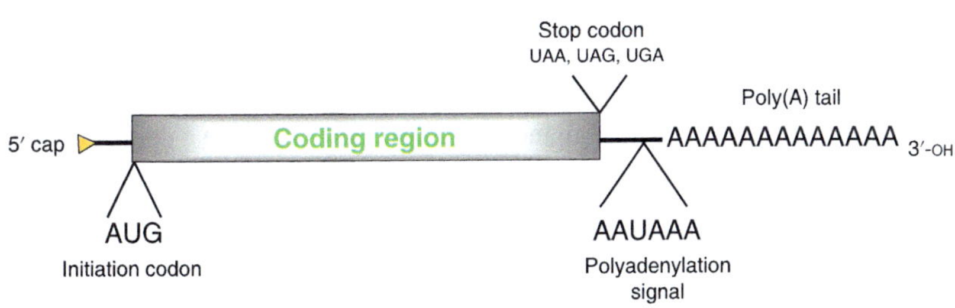

FIGURE 1.11 Structural features of a typical eukaryotic mRNA molecule.

7-methylguanosine (m^7G) nucleotide, the polarity of which effectively seals the 5′ end of the transcript. In the realm of RNA biogenesis, 5′ capping is the first example of a posttranslational modification associated with gene expression and is observed in virtually all eukaryotic mRNAs and is likewise a feature of most eRNA molecules.

The formation of the 5′ cap begins with the addition of the terminal guanosine nucleotide (G) to the 5′ end of the nascent transcript. This occurs in the nucleus and first requires the removal of the γ-phosphate from the 5′ end of the hnRNA by RNA 5′ triphosphatase (RTPase). Subsequently, the terminal "G" is ligated by the enzyme RNA guanylyl transferase (GTase) and results in the structure

5′ GpppRNNN.....3′

The new terminal guanosine nucleotide is joined 5′ → 5′ to what was the first transcribed nucleotide via what is best thought of as an inverted linkage. The resulting 5′ cap structure is then subjected to one or more methylation events (methyl group donor is S-adenosylmethionine; SAM), the first of which is directed toward the number 7 position of the terminal guanine, courtesy of the enzyme mRNA (guanine-N7)-methyltransferase (MTase) and represented as

m^7G(5′)ppp(5′)RNNN.....3′

Most eukaryotic mRNAs subsequently experience additional methylation directed toward the 2′ oxygen in the ribose of the second nucleotide, catalyzed by the enzyme 2′-O-methyl-transferase. Depending on the mRNA, as many as four methylation events can occur involving the 5′ guanine nucleotide, the sugar of the penultimate- and subpenultimate nucleotides, and may also include the formation of N^6-methyladenosine (if the second nucleotide contains adenine). The extent of methylation is a function of mRNA species, and higher organisms usually have more extensively methylated caps. The formation of various methylated RNA nucleotides is part of the epitranscriptome (described in Chapter 23: Transcriptomes and Bioinformatics); once thought to be a rare occurrence, these and other modified RNA nucleotides are widely distributed across the transcriptome. Of note, alternative noncanonical cap-like structures of unclear function have also been observed in small nuclear RNA species and in prokaryotic organisms, including AMP-capped RNA, 5′ methyl phosphate caps, 5′ trimethylguanosine caps, and 5′ NAD^+-capped RNA (Mattay, 2022; Wiedermannová et al., 2021).

The capping reaction associated with the 5′ terminus also has a role regulating transcription itself. It appears that the RNA polymerase slows down or altogether pauses when the nascent hnRNA is no more than 30 nts in length, facilitating the recruitment of the capping enzymes. This might be thought of as a quality control checkpoint, as transcription resumes only when the capping process has been completed.

The presence of the 5′ cap is required to support the initiation of translation in eukaryotes. Eukaryotic ribosomes distinguish mRNAs from non-mRNAs by the presence of the 5′ cap. Succinctly, rRNA and tRNA are not translated because they are not capped. When produced by in vitro transcription, RNAs must be subjected to an in vitro capping reaction if the transcripts are to be expected to support the synthesis of the encoded protein. The formation of the translation apparatus is initiated in part by cap-binding proteins, particularly eIF4E, followed by the assembly of the ribosomal subunits is mediated by initiation factors.

Capping also confers transcript stability by protecting against phosphatase attack and 5′→3′ exonucleolytic degradation. In contrast, prokaryotic mRNAs, which naturally lack a 5′ cap structure, are degraded exonucleolytically from the 5′ end even while translation is ongoing downstream. Mitochondrial and chloroplast mRNAs are not capped, while most animal viruses that replicate in eukaryotic cells manifest 5′ capped mRNAs, a noteworthy exception being poliovirus.

Finally, the 5′ cap has a role in nuclear egress. Export of mRNA into the cytoplasm occurs when the 5′ cap engages the heterodimeric nuclear cap-binding complex (CBC). Once through the nuclear pore, eIF4E replaces CBC; this substitution is needed to ensure efficient translation, as noted above.

5′ Untranslated region (leader sequence)

The first nucleotides immediately 3′ to the eukaryotic cap structure and upstream of the initiation codon constitute the 5′ untranslated region (5′ UTR), also known more casually as the (nontranslated) leader sequence. In eukaryotic cells, the typical 5′ UTR is 30–50 bases long, though 5′ UTRs as long as hundreds of bases long or as short as three bases have been observed. As its name implies, this variable-length region is not translated, though complex intramolecular base pairing with a role in translation regulation commonly forms in this region. According to the ribosome scanning model (Kozak, 1978; Kozak, 1989), the 40S ribosome subunit binds to the 5′ cap and then travels along the 5′ UTR, pausing only when reaching the first AUG codon, which usually (but not always) demarcates the boundary between the leader sequence and the coding region. The initial association between the 5′ cap and the 40S subunit is mediated by cap binding proteins while the eventual assembly of the functional ribosome is mediated by eukaryotic initiation factors (eIFs) and other mRNA binding proteins.

Coding region

The structural portion of the mRNA, that is, the coding region, begins with an initiation codon, most often the triplet AUG. Far less frequently, translation in eukaryotic cells may be initiated from non-AUG start codons (Hann et al., 1988; Florkiewcz and Sommer, 1989), including GUG, ACG, and CUG. Curiously, these noncanonical initiation codons differ from AUG by only one base. Prokaryotes, in contrast, seem to like both GUG and AUG for the purpose of translation initiation.

In order for the first AUG codon to function as the initiation codon, it must be in "good context," meaning that the surrounding nucleotides are just as important (Kozak, 1986; Kozak, 1987). The critical landmarks are adenine three bases upstream (−3), and guanine four bases downstream (+4) from the putative AUG initiation codon. These flanking structural elements are presumed to be indicators pointing to the initiation codon. For example, in higher animals the consensus sequence GCCGCCRCCAUGG is well known, and two slightly different consensus sequences are observed among monocots and dicots, respectively, in the plant kingdom (Joshi et al., 1997). An AUG codon found not to be in good context, including being within 10 nts of the 5′ cap, is ignored and the ribosomal 40S subunit will proceed downstream, scanning for a proper initiation codon.

An AUG codon at the beginning of the coding region results in the placement of the amino acid methionine at the amino terminus of eukaryotic cytoplasmic proteins, and the placement of formyl-methionine at the amino terminus of proteins in prokaryotic cells and in eukaryotic organelles. Each group of three nucleotides beyond the initiation codon in turn specifies the placement of another amino acid into the nascent polypeptide. Elongation continues in this manner until a nonsense (stop) codon (UAA, UAG, or UGA) is encountered which, in the absence of tRNA suppressor mutations, is the normal signal to halt translation. Thus, the 5′ end of the mRNA coding region corresponds to the amino terminus, and the 3′ end corresponds to the carboxy terminus of the primary polypeptide that results from translation. The region of the mRNA sandwiched between the initiation codon and the stop codon is also described as an ORF because this uninterrupted sequence of codons has the ability to encode the cognate polypeptide. Typical eukaryotic mRNA coding regions range from 200 to 1500 nts long, though extremely short and extremely long ORF coding regions have been identified.

3′ Untranslated region (trailer sequence)

Beyond the stop codon lies another nontranslated area is known as the 3′ UTR, also known as the trailer sequence. This sequence usually ranges from 50 to 150 nucleotides long, although trailer sequences several hundred bases long have been reported. Further, different mRNA molecules encoding the same polypeptide may differ only with respect to the length of their 3′ UTR, of which dihydrofolate reductase mRNA is one example (Kubo et al., 2006). Characteristically, AU-rich elements (AREs) are observed in the 3′ UTR and are major *cis*-regulatory elements related to mRNA stability, doing so in concert with ARE binding proteins. The slightly different classes of AREs are typically spatial organizational variations of AUUUA and related U-rich sequences.

Most 3′ UTR trailer regions also manifest one or several copies of the highly conserved AAUAAA hexamer within 10–30 nucleotides of the location where the synthesis of the poly(A) tail will begin; mutations within this hexamer or spatial rearrangements interfere profoundly with the formation of poly(A)$^+$ mRNA. The polyadenylation signal, while highly conserved, is not a universal signal. For example, the polyadenylation signal AUUAAA is observed in 5%–10% of eukaryotic mRNAs, and the sequence AAUAUA has likewise been observed in a minor form of pancreatic α amylase mRNA (Tosi et al., 1981). The presence of multiple polyadenylation signals is widespread among eukaryotes (Tian et al., 2005; Shen et al., 2008), a strategy by which multiple transcripts can be produced from a single genetic locus. To ensure the cessation of translation, however, there are often stop codons in all three reading frames in the 3′ UTR, presumably to control a phenomenon known as "stop codon readthrough."

Beyond regulating polyadenylation, the 3′ UTR attracts other proteins which influence transcript stability. In addition, microRNA (miRNA) molecules hybridize into miRNA response elements (MREs) which often reside in the 3′ UTR. In so doing, miRNAs are able to influence gene expression by downregulating translatability. In the realm of target gene silencing by RNA interference (RNAi), miRNAs are often designed to hybridize to this region of the mRNA (Chapter 12: miRNA and Other Noncoding RNAs).

Poly(A) tail

Most eukaryotic mRNA species are characterized by a 50–250 nucleotide tract of polyadenylic acid at the 3′ end of the molecule. This structure is known as the poly(A) tail, the precise length of which is a function of mRNA species, age, and translational status in the cell (reviewed by Stewart, 2019). The poly (A) tail is added enzymatically to precursor hnRNA by the nuclear enzyme polyadenylate polymerase (polynucleotide adenylyltransferase) or simply poly (A) polymerase, very soon after transcription is completed, in concert with numerous proteins. Polyadenylation, as the process is known, may occur even before splicing (i.e., intron removal) is complete. The poly(A) tail is not encoded by the corresponding gene locus: there is no poly(T) tract found on the template strand of the gene. The poly(A) tail is added efficiently in a processive, nontemplate-dependent manner, the overall length of which is partially controlled by the action of the nuclear poly(A) binding protein, *PABPN1*.

Polyadenylation does not begin at the site where transcription ends, but rather as many as 1000 bases upstream, and was first observed in adenovirus (Nevins and Darnell, 1978; Fraser et al., 1979). Splicing of the primary transcript occurs immediately 3′ to the first CA dinucleotide located within 30 nts downstream of the polyadenylation signal described in the previous section. This event, which is stringently regulated, produces a polyribonucleotide with a free 3′-OH to which AMP nucleotides are sequentially attached. To be clear, the poly(A) signal is completely different from the poly(A) site. A GU-rich sequence, another conserved motif, is located downstream from the polyadenylation start site in animals (upstream from the polyadenylation site in plants); in cooperation with other sequences upstream of the polyadenylation signal it may actually enhance the efficiency of the polyadenylation process. It is the precise spatial relationship of these sequences and the degree of sequence conservation that direct the cleavage of an hnRNA molecule downstream from the polyadenylation signal and subsequent poly(A) polymerization; these events are mediated by small nuclear ribonucleoproteins (snRNPs) (Birnstiel et al., 1985; Berget and Robberson, 1986; Black and Steitz, 1986; Hashimoto and Steitz, 1986). These landmarks, or subtle variants thereof, are widely observable among the eukaryotes, with the exception of yeast, in which polyadenylation enhancer elements are uniquely located upstream of the poly(A) site. Alternative 3′ end processing is also a distinct possibility when an hnRNA molecule contains multiple polyadenylation signals which may involve as many as 70% of all human genes (Derti et al., 2012). Variations in the splicing and polyadenylation patterns at the 3′end of transcripts from a single gene locus may well result in differential protein synthesis in a tissue-specific manner.

It is well documented that mRNA polyadenylation correlates well with the capacity of the transcript for nuclear export. This process is best thought of as another quality control measure that prevents improperly formed mRNAs from traveling to the cytoplasm where, if translation were to occur, a dysfunctional peptide might result. When polyadenylation is inhibited, for example with cordycepin (3′-deoxyadenosine), then normally adenylated mRNAs fail to emerge from the hnRNA pool (Zeevi et al., 1982), all the while there is no other apparent effect on transcription in general or on the synthesis of hnRNA. As noted above, nucleocytoplasmic transport depends on other features of the mRNA as well.

The mRNA fraction that exhibits the polyadenylation feature is collectively described as poly(A)$^+$ mRNA, and nearly all ($>97\%$) types of eukaryotic mRNAs are believed to exhibit the poly(A) structure. In contrast to the great variety of mRNA species that are polyadenylated, fully one-third of the total *mass* of cellular mRNA lacks a 3′ poly(A) tail, of which the overwhelming component is the nonadenylated histone mRNAs, which encode major chromosomal structural proteins. The highly conserved nature of the polyadenylation signal and the abundance of adenylated eukaryotic transcripts suggest a central regulatory role in gene expression. The poly(A) tail is important for mRNA stability as well as translational efficiency. For example, as with the 5′ cap, the specialized poly(A) feature at the 3′ end protects eukaryotic mRNAs from 3′→5′ exonuclease digestion, or at least stalls it. In the cytoplasm, polyadenylate-binding protein (PABP) appears to bind to every 15–20 bases within the poly(A) tail the length, correlating well with the stability of the mRNA. An mRNA may, or may not, be translated, depending on local conditions.

Organellar messenger ribonucleic acids

It has been recognized for some time that both mitochondria and chloroplasts have their own chromosome(s), circular in form, which are inherited independently of nuclear chromatin and in a non-Mendelian fashion. The genomes of these organelles are referred to as mtDNA and ctDNA, respectively. The genes found in mtDNA and ctDNA encode proteins that are uniquely localized in these organelles, though mitochondria and chloroplasts each import proteins encoded by nuclear genes that are needed to support normal organellar physiology.

Although the mechanics of transcription and translation are highly conserved, certain structural differences distinguish mRNA species confined to organelles. For example, neither mitochondrial nor chloroplast mRNAs exhibit a 5′ cap structure. Most mitochondrial transcripts are constitutively polyadenylated at the 3′ terminus, while most chloroplast mRNAs are not polyadenylated. In the case of both organelles, the poly(A) tail, when present, is usually quite short by comparison with adenylated cytoplasmic transcripts. Curiously, the effect of polyadenylation on transcripts in the chloroplast is destabilizing (Lisitsky et al., 1997; Gagliardi and Leaver, 1999; reviewed by Hayes et al., 1999; Schuster et al., 1999), rather than stabilizing as seen in cytoplasmic mRNAs. Mitochondrial mRNAs frequently utilize the noncanonical AUA and AUU translation start codons, and these are commonly observed very close to the 5′ terminus. Interestingly, the mRNAs found in the mitochondria of yeast are more closely related to typical cytoplasmic mRNAs, structurally speaking, than to mitochondrial mRNAs observed in other cell types.

Messenger ribonucleic acid—stability and turnover

A fundamental regulator of gene expression in all cell types and within subcellular compartments is the stability of translatable transcripts. In general, mRNAs do not have long half-lives (Table 1.7), presumably to prevent the over-production of a normal protein which could, in turn, disrupt homeostasis or give rise to a disease state. As a cell must be flexible enough to respond quickly to changes in its environment or in response to both intracellular and extracellular cues, mRNAs with relatively short half-lives favor a

TABLE 1.7 Comparison of transcription and translation in prokaryotic and eukaryotic cells.

		Prokaryotes	Eukaryotes
Transcription	Gene organization	Mostly polycistronic (operon model); some are monocistronic	Nearly all are monocistronic; some polycistronic and bicistronic mRNAs observed in plants
	Location	Single cellular compartment	Nucleus
	mRNA half-life	2 min; range 1–10 min	10 h; range 1–24 h
	Ancillary factors	Sigma factor	Transcription factors
	Transcription rate	40 nts/s	40 nts/s
	Master enzyme	RNA polymerase	RNA polymerases (various)
	Posttranscriptional modifications	Limited-to-none	Extensive, including 5′ capping, intron removal, and polyadenylation
	mRNA average size	0.9–1.5kb	1.9–2.2kb
Translation	Intact ribosomes	70S	80S
	Ribosomal subunits	50S, 30S	60S, 40S
	rRNA species	23S, 16S, 5S	28S, 18S, 5.8S, 5S
	Location	Single cellular compartment	Cytoplasm
	Ancillary factors	Needed for initiation, elongation, and termination	Needed for initiation, elongation, and termination
	Translation rate	15 amino acids/s	5 amino acids/s
	Master enzyme	Peptidyl transferase	Peptidyl transferase
	Posttranslational modifications	Limited-to-none	Extensive

This table does not include specific transcription and translation information associated with mitochondrial or chloroplast genes.

rapid cellular response. At the same time, mRNAs must remain stable for a sufficient amount of time to be recognized and engaged by the translation apparatus. As noted above, the eukaryotic 5′ cap not only identifies a transcript specifically as an mRNA but also stabilizes the 5′ end against nuclease assault. At the other end of the molecule, it has long been known that the length of the poly(A) tail plays a role in mRNA stability, as a shortening of the poly(A) tail results in destabilization of cytoplasmic transcripts and is part of the natural decay process. In fact, the length of the poly(A) tail as a marker for degradation may be thought of as the nucleic acid counterpart of ubiquitination as a marker for the destruction of proteins.

If large quantities of a protein are to be produced in a cell, one may expect that the corresponding gene will be transcribed with a greater frequency compared to other genes with, for example, housekeeping functions. There is, however, another strategy. Translational efficiency, which is generally taken to mean the number of peptides molecules synthesized on each mRNA template, is more likely to account for protein levels in the cell (Marguerat et al., 2012; Vogel and Marcotte, 2012; Ly et al., 2014). The ability of the cell to perform efficiently in this manner is a direct function of the balance that exists between mRNA synthesis and turnover. In order to further stabilize a typical mRNA, these transcripts commonly assume a circular configuration (Fig. 1.12) in order to further protect their 5′ and 3′ ends. In so doing, enhanced transcriptional efficiency is favored.

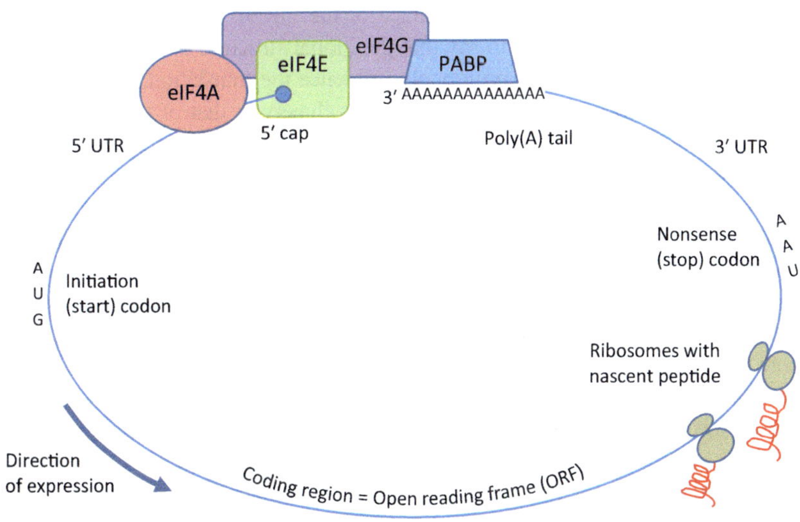

FIGURE 1.12 Circularized mRNA configuration enhances stability and translation. Transient circularization is mediated through a bridging function involving translation elongation factor eIF4G, as well as translation-associated factors eIF4A and eIF4E, the latter of which also acts as a 5′ cap-binding protein; poly(A) binding protein (PABP) is responsible for the involvement of the poly(A) in "closing the loop." This structure is possible as long as the 5′ and 3′ end remain intact.

There are various distinct degradative pathways that directly impact levels of mRNA in eukaryotic cells. In each case, the degradative strategy involves the recognition of an intrinsic structural element associated with the products of RNA polymerase II. In one example, the occurrence of a point mutation that changes an amino acid-specific codon into a stop or nonsense codon upstream of a naturally occurring stop codon induces a surveillance-type system which usually leads to mRNA destruction (Le Hir et al., 2000; Lykke-Andersen et al., 2000; Frischmeyer et al., 2002). Nonsense-mediate decay (NMD), as this mechanism is widely known, is a well-characterized quality control mechanism believed to be triggered in all eukaryotic cells upon the formation of these so-called premature termination codons and which leads to elimination of these mRNAs. It has been hypothesized that, by ensuring that only open reading frames (ORFs; uninterrupted coding regions) are available to the translation apparatus, NMD may be the reason for the relatively low observed incidence of recessive disorders.

Another degradation strategy involves the presence of multiple copies of the pentanucleotide motif AUUUA in the 3′ UTR, known as an ARE or AU-rich element. AREs which mediate destabilization in a step-wise manner. In the first step, deadenylation, that is, removal of the poly(A) tail, is promoted. In the second step, the ensuing degradation which ensues involves exonucleolytic molecular dismantling at both the 5′ and 3′ ends, facilitated by the concomitant removal of the 5′ cap. With the poly(A) tail and 5′ cap removed, the natural protective characteristic closed-loop shape of the mRNA is lost, thereby accelerating degradation. It is safe to say that the global structure of mRNA, particular the 5′ end and the 3′ end, profoundly influence the translation.

Perhaps the most intriguing degradation machinery is that of the exosome, a multiprotein complex found in eukaryotes (reviewed by Wasmuth and Lima, 2012); structurally similar, but not identical, machinery has been identified in the Archaea. Eukaryotic exosome substrates include mRNA, rRNA, and several types of noncoding transcripts which are dismantled $3' \rightarrow 5'$ both *exo*- and endoribonucleolytically. It seems that the role of exomes is to eliminate unstable transcripts as well as improperly processed transcripts that manage to evade the nuclear RNA surveillance system (Schneider et al., 2012; Szczepińska et al., 2015). This includes the degradation of enhancer RNA (eRNA) and presumably other molecules that could be classified generically as transcriptional noise. In higher eukaryotes, exosomes are distributed throughout the cell and are often prevalent in the nucleolar region of the nucleus. As has been a central theme is this chapter, exosome functionally is another extension of the quality control arm of the cell.

Bicistronic messenger ribonucleic acids

The one mRNA, one polypeptide relationship is widespread among eukaryotes, although bicistronic mRNAs have been identified in certain organisms. A bicistronic mRNA is a single transcript capable of directing the synthesis of two different proteins. One might think of biscistronic mRNAs as the eukaryotic answer to the polycistronic mRNAs that are nearly universally observed among prokaryotes. Taking this a step further, functional tricistronic mRNAs, encoding three different polypeptides, are in use in certain in vitro applications, most notably when cloning into lentiviral expression vectors. Due to the peculiarities associated with translation in eukaryotes, the first (upstream) encoded protein is synthesized in the 5′-cap dependent manner usually associated with translation of monocistronic mRNAs, while initiation of the synthesis of the second (downstream) polypeptide is under the control by an internal ribosome entry site (IRES; also known as ribosome binding sites, RBSs) which allows ribosome assembly in a noncap-dependent manner. This translation strategy is widespread among eukaryotic viruses and, while still considered a rarity in higher animal cells, there are reports of bicistronic mRNAs in plants, including the tomato tom*Pro1* locus and in *Arabidopsis*. It is also possible for a single-reading-frame mRNA to produce two or more polypeptides by cleavage of large precursor protein, as is the case with zymogens. These observations have lead investigators to rethink the entire process of the regulation of gene expression and to analyze gene expression data circumspectly.

Prokaryotic messenger ribonucleic acids

The goals of transcription and translation in the prokaryotic cell are similar to these processes in eukaryotes, namely the manifestation of gene expression. There are, however, structural differences that are readily observable when comparing transcripts, and the transcription and translation machinery, from these two cell types. As described above, prokaryotic RNAs are all produced by a single type of RNA polymerase. The resulting mRNA subpopulation is characterized by molecules that are characteristically polycistronic, encoding more than one polypeptide. These transcripts often experience polysome-associated translation and simultaneous 5′ degradation even prior to the completion of transcription at the 3′ end. While endoribonucleases are at work destroying RNA from within the molecule, a process initiated by RNase E, exoribonucleases are also at work disassembling the RNase E cleavage products one base at a time from the free 3′ end of each fragment (reviewed by Mackie, 2013).

Remarkably, polyadenylation in prokaryotes, which was discovered in the early 1960s (August et al., 1962) was largely ignored for the next three decades. It is now known that this process plays a role in the lifecycle of RNA in bacteria and "tailing" does not appear to be limited to mRNAs; tailed rRNA and tRNA species have been observed. It appears that the prokaryotic counterparts of poly(A) polymerase add adenine nucleotides to some RNAs, though to a much lesser extent than is observed in the eukaryotic nucleus. Not only are the tails shorter (10–20 nts) and more dynamic in terms of their stability, it appears that many tails consist predominantly, but not exclusively, of tracts of adenine nucleotides; the bases cytosine and uracil are occasionally observed interspersed (reviewed by Mohanty and Kushner, 2022). The criteria by which one nucleotide is selected over another, giving rise to heteropolymeric tails, is poorly understood. Moreover, the quantity of mRNA that is poly(A) tailed in bacterial varies widely from bacterial species to the next and many of the difficulties associated with the study of polyadenylated RNA is due to the fact that the tails often consist of twenty or fewer nucleotides. While the role of polyadenylation in bacteria remains something of a mystery, the consensus is that it has a fundamental, elusive role in overall RNA metabolism and most likely orchestrates transcript obliteration rather than promoting stability (O'Hara et al., 1995).

Another prokaryotic RNA phenomenon is known as riboswitch. Riboswitches are ligand-binding transcripts that act as modulators of gene expression (Nahvi et al., 2002; Winkler et al., 2002a,b). Formed by intramolecular base-pairing, a riboswitch is a double-stranded region of an RNA molecule that is able to bind a small metabolite or ion. When this occurs, the RNA experiences a conformational change that influences subsequent steps in a biochemical pathway. The "sensing" of cellular metabolites in this manner directly regulates the activity of genes whose products are often associated with the synthesis or transport of the metabolite itself. The regulatory influence of riboswitches is exerted on both transcription and translation, and may profoundly impact transcript stability. Although most riboswitches are prokaryotic, at least one type of riboswitch has been found in plants, fungi, algae, and in the archaea (reviewed by Serganov and Nudler, 2013).

Messenger ribonucleic acid sequence and structure affect translation

The translation is the culmination of variegated biochemical events acting concertedly to sustain the production of proteins. In the cytoplasm, protein synthesis is mediated either by membrane-bound or free ribosomes; in the case of the latter, the proteins are most often directed to another subcellular compartment, for example, mitochondria, chloroplasts, or the nucleus, rather than being retained in the cytoplasm. The initiation, elongation, and termination stages of translation are sequence-dependent (e.g., the initiation and nonsense codons). Translation ends when a stop codon enters the aminoacyl (A) site of the ribosome, thereby triggering the termination mechanism. Mutations therein can have strong negative consequences on the final manifestation of gene expression.

In the case of bicistronic or polycistronic mRNAs, an IRES can initiate translation by promoting ribosome binding within the body of the transcript. An IRES can initiate translation of the second or subsequently encoded polypeptide in a 5′ cap-independent manner, that is, without requiring the ribosome to travel from the 5′ end of the transcript (reinitiation; Kozak, 1984).

This strategy reduces translation initiation complexity. Though there is very limited sequence homology among IRESs, most are characterized by an AUG initiation codon at the boundary of their upstream-most region to which the small ribosomal subunit binds to form the translation apparatus. It is important to note, however, that while key features of the transcript sequence are intimately linked to functionality, the three-dimensional shape(s) that a transcript can form also influence its function profoundly. That higher-order folding of RNA imparts function comes as no surprise; the cloverleaf form of tRNA molecules is an example *par excellence*.

RNA molecules exhibit a variety of single- and double-stranded domains which together give rise to the formation of tertiary or three-dimensional architecture. As with proteins, a great deal of what an RNA molecule does, including binding of proteins and recognition of other molecules, splicing, and other catalytic activities, is related to its shape. Despite the comparatively enhanced stability of dsRNA, secondary and tertiary domains are transient and subject to dynamic remodeling ranging from subtle alterations to dramatic refolding. When any portion of an RNA molecule does acquire a secondary structure, the resulting double-stranded region is an α-helix (favored by the 2′-OH); in contrast, the β-helix conformation (B-form) is normally found in well-hydrated dsDNA. For this reason, dsRNA domains are sometimes referred to as A-RNA; other types of secondary structures are also observed (refer back to Fig. 1.4). The secondary structure(s) that an individual RNA forms strongly influences the overall protein content of the resulting mRNP and, concomitantly, impacts its ability to move through the nuclear pores and into the cytoplasm in the first place. It should be noted, however, that large dsRNA molecules made from two complementary transcripts are very unusual in higher eukaryotes and are poorly tolerated.

The formation of secondary structure(s) involving the 5′ leader sequence of the mRNA can limit or altogether suppress translation. Moderately stable hairpins upstream of an AUG codon and close to the 5′ end of a transcript are known to diminish translation efficiency, often dramatically, presumably by impacting the ability of ribosomes to load onto the mRNA. This mode of gene regulation has been observed in both plant and animal models, presumably to prevent the over-expression of a specific protein (reviewed by Merrick and Pavitt, 2018). In order to overcome this translation impediment, eIF4E localizes to the 5′ UTR region and, in so doing, recruits eIF4A helicase to relax the secondary structures (Feoktistova et al., 2013).

Not unexpectedly, mutation and deletion experiments that reduce hairpin stability usually result in an increase in the amount of the encoded protein. In contrast, a hairpin up to 15 nts downstream from an AUG codon can actually enhance translation initiation, perhaps by causing the ribosome to pause directly over the start codon. This is another quality checkpoint in that it optimizes translation. As noted above, eukaryotic mRNAs acquire a circular conformation by bringing the 3′ end of the transcript close to the 5′ cap and the 5′-UTR, thereby influencing overall translation efficiency. This translation regulatory mechanism appears to be widespread in animals, plants, and plant viruses (see Biziaev et al., 2022).

A final note regarding the role of secondary structure. The cessation of transcription in prokaryotes is often mediated by the formation of a hairpin-like structure in the RNA which is being synthesized; an ensuing pause of the RNA polymerase results in the breaking of the hydrogen bonds between the nascent RNA and the DNA template, and the disengagement of the polymerase from the gene.

Alternative splicing of messenger ribonucleic acid from a single genetic locus

Alternative splicing refers to variation in the way that transcripts from a single gene locus are processed posttranscriptionally. This phenomenon is well-characterized in eukaryotic cells. Understanding that higher plant and animal cells can differentially process a particular species of pre-mRNA molecules goes a long way toward explaining how a cell with approximately 20,000 different genes can encode greater numbers of transcripts which, in turn, may be able to encode several hundred thousand (or more) functional proteins. By joining together various combinations of exons and even changing the functionality of introns, a remarkable diversity of proteins may result from a seeming economy of genomic sequences. The most direct method for the detection of subtle, and sometimes not so subtle, changes in transcript identity is through RNA-seq which provides a global snapshot of the structure of the transcriptome

The more common strategies used by cells to accomplish alternative splicing are known as exon skipping, intron retention, the manifestation of cryptic introns, and the use of alternative TSSs, to name a few examples. These phenomena can occur within a single cell in response to the environmental stimuli or may occur in a tissue-specific manner in order to support the physiology of the organism. Alternative processing of transcripts may also involve modulating the addition of the poly(A) tail associated with the 3′ end of a transcript through the use of alternative polyadenylation sites. Such poly(A) variants may influence the stability of the transcript in the cell as well as the precise nature of alternative splicing that ensues (Li et al., 2017).

Upon completion of the human genome project, the apparent paucity of genes came as a great surprise, given the complexity of the proteome. In the context of alternative transcript processing and posttranslational modification of proteins, the relatively small number of genes in most species is no longer inconsistent with the vast biochemical complexity that they produce in the cell.

Levels of gene regulation

Controlling gene expression requires modulating either transcription or translation directly, or the postsynthesis products of these biochemical processes. The number of potential points of gene regulation in living cells is virtually infinite. In the *broadest* sense, these myriad potential regulatory points may be classified under one of four main headings (Fig. 1.13):

1. Regulation at the transcriptional level,
2. Regulation at the posttranscriptional level,
3. Regulation at the translational level, or
4. Regulation at the posttranslational level.

RNA molecules are produced by transcription. This is the primary regulation level in the cell: either a gene is transcribed or it is not. In eukaryotes,

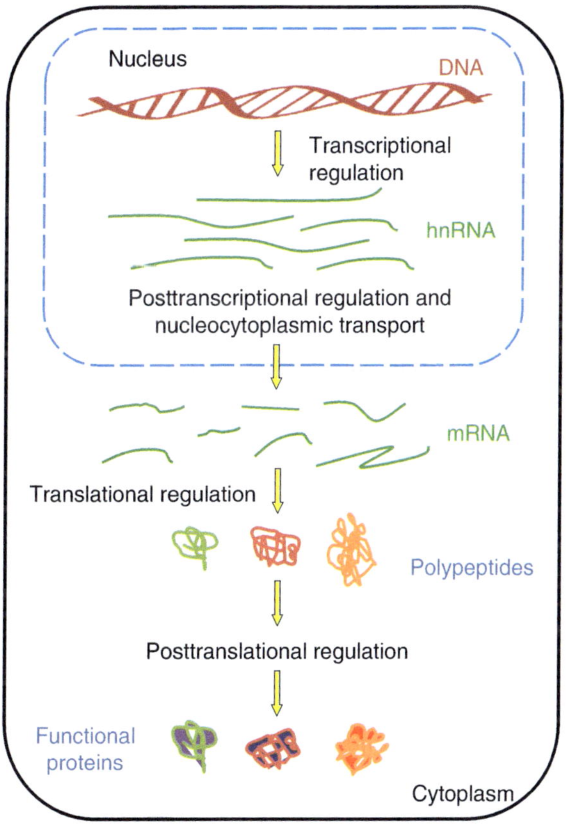

FIGURE 1.13 Gene expression is subject to regulation at key control points within the cell, beginning with the information encoded in the genome of the cell, through mRNA biogenesis and resulting in final manifestation as a functional protein. See text for details.

subsequent processing of the primary transcript is a continuation of what is known as mRNA biogenesis. Mature RNA molecules, which have been appropriately modified, and not the intron sequences that have been removed by excision, may be exported from the nucleus into the cytoplasm via a discriminating nucleocytoplasmic transport mechanism, involving movement to and then through nuclear pore complexes. Once in the cytoplasm, mature messenger RNA may or may not become associated with the protein translational apparatus. While some of the mRNA is engaged in polysome complexes, the nonpolysome-engaged mRNA exists as an mRNP particle; the now obsolete term "informosome" (Preobrazhensky and Spirin, 1978) was once used describe these mRNA-protein complexes. A new type of RNP structure, the L-body, is a membraneless storage granule where, among other reasons, mRNA can be localized presumably to support functions in cells where polarity or spatial organization is critical, for example, in oocytes (Neil et al., 2021). As the mRNA is biologically and chemically competent, sequestering the molecules in this way is an example of a posttranscriptional regulatory strategy.

When a eukaryotic mRNA molecule does undergo translation, then the primary product of translation, a polypeptide, is usually subjected to extensive posttranslational modifications, examples of which include removal of the peptide signal sequence, proteolytic cleavage, acylation, methylation, prenylation, carboxylation, glycosylation, phosphorylation, acetylation, and hydroxylation. Addition or removal of one or more of these functional groups greatly contributes to the observed diversity and dynamic nature of the proteome.

Any variable that influences the efficiency of transcription or prevents transcription from occurring altogether is said to act at the transcriptional level (transcriptional regulation). This may include epigenetic events, such as direct modification of chromatin by DNA methylation or by covalent modification of histone proteins (e.g., methylation of lysine 9 in histone 3) that facilitates RNA polymerase accessibility to gene promoter elements resident within the nucleosome by changing the actual shape of the chromatin. Following transcription, any event that influences the processing of hnRNA, its stability in the nucleus, nucleocytoplasmic transport, or the very stability of mRNA itself in the cytoplasm is said to act posttranscriptionally (posttranscriptional regulation) and these events are indeed multifarious. For example, while as much as one-half of the mRNA in mammalian cells propagated in vitro may have a half-life of several hours, many mRNA species have half-lives of less than one hour, the brevity of which is a natural modulator of gene expression. Half-lives are also subject to change, often in response to environmental stimuli. Stating that gene regulation occurs at a translational or posttranslational level usually implies that a mature mRNA in the cytoplasm is somehow experiencing translational interference at the level of protein synthesis or beyond.

The biochemical profile of a population of cells or a tissue sample is often qualitatively and quantitatively defined as a function of mRNA complexity at the time of cellular disruption. For example, an investigator may wish to determine whether the observed modulation of gene expression in a model system is regulated transcriptionally or by a posttranscriptional event(s). In such a case, the method of cellular disruption and subsequent RNA purification and analysis must permit examination of the nuclear abundance of salient gene transcripts independently of the cytoplasmic abundance of the same. In addition, transcription rate studies require the isolation of intact nuclei that are able to support elongation of, and label incorporation into, initiated transcripts (see nuclear run-on assay, Chapter 19: Analysis of Nuclear RNA). Therefore, the conditions under which the RNA is purified from biological sources must support the chemical stability of the RNA and ensure RNase inactivation throughout the procedure. Scrupulous, if not compulsive, attention to these details is often the deciding factor in the success or failure of an RNA isolation procedure and subsequent analysis.

References

Andersson, R., et al., 2014. An atlas of active enhancers across human cell types and tissues. Nature 507, 455–461.

August, J., Ortiz, P.J., Hurwitz, J., 1962. Ribonucleic acid-dependent ribonucleotide incorporation. I. Purification and properties of the enzyme. J. Biol. Chem. 237, 3786–3793.

Baltimore, D., 1970. Viral RNA-dependent DNA polymerase. Nature 226, 1209–1211.

Benne, R., van den Berg, J., Brakenhoff, J.P., Sloof, P., Van Boom, J.H., Tromp, M.C., 1986. Major transcript of the coxII gene from trypanosome mitochondria contains four nucleotides that are not encoded in the DNA. Cell 46, 819–826.

Berget, S.M., Robberson, B.L., 1986. U1, U2, and U4/U6 small nuclear ribonucleoproteins are required for in vitro spicing but not for polyadenylation. Cell 46, 691–696.

Birnstiel, M.L., Busslinger, M., Strub, K., 1985. Transcription termination and 3' end processing: the end is in site!. Cell 41, 349–359.

Biziaev, N.S., Egorova, T.V., Alkalaeva, E.Z., 2022. Dynamics of eukaryotic mRNA structure during translation. Mol. Biol. 56, 382–394. Available from: https://doi.org/10.1134/S0026893322030037.

Black, D.L., Steitz, J.A., 1986. Pre-mRNA splicing in vitro requires intact U4/U6 small nuclear ribonucleoprotein. Cell 46, 697–704.

Brenner, S., Jacob, F., Meselson, M., 1961. An unstable intermediate carrying information from genes to ribosomes for protein synthesis. Nature 190, 576–581.

Cajal, S., 1903. Un sencillo metodo de coloracion selective del reticulo protoplasmico y sus efectos en los diversos organos nerviosos de vertebrados e invertebrados. Trab. Lab. Invest. Biol. 2, 129–221.

Cech, T.R., Zaug, A.J., Grabowski, P.J., 1981. In vitro splicing of the ribosomal RNA precursor of tetrahymena: involvement of a guanosine nucleotide in the excision of the intervening sequence. Cell 27, 487–496.

Derti, A., Garrett-Engele, P., Macisaac, K.D., Stevens, R.C., Sriram, S., Chen, R., et al., 2012. A quantitative atlas of polyadenylation in five mammals. Genome Res. 22, 1173–1183. Available from: https://doi.org/10.1101/gr.132563.111.

Djebali, S., et al., 2012. Landscape of transcription in human cells. Nature 489, 101–108.

Eisenberg, E., Levanon, E.Y., 2018. A-to-I RNA editing-immune protector and transcriptome diversifier. Nat. Rev. Genet. 19, 473–490. Available from: https://doi.org/10.1038/s41576-018-0006-1.

Farrell, R.E., Bassett, C.L., 2007. Multiple transcript initiation as a mechanism for regulating gene expression. In: Bassett, C.L. (Ed.), Regulation of Gene Expression in Plants. Springer, New York, pp. 44–50.

Feoktistova, K., Tuvshintogs, E., Do, A., Fraser, C.S., 2013. Human eIF4E promotes mRNA restructuring by stimulating eIF4A helicase activity. Proc. Natl. Acad. Sci. 110, 13339–13344.

Field, A., Adelman, K., 2020. Evaluating enhancer function and transcription. Ann. Rev. Biochem. 89, 213–234. Available from: https://doi.org/10.1146/annurev-biochem-011420-095916.

Flavell, A.J., Cowie, A., Legon, S., Kamen, R., 1979. Multiple 5' terminal cap structures in late polyoma virus RNA. Cell 16, 357–371.

Florkiewcz, R.Z., Sommer, A., 1989. Human basic fibroblast growth factor gene encodes four polypeptides: three initiate translation from non-AUG codons. Proc. Natl. Acad. Sci. 86, 3978–3981.

Fraser, N.W., Nevins, J.R., Ziff, E., Darnell, J.E., 1979. The major late adenovirus type-2 transcription unit: termination is downstream from the last poly(A) site. J. Mol. Biol. 129, 643–656.

Frischmeyer, P.A., van Hoof, A., O'Donnell, K., Guerrerio, A.L., Parker, R., Dietz, H.C., 2002. An mRNA surveillance mechanism that eliminates transcripts lacking termination codons. Science 295, 2258–2261.

Gagliardi, D., Leaver, C.J., 1999. Polyadenylation accelerates the degradation of the mitochondrial mRNA associated with cytoplasmic male sterility in sunflower. EMBO J. 18, 3757–3766.

Ghosh, P.K., Reddy, V.B., Swinscoe, J., Lebowitz, P., Weissman, S.M., 1978. Heterogeneity and 5' terminal structures of the late RNAs of simian virus 40. J. Mol. Biol. 126, 813–846.

Gros, F., Hiatt, H., Gilbert, W., Kurland, C.G., Risebrough, R.W., Watson, J.D., 1961. Unstable ribonucleic acid revealed by pulse labelling of *Escherichia coli*. Nature 190, 581–585.

Grünwald, D., Singer, R.H., Rout, M., 2011. Nuclear export dynamics of RNA-protein complexes. Nature 475, 333–341.

Guerrier-Takada, C., Altman, S., 1984. Catalytic activity of an RNA molecule prepared by transcription in vitro. Science 223, 285–286.

Hall, B.D., Spiegelman, S., 1961. Sequence complementarity of T2 DNA and T2-specific RNA. Proc. Natl. Acad. Sci. 47, 137–146.

Hann, S.R., King, M.W., Bentley, D.L., Anderson, W.C., Eisenman, R.N., 1988. A non-AUG translational initiation in c-myc exon 1 generates an N-terminally distinct protein whose synthesis is disrupted in Burkitt's lymphomas. Cell 52, 185–195.

Hashimoto, C., Steitz, J.A., 1986. A small nuclear ribonucleoprotein associates with AAUAAA polyadenylation signal in vitro. Cell 45, 581–591.

Hayes, R., Kudla, J., Gruissem, W., 1999. Degradation of chloroplast mRNA: the role of polyadenylation. Trends Biochem. Sci. 24, 199–202.

Henras, A.K., Plisson-Chatang, C., O'Donohue, M., Chakraborty, A., Gleizes, P., 2015. An overview of pre-ribosomal RNA processing in eukaryotes. WIREs RNA 6, 225–242.

Herr, A.J., Jensen, M.B., Dalmay, T., Baulcombe, D.C., 2005. RNA polymerase IV directs silencing of endogenous DNA. Science 308, 118–120.

Jacob, F., Monod, J., 1961. Genetic regulatory mechanisms in the synthesis of proteins. J. Molec. Biol. 3, 318–356.

Joshi, C.P., Zhou, H., Huang, H., Chiang, V.L., 1997. Context sequences of translation initiation codon in plants. Plant Mol. Biol. 35, 993–1001.

Kanno, T., Huettel, B., Mette, M.F., Aufsatz, W., Jaligot, E., Daxinger, L., et al., 2005. Atypical RNA polymerase subunits required for RNA-directed DNA methylation. Nat. Genet. 37, 761–765.

Kozak, M., 1978. How do eukaryotic ribosomes select initiation regions in messenger RNA? Cell 15, 1109–1123.

Kozak, M., 1984. Selection of initiation sites by eukaryotic ribosomes: effect of inserting AUG striplets upstream from the coding sequence for preproinsulin. Nucleic Acids Res. 12, 3873–3893.

Kozak, M., 1986. Point mutations define a sequence flanking the AUG initiator codon that modulates translation in eukaryotes. Cell 44, 283–292.

Kozak, M., 1987. At least six nucleotides preceding the AUG initiator codon enhance translation in mammalian cells. J. Mol. Biol. 196, 947–950.

Kozak, M., 1989. The scanning model for translation: an update. J. Cell Biol. 108, 229–241.

Kruger, K., Grabowski, P.J., Zaug, A.J., Sands, J., Gottschling, D.E., Cech, T.R., 1982. Self-splicing RNA: autoexcision and autocyclization of the ribosomal RNA intervening sequence of tetrahymena. Cell 31, 147–157.

Kubo, T., Wada, T., Yamaguchim, Y., Shimizu, A., Handa, H., 2006. Knock-down of 25 kDA subunit of cleavage factor IM in HeLa cells alters alternative polyadenylation within 3'-UTRs. Nucleic Acids Res. 34, 6264–6271.

Le Hir, H., Moore, M.J., Maquat, L.E., 2000. Pre-mRNA splicing alters mRNP composition: evidence for stable association of proteins at exon-exon junctions. Genes Dev. 14, 1098–1108.

Lewin, B., 1997. Genes VI. Oxford Universiity Press.

Li, Q.Q., Liu, Z., Lu, W., Liu, M., 2017. Interplay between alternative splicing and alternative polyadenylation defines the expression outcome of the plant unique OXIDATIVE TOLERANT-6 gene. Sci. Rep. 7, 2052. Available from: https://doi.org/10.1038/s41598-017-02215-z.

Lisitsky, I., Klaff, P., Schuster, G., 1997. Blocking polyadenylation of mRNA in the chloroplast inhibits its degradation. Plant J. 12, 1173–1178.

Lu, Z., Gong, J., Zhang, Q.C., 2018. PARIS: psoralen analysis of RNA interactions and structures with high throughput and resolution. In: Gaspar, I. (Ed.), RNA Detection. Methods in Molecular Biology, 1649. Humana Press, New York. Available from: https://doi.org/10.1007/978-1-4939-7213-5_4.

Ly, T., Ahmad, Y., Shlien, A., Soroka, D., Mills, A., Emanuele, M., et al., 2014. A proteomic chronology of gene expression through the cell cycle in human myeloid leukemia cells. Elife 3, e01630.

Lykke-Andersen, J., Shu, M.D., Steitz, J.A., 2000. Human Upf proteins target mNRA for nonsense-mediated decay when bound downstream of a termination codon. Cell 103, 1121–1131.

Mackie, G.A., 2013. RNase E: at the interface of bacterial RNA processing and decay. Nat. Rev. Microbiol. 11 (1), 45–57.

Marguerat, S., Schmidt, A., Codlin, S., Chen, W., Aebersold, R., Bähler, J., 2012. Quantitative analysis of fission yeast transcriptomes and proteomes in proliferating and quiescent cells. Cell 151, 671–683.

Mattay, J., 2022. Noncanonical metabolite RNA caps: classification, quantification, (de)capping, and function. WIRES RNA e1730. Available from: https://doi.org/10.1002/wrna.1730.

Merrick, W.C., Pavitt, G.D., 2018. Protein synthesis initiation in eukaryotic cells. Cold Spring Harb. Perspect. Biol. 10 (12), a033092. Available from: https://doi.org/10.1101/cshperspect.a033092.

Mohanty, B.K., Kushner, S.R., 2022. Regulation of mRNA decay by *E. coli*. Crit. Rev. Biochem. Mol. Biol. 57 (1), 48–72. Available from: https://doi.org/10.1080/10409238.2021.1968784.

Morais, P., Adachi, H., Yu, Y.-T., 2021. Splicesomal snRNA epitranscriptomics. Front. Genet. 12, 652129. Available from: https://doi.org/10.3389/fgene.2021.652129.

Nahvi, A., Sudarsan, N., Ebert, M.S., Zou, X., Brown, K.L., Breaker, R.R., 2002. Genetic control by metabolite binding mRNA. Chem. Biol. 9, 1043–1049.

Neil, C.R., Jeschonek, S.P., Cabral, S.E., O'Connell, L.C., Powrie, E.A., Otis, J.P., et al., 2021. L-bodies are RNA-protein condensates driving RNA localization in Xenopus oocytes. Mol. Biol. Cell. 32 (22), ar37. Available from: https://doi.org/10.1091/mbc.E21-03-0146-T.

Nevins, J.R., Darnell, J.E., 1978. Steps in the processing of Ad2 mRNA: poly(A)$^+$ nuclear sequences are conserved and poly (A) addition precedes splicing. Cell 15, 1477–1493.

O'Hara, E.B., Chekanova, J.A., Ingle, C.A., Kushner, Z.R., Peters, E., Kushner, S.R., 1995. Polyadenylation helps regulate mRNA decay in E. coli. Proc. Natl. Acad. Sci. 92, 1807–1811.

Onodera, Y., Haag, J.R., Ream, T., Nunes, P.C., Pontes, O., Pikaard, C.S., 2005. Plant nuclear RNA polymerase IV mediates siRNA and DNA methylation-dependent heterochromatin formation. Cell 120, 613–622.

Park, S.V., Yang, J.-S., Jo, H., Kang, B., Oh, S.S., Jung, G.Y., 2019. Catalytic RNA, ribozyme, and its applications in synthetic biology. Biotechnol. Adv. 37 (8), 107452. Available from: https://doi.org/10.1016/j.biotechadv.2019.107452.

Péterfy, M., Gyuris, T., Takács, L., 2000. Signal-exon trap: a novel method for the identification of signal sequences from genomic DNA. Nucleic Acids Res. 28 (7), e26. Available from: https://doi.org/10.1093/nar/28.7.e26.

Pontier, D., Yahubyan, G., Vega, D., Bulski, A., Saez-Vasquez, J., Hakimi, M.A., et al., 2005. Reinforcement of silencing at transposons and highly repeated sequences requires the concerted action of two distinct RNA polymerases IV in Arabidopsis. Genes. Dev. 19, 2030–2040.

Preobrazhensky, A.A., Spirin, A.S., 1978. Informosomes and their protein components: the present state of knowledge. Prog. Nucleic Acid. Res. Mol. Biol. 21, 1–38.

Roeder, R.G., 1976. RNA polymerase. In: Losick, R., Chamberlain, M. (Eds.), RNA Polymerase. Cold Spring Harbor Laboratory Press, Cold Spring Harbor, NY, pp. 285–329.

Rymen, B., Ferrafiat, L., Blevins, T., 2020. Non-coding RNA polymerases that silence transposable elements and reprogram gene expression in plants. Transcription 11 (3–4), 172–191. Available from: https://doi.org/10.1080/21541264.2020.1825906.

Schneider, C., Kudla, G., Wlotzka, W., Tuck, A.C., Tollervey, D., 2012. Transcriptome-wide analysis of exosome targets. Mol. Cell 48, 422–433.

Schuster, G., Lisitsky, I., Klaff, P., 1999. Update on chloroplast molecular biology: polyadenylation and degradation of mRNA is the chloroplast. Plant Physiol. 120, 937–944.

Serganov, A., Nudler, E., 2013. A decade of riboswitches. Cell 152 (1–2), 17–24. Available from: https://doi.org/10.1016/j.cell.2012.12.024.

Shen, Y., Ji, G., Haas, B.J., Wu, X., Zheng, J., Reese, G.J., et al., 2008. Genome level analysis of rice mRNA 3′-end processing signals and alternative polyadenylation. Nucleic Acids Res. 36 (9), 3150–3161.

Stewart, M., 2019. Polyadenylation and nuclear export of mRNAs. J. Biol. Chem. 294 (9), 2977–2987. Available from: https://doi.org/10.1074/jbc.REV118.005594.

Suzuki, T., 2021. The expanding world of tRNA modifications and their disease relevance. Nat. Rev. Mol. Cell Biol. 22, 375–392. Available from: https://doi.org/10.1038/s41580-021-00342-0.

Szafranski, K., Schindler, S., Taudien, S., Hiller, M., Jahn, N., Schreiber, S., et al., 2007. Violating the splicing rules: TG dinucleotides function as alternative 3′ splice sites in U2-dependent introns. Genome Biol. 8 (8), R154. Available from: https://doi.org/10.1186/gb-2007-8-8-r154.

Szczepińska, T., Kalisiak, K., Tomecki, R., Labno, A., Borowski, L.S., Kulinski, T.M., et al., 2015. DIS3 shapes the RNA polymerase II transcriptome in humans by degrading a variety of unwanted transcripts. Genome Res. 25, 1622–1633.

Taub, R., Kirsch, I., Morton, C., Lenoir, G., Swan, D., Tronick, S., et al., 1982. Translocation of the c-myc oncogene into the immunoglobulin heavy chain locus in human Burkitt lymphoma and murine plasmacytoma cells. Proc. Natl. Acad. Sci. 79, 7837–7841.

Temin, H., Mizutani, S., 1970. RNA-dependent DNA polymerase in virons of Rous sarcoma virus. Nature 226, 1211–1213.

Tian, J., Hu, J., Zhang, H., Lutz, C.S., 2005. A large-scale analysis of mRNA polyadenylation of human and mouse genes. Nucleic Acids Res. 33 (1), 201–212.

Tosi, M., Young, R.A., Hagenbuchle, O., Schibler, U., 1981. Multiple polyadenylation sites in a mouse α-amylase gene. Nucleic Acids Res. 9, 2313–2323.

Vogel, C., Marcotte, E.M., 2012. Insights into the regulation of protein abundance from proteomic and transcriptomic analyses. Nat. Rev. Genet. 13, 227–232.

Volkin, E., Astrachan, L., 1956. Phosphorus incorporation of E. coli ribonucleic acid after infection with bacteriophage T2. Virology 2, 146–161.

Wasmuth, E.V., Lima, C.D., 2012. Structure and activities of the eukaryotic RNA exosome. Enzymes 31, 53–75.

Wiedermannová, J., Julius, C., Yuzenkova, Y., 2021. The expanding field of non-canonical RNA capping: new enzymes and mechanisms. R. Soc. Open. Sci. 8 (5), 201979. Available from: https://doi.org/10.1098/rsos.201979.

Winkler, W., Nahvi, A., Breaker, R.R., 2002a. Thiamine derivatives bind messenger RNAs directly to regulate bacterial gene expression. Nature 419, 952–956.

Winkler, W.C., Cohen-Chalamish, S., Breaker, R.R., 2002b. An mRNA structure that controls gene expression by binding FMN. Proc. Natl. Acad. Sci. 99 (25), 15908–15913.

Zamenhof, S., Brawerman, G., Chargaff, E., 1952. On the desoxypentose nucleic acids from several microorganisms. Biochim. Biophys. Acta 9, 402–405.

Zeevi, M., Nevins, J.R., Darnell Jr., J.E., 1982. Newly formed mRNA lacking polyadenylic acid enters the cytoplasm and the polyribosomes but has a shorter half-life in the absence of polyadenylic acid. Mol. Cell Biol. 2, 517–525.

Zhang, X., Henderson, I.R., Luc, C., Green, P.J., Jacobsen, S.E., 2007. Role of RNA polymerase IV in plant small RNA metabolism. Proc. Natl. Acad. Sci. 104, 4536–4541.

Zhang, M., Li, K., Bai, J., Velema, W.A., Yu, C., van Damme, R., et al., 2021. Optimized photochemistry enables efficient analysis of dynamic RNA structuromes and interactomes in genetic and infectious diseases. Nat. Commun. 12, 2344. Available from: https://doi.org/10.1038/s41467-021-22552-y.

Further reading

Breaker, R.R., 2018. Riboswitches and translational control. Cold Spring Harb. Perspect. Biol. 10 (11), a032797. Available from: https://doi.org/10.1101/cshperspect.a032797.

Chandrasekaran, V., Juszkiewicz, S., Choi, J., Puglisi, J.D., Brown, A., Shao, S., et al., 2019. Mechanism of ribosome stalling during translation of a poly(A) tail. Nat. Struct. Mol. Biol. 26, 1132–1140. Available from: https://doi.org/10.1038/s41594-019-0331-x.

Crick, F., 1957. On protein synthesis. Paper presented to the Society for Experimental Biology.

Dueck, H., Khaladkar, M., Kim, T.K., Spaethling, J.M., Francis, C., Suresh, S., et al., 2015. Deep sequencing reveals cell-type-specific patterns of single-cell transcriptome variation. Genome Biol. 16 (1), 122. Available from: https://doi.org/10.1186/s13059-015-0683-4.

Huang, Y., Kendall, T., Forsythe, E.S., Dorantes-Acosta, A., Li, S., Caballero-Pérez, J., et al., 2015. Ancient origin and recent innovations of RNA polymerase IV and V. Mol. Biol. Evol. 32 (7), 1788–1799.

Hunt, A.G., 2007. Messenger RNA 3'-end formation and the regulation of gene expression. In: Bassett, C.L. (Ed.), Regulation of Gene Expression in Plants. Springer, New York, pp. 101–122.

Julius, C., Yuzenkova, Y., 2018. Noncanonical RNA-capping; discovery, mechanism, and physiological role debate. WIRES RNA 10, e1512. Available from: https://doi.org/10.1002/wrna.1512.

Kozak, M., 1999. Initiation of translation in prokaryotes and eukaryotes. Gene 234, 187–208.

Rybak-Wolf, A., Stottmeister, C., Glažar, P., Jens, M., Pino, N., Giusti, S., et al., 2015. Circular RNAs in the mammalian brain are highly abundant, conserved, and dynamically expressed. Mol. Cell 58 (5), 870–885.

Sachs, A.B., Sarnow, P., Hentze, M.W., 1997. Starting at the beginning, middle, and end: translation initiation in eukaryotes. Cell 89, 831–838.

Sun, M., Gao, A.X., Li, A., Liu, X., Wang, R., Yang, Y., et al., 2021. Bicistronic design as recombinant expression enhancer: characteristics, applications, and structural optimization. Appl. Microbiol. Biotechnol. 105, 7709–7720. Available from: https://doi.org/10.1007/s00253-021-11611-5.

You, X., Vlatkovic, I., Babic, A., Will, T., Epstein, I., Tushev, G., et al., 2015. Neural circular RNAs are derived from synaptic genes and regulated by development and plasticity. Nat. Neurosci. 18 (4), 603–610.

Chapter 2

Creating a ribonuclease-free environment

Rationale

The difficulties associated with the isolation of full-length, naturally labile ribonucleic acid (RNA) are further compounded by the ubiquity of ribonuclease (RNase) activity. RNases are a family of enzymes, present in virtually all living cells. They exist as a means of recycling RNA and as a natural defense against RNA viral attacks. RNases are able to degrade RNA molecules through often aggressive endonucleolytic and exonucleolytic activity. In the lab, this can be useful or troublesome: RNases are useful in procedures such as the preparation of RNA-free DNA preps or for eliminating residual RNA molecules in the RNase Protection Assay (Chapter 18: Quantification of Specific mRNAs by Nuclease Protection). In contrast, RNase activity is bad when an investigator is attempting to isolate intact RNA because even trace amounts of these highly destructive enzymes can wreak havoc during isolation procedures, in previously purified samples, and even in RNA sequencing facilities. Thus, their control is of paramount importance.

RNases are small, remarkably stable enzymes that have minimal cofactor requirements and are active over a wide pH range. Their tertiary configurations are maintained by virtue of as many as four disulfide bridges (e.g., RNase A) that allow these enzymes to refold following treatment with many denaturants (reviewed by Raines, 1998), and even after boiling. Particularly resilient are RNases of the pancreatic variety (RNase A-like enzymes); the investigator often has seconds, literally, to inhibit RNase by flash freezing fresh tissue, by homogenizing it in chaotropic lysis buffer (Chapter 4: RNA Isolation Strategies), or by chemically stabilizing the RNA within the tissue. This caveat is likewise true for many other tissues. It is incumbent upon the investigator to ensure that both equipment and reagents are purged of nucleases, including those intended for downstream applications, from the onset of the experiment. For most RNA-minded molecular biologists, saying that a reagent or apparatus is sterile is often a statement that it is RNase sterile, that is, RNase-free. Chemicals and premade reagents and solutions marketed as molecular biology grade are generally certified as being RNase- and

RNA Methodologies. DOI: https://doi.org/10.1016/B978-0-323-90221-2.00027-8

DNase-free; while often more expensive than standard American Chemical Society grade products, they are well worth the investment.

Many investigators often worry less about RNase degradation of their samples than in the past. This is mainly due to the ready availability of reagents that favor expedient isolation of RNA from biological sources of all types. As such, attentiveness to many of the traditional techniques for the control of nuclease activity is often minimized and RNase is not viewed as a concern of central importance. Such an attitude may unwittingly compromise the reliability of the sample as well as data reproducibility. Many of the products used for RNA isolation assume that the user has received appropriate instruction in the art of creating and maintaining an RNase-free environment. Thus, the investigator needs to always make a conscious effort to prevent the accidental introduction of RNase from sources within the lab (RNase A is the most prevalent enzyme) and, with the same zeal, suppress endogenous (cell and tissue) RNase activity. The most straightforward procedure to this end is for the lab to develop a detailed standard operating procedure (S.O.P.), including a checklist, which each person in the lab is obligated to follow. An important part the S.O.P. involves working in a designated low-traffic "RNA only" area and using a dedicated set of micropipettors and ultrapure reagents, as described below. This strategy is strongly recommended.

Elimination of resilient ribonucleases

Controlling RNase activity must be considered from two perspectives, namely, external (lab) and endogenous (from the biological sample) sources. First, potential external sources of RNase contamination must be identified and neutralized from the onset of the experiment and all reagents must be maintained RNase-free at all times. As shown in Table 2.1, external sources include but are not limited to, bottles and containers in which chemicals are packaged, nonnuclease-free water, disposal laboratory plasticware, gel boxes and combs, and even microbial and fungal contamination of buffers. Many types of RNase are known and each of which has a unique sequence specificity and degradation pattern. It is worth noting that the presence of a 5′ cap is no guarantee of mRNA stability. The single greatest external contamination source, however, is the RNase "cocktail" found on the surface of the skin, particularly the fingertips, dried skin cells, and even hair. Consequently, rule number one when working with RNA is to wear gloves!

Second, endogenous RNase activity must also be controlled. Beyond the potential for accidental contamination of an RNA preparation with RNases from sources in the lab, intracellular RNases normally sequestered within the cell are liberated upon cellular lysis. Unless they are inhibited without delay, RNases are free to initiate degradation of the very RNA that the investigator is attempting to isolate. As endogenous RNase activity varies tremendously from one biological source to the next, the degree to which action must be

TABLE 2.1 Common sources of RNase activity.

Source	Remedy
Ungloved hands	Wear gloves and change periodically.
Saliva	Don't talk or sing while working with RNA.
Contaminated solutions	Visual inspection; pay attention to expiration dates.
Dust (noun)	Dust (verb).
Sweat	Remain calm; turn on the air-conditioning.
Lab equipment	Pretreatment; reserve exclusively for RNA work as much as possible.
Appliances (refrigerator and microwave handles, draw handles, doorknobs)	Change gloves before returning to the bench.
Devices (cell phones, tablets, keyboards, earbuds, headphones)	Don't use until bench work is complete and the sample is secure.

taken to inhibit nuclease activity is a direct function of cell or tissue type. Knowledge of the extent of intrinsic nuclease activity is derived from two primary sources: salient literature and personal experience. The good news is that, while RNases can function in a variety of chemical environments, they really don't like cold temperatures; this is the reason some protocols stress the importance of keeping RNA on ice.

The method selected for controlling RNase activity must, first and foremost, demonstrate compatibility with the cell lysis procedure. For example, one objective of a study might be the examination of nuclear RNA independently of cytoplasmic RNA, often performed to assess the level(s) of gene regulation. In this case, nuclease inhibitors are added to relatively gentle lysis buffers (Chapter 4: RNA Isolation Strategies) when nuclei and other subcellular organelles must remain intact. Further, the method of nuclease inhibition must support the integrity of the RNA throughout the subsequent fractionation or purification steps, some of which can be time-consuming. Finally, the reagents used to inhibit RNase activity must be easily removed from purified RNA preparations so as not to interfere with subsequent procedural steps. Laboratories routinely isolating high-quality RNA almost certainly maintain a "private stock" of RNA reagents and supplies that are not used for any purpose other than RNA recovery and subsequent manipulation. Such materials should definitely not be in general circulation in the laboratory. In all cases, actions taken to control nuclease activity need to be aggressive. Failure to do so is likely to yield a useless sample of degraded RNA.

While no one would question the RNase inhibitory properties of many commonplace reagents used to purify RNA, of even greater concern is the stability of the RNA once the isolation reagent has been removed. For example, purified RNA is never more susceptible to nuclease attack than when stored in an aqueous buffer, such as TE (10 mM Tris, 0.1 mM EDTA, pH 7.5), or water. Unfortunately, losing a valuable sample is often a pivotal lesson on the need to maintain the stability of the RNA before, during, and after its isolation from the cell. The correct way to store purified RNA is described in Chapter 4, RNA Isolation Strategies.

Latent RNase contamination issues

The term "latent RNase" refers to RNase that is (transiently) inactivated because it is complexed with an RNase inhibitor or otherwise finds itself in an inhibitory chemical environment. The reason that latent RNase is a concern is because, upon removal, inactivation, or dissociation of the inhibitor, the RNase can become active and proceed to decimate the RNA. This commonly occurs when an RNase-RNase inhibitor complex is exposed to elevated temperatures (usually >65°C), strongly denaturing conditions (e.g., guanidinium salts >3 M or urea >6 M), pH extremes (pH <5.5 or pH >9.0), or upon oxidation. The different types of inhibitors described below each have a unique pH, salt, and temperature range over which they are functional, and it is wise to heed those parameters. A standard low-cost test for latent RNase contamination is outlined in Fig. 2.1. In one tube, RNA known to be of high quality

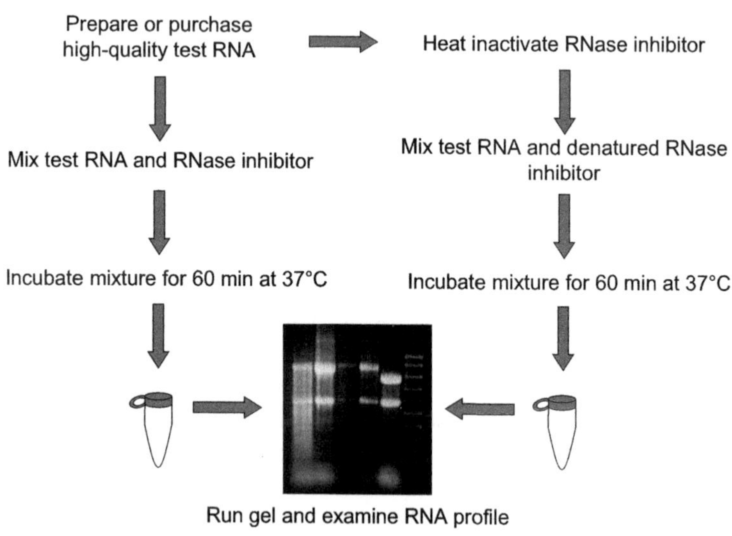

FIGURE 2.1 Detection of latent RNase activity. Inactivation of RNase-inhibitors will release activated RNase that may compromise the integrity of a sample. See text for details.

TABLE 2.2 General strategies for RNase control.

Do this	How
Prevent it	Wear gloves and follow SOPs
Oxidize it	H_2O_2
Crosslink it	UV light
Chemically destroy it	Bleach, commercial products, or DEPC
Heat it	Boil, bake, or autoclave
Bind it	RNasin, nitrocellulose filtration, or VDR

Note: The best way to control RNase is to keep it out from the start. Refer to text for details.

(test RNA) is dissolved in water or another nondenaturing buffer (TE, for example) and then incubated for 1 hour at 37°C. In a second tube, heat-inactivated (15 minutes at 70°C) RNase inhibitor is mixed with test RNA for 1 hour at 37°C; some RNase-inhibitors can be intentionally inactivated by incubation with oxidized glutathione or by withholding DTT. At the conclusion of the incubation period, the contents of both tubes are electrophoresed under denaturing conditions to look for signs of degradation (Chapter 8: Quality Control for RNA Preparations). This incubation can be extended to 2–16 hours (overnight) if very low levels of RNase are suspected. In well-funded laboratories, a precision integrity profile of an RNA sample can be generated by Bioanalyzer analysis. Standard analysis strategies for controlling RNase are described in the following sections and summarized in Table 2.2. Regardless of the method used, the integrity of an RNA sample must be demonstrated if data generated in downstream applications are to be believed. A good laboratory SOP involves the removal of "strategic" aliquots to test for the presence of RNase.

Types of ribonuclease inhibitors

Compounds used for control of RNase activity fall into two broad categories, specific and nonspecific RNase inhibitors. Specific inhibitors compromise the activity of RNases, while nonspecific inhibitors eliminate any type of enzyme activity, including DNase, RNase, and protease. Some RNase inhibitors are added directly to lysis buffers and reaction mixtures to control RNase activity, whereas others are used in the standard preparation of reagents, equipment, and other lab implements intended for RNA work. The effectiveness of an RNase inhibitor in a particular system can be ascertained by a short incubation of RNase A, test RNA, and the putative inhibitor. The postincubation products may then be viewed on a simple agarose gel (see Chapter 8: Quality Control for RNA Preparations, Fig. 8.5). If there is evidence of RNA degradation, a change in protocol is clearly indicated. The biotech industry has fueled the demand for both broad-spectrum and application-specific nuclease inhibitors and, fortunately, many such products are widely available.

Specific inhibitors

RNasin

Perhaps the best-known RNase inhibitor is the protein RNasin (Promega, Madison, WI). Although it is a trademark-protected name, in many circles "RNasin" is used in a generic sense to describe any of a variety of excellent RNase inhibitors from other well-respected manufacturers. Originally purified from the human placenta (native RNasin is still available), there is an ample supply of the recombinant form of this very effective RNase inhibitor. RNasin is a 50 kDa protein that inactivates RNase by competitive, noncovalent binding (1:1) to the enzyme with a binding constant $K_i \approx 10^{-14}$ (Blackburn et al., 1977; Blackburn, 1979; reviewed by Lee and Vallee, 1993). Mammalian, nonhuman RNasin and related inhibitors are also available; these are quite suitable for some clinical and diagnostic research involving RNA from *Homo sapiens* useful because they present no risk of human DNA contamination. One additional note: inasmuch as some RNase inhibitor proteins are DTT-dependent, one should follow the manufacturer's guidelines carefully because excessive DTT can be detrimental to ensuing procedural steps.

RNasin is routinely used at a concentration of 250–1000 U/mL and is effective against RNase A, RNase B, RNase C, and human placental RNase; it does not inhibit RNase T_1, S1 nuclease, RNase H, or RNase from *Aspergillus*. Other formulations of nonspecific nuclease inhibitors are also available. For example, several formulations of nonhuman, non-DTT-dependent proteins which offer broader protection against RNase, including RNase T_1 are widely available. In this regard, the common generic abbreviations in the literature are ribonuclease inhibitor (RI), ribonuclease inhibitor protein (RIP) and human placental ribonuclease inhibitor (hPRI).

A major advantage of the use of protein inhibitors as a means of controlling nuclease activity is compatibility with a great variety of in vitro reactions, including protection of template messenger RNA (mRNA) during the synthesis of cDNA, in vitro transcription, and in vitro translation. As noted above, care must be taken to avoid powerful denaturing conditions which will cause the uncoupling of RNase–RNasin complexes and accompanying restoration of RNase degradative activity. Heating samples above 50°C also results in fairly rapid RNase inhibitor inactivation, although some newer RNase inhibitor formulations are able to withstand temperatures as high as

70°C for up to 15 minutes. This is particularly handy when RNAs with significant secondary structures need to be denatured. It is less common to put RNasin into aqueous lysis buffers; RNasin is typically used as a protectant of previously purified RNA in small reaction volumes.

Vanadyl ribonucleoside complexes (VDR; VRC)

Vanadyl ribonucleoside complexes (VDR) were developed in the mid-1970s as a means of controlling RNase activity when using relatively gentle methods of cellular lysis. VDR is not widely used any longer for three reasons: (1) widespread use of chaotropic lysis reagents that are efficient inhibitors of RNase, (2) VDR can interfere with cDNA synthesis by competing with dNTPs when present at low concentrations, and (3) VDR in trace quantities can inhibit in vitro translation of purified mRNA. Therefore, VDR is contraindicated for RT-PCR and the numerous permutations thereof, and its use is not recommended. VDR was previously used in this laboratory with gentle NP-40 lysis buffers, though we have found that keeping all reagents ice-cold and working out of an ice bucket throughout the isolation procedure is equally as effective. There are examples of VDR being used successfully as an RNase inhibitor in sucrose gradients used for polysome profiling (Del Prete et al., 2007).

VDR consists of complexes formed between the oxovanadium ion and any or all of the four ribonucleosides (Leinhard et al., 1971; Berger and Birkenmier, 1979) in which vanadium takes the place of phosphate. These complexes then function as transition-state analogs. In the absence of VDR, RNase-mediated cleavage of the phosphodiester backbone of RNA results in the transient formation of a dicyclic transition-state intermediate, which is subsequently opened up by reaction with a water molecule. In its capacity as an RNA analog, the vanadium complex forms a highly stable dicyclic species to which the enzyme remains irreversibly bound, locking the RNase and its "pseudo-substrate" in the transition state. Succinctly, VDR inhibits RNase by binding irreversibly to a broad spectrum of cellular RNases.

When used, VDR is routinely prepared as a 200 mM stock solution and used at a final concentration of 5–20 mM, depending on the anticipated degree of RNase activity. A protocol for VDR synthesis is provided at the end of this chapter, though VDR can still be purchased from a handful of vendors. VDR is well known to suppress RNase activity in cells with significant levels of endogenous RNase, including pancreatic RNase A and RNase T_1, but not RNase H. It is also inhibitory against the family of enzymes commonly known as the poly(A) polymerases, which may or may not be significant in the context of the types of experiments being undertaken. After it has neutralized RNase activity, VDR can be removed from an aqueous lysate by repeated, systematic extraction with an organic extraction buffer containing 0.1% 8-hydroxyquinoline (w/v), which is itself a partial inhibitor of RNase. Alternatively, VDR can be removed by the addition of 10 equivalents of

EDTA followed by ethanol precipitation. The investigator is cautioned, however, that 1 equivalent of EDTA is more than sufficient to dissociate the VDR complex. Thus, when VDR is present, EDTA-containing buffers are to be avoided until another method of controlling RNase activity has been employed. Moreover, excess EDTA carrying over can inhibit PCR dramatically (discussed in Chapter 10: RT-PCR: A Science and An Art Form and Chapter 11: Quantitative PCR Techniques). While the habitual use of VDR is one of the more cost-effective means of controlling nuclease activity, in the world of RT-PCR it is not an appropriate choice. Fortunately, much better options for nuclease control are available.

Nonspecific inhibitors

The use of a variety of different types of nonspecific inhibitors has been widely reported. These compounds function by eliminating all intrinsic enzymatic activity and have been incorporated into RNA isolation procedures with varying degrees of success. None of these has been found to be completely satisfactory for every application; one or more of these compounds, however, may provide a means of controlling RNase activity that would otherwise not be possible. Their persistence in the literature, however, mandates at least mentioning them here. These nonspecific inhibitors include heparin iodoacetate, polyvinyl (dextran) sulfate, Catrimox-14 cationic surfactant, and the clays macaloid and bentonite. Historical note: In addition to its biotech and pharmaceutical applications, the use of bentonite in winemaking is nearly universal, dating back to the mid-1930s (Saywell, 1934).

More contemporary methods include dilute solutions (3%) of hydrogen peroxide (H_2O_2), a mixture of 0.1 N NaOH and 1 mM EDTA, and a solution of 0.1% sodium dodecyl sulfate (SDS). For those investigators who do not wish to prepare RNase remedies in-house, several commercial preparations are readily available. These proprietary formulations include ELIMINase, RNase *Away*, RNaseZap, RNAse-OFF, ProtectRNA, and others, and provide very satisfactory removal of RNase from a variety of laboratory surfaces, per the manufacturers' instructions. Alternatively, some laboratories wipe down benchtops and countertops before and after RNA work with a solution of 2.5%–5% household bleach; other labs find 70% ethanol to be completely satisfactory. If used to treat laboratory implements, it is imperative that these formulations be rinsed away completely because even small residual amounts may have a negative effect on downstream applications. Finally, be sure to observe appropriate laboratory safeguards when handling bleach or any other materials.

Preparation of equipment and reagents

As delineated above, rule number one when working with RNA is to wear gloves (powder-free work best) during the preparation of reagents and equipment and during the actual RNA extraction procedure. Finger greases, especially those of overly zealous lab technicians, are notoriously rich in RNase and are generally accepted as the single greatest source of RNase contamination. Further, one should not hesitate to change gloves several times during the course of an RNA-related experiment, since doorknobs, micropipettors, stir bars, refrigerator door handles, telephone receivers, cell phones, tablets, computer keyboards, and even animals are also potential sources of nuclease contamination. It is the strict policy of this Author to remove his gloves when stepping away from the bench and don a fresh pair when returning to the bench and the RNA protocol.

With respect to laboratory plasticware, individually wrapped serological pipettes are always preferred for RNA work as no special pretreatment is required. Plasticware that is certified as tissue culture sterile, including both polypropylene and polystyrene conical 15 and 50 mL tubes, is not likely to be a source of nuclease activity unless subsequently contaminated (be certain to wear gloves when handling these items for RNA work). Bulk-packed polypropylene products are potential sources of nuclease contamination, mainly due to handling and distribution from a single bag. This pertains to microfuge tubes and polypropylene micropipette tips because they can become contaminated and, in turn, contaminate stock solutions. Thus, any nonsterile plastic product that comes into contact with an RNA sample at any time, either directly or indirectly, and which can withstand autoclaving, should be so treated. Bulk packages of microfuge tubes are best distributed with gloved hand into some type of autoclavable receptacle. Alternative, place the tubes into several 500 mL glass beakers and cover with heavy duty aluminum foil. Then, autoclave microfuge tubes for no more than 15–20 minutes. These microfuge tubes are then handled only with gloves and set aside exclusively for RNA work.

When glassware must be used, as may be required for the manipulation of organic (phenol and chloroform) extraction buffers, individually wrapped borosilicate glass pipettes are strongly preferred. Glassware that must be reused most certainly should be reserved exclusively for RNA work and should not be in general circulation in the laboratory. Contrary to popular belief, however, the temperature and pressure generated during the autoclaving cycle are usually not sufficient to eliminate all RNase activity. However, RNases can be destroyed quite effectively by baking in a dry-heat oven. Glassware should be cleaned scrupulously, rinsed with RNase-free water (discussed below), and then baked for 2–3 hours at approximately 200°C. This is a very effective method for purging glassware of RNase activity. It is important to note that not all laboratory implements can withstand the heat generated in a dry-heat oven. If there is a question about a particular type of plastic or other material, be certain to check with the manufacturer before baking. Finally, all stock- and working solutions in the lab should be labeled with a predetermined expiration date, in addition to the date of preparation. The use of out-of-date or, even worse, "ancient" solutions may have the unintended consequence of introducing RNase from microorganisms that may have taken up residence in these bottles. Keep the QA/QC people happy: pay attention to expiration dates. Finally, it is prudent to maintain separate containers of chemicals as well as stock solutions for exclusive use as RNA reagents. This is most easily accomplished by aliquoting sterile reagents into suitable volumes, rather than drawing repeatedly from the stock bottles. Aliquots used once should be discarded. Although such actions may at first seem excessive, they may well preclude the accidental introduction of RNases and facilitate recovery of the highest possible quality RNA. Finally, don't forget that chemical solids should be weighed with an RNase-free spatula. It is unfortunate that some investigators fail to pay heed to these guidelines, citing "no measurable problems with RNase activity at present."

UV light

Among the most efficient and simplest methods for ensuring RNase inactivation is to irradiate exposed surfaces to ultraviolet (UV) light. The energy imparted by short wavelength irradiation very effectively crosslinks organic molecules, thereby rendering them permanently nonfunctional. An added benefit of UV irradiation beyond RNase inactivation is that DNA molecules from previous experiments that may be lingering will likewise be crosslinked and will no longer be accessible as possible contaminants in future applications such as the polymerase chain reaction (Chapter 10: RT-PCR: A Science and An Art Form). It cannot be stressed enough that, when working with a UV light source, great care must be taken to prevent exposure of the skin and eyes to UV light due to the severe, permanent tissue damage that would occur very quickly without protection.

It is not necessary to use a calibrated UV light source for RNase inactivation, though doing so would allow the investigator to develop an SOP that everyone can follow. In this Author's lab, portable lab implements are placed in a laminar flow biological safety cabinet with a factory-installed UV light and irradiated for five minutes. A new generation of high-intensity UV LED light instruments, emitting at one or more wavelengths, are particularly effective for smaller surface irradiation (Fig. 2.2), with direct applicability for RNA library preparation, various RT-PCR methods, and next-generation sequencing. Again, UV light should only be turned on when everyone is out of the lab or otherwise safely protected from exposure.

Sterile water options

Water certified as being nuclease-free water is also excellent for making dilutions of stock solutions. For those labs affiliated with hospitals, note that

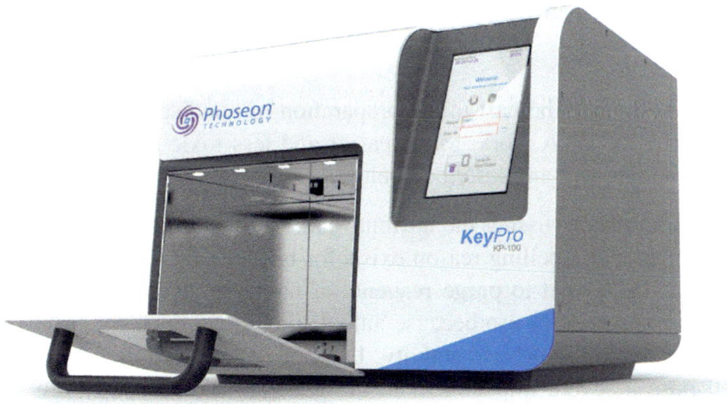

FIGURE 2.2 KeyPro decontamination system. A proprietary UV LED light system is effective for the deactivation of RNase as well as having antimicrobial and antiviral properties. *Image courtesy Phoseon Technology (http://www.phoseon.com).*

water labeled "sterile water for irrigation" or "sterile water for injection" meets the United States Pharmacopeia monograph for sterility, is free of contaminants, and is good for RNA work! Better still for RNA work is water that has been UV irradiated. Many water purification systems contain this feature; after the water is cartridge deionized it is transferred to a holding tank outfitted with a UV light source. Periodic UV irradiation crosslinks RNase, if present, thereby inhibiting it. Water prepared in this manner is ready for applications requiring nuclease-free sterility without any additional treatment; one should, however, periodically test the water for added assurance. It is well worth the investment.

An older and much less desirable method of preparing water for RNA applications is using diethyl pyrocarbonate (DEPC), a potent nonspecific chemical inhibitor of RNase. DEPC is also carcinogenic and must be handled with care and with strict attention to the safety recommendations of the manufacturer, including working in a chemical fume hood. Appropriately, there remains a fair amount of resistance to using DEPC in the contemporary molecular biology laboratory. It is the opinion of this Author that there are four legitimate reasons for avoiding DEPC:

1. Certified nuclease-free solutions, including water, can be purchased from one's preferred vendor and represent a very worthwhile investment. This alone justifies eliminating DEPC from the lab.
2. DEPC is toxic and not appropriate in a lab where it cannot be handled properly. This concern has merit, as improper handling may present a serious health risk.

3. DEPC is too difficult to remove from reagents and interferes with PCR and other reactions. This is often a direct result of adding too much DEPC.
4. DEPC was used in the past in the preparation of nondisposable glassware prior to use for RNA work. Much easier and less toxic methods are currently available and are described below.

Still, it is worth discussing the attributes and shortcomings of DEPC here in the event that a compelling reason exists for bringing it into the lab.

DEPC has been used to purge reagents of nuclease activity from certain aqueous solutions prior to use because autoclaving alone does not ensure the complete elimination of RNase activity. Lack of attention to this detail may very well pose a severe impediment to the generation of data that are both representative and reproducible. If the decision is made to use DEPC, it should be added to a final concentration of 0.05% (v/v). Whereas previous reports called for the use of DEPC at 0.1%, a reduction to 0.05% works just as well. Any concentration above 0.1% will be difficult to purge from the solution; in various consulting sorties over the years, laboratory personnel have confessed to this Author to using DEPC at concentrations ranging from 0.5%–5.0% (with many ensuing problems)!

Water to which DEPC is added is then shaken for several hours on an orbital shaker platform or stirred vigorously with a magnetic stirrer for 20–30 minutes, after which DEPC must be destroyed completely. This is most commonly accomplished by autoclaving. At autoclaving temperature and pressure, DEPC degrades into carbon dioxide and ethanol, both of which are quite volatile under these conditions. Alternative approaches include (1) maintaining certain solutions at 60°C overnight, a process that should be conducted in a chemical fume hood with the bottle cap slightly loosened and (2) removing DEPC from large volumes of DEPC-treated water by boiling it for 1 hour in a fume hood. Complete removal of DEPC can be further promoted by rapidly stirring the hot solutions with a nuclease-free magnetic stir bar. Important: Do not autoclave or otherwise heat solutions that would not ordinarily be so treated, such as those containing SDS, NP-40 (Igepal CA-630), or NaOH. Start with water that is already nuclease-free and then add the nonautoclavable components to it.

Not only is DEPC carcinogenic, which alone would justify its complete removal and careful handling, but it also has a strong affinity for adenosine nucleotides. Even trace amounts of residual DEPC will result in chemical modification of the base adenine (Henderson et al., 1973; Ehrenberg et al., 1976), thereby changing the physical properties of RNA and compromising its usability. This is also the reason that it has never been recommended that DEPC be added directly to cell suspensions or lysates containing RNAs to be purified. When the autoclaving time is inadequate, one notices the distinctive odor of residual DEPC. In days gone by in this laboratory, solutions were routinely kept in the warm autoclave chamber for approximately 30 minutes after completion of the autoclaving cycle to purge them completely of DEPC.

Important considerations if DEPC is to be used in the lab:

1. Do not add DEPC to any buffer containing mercaptans or 1° amine groups, with which DEPC is reactive. The most common buffers to which DEPC exposure is to be avoided are the Tris (*tris*(hydroxymethyl) aminomethane) buffers. Instead, use DEPC-treated water (add DEPC to water; autoclave to remove DEPC) to make up the Tris-containing solution and then autoclave the solution again.
2. Buffers consisting of chemicals that are known (or suspected) to have DEPC incompatibility can be twice filtered through a nitrocellulose membrane to remove RNase activity and other trace proteins.
3. The electrophoresis unit itself, including the gel comb, casting tray, and interior of the gel box can be treated to remove RNase contamination by rinsing and soaking in DEPC-treated water only *after* autoclaving the water. *Never* expose the electrophoresis unit to DEPC because acrylic is not resistant to DEPC. The "Hydrogen Peroxide" section, below, offers a much safer alternative approach of the purging of RNase from electrophoresis equipment.

Hydrogen peroxide

The need to eliminate RNase from equipment and supplies in the lab, including gel combs, casting trays, electrophoresis chambers, and graduated cylinders is unquestioned. Soaking such implements in a 3% solution of hydrogen peroxide (H_2O_2), commonly available over the counter, is a very effective and inexpensive measure. H_2O_2 is a powerful oxidizing agent that can render a surface nuclease-free by soaking for 20–30 minutes, followed by rinsing with copious amounts of water that, at the very least, has been autoclaved. *Do not* use the more concentrated forms of H_2O_2 (e.g., 30%) commonly available from standard chemical supply companies. The higher concentration H_2O_2 is extremely dangerous and may well cause irreparable damage to acrylic gel box components and other equipment, and tissue damage to the investigator. Also, be sure to avoid old solutions of H_2O_2, as they may no longer be solutions of H_2O_2!

NaOH and sodium dodecyl sulfate

An older method for the removal of RNase activity from laboratory surfaces is to soak resistant laboratory implements or wipe laboratory surfaces, with a solution containing 0.1 N NaOH and up to 0.1% SDS. The reader is cautioned to handle such a formulation with care, including skin and eye protection. After treatment, subsequent extensive washing is required to remove the alkali. Failure to do so could pose a health hazard to others working in the lab, not to mention the fact that residual NaOH could indirectly create a chemical environment favoring alkaline hydrolysis of an RNA sample. Even though thorough rinsing with water precludes this potential difficulty, it is somewhat messy and not as cost-effective as H_2O_2, especially when NaOH and SDS are purchased premixed. This approach to eliminating RNase is best avoided.

Other compounds used to control nuclease activity

The following compounds and reagents are commonly associated directly or indirectly with the purification of RNA. They are often components of popular RNA isolation buffers. What they all have in common is the ability to minimize or altogether eliminate RNase activity. In the absence of a strong biologically disruptive environment, RNase inhibitors, as described above, should be added to the lysis buffer just prior to use.

Polyvinylsulfonic acid

Polyvinylsulfonic acid (PVSA) is a branched anionic polymer that is under investigation as a possible broad-spectrum RNase inhibitor (Earl et al., 2018). It has been demonstrated to protect purified RNA and is compatible with in vitro transcription reactions; it is, however, not compatible with in vitro translation reactions. More recently, PVSA has been explored as a component in RNA-extraction-free COVID-19 testing by RT-PCR (Smyrlaki et al., 2020). As a nonbiologic, it has the advantage of being used under conditions that would not be compatible with the use of protein-type RNase inhibitors. Its low cost makes PVSA a potentially attractive means of protection RNA, particularly when immediate processing is not possible, in the ongoing war to control RNase.

Guanidinium salts

Solutions containing guanidine hydrochloride (G · HCl), or guanidine thiocyanate (GTC) are strong ionic denaturants and are often described as "chaotropic" because of their biologically disruptive nature. Cellular and organellar membranes are disrupted on contact with solutions containing G · HCl (4–6 M) and GTC (4 M); the unfolding and solubilization of proteins which ensues make these compounds ideal inhibitors of RNase activity during purification of nucleic acids from a variety of biological sources. GTC is a stronger protein denaturant than G · HCl and is a denaturant of choice for the preparation of RNA from sources enriched in RNase activity, especially pancreatic tissue. It is routinely used in conjunction with a reducing agent (e.g., β-mercaptoethanol; β-ME) and an ionic detergent such as sarcosyl (*N*-laurylsarcosine).

For cells enriched in RNase, and even those that are not, homogenization in guanidinium lysis buffers is widely accepted as the method of choice for consistency from sample to sample. These compounds are well-suited for disruption of many tissue types as well as cells grown in culture and are the key component of several commercial RNA isolation products. A noteworthy disadvantage to lysing cells directly in guanidinium-containing buffers is the intermixing of nuclear and cytoplasmic message, as described in Chapter 4, RNA Isolation Strategies. Other excellent methods for RNA isolation which preclude this difficulty are also presented throughout this laboratory guide.

Sodium dodecyl sulfate

SDS, also known as lauryl sulfate, is an ionic detergent that is useful for the rapid disruption of biological membranes as well as the breakdown of tissue architecture. Concurrently, SDS is a powerful inhibitor of both RNase and DNase activity. SDS is commonly prepared as either a 10% or a 20% (w/v) stock solution and is used most often at a working concentration of 0.1% to 0.5%. SDS is easily precipitable in the presence of potassium salts and generally is not added to guanidinium buffers because it has very low solubility in high-salt, chaotropic solutions.

N-Laurylsarcosine

N-Laurylsarcosine, also known as sarkosyl, is an ionic detergent which is similar to SDS. Unlike SDS, sarkosyl exhibits excellent solubility in high-salt, chaotropic solutions and is therefore the detergent of choice in guanidinium-based lysis buffers. It disrupts tissue and cell structure, inhibits nuclease activity, and facilitates the disaggregation of proteins and nucleic acids in a cellular lysate.

Phenol:chloroform:isoamyl alcohol

Caution: Phenol, chloroform, and combinations thereof are caustic, carcinogenic reagents that must be handled with great care. Be sure to observe all safety precautions indicated by the manufacturer, including eye and skin protection.

Phenol and chloroform are organic solvents that efficiently denature and cause the precipitation of proteins. Molecular biology-quality phenol is itself a good protein denaturant, although it is quite unstable. Phenol oxidizes rapidly into quinones, which impart a pinkish tint to phenol solutions; if observed, pink-tinted phenol solutions should be discarded immediately. Quinones form free radicals that break phosphodiester linkages and crosslink nucleic acids. Phenol reagents are usually prepared by saturating the phenol with an aqueous buffer and then adding an equal volume of chloroform. The addition of chloroform stabilizes the phenol, imparts a greater density to this organic extracting material, improves the efficiency of deproteinization of the sample, and also facilitates the removal of lipids from the lysate. Very often, isoamyl alcohol is used in conjunction with chloroform or with mixtures of phenol and chloroform. Isoamyl alcohol reduces the foaming of proteins that would normally be generated by the mechanics of the extraction procedure. The most common organic extraction buffer formulations consist in part of phenol:chloroform:isoamyl alcohol in a 25:24:1 ratio. In the presence of these solvents, RNase activity is inhibited. A more thorough discussion of the preparation of phenol:chloroform:isoamyl alcohol is presented in Chapter 4, RNA Isolation Strategies.

8-Hydroxyquinoline

8-Hydroxyquinoline is a partial inhibitor of RNase and is occasionally added to organic extraction buffers that contain phenol. As an antioxidant, 8-hydroxyquinoline stabilizes phenol and retards the formation of quinones (phenol oxidation products). It is usually added to a final concentration of 0.1% (w/v). 8-Hydroxyquinoline imparts a bright yellow color to the phenol: chloroform to which it is added, thereby helping the investigator keep track of the organic and aqueous phases during the nucleic acid purification process.

8-Hydroxyquinoline also chelates heavy metals, making it especially useful for removing VDR, another RNase inhibitor, following cell lysis. Upon binding VDR, the 8-hydroxyquinoline changes from yellow to dark green, and repeated extractions with 8-hydroxyquinoline-containing phenol:chloroform are often necessary to fully remove it. When the phenol-containing phase of the extraction buffer remains yellow, all VDR has been removed. The inclusion of 8-hydroxyquinoline in phenol formulations may also be advantageous even when VDR is not used because heavy metals can cause RNA degradation when they are present with RNA for extended periods. Needless to say, all reagents should have been prepared using high-purity, nuclease-free water.

Cesium salts

Cesium chloride (CsCl) is a dense salt (3.99 g/cm^3) that was once a standard centrifugation medium for both analytical and preparative separation of nucleic acids. Its use for this application has declined significantly in favor of newer resins and silica filters, the use of either of which precludes the requirement for density gradient (isopycnic) centrifugation (discussed in Appendix I: Centrifugation as a mainstream tool for the molecular biologist).

If it is to be used, only ultrapure nuclease-free preparations of CsCl should be purchased because CsCl has a limited ability to inhibit RNase activity. If necessary, impure, solid CsCl may be baked at 200°C for 4–6 hours to remove residual RNase activity prior to exposure to RNA. The use of cesium sulphate ($CsSO_4$) also appears in some older protocols involving ultracentrifugation as a means of purifying RNA.

Cesium trifluoroacetate (CsTFA) is a highly soluble salt that solubilizes and dissociates proteins from nucleic acids without the use of detergents. Thus, CsTFA is an excellent inhibitor of RNase, and its use precludes removal of proteins from a sample by more traditional methods such as phenol:chloroform extraction or incubation with proteinase K. CsTFA is more chaotropic than CsCl, inhibits RNase to a greater extent than CsCl and shows greater solubility in ethanol, which expedites its removal following isopycnic centrifugation of RNA (Carter et al., 1983). As desired, RNA can be banded or pelleted in CsTFA because solution densities up to 2.6 g/mL are possible using this reagent. A major disadvantage of this approach, as with CsCl, is the requirement for ultracentrifugation and for this reason, it is rarely used in the contemporary molecular biology laboratory.

Proteinase K

Proteinase K is a proteolytic enzyme (a serine protease) naturally found in the mold *Tritirachium album*. In solution, it is stable over a pH range of 4.0–12.5 with an optimum of pH 8.0, and a temperature range of 25°C–65°C. The enzyme has two binding sites for Ca^{++}, though in the absence of this divalent cation some catalytic activity is retained. Maximum proteinase K activity is observed with the inclusion of 1 mM Ca^{++} in the reaction buffer. Proteinase K digestion is routinely performed at 50°C, and occasionally in the presence of EDTA to inhibit labile, Mg^{++}-dependent nucleases. Proteinase K is prepared commonly as a 20 mg/mL stock solution in sterile water (stable for 1 year at −20°C) or in a solution of 50 mM Tris, pH 8.0, 1 mM $CaCl_2$ (stable for months at 4°C). It is generally used at a working concentration of up to 50 μg/mL in any of a number of buffer formulations, including those that contain as much as 0.5% SDS.

Pronase (*Streptomyces griseus*) can be used in place of proteinase K. If prepared in-house, it may be necessary to perform a pronase self-digestion to eliminate contaminating RNase and DNase activity. If necessary, this is easily accomplished by incubation of the pronase stock (20 mg/mL in 10 mM Tris-Cl, pH 7.5; 10 mM NaCl) at 37°C for 1 hour. This extra task can usually be avoided by the purchase of predigested pronase, readily available from many vendors. In either case, suitable aliquots of pronase are stored at −20°C. Reaction conditions for pronase are identical to those for proteinase K except that the recommended working concentration for pronase is about 1 mg/mL.

RNA*later*

RNA*later* (Thermo Fisher Scientific) is a proprietary aqueous formulation that is well known in RNA circles for its ability to prevent chemical and RNase-mediated deterioration of RNA in cells, tissues, and organs until such time as it is convenient for the investigator to process the biological material to recover the RNA. RNA*later* permeates tissues rapidly, offering several strategic advantages with respect to the collection of samples, especially when working out in the field. RNA*later* is compatible with plant and animal cells and tissues, as well as *E. coli*. RNA*later* preserves histological organization of tissue comparable to formalin-fixed and frozen sectioned tissues (Florell et al., 2001). Similar outstanding RNA stabilization is achievable with the use of RNAprotect (Qiagen).

Protocol: synthesis of vanadyl ribonucleoside complexes

The protocol described here is a modification of the procedures described by Leinhard et al. (1971) and Berger and Birkenmier (1979). Although one may purchase VDR (e.g., New England Biolabs), this protocol was used in days gone by in this laboratory to prepare VDR in-house. Synthesis of VDR is performed by mixing all reagents in a three-neck flask (Fig. 2.3) submerged in a hot oil bath. The three-neck configuration permits access to a

FIGURE 2.3 The three-neck flask configuration permits easy access to the reaction components and mounting of a pH electrode, thermometer, and N_2 stream. *Photograph courtesy Corning Glass, Corning, New York.*

thermometer and pH probe, while allowing the dropwise addition of NaOH for titration. The volumes and masses indicated may be increased proportionately to accommodate larger-scale preparations.

1. Mix 0.5 mmol each of adenosine, guanosine, cytidine, and uridine in a total volume of 8 mL of nuclease-free H_2O.
2. Heat this mixture in a heating mantle, boiling water bath, or hot oil bath until all solids are dissolved.

   Caution: Perform this procedure in a chemical fume hood and be sure to seek the guidance of someone who is familiar with the proper handling of a hot oil bath or heating mantle. Improper or careless handling can cause serious, permanent injury.
3. While flushing or bubbling the solution with a steady stream of nitrogen gas, add 1 mL of 2 M vanadium sulfate ($VOSO_4$).

   CAUTION: Be sure to observe all the guidelines pertaining to the safe handling of poisonous vanadium sulfate.
4. Increase the pH of the solution to about 6.0 with the dropwise addition of 10 N NaOH.
5. Increase the pH of the solution to 7.0 with the dropwise addition of 1 N NaOH.

   Note: Be sure to change to 1 N NaOH between pH 6.0 and pH 7.0 because the pH of the solution changes very rapidly in this interval.
6. At pH 7.0, a heavy precipitate of oxovanadium (IV) is observed. Accompanying the formation of this precipitate is an abrupt change of color from bright blue to a characteristic green-black appearance.

   Note: In the absence of a nitrogen atmosphere (step 3, above), rapid oxidation of the oxovanadium (IV) species to the oxovanadium (V) species occurs above pH 3.5, the latter being a less efficient inhibitor of RNase.
7. Increase the final volume of the mixture to 10 mL with H_2O.

   Optional: A by-product of this synthesis reaction, Na_2SO_4, can now be removed by brief centrifugation of the newly synthesized VDR, or just prior to use.
8. The product of this synthesis reaction is a 200 mM stock solution of vanadyl ribonucleoside complexes. Store small aliquots of this preparation under N_2 at $-20°C$ for 1–2 months. At $-80°C$ these complexes are stable for at least 1 year.

References

Berger, S.L., Birkenmier, C.S., 1979. Inhibition of intractable nucleases with ribonucleoside–vanadyl complexes: isolation of messenger ribonucleic acid from resting lymphocytes. Biochemistry 18 (23), 5143–5149.

Blackburn, P., 1979. Ribonuclease inhibitor from human placenta: rapid purification assay. J. Biol. Chem. 254, 12484–12487.

Blackburn, P., Wilson, G., Moore, S., 1977. Ribonuclease inhibitor from placenta. Purification and properties. J. Biol. Chem. 252, 5904–5910.

Carter, C., Britton, V.G., Haff, L., 1983. CsTFA: a centrifugation medium for nucleic acid isolation. BioTechniques 1, 142–147.

Del Prete, M.J., Vernal, R., Dolznig, H., Mullner, E.W., Garcia-Sanz, J.A., 2007. Isolation of polysome-bound mRNA from solid tissues amenable for RT-PCR and profiling experiments. RNA 13, 414–421.

Earl, C.C., Smith, M.T., Lease, R.A., Bundy, B.C., 2018. Polyvinylsulfonic acid: a low-cost inhibitor for enhanced RNA preservation and cell-free protein translation. Bioengineered 9, 90–97. Available from: https://doi.org/10.1080/21655979.2017.1313648.

Ehrenberg, L., Fedorcsak, I., Solymosy, F., 1976. Diethyl pyrocarbonate in nucleic acid research. Prog. Nucleic Acid. Res. Mol. Biol. 16, 189–262.

Florell, S.R., Coffin, C.M., Holden, J.A., Zimmerman, J.W., Gerwels, J.W., Summers, B.K., et al., 2001. Preservation of RNA for functional genomic studies: a multidisciplinary tumor bank protocol. Mod. Pathol. 14, 116–128.

Henderson, R.E.L., Kirkegaard, L.H., Leonard, N.J., 1973. Reaction of diethyl pyrocarbonate with nucleic acid components. Adenosine-containing nucleotides and dinucleoside phosphates. Biochim. Biophys. Acta 294, 356–364.

Lee, F.S., Vallee, B.L., 1993. Structure and action of mammalian ribonuclease (angiogenin) inhibitor. Prog. Nucleic Acid. Res. Mol. Biol. 44, 1–30.

Leinhard, G.E., Secemski, I.I., Koehler, K.A., Lindquist, R.N., 1971. Enzymatic catalysis and the transition state theory of reaction rates: transition state analogs. Cold Spring Harb. Symp. Quant. Biol. 36, 45–51.

Raines, R.T., 1998. Ribonuclease A. Chem. Rev. 98, 1045–1065.

Saywell, L.G., 1934. Clarification of wine. Ind. Eng. Chem. 26, 981–982.

Smyrlaki, I., Ekman, M., Lentini, A., de Sousa, N.R., Papanicolaou, N., Vondracek, M., et al., 2020. Massive and rapid COVID-19 testing is feasible by extraction-free SARS-CoV-2 RT-PCR. Nat. Commun. 11, 4812. Available from: https://doi.org/10.1038/s41467-020-18611-5.

Further reading

Cuthbert, B.J., Burley, K.H., Goulding, C.W., 2018. Introducing the new bacterial branch of the RNase A superfamily. RNA Biol. 15 (1), 9–12. Available from: https://doi.org/10.1080/15476286.2017.1387710.

Jones, S.P., Goossen, C., Lewis, S.D., Delaney, A.M., Gleghorn, M.L., 2022. Not making the cut: Techniques to prevent RNA cleavage in structural studies of RNase–RNA complexes. J. Struct. Biol. X 6. doi.org/10.1016/j.yjsbx.2022.100066.

Chapter 3

Stringency: conditions that influence nucleic acid structure

Rationale

One of the principal tools of the molecular biologist is hybridization technology, that is, the ability to promote the formation of double-stranded molecules from two complementary single-stranded nucleic acid molecules, as well as the ability to reverse the process. This technology allows the identification, enrichment, recovery, and/or amplification of unique sequences from a mixed population of DNA or RNA molecules and is at the very heart of classical and contemporary molecular biology. Depending on the application, when two complementary polynucleotides form hydrogen bonds between complementary bases, the process is known as annealing, hybridization, strand reassociation, duplex formation, zippering, or just plain old base-pairing. When a double-stranded molecule separates into its constituent single strands, the process is known as melting, unzippering, denaturation, or strand dissociation. There are several adjustable parameters that promote or prevent, the formation of double-stranded molecules, as needed. Thus, successful hybridization requires accurate base sequence information because, more than anything else, the process is guided by Watson-Crick (canonical) base-pairing. To that end, conditions can be created that will also promote noncanonical base-pairing, assuming that it is permissible within the experimental design.

Types of double-stranded molecules

The formation of a double-stranded molecule is solely a function of the length and base composition, and not the nature of the molecules' backbone (polyribonucleotide or polydeoxyribonucleotide). Thus, conditions are easily created that will promote hybridization between any two complementary polynucleotide strands, resulting in the types of double-stranded molecules shown in Table 3.1, albeit with different thermodynamic stabilities. Keep in mind also that, although RNA molecules exist primarily in a single-stranded form in vivo, the formation of extensive intramolecular base-pairing occurs is well-documented and requires heat denaturation prior to use in many applications.

RNA Methodologies. DOI: https://doi.org/10.1016/B978-0-323-90221-2.00001-1

TABLE 3.1 Types of double-stranded molecules.

Type of molecule	Symbol	Relative thermostability
Double-stranded DNA	dsDNA or DNA:DNA	Least
DNA:RNA hybrid	DNA:RNA	Intermediate
Double-stranded RNA	dsRNA or RNA:RNA	Greatest

Note: The thermostability of double-stranded molecules is a function of the base composition and the length of the molecule, in addition to the nature of the participating molecules (RNA or DNA).

Regardless of base composition, double-stranded DNA (dsDNA) molecules are the least stable of the three possible types of double-stranded molecules. DNA:RNA hybrids are more stable than dsDNA; the RNA 2′-OH promotes additional hydrogen bond formation involving the backbone of the molecule. For the same reason, the most stable type of double-stranded molecule forms upon base-pairing of two complementary strands of RNA. In an experimental context, the formation of double-stranded RNA molecules permits very stringent hybridization and wash conditions and are therefore often desirable as a means of increasing the signal-to-noise ratio in an experiment. If not carefully controlled, however, RNA molecules, labeled or not, can participate in extensive and highly problematic nonspecific base-pairing which will produce extensive background, thereby interfering with accurate interpretation of the data.

Importance of controlling stringency

The individual strands of a double-stranded nucleic acid molecule are dynamic indeed. Several variables, all of which can be manipulated in the laboratory, affect the ability of two single-stranded nucleic acid molecules to form stable hydrogen bonds between their bases. Many molecular biology applications require the investigator to be able to promote or prevent the formation of a double-stranded molecule, and it is commonplace to alternate between the double-stranded and single-stranded forms of a molecule in the course of a single experiment. The term "stringency" describes a defined set of conditions that are used to control the "strandedness" of nucleic acid molecules (Fig. 3.1).

Stringency is a measure of the likelihood that a double-stranded nucleic acid molecule will dissociate into its constituent single strands. Thus, stringency is all about the behavior of complementary nucleic acid molecules under what might be considered fairly hostile conditions compared to the cellular milieu. In practice, higher-stringency conditions favor stable hybridization only between nucleic acid molecules with a high degree of complementarity, the so-called "exact-match" sequences. As the stringency in a system is lowered, a proportional

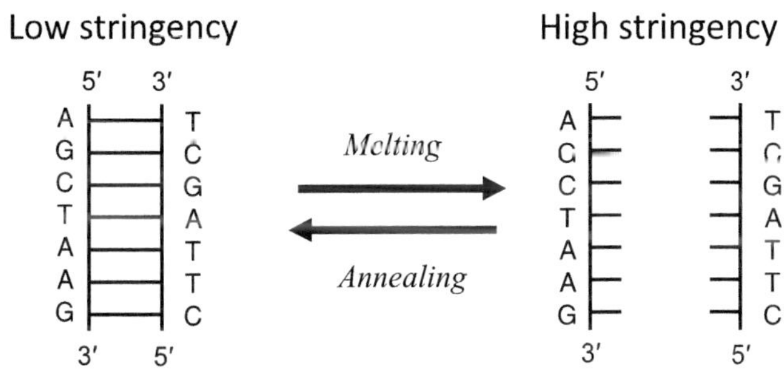

FIGURE 3.1 Effect of stringency on a nucleic acid structure. High stringency conditions favor single-strandedness while low stringency conditions promote base-paring between complementary molecules. Annealing and melting are repeatable, reversible processes. The formation of all double-stranded nucleic acid molecules occurs in an antiparallel configuration resulting in an orderly helical structure.

increase in nonspecific hybridization is tolerated, meaning that the base on one molecule is not complementary to the base directly across from it in a double-stranded molecule. These so-called "mismatches" cause distortions of the helix architecture.

The conditions that are commonly manipulated to either promote or prevent nucleic acid hybridization include pH, ionic strength, temperature, and the inclusion of formamide and urea (Table 3.2). Each of these acts as a function of time, meaning that the conversion from the double-stranded to the single-stranded form, and vice versa, while often rapid, is not instantaneous. The base composition (particularly the G + C content), the length of the molecule (number of bases), and the number of mismatches (if any) all influence the stability of the duplex at any given level of stringency. Practically speaking, manipulating these variables is really fine-tuning the ability of nucleic acid molecules to discriminate between other molecules that have a high degree of complementarity and those that exhibit a lesser degree of complementarity.

Frequently, the temperature required to create a particular high-stringency environment is so great that the chemical stability of the RNA or DNA molecule itself would be at risk. To preclude this potentially serious difficulty, formamide can be used to create high stringency conditions at lower temperatures such that the integrity of the molecules is unlikely to be compromised. The function of formamide is to lower the melting temperature (T_m) of base-paired molecules by interfering with hydrogen bonding. The T_m of a double-stranded molecule is the best thought of as an equilibrium point, that is, that temperature at which 50% of the double-stranded molecules will dissociate into their constituent single-strands; high-stringency hybridizations are often performed only a few degrees below the calculated T_m to ensure complete hybridization wherever possible. The net result of the inclusion of formamide is a favorable reduction in the

TABLE 3.2 Factors that influence nucleic acid structure.

Lower stringency	Higher stringency
High salt	Low salt
Slightly acidic pH	Alkaline pH
Low temperature	High temperature
No formamide	High formamide [50%–80%]
No urea	High urea [2–4 M]

Notes: The formation of double-stranded molecules is promoted in a low-stringency environment, while dissociation into constituent single strands is favored as the stringency is increased. The use of formamide or urea permits maintaining high-stringency conditions at temperatures lower than would otherwise be required. Both formamide and urea also have the added benefit of inhibiting RNase.

temperature needed to perform high-stringency hybridizations, thereby affording the investigator greater latitude with respect to experimental parameters. Similar results are often possible through the use of urea, instead of formamide.

The inclusion of formamide or urea as a means of modulating stringency can be complicated. In some cases, a stringency reduction may be achievable by simply raising the salt concentration and/or lowering the temperature of the system. The take-home lesson: low-stringency conditions favor or promote double-strandedness while high-stringency conditions favor or promote single-strandedness in a mixture of nucleic acid molecules. This is expanded upon in Chapter 17, Nucleic Acid Probe Technology.

Effect of salt on stringency

In an experimental setting the natural electrostatic repulsion that would prevent two complementary, negatively charged phosphodiester strands from base-pairing can be neutralized with low, moderate, or high concentrations of a monovalent cation, most often Na^+, K^+, or Li^+. Referred to as counterions, the positive charge that each contributes will reduce the tendency for natural electrostatic strand repulsion. At low concentrations or in the absence of salt, electrostatic repulsion will easily exceed the strength of the hydrogen bonds that could otherwise form between complementary base pairs, resulting in the dissociation of double-stranded molecules, particularly at elevated temperatures. Within the cell, natural cationic proteins neutralize the negative charge associated with nucleic acids, thereby preventing inappropriate strand denaturation. Physiological ionic strength, which varies by cell type, averages around 155 mM and ensures that appropriate base-pairing is tolerated.

Effect of pH on stringency

The degree of ionization of a nucleic acid molecule is strongly influenced by pH. At physiological (near neutral) pH, nucleic acids are slightly ionized; hybridization can occur between complementary polynucleotides even though the individual molecules demonstrate a net negative charge. Just below physiological pH, a slightly acidic environment will neutralize the phosphates, thereby favoring greater base-pairing between complementary strands. In vitro, however, a concern in a stronger acidic environment (pH <5.5) is nucleic acid depurination (read: degradation) which will occur in fairly short order. In contrast, a more alkaline pH will increase the ionization of the phosphate groups to the extent that electrostatic repulsion can easily exceed the thermodynamic stability of a double-stranded molecule, resulting in strand dissociation. Although using NaOH to denature double-stranded DNA is convenient, when working with RNA it is of paramount importance to remember the exquisite susceptibility of these molecules to degradation by alkaline hydrolysis. Do not expose RNA to a strongly alkaline environment (pH >8.6) unless the intention is to destroy it.

Effect of temperature on stringency

In general, the effect of temperature is opposite to that of salt concentration. By applying more heat to the system, one will eventually exceed the T_m of the duplex, causing dissociation of the strands. The shorter a nucleic acid molecule, the more sensitive it is to even subtle temperature differences. Any molecular biologist can attest to the fact that oligonucleotides used as hybridization probes or as primers for PCR often work (or don't work!) precisely because of what this Author refers to euphemistically as a thermodynamic miscalculation: one may easily promote or prevent hybridization by changing the temperature by as little as one degree. The thermodynamic stability of a duplex is a function not only of the number of bases involved but also of the guanine and cytosine (G + C) content: remember that three hydrogen bonds naturally occur between the bases guanine and cytosine (G:::C) and only two hydrogen bonds exist between adenine and thymine (A::T) and between adenine and uracil (A::U). The exact experimental conditions that define stringent hybridization and posthybridization washes are addressed in Chapter 17, Nucleic Acid Probe Technology, while the use of oligonucleotides as PCR primers is discussed in Chapter 10, RT-PCR: A Science and an Art Form, and Chapter 11, Quantitative PCR Techniques.

Effect of formamide on stringency

As noted above, the T_m of a double-stranded molecule is largely a function of the number of base pairs involved and the G + C content. As the length of the molecule decreases, the impact of G + C content on the thermodynamic behavior of the molecule becomes more pronounced. With longer probes, the hybridization temperature needed to maintain a comparable level of stringency would simply be too high to permit base pairing under any circumstances. When this occurs, the investigator may choose to incorporate formamide into a hybridization recipe. The reason for including formamide is to lower the T_m while maintaining high stringency: a high stringency environment is created at a temperature that is low enough to permit base pairing to occur while maintaining the specificity of the hybridization events. When performed correctly, high-fidelity hybridization is strongly favored without the concerns associated with temperature extremes. As a general rule, one may expect to see the T_m lowered by approximately 0.75°C for each 1% formamide added. While an effective additive for the T_m reduction, formamide is also a potent, toxic compound that tends to vaporize at the elevated temperatures commonly associated with high-stringency manipulations; it should therefore be handled with great care, following the containment recommendations of the manufacturer. In some settings, the substitution of urea may be an excellent alternative to formamide for controlling stringency.

Effect of urea on stringency

Urea has been widely used with polyacrylamide gel electrophoresis of single-stranded RNA and DNA molecules, as well as proteins. It has the desirable qualities of being a duplex destabilizing compound and is considerably less toxic than formamide. Thus, urea has been used successfully at concentrations ranging from 2 to 4 M in place of 50% formamide (Simard et al., 2001). At concentrations greater than 4 M it is believed that the concomitant increase in viscosity may be responsible for the diminished sensitivity of the assay. Urea is able to participate in hydrogen bond formation and therefore is able to destabilize double-stranded polynucleotide formation by providing alternative hydrogen bonding sites for the nitrogenous bases. As a direct result of these interactions, the T_m of a double-stranded molecule is reduced, making it easier to melt apart at lower temperatures. A protocol for the use of urea in agarose gel electrophoresis is presented in Chapter 14, Electrophoresis of RNA.

References

Simard, C., Lemieux, R., Côté, S., 2001. Urea substitutes toxic formamide as destabilizing agent in nucleic acid hybridizations with RNA probes. Electrophoresis 22, 2679–2683.

Further reading

Wittwer, C.T., Makrigiorgos, G.M., 2018. Nucleic acid techniques. In: Rifai, N., et al., (Eds.), Principles and Application of Molecular Diagnostics. Elsevier, pp. 47–86.

Chapter 4

RNA isolation strategies

Rationale

RNA is more labile than DNA, chemically and biologically, particularly at elevated temperatures ($>65^\circ C$) and in the presence of alkali. These intrinsic handling difficulties are further compounded by the aggressive activity of a variety of resilient ribonucleases (RNase), the apparent ubiquity of which is undisputed. These factors mandate the rapid recovery of RNA with reagents that are free of RNase contamination. Likewise, extraction tactics must be formulated to include sufficient management of endogenous RNase activity upon disruption of the organelles and vacuoles that normally sequester these enzymes. Failure to scrupulously observe procedures for eliminating potential sources of RNase contamination will almost always yield a useless sample of degraded RNA. Further, even trace quantities of contaminants carried over from some isolation procedures may compromise the efficiency of subsequent reactions. Due to their naturally short half-life, experiments involving RNA are most judiciously planned around the availability of cell cultures or tissue samples and the actual date of RNA isolation.

Efficient methodologies have been empirically derived to accommodate the expedient isolation of RNA, techniques that should be scrutinized and refined continuously. In general, these methods yield cytoplasmic RNA, nuclear RNA, or mixtures of both, commonly known as total or cellular RNA. Protocols for the isolation of RNA begin with cellular lysis mediated by buffers that typically fall into one of two categories: (1) those consisting of harsh chaotropic agents such as one of the guanidinium salts (guanidinium thiocyanate, GTC; guanidinium hydrochloride, G · HCl) sodium dodecyl sulfate (SDS), lithium dodecyl sulfate (LiDS), N-laurylsarcosine (sarcosyl), urea, phenol, or chloroform, all of which disrupt the plasma membrane and subcellular organelles and which simultaneously inactivate RNase; and (2) those that gently solubilize the plasma membrane while maintaining the integrity of the nucleus and other organelles, which are characteristically nonionic, hypotonic lysis buffers; organelles and cellular debris are then removed from the lysate by differential centrifugation. The best known and most efficient lysis buffers in this category contain Nonidet P-40 (NP-40), also sold as IGEPAL CA-630 and which is chemically indistinguishable from NP-40. While osmotic lysis is the most gentle method for cell disruption, it does nothing when carbohydrate-rich cell walls are present, as with certain

RNA Methodologies. DOI: https://doi.org/10.1016/B978-0-323-90221-2.00042-4

bacteria, fungi, and plant cells. Appropriate steps must be taken to break through the cell wall in order to access the cellular contents. It is worth noting that a seemingly endless list of permutations on a few fundamental RNA extraction techniques exist; for example, some techniques support the isolation of poly(A)$^+$ material directly from a cellular lysate without prior purification of total RNA. As far as which RNA extraction procedure to use, it is incumbent upon the investigator to always think two steps ahead and ask, "What is to be done with this RNA after it has been purified?"

RNA isolated from the nucleus or cytoplasm by direct cell lysis is known as *steady-state* RNA; it represents the final accumulation of the RNA in the cell or a subcellular compartment. Whereas the abundance of specific steady-state RNA species may certainly be interpreted as an indication of how specific genes are modulated, it furnishes no information about the *rate* at which these RNA molecules are transcribed or about their stability or half-life in the cell. This is the major disadvantage of evaluating the steady-state RNA alone. Cells respond to environmental change not only by altering the rate of transcription, but also through the processing efficiency of precursor mRNA, the efficiency of nucleocytoplasmic transport, the stability of coding and noncoding RNA in the cytoplasm, and the translatability of salient messages. The protocols presented in this chapter are directed at the recovery of steady-state transcripts primarily from cultured cells. Protocols for the isolation of RNA from animal and plant tissue are presented in Chapter 6, The truth about tissues, and Chapter 7, Going green: RNA and the molecular biology of plants, respectively. Protocols for transcription rate assays and the isolation of nuclear RNA, in general, are presented in Chapter 19, Analysis of nuclear RNA. The subpopulations of RNA that are common of experimental interest are listed in Table 4.1. It is incumbent upon the investigator to consider which population of RNA will result from a particular isolation procedure before picking up the micropipettors.

Goals in the purification of ribonucleic acid

Any worthwhile strategy for RNA purification must achieve specific goals if data derived from the final RNA preparation are to be meaningful and trustworthy. There are many procedures for isolating RNA from various biological sources, and each procedure in some way accomplishes these goals in order to produce a representative, high-quality sample.

Goal 1: Select an appropriate method of membrane solubilization.

The first decision to factor into the design of an RNA isolation strategy is the method of cellular disruption, and it is based on which population of RNA or subcellular compartment is to be investigated. For example, is the observed modulation of gene expression in an experimental system regulated transcriptionally or by some posttranscriptional event(s)? In such an instance,

TABLE 4.1 Categories of eukaryotic RNA.

RNA population	Membership
Total cellular RNA	All RNA from the cell
Total cytoplasmic RNA	Cytoplasmic RNA, excluding mtRNA
Nuclear RNA	Transcripts (mature and immature) isolated directly from intact nuclei
mRNA	Transcripts that have matured and been exported to the cytoplasm. This subpopulation includes both poly(A)$^+$ and poly(A)$^-$ transcripts
miRNA	MicroRNA (all types)
Cellular Poly(A)$^+$ RNA	Nuclear and cytoplasmic polyadenylated transcripts
Cytoplasmic Poly(A)$^+$ RNA	Cytoplasmic polyadenylated mRNA
Poly(A)$^-$ RNA	Nonpolyadenylated RNA (mostly rRNA, tRNA, and several noncoding RNA species)
mtRNA	Mitochondrial RNA (all types)
cpRNA	Chloroplast RNA (all types; plants only)

selection of the method of cellular disruption and subsequent RNA isolation must permit analysis of the nuclear abundance of salient transcripts independently of the abundance in the cytoplasm.

The method of cell lysis will determine the extent of subcellular disruption of the sample. For example, a lysis buffer that is used successfully with tissue culture cells may be entirely inappropriate for whole tissue samples. Just as important, the method by which membrane solubilization is accomplished will dictate whether additional steps will be required to remove DNA from the RNA preparation, and whether or not it will be possible to separate nuclear RNA and cytoplasmic RNA species independently of one another, postlysis. For example, DNA can be purged from an RNA preparation with minimal fanfare, though it is not possible to determine the relative contribution of RNA from the nucleus and the cytoplasm once RNA from these two subcellular compartments have copurified. A particular lysis procedure must likewise demonstrate compatibility with ensuing protocols once the RNA has been recovered. Always think two steps ahead; the proper method of solubilization is dependent on the plans for the RNA after purification and the questions being asked in a particular study.

Goal 2: Ensure total inhibition of nuclease activity.

The imperative for controlling nuclease activity should be abundantly clear, if not from personal experience, then from the discussion in Chapter 2, Creating a ribonuclease-free environment. This includes purging RNase from reagents, glassware, consumable plasticware, and equipment, controlling RNase in a cell lysate or tissue homogenate, and in all downstream applications. Although some lysis reagents inhibit nuclease activity in their own right, other lysis reagents require additional nuclease inhibitors to safeguard the RNA during the isolation procedure. Steps for the inhibition or elimination of RNase activity must, first and foremost, demonstrate compatibility with the lysis buffer. The isolation of RNA from biological sources enriched in RNase is challenging but conquerable.

Goal 3: Keep sample volumes as small as possible.

The loss of RNA will be minimized when isolation procedures and, if required, ensuing precipitation steps, involve small volumes. At higher nucleic acid concentrations, precipitation will be as efficient as possible. In addition, it is much easier to dilute a more concentrated sample (one step) than to try to increase the concentration of samples that are too dilute (several steps), which could potentially compromise sample integrity. Of course, it will be necessary to increase volumes with a larger mass of starting material. In this case, high-strength glass centrifuge tubes (e.g., Corex glass tubes or the equivalent) will be needed for the required high-speed centrifugation steps to collect the RNA precipitate. Smaller volumes in smaller tubes usually allow protocols to be completed in less time. Given the improvements in RNA isolation kits and procedures, working with volumes small enough to the contained in microcentrifuge tubes, and working out of an ice bucket, is always the much-preferred scenario.

Goal 4: Select a method for deproteinization of the sample.

The complete removal of protein from a cellular lysate or tissue homogenate is of paramount importance in the isolation of both RNA and DNA. Meticulous attention to this detail is required for accurate quantification, precision hybridization, and compatibility with ensuing processes. For example, the restriction of double-stranded DNA (dsDNA) and the action of DNA- and RNA ligase can be inhibited by carry-over protein, especially histone proteins. Further, reverse transcription (Chapter 9: cDNA: a permanent biochemical record of the cell) and amplification by PCR (Chapter 10: RT-PCR: a science and an art form) are inhibited strongly by the use of "dirty" RNA. Removal of proteins may be accomplished by:

a. digestion of the sample with the enzyme proteinase K,
b. repeated extraction with mixtures of organic solvents such as phenol and chloroform,
c. solubilization in guanidinium buffers,
d. physical separation by (silica) column chromatography,
e. salting-out of proteins,
f. any combination of the above.

Any procedure for deproteinization is itself a means of controlling RNase activity. It is important to remember that the RNA in the sample is subject to nuclease degradation once the protein denaturant or inhibitor is removed. RNA is never more susceptible to nuclease attack than when it is dissolved in a simple, nondenaturing aqueous buffer such as TE buffer or nuclease-free water.

Goal 5: Select a method for nucleic acid concentration.

This is the final step in most RNA purification schemes. The most versatile method for concentrating nucleic acids is precipitation using various combinations of salt and alcohol (Table 4.2). For example, one common method is to add 0.1 vol of 3 M sodium acetate (pH 5.2) to a nucleic acid sample, followed by the addition of 2.5 vol of 95%–100% ethanol. Nucleic acids and the salt that drives their precipitation to form complexes that have greatly reduced solubility in ethanol and in isopropanol. Further, the rate of precipitation using various salt and alcohol combinations is temperature-dependent. The dramatic precipitation of genomic DNA observed in ethanol is a direct function of genome size and complexity. In contrast, the

TABLE 4.2 Salt and alcohol combinations for nucleic acid[a] precipitation.

A.	Salt	Typical stock concentration	Amount required	Final concentration
	NaOAc	3 M, pH 5.2	0.1 vol	300 mM
	NaCl	1 M	0.1 vol	200 mM
	LiCl[b]	8 M	0.1 vol	800 mM
	NH_4OAc[c]	10 M	0.2 vol	2 M
	KOAc	2.5 M	0.1 vol	250 mM
	Glycogen[d]	10 mg/mL	0.01 vol	100 μg/mL
B.	Alcohol		Volume required after addition of salt	
	Ethanol (95%–100%)		2.2–2.5 vol	
	Isopropanol (100%)		0.6–1.0 vol	

[a]*Precipitation thresholds: DNA 50 ng/mL, RNA 100 ng/mL. If the concentration falls below the threshold, precipitation will be very inefficient unless a carrier is added.*
[b]*Lithium chloride does not coprecipitate with nucleic acids. As with NaCl, it is very soluble in ethanol. Avoid LiCl if RNA is to be reverse transcribed unless the RNA will be double-precipitated in order to remove the excess LiCl.*
[c]*Preferred salt for precipitating cDNA after synthesis is ammonium acetate because it does not precipitate unincorporated dNTPs.*
[d]*The judicious use of glycogen is recommended because excessive quantities may inhibit subsequent* in vitro *reactions.*

precipitation of RNA is much more dignified, often requiring longer incubation periods at −20°C to ensure complete recovery, especially when using the nonionic lysis procedure described below, and owing to the relatively low complexity of cellular RNA. Other procedures for concentrating nucleic acid samples, including the use of commercially available concentrating devices, dialysis, and centrifugation under vacuum, are not as frequently employed because the degree of success is much more dependent on the skill of the investigator than with simple salt and alcohol concentration. An older procedure for the concentration of nucleic acids, repeated extraction with butanol, is not recommended because of the inherent difficulties associated with the complete removal of butanol. Finally, the use of the newer silica-based technologies often precludes the need to add any salt or alcohol to precipitate the sample because the RNA can be eluted from the column in concentrated form and immediately ready for the next step. Silica spin columns are in widespread use and yield very pure and high-quality nucleic acid preps.

Goal 6: Select the proper storage conditions for purified samples of RNA.

Because of the naturally labile character of RNA, improper storage of excellent RNA samples will often result in degradation in a short time. There are many opinions as to the proper temperature, buffer, and storage form for both RNA and DNA. A discussion of storage considerations and options appears at the end of this chapter.

The word on kits

The investigator who prepares in-house reagents "from scratch" for RNA isolation is something of a rarity now because so many vendors are marketing products that save both time and money. It is certainly true that kits can increase productivity, though it is essential for the investigator to ascertain that a kit which looks and sounds attractive (1) produces RNA that will be compatible with downstream applications, and (2) yields the population of RNA that is needed (e.g., total cellular, cytoplasmic, poly(A)$^{+}$), or at least enrich in favor of that subpopulation. Don't be afraid to pick up the telephone and call the local sales rep or even the national sales office to politely request an evaluation kit (read: free kit). If you are genuinely interested in a particular product, most vendors will be happy to provide you with a one-time sample, or at least offer an attractive discount.

The more desirable kits minimize or altogether eliminate the use of the organic solvents phenol and chloroform, and do so mostly through the use of silica columns or a magnetic separation platform. If a kit or reagent does utilize either or both of these hazardous compounds, the isolation procedure has been optimized to require the smallest possible volumes. Be sure that residual organic materials, regardless of the volume, are disposed of in accordance with in-house policies. Perhaps the best innovation in "kit technology" in recent years has been the development of products that are designed to recover RNA molecules that range between 20 and 200 bases, rather than discarding them as "undesirables." The exigency of this technological improvement is apparent when one considers the tremendous importance of miRNA and other small molecules, both coding and noncoding, in the regulation of gene expression.

Finally, there are many options in terms of instrumentation designed for automated RNA isolation or for DNA isolation, or both. The notion of automation for RNA extraction is attractive, but the instrumentation, commonly referred to as robotic liquid handling, is found mostly in research lab settings that would be considered high-throughput enterprises. Beginning with the coronavirus disease 2019 (COVID-19) pandemic, however, there was a sudden and sustained need for the ability to process very large numbers of COVID-19 sputum samples as well as specimens collected as nasal swabs. Methodologies often include automated magnetic separation, a manual version of which is described in the next chapter for poly(A)$^+$ mRNA selection. RNA extraction machines are compatible with both tissue samples and cell samples, though the protocols differ. Given the number of samples that nucleic acid extraction systems typically process, many are outfitted with a UV light for both decontamination and disinfection purposes.

Silica technology

An important improvement in the area of nucleic acid isolation was the development of silica filter columns that are small enough for use with a standard microcentrifuge. These filters consist of glass microfibers that are positioned at the bottom of a small plastic column that fits inside a standard 1.5 mL microcentrifuge tube (Fig. 4.1). Not only are these products used to purify RNA from biological sources, but they can also be used to clean up RNA or DNA after various manipulations, such as restriction enzyme digestion, ligation reactions, cDNA synthesis, and PCR amplifications. Silica filter-based products are readily available from any number of vendors. Best of all, nucleic acid purification and cleanup are performed in a remarkably short time and in convenient small volumes.

RNA and DNA bind to silica in the high-salt, chaotropic environment produced by diluting a nucleic acid sample in guanidinium thiocyanate. Because of the affinity to silica, nucleic acid isolation in this manner is referred to as solid-phase extraction. Following a series of washes, the purified material is eluted from the silica matrix under very low salt conditions and often in a sufficiently small volume such that precipitation with salt and alcohol is not necessary, which is always desirable. One common mistake that investigators make when using the silica column approach is the incomplete removal of the wash buffer at the end of each centrifugation step.

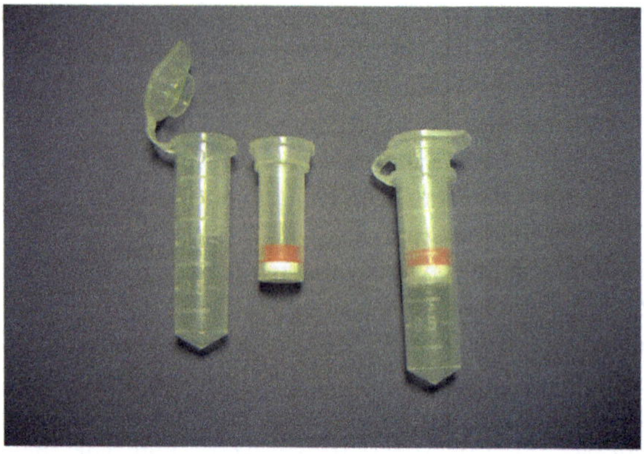

FIGURE 4.1 Silica filter for nucleic acid isolation. Under chaotropic, high-salt conditions, nucleic acids bind to the silica filter, while lysis buffer components and other cellular macromolecules are washed away. The purified sample is then eluted under low-salt conditions. The most commonly used filter format used in the laboratory is small enough to fit into a standard 1.7 mL microcentrifuge tube or 2.2 mL microcentrifuge tube (shown). *Courtesy of Sigma-Aldrich.*

The wash buffers usually contain significant amounts of ethanol, which retains RNA on silica columns. More often than not, a small droplet of residual ethanol remains at the bottom of the column at the point of egress. It is also common to find residual wash buffer on the outer wall of the silica column. Inasmuch as the column is removed from the microcentrifuge waste collection tube so that the flowthrough material can be discarded, it is wise to wipe the outside of the column with a Kimwipe before reinserting the column into a fresh microcentrifuge collection tube for the elution step. If any of the ethanol-containing wash buffer carries over with the eluted RNA, the sample will be completely useless for electrophoresis (the sample will float out of the well, even after the addition of loading buffer) and any enzymatic manipulations will be inhibited. Ethanol-tainted samples should be reapplied to a fresh silica column, in lysis buffer, and then eluted properly.

Isolation of cytoplasmic RNA on a silica column

Many kits have been developed for the isolation of DNA and RNA since all nucleic acids have a strong binding affinity for silica filters under high-salt (chaotropic) conditions. Thus, the preliminary handling of the biological material and the lysis buffer formulation determine whether DNA or RNA will be recovered at the end of the procedure. For example, cell lysis might be performed with a quasialkaline lysis buffer to which RNase may have been added, resulting in DNA isolation and RNA obliteration. Alternatively, incubation of an RNA prep with RNase-free DNase will usually render a very respectable RNA yield, assuming that standard RNase-free techniques are observed throughout. One might even consider gentle nonionic cell lysis to remove the nucleus (below), after which the cytosol is processed as if working with a whole-cell suspension. This will render only cytoplasmic RNA, rather than cellular RNA, and has all of the advantages associated with silica column chromatography. This approach to nucleic acid isolation is very popular because it can be used with small amounts of tissue or cells, can be performed on the bench quickly, easily, and economically with standard equipment, and eliminates the need for any organic solvents. However, these systems often contain guanidinium buffers, which are hazardous, and should be handled according to the manufacturer's recommendations and the material safety data sheet (SDS; *formerly*, MSDS).

Affinity matrices

In addition to the approaches described above for the isolation of total cellular RNA or total cytoplasmic RNA, there are products available which capture polyadenylated transcripts. For example, mRNA isolation kits feature tracts of oligo(dT) that have been covalently linked to a solid support, such that polyadenylated transcripts are captured through canonical base-pairing between the poly(A) tail and the oligo(dT) in a high-salt environment. Some kits feature columns that are prepacked with oligo(dT) linked to cellulose, while in other cases the oligo(dT) is linked to synthetic beads. Affinity selection in this manner is similar to the magnetic separation methods described in Chapter 5, Isolation of polyadenylated RNA. The benefit associated with affinity selection is the enrichment of a nucleic acid sample in favor of mRNA by minimizing, if not altogether eliminating, carryover of rRNA and tRNA; enrichment in this manner may increase the sensitivity of an assay. Affinity selection in which poly(A)$^+$ mRNA was affinity-captured using a column packed with poly(U) linked to sepharose beads is no longer in use because of the relatively poor binding capacity of poly(U) matrices and the fact that quantitative recovery of RNA from a poly(U) matrix usually required formamide-based elution buffers.

Another variant of the affinity matrix approach is designed to study nucleic acid–protein interactions by passing a heterogeneous protein mixture over a column packed with either RNA or DNA oligonucleotides to capture proteins with some level of binding affinity to the sequences on the column. The nucleic acid species on the column is often referred to as the bait, while the proteins that can bind to it are known as the prey. This is sometimes referred to as a pull-down method and is popular for the characterization of RNA or DNA binding proteins. Finally, chromatin immunoprecipitation as a means of enriching for nucleic acid sequences associated with the binding of specific proteins is described in Chapter 23, Transcriptomes and bioinformatics.

Lysis buffer formulations

Everyone has a favorite method for the isolation of the RNA, and there are numerous permutations on the protocols that accompany many of the commercially available products for RNA isolation. In general, however, all of these methodologies fall into one of two major categories, depending on the degree of cellular disruption and capacity for RNase inhibition. These categories are referred to herein as gentle lysis buffers and harsh lysis buffers. Each has a role in the investigation of gene expression, and each has limitations, as described in the following sections. The most important thing to remember is that there is no one right way to extract RNA from cells and tissues and the only wrong way of doing things is to permit the introduction of RNase activity.

Gentle lysis buffers

Cellular lysis mediated by nonionic, hypotonic buffer is not disruptive to most subcellular organelles. The inclusion of nonionic detergents such as NP-40 (Favaloro et al., 1980) and $MgCl_2$ in the lysis buffer facilitates plasma membrane solubilization while maintaining nuclear integrity. Thus, lysis of cells with NP-40-containing buffer precludes the mixing of nuclear and cytoplasmic RNA from the onset. In such protocols, intact nuclei, large organelles, and cellular debris are easily removed by differential centrifugation. The resulting cytosolic supernatant is rich in cytoplasmic RNA and proteins, the latter being easily removed by a series of extractions with mixtures of organic solvents such as phenol and chloroform or, preferably, by silica column chromatography. The obvious advantage of this isolation strategy is that the RNA precipitated at the end of the procedure represents only the cytoplasmic population. This may be especially helpful when attempting to assign gene regulation as transcriptional or posttranscriptional in nature.

A disadvantage of using this approach is that the lysis buffer alone is not sufficiently chaotropic to fully inhibit RNase activity. Keep in mind that upon cellular lysis, normally sequestered RNases are suddenly liberated, and their activity will greatly compromise the integrity of the RNA even as the investigator is working diligently to purify it. It may be helpful, but it is not necessary, to add some type of RNase inhibitor to the lysis buffer *just prior to use* in order to control nuclease activity. In this laboratory, rather than adding a nuclease inhibitor to gentle lysis buffers, very high-quality RNA is recovered when the sample is maintained on ice at all times unless the protocol specifically dictates otherwise. It is also very helpful to use ice-cold reagents and tubes that have been prechilled. Should one wish to take advantage of the disruptive nature of the guanidinium-based isolation procedures discussed below and still maintain the subcellular compartmentalization of RNA, one worthwhile strategy is to begin the isolation procedure with a nonionic lysis buffer, recover intact nuclei, and then lyse the nuclei with guanidinium buffer; one may proceed to purify the nuclear (or cytoplasmic) RNA as if working with intact cells. This approach is particularly suited for the isolation of nuclear RNA. Protocols describing the use of guanidinium buffers are described here; for a more complete discussion of the isolation of nuclear RNA, see Chapter 19, Analysis of nuclear RNA.

Protocol: isolation of cytoplasmic RNA by gentle hypotonic lysis

This procedure is based on nonionic, hypotonic solubilization of the plasma membrane while maintaining nuclear integrity; the investigator may wish to optimize empirically the lysis conditions because cells and organelles exhibit different sensitivities to detergents, salts, and RNase inhibitors. It is used for the isolation of cytoplasmic RNA as described here and is likewise used for the isolation of intact nuclei and accompanying purification of nuclear RNA. The procedure is utilized routinely in this lab with ice-cold reagents and without any RNase inhibitors, and renders very high-quality RNA. An in-house modification of this valuable technique is presented here (Fig. 4.2). If RNase inhibitors are to be used they must remain with the lysate until the phenol:chloroform mixtures are introduced into the system to remove the protein. Finally, the RNA is recovered by salt and alcohol precipitation of the RNA, a widely used preparative technique. Whereas this method is faster than the older guanidinium-based methods requiring ultracentrifugation, it may appear to be more time-consuming than the acid–phenol, and nonphenol, methods described below. This is due, in part, to the requirement for gentle handling of the sample to preclude nuclear breakage, which, to a limited extent, *may* put the RNA molecules at slightly greater risk of RNase degradation. RNA prepared according to this method is quite adequate for most standard applications. This procedure works best when starting with cells grown in tissue culture; the lysis buffer does not disrupt whole tissue samples as efficiently as cells harvested from tissue culture. The protocol may be adapted for whole tissue samples when combined with gentle Dounce homogenization to prevent serious damage to organellar membranes.

In advance: preparation of extraction buffer

An adequate volume of extraction buffer should be freshly prepared before each use, as the oxidation of phenol can greatly compromise the quality of an RNA preparation (Fig. 4.3). Be sure to handle all organic solvents with proper skin and eye protection, in a chemical fume hood, and be sure to dispose of organic waste according to departmental regulations.

1. Melt redistilled (molecular biology grade) phenol at 60°C and, as indicated in Fig. 4.3A, saturate it by mixing the phenol with an equal

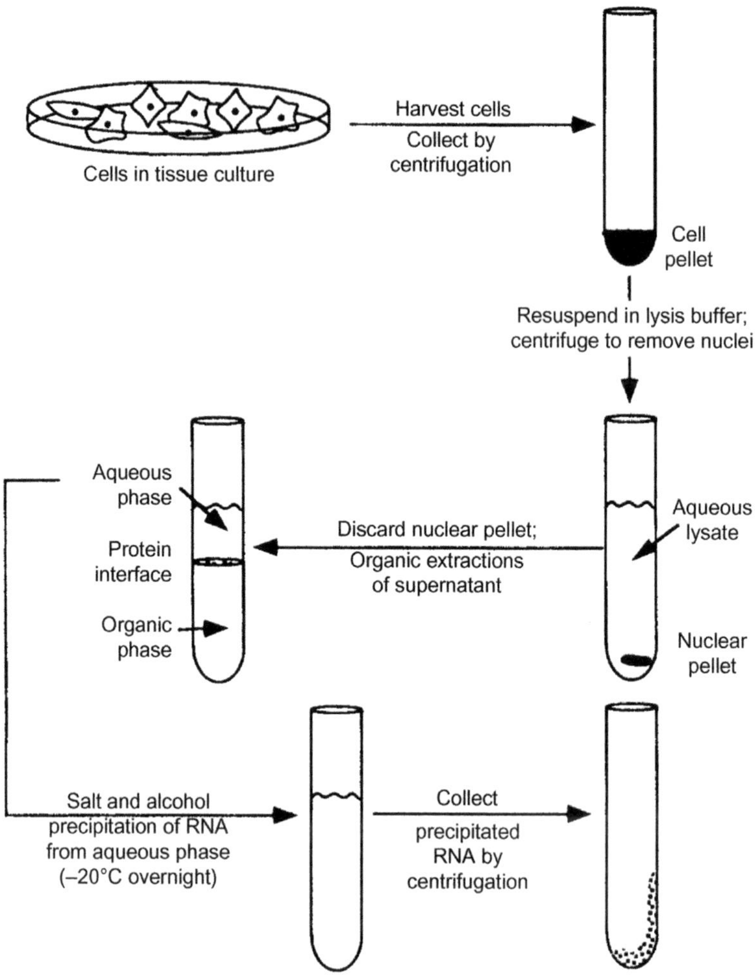

FIGURE 4.2 Preparation of total cytoplasmic RNA by gentle hypotonic lysis. This method is best suited for cells grown in culture; soft tissue samples may require a gentle, nonaggressive homogenization step.

volume of Tris-SDS buffer:

100 mM NaCl,

1 mM EDTA,

10 mM Tris, pH 8.5,

0.5% SDS.

Note: The inclusion of SDS ensures the quantitative recovery of poly $(A)^+$ RNA in the aqueous phase during extraction with organic buffer. Saturation of phenol with Tris-SDS buffer is desirable also because this partitioning phenomenon in the presence of SDS is pH-independent.

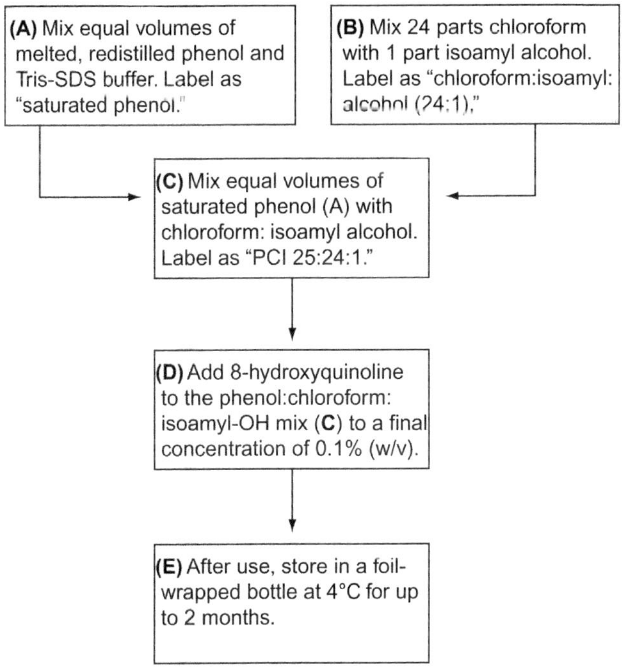

FIGURE 4.3 Method of preparation of the organic extraction mixture used for protein removal. Saturation of phenol with Tris/SDS buffer, rather than Tris alone, and the inclusion of 8-hydroxyquinoline renders a superior formulation for the deproteinization of the cytosol. This buffer can be substituted in most applications when the use of phenol:chloroform:isoamyl alcohol is dictated, and can be used in both DNA and RNA isolation procedures. The extraction buffer is assembled stepwise (A – E); refer to protocol in the text for details.

Be sure to autoclave this buffer, or at least the stock solutions, before adding SDS to complete this solution.

2. Mix the resulting Tris-SDS/phenol with an equal volume of a mixture of chloroform:isoamyl alcohol (24:1) (Fig. 4.3B) in a sterile glass bottle. The resulting reagent is known as the extraction buffer (Fig. 4.3C).
3. Add 8-hydroxyquinoline to the extraction buffer (Fig. 4.3D) to a final concentration of 0.1% (w/v). Allow the phases to separate at 4°C. Ideally, the bottle should be wrapped in foil when not in use, especially if the refrigerator is opened and closed continuously throughout the day.

Example:: Mix 20 mL of melted, redistilled phenol with 20 mL (an equal volume) of Tris/SDS buffer. To the resulting 40 mL of Tris/SDS-saturated phenol, add 40 mL (an equal volume) of chloroform:isoamyl alcohol, previously mixed in a ratio of 24:1. This is the extracting buffer, the stability of which is enhanced by the addition of 0.08 g of 8-hydroxyquinoline (final concentration of 0.1% w/v). This buffer is conveniently prepared and stored in a 100 mL glass bottle and has a shelf life of about 4–8 weeks at 4°C.

RNA isolation

1. Wear gloves and appropriate protective eyewear!
2. Harvest cells from tissue culture and collect by centrifugation in a suitable tube, prechilled on ice.
3. Gently, and completely, resuspend the cell pellet in ice-cold lysis buffer:

   140 mM NaCl,
   1.5 mM MgC12,
   10 mM Tris-Cl, pH 8.5,
   0.5% NP-40,
   Use at least 100 μL lysis buffer/10^6 cells.

   Note 1: An easy way to disperse the cell pellet is to drag the tube with the pelleted cells across the surface of a microcentrifuge tube rack, the action of which will help disperse the cell when the lysis buffer is added.

   Note 2: Autoclave the lysis buffer before adding the NP-40 needed to complete this reagent.
4. Incubate on ice for 5 minutes. Periodic *gentle* inversion of the tube may facilitate cell lysis. Incubate on ice for 5 minutes.

   Note 1: The extent of cell lysis may be assessed by examining a 1-μL aliquot of the lysate on a microscope slide for the presence of intact cells vs free nuclei.

   Note 2: In older versions of this protocol, an optional component of the lysis buffer was 10 mM vanadyl ribonucleoside (VDR) complexes. VDR oxidizes very rapidly and therefore should be added to the lysis buffer just before use. It is usually prepared/supplied as a 200 mM stock solution and should be used at a working concentration of 5–20 mM. The use of VDR has fallen out of favor, however, because of its strong inhibitory effect on reverse transcriptase and other enzymes. See Chapter 2, Creating a ribonuclease-free environment, for the discussion of advantages and disadvantages of VDR, and for a synthesis protocol.

   Note 3: As an alternatively to VDR, RNasin (Promega) can be added to the lysis buffer (250–1000 U/mL) to inhibit Rnase activity. Beyond this point in the protocol, the tube containing the RNA should be maintained on ice unless otherwise indicated.
5. Pellet the nuclei and other cellular debris by differential centrifugation for 5 minutes at 4°C. Centrifuge at 5000 × *g* in a microfuge, or at a suitable *g*-force if another type of tube is used.

   Note 1: Although nuclei can be efficiently pelleted with as little as 500 × g, the added force is useful for the more complete removal of large organelles and the tight packing of the nuclei. This centrifugation will result in the removal of genomic DNA as well as the nuclear hnRNA species and will facilitate the analysis of extranuclear RNA. Some nuclear breakage is inevitable and contaminating genomic DNA can be easily removed by incubation with DNase.

Note 2: If the nuclei are to be used in a later application, such as the isolation of hnRNA or genomic DNA, then collect the nuclei at a maximum of 500 × g and extend the centrifugation time as need. Additional details may be found in Chapter 19, Analysis of nuclear RNA.

6. Transfer the supernatant (cytosol) to a fresh, RNase-free tube and add EDTA (pH 8.0) to a final concentration of 10 mM.

Note: Na_2-EDTA, routinely prepared as a 500 mM stock solution, is added to prevent the magnesium-mediated aggregation of nucleic acids with proteins and with each other. Adding EDTA to the lysate prior to this step would have caused nuclear rupture and contamination of cytoplasmic RNA with nuclear transcripts as well as DNA.

7. Add an equal volume of the organic phase (lower phase; yellow) of the extracting buffer to the aqueous cytosol (lysate). Using great care, mix thoroughly by inversion for 15 seconds or longer. Incubate the sample at 55°C for 5 minutes.

Note 1: If the prep is being done in a microfuge tube, hold the tube between the thumb and index finger and invert the tube several times. This technique will give the most complete intermixing of the phases, the most complete denaturation of proteins, and minimize the likelihood of accidental spillage.

Note 2: The heating of the sample will facilitate a more thorough removal of proteins. Some variations of this procedure direct the investigator to do a series of organic extractions without heating the sample. The approach recommended here reduces the volume of organic waste.

Note 3: Upon contact with VDR, the initially yellow 8-hydroxyquinoline becomes dark green-black. This is because 8-hydroxyquinoline will chelate heavy metals (vanadium). In the absence of 8-hydroxyquinoline (or VDR), no color change is expected.

8. Incubate the sample on ice or in an ice water bath for 5 minutes.

9. Centrifuge the sample at 500–2000 at 4°C or at top speed in a microfuge to separate the phases.

Note: The time required to complete the separation will be a direct function of the volume and the size of the tube. A maximum of 5 minutes is usually adequate. Although a refrigerated microfuge is useful, it is not required.

10. Carefully transfer the upper aqueous phase to a fresh, RNase-free tube. Use great care to avoid disturbing the protein interface which appears as a thin white band between the upper and lower phases. It is much better to leave a few microliters of aqueous material behind than to risk transfer of any material at the interface.

11. Add a fresh aliquot (equal volume) of extracting buffer to the aqueous phase and mix carefully and thoroughly by inversion. Centrifuge at 500–2000 × g at 4°C or at top speed in a bench-top microfuge to separate the phases. It is not necessary to heat and cool the sample again, as described in steps 7–8.

12. Carefully transfer the upper aqueous phase to a fresh RNase-free tube. If a protein interface is apparent or if the aqueous phase remains cloudy, repeat the organic extraction described in Step 11 until no protein appears at the interface and the aqueous (upper) phase is clear.

Note: Additional phenol extractions are usually necessary when extracting a large number of cells in a relatively small volume. Using 100 μL lysis buffer per 10^6 cells at the beginning of this protocol usually precludes this difficulty.

13. Extract the aqueous material one last time with an equal volume of chloroform or chloroform:isoamyl alcohol (24:1) and mix carefully and thoroughly by inversion. Centrifuge for 30 seconds to separate the phases.

Note: Because phenol is very soluble in chloroform, a final chloroform extraction will remove any traces of phenol from the RNA-containing aqueous phase. This should be a standard technique whenever phenol is involved. Remember that even trace amounts of phenol will oxidize into quinones which will compromise the quality of a nucleic acid sample.

14. Transfer the aqueous phase (upper) to a fresh, RNase-free tube and add 0.1 vol 3 M sodium acetate, pH 5.2 and 2.5 vol ice-cold 95% ethanol. Mix thoroughly by inversion.

15. Precipitate the RNA overnight at −20°C.

Note: To ensure complete RNA recovery, precipitate the samples overnight at -20°C.

16. Collect the precipitated RNA by centrifugation at 9000 × *g* at 4°C for 30 minutes if using Corex glass tubes or the equivalent (be sure to use a rotor adapter to support glass tubes) or at 12,000 × g in a microfuge for 10 minutes at 4°C (preferred) or at room temperature.

17. Carefully decant the ethanol supernatant. Excess coprecipitated salt should be removed by washing the RNA pellet 2–3 times with 70%–75% ethanol, into which the salt will dissolve (sodium acetate is particularly soluble in 70%–80% ethanol). This concentration of ethanol is not sufficiently aqueous, however, to redissolve precipitated nucleic acids.

18. Briefly air dry the tube(s). If desired, a final wash with 95% ethanol will facilitate the drying of the sample. The RNA is best stored at −80°C as an ethanol precipitate or, following quantification, stored in suitable aliquots at −80°C so as to avoid repeated freezing and thawing. Depending on the cell type, typical yields range from 75 to 100 μg cytoplasmic RNA per 10^7 cells.

Chaotropic lysis buffers

There is probably no better way to deal with seemingly recalcitrant RNases than to disrupt cells in guanidinium lysis buffer (Chirgwin et al., 1979). Guanidinium buffers efficiently denature and solubilize proteins, including RNase, which inhibits their degradative prowess. This effect is enhanced by the inclusion of the ionic detergent sarkosyl. It is not necessary to add additional RNase inhibitors to such lysis buffers, and RNA isolation procedures conducted under these harsh conditions can be performed with room temperature reagents. Because these buffers are so chaotropic (biologically disruptive), organelle lysis also accompanies disruption of the plasma membrane. This occurrence liberates heterogeneous nuclear RNA (hnRNA) and genomic DNA from the nucleus as well as mitochondrial DNA and RNA, all of which copurify with the cytoplasmic RNA. Therefore, further purification is required to remove DNA from the sample. Formerly, the most prevalent of these methods was a technique known as isopycnic centrifugation through a cesium chloride (CsCl) gradient (Glišin et al., 1974; Ullrich et al., 1977) or a cesium trifluoroacetate (CsTFA) gradient (Carter et al., 1983; Zarlenga and Gamble, 1987).

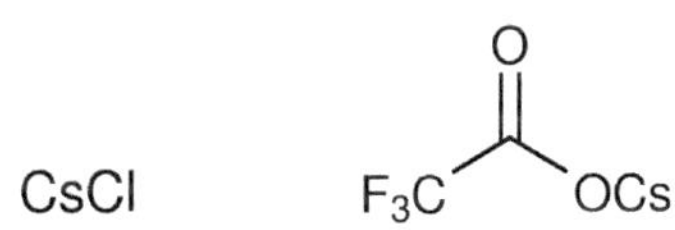

CsCl
Cesium Chloride

CsTFA
Cesium trifluoroacetate

Isopycnic separation is possible because of the differing buoyant densities of DNA (1.5−1.7 g/mL) and RNA (1.8−2.0 g/mL). For the convenience of the reader, a review of the fundamentals of centrifugation as a tool in molecular biology is presented in Appendix I.

Subsequently, the differential partitioning of DNA, RNA, and protein by guanidinium−acid−phenol extraction was described (Chomczynski and Sacchi, 1987). Upon publication, this method eliminated the need for the ultracentrifugation step previously associated with the use of guanidinium lysis buffers. It is now one of the most commonly used techniques for the rapid purification of RNA. In this method, the salient chemical differences between RNA, protein, and DNA are exploited by creating an acidic pH environment in the presence of the organic solvents phenol and chloroform, hence the nomenclature as the "GTC-acid-phenol" method. The convenience and simplicity of this method led to the development of several products and related methodologies for the rapid, efficient purification of both RNA and DNA (and protein) from the same biological source (e.g., Majumdar et al., 1991; Chomczynski, 1993). Currently, the use of glass fiber filters (a.k.a. silica binding technology) is the most popular method for small-scale RNA isolation and is used in conjunction with guanidinium-based cell lysis. As noted above, the principal drawback of all of these chaotropic procedures, however, is that the investigator is not able to easily discriminate between cytoplasmic and nuclear RNA, as there is no method for separating hnRNA from spliced messenger RNA (mRNA) once these two populations have been mixed (size fractionation may result in a partial separation, but is not at all definitive).

It is unfortunate that some seasoned investigators begin to show signs of sloppiness with respect to keeping buffers and equipment nuclease free when routinely working with guanidinium buffers. Although it is true that RNA is safe from nuclease degradation in the presence of these agents, purified RNA is once again susceptible to nuclease degradation when the denaturants have been removed. Therefore, it is necessary to guard against the RNase peril consistently, according to the procedures and guidelines described in the previous chapter.

Isolation of RNA with guanidinium buffers

RNA lysis buffers that contain guanidinium thiocyanate or guanidinium–HCl reproducibly yield very high-quality RNA samples. This is true because of the extremely chaotropic nature that these chemicals exhibit; they are among the most effective protein denaturants. As such, guanidinium is a dependable reagent for RNA extraction from whole tissue samples, particularly those enriched in RNase. The nuclear and organelle disruption that accompanies plasma membrane solubilization liberates nuclear RNA and genomic DNA, both of which copurify with cytoplasmic RNA species. The efficiency of protein denaturation, including disruption of RNases, may be enhanced by the inclusion of β-mercaptoethanol (β-ME). This reducing agent acts to break intramolecular protein disulfide bonds. The addition of exogenous inhibitors of RNase is not required. Another common reducing reagent, dithiothreitol (DTT), should be avoided in this particular application because it is chemically reactive with guanidinium.

The use of guanidinium-based lysis buffers mandates a procedure for the partitioning of RNA, DNA, and protein in the resulting lysate. There are currently three basic approaches for accomplishing this task: examples of each are presented here. These approaches involve whole cell lysis followed by isopycnic ultracentrifugation, acid–phenol extraction, or binding to glass fiber filters (Fig. 4.4). Although the RNA is most often precipitated directly from a guanidinium-containing lysate, ultracentrifugation through CsCl

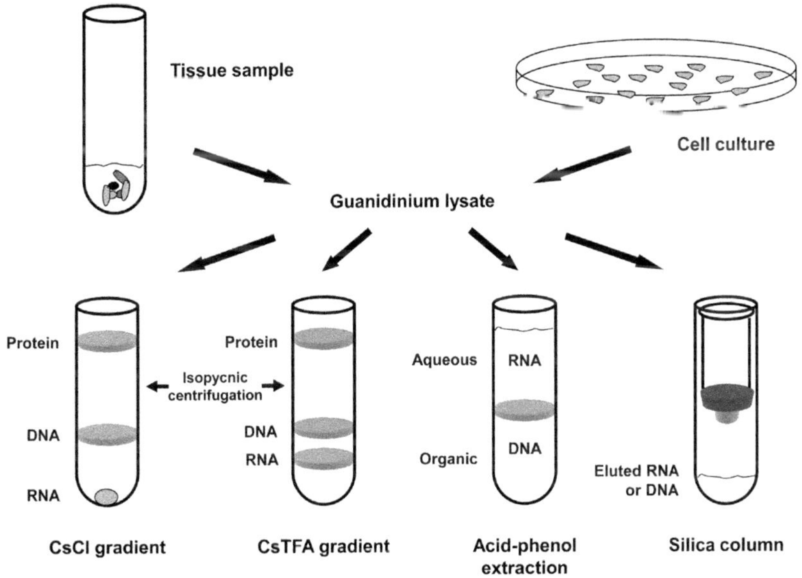

FIGURE 4.4 Methods for isolation of total cellular RNA following cellular disruption with guanidinium buffer: CsCl or CsTFA ultracentrifugation, acid-phenol extraction, or binding to a glass fiber filter. See text for details pertaining to the advantages and disadvantages of each.

(or CsTFA), though cumbersome, is useful when ultrapure RNA is required. Running gradients is generally a time-consuming procedure that is usually no longer required for mainstream molecular biology applications.

In contrast to the gentle lysis methods, which facilitate the isolation of total cytoplasmic RNA, procedures involving the use of guanidinium-containing buffers are used in the isolation of total cellular RNA (nuclear and cytoplasmic species). The RNA size distributions from total cytoplasmic and total cellular isolation procedures are compared in Fig. 4.5.

Guanidinium–acid–phenol extraction techniques

The highest quality RNA indisputably results from the extraction of RNA mediated by chaotropic lysis buffers, and guanidinium-containing buffers are among the most effective. As described above, these isolation procedures have the advantage of disrupting cells grown in culture or whole tissue samples rapidly and completely, simultaneously inactivating RNase activity, even when it is present in great abundance. The original procedures were labor-intensive because of the inclusion of an ultracentrifugation step to separate RNA from DNA based on differences in buoyant density. In contrast, the acid–phenol method describes the isolation and purification of undegraded RNA by treatment of cells with guanidinium thiocyanate-containing buffer formulations, but

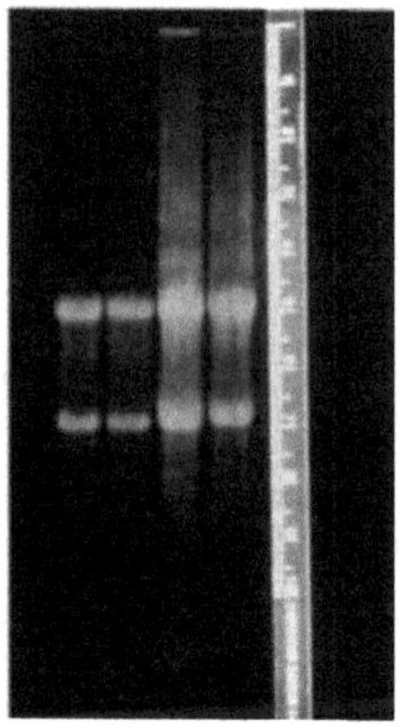

FIGURE 4.5 Comparison of size distribution of total cellular RNA and total cytoplasmic RNA. Lanes a and b: 20 μg total cytoplasmic RNA isolated with NP-40 lysis buffer and subsequent deproteinization with phenol:chloroform, as described. Lanes c and d: 25 μg of total cellular RNA isolated by guanidinium-acid-phenol extraction. The higher molecular weight species (hnRNA) are clearly visible above the 28S rRNA in lanes c and d. The obvious fluorescence in the wells of lanes c and d suggests a small amount of contaminating DNA. Fluorescent ruler courtesy of Diversified Biotech, Inc.

without the need for subsequent CsCl ultracentrifugation of the sample. In this and related procedures, the RNA is isolated in a very short time by extraction of a guanidinium cell or tissue lysate with an acidic phenol solution, followed by the addition of chloroform to facilitate partitioning of the aqueous and organic material. The required components for this type of RNA extraction procedure are easily prepared in the molecular biology laboratory. Alternatively, very popular premixed monophasic formulations of guanidinium thiocyanate and phenol are readily available [e.g., TRI Reagent (Molecular Research Center), TriPure (Roche), TRIzol (ThermoFisher Scientific)]. These products represent modifications of and improvements to the original published protocols. Upon phase separation, RNA is retained in the aqueous phase, while DNA and proteins partition into the organic phase; a protein interface is generally not observed due to protein solubility in guanidium solutions. After transferring the aqueous phase to a separate tube, the RNA is then recovered by precipitation with isopropanol and collected by centrifugation. In some cases, these commercial formulations are used in conjunction with a spin column or in magnetic bead format for ease in processing and to maximize purity [e.g., Direct-Zol (Zymo Research), QIAzol + RNeasy (Qiagen)]. In general, RNA can be efficiently isolated from less than 1 mg of tissue or 10^5 cells, usually in less than 1 hour, using these procedures. The following protocol is a modification of the procedure of Chomczynski and Sacchi (1987).

Protocol: guanidinium–acid–phenol extraction

1. Wear gloves and appropriate protective eyewear!
2. Collect harvested cells by centrifugation and resuspend cell pellets in 100 μL of the following formulation ("solution D") per 10^6 cells:
   4 M guanidinium thiocyanate,
   25 mM sodium citrate, pH 7.0,
   0.5% sarcosyl,
   100 mM β-mercaptoethanol.
   Micropipette up-and-down, or gently vortex in order to break up the cell pellet.
   Note: Cells may be lysed directly by the addition of 1.0–1.5 mL of solution D per 100 mm tissue culture dish. Lysis in the culture dishes and flasks usually mandates the use of larger reagent volumes, and it may prove difficult to recover all of the lysate due to its tremendous viscosity. In such an event, a sterile cell scraper (e.g., Corning #3010) may be helpful for the mechanical harvesting of cells from culture dishes of flasks. Scaled down, the entire procedure can be performed in a 2.2 mL microfuge tube.
3. Transfer the lysate to a polypropylene tube. For each 1 mL of solution D lysis buffer used in step 2, add:
   0.1 mL 2 M sodium acetate, pH 5.2,
   1.0 mL water-saturated phenol (molecular biology grade),
   0.2 mL chloroform:isoamyl alcohol (49:1).
   Cap tube and mix carefully and thoroughly by inversion following the addition of each reagent and invert vigorously for an additional 30 seconds after all reagents have been added.
4. Cool sample on ice for a minimum of 15 minutes; centrifuge at 4°C to separate the phases.
5. Transfer the aqueous (upper) phase containing the RNA to a fresh tube and mix with 0.75 vol of ice-cold isopropanol. Store at −20°C for at least 1 hour to precipitate RNA.
   Note: For larger-scale preparations, Corex glass tubes (or the equivalent), placed in the correct adapters, can be used for precipitation and recovery of RNA.
6. Collect precipitate by centrifugation at 10,000 × *g* for 20 minutes at 4°C. Carefully decant and discard supernatant.
   Caution: *Do not exceed the manufacturer's recommended maximum g-force for any of the tubes used in this protocol.*
7. Completely dissolve RNA pellet in 300 μL of solution D (see step 2) and then transfer it to an RNase-free 1.7 mL microfuge tube.
8. Reprecipitate the RNA by the addition of 0.75 vol of ice-cold isopropanol and store at −20°C for 1 hour.

9. Collect precipitate at 12,000 × *g* in a microcentrifuge for 10 minutes at 4°C. Carefully decant and discard supernatant.
10. Wash pellet 3 × with 500 μL 70% ethanol per wash, followed by a final wash with 500 μL 95% ethanol. If the RNA does not dislodge during these washes, there is no need to re-centrifuge. Allow tubes to air-dry to remove residual ethanol. Store the RNA as an ethanol precipitate until it is to be used.

    Note: The final wash with 95% ethanol will accelerate the drying process.
11. Redissolve RNA in the smallest possible volume of TE buffer or nuclease-free H_2O. Incubation at 65°C for 10 minutes may facilitate solubilization, though this is unnecessary if the RNA did not dry out completely following the ethanol washes. Following the determination of concentration, store the RNA in suitable aliquots at −80°C. It is important to avoid repeated freezing and thawing.

Density gradient centrifugation

The original protocols and several current procedures for the isolation of RNA from biological sources enriched in RNase described cell and tissue disruption with guanidinium buffer. The resulting intermixing of subcellular components mandates the separation of these biochemical macromolecules from each other, with particular regard to the removal of DNA from RNA preparations. Subtle, though measurable, differences in the densities of DNA, RNA, and protein (Table 4.3) allow partitioning by banding or pelleting them in a density gradient.

Isopycnic centrifugation, a type of density gradient centrifugation, is a technique in which macromolecules move through a density gradient until they reach a density equal to their own. Macromolecules, therefore, accumulate at that position in the gradient, floating there until the end of the centrifugation run. In some cases, RNA is of greater density than any position in the gradient and therefore accumulates as a pellet at the bottom of the tube.

TABLE 4.3 Typical buoyant density ranges of DNA, RNA, and protein.

Molecule	Density (g/mL)
Protein	1.2–1.5
DNA	1.5–1.7
RNA	1.7–2.0

Note: Buoyant density changes with G + C content, methylation and other chemical modifications, as well as molecular configuration, for example, single- versus double-stranded, linear vs. covalently closed circle.

Gradients of greater maximum density may permit the banding of RNA, much as plasmid DNA is banded for purification. The classical gradients for this type of separation consist of CsCl, CsTFA, and cesium sulfate (Cs_2SO_4). Other materials such as glycerol, ficoll, metrizamide, and sucrose also have specific applications; they are readily available at low cost and high purity. Nucleic acid resolution based on differences in density is best performed using one of the cesium salts.

Cesium chloride

Cesium chloride is a dense salt that exhibits the ability to form a linear gradient when ultracentrifugation g-forces are applied to a homogeneous cell lysate suspension. There is no need to preform the gradient. CsCl gradients are generally steep, and the maximum density within the gradient usually exceeds all but the densest material, meaning that higher density molecules such as RNA will collect as a pellet at the bottom of the tube. CsCl gradients are subjected to ultracentrifugation *g*-forces for a period sufficiently long to complete the migration of all macromolecules into their equilibration positions. The actual time required is a function of the rotor size; traditional RNA preps in floor model ultracentrifuges require overnight centrifugation,

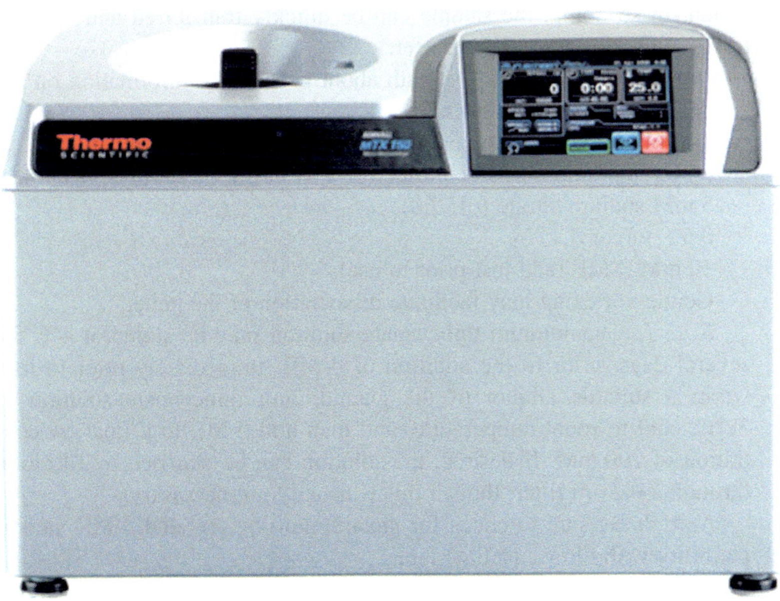

FIGURE 4.6 Sorvall MTX 150 Micro-Ultracentrifuge. Smaller radius rotors permit higher g-forces which, in turn, permit isopycnic preparation of RNA in 3 h or less. *Reproduced with permission of Thermo Fisher Scientific.*

while only a few hours are needed to accomplish the same type of separation in a tabletop micro-ultracentrifuge, for example, the Sorvall MTX 150 (Fig. 4.6).

Protocol: cesium chloride gradients

***Caution**: If you are not familiar with the proper operation of an ultracentrifuge or the procedure for derating a rotor be sure to get proper technical assistance. If not used properly, a rotor can be lethal.*

The RNA isolation protocol presented here is a modification of the procedures described by Glišin et al. (1974); Ullrich et al. (1977), and Chirgwin et al. (1979).

1. Wear gloves and appropriate protective eyewear!

2a. For tissue culture: collect harvested cells by centrifugation at $500 \times g$ for 5 minutes (at 4°C if possible). Decant as much of the supernatant as possible. Wash cells once with ice-cold PBS and then decant as much PBS as possible.

2b. For tissue: *rapidly* harvest the tissue and mince/dice in ice-cold saline solution. Quickly transfer tissue into a suitable homogenization vessel that already contains an aliquot of lysis buffer (step 3). If tissue was flash-frozen in liquid nitrogen, it may be helpful to shatter it (while wrapped in aluminum foil and in a Ziploc bag) by impact upon the lab bench. In so doing, the sample can be quickly transferred and allowed to thaw directly in the lysis buffer.

Note: See Chapter 6, The truth about tissues, for information on the subtleties of working with tissue.

3. Add 5 vol (the size of the pellet) of a solution consisting of:

6 M guanidinium hydrochloride (or 4 M guanidinium thiocyanate),
5 mM sodium citrate (pH 7.0),
0.5% sarcosyl,
10 mM β-ME (add just prior to use).
Gentle vortexing may facilitate dissociation of the pellet.

Note 1: Guanidinium thiocyanate solution may be stored at 4°C for several days, prior to the addition of β-ME. Immediately prior to use, warm a suitable aliquot of the guanidinium thiocyanate solution to 37°C, cool to room temperature, and then add β-ME to a final concentration of 200 mM. If desired, the solution can be clarified by filtration through a 0.2 μm filter, though this is usually not necessary.

Note 2: Be sure to check for guanidinium buffer- and β-ME incompatibility with ultracentrifuge tubes.

4. Disrupt the cells or tissue at room temperature by either manual homogenization (repeatedly drawing the lysate through a 19-gauge needle; several strokes using a Dounce homogenizer) or mechanical homogenization (e.g., Polytron).

Note 1: Shearing forces physically break up genomic DNA so as to preclude the formation of an impenetrable web of high-molecular-weight DNA that would otherwise impede the sedimentation of RNA.

Note 2: For optimal recovery of high quality RNA, avoid generating excess heat or foam during this step. Short bursts of homogenization are usually more effective than one prolonged application.

5. If desired, add 1 g of solid CsCl for every 2.5 mL of lysate. This is required only when other molecules are to be copurified from the same gradient along with the RNA.
6. Pipette a cushion of 5.7 M CsCl (dissolved in 100 mM EDTA, pH 7.5) into an ultracentrifuge tube, typically cellulose nitrate or polyallomer, compatible with the swinging bucket rotor to be used. The 5.7 M CsCl cushion should occupy about 25% of the total volume of the tube.
7. Gently layer the lysate on top of the CsCl cushion until the ultracentrifuge tube is filled. Be sure to weigh each centrifuge tube along with its swinging bucket and corresponding top. Each sample (bucket, screw top, centrifuge tube containing the sample) should weigh within 0.05 g of the other samples to be loaded onto the rotor. Follow the manufacturer's instructions for sealing the tubes and be sure to weigh each sealed sample again before loading the rotor. Failure to precisely balance the rotor buckets according to the specifications of the centrifuge manufacturer will have catastrophic consequences.
8. Centrifuge the preparation in a swinging bucket rotor. Floor model ultracentrifuges generally require an overnight run, while micro-ultracentrifuge models require only 3–4 hours. Each rotor instruction guide will delineate the correct relative centrifugal force (RCF) needed to purify RNA safely using this type of centrifugation.

   Note: Perform this type of centrifugation at ambient temperature. Cooling the centrifuge chamber at any time during the run will result in precipitation of the CsCl and ruination of the gradient.
9. At the conclusion of the run, carefully remove the supernatant by aspiration (sterile Pasteur pipettes, or sterile 5 mL pipettes with the tapered tip removed, work well). Use great care not to disrupt the pellet at the bottom of the tube. Many investigators prefer to remove 80% of the gradient by pipetting, and the remainder by decanting. After the supernatant has been removed, it is useful to cut off the top portion of the ultracentrifuge tube just above the RNA pellet. In this lab, a sterile razor blade is clamped in a hemostat, briefly heated with a Bunsen burner, and used to slice through the emptied centrifuge tube. In so doing, the translucent RNA pellet will be easily accessible in what looks like a giant contact lens-shaped cradle (Fig. 4.7).

   Note: This technique will help prevent contamination of the RNA pellet with protein and/or DNA that may be clinging to the side of the ultracentrifuge tube, even after removal of the gradient.

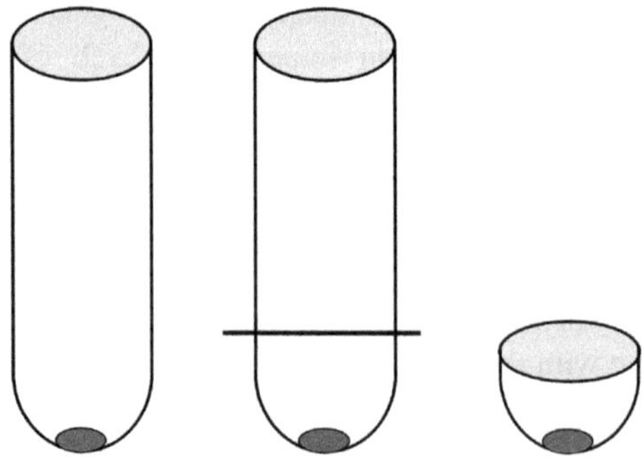

FIGURE 4.7 Method for recovery of RNA following isopycnic centrifugation. (A) Appearance of RNA pellet after removal of gradient. (B) Ultracentrifuge tube is cut just above the RNA. (C) RNA rests in contact lens-shaped cradle. This approach favors recovery of RNA free of DNA, protein, and other contaminates clinging to the inside wall of the tube.

10. Dissolve the RNA pellet completely in the smallest possible volume of 10 mM Tris, pH 7.4, 5 mM EDTA, 0.1% SDS (optional) and then transfer the sample to a sterile microfuge tube. For large-scale RNA extractions, transfer the RNA suspension to a conical 15 mL polypropylene tube.

    Note: The RNA pellet is often difficult to dissolve in aqueous buffer after this type of purification, and the larger the sample, the more difficulty will be encountered. Freezing the pellet at −20°C and then thawing usually facilitates dissolving it and removal from the centrifuge tube. Limited vortexing may help.
11. Extract the sample once with 2 vol of chloroform:isoamyl alcohol (4:1, v/v) or with 2 vol of chloroform:butanol (4:1, v/v). Pulse centrifuge to separate the two phases. Transfer the upper aqueous phase to an autoclaved microfuge tube.
12. Re-extract the organic phase and interface with an equal volume of 10 mM Tris, pH 7.4, 5 mM EDTA, 0.1% SDS (optional) and combine the two aqueous phases.

    Note: These chloroform extractions are needed to remove residual contaminants from the RNA preparation.
13. Precipitate the RNA for several hours at −20°C, with the addition of 0.1 vol 3 M sodium acetate (pH 5.2) and 2.5 vol of ice-cold 95% ethanol. Depending on the volumes involved, it may be necessary to split the sample into two or more tubes or a 15 mL Corex other high-strength glass tubes.
14. If the RNA was precipitated in a microfuge tube, collect the precipitate at 12,000 × *g* for 10 minutes, at 4°C if possible. If precipitation was

carried out in a Corex glass tube (or the equivalent), centrifuge in a suitable adapter for at least 20 minutes at 9000 × *g*.

Caution: *Be certain not to exceed the maximum g-force rating for the rotor or the type of tube in which the RNA was precipitated.*

15. Carefully decant the supernatant and wash the pellet at least once with 70% ethanol. Decant the 70% ethanol wash and, if desired, wash the sample once with 95% ethanol to accelerate the drying of the sample. Do not allow the pellet to dry out completely.

16. The RNA is best stored at −80°C as an ethanol precipitate or, following quantification, in suitable aqueous aliquots at −80°C to avoid repeated freezing and thawing. Depending on the cell type, typical yields range from 50 to 75 μg of very high-quality cellular RNA per 10^6 cells.

Cesium trifluoracetate

Cesium trifluoracetate (CsTFA) is an excellent CsCl alternative for density gradient separation of RNA and DNA by isopycnic centrifugation. At ultracentrifugation RCFs, CsTFA will likewise self-form a gradient (Fig. 4.8).

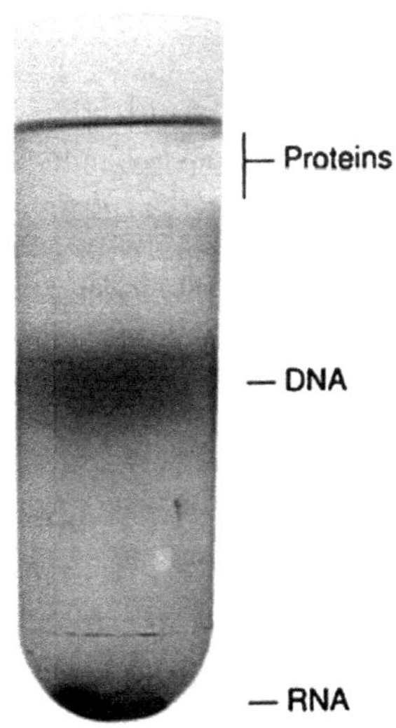

FIGURE 4.8 CsTFA gradient. Isopycnic gradient separation of calf liver macromolecules extracted with guanidinium thiocyanate. A guanidinium thiocyanate extract of calf liver (18 mL) was layered on top of a 19 mL cushion of CsTFA ($\rho = 1.51$ g/mL) and centrifuged at 125,000 × *g* for 16 h. Approximately 3.5 mg of total RNA was pelleted at the bottom of the tube ($\rho \approx 1.62-1.9$ g/mL). This RNA contained less than 0.4% DNA as measured with a DNA probe. DNA forms a band near the original interface of the cushion and the sample solutions ($\rho \approx 1.6$ g/mL). Protein in the sample collects in the upper portion of the gradient ($\rho \approx 1.2-1.5$ g/mL). *Courtesy of Amersham Biosciences (now part of Cytiva).*

The trifluoracetate anion gives CsTFA properties that result in very high-quality nucleic acid preparations of extremely high quality, including (1) more ready dissociation of proteins from nucleic acids; (2) the ability to band, rather than pellet RNA, because of high solution density; (3) simultaneous banding and recovery of DNA, RNA, and protein from the same CsTFA gradient (Zarlenga and Gamble, 1987); (4) improved yields of pure RNA due to the salting-in effect of the trifluoroacetate ions, in contrast to the salting-out effect of the chloride and sulfate anions which causes protein precipitation and concomitant loss of nucleic acids during isolation; (5) solubility in polar solvents, such that nucleic acids may be recovered directly from the gradient by precipitation with ethanol and without prior dialysis or extraction to remove the salt; and (6) excellent inhibition of RNase.

Among the disadvantages of such density, gradients are the cost of the density gradient material (e.g., CsCl, CsTFA), the requirement for a very expensive ultracentrifuge, the relatively long period of centrifugation, the labor-intensive clean-up of gradient-purified RNA, and the limited ability to prep more than a few samples at a time. Thus, isopycnic centrifugation has become a rarity in the isolation of RNA, in favor of newer reagents and kits. In some applications, there remains the need to resolve RNA in a density gradient and for this reason, the CsCl protocol (above) and the following CsTFA protocol are included here.

Protocol: cesium trifluoroacetate gradients

This protocol is a modification of the procedure originally described by Okayama et al. (1987). As presented, the RNA will form a pellet at the bottom of the tube, rather than banding within the gradient. It is recommended for the preparation of total RNA from a variety of tissues and can be modified to accommodate as little as 25 mg or as much as 6 g of starting material. Do not use cellulose nitrate tubes when running a CsTFA gradient, nor should one use cellulose nitrate to attempt to filter-sterilize CsTFA.

Advance preparation of cesium trifluoroacetate

CsTFA is usually provided at a density of 2.0 ± 0.05 g/mL, and prepared as a working solution with a density of 1.51 ± 0.01 g/mL, in 100 mM EDTA, pH 7.0. To prepare 100 mL of this solution, mix $51/(\rho-1)$mL of CsTFA (where $\rho =$ density of the product as received) with 40 mL of a solution of 100 mM Tris (pH 8.0), 100 mM KCl, 1 mM EDTA, and $60-[51/(\rho-1)]$mL of sterile distilled water. The density of the working solution of CsTFA may be altered to fit a particular experimental protocol and additional formulae for preparing precisely defined densities of CsTFA are provided in the instructions that accompany this product. Densities as great as 2.6 g/mL ($\sim$10 M) are possible with CsTFA.

1. Wear gloves and appropriate protective eyewear!

2a. For tissue culture: Collect harvested cells by centrifugation at 500 × *g* for 5 minutes. Decant as much of the supernatant as possible. Wash cells once with PBS. Decant as much PBS as possible. If the RNase is likely to be a major challenge, perform centrifugations at 4°C, and wash cells with ice-cold PBS.

2b. For tissue: *rapidly* harvest the tissue and mince in an ice-cold saline solution. Quickly transfer tissue into a suitable homogenization vessel that already contains an aliquot of lysis buffer (step 3). If tissue was flash-frozen in liquid nitrogen, it may be helpful to shatter it (while wrapped in aluminum foil in a Ziploc bag) by impact upon the lab bench. In so doing, the sample can be quickly transferred and allowed to thaw directly in the lysis buffer.

Note: See Chapter 6, The truth about tissues, for a review of the subtleties of working with tissue.

3. For each gram of tissue or cell pellet, add 18 mL of 5.5 M guanidinium thiocyanate solution (5.5 M guanidinium thiocyanate, 25 mM sodium citrate, 0.5% sarcosyl; adjust pH to 7.0 with NaOH; add 200 mM β-ME just prior to use).

Note 1: Guanidinium thiocyanate solution may be stored at 4°C for several days, prior to the addition of β-ME. Immediately prior to use, warm a suitable aliquot of the guanidinium thiocyanate solution to 37°C, cool to room temperature, and then add β-ME to a final concentration of 200 mM. If desired, the solution can be clarified by filtration through a 0.2-μm filter, though this is usually not necessary.

Note 2: Be sure to check for guanidinium buffer, CsTFA, and β-ME incompatibility with ultracentrifuge tubes prior to use.

4. Disrupt the cells or tissue at room temperature by either manual homogenization (repeated shearing through a 19-gauge needle; several strokes using a Dounce homogenizer) or mechanical (e.g., Polytron) homogenization.

Note 1: Shearing forces physically break up genomic DNA so as to preclude the formation of an impenetrable web of high-molecular-weight DNA that would otherwise impede the sedimentation of RNA.

Note 2: For optimal recovery of high-quality RNA, avoid generating excess heat or foam during this step.

Short bursts of homogenization are usually more effective than a single, prolonged application.

5. Carefully transfer the lysate to disposable conical centrifuge tubes (e.g., 50 mL size). Centrifuge at 1500 × *g* for 5 minutes to sediment insoluble materials.

6. Without disturbing the pelleted material, transfer the clarified lysate to a fresh tube.

7. If the lysate still appears viscous, shear the DNA by once again drawing it through a 19-gauge needle repeatedly until the viscosity decreases. Between 10 and 12 passes into and out of the syringe is usually adequate.
8. Centrifuge at $5000 \times g$ for 20 minutes at 15°C to remove the cell debris. Decant and retain the supernatant, taking great care not to dislodge the pelleted debris.

   Caution: *Do not exceed the maximum g-force recommended for any of the centrifuge tubes used in this protocol.*
9. If necessary, adjust the volume of the supernatant with 5.5 M guanidinium thiocyanate solution (see step 3).
10. Select a rotor suitable for the number and volume of samples. DO NOT use cellulose nitrate tubes with CsTFA. Pipette cushions of CsTFA into each tube, and then overlay aliquots of the sample onto these cushions, filling each tube to the very top. The CsTFA cushion should occupy about 25% of the total volume of the tube.
11. Be sure to weigh each centrifuge tube along with its swinging bucket and corresponding top. Each sample (bucket, screw top, and centrifuge tube containing the sample) should weigh exactly the same as the other samples to be loaded onto the rotor, within 0.05 g of each other. Follow the manufacturer's instructions for sealing the tubes and be sure to weigh each sealed sample again before loading the rotor. Failure to precisely balance the rotor buckets according to the specifications of the centrifuge manufacturer will have catastrophic consequences.
12. Centrifuge at $125{,}000 \times g$ in a swinging bucket rotor for 16–20 hours at 15°C. Under these conditions RNA will pellet at the bottom of the tube and DNA will collect in a band in the lower third of the gradient.
13. After centrifugation, carefully aspirate off most of the liquid in each tube, stopping as soon as the DNA band has been removed. If the DNA band is not visible, discontinue aspiration within 1 cm from the bottom of the tube. Decant the remaining liquid, taking care not to disturb the RNA pellet. Invert the tubes and allow them to drain onto a stack of 3–4 Kimwipes for 5 minutes.
14. Carefully cut off the bottom of each tube with a sterile razor blade or scalpel, as shown in Fig. 4.7. Leave a sufficient portion of the sides of the tube to form a small cup.
15. Dissolve RNA pellets directly in the tube bottoms using two aliquots of an appropriate buffer, either nuclease-free H_2O or TE, pH 7.4 (10 mM Tris-Cl, pH 7.4; 0.1 mM EDTA).

    Note: At least separate aliquots (100–200 μL) of buffer are recommended to recover all of the RNA samples from the tubes. Repeated pipetting may be necessary for complete recovery.

As always keep the sample as concentrated as possible so that all manipulations, including reprecipitation if necessary, can be carried out in a microfuge tube.

16. Combine both aliquots of RNA and vortex. Heat tubes to 65°C for 10 minutes and vortex again.

17. Spin tubes briefly and at low speed to remove any insoluble material.

18. Determine RNA concentration, purity, and yield (Chapter 8: Quality control for RNA preparations). If desired, RNA can be precipitated to concentrate the RNA sample and/or change buffers.

19. Store RNA in suitable aliquots at −80°C until they are ready for use. Avoid multiple freeze-thaw cycles.

Isolation of RNA and DNA from the same source

Many noteworthy procedures representing varying degrees of complexity have been described for the isolation of RNA from cells and tissues. Most biotech companies now offer devices, reagents, solutions, or resins for the expedient isolation of RNA from various types of biological sources. These products vary widely in their simplicity, speed, cost, and effectiveness. Several procedures have also been described for the simultaneous recovery of RNA, DNA, and protein from the same biological source and some of these, too, are available for purchase.

The method presented here has the advantage of requiring no specialized equipment other than what is found even in the most modest of molecular biology labs. This protocol is a modification of the procedure of Majumdar et al. (1991) and is based on the selective partitioning of RNA and DNA by adjusting the pH of a phenolic lysate (Chomczynski and Sacchi, 1987; Chomczynski, 1993) but without the use of guanidinium reagents. Succinctly, the RNA is recovered in the aqueous phase by first extracting the sample at acidic pH; DNA in the sample is retained at the interface and in the organic phase. Following the transfer of the RNA-containing aqueous phase to a fresh tube, the DNA is eluted from (i.e., drawn out of) the organic phase by establishing an alkaline pH. This is known as reverse extraction.

Cells may be lysed directly in the tissue culture vessel or harvested first. Cell pellets should be gently dispersed *prior* to addition of the lysis buffer because shearing of the DNA is likely to occur when an aggressive attempt is made to resuspend a cell pellet *after* addition of the lysis buffer. In this regard the investigator is strongly urged not to succumb to the temptation to vortex this or any lysate from which genomic DNA is to be purified, as the enormous shearing forces created will damage DNA in a very short time. The size distribution of the resulting fragments is likely to be unacceptably low for some applications.

Protocol: simultaneous isolation of RNA and DNA

1. Wear gloves and appropriate protective eyewear!
2. In advance: Prepare phenol:chloroform:isoamyl alcohol extraction buffer as described in Fig. 4.3, with the exception that the phenol should be water-saturated 2–3 times, rather than being Tris-saturated.
3. For tissue culture: Remove growth medium by aspiration and wash the monolayer twice with ice-cold PBS. If desired, it may be helpful to prepare the cell cultures for lysis by cooling them directly on a bed of ice for 15 minutes. Remove, by aspiration, as much PBS as possible. Add 2 mL of lysis buffer (10 mM EDTA, pH 8.0; 0.5% SDS) per 75 cm^2 flask containing up to 5×10^6 cells (for higher density cell cultures, use 100 μL lysis buffer/cm^2). Rock the flask gently from side to side to ensure lysis of all cells. Incubate on ice for 3 minutes. If desired, the lysate can be transferred to lyse cells in a second flask, followed by incubation for an additional 3 minutes. Gentle agitation throughout may help. Do not use an aliquot of lysis buffer for more than two tissue culture flasks; excessive DNA/RNA/protein concentrations in the lysate will compromise partitioning efficiency. Cells that were harvested from tissue culture and pelleted should be lysed using 200 μL lysis buffer per 10^6 cells.

   Note: Cooling reagents and cells on ice is an excellent method for controlling RNase activity. It is the responsibility of the investigator, however, to empirically determine whether a 15-minute incubation on ice prior to lysis will change the cellular biochemistry enough to influence the outcome of the study.
4. For tissue: mince freshly isolated tissue (up to 100 mg) on ice. For each milligram of tissue, add 250 μL lysis buffer (10 mM EDTA, pH 8.0; 0.5% SDS) and gently Dounce homogenize at room temperature. Do not use a Polytron to homogenize, as this will cause unacceptable shearing of the genomic DNA to be recovered later in this protocol.

   Note: See Chapter 6, The truth about tissues, for a review of the subtleties of working with tissue.
5. Transfer the lysate into a prechilled 15 mL polypropylene tube. When scaling down, a polypropylene microfuge tube may be satisfactory.

   Note: Due to the viscous nature of liberated chromatin, a sterile cell scraper may be useful for the quantitative recovery of the lysate.
6. Add an equal volume of cold organic extraction buffer, prepared as described in step 2. Be sure to check for centrifuge tube compatibility with phenol and chloroform.
7. Carefully mix the tube(s) by gentle inversion several times. Do not vortex! The yield will improve dramatically if the tubes are maintained on ice at all times when not mixing.

8. Centrifuge samples in an appropriate rotor in order to separate the aqueous material from the organic material.

   Caution: Be careful not to exceed the recommended rotor speed or the maximum g-force recommended for any of the tubes used in this protocol.

Recovery of RNA

9. Transfer the aqueous (upper) phase to a fresh tube, prechilled and resting on ice. Take care to avoid disrupting the protein interface. Recovery of DNA (step 21), which must be eluted from the interface and organic phase, can be delayed until all steps pertaining to RNA recovery have been completed. Leave tubes on ice until then.
10. Extract the aqueous phase with a fresh, equal-volume aliquot of the cold organic mixture described in step 2.
11. Centrifuge as described above. Carefully recover the aqueous phase and transfer to a nuclease-free Corex or another high-strength glass tube, taking care to avoid disrupting the interface or organic material. If the volumes involved are sufficiently small, microcentrifuge tubes should be used instead.
12. Add 500 μL of ice-cold 1 M Tris-Cl, pH 8.0 and 200 μL of 5 M NaCl for every 4 mL of RNA-containing aqueous material. Mix carefully and thoroughly. Add 2.5 vol of ice-cold 95% ethanol and mix thoroughly. Store at −20°C for at least 1 hour.
13. Collect RNA precipitate by centrifugation at 9000 × *g* for 10 minutes in a suitable tube. Centrifugation at 4°C is preferred, though not essential.

    Caution: *Do not exceed the recommended maximum RCF for the centrifuge tube. If using high-strength glass tubes, be sure to support them with a suitable rotor adapter.*
14. Carefully decant and discard the supernatant.
15. Wash pellet 2–3 times with 70% ethanol. Centrifuge again to recover the precipitate. Carefully decant and discard supernatant.
16. Allow sample to air-dry. If desired, a final wash with 95% ethanol may accelerate air drying.
17. Dissolve the RNA pellet in 200 μL of ice-cold TE buffer (10 mM Tris-Cl, pH 8.0; 0.1 mM EDTA). Incubate on ice for 1 hour and then transfer to a fresh microfuge tube. A second aliquot of ice-cold TE buffer may be used to rinse the centrifuge tube and should be combined with the first aliquot.
18. Add 6 μL of 5 M NaCl and 800 μL of ice-cold ethanol for every 300 μL RNA solution. Incubate on ice for 5–10 minutes to precipitate the RNA. The sample can now be stored as an ethanol precipitate at −80°C or used immediately.

19. Collect RNA precipitate by centrifugation at 12,000 × *g* in a microfuge for 8 minutes, preferably at 4°C. Carefully decant the supernatant and dry to remove excess ethanol as described in steps 14–15.
20. Resuspend the RNA pellet(s) in a minimum volume of the desired buffer (e.g., nuclease-free water) and allow 15 minutes on ice for the sample to dissolve. Depending on the degree to which the sample has dried, longer incubation on ice may be necessary to dissolve the sample completely. Recommended: incubate the sample with RNase-free DNase (Appendix C) to remove any contaminating DNA.
21. Calculate the concentration, purity, and yield of the RNA. Store hydrated samples in suitable aliquots at −80°C.

Recovery of DNA

22. Return to the tubes containing organic material saved from step 8. Remove any remaining aqueous material as completely as possible. Any remaining RNA will be hydrolyzed in the steps that follow. Alternatively, the purified DNA may be incubated with RNase (Appendix D) to remove all contaminating RNA at the end of the procedure.
23. To the remaining organic/interface material add an equal volume of 1 M Tris base solution. This solution should not be pH-adjusted (pH ≈ 10.5). Mix thoroughly by gentle shaking. Do not vortex.

    Note: This step is known as reverse extraction; by creating a strongly alkaline environment the DNA will be drawn into the aqueous phase.
24. Centrifuge samples at 2000 × *g* for 15 minutes, at 4°C if possible, in order to separate the aqueous material from the organic material. Carefully transfer the aqueous (upper) phase to a fresh tube.
25. Repeat steps 21 and 22, but do not mix the two aqueous phases. Keep them in separate centrifuge tubes.
26. To each Tris:DNA aqueous sample, add an equal volume of chloroform:isoamyl alcohol (24:1). Mix carefully and thoroughly. Pulse centrifuge samples for 30 seconds to separate the phases. Carefully transfer the aqueous (upper) phase from each tube to fresh tubes.
27. Precipitate the DNA with the addition of 0.1 vol of 3 M sodium acetate, pH 5.2, and 2.5 vol of 95% ethanol. Mix thoroughly, first by gentle swirling and then by thorough inversion. Do not vortex.

    Note: Adding ethanol should cause the immediate precipitation of the genomic material, which, characteristically, should appear quite stringy. The DNA can be collected immediately by low-speed centrifugation (500 × *g*) for 1–2 minutes. Alternatively, genomic DNA can be "fished" out of the ethanol solution using a silanized Pasteur pipette

(Appendix H) such that there is minimal sticking of the DNA to the glass.

28. Transfer the precipitated DNA to a fresh microfuge tube. Wash the DNA 3 × with 500 μL aliquots of 70% ethanol per wash to remove excess salt. A final wash with 95% ethanol will accelerate the drying process. Allow DNA to redissolve in a suitable buffer (e.g., TE buffer, pH 8.0) and then determine DNA concentration.

Other methods

The following procedure, a modification of Peppel and Baglioni (1990), Salvatori et al. (1992), and Zolfaghari et al. (1993), takes advantage of the ability of the ionic detergent SDS to inhibit RNase activity. It is a very rapid method for isolating total cellular RNA that does not require guanidinium solutions, phenol, or isopycnic centrifugation, and was the basis for the development of a number of popular nonphenol-based kits for RNA isolation. It is suitable for the isolation of both eukaryotic and prokaryotic RNA. When RNA is to be isolated from tissue, it is best to couple this method with some form of mechanical disruption. Most of the reagents required are commonly found in labs performing even the most basic molecular biology procedures. The volumes can be scaled up to accommodate larger extractions, and one may expect to harvest approximately 100 μg RNA per 10^6 cells.

Protocol: rapid isolation of RNA with SDS and potassium acetate reagents

1. Wear gloves and appropriate protective eyewear!

2a. For tissue culture: Remove growth medium from the tissue culture vessel containing cells of interest and rinse with PBS. Remove PBS by aspiration and discard. Lyse cells were directly on tissue culture plastic with lysis buffer (2% SDS; 200 mM Tris-Cl, pH 7.5, 0.5 mM EDTA), using 50 μL lysis buffer/cm^2. Gently agitate the culture vessel on the bench top or on an orbital platform for 2 minutes to ensure complete lysis. This lysis buffer can be prepared ahead of time and stored at room temperature.

2b. For tissue: *Rapidly* harvest the tissue and mince in an ice-cold saline solution. Quickly transfer tissue into a suitable homogenization vessel that already contains an aliquot of lysis buffer (step 2a). If tissue was flash-frozen in liquid nitrogen, it may be helpful to shatter it (while wrapped in aluminum foil in a Ziploc bag) by impact upon the lab bench. In so doing, the sample can be quickly transferred and allowed to thaw in the lysis buffer. Use 1 mL lysis buffer for each milligram of tissue. Disrupt the cells or tissue at room temperature by either manual homogenization (repeated shearing through a 19-gauge needle;

Dounce homogenizer) or mechanical homogenization (e.g., Polytron). Several short bursts with an electronic homogenize, rather than one sustained application, will quite effectively avoid generating excess heat or foam during this step.

3. Transfer the lysate to a sterile microfuge tube (1.7- or 2.2-mL size). Cap and invert tube sharply 15–20 times.
4. For each 500 μL lysis buffer used in step 2, add 150 μL of potassium acetate solution (50 g potassium acetate; 11 mL glacial acetic acid; H_2O to 100 mL). This solution can be prepared ahead of time and stored at −20°C.
5. Carefully invert the tube sharply 15–20 times to ensure thorough mixing.
6. Incubate on ice for at least 3 minutes but no longer than 5 minutes.
7. Centrifuge sample for 5 minutes at room temperature at maximum speed in a microcentrifuge.
8. Carefully recover the supernatant and transfer it to a fresh tube.
9. Extract the supernatant with 300 μL of a mixture of chloroform:isoamyl alcohol (24:1). Centrifuge for 30 seconds to separate the phases. Transfer the aqueous (upper) phase to a fresh tube without disturbing the protein interface and then repeat the extraction.
10. Transfer the aqueous (upper) phase to a fresh microfuge tube prechilled on ice.
11. Precipitate the RNA with the addition of an equal volume (~650 μL) ice-cold isopropanol. Incubate at −20°C for 30 minutes.
12. Collect the RNA precipitated by centrifugation for 5 minutes at room temperature.
13. Carefully decant the supernatant; wash the cell pellet 2–3 times with 70% ethanol. If the pellet becomes dislodged, centrifuge the sample again before attempting to remove the ethanol. Decant and discard the ethanol and allow the sample to air-dry. If desired, a final wash with 95% ethanol may facilitate air drying.
14. Dissolve RNA in the smallest possible volume of TE buffer or nuclease-free H_2O. Determine RNA concentration and store sample in suitable aliquots at −80°C until further use. *Optional*: Incubate purified sample with RNase-free DNase (Appendix C), to remove any contaminating DNA.

Protocol: isolation of prokaryotic RNA

This rapid and efficient protocol for the isolation of bacterial RNA is a modification of the procedures of Peppel and Baglioni (1990), Salvatori et al. (1992), and Zolfaghari et al. (1993). Be sure to use the lysis buffer described below; *never* use the standard NaOH/SDS lysis buffer used to disrupt bacterial cells for plasmid preps, as this buffer will cause rapid and extensive hydrolysis of the RNA!

In advance: NP-40 buffer and potassium acetate buffer preparation

1. Prepare the NP-40 lysis buffer: 250 mM sucrose; 20 mM EDTA, pH 8.0; 0.75% NP-40 (use IGEPAL CA-630 if NP-40 is not available). Important: do not add the NP-40 detergent until after the other two components have been mixed and autoclaved. Store at 4°C until ready to use.
2. Prepare potassium acetate buffer: Prepare 3 M potassium acetate (KCH_3O_2) in 80% of the intended final volume. Adjust the pH to 5.5 with glacial acetic acid. Add remaining H2O. Add 100 g guanidinium thiocyanate (GTC) per 100 mL. Mix carefully and thoroughly. Store at 4°C until ready to use.

Note: Although these buffers have a shelf life of several months, they should be inspected prior to use for the growth of opportunistic prokaryotes. This is especially important because this protocol is for the isolation of prokaryotic RNA.

RNA isolation

1. Wear gloves and appropriate protective eyewear!
2. Harvest bacterial cells from a 100 mL culture by centrifugation. Completely remove supernatant and dispose of as biological waste according to departmental guidelines.
3. Combine 1.5 mL of ice-cold lysis buffer with 2 mL of ice-cold potassium acetate buffer, above, and then gently resuspend the cell pellet in this mixture. Allow lysate to incubate on ice for 10 minutes. Divide lysate into two or more tubes, as needed.
4. Centrifuge at 10,000 × *g* for 10 minutes. Be sure to use centrifuge tubes rated for this *g*-force.
5. Transfer the supernatant to a fresh tube, taking care not to transfer any of the precipitated solids. It is far better to leave a small amount of supernatant behind than to risk carryover of cell debris. Incubate supernatant on ice for an additional 15 minutes.
6. Centrifuge at 10,000 × *g* for 10 minutes, at 4°C if possible.
7. Transfer the supernatant to a fresh polypropylene tube, and then extract with an equal volume of chloroform:isoamyl alcohol (24:1). Mix *carefully* and thoroughly. Centrifuge for 3 minutes to separate the phases. If a protein band appears between the upper and lower phases, repeat the extraction with a fresh aliquot of chloroform:isoamyl alcohol (24:1).
8. Transfer the upper aqueous phase to a fresh tube, and then add 0.6 vol of ice-cold isopropanol. Store sample at −20°C for 20 minutes to precipitate the RNA.

9. Collect RNA by centrifugation at 10,000 × *g* for 10 minutes, at 4°C if possible.
10. *Carefully* decant supernatant and then wash pellet with several aliquots of 70% ethanol. A final wash with 95%–100% ethanol will accelerate the drying process.
11. Allow the sample to air-dry, and then resuspend in TE buffer or nuclease-free H_2O. After determination of RNA concentration and purity, store in suitable aliquots at −80°C. If RNA is not to be used right away, store it as an ethanol precipitate at −80°C.

Protocol: isolation of RNA from yeast

Yeast cells produce a cell wall that acts as a formidable impediment to the rapid recovery of RNA from these eukaryotic cells. Many of the classical methods for the isolation of RNA from yeast involve vortexing yeast cultures in the presence of glass beads, a technique intended to compromise severely the ultrastructure of these cells in a nonenzymatic manner. This type of approach is tedious and, when scaled down to the mini-prep level, often results in poor yields. Alternatively, incubation of yeast cells with the enzyme zymolyase results in degradation of the cell wall and concomitant spheroplast formation. This approach causes gentle cellular disruption and favors the recovery of high-molecular-weight DNA. Zymolyase predigestion of the cell wall can also be used for RNA isolation and in conjunction with other extraction methods.

The following RNA isolation procedure is a modification of the method of Schmitt et al. (1990). It has the advantages of being a rapid isolation procedure and having minimum cell culturing requirements. The method exploits the chaotropic nature of phenol and SDS, enhanced by heating, freezing, and then thawing the sample. As the sample freezes, phenol crystals form which pierce the yeast cell, thereby liberating its contents. DNA is separated from RNA by phenol extraction under acidic conditions, a frequent tactic for RNA purification when chaotropic reagents are utilized.

In advance: Inoculate 0.5 mL of yeast culture medium (YPD) in a sterile 15 mL tube. Grow culture overnight at 30°C with gentle shaking. The following morning dilute the culture 1:40 with fresh YPD medium. Continue to incubate for 4 hours or until the culture is in an exponential growth phase.

Note 1: Ensure that the tubes are properly covered, not only to prevent evaporation but also to prevent infection by opportunistic prokaryotes.

Note 2: YPD medium (per liter): Mix 10 g yeast extract, 20 g peptone, 20 g dextrose; autoclave. After autoclaving, YPD medium acquires a dark brown appearance due to the "caramelizing" of the dextrose at high temperatures.

1. Wear gloves and appropriate protective eyewear.
2. Remove a 5 mL aliquot of the culture and collect cells by centrifugation (500 × *g* for 5 minutes). Remove the supernatant as completely as possible and discard it.
3. Resuspend cell pellet in 400 μL of AE buffer (50 mM sodium acetate, pH 5.2; 10 mM EDTA). If the cells do not resuspend quickly and thoroughly, then add more AE buffer.

   Note: If desired, an appropriate RNase inhibitor may be added to the solution just prior to use, though this is usually not necessary.
4. Transfer the cell suspension to a microfuge tube and add 40 μL of 10% SDS. Invert the tube several times to mix. *If the volume of AE buffer was increased in step 4, be sure to increase the volumes of all other reagents proportionally.*
5. Add an equal volume (~1 mL) of molecular biology-grade phenol that has been equilibrated with AE buffer.

   Important: Do not adjust the pH of the equilibrated phenol.
6. Mix the sample carefully and thoroughly and then incubate at 65°C for 5 minutes.
7. Rapidly cool the lysate on dry ice until phenol crystals appear. If an ethanol/dry ice bath is not available, the sample may be incubated at −20°C for 30 minutes.
8. Centrifuge the sample at 12,000 × *g* (top speed) in a microfuge for 2 minutes to separate the phases.

   Note: It is very common to observe a large amount of precipitate, consisting of cellular debris, at the bottom of the tube in addition to a protein interface between the upper aqueous phase and the lower organic phase.
9. Carefully recover the upper (aqueous) phase, taking care not to disturb either the protein interface or the organic phase. Transfer the aqueous material to a fresh microfuge tube and then add an equal volume of room temperature modified phenol:chloroform:isoamyl alcohol (25:24:1), prepared as described in Fig. 4.3. Mix carefully and thoroughly by inversion for 2–3 minutes.
10. Centrifuge the sample at 12,000 × *g* in a microfuge for 2 minutes to separate the phases.
11. Carefully recover the upper (aqueous) phase, taking care not to disrupt either the protein interphase (if present) or the organic phase. Transfer the aqueous material to a new microfuge tube.
12. Extract the aqueous material once with an equal volume of chloroform or chloroform:isoamyl alcohol (24:1). Pulse centrifuge for 30 seconds to separate phases and then carefully transfer the aqueous phase to a fresh tube.
13. Add 0.1 vol 3 M sodium acetate, pH 5.2, and 2.5 vol of 95% ethanol. Store sample overnight (or at least for a few hours) at −20°C to precipitate RNA.
14. Collect the precipitate by centrifugation at 12,000 × *g* for 8 minutes and do so at 4°C if possible.

15. Decant ethanol. Wash the pellet 3 × with 500 μL aliquots of 70% ethanol. Air-dry the pellet, but do not allow the pellet to dry completely. If desired, a final wash with 95% ethanol may accelerate air-drying of the RNA pellet. Store the sample as an ethanol precipitate at −80°C until ready to use.
16. Resuspend the pellet in 25−50 μL nuclease-free H_2O or TE (10 mM Tris, 0.1 mM EDTA; pH 7.5).
17. Determine RNA concentration, purity, and electrophoretic profile as described in Chapter 8, Quality control for RNA preparations. Store the sample in suitable aliquots at −80°C. The expected yield is about 150 μg total RNA from an exponential 10 mL yeast culture.

RNA isolation from fluid matrices

miRNA and other small RNA species have already been identified as reliable biomarkers for many disorders and for the preclinical detection of the disease state. miRNAs are readily found in blood (serum and plasma), sputum, cerebrospinal fluid, urine, and other biofluids, meaning that in many cases they can be collected in a noninvasive or minimally invasive manner. As with other RNA isolation methods, meticulous attention to details pertaining to expedient purification and RNase control are essential. Given the unique composition of each type of biofluid, special attention must also be given to avoid or at least minimize the copurification of compounds (e.g., heparin in blood samples) that interfere with downstream applications.

Kit procedures for miRNA isolation are similar in format to kits used for the isolation of larger cellular RNA species, often purifying cellular RNA first followed by recovery of the miRNA subpopulation. These procedures commonly utilize guanidinium isothiocyanate-based lysis buffers and phenol as a monophasic suspension which require the addition of chloroform for phase separation (do not use chloroform with isoamyl alcohol added). They work by trapping RNA recovered from the aqueous phase on spin columns that consist of silica beads or glass-fiber filters, followed by column washing and then elution of the RNA in a minimum volume.

When evaluating a new kit or procedure for the first time, or when comparing two or more protocols, one strategy to determine how well the procedure or kit works is to spike the biofluid (or at least the aqueous phase material following the initial phase separation) with a known quantity of one or more exogenous miRNAs; optimized kits specifically for spiking the sample as a quality control measure are widely available. These molecules are then subjected to the same physical conditions as endogenous miRNA during the remainder of the isolation procedure. The abundance of recovered spiked-in miRNA (e.g., real-time PCR) is indicative of the efficiency of the purification procedure (Fig. 4.9).

It is worth noting that the retention of small RNAs, including miRNA, is highly ethanol-dependent. A minimum of 50% ethanol is needed to retain

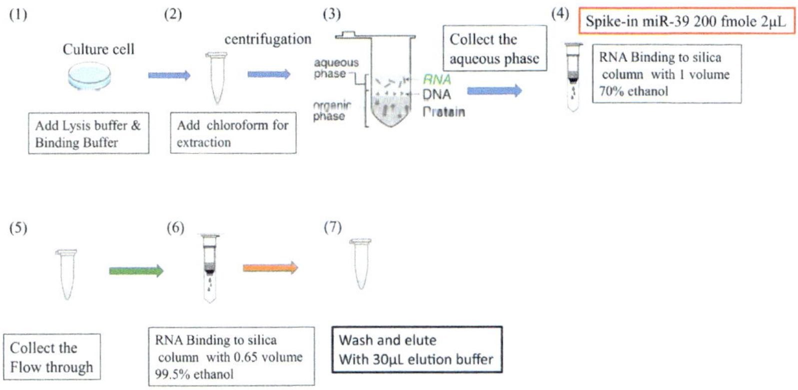

FIGURE 4.9 Assessing the efficiency of miRNA isolation. Spiking the aqueous phase with exogenous miRNA(s) following the organic extraction subjects all RNA molecules in the sample to the same physical conditions during the remainder of the protocol. Quantification of the recovered spiked-in miRNA points to the efficiency of the procedure (Hu et al., 2020). *Courtesy of Hu, W.P., Chen, Y.C., Chen, W.Y., 2020. Improve sample preparation process for miRNA isolation from the culture cells by using silica fiber membrane. Sci Rep 10, 21132. doi: 10.1038/s41598-020-78202-8. Reproduced with permission under the Creative Commons Attribution 4.0 International License (CC BY 4.0; creativecommons.org/licenses/by/4.0/).*

molecules less than 200 nts on the column, whereas these small RNAs (20–200 nts) elute readily at an ethanol concentration of 35% or lower. In terms of a practical application, the subpopulation of RNA to be purified depends in large measure on the concentration of ethanol in the wash buffers. In order to provide the investigator with the greatest latitude, some RNA isolation products feature a two-column system through which the larger RNAs can be purified independently from smaller RNAs from the same biological source. Still, other products are based on magnetic bead separation. Product enhancements may include altered wash and elution buffers as well as the use of fractionation-type columns or devices specifically designed to prevent small RNA flow-through.

In cases where exosomes (and the RNA they carry) are desired, the isolation procedure is similar in terms of RNA recovery, though the RNA isolation protocol is preceded by (a) biofluid filtration which eliminates particles larger than approximately 1 μm, and (b) exosome column binding steps that separate exosomes of all sizes from cellular debris. These steps enrich vesicle-bound nucleic acids by sequestering them from most blood plasma components before lysing them. When collected from blood, it is also important to avoid the anticoagulant heparin, commonly found in the blood collection tubes used by phlebotomists, because the reverse transcriptase step may be compromised if the sample becomes heparin-tainted. Heparin that is naturally present in an RNA prep is easily degraded enzymatically by incubating the purified RNA with heparinase I (Roest et al., 2016). Other large structures, including protein complexes, are likewise removed during the exosome disruption, protein denaturation, and the nucleic acid binding and washing steps that follow.

Short- and long-term storage of purified RNA

The correct storage conditions for RNA samples are often a hotly discussed topic within the laboratory. Improper storage, over a period of merely a few hours or as long as several months, is likely to have a profound negative impact on the probable utility of a sample of purified RNA. The key issues are as follows:

a. Has the RNA been purified from the cellular or tissue architecture?
b. Is −80°C storage available in the lab?
c. Will the RNA be used within the next 5–7 days?
d. How often will aliquots be withdrawn from a stock tube of RNA?
e. Has the RNA been redissolved in an aqueous solution (nuclease-free water or TE buffer)?

Tissues and cell pellets may be harvested and then flash-frozen for storage in liquid nitrogen or at −80°C. When stored as such, RNA can be purified up to a year later, and sometimes longer. In general, the frozen material is thawed in lysis buffer, accompanied by immediate mechanical disruption. Other investigators will homogenize fresh tissue in guanidinium buffer immediately upon receipt and then freeze the homogenate at −80°C, continuing the RNA isolation procedure at a later date.

Purified RNA is most stable when stored as an ethanol precipitate at −80°C. Under these conditions RNA can be stored confidently for several months or even longer, the half-life being a direct function of the biological source. RNA purified from certain sources (plants, for example) is remarkably stable for years when maintained as an ethanol precipitate at −80°C. These RNA species also show enhanced stability when stored for weeks in the hydrated form at −20°C. If the sample will surely be used within the next week, the RNA may be stored stably at −20°C, again as an ethanol precipitate. The "clock begins ticking" when a purified sample of RNA is dissolved in an aqueous buffer, either sterile water or modified TE buffer (10 mM Tris; 0.1 mM EDTA; overall pH 7.5). TE buffer has the advantage of being able to chelate the Mg^{2+} that is often carried over from certain isolation procedures, and the low EDTA concentration is unlikely to cause havoc in downstream PCR procedures (Chapter 10: RT-PCR: a science and an art form and Chapter 11: Quantitative PCR techniques). Chelation is advantageous because magnesium and other heavy metals are known for their ability to catalyze nonspecific degradation of RNA, particularly at elevated temperatures.

RNA stored in nuclease-free H_2O, nuclease-free H_2O with 0.1 mM EDTA added, or TE buffer should be maintained on the ice during handling unless a protocol specifically dictates otherwise. The most prudent course of action is to determine the RNA concentration and then store the remaining RNA sample in suitable aliquots at −80°C. This will preclude repeated freezing and thawing of any one tube. Some investigators add RNase inhibitors to RNA samples during long-term storage, an action that is usually unnecessary and may even be counterproductive. It is incumbent upon the investigator to ensure that any RNase inhibitors added to the prep at any time will not interfere with any subsequent manipulations and/or reactions involving the RNA. Historically, RNA has been stored in widely varying concentrations of SDS and VDR. Unfortunately, the incomplete removal of these "protectant" compounds prior to use will severely reduce the utility of the RNA in most downstream applications.

Other options for long-term storage of RNA include (1) highly purified 100% formamide (Chomczynski, 1992); (2) a commercially available stabilized form of formamide known as FORMAzol (Molecular Research Center, Cincinnati, OH), or (3) RNA*later* (Thermo Fisher Scientific), which is a proprietary formulation for stabilizing RNA within tissue samples and cell pellets prior to isolation.

RNA is stable for up to two years at −20°C when stored in FORMAzol, and this reagent has been used very successfully in this lab. The RNA can then be used directly for denaturing gel electrophoresis (Chapter 14: Electrophoresis of RNA). If required for cDNA synthesis or RT-PCR, either formamide or FORMAzol are easily removed by precipitating the RNA with the addition of 4 vol of 95% ethanol; at dilute concentrations, the addition of 0.2 M NaCl may also be required for efficient recovery of the RNA. We have also used RNA*later* extensively and with great success for a number of applications.

There are many considerations with respect to the correct storage of RNA. If ever in doubt, store purified RNA samples, in whatever form, at −80°C.

References

Carter, C., Britton, V.G., Haff, L., 1983. CsTFA: a centrifugation medium for nucleic acid isolation. Biotechniques 1, 142–147.

Chirgwin, J.M., Przybyla, A.E., MacDonald, R.J., Rutter, W.J., 1979. Isolation of biologically active ribonucleic acid from sources enriched in ribonuclease. Biochemistry 18, 5294–5299.

Chomczynski, P., 1992. Solubilization in formamide protects RNA from degradation. Nucl. Acids Res. 20, 3791–3792.

Chomczynski, P., 1993. A reagent for the single-step simultaneous isolation of RNA, DNA, and proteins from cell and tissue samples. Biotechniques 15, 532–536.

Chomczynski, P., Sacchi, N., 1987. Single-step method of RNA isolation by acid guanidinium thiocyanate-phenol-chloroform extraction. Anal. Biochem. 162, 156–159.

Favaloro, J., Triesman, R., Kamen, R., 1980. Transcriptional maps of polyoma virus-specific RNA: analysis by two-dimensional S1 gel mapping. Methods Enzymol. 65, 718–749.

Glišin, V., Crkvenjakov, R., Byus, C., 1974. Ribonucleic acid purified by cesium chloride centrifugation. Biochem. 13, 2633–2637.

Hu, W.P., Chen, Y.C., Chen, W.Y., 2020. Improve sample preparation process for miRNA isolation from the culture cells by using silica fiber membrane. Sci. Rep. 10, 21132. Available from: https://doi.org/10.1038/s41598-020-78202-8.

Majumdar, D., Avissar, Y.J., Wyche, J.H., 1991. Simultaneous and rapid isolation of bacterial and eukaryotic DNA and RNA: a new approach for isolating DNA. Biotechniques 11 (1), 94–101.

Okayama, H., Kawaichi, M., Brownstein, M., Lee, F., Yokata, T., Arai, K., 1987. High efficiency cloning of full length cDNA: construction and screening of cDNA expression libraries from mammalian cells. Methods Enzymol. 154, 3–28.

Peppel, K., Baglioni, C., 1990. A simple and fast method to extract RNA from tissue and other cells. Biotechniques 9, 711–713.

Roest, H.P., Verhoeven, C.J., de Haan, J.E., de Jonge, J., IJzermans, J.N., van der Laan, L.J., 2016. Improving accuracy of urinary miRNA quantification in heparinized patients using heparinase I digestion. J. Mol. Diagnost. 18 (6), 825–833. Available from: https://doi.org/10.1016/j.jmoldx.2016.06.006.

Salvatori, R., Bockman, R.S., Guidon Jr., P.T., 1992. A simple modification of the Peppel/Baglioni method for RNA isolation. Biotechniques 13, 510–511.

Schmitt, M.E., Brown, T.A., Trumpower, B.L., 1990. A rapid and simple method for preparation of RNA from *Saccharomyces cerevisiae*. Nucl. Acids Res. 18 (10), 3091–3092.

Ullrich, A., Shine, J., Chirgwin, J., Pictet, R., Tisher, E., Rutter, W.J., et al., 1977. Rat insulin genes; construction of plasmids containing the coding sequences. Science 196, 1313–1319.

Zarlenga, D.S., Gamble, H.R., 1987. Simultaneous isolation of preparative amounts of RNA and DNA from *Trichinella spiralis* by cesium trifluoroacetate iso-pycnic centrifugation. Anal. Biochem. 162, 569–574.

Zolfaghari, R., Chen, X., Fisher, E.A., 1993. Simple method for extracting RNA from cultured cells and tissues with guanidine salts. Clin. Chem. 39, 1408–1411.

Further reading

Baronti, L., Karlsson, H., Marušič, M., Petzold, K., 2018. A guide to large-scale RNA sample preparation. Anal. Bioanalyt. Chem. 410, 3239–3252. Available from: https://doi.org/10.1007/s00216-018-0943-8.

Das, S., Ansel, K.M., Bitzer, M., Breakefield, X.O., Charest, A., Galas, D.J., et al., 2019. The extracellular RNA communication consortium: establishing foundational knowledge and technologies for extracellular RNA research. Cell 177 (2), 231–242.

Drygin, Y.F., Butenko, K.O., Gasanova, T.V., 2021. Environmentally friendly method of RNA isolation. Anal. Biochem. 620, 114113. Available from: https://doi.org/10.1016/j.ab.2021.114113.

Roest, H.P., IJzermans, J.N.M., van der Laan, L.J.W., 2021. Evaluation of RNA isolation methods for microRNA quantification in a range of clinical biofluids. BMC Biotechnol. 21, 48. Available from: https://doi.org/10.1186/s12896-021-00706-6.

Scholes, A.N., Lewis, J.A., 2020. Comparison of RNA isolation methods on RNA-seq: implications for differential expression and meta-analyses. BMC Genomics 21, 249. Available from: https://doi.org/10.1186/s12864-020-6673-2.

Xiao, M.-S., Wilusz, J.E., 2019. An improved method for circular RNA purification using RNase R that efficiently removes linear RNAs containing G-quadruplexes or structured 3′ ends. Nucl. Acids Res. 47 (16), 8755–8769. Available from: https://doi.org/10.1093/nar/gkz576.

Chapter 5

Isolation of polyadenylated RNA

Rationale

Why purify mRNA? The purpose for purifying mRNA, that is, removal of the rRNA and tRNA species, is to increase the statistical representation of one, several, or all of the mRNAs in a sample. This may be an important assay design consideration because only a small fraction (1%–3%) of all mature transcripts are mRNA. The mRNA selection process is directed at the unique 3′ feature exhibited by most eukaryotic mRNAs, namely a tract of adenosine nucleotides that not only stabilizes the molecules so endowed but also contributes to their ability to support translation. Polyadenylated molecules, as these mRNAs are classified, are collectively referred to as poly $(A)^+$. Sequestration of poly$(A)^+$ is useful for increasing assay sensitivity (e.g., RT-PCR, nuclease protection, Northern analysis), for producing a representative (more full-length) cDNA pool for cloning, for RNA-seq applications, or for recovery of RNAs produced by in vitro translation. This is known as an enrichment strategy, the goal of which is to favor the presence of one or more specific transcript types. Enrichment may also be promoted by pretreatment of the biological material prior to cellular disruption to superinduce, or accumulate, certain types of transcripts in the cell.

Eukaryotic mRNAs are categorized as falling into an abundance class based on prevalence in the cell. High-abundance mRNAs are those that are present in hundreds of copies per cell. When a cell produces large amounts of a particular protein, it is sometimes (but not always) the case that the cell transcribes a correspondingly elevated amount of the cognate mRNA. Reticulocytes, for example, produce very large amounts of β-globin mRNA. Certain non-mRNA transcripts are also classified as being highly abundant in the cell, namely rRNA, tRNA, and some miRNA species. Medium abundance mRNAs are those present in dozens of copies per cell. Many of the standard housekeeping genes that are assayed frequently as transcription control or reference genes (e.g., histone, β-actin, GAPDH) produce transcripts of sufficient number to be members of this class. Low abundance mRNAs are those transcripts that are generally present in ten or fewer copies per cell; these are the so-called "rare" mRNAs that were difficult to detect until the advent of PCR. It is believed that fully 30% of cellular mRNAs fall in the low abundance category. Further, this Author's definition of *very low*

RNA Methodologies. DOI: https://doi.org/10.1016/B978-0-323-90221-2.00020-5

abundance transcripts are those present, on average, in fewer than one copy per cell and which at one time were referred to as the "hard to clone" genes!

It is important to understand that given the refinements in reagents, techniques, and instrumentation, it may no longer be necessary or even desirable to work with the mRNA fraction because the mechanics of mRNA selection may further underrepresent low abundance transcripts and alter the interpretation of the data. In this laboratory, nearly all of the transcription assays performed, including quantitative and semiquantitative PCR, begin with total cellular or total cytoplasmic RNA as template material. Owing to assay sensitivity, mRNA preselection may actually be counterproductive.

Polyadenylation

The basis of selecting mRNA from a more heterogeneous population of transcripts is the well-known poly(A) tail, a tract of adenosine nucleotides that characterizes the 3′ end of most types of eukaryotic mRNA molecules. Polyadenylation, the process by which this structure is added during mRNA biogenesis, is not an integral part of transcription. Rather, it is a rapid post-transcriptional regulatory event catalyzed by the nuclear enzyme poly(A) polymerase in concert with several ancillary proteins. The length of the poly(A) tail is variable among mRNAs, even among those transcribed from the same locus. The typical mammalian poly(A) tail ranges from 50 to 300 nts long. All RNA molecules that possess a poly(A) tail are referred to collectively as poly(A)$^+$ mRNA and are correctly said to be polyadenylated.

The addition of the 3′ poly(A) tract is actually the product of two sequential enzymatic activities, namely a cleavage reaction of newly transcribed precursor RNA (hnRNA) followed by the polyadenylation event itself; prior to cleavage, hnRNA transcripts may extend up to one thousand or more bases beyond the polyadenylation signal. The precise site at which the poly(A) tail is added is formed by endonucleolytic cleavage of the hnRNA molecule, and not by exonucleolytic degradation from the 3′ end of the molecule generated upon transcription termination. This cleavage event is then followed by the sequential addition of the adenylate residues that ultimately constitute the poly(A) tail.

The cleavage that precedes polyadenylation is not a random or haphazard occurrence. Among the higher eukaryotes, virtually all polyadenylated mRNA molecules exhibit the highly conserved hexanucleotide polyadenylation signal motif 5′...AAUAAA...3′, or a closely related variant, which is typically found with fewer than 30 bases upstream from the polyadenylation start site. Polyadenylation begins immediately 3′ to the first CA dinucleotide after the polyadenylation signal, and the efficiency of the process is influenced by the precise juxtaposition of a second element, a GU-rich cluster found approximately 20–40 nucleotides downstream from the actual cleavage site (Tian and Graber, 2012, review).

At least one role for the poly(A) tail is regulating the stability and functionality of the mRNA in the cell (reviewed by Hunt, 2007). In vivo, the poly(A) tract associates with poly (A)-binding proteins (PABPs), which bind stoichiometrically to the poly(A) tail in characteristic 27-nucleotide heteromeric repeat patterns in vertebrates; these important proteins associate with and modulate miRNA functions as well (for review, see Goss and Kleiman, 2013). The association of the poly(A) tail with PABPs extends the life of the transcript by resisting 3′ → 5′ nuclease attack. Polyadenylation is also considered to be a cellular quality control measure: immature nonpolyadenylated transcripts in the nucleus are blocked from premature export into the cytoplasm, where translational havoc could ensue. Deadenylation, in contrast, which is the gradual shortening of the poly(A) tail, is experienced by aging cytoplasmic poly(A)$^+$ transcripts. As this natural process occurs, the mRNA is translated less frequently and is targeted for degradation. Moreover, the presence of miRNAs, which are powerful posttranscriptional regulators of gene expression, may accelerate the mRNA destabilization process.

Selection of polyadenylated molecules: how it works

The sequestering of molecules with a poly(A) tract from nonpolyadenylated RNA species is possible by taking advantage of the naturally occurring hydrogen bonding affinity between the bases adenine and thymine. In so doing, the relatively small poly(A)$^+$ mRNA fraction can be physically separated from total cytoplasmic or total cellular RNA. Sample fractionation in this manner is an example of affinity chromatography, that is, a separation or chromatography based on a biological activity or structure, in this case the poly(A) tail. The separation is performed using oligodeoxythymidylate residues [oligo(dT) or poly(T)] linked to any of a variety of support matrices. In the original descriptions of the technique, oligo(dT) molecules were covalently bound by their 5′ phosphate to cellulose beads (Gilham, 1964; Aviv and Leder, 1972), hence the nomenclature "oligo(dT)-cellulose." During the isolation procedure the oligo(dT) forms hydrogen bonds with the poly(A) tail, thereby retaining it on the matrix. While oligo(dT)-cellulose has proven to be fairly effective for the capture of poly(A)$^+$ mRNA and remains available for purchase, newer methods described below, which require minimal amounts of starting material, are now preferred and are in widespread use. These contemporary poly(A) selection procedures remain variations on the theme of affinity separation and have streamlined the somewhat cumbersome mechanics of what was once known as "running an oligo(dT) column." Alternative approaches include the following:

1. Using paramagnetic beads to which oligo(dT) has been covalently linked. Following solution hybridization, the beads with the bound poly(A)$^+$ mRNA are sequestered using a magnet.

2. Using oligo(dT) linked to latex and/or polystyrene beds in a packed column format to support chromatography.
3. Exploiting the natural affinity between biotin and avidin. Biotin can be conjugated to nucleic acids and a variety of other molecules without changing their biological activities. The tight binding of biotin by streptavidin (K_d 10^{-15}M) makes many applications possible. In one format, biotinylated oligo(dT) molecules base-pair with poly(A)$^+$ mRNA and are then precipitated as a biotin/streptavidin complex. In another format, the hybrids may also be recovered using streptavidin linked to a paramagnetic bead and eluted in small, very convenient volumes without the need for further precipitation. Many permutations of this general approach are available as kits for mRNA purification.
4. Mixing oligo(dT)-cellulose or microcrystalline oligo(dT)-cellulose and total RNA together directly in a microfuge tube without any prior formation of some type of matrix geometry.

Regardless of the methodology, the key to successful poly(A)$^+$ selection is to keep the volumes involved as small as possible, thereby keeping the sample as concentrated as possible.

Methodologically, techniques for the purification of poly(A)$^+$ mRNA have at least one thing in common: poly(A)$^+$ mRNA anneals to an affinity matrix under high ionic strength (high salt) conditions. This is necessary to eliminate the electrostatic repulsion that would occur naturally between the target RNA and the thymidylate residues which are linked to the matrix. Under high-salt conditions, a monovalent cation (usually Li^+ or Na^+) acts as a counter-ion, neutralizing the net negative charge intrinsic to the phosphodiester backbone of polynucleotides. In the absence of a high-salt milieu, the electrostatic repulsion between the oligo(dT) and the poly(A) tail would be of sufficient magnitude to prevent base-pairing and, consequently, the purification of poly(A)$^+$ mRNA in this manner.

The binding capacity of oligo(dT)-matrices increases with salt concentration up to about 500 mM LiCl, NaCl, or even KCl, though the binding capacity is greater when KCl or LiCl is used in place of an equivalent concentration of NaCl (Mercer and Naora, 1975); see manufacturer's package insert for the particulars of each grade and lot of oligo(dT) separation matrix. This increase in binding capacity, however, may be accompanied by an increase in nonspecific hybridization to the matrix. Once the nonhybridized, nonpoly(A)$^+$ material is removed by washing the matrix with a fairly high-salt buffer, the hybridized species of poly(A)$^+$ RNA are recovered from the matrix by changing to a very low-salt buffer (0–5 mM salt). Removal of the monovalent cation from the system reestablishes electrostatic repulsion; rapid dissociation of the hybridized RNA from the matrix in a small volume is favored by preheating the elution buffer to 55°C. Although usually unnecessary, the now separated poly (A)$^+$ and poly(A)$^-$ material can be concentrated further by salt and alcohol precipitation. The poly(A)$^-$ RNA should be saved because it makes an excellent negative control in hybridization analyses as well as in the synthesis of cDNA (Chapters 15: Photodocumentation and image analysis and 7: Going green: RNA and the molecular biology of plants, respectively) because samples which have been depleted of poly(A)$^+$ RNA should fail to produce a signal.

The poly(A) caveat

What is the significance of the mRNA subpopulation that is purified by affinity chromatographic selection directed toward the poly(A) tail? Although most eukaryotic mRNAs are polyadenylated, a small percentage is not. Thus, populations of poly(A)-selected mRNAs molecules do not include nonpolyadenylated mRNAs, an element that may or may not be meaningful in the context of a particular investigation. The reader is cautioned that kits that purport to facilitate the selection of "total mRNA" through poly(A) selection will fail to yield poly(A)$^-$ mRNA. Further, because the poly(A) tail is added in the nucleus, the selection of poly(A)$^+$ RNA from a whole cell lysate will unquestionably render polyadenylated cytoplasmic RNA (mRNA) as well as polyadenylated nuclear RNA (hnRNA), which may or may not be meaningful. Note that the mixture of the two cannot be construed as representing functional (translatable) mRNA; natural, elaborate posttranscriptional regulatory circuits orchestrate the degradation of a great deal of hnRNA in the nucleus, thereby preventing it from ever maturing and entering the cytoplasm. Again, this may or may not be meaningful in the interpretation of the data.

cDNA synthesis considerations

A common approach to the synthesis of first-strand cDNA for RT-PCR analysis and other applications involves annealing an oligo(dT) primer to the poly(A) tail on the mRNA template. This positions the requisite 3′-OH group from the primer in the proper antiparallel orientation relative to the template. The net effect is that reverse transcriptase (RNA-dependent DNA polymerase) initiates first-strand cDNA synthesis at or near the 3′ end of poly(A)$^+$ mRNA, while poly(A)$^-$ mRNA is excluded. In order to circumvent the potential shortcomings of oligo(dT)-primed cDNA populations, more inclusive cDNA pools are prepared by random-primed synthesis of the first strand cDNA, and sometimes generated through the use of random primers *and* oligo(dT) in the same reaction tube, which is a standard procedure in this laboratory. Briefly, a number of nucleotide hexamers or nonomers of random nucleotide sequence are allowed to anneal to heat-denatured RNA. This annealing effectively supports cDNA synthesis regardless of whether the 3′ terminal structure of the molecule is poly(A)$^+$ or poly(A)$^-$. cDNA synthesized by random priming alone will not prime from within the poly(A) tail unless an oligo(dT) random primer had been included in the primer mixture. Moreover, random priming generally produces a greater number of full-length cDNA molecules. This strategy and many others are described in detail in Chapters 9, cDNA: A permanent biochemical record of the cell and 10, RT-PCR: a science and an art form. Historical note: When screening a library that may have been synthesized some years ago (they still exist in the recesses of −80°C freezers), knowledge of the method of cDNA priming may influence the interpretation of the outcome of the experiment. Also appearing in older literature are reports of treating total RNA with poly(A) polymerase to polyadenylate all RNA prior to cDNA synthesis! This is neither necessary nor recommended.

In order to focus more closely on only mature poly(A)$^+$ mRNA being produced by the cell, one could lyse the cells gently with hypotonic, nonionic buffer (Chapter 4: RNA isolation strategies), remove the intact nuclei, and then proceed to select the polyadenylated RNA from the cytosol. Alternatively, to give even clearer definition to those specific mRNAs destined to be translated, polysome profiling is a suitable approach for determining which RNAs are actually engaged by the ribosomes (Seimetz et al., 2019). The inclusion of high concentrations of cycloheximide arrests translation by blocking the translocation step, thereby stabilizing the polysomes during processing. Active polysomes are recovered by fractionation in sucrose gradients and the bound mRNA is then recovered and subject to deep sequencing in order to identify them; poly(A)$^+$ selection can increase sensitivity and resolution. This and other methods are very useful for discerning the active mRNA population based on ribosome occupancy.

Assay sensitivity considerations

Poly(A)$^+$-selected mRNA could not be utilized to derive qualitative or quantitative information about cellular or cytoplasmic poly(A)$^-$ species, which includes the abundant histone mRNAs. The situation becomes more complicated when one considers the possibility that alternative posttranscriptional processing pathways might yield a collection of polyadenylated and nonpolyadenylated transcripts from the same locus. The advantages and disadvantages to selecting poly(A)$^+$ RNA are highlighted in Table 5.1. One must weigh the relative merits of working with poly(A)$^+$ instead of total RNA in the context of the required level of sensitivity, the nature of the assay, the type of data required, and the number of cells or mass of tissue available. In many cases, it may be worth running experiments with total cellular RNA first, before incurring the added expense and the added risk associated with the selection of poly(A)$^+$ mRNA.

Magnetic bead technology for poly(A)$^+$ purification

The covalent linkage of a variety of ligands to what are essentially coated iron beads has become an invaluable tool for the resolution of a variety of

TABLE 5.1 **Poly(A)$^+$ purification: advantages and disadvantages.**

Advantages
• Poly(A)$^+$ purification increases mRNA as a percentage of all RNA in the sample.
• Poly(A)$^+$ purification greatly reduces interference from rRNA and tRNA.
• Poly(A)$^+$ purification increases assay sensitivity.
• Magnetic methods for poly(A) capture are able to concentrate the sample dramatically without the need to reprecipitate the sample. Purified poly(A)$^+$ mRNA is ready for immediate use.
• The poly(A)$^-$ fraction, which can be collected and precipitated, is an excellent negative control, especially for blot analysis.
Disadvantages
• Poly(A)$^+$ selection excludes poly(A)$^-$ mRNAs.
• Loss of poly(A)$^+$ during the purification process may further underrepresent rare mRNA, especially if starting with a previously isolated sample of total RNA.
• Depending on the cell lysis method, poly(A)$^+$ purification may also yield polyadenylated hnRNA.
• Oligo(dT) matrices are expensive.
• Given the improvements in cloning and RT-PCR efficiency, poly(A)$^+$ selection may not be necessary or may even be counterproductive.

compounds from complex mixtures. Ligands include antibodies, which can be used to recognize and bind cell surface antigens and, in the context of this resource, oligothymidylate, for the efficient recovery of polyadenylated transcripts. Several companies produce magnetic bead-based products; Dynabeads (ThermoFisher Scientific), for example, have been used in this lab for many years when application required poly(A)$^+$ mRNA. The Dynabeads used for mRNA capture are uniform superparamagnetic beads, 2.8 μm in diameter, with a polystyrene coating that encloses iron oxide (Fe_2O_3) particles. These particles form a magnetic dipole when exposed to a magnetic field. Chains of deoxythymidylate, 25 nucleotides long, have been covalently attached to the surface of the beads via a 5′ linker group. A mixture of total RNA added to the beads under conditions of high ionic strength will favor hybridization between polyadenylated RNA and the oligo(dT) tract linked to the beads (Fig. 5.1). Hybridization kinetics are similar to those found in free solution, with complete hybridization being observed within five minutes. The magnetic beads and hybridized poly(A)$^+$ RNA are then collected by placing the sample tube in the vicinity of a magnet (Fig. 5.2). Washing is performed, followed by elution using a standard combination of elevated temperature and very low-salt buffer. Isolation of polyadenylated RNA by magnetic capture is the

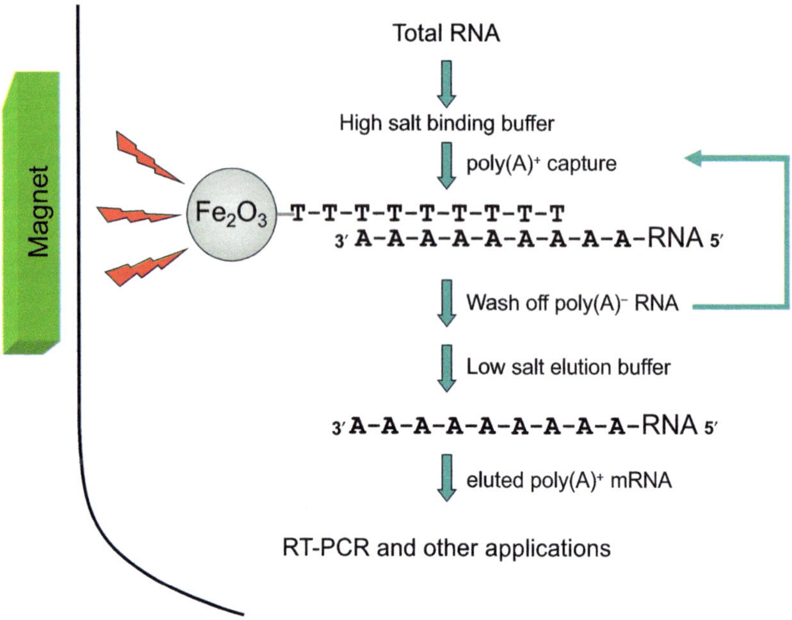

FIGURE 5.1 Selection of poly(A)$^+$ RNA using magnetic bead technology. Under high-salt conditions previously purified (raw) RNA or a whole cell lysate is mixed in a microfuge tube with oligo(dT) linked to a paramagnetic bead. After annealing of the poly(A) tail and the oligothymidylate, polyadenylated mRNA is sequestered to the inside wall of the tube by the tube next to a magnet. After washing away nonadenylated transcripts, the purified mRNA is eluted in a very small volume of low-salt buffer or sterile water. This approach is compatible with polyadenylated RNA from any source isolated by any procedure.

strongly preferred approach to poly (A)$^+$ selection because it precludes the requirement for traditional column affinity chromatography, thereby rendering a new level of resolution in the chromatography of nucleic acids. The principal advantage of this technique is the rapid isolation of pure, intact poly(A)$^+$ RNA in concentrated form. As the purified poly(A$^+$) RNA is already highly concentrated, there is no need for reprecipitation, which often results in the further loss of unique or low-abundance transcripts. Transcripts that are affinity-purified in this manner are ready for immediate use. The average binding capacity of these beads is 2 μg poly(A)$^+$ RNA per milligram of beads.

There are numerous protocols that accompany magnetic beads for nucleic acid purification; because protocols are product-specific, they have not been included here. The focus of this chapter is to inform the reader of the available options for improving assay sensitivity. Conveniently enough, there is wide latitude with respect to the selection of starting materials for poly(A)$^+$ selection, including previously purified total cellular or cytoplasmic RNA, cells growing in culture, and solid tissue. It is very important to realize that in cases where there is a limited amount of starting material it may be

(A) (B)

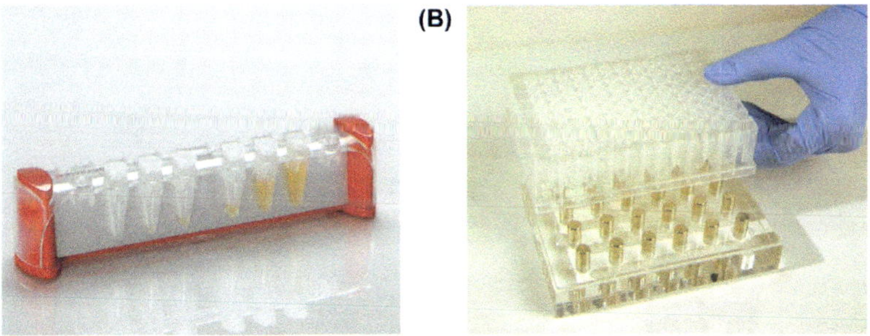

FIGURE 5.2 Magnetic capture of poly(A)$^+$ RNA after base-pairing with oligo(dT) linked to superparamagnetic particles. The beads can be sequestered in (a) microfuge tubes or (b) in specially designed 96-well plates. *Image courtesy of Invitrogen.*

prudent to add the beads directly to a whole cell lysate or tissue homogenate for the direct harvest of poly(A)$^+$ RNA. In the context of assigning gene regulation as either transcriptional or posttranscriptional, the investigator is reminded that thymidylate residues recognize and capture *any* transcript with a poly(A) tract, both mRNA *and* hnRNA, and will fail to capture nonpoly(A) mRNA. Regardless of the product, the success of the magnetic separation technique requires the complete removal of all buffers from the tube containing the magnetic beads before moving on to the next step to avoid creating salt gradations in the tube. Most importantly, *never* remove anything from the tube containing the magnetic beads unless the tube is in the magnet!

Finally, note that other formats for poly(A)$^+$ selection are available for purchase, including magnetic-based separation products and spin column configurations, all of which in some way target the poly(A) tail. Magnetic separation technology has also been expanded for immunocapture of proteins, cell separation, genomic DNA, total cellular RNA, and other applications.

Oligo(dT) affinity chromatography

Running an oligo(dT) column was standard fare in the 1980s and early 1990s. Preparing the matrix and attending to the column was a time-consuming and somewhat monotonous process and the success of which was highly user dependent. A protocol for the use of oligo(dT)-cellulose is presented here, though its availability continues to decline in favor of other matrices such as oligo (dT)-agarose beads (MCLAB Products), magnet bead capture, or even automated instrumentation. For larger-scale purification, as would be observed in manufacturing processes, POROS oligo(dT)$_{25}$ affinity resin or a prepacked column (Fig. 5.3; ThermoFisher) in conjunction with an absorbance detection monitor (Fig. 5.4) are appropriate matrix options

The selection of poly(A)$^+$ RNA can also be accomplished without packing a column. In fact, relatively small masses of starting material (typically

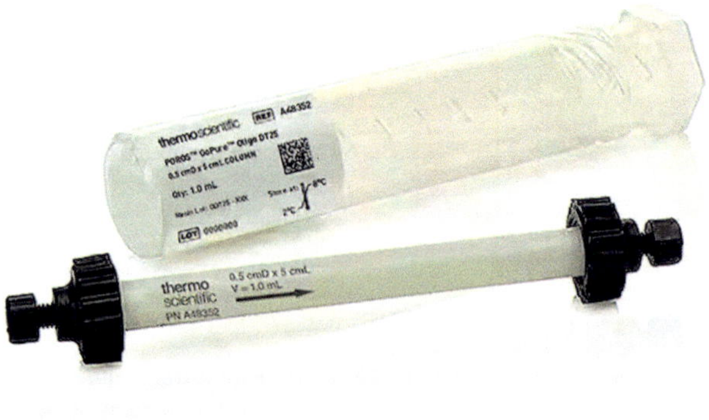

FIGURE 5.3 Prepacked column for poly(A)$^{+}$ selection. This 1 mL POROS GoPure Oligo (dT)$_{25}$ column contains an oligo(dT) matrix that is appropriate for use in manufacturing-scale polyadenylated RNA recovery. The entrance and exit ports located at the ends of the column facilitate efficient buffer movement into and out of the column via sterile tubing. *Image © Life Technologies Corporation, owned by and courtesy of Thermo Fisher Scientific (www.thermofisher.com).*

less than 200 μg total RNA) mandate a method of separation other than a column format. This is mainly because of the dilution of RNA that inevitably occurs and the very low percentage of the total mass that the poly(A)$^{+}$ species represent, not to mention the labor-intensive nature of preparing and running a homemade column. If the yield is extremely small or the sample is too dilute, it will be difficult to precipitate the purified RNA without the addition of carrier molecules such as tRNA or glycogen. Of course, such a strategy defeats the purpose of poly(A)$^{+}$ selection in the first place, and excessive glycogen carryover is incompatible with several subsequent RNA applications.

Alternative matrix configurations include the use of magnetic beads, described above, the use of standard grade or microcrystalline oligo(dT)-cellulose, or oligo(dT)-agarose beads for the selection of poly(A)$^{+}$ mRNA in a noncolumn format. Microcrystalline oligo(dT)-cellulose generally can bind in excess of 100 OD (A_{260} units) per gram of matrix. By comparison, the typical binding capacity of the oligo(dT)-cellulose under standard assay conditions generally ranges from 50 to 80 OD/g of matrix, depending on the manufacturer, the grade of refinement, and the monovalent cation used to prepare the binding buffer: Li^{+} is better for this application than either K^{+} or Na^{+}. The selection of poly(A)$^{+}$ RNA is best accomplished by repeated suspension of the matrix in traditional high-salt buffers, followed by

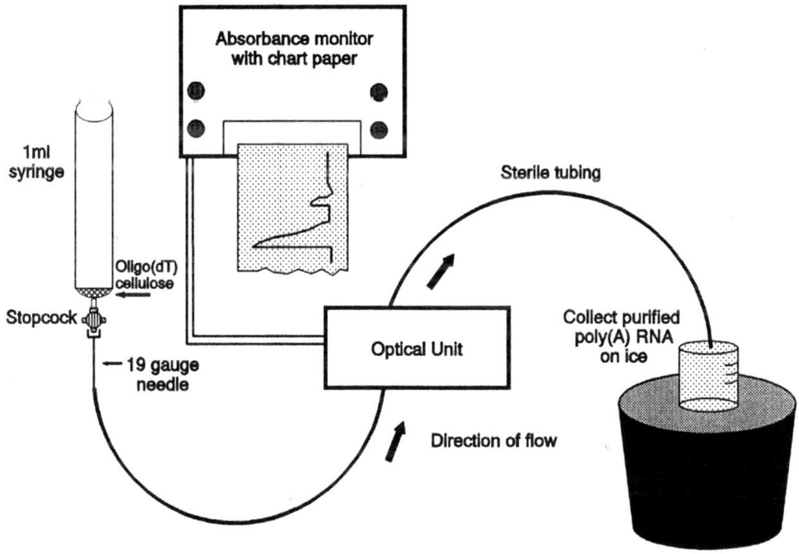

FIGURE 5.4 Method for purification of large quantities of poly(A)$^+$ material. An oligo(dT) matrix may be purchased prepacked or may be prepared in a syringe. In the example shown, the flow rate is regulated by an autoclavable (RNase-free) stopcock placed beneath the syringe; ideally, the flow through the cell should be slow and uniform. The needle below the stopcock channels the eluate into the tubing which leads to the optical unit of the UV detection monitor. The optical unit is outfitted with a 254 nm absorbance filter designed for nucleic acid chromatography. The concentration of nucleic acid flowing through the optical unit is interpreted by the magnitude of the deflection of the pen mounted on the monitor, thereby providing the user with either a physical or electronic record of the binding and elution profiles for each sample. The eluate is then directed out of the optical unit and is collected in one or more sterile tubes, usually resting on ice. Sudden increases or decreases in absorbance are indicative of when portions of the sample passing through the flow cell detector should be collected as separate fractions.

centrifugation. Using this approach, one avoids the excessively slow flow rates commonly associated with column chromatography; note that microcrystalline oligo(dT)-cellulose is not recommended for use in column format unless prepacked by the manufacturer. If scaled down, the entire procedure can be carried out in a microfuge tube. The eluted RNA thus remains in a relatively small volume and at a concentration that permits precipitation with salt and alcohol alone, without an added carrier.

The following protocol is one variation of several procedures for microcrystalline oligo(dT)-cellulose. Although this type of separation can also be used with standard oligo(dT)-celluloses, the microcrystalline grade maximizes recovery. Note that all requirements for controlling RNase activity must be satisfied for quantitative and qualitative recovery. In all cases, it is important to adhere closely to the recommendations of the manufacturer of the oligo(dT) product for optimum performance.

Protocol: noncolumn poly(A)$^+$ RNA purification

1. Wear gloves and appropriate protective eyewear.
2. Weigh out an appropriate mass of matrix material. Use 100 mg (dry weight) for every 150 μg of total RNA starting material. Place matrix in a nuclease-free microcentrifuge tube.

   Note: Oligo(dT)-cellulose should be weighed out under RNase-free conditions. While autoclaved or baked metal spatulas may be used (see Chapter 2: Creating a ribonuclease-free environment, for details), it is better to gently tap the bottle of oligo(dT)-cellulose to remove a suitable mass.
3. Add 500 μL binding buffer (500 mM LiCl; 20 mM Tris-Cl, pH 7.5; 1 mM EDTA) to the oligo(dT)-cellulose. Allow the beads to equilibrate in this high-salt buffer for 5 minutes.
4. Centrifuge the tube containing the equilibrate oligo(dT)-cellulose at 200 × *g* for 30 seconds. Carefully remove and discard the supernatant without disturbing the oligo(dT)-cellulose pellet. Set tube aside.
5. In a separate tube, dissolve RNA in nuclease-free H_2O in a final volume of 100 μL and heat to 65°C for 5 minutes.

   Note: The heating step in this and related protocols is necessary to denature RNA molecules, thereby making the poly(A) tail more available for base-pairing to the matrix.
6. Add an equal volume of binding buffer (step 3, above) to the heat-denatured RNA and allow it to cool to room temperature. Add the mixture to the oligo(dT)-cellulose that was set aside in step 4.
7. Place the tube in a microfuge tube rack and allow the sample to stand at room temperature for 5 minutes. Agitate the tube gently by hand once or twice during this brief incubation.
8. Centrifuge at 1500 × *g* for 5 minutes, at 4°C if possible. Carefully transfer the supernatant to a fresh tube and label as "poly(A)$^-$ RNA." Set tube aside on ice.

   Note: The poly(A)$^-$ fraction can be extremely useful as a negative control in a variety of applications or can itself be used as carrier RNA in future experiments.
9. Wash the oligo(dT)-cellulose up to five times with 500 μL aliquots of binding buffer. Centrifuge at 1500 × *g* after each wash and then carefully remove the supernatant between each wash.
10. Elute the poly(A)$^+$ RNA from the matrix with one or more 50 μL aliquots of elution buffer (10 mM Tris-Cl, pH 7.5; 0.1 mM EDTA). Alternatively, RNA may be eluted in nuclease-free H_2O.

    Note: The number of elution buffer aliquots needed to recover all of the bound material will be a function of poly(A)$^+$ mass and the volume of elution buffer used. The first aliquot of elution buffer will contain a great majority of the poly(A)$^+$ material. Logistically speaking, it may be useful to elute the poly(A)$^+$ in empirically determined smaller volumes. As needed, the RNA should be reprecipitated in one or more microfuge tubes. This is easily accomplished by adding 0.1 vol 3 M NaOAc (pH 5.2) and 2.5 volumes ice-cold 95% ethanol. Store sample(s) on dry ice for 20 minutes or at −20°C for at least 4 hours. Following precipitation and centrifugation, carefully decant the ethanol supernatant. Wash the RNA pellet two or three times with 500 μL aliquots of 70% ethanol (prepared in sterile H_2O). A final wash with 95% ethanol may facilitate air drying. Resuspend RNA in a minimum volume of TE buffer, pH 7.5 or sterile H_2O.

11. Determine poly(A)$^+$ concentration by measuring A_{260}, as described in Chapter 8, Quality control for RNA preparations.

12. Store the RNA is suitable aliquots at −80°C until further use.

References

Aviv, H., Leder, P., 1972. Purification of biologically active globin messenger RNA on oligothymidylic acid-cellulose. Proc. Natl. Acad. Sci. 69, 1408−1412.

Gilham, P.T., 1964. Synthesis of polynucleotide celluloses and their use in fractionation of polynucleotides. J. Am. Chem. Soc. 86, 4982−4985.

Goss, D.J., Kleiman, F.E., 2013. Poly(A) binding proteins: are they are created equal? WIREs RNA 4, 167−179.

Hunt, A.G., 2007. Messenger RNA 3′-end formation and the regulation of gene expression. In: Bassett, C.L. (Ed.), In Regulation of Gene Expression in Plants. Springer, New York, pp. 101−122.

Mercer, J.F.B., Naora, H., 1975. A comparison of the chromatographic properties of various polyadenylate binding materials. J. Chromatogr. 114, 115−128.

Seimetz, J., Arif, W., Bangru, S., Hernaez, M., Kalsotra, A., 2019. Cell-type specific polysome profiling from mammalian tissues. Methods 155, 131−139. Available from: https://doi.org/10.1016/j.ymeth.2018.11.015.

Tian, B., Graber, J.H., 2012. Signals for pre-mRNA cleavage and polyadenylation. WIREs RNA 3, 385−396. Available from: https://doi.org/10.1002/wrna.116.

Further reading

Gruber, A.J., Zavolan, M., 2019. Alternative cleavage and polyadenylation in health and disease. Nat. Rev. Genet. 20, 599−614. Available from: https://doi.org/10.1038/s41576-019-0145-z.

He, S.L., Green, R., 2013. Polysome analysis of mammalian cells. Meth. Enzymol. 530, 183−192. Available from: https://doi.org/10.1016/B978-0-12-420037-1.00010-5.

Viero, G., Lunelli, L., Passerini, A., Bianchini, P., Gilbert, R.J., Bernabò, P., et al., 2015. Three distinct ribosome assemblies modulated by translation are the building blocks of polysomes. J. Cell Biol. 208 (5), 581−596.

Yuan, F.W., Hankey, E.J., Wagner, W., Li, Wang, Q., 2021. Alternative polyadenylation of mRNA and its role in cancer. Genes & Diseases 8 (1), 61−72. Available from: https://doi.org/10.1016/j.gendis.2019.10.011.

Chapter 6

The truth about tissues

Rationale

There is much to be said in favor of a good in vitro model for both basic and applied research. However, it is nearly impossible to extrapolate results obtained in vitro to describe in vivo phenomena completely and accurately. Unquestionably, the ability to detect the many subtle and not-so-subtle biochemical changes that occur in tissues is lost when working with cell monolayers. In contrast, working with cells grown in culture presents fewer logistical difficulties than the acquisition of tissue samples or working with whole organism models, both animals and plants. While there is much to be said in favor of developing a cell culture model, it is not without its shortcomings. This chapter addresses issues unique to the isolation of RNA from animal tissue; the isolation of RNA from plant tissue is described in Chapter 7, Going Green: RNA and the Molecular Biology of Plants.

Tissue culture or tissue?

The isolation of high-quality RNA from cells grown in culture is generally far easier to accomplish than is an attempted RNA isolation from tissue. The primary reason is the three-dimensional architecture of the tissue must be dismantled and removed compared to the disruption of cells grown on tissue culture plastic. Frequently, aggressive methods of homogenization in the presence of rather harsh lysis buffer formulations are required to provide a satisfactory level of tissue disruption so as to render an RNA yield of acceptable quantity and purity. Further, some tissues are notorious for their legendary levels of RNase and reluctance to give up their RNA. Thus, the aggressive methods needed to compensate for the unique character of tissue often renders total cellular RNA, as opposed to an RNA subpopulation, for example, cytoplasmic *or* nuclear transcripts. The good news is that tissues can be snap-frozen, or otherwise stabilized and stored at −80°C immediately after harvesting so that RNA isolation may be performed at a more convenient time.

RNA isolation from tissue is a nonnegotiable when studying diseased tissue, when attempting to ascertain a tissue-specific response to an experimental challenge or other insults, or when trying to characterize the molecular physiology of normal tissue. In addition to some of the RNA isolation considerations described

RNA Methodologies. DOI: https://doi.org/10.1016/B978-0-323-90221-2.00008-4

previously, the tissue sample itself may be difficult to acquire especially when the "phenomenon" is rare or when patient consent is required. Moreover, investigators working with human samples know all too well of the inherent dangers of HIV-, hepatitis-, or COVID-19-tainted samples, to name but three prevalent and very dangerous infectious agents. Not only are the mechanics of RNA isolation from tissue samples modified, compared to RNA isolation from cell culture samples, so is the level of difficulty in the storing and handling of the sample so as to maintain the safety of all parties involved.

The following is a summary of the advantages and disadvantages associated with the use of cell culture and tissue samples, respectively.

Advantages of cell culture

1. The total exposed surface area of cells grown in culture is much greater than is observed in a tissue sample containing an equivalent number of cells. Consequently, it is easier to ensure complete cell lysis. In contrast, some tissues are extremely difficult to disrupt, while other tissues are quite reluctant to give up their RNA.
2. Many primary cells and established cell lines are easy to obtain and can be propagated economically.
3. Cells growing in culture are available on an as-needed basis. In contrast, it is often difficult to obtain tissue samples or biopsies from animals on a regular basis and, especially with humans, nearly impossible to acquire multiple tissue specimens from the same organism.
4. Isolation of RNA from cells in culture is generally more expedient than isolating the same from a tissue sample.
5. Unlike cultured cells, disruption of tissue samples generally requires fairly aggressive homogenization, including Polytron disruption or Dounce homogenization. In the case of the former method, excessive heat will compromise the stability of the RNA.
6. The mechanical shearing forces that accompany tissue disruption usually disrupt membrane-bound organelles, including the nucleus. Thus, tissue homogenization should be expected to yield routinely significant quantities of nuclear RNA, genomic DNA, and mitochondrial and/or chloroplast DNA and transcripts. This often necessitates additional steps to clean up the RNA prior to use. In contrast, it is very easy to isolate RNA from specific subcellular compartments when working with cell culture.
7. Cells grown in vitro, once removed from the culture vessel, generally dissociate from each other with minimal difficulty. However, cells that constitute a tissue are tethered together via various types of intercellular junctions as well as the components of the extracellular matrix (ECM) that contribute to the heavy debris that commonly accompanies tissue disruption. This necessitates one or more additional steps to remove the debris completely to recover RNA of sufficient quantity and purity to support any of a variety of downstream

applications. These steps are not technically challenging, though they do constitute additional manipulations and, therefore, more opportunities to lose or compromise the sample.

8. Cells grown in culture usually represent a homogeneous population, even though the cell population may be aneuploid. In contrast, even a very small tissue snippet may contain many types of cells, making it difficult to pinpoint the specific cells responsible for any observed modulation of gene expression.

Advantages of tissue samples

1. The three-dimensional architecture of the tissue is preserved. This permits thin sectioning of flash frozen tissues with a cryostat or with a microtome in the case of formalin-fixed samples for localization of gene expression by in situ hybridization. Alternatively, laser-capture microdissection (reviewed by Bevilacqua and Ducos (2018); Datta et al. (2015)) can be applied to tissue samples for the isolation of RNA from a very small number of specific cells within the tissue.
2. The natural molecular physiology of the tissue is preserved, whether normal, wild type, untreated, treated, mutant, or diseased.
3. Relatively small tissue samples (100 mg range) can yield sufficient RNA for multiple applications.
4. Depending on the mass of a tissue sample, a very large quantity of RNA can be isolated because of the enormous numbers of cells involved. Storing RNA from large-scale isolation ensures the availability of identical aliquots of RNA for multiple experimental replicates.
5. The stability of RNA in a tissue sample is excellent when the tissue is flash-frozen in liquid nitrogen and then stored at −80°C, even for extended periods. Ultracold tissue samples can be fractured into smaller pieces or pulverized for later use. When ready to perform RNA isolation, thawing the tissue directly in the presence of the lysis buffer has always worked very well in this lab.
6. The results of transcription-based assays when starting with tissue samples reflect the pharmacology of a drug on a specific organ or system.
7. Isolation of RNA from tissue samples offers a glimpse of gene expression of biochemically different organisms, which has profound ramifications in the context of personalized medicine.

Homogenization methods

A common mistake that investigators make when working with tissue for the first time is to begin with too much tissue at the onset of the RNA isolation procedure. If accustomed to working with cell culture, it is very easy to underestimate the number of cells present even in a very small amount of tissue. If an insufficient amount of lysis buffer is used, then the purity of the RNA is likely to suffer, and the overall yield will plummet. If, on the other hand, the volume of lysis buffer is increased, it may be necessary to split the homogenate into multiple tubes, making processing of the sample cumbersome at best, and reducing the number of different samples that can be handled simultaneously.

Two basic approaches to homogenization are presented here, the choice of which depends upon the difficulty encountered in disrupting a particular tissue type. Keep in mind that animal tissue is generally easy to disrupt because of the absence of the cell wall that is associated with plants, bacteria, and yeast. As a general rule, one should use the least aggressive means necessary for creating the homogenate. The idea behind homogenization is simply to increase the surface area of the biological material to afford the lysis buffer a chance to do its job: lyse cells. Overly zealous investigators endeavoring to disrupt every single cell by mechanical means run the risk of damaging the RNA to the point where it will no longer support downstream analyses.

Motorized homogenizers

The term *polytron* enjoys commonplace usage in many research laboratories. Polytron (Kinematica AG) is the trademark-protected name of a motorized device used to rapidly homogenize all types of plant and animal tissue, as well as bacterial cultures, in large and small volumes (Fig. 6.1). Attached to the motor is a changeable stainless-steel probe, properly known as a

FIGURE 6.1 Polytron homogenizer. Models range from hand-held units (shown) to those with much higher throughput dispersing capacity. *Courtesy Kinematica, Inc.*

"generator." The end of the generator is characterized by very sharp, knife-like, stainless-steel saw-tooth blades that are designed to shred tissue efficiently. The variable-speed motor allows the user to control the degree of tissue disruption while minimizing the amount of foaming and heat generation. It is very important for the user to observe all safety procedures recommended by the manufacturer, as intentional misuse or accidents may result in serious, permanent injury. Motorized homogenization devices are also available from other companies such as Omni International (Kennesaw, GA). The use of hand-held and stationary homogenizers for nucleic acid work has replaced the use of the Waring blender, one time a laboratory staple, for many applications.

Although stainless steel generators are durable and may be required to homogenize recalcitrant tissues, disposable generators are also available (Fig. 6.2). The use of disposable generators dispenses the need to clean a single stainless-steel probe when multiple samples must be processed. Disposable generators also eliminate the potential for cross-sample contamination and are particularly helpful in the control of RNase activity, thereby favoring the maximum stability of the RNA. Disposable homogenizer generators have been used extensively and with great success in this lab for soft-tissue homogenization, including liver, brain, and cell suspensions. In the case of human tissues, the use of disposable generators also reduces the potential for the spread of infectious agents.

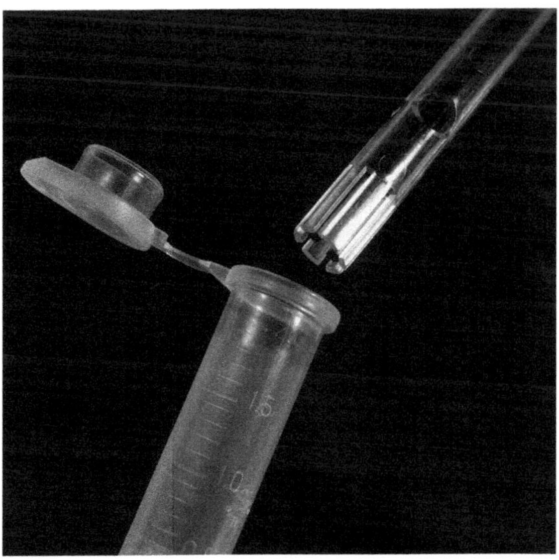

FIGURE 6.2 Hand-held motorized homogenizer showing disposable generator tips. Tips are discarded after use, minimizing the downtime required to clean the generator between each sample. Tip disposal also eliminates at least one potential source of cross-sample contamination and accidental introduction of RNase activity. *Courtesy Omni International.*

Dounce homogenization

The Dounce homogenizer is a hand-held glass implement that resembles a test tube (Fig. 6.3). Generally, the open end of the instrument is enlarged to accommodate the introduction of biological material. Inserted into the glass tube is a stylized tightly-fitting round glass or Teflon pestle with a grip at one end. This design allows the user to apply the desired number of up-and-down strokes while twisting the pestle in the process; this motion maximizes the shearing forces; tissue disruption occurs as the sample is forced through a narrow space, specifically between the inner wall of the homogenizer and the pestle. In contrast to some of the physical (mechanical) forms of cell and/or tissue disruption that often require the use of expensive equipment, Dounce homogenization is a much more gentle, controlled, and economical means of breaking up tissue. Dounce homogenization, considered a form of liquid homogenization, is ideal for the isolation of intact nuclei and other organelles that would otherwise be compromised by aggressive mechanical disruption.

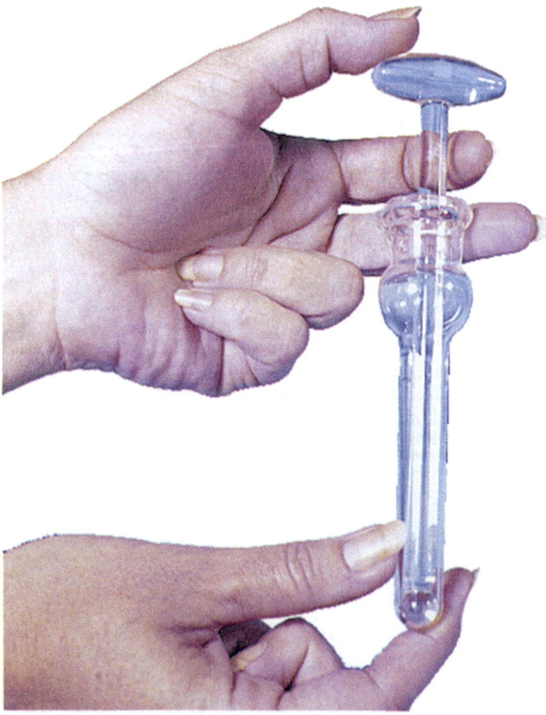

FIGURE 6.3 Dounce homogenizer. This handheld instrument is an effective, gentle homogenization device for use with many types of tissue. With the proper lysis buffer formulation, nuclei can be collected intact following whole-cell disruption, a useful approach for the analysis of nuclear transcripts. See text for details. *Courtesy Bellco Glass Company.*

The traditional Dounce homogenizer is designed to be re-used repeatedly. In this laboratory, glass homogenizers are thoroughly scrubbed and rinsed, after use, and then soaked in 3% H_2O_2 (see Chapter 2: Creating a Ribonuclease-Free Environment) for 20 minutes before being rinsed with copious amounts of autoclaved water. After air-drying, each homogenizer and its pestle are individually wrapped in foil and autoclaved to ensure that the homogenizer is nuclease-free the next time it is used.

One may also purchase disposable single-use conical tissue grinders. These devices are Dounce-like homogenizers that consist of a plastic pestle mounted in the cap of a conical 50 mL or 15 mL tube (Fig. 6.4); small versions are also available for use with 1.5 mL microfuge tubes. These devices are extremely handy when large numbers of samples need to be processed in a short span or to control the dispersal of infectious agents that could potentially be in the sample.

BeadBeater homogenization

The BeadBeater product line (BioSpec Products, Inc., Bartlesville, OK) is innovative instrumentation in widespread use for the disruption of cells, particularly those that have a cell wall. This is an economical method that can be used when chemical lysis is inefficient or when chemical lysis yields are unacceptably low. The BeadBeater (Fig. 6.5) approach is a mechanical lysis method in which either smooth or sharp small beads are used to disrupt the cells rapidly and in various types of lysis buffer. The beads are commonly made of glass, ceramic, or stainless steel and the bead size, typically ranging from a diameter of 0.1 mm to more than 2.0 mm, is selected based on the

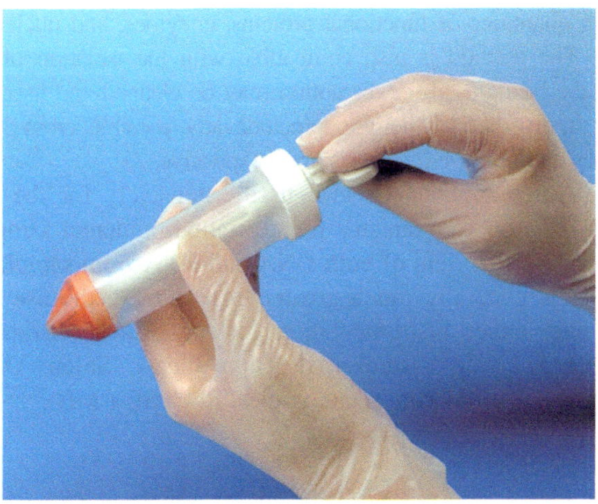

FIGURE 6.4 Disposable closed-system tissue grinder. Presterilized and designed for single use, this homogenizer is ideal for preventing intersample contamination and for controlling the spread of infectious agents. *Courtesy Kimble Chase Life Science.*

FIGURE 6.5 Mini-BeadBeater-16. This is a high-energy cell disruptor that can process up to 16 samples simultaneously. *Courtesy BioSpec Products, Inc.*

type of material to be disrupted. Other bead materials and densities are also available, depending on the nature of the biological material.

The method of cell lysis is based on the crushing action that occurs when the beads collide with the sample, particularly at high speed. By limiting the mechanical disruption to short bursts of the instrument and when used in conjunction with the accompanying cooling jacket, the BeadBeater method facilitates high-yield macromolecule recovery. Through the use of relatively gentle lysis buffers, for example, large quantities of functional proteins, enzymes, and nucleic acids may be recovered. The method is also compatible with the isolation of small organelles. While the beads used for disruption may be cleaned and used again, most investigators simply discard them to preclude any possible cross-contamination among samples and because the beads are inexpensive.

The BeadBeater approach is also used extensively for the isolation of nucleic acids from microorganisms found in environmental samples, including soil and water. The yield of both DNA and RNA from samples processed in this manner is generally much greater than through the use of chemical extraction alone. Other instrumentation for mechanical disruption of this nature are the FastPrep system (MP Biomedicals; Solon, OH) and the Precellys homogenizer (Bertin Technologies, Rockville, MD).

RNA isolation strategies for various organs and tissues

There is no single fool-proof method for the expedient isolation of RNA from tissue samples. Each tissue type has qualities that make the isolation methodology unique to that sample. The tissue mass, the cell type(s) involved, the age of the tissue, and the status of the tissue (fresh, flash-frozen, formalin-fixed, animal or plant) are all important considerations. No matter what the source, the key concern is the stability of the RNA in the tissue during the interval between specimen collection time and the commencement of the RNA isolation procedure.

One guiding principle when planning to isolate RNA from tissue samples: it is *not* necessary to accomplish complete tissue disruption in order to recover acceptable quantities of usable RNA. Being overly enthusiastic, that is, trying to break up every single clump of tissue, maybe to the detriment of the resulting RNA in terms of both quality and quantity. If the tissue breaks up quickly and efficiently, fine. If the tissue does not dissociate efficiently, remember that even moderate disruption results in an increased surface area on which the lysis buffer can act. A sufficient incubation time in lysis buffer may be all that is necessary to isolate enough RNA for the intended application. Conversely, leaving a partial homogenate in the presence of the lysis reagent for too long at room temperature can be completely counterproductive, meaning that the RNA quality declines with increased exposure to the lysis conditions. In some laboratories tissue samples are homogenized on receipt in one of the guanidinium-based buffers, followed by storing the homogenate at −80°C without any further purification. This approach is generally acceptable for most tissues, though some labs have experienced difficulty recovering high-quality RNA after the homogenate was thawed and the RNA isolation procedure completed. There really are no hard-and-fast rules, as the success of this strategy is both tissue- and user-dependent. The bottom line is that when working with tissues, timing is everything.

Another common mistake made when working with tissue samples for the first time involves the actual mass of tissue that is used. In general, most of the protocols that appear in the literature or accompany commercial reagents for RNA isolation are optimized for 100−1000 mg of tissue. When too much tissue is used without scaling up reagent volumes appropriately the RNA yield almost always declines. Unfortunately, scaling-up often results in a final volume that is too large for a single tube, resulting in the extreme inconvenience of having to perform multiple repeated steps. At the same time, the tissue should be in small enough pieces so as to be able to interact with the lysis buffer and the homogenizer itself. Large chunks of tissue, including soft tissue, do not homogenize well and often clog the homogenizer, including motor-driven, hand-held homogenizers.

Successful RNA isolation from tissue samples can render very large quantities of RNA, depending on the input mass and the type of tissue. This may require as much as 500−1000 μL of sterile buffer to dissolve the RNA. For investigators accustomed to working with RNA from cell culture, this required large volume may be quite unsettling; even small tissue samples consist of a surprisingly large number of cells, yielding a proportional mass of RNA. If the RNA is not dissolved completely, it will appear electrophoretically as a massive smear (Fig. 6.6), indicating that more buffer is required. In such an instance, it is wise to take a small, representative aliquot from the master tube, and then dilute only the aliquot until the 28S and 18S rRNAs become plainly visible upon denaturing electrophoresis, as described in Chapter 14, Electrophoresis of RNA. After the concentration of the now-diluted aliquot has been determined spectrophotometrically, the remaining RNA can be diluted down to a predetermined standard storage concentration and then, if desired, reprecipitated in convenient aliquots for long-term storage at −80°C. At some point prior to the completion of this procedure, it is very important to perform some of the quality control methods described in Chapter 8, Quality Control for RNA Preparations, because of the added complexity of many RNA isolation procedures involving tissue, compared to cell culture.

Fresh tissue

Uncontrolled RNase remains a viable weapon of mass destruction after a tissue sample has been removed from the donor organism. It remains active following cell disruption as long as conditions persist that support the function of these recalcitrant enzymes. The idea is to disrupt the tissue as quickly as possible and, in so doing, permanently inactivate problematic RNase. Immediately upon harvesting, the tissue should be rinsed in ice-cold phosphate-buffered saline (PBS), weighed, and then rapidly minced. Tissue mincing is ideally performed over no more than 4–5 minutes in a precooled weigh boat that was prerinsed with H_2O_2. Use a pair of new nuclease-free

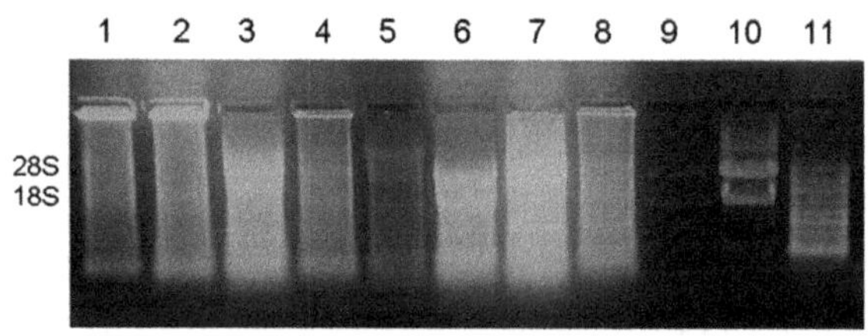

FIGURE 6.6 Appearance of incompletely dissolved RNA. Lanes 1–8: RNA was isolated from various human tissues and precipitated. The resulting clear, jelly-like pellet was diluted in 500 μL TE buffer, incubated at 60°C for 10 min, and then stored on ice for 2 h. RNA samples were electrophoresed through a denaturing formaldehyde-agarose gel. Since the amount of RNA recovered from the tissue was so large, 500 μL was not enough to completely dissolve it, which also compromised the accuracy of spectrophotometric concentration measurements. Aliquots of the RNA samples were removed from the master tube and diluted further until the 28S and 18S rRNAs were plainly visible upon electrophoresis. Lanes 9–10: RNA that was isolated from human tissue at an earlier date and dissolved properly. Lane 11: RNA molecular weight standard (Promega). The samples shown in lanes 1–8 eventually demonstrated properly formed 28S and 18S rRNA bands after appropriate dilution.

scalpels; avoid reusing scalpel blades, especially if the history of the scalpel is unknown. The resulting tissue chunks should not exceed 3–4 mm pieces and, when performed properly, mincing produces cut-up tissue that is finer than either chopping or dicing (chopping > dicing > mincing). The minced tissue is then dropped into lysis buffer as quickly as possible, taking care to avoid transferring any of the PBS; if the lysis buffer is diluted too much it will lose its ability to disrupt the architecture of the tissue and, even worse, to inhibit RNase efficiently.

Frozen tissue

Tissue samples that are not processed immediately can also be frozen for analysis at a later date, usually by snap- or flash-freezing in liquid nitrogen as quickly as possible after harvesting and then maintaining at −80°C or on dry ice until ready for use. Assuming that tissue samples are properly handled, there should be no difficulties isolating RNA of sufficient quantity and quality to support contemporary methods of analysis. Fresh, frozen tissue is then pulverized and added to chaotropic lysis buffer or, if the sample is small enough, allowed to simply thaw in a suitable aliquot of lysis buffer and then mechanically disrupted. However, if the tissue is to be sectioned on a cryostat for histological examination or in situ hybridization, most tissues section much better when they are embedded in Tissue-Tek O.C.T. (optimal cutting temperature) compound and then frozen in the cryostat itself rather than being flash-frozen. O.C.T. is a standard material used to prepare tissue samples for thin sectioning on a cryostat. It is also used to attach the frozen tissue to the chuck, holding it in place while it is advanced toward the cryostat blade. Fresh tissue samples are generally cooled gradually in O.C.T. medium within the cryostat chamber at −22°C. Ensuring that the spatial distribution of the RNA is maintained during tissue sectioning and in all subsequent steps is an absolute requirement if in situ hybridization experiments are to be meaningful. Flash-freezing, in contrast, often distorts the spatial arrangement of certain tissue regions, thereby rendering unreliable data with respect to the local distribution of the transcripts of interest within specific cells or regions of the tissue.

Once the tissue has been frozen, by any method, it is generally advisable to avoid thawing the tissue more than once. Freezing also makes subdividing or mincing the tissue extremely difficult without compromising the stability of the RNA. In this lab, small-to-medium size samples of flash frozen tissue (0.5–2 g) are twice wrapped in heavy-duty foil and stored in Ziploc plastic bags at −80°C. When ready to perform the RNA isolation, the sample is removed from ultracold storage and immediately whacked onto the surface of the lab bench while still frozen and wrapped. Be sure to wear safety glasses, gloves, and be gowned according to the departmental policy before performing this technique and do be sure to check first to determine if it is allowable. Moreover, human specimens that could contain infectious agents should be handled in a suitable containment area. This caveat applies to other tissue as well. When performed, the resulting impact generally shatters the tissue quite satisfactorily. Upon rapid unwrapping, shards of the frozen tissue are quickly transferred to a weigh boat that was precooled by placing it on dry ice a few minutes before use. Thus, the mass of the biological material is measured and recorded prior to thawing the sample which, ideally, should be done in lysis buffer. It is usually a big mistake to allow frozen tissue to thaw prior to submersion in lysis buffer. Because the lysis buffer itself is not frozen, the immersed tissue will thaw very quickly in a strongly denaturing environment and mechanical disruption can commence immediately.

Fixed tissue

A well-established method for preserving tissue samples is fixation in 10% buffered formalin or 4% paraformaldehyde. While formalin fixation is commonplace, it has the undesirable consequence of making RNA recovery and the analysis extremely difficult and inefficient because formalin creates extensive protein-protein and protein-nucleic acid crosslinked structures and because of the ensuing limited solubility of formalin-fixed tissues in guanidinium-based reagents. Following paraffin removal, aggressive proteinase K digestion is an effective method for solubilizing formalin-fixed tissues, thereby facilitating both RNA and DNA recovery (Coombs et al., 1999).

Regardless of the exact procedure, the four integral steps for nucleic acid recovery of formalin-fixed paraffin-embedded (FFPE) samples include:

1. Deparaffinization, by which the wax embedding the tissue sample is removed by one or more incubations with xylene and with ethanol (crushing or grinding the sample on dry ice may enhance nucleic acid recovery).
2. Protease digestion, by which the histological architecture of the tissue is broken down, usually with proteinase K or pronase, thereby liberating the trapped nucleic acids.
3. Recovery of all varieties of nucleic acid, by which DNA and RNA are captured, washed, and then eluted in purified form from a silica glass matrix, usually assembled as a microcentrifuge spin column.
4. Nuclease digestion, by which either DNA is destroyed (by incubation with DNase) or RNA is destroyed (by incubation with RNase), followed by spin column purification or, less frequently, by direct salt and ethanol-mediated precipitation.

Procedural note: if the recovery procedure is intended to capture the miRNA population, it is highly recommended that the investigator works with tissue sections that are greater than 10–12 μm in thickness. When sliced thinner, it is highly probable that the monolayer(s) of cells represented in that tissue section will have ruptured during the sectioning procedure, resulting in the loss of some or most of the RNA content.

FFPE-purified RNA is able to support amplification PCR after the removal of the mono-methylol groups ($-CH_2OH$) that are added to all four bases during the formalin fixation process. These groups can be removed by incubating formalin-modified RNA in TE buffer (Masuda et al., 1999) or using a three-day lysis method at 65°C (Chung et al., 2006); in the case of the latter, amplicons in the range of 300 bp are routinely generated because the purity of the RNA recovered from FFPE samples is good enough to support efficient reverse transcription (Cronin et al., 2004; Bonin et al., 2010). cDNA synthesized from RNA that was previously trapped in fixed tissue is usually smaller, compared to the situation where the RNA template is isolated from fresh- or flash-frozen tissue, though improvements in cDNA synthesis and PCR primer design have enhanced the overall efficiency of RT-PCR when beginning with FFPE RNA as a template (Specht et al., 2001; Mikhitarian et al., 2004). As a general rule, it is much easier to extract and work with RNA from fresh or frozen tissue compared to tissue that has been fixed.

The ability to recover RNA from archival tissue offers an unparalleled opportunity to study unique samples that have been stored for perhaps as long as several years. This prospect is particularly tantalizing because of the substantial archives found in most clinical settings. This ability may provide new insight into disease initiation and progression, not to mention the ability to examine the genomes of both DNA and RNA viruses that may have been present in a tissue sample at the moment it became formalin-fixed. Because some viruses are known to mutate at an astonishing rate, one might be able to generate a genetic timeline with which to compare contemporary viral strains with their predecessor. In addition to the protocol shown below, there are several kits, based on glass-fiber filter purification, available for streamlined recovery of FFPE RNA.

Protocol: LiCl–urea method for RNA isolation from tissue

The following is a modification of the procedures of Auffray and Rougeon (1980) and Kato et al. (1987). It has been used successfully for the isolation of RNA from a variety of tissues including breast, pancreas, liver, prostate, spleen, and brain. The key to success in this procedure is to homogenize the tissue in short bursts totaling no more than 60 seconds of homogenization time and do so while submerging the tube containing the tissue and lysis buffer in an ice water bath (Fig. 6.7). An earlier version of this protocol included incubation with proteinase K, which has been omitted from this protocol. In this laboratory, the LiCl–urea method has produced large quantities of very pure RNA which often exceed the quality of RNA produced using guanidinium-based methods.

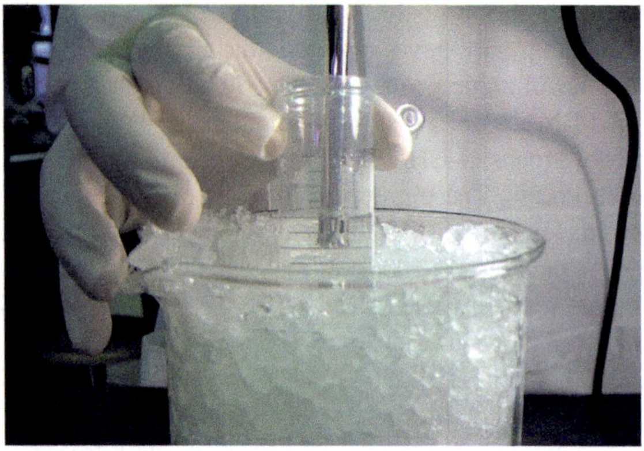

FIGURE 6.7 Homogenization of tissue using the LiCl-urea method. The conical 50 mL tube containing the tissue sample and the LiCl-urea homogenization buffer is held in an ice bath during the homogenization process. This is an excellent way to avoid generating heat caused by the action of an electronic homogenizer. Short bursts of homogenization are applied, totaling no more than 60 s, as described in the text. The tube is maintained in the ice bath between applications of the homogenizer.

In advance: Make sure that all buffers and tubes are prechilled on ice in order to minimize nuclease activity and stabilize the RNA.

1. Place a weigh boat on dry ice for a few minutes to cool it down. This will allow accurate weighing of tissue without the sample thawing.
2. Rapidly transfer a suitable mass of the tissue into a prechilled 50 mL tube and then add 10 vol of LiCl buffer (3 M LiCl; 6 M urea; 50 mM Tris, pH 7.4; 1 mM EDTA; 100 mM β-mercaptoethanol, 0.05% sarcosyl).

   Note: Add the β-mercaptoethanol just prior to use.
3. Immerse the tube in an ice water bath and homogenize the sample with several short bursts with a motorized tissue homogenizer. The total homogenization time should be no more than 60 seconds.

   Note 1: It is very important to minimize foaming and the generation of heat during the homogenization process.

   Note 2: Be sure to observe all safety precautions throughout the procedure when using a motorized homogenizer.
4. Transfer the homogenate to a prechilled polypropylene conical 15 mL tube (no more than 12 mL/tube). Place the tubes containing the homogenate at −20°C overnight.
5. The following morning, place the tubes into a precooled rotor and centrifuge for 75 minutes at 5000 × *g* at 4°C.

   Note: Be sure to check the maximum relative centrifugal force (RCF) and chemical compatibility of the tubes being used in this step.
6. At the conclusion of the run, carefully decant and discard the supernatant, and then rapidly blot each tube dry with a Kimwipe. Observe the

tube for the presence of solidified lipid clinging to the inside wall and, if present, wipe it away with a sterile cotton swab. Do not use a Kimwipe to try to remove anything from the inside walls of the tube. All tubes should be maintained on the ice during this manipulation.

7. Add 8–10 mL cold 3 M LiCl to each tube, and gently vortex for 30 seconds.

   Note: The vortexing step should be only aggressive enough to disperse the pelleted material.

8. Centrifuge the samples for 45 minutes at 5000 × *g*, at 4°C.

   Note: Be sure to check the maximum RCF and chemical compatibility of the tubes being used in this step.

9. Repeat steps 6 to 8.
10. Resuspend the pellet in 2.5 mL of TE/SDS buffer (10 mM Tris, pH 7.5; 0.1 mM EDTA, 0.5% SDS). Vortex as needed to resuspend the pellet. If necessary, use a sterile serological pipette to disperse the pelleted material.
11. Add 150 μL 5 M NaCl and mix by gentle vortexing.
12. If the material from step 11 is in a 15 mL polypropylene tube, then add an equal volume of modified phenol:chloroform:isoamyl alcohol (see note below). If the sample from step 11 is not in a polypropylene tube, it must be transferred to a tube that is compatible with these organic solvents. Secure the cap and then carefully mix the sample by vortexing.

    Note: Modified phenol:chloroform:isoamyl alcohol is prepared by saturation molecular biology grade phenol with an equal volume of Tris-SDS buffer (100 mM NaCl; 1 mM EDTA; 10 mM Tris, pH 8.5; 0.5% SDS) and then adding a volume of chloroform:isoamyl alcohol (24:1) equal to the volume of the phenol/Tris buffer mixture. Finally, add 8-hydroxyquinoline to a final concentration of 0.1%.

13. Centrifuge samples at 1000 × g for 10 minutes at 4°C.
14. Transfer the upper aqueous phase to a fresh polypropylene tube. Be careful to avoid the protein interface.
15. Add an equal volume of modified phenol:chloroform:isoamyl alcohol. Carefully mix the sample by vortexing.
16. Centrifuge the sample at 1000 × g for 10 minutes at 4°C.
17. Transfer the upper aqueous phase to a fresh polypropylene tube. Be careful to avoid the protein interface, if any.

    Note: If a band of protein persists between the aqueous and organic phases, it would be wise to repeat the organic extraction (steps 15–17) until all of the protein has been removed.

18. Perform one final organic extraction by the addition of an equal volume of chloroform alone or a mixture of chloroform:isoamyl alcohol (24:1). Centrifuge the sample at 1000 × *g* for 3 minutes at 4°C.
19. Transfer the aqueous phase to an RNase-free Corex glass tube that has been prechilled on ice. Add 3 vol 100% ethanol. Cover each tube with

Parafilm. If it does not appear that the ethanol and aqueous solution have mixed, then carefully and gently vortex the tube for a few seconds. Do not invert the tube to mix.

Note: Be sure to use only those high-strength tubes that can withstand the required RCF. If Corex tubes are not available, then equivalent Kimble brand high-strength glass tubes can be purchased from DWK Life Sciences. Use catalog number 45500-15 for 15 mL tubes and 45500-30 for 30 mL tubes. High-strength tubes must be placed in a suitable rubber adapter sleeve for centrifugation; otherwise, the tubes will shatter.

20. Place each tube at −20°C and store overnight in order to maximize the recovery of RNA.

21. Place each Corex tube into a suitable rubber adapter sleeve in a pre-cooled rotor and then centrifuge at 9000 × *g* for 30 minutes at 4°C.

22. At the conclusion of the run, carefully decant and discard the supernatant, taking care not to lose any of the precipitated material on the side or bottom of the tube.

23. Wash RNA with 75% ethanol. Use 5 mL to wash RNA that was precipitated in a 15 mL Corex tube or 10 mL to wash RNA that was precipitated in a 30 mL Corex tube.

24. Vortex the tube(s) gently, and then centrifuge at 9000 × *g* for 15 minutes at 4°C.

25. Repeat steps 22–24 twice, for a total of three washes.

26. Wash RNA once with 5 mL of 95% ethanol. Vortex gently, and then centrifuge at 9000 × *g* for 10 minutes at 4°C.

Note: Washing the RNA in 100% ethanol will expedite the ethanol evaporation process and favor maximum recovery of RNA. The very high-quality RNA that will result should perform extremely well in any of a variety of downstream applications.

27. Allow the RNA in the tube to air dry but take care to avoid drying out the sample completely. Having washed the precipitate with 100% ethanol, the sample should be dry within 5 minutes. The drying process can be further accelerated by placing the tube on a downward angle in a 37°C incubator for no more than 2–3 minutes. Be sure that the tube is uncovered so residual ethanol can drain out; also ensure that the opening of the tube is resting on a stack of 3–4 Kimwipes, and not directly on the metal inside the incubator.

28. Resuspend the RNA in the smallest possible volume of nuclease-free H_2O or modified TE buffer (10 mM Tris, pH 7.5; 0.1 mM EDTA). Depending on the starting mass of the tissue, completely dissolving the RNA may require a volume of as much as 500–1000 μL or more.

29. Remove an aliquot of the sample for spectrophotometric determination of the concentration of the RNA (Chapter 8: Quality Control for RNA Preparations). Next, run an aliquot on a denaturing agarose gel (Chapter 14: Electrophoresis of RNA) to assess sample quality.

Note: Smearing on the gel long the entire length of the lane may be an indicator that the RNA has not completely dissolved (Fig. 6.6).

30. Distribute RNA into convenient aliquots for storage at −80°C. RNA can be stored in hydrated form or can be reprecipitated for long-term storage.

Protocol: RNA isolation from lipid-enriched tissue

The isolation of RNA from adipose tissue is a simple matter, accomplished most easily through the use of guanidinium buffers to disrupt the tissue architecture. Perhaps the most cumbersome issue related to this method is the removal of the preponderance of lipids stored in adipose and other types of tissue, including brain and breast tissue. In general, cells or tissues are homogenized in guanidinium buffer (or LiCl−urea homogenization buffer), followed by brief, low-speed centrifugation during which the lipid floats to the top of the tube, from which it may then be aspirated away. This low-speed centrifugation is repeated as needed. Alternatively, one may begin with the addition of chloroform directly to a guanidinium lysate, in order to draw the lipids into the organic phase, which is then removed. RNA isolation then continues as usual with the addition of acidic phenol and chloroform, followed by the purification of the RNA. While the following procedure (a modification of Chomczynski and Sacchi, 1987) assumes the investigator will be preparing in-house reagents, one may confidently substitute commercial RNA isolation reagents such as Trizol or TRI Reagent.

1. Wear gloves!
2. Mince freshly isolated tissue (up to 100 mg) on ice. For each 0.1 g of tissue add 1 mL of solution D:
   4 M guanidinium thiocyanate,
   25 mM sodium citrate, pH 7.0,
   0.5% sarcosyl,
   100 mM 2-mercaptoethanol (add just prior to use)
   and homogenize at room temperature. Add an equal volume of chloroform and then pulse centrifuge to separate the phases. Remove and discard the chloroform (lower) phase. Add a fresh volume of chloroform, pulse centrifuge to separate the phases, and then transfer the upper aqueous phase to a fresh polypropylene tube.
3. For each 1 mL of solution D lysis buffer used in step 2, add:
   0.1 mL 2 M sodium acetate, pH 5.2,
   1.0 mL water-saturated phenol (molecular biology grade),
   0.2 mL chloroform:isoamyl alcohol (49:1).
4. Cap tube and mix carefully and thoroughly by inversion following the addition of each reagent and invert vigorously for an additional 30 seconds after all reagents have been added.

5. Cool sample on ice for 15 minutes.
6. Centrifuge to separate the phases; for microfuge tubes use $12{,}000 \times g$ at 4°C, if possible.
7. Transfer the aqueous phase (containing RNA) to a fresh tube and mix with 0.75 vol of ice-cold isopropanol. Store at −20°C for at least 1 hour to precipitate RNA. Note: For larger scale preparations, Corex glass tubes, placed in the correct adapters, can be used for precipitation and recovery of RNA.
8. Collect precipitate by centrifugation at $9000 \times g$ for 10 minutes at 4°C. Carefully decant and discard supernatant. CAUTION: Do not exceed the maximum g-force for any of the tubes used in this protocol.
9. Completely dissolve RNA pellet in 300 μL of solution D (see step 2) and then transfer it to an RNase-free 1.7 mL microfuge tube.
10. Reprecipitate the RNA by the addition of 0.75 vol of ice-cold isopropanol and store at −20°C for 1 hour.
11. Collect precipitate at top speed in a microcentrifuge for 10 minutes at 4°C. Carefully decant and discard supernatant.
12. Wash pellet 3–4 times in 500 μL 70% ethanol. If the RNA does not dislodge during these washes, there is no need to re-centrifuge. Allow tubes to air-dry to remove residual ethanol. Wash the sample once more with 95% ethanol to accelerate the ethanol evaporation process.
13. Dissolve the RNA in the smallest possible volume of TE buffer or nuclease-free H_2O. Incubation at 65°C for 10 minutes may facilitate solubilization, though this is unnecessary if the RNA did not dry out completely following the ethanol washes. Store RNA as an ethanol precipitate until ready to use. Following the determination of concentration, store the RNA in suitable aliquots at −80°C. Avoid repeated freezing and thawing.

Purification of polysome- and protein-engaged mRNA

mRNA is an excellent parameter of gene expression, but it is not the only one. Standard RNA isolation techniques, even when coupled to PCR, reveal information about the steady-state abundance of certain RNAs in the cell at the moment of cell lysis. These methods reveal nothing, however, about the translational fate of the transcripts under investigation; recall that the gene expression is also subject to regulation at the translational and posttranslational levels. Not unexpectedly, then, translational dysregulation is closely tied to the initiation and progression of a variety of pathological conditions.

A clearer definition of the translational aspect of gene expression may be gained by collecting and analyzing the mRNA fraction that has engaged the translational machinery. The polysome fraction of the cell, meaning all of the mRNAs in the act of being translated by one or multiple ribosomes, is a reliable indicator of the proportion of the mRNA population that has advanced to the translational level along the gene expression pathway. It is important to keep in mind that there is often a significant disconnect between transcriptional activity in a cell and the abundance of the cognate protein (for reviews, see Buccitelli and Selbach, 2020; Hershey et al., 2019). Thus, the study of polysome-engaged mRNA to profile gene expression at the translational level offers a different perspective on gene expression. Such data are useful for understanding the dynamic foundation of the proteome because functional proteins, and hence the imparted cellular phenotype, are observed only after experiencing posttranslational modification(s).

In the cell, some polysomes are associated with the endoplasmic reticulum (ER), synthesizing secreted proteins, mitochondrial proteins, and proteins to be embedded in the membrane, while non-ER-bound, free polysomes synthesize proteins that will remain in the cytoplasm or move into the nucleus. The standard isolation procedure, high-speed centrifugation through a linear 5%−50% sucrose gradient, facilitates the partitioning of heavy polysomes (higher translational efficiency; four or more ribosomes per transcript) from light polysomes (lesser translational efficiency; one−three ribosomes per transcript). Succinctly, polysomal RNA purified from cell culture or from tissue can be analyzed to determine which mRNAs are being translated and with what efficiency.

Methods describing the immunological precipitation of specific polysomes first appeared in the literature in the early 1970s (Palacios et al., 1972; Schechter, 1973; Palmiter, 1974). Methodologies evolved and were improved upon by restructuring the mechanics of the technique to the affinity column format, originally based on *p*-aminobenzyl (PAB)-cellulose (Schultz et al., 1977) and later proteins A and G. Noteworthy examples of the efficacy of this method as an RNA enrichment technique include the isolation of the mRNA encoding the heavy chain of the human HLA-Dr histocompatibility antigen using epitope-specific monoclonal antibodies (Korman et al., 1982) and trypanosome variable surface antigen (Shapiro and Young, 1981). Shortly thereafter, polyvalent sera were used successfully for the isolation of mRNAs encoding cystathione β-synthase from rat liver (Kraus and Rosenberg, 1982) and the low-density-lipoprotein receptor from bovine adrenal cells (Russell et al., 1983). In general, the use of antibodies for polysome immunopurification was difficult to perform and offered only limited success because many of the well-characterized antibodies available at the time were specific for antigens that were glycosylated or had otherwise matured, a characteristic not associated with nascent proteins.

Newer approaches involve the precipitation of all polysomes for direct global analysis of the translatome. Ribosome profiling (Ingoglia et al., 2009; Ingolia et al., 2018), also known as ribo-seq, involves deep sequencing of mRNAs that were engaged with ribosome complexes following pull-down of that subpopulation of mRNA that had actively engaged the ribosomes (reviewed by Kage et al., 2020; Wang et al., 2020); compare this approach to the isolation of one or more specific messages by antibody recognition as described above. Expression profiling analysis in this manner is a rather accurate indicator of translational efficiency because of the high correlation between the formation of the translation apparatus and the emergence of the mRNA-encoded primary polypeptide. Once free from the ribosomes and ancillary translation factors, purified polysomal mRNA can be reverse transcribed to synthesize probes for microarray analysis (Johannes et al., 1999), a standard approach that was later refined to RNP immunoprecipitation-microarray (RIP-Chip) analysis (Keene et al., 2006; Townley-Tilson et al., 2006).

Much greater attention is now given to understanding why certain RNA binding proteins are attracted to certain RNA nucleotide sequences, often known as RNA recognition elements. One of several contemporary techniques used to study macromolecular interactions of this nature is known as (photoactivatable ribonucleoside-enhanced crosslinking and immunoprecipitation, PAR-CLIP), a pull-down method by which the RNA molecules associated with certain proteins or protein complexes are sequencing and quantified (Hafner et al., 2010; Garzia et al., 2017). The method involves immunoprecipitation of a ribonucleoprotein complex of interest, recovery of the engaged RNA followed by partial digestion, conversion into cDNA, sequencing, and bioinformatics-based identification (Spitzer et al., 2014). Crosslinking of the ribonucleoprotein complex, prior to immunoprecipitation and in the presence of the photoreactive nucleosides, ensures recovery of the engaged RNA(s) with little or no interference from noncrosslinked background transcripts.

Protocol: isolation of polysomal mRNA

The method which follows is a variant of the magnesium precipitation procedure described by Palmiter (1974). It has the advantage of not requiring ultracentrifugation and is compatible with tissue and cells were grown in vitro. From an RNase control perspective, heparin is a competitive inhibitor of RNase as is magnesium at the elevated concentration used in the precipitation and polysome buffers.

1. Wear gloves! Be sure to observe RNase-free technique throughout this procedure and perform all centrifugations at 4°C unless otherwise noted.
2. *Tissue samples*: For each mg tissue add 100 μL of grinding buffer (25 mM NaCl; 5 mM $MgCl_2$; 25 mM Tris, pH 7.5; 1.5% Triton X-100; 0.5 mg/mL heparin). Homogenize the tissue on ice in a suitably sized Dounce homogenizer with a loosely fitting pestle. Add small volumes of additional grinding buffer, as needed. Proceed to step 3.

   Note: Tissue can be flash-frozen in liquid N_2 and then pulverized with a precooled, nuclease-free mortar and pestle prior to the addition of the grinding buffer.

3. *Cell culture*: Harvest the cells according to in-house procedures. For each 10^6 cells add 100 μL of grinding buffer (25 mM NaCl; 5 mM $MgCl_2$; 25 mM Tris, pH 7.5; 1.5% Triton X-100; 0.5 mg/mL heparin).

   Note 1: RNasin, at a final concentration of 200 U/mL, can be used in addition to heparin as an RNase inhibitor. If heparin is used, it will be necessary to remove all traces of it (below) if the RNA is to be reverse transcribed.

   Note 2: In order to maintain polysome integrity throughout the isolation procedure, the cell culture growth medium can be supplemented with up to 50 μg/mL cycloheximide for no more than 15 minutes just prior to harvesting. The idea is to prevent polysome dissociation prior to mRNA capture and this treatment will, in fact, maximize polysome recovery.
4. Transfer the homogenate into a suitable RNase-free centrifuge tube. For larger volumes, use a silanized Corex glass tube, and don't forget to place the tube(s) into a suitable rubber rotor adapter sleeve. For smaller volumes, use an RNase-free microfuge tube. In either case, centrifuge for 5 minutes at 10,000 × *g* in a prechilled rotor (be sure not to exceed the *g*-force rating for the tube).
5. While the sample is being centrifuged, prechill an appropriate number of sterile conical 50 mL tubes on ice.
6. At the conclusion of the centrifugation in step 3, carefully decant the supernatant into the prechilled tubes. Depending on the original volume of the homogenate, it may be necessary to split the supernatant into two tubes in order to accommodate the subsequent required reagents.
7. Add 1.2 vol of precipitation buffer (25 mM NaCl; 200 mM MgC12; 25 mM Tris, pH 7.5; 1.5% Triton X; 0.5 mg/mL heparin or 200 U/mL RNasin). Incubate on ice for 2 hours.
8. At the conclusion of the incubation period, add 3.5 mL polysome buffer (25 mM NaCl; 200 mM MgC12; 25 mM Tris, pH 7.5; 1 M sucrose) into sterile, silanized 15 mL Corex glass tube(s). Then, carefully overlay 10 mL of the mixture from step 6, above.

   Note: See Appendix H for silanization procedure.
9. Centrifuge for 15 minutes at 10,000 × *g* at 4°C in a prechilled rotor (swinging bucket rotor is best).
10. Carefully decant and discard the supernatant, taking care not to dislodge the pellet. If the pellet appears loose, then remove the supernatant by gentle aspiration.
11. Resuspend the polysome pellet in 1 mL of room temperature HEPES-SDS buffer (0.1 M NaOAc; 10 mM HEPES, pH 7.5; 0.1% SDS). More of this buffer can be added to completely redissolve the pellet as needed.
12. If the volume of the resuspended pellet from step 10 is 1 mL or less, transfer all of it to a 2.2 mL microfuge tube. If the volume of the

resuspended pellet from step 10 is more than 1 mL, it will be necessary to either split the volume into two microfuge tubes.

13. Add an equal volume of phenol:chloroform:isoamyl alcohol (25:24:1) to each tube. Secure the cap and then mix carefully. Centrifuge for 5 minutes at 10,000 × *g* to separate the phases.

    Note: If larger tubes are used for the extraction, be sure not to exceed the maximum recommended RCF for that tube.
14. Transfer the aqueous (upper) phase to a fresh tube, and then repeat step 12.
15. Transfer the aqueous (upper) phase to a fresh tube and then add an equal volume of chloroform or chloroform:isoamyl alcohol (24:1). Centrifuge for 30 seconds at 10,000 × *g* to separate the phases.
16. Transfer the aqueous material to a larger volume silanized Corex glass tube. If any samples were split into multiple tubes in step 11, they can now be recombined.
17. Add 0.08 vol 3 M NaOAc, pH 5.2 and 2.5 vol 100% ethanol. Store at −20°C for a minimum of 2 hours to precipitate the RNA; an overnight incubation at −20°C will maximize RNA recovery.
18. Collect the RNA by centrifugation at 10,000 × *g* for 20 minutes at 4°C.
19. Carefully decant the ethanol, taking care not to lose the precipitated RNA.
20. Wash to RNA twice with 75% ethanol.
21. Wash the RNA once with 95% ethanol to facilitate rapid ethanol evaporation. Do not allow the RNA to dry out completely.
22. Resuspend the RNA in the smallest possible volume of nuclease-free H_2O. Gently vortexing the sample will help dissolve the material in the tube. If necessary, heat the tube in a 65°C water bath for no more than 2 minutes to help dissolve the sample and then vortex gently again.
23. Optional but recommended: precipitate the RNA in 2 M LiCl overnight at 4°C to remove all traces of heparin. Collect the precipitate at 10,000 × *g* for 15 minutes at 4°C in a suitable centrifuge tube. Wash pellet twice with 70% ethanol to wash away residual salt and then wash once with 95% EtOH to expedite ethanol evaporation. Dissolve the RNA in a suitable aliquot of sterile water or TE buffer (pH 7.5).
24. Determine the concentration and purity of the RNA as described in Chapter 8, Quality Control for RNA Preparations. Aliquots of this RNA should be stored at −80°C until ready for use.

Collecting samples in the field

A very important compound that has revolutionized the way samples are stored for RNA analysis at a later time, date, or location is RNA*later* (Ambion; now part of Thermo Fisher). This product is a proprietary, ammonium compound-based aqueous formulation that is well known in RNA circles for its ability to prevent the deterioration of RNA in tissues and organs, until such time as it is convenient for the investigator to process the biological material and recover high-quality RNA for any of several uses. RNA*later* quickly permeates cells and tissue samples thereby allowing safe storage of tissue for up to a week at room temperature, for as long as a month at 4°C, and indefinitely at −20°C or colder (product specification sheet).

The chemical attributes of RNA*later* offer several strategic advantages with respect to the collection of samples and subsequent processing, especially when working out in the field. First, it is generally not practical to transport even a small dewar of liquid nitrogen for the purpose of flash freezing, especially if the investigator is planning to be in the field for several days. Second, RNA in a tissue sample is immediately stabilized when placed in RNA*later*. As such, tissue samples and their RNA contained therein are stabilized without risk of degradation. Third, RNA*later* preserves the histological architecture of tissues, favoring detailed microscopic examination later on. Fourth, RNA*later* is effective for the storage of Gram-positive and Gram-negative bacteria, yeast, plant tissue, animal tissue, and tissue culture cells. It is less effective when used with plant tissue with a waxy cuticle or with bone. This gives the investigator much greater latitude in terms of when the RNA can actually be isolated from the cells. Fifth, the use of RNA*later* is compatible with most RNA isolation procedures associated with various downstream applications. Finally, because of its nontoxic composition, RNA*later* can be disposed of according to local regulations or institutional policy; in some jurisdictions sink disposal, accompanied by thorough flushing with water, is permitted.

RNA "clean-up" methods

No matter which procedure is used to isolate RNA from fresh-, frozen-, or fixed archival tissue, the RNA is often contaminated with small amounts of residual protein or compounds used as part of the isolation procedure. These contaminants can have a negative impact on the utility of the RNA in most popular applications. If the electrophoretic appearance of newly isolated RNA is good but the RNA does not support conversion into cDNA or function in other types of assay, then it may be necessary to clean up the RNA. This can be performed in any number of ways. For example, one may increase the volume of the dissolved RNA a bit with sterile water or TE buffer and then extract it with an equal volume of phenol:chloroform:isoamyl alcohol (25:24:1), as described in this chapter and in Chapter 4, RNA Isolation Strategies. It is wise to keep the combined volume of RNA and the organic material under 1500 μL so that the extraction can be performed in a microfuge tube. Upon centrifugation, the appearance of a white band between the aqueous- and organic phases is a sure indicator that all of the protein was not removed. The extraction should be repeated until all of the protein is removed, as demonstrated by the absence of a band between the upper and lower phases. Then, the aqueous phase should be extracted once with equal volume of chloroform *or* chloroform:isoamyl alcohol (24:1) to remove all traces of phenol that could oxidize into very damaging quinones. Next, the RNA can be salt and alcohol precipitated and quality control assayed again.

Alternatively, there are a number of silica-based spin columns that bind RNA under high-salt chaotropic conditions, as described previously. These products, available from nearly all vendors of molecular biology wares, allow the RNA to be sequestered from all of the other chemicals and compounds in a tainted sample. Purified RNA is then eluted from the column in a low salt buffer or in sterile water. This approach is rapid and, although it often utilizes guanidinium thiocyanate as the chaotropic agent, it does not involve the use of any organic solvents. In this lab, we have successfully cleaned up suspect RNA samples by redissolving them in the RNA lysis buffer that accompanies spin column RNA isolation kits. This sample is then processed as if working with a whole cell lysate. The results of this type of cleanup approach are remarkable, allowing the RNA to be recovered from the spin column in a minimum volume.

RNA clean-up can also mean the removal of contaminating genomic or mitochondrial DNA from the sample. Recall that the more aggressive methods of RNA isolation disrupt all organelles, thereby liberating DNA into the homogenate. It is a standard operating procedure in most labs to treat RNA isolated from tissue with DNase I, as described in Appendix C. In so doing, the investigator will not need to worry about competition between cDNA and genomic DNA during PCR. Nor will contaminating DNA produce unwelcome bands during blot analysis.

Finally, it is important to recognize the fact that certain tissues are characterized by excessive quantities of natural cell products that may copurify with the RNA. Noteworthy examples include the copurification of glycogen from liver tissue and the copurification of lipid from adipose tissue. In the case of the former, the addition of 3 vol of 4 M sodium acetate, pH 7.0 (*not* pH 5.2) followed by incubation overnight at −20°C will precipitate the RNA while leaving the glycogen behind in the solution (Sambrook et al., 1989). The RNA is covered by centrifugation at a minimum of $8000 \times g$ (depending on the centrifuge tube) for 15 minutes at 4°C. Alternatively, adding 4 M LiCl to a tube containing precipitated RNA, and breaking up the pellet, will move the contaminating glycogen away from the RNA (Puissant and Houdebine, 1990); the RNA is then recovered by centrifugation for 10 minutes at $3000-5000 \times g$, depending on the tube. In the case of the latter, strategies for dealing with excessive quantities of lipid are described above.

The take-home lesson: be attentive to the details of the various isolation procedures in order to generate the highest quality RNA possible. It is wise to minimize the number of manipulations to which the RNA must be subjected so as to favor maximum stability of the sample while minimizing the risk of accidental introduction of RNase. RNA has a low tolerance for repeated precipitations that might be required if the sample was not properly purified the first time.

Troubleshooting RNA isolation from tissue

When RNA yield, quantity, or quality from tissue samples is marginal or fails altogether, the reason(s) may be due to any or all of the following:

1. **Incomplete tissue disruption**. This is an especially important concern because, unlike the lysis of cells propagated in culture, the three-dimensional architecture of the tissue is formidable due to the presence of resilient proteins in the ECM as well as myriad intercellular junctions. The lysis conditions may need to be modified, and homogenization by mechanical means, rather than chemical disruption, may be necessary. Some tissues are notoriously difficult to break up. Pulverizing (also known as powdering) snap-frozen tissue is an ideal strategy.
2. **Overly zealous homogenization**. The forces associated with tissue disruption are usually needed to break apart challenging tissues. In practice, short bursts with an electronic homogenizer or fewer, rather than more, strokes with a Dounce homogenizer will usually yield RNA of the best possible quality. Limit the use of the homogenization device to only what is necessary to do the job.
3. **Failure to control RNase activity**. Some animal organs and tissues contain abundant levels of RNase. Degradation of RNA can begin after removal of the tissue sample from its biological source if not frozen immediately or otherwise stabilized. To address this, modify lysis buffers with the addition of one or more nonspecific inhibitors of RNase, or use buffers that are highly chaotropic.
4. **Incomplete removal of protein**. The methods described herein and elsewhere for RNA isolation from tissue, when applied correctly, usually produce high-purity RNA as assessed by A_{260} and A_{280} measurements. When the A_{260}/A_{280} ratio is below 1.8 (a ratio of 2.0 is ideal), one should consider dissolving the RNA in a fresh aliquot of lysis buffer and then proceed to purify it again as if starting with a sample of cells grown in culture (re-homogenization not necessary). Paying careful attention to the lab protocol and the recommended volumes will generally preclude this difficulty.
5. **Too much tissue in too little buffer**. A common mistake among those investigators working with tissue, particularly for the first time, is the failure to use enough lysis buffer for the mass of the tissue sample. The rule of thumb when working with cells produced in culture is to keep the sample as concentrated as possible in order to facilitate RNA

precipitation at the end of the procedure. It is easy to underestimate the number of cells in a milligram or more of tissue; using too little lysis buffer will surely result in incomplete partitioning of macromolecules. Even when the proportion of lysis buffer to tissue is correct, investigators often fail to use a suitable amount of sterile water, TE, or other buffers to dissolve the purified RNA. This makes accurate spectrophotometric measurement difficult and electrophoretic profiling all but impossible. Remember that tissue samples often yield very large quantities of RNA and the volume of the buffer used to dissolve the purified RNA needs to be scaled up appropriately.

References

Auffray, C., Rougeon, F., 1980. Purification of mouse immunoglobulin heavy-chain messenger RNAs from total myeloma tumor RNA. Eur. J. Biochem. 107, 303–314.

Bevilacqua, C., Ducos, B., 2018. Laser microdissection: a powerful tool for genomics at cell level. Mol. Aspects Med. 59, 5–27.

Bonin, S., Hlubek, F., Benhattar, J., Denkert, C., Dietel, M., Fernandez, P.L., et al., 2010. Multicentre validation study of nucleic acids extraction from FFPE tissues. Virchows Arch. 457, 309–317.

Buccitelli, C., Selbach, M., 2020. mRNAs, proteins and the emerging principles of gene expression control. Nat. Rev. Genet. 21, 630–644. Available from: https://doi.org/10.1038/s41576-020-0258-4.

Chomczynski, P., Sacchi, N., 1987. Single-step method of RNA isolation by acid guanidinium thiocyanate-phenol-chloroform extraction. Analyt. Biochem. 162, 156–159.

Chung, J.Y., Braunschweig, T., Hewitt, S.M., 2006. Optimization of recovery of RNA from formalin-fixed, paraffin-embedded tissue. Diagnostic Mol. Pathol. 15, 229–236.

Coombs, N.J., Gough, A.C., Primrose, J.N., 1999. Optimization of RNA and RNA extraction from archival formalin-fixed tissue. Nucleic Acids Res. 27 (16), e12–e14.

Cronin, M., Pho, M., Dutta, D., Stephans, J.C., Shak, S., Kiefer, M.C., et al., 2004. Measurement of gene expression in archival paraffin-embedded tissues: development and performance of a 92-gene reverse transcriptase-polymerase chain reaction assay. Am. J. Pathol. 164 (1), 35–42.

Datta, S., Malhotra, L., Dickerson, R., Chaffee, S., Sen, C.K., Roy, S., 2015. Laser capture microdissection: big data from small samples. Histol. Histopathol. 30 (11), 1255.

Garzia, A., Meyer, C., Morozov, P., Sajek, M., Tuschl, T., 2017. Optimization of PAR-CLIP for transcriptome-wide identification of binding sites of RNA-binding proteins. Methods 118–119, 24–40. Available from: https://doi.org/10.1016/j.ymeth.2016.10.007.

Hafner, M., Landthaler, M., Berger, L., Khorshid, M., Hausser, J., Berninger, P., et al., 2010. PAR-CLIP – transcriptome-wide identification of RNA targets and binding sites of RNA-binding proteins. Cell 141, 129–141.

Hershey, J.W., Sonenberg, N., Mathews, M.B., 2019. Principles of translational control. Cold Spring Harb. Perspect. Biol. 11 (9), a032607.

Ingoglia, N.T., Ghaemmaghami, S., Newman, J.R., Weissman, J.S., 2009. Genome-wide analysis in vivo of translation with nucleotide resolution using ribosome profiling. Science 324, 218–223.

Ingolia, N.T., Hussmann, J.A., Weissman, J.S., 2018. Ribosome profiling: global views of translation. Cold Spring Harb. Perspect. Biol. Available from: https://doi.org/10.1101/cshperspect.a032698.

Johannes, G., Carter, M.S., Eisen, M.B., Bowen, P.O., Sarnow, P., 1999. Identification of eukaryotic mRNAs that are translated at reduced cap binding complex eIF4F concentrations using a cDNA microarray. Proc. Natl. Acad. Sci. 96, 13118–13123.

Kage, U., Powell, J.J., Gardiner, D.M., Kazan, K., 2020. Ribosome profiling in plants: what is not lost in translation? J. Exp. Bot. 71 (18), 5323–5332. Available from: https://doi.org/10.1093/jxb/eraa227.

Kato, N., Pfeifer-Ohlsson, S., Kato, M., Larsson, E., Rydnert, J., Ohlsson, R., et al., 1987. Tissue-specific expression of human provirus ERV3 mRNA in human placenta: two of the three ERV3 mRNAs contain human cellular sequences. J. Virol. 61 (7), 2182–2191.

Keene, J.D., Komisarow, J.M., Friedersdorf, M.B., 2006. RIP-Chip: the isolation and identification of mRNAs, microRNAs and protein components of ribonucleoprotein complexes from cell extracts. Nat. Protoc. 1, 302–307.

Korman, A.J., Knudsen, P.J., Kaufman, J.F., Strominger, J.L., 1982. cDNA clones for the heavy chain of HLA-DR antigen obtained after immunopurification of polysomes by monoclonal antibody. Proc. Natl. Acad. Sci. 70, 1844–1848.

Kraus, J.P., Rosenberg, L.E., 1982. Purification of low-abundance messengers RNAs from rat liver by polysome immunoadsorption. Proc. Natl. Acad. Sci. 79, 4015–4019.

Masuda, N., Ohnishi, T., Kawamoto, S., Monden, M., 1999. Analysis of chemical modification of RNA from formalin-fixed samples and optimization of molecular biology applications for such samples. Nucleic Acids Res 27 (22), 4436–4443.

Mikhitarian, K., Reott, S., Hoover, L., Allen, A., Cole, D.J., Gillanders, W.E., et al., 2004. Enhanced detection of RNA from paraffin-embedded tissue using a panel of truncated gene-specific primers for reverse transcription. Biotechniques 36 (3), 474–478.

Palacios, R., Palmiter, R.D., Schimke, R.T., 1972. Identification and isolation of ovalbumin-synthesizing polysomes. I. Specific binding of ^{125}I-anti-ovalbumin to polysomes. J. Biol. Chem. 247, 2316–2321.

Palmiter, R.D., 1974. Magnesium precipitation of ribonucleoprotein complexes. Expedient techniques for the isolation of undegraded polysomes and messenger ribonucleic acid. Biochemistry 13 (7), 3606–3615.

Puissant, C., Houdebine, L., 1990. An improvement of the single-step method of RNA isolation by acid guanidinium thiocyantate-phenol-chloroform extraction. Biotechniques 8 (2), 148–149.

Russell, D.W., Yamamoto, T., Schneider, W.J., Slaughter, C.J., Brown, M.S., Goldstein, J.L., 1983. cDNA cloning of the bovine low density lipoprotein receptor: feedback regulation of a receptor mRNA. Proc. Natl. Acad. Sci. 80, 7501–7505.

Sambrook, J., Fritsch, E.F., Maniatis, T., 1989. Molecular Cloning: A Laboratory Manual, second ed. Cold Spring Harbor Laboratory Press, Cold Spring Harbor, NY, p. 7, 22.

Schechter, I., 1973. Biologically and chemically pure mRNA coding for a mouse immunoglobulin L-chain prepared with the aid of antibodies and immobilized oligothymidine. Proc. Natl. Acad. Sci. 70, 2256–2260.

Schultz, G., Kieval, S., Groner, B., Sippel, A.E., Kurtz, D.T., Feigelson, P., 1977. Isolation of specific messenger RNA by adsorption of polysomes to matrix-bound antibody. Nucleic Acids Res. 4, 71–84.

Shapiro, S.Z., Young, J.R., 1981. An immunochemical method for mRNA purification. Application to messenger RNA encoding trypanosome variable surface antigen. J. Biol. Chem. 256, 1495–1498.

Specht, K., Richter, T., Muller, U., Walch, A., Werner, M., Hofler, H., 2001. Quantitative gene expression analysis in microdissected archival formalin-fixed and paraffin-embedded tumor tissue. Am. J. Pathol. 158 (2), 419–429.

Spitzer, J., Hafner, M., Landthaler, M., Ascano, M., Farazi, T., Wardle, G., et al., 2014. PAR-CLIP (photoactivatable ribonucleoside-enhanced crosslinking and immunoprecipitation): a step-by-step protocol to the transcriptome-wide identification of binding sites of RNA-binding proteins. Methods Enzymol. 539, 113–161.

Townley-Tilson, W.H.D., Pendergrass, S.A., Marzluff, W.F., Whitfield, M.L., 2006. Genome-wide analysis of mRNAs bound to the histone stem–loop binding protein. RNA 12 (10), 1853–1867.

Wang, Y., Zhang, H., Lu, J., 2020. Recent advances in ribosome profiling for deciphering translational regulation. Methods 176, 46–54. Available from: https://doi.org/10.1016/j.ymeth.2019.05.011.

Further reading

Brown, R.A.M., Epis, M.R., Horsham, J.L., Kabir, T.D., Richardson, K.L., Leedman, P.J., 2018. Total RNA extraction from tissues for microRNA and target gene expression analysis: Not all kits are created equal. BMC Biotechnol. 18 (16). Available from: https://doi.org/10.1186/s12896-018-0421-6.

Chassé, H., Boulben, S., Costache, V., Cormier, P., Morales, J., 2017. Analysis of translation using polysome profiling. Nucleic Acids Res. 45 (3), e15.

Gaffney, E.F., Riegman, P.H., Grizzle, W.E., Watson, P.H., 2018. Factors that drive the increasing use of FFPE tissue in basic and translational cancer research. Biotech. Histochem. 93 (5), 373–386.

Chapter 7

Going green: RNA and the molecular biology of plants

Rationale

As eukaryotic organisms, plants exhibit complex gene expression pathways and regulatory controls that are often similar to those observed in animals. Consequently, techniques that are commonly used to assay gene expression in animal cells may be used to assay the same in Plantae. The increased awareness and interest in agricultural biotechnology and applications thereof have created a need for expedient isolation of high-quality RNA from vastly different plant tissues, such as leaf, root, bark, floral tissue, fruit, and cotyledon. The fact that plants are unable to move when local conditions become unfavorable makes RNA profiling of plant tissue particularly fascinating, and much is to be gained by understanding the biochemistry of these sessile organisms. Contemporary lines of research include identifying the genes responsible for cold-hardiness, discerning plant molecular responses to the biotic (e.g., pathogen) and abiotic (e.g., drought) stress, identifying genes that may lead to the cost-effective synthesis of biofuels, and plant bioremediation, to name but a few examples. The good news is that plant RNA when stored properly, seems to be remarkably stable when compared to the animal cell counterpart, meaning that there are often much greater opportunities for long-term storage and analysis.

RNA isolation and the peculiarities of plants

The most obvious ultrastructural features of plant cells that distinguish them from animal cells are (1) the presence of a cell wall, (2) the presence of a large fluid-filled vacuole, and (3) the presence of chloroplasts. While it is true that RNA is chemically the same type of molecule regardless of its biological source, the rather fibrous nature of plant tissue often requires far more aggressive measures to disrupt cells than are required for an equal starting mass of animal tissue. Plant cell lysates and tissue homogenates are characteristically rich in polyphenols, lipids, polysaccharides, and a variety of secondary metabolites. As such, reliable and efficient RNA isolation

RNA Methodologies. DOI: https://doi.org/10.1016/B978-0-323-90221-2.00003-5

procedures must incorporate steps to eliminate the potential contaminants from copurifying with RNA, DNA, or both.

A number of isolation procedures have been described, and excellent kits developed, to deal with the unique qualities of individual plant species, tissues, and organs. There is no standard protocol for RNA isolation from all plant species; as with other types of organisms, plants exhibit multiple tissue types, each of which poses specific challenges with regard to RNA isolation. Examples of strategies used to isolate RNA from various plant structures are presented here; procedures best suited for one's own research will have to be determined empirically.

The plant cell wall is obviously structural in nature, contributing to defense at both the cell and organism levels and regulating the distribution of water, minerals, and growth-associated compounds. Its rigid structure likewise contributes to the overall architecture of the plant. From an experimental perspective, the plant cell wall represents a challenge for getting macromolecules out of cells as well as for getting things into cells, an example of which is the introduction of new gene constructions or siRNA molecules for gene silencing (Chapter 13: RNA Interference and Gene Editing).

The major components of what is known as the primary cell wall are cellulose microfibrils, hemicellulose, pectins, and glycans, the latter of which have a crosslinking role in the architecture of this remarkable structure. As the cell and the wall surrounding it mature, the formation of a secondary cell wall is commonly observed. While there are often considerable biochemical similarities between the primary and secondary cell walls, compounds that are most often associated with the secondary cell wall include lignin, which is a composite of aromatic alcohol polymers, as well as cutin, suberin, and other materials. In a manner very much like the extracellular matrix in animal cells, the plant cell wall tethers cells together resulting in tissue formation. Since the cell wall is a barrier between the investigator and the transcripts, it must be disrupted, dismantled, or otherwise penetrated.

Mature plant cells characteristically have a rather large fluid-filled central vacuole in the cytoplasm. Bound by a dynamic membrane known as the tonoplast, the vacuole is filled with a variety of dissolved organic and inorganic compounds, some of which are distasteful to insects and other plant-eating animals. Among its many functions, the vacuole maintains turgor pressure, which refers to the outward pressure exerted against the cell wall by the water in the central vacuole. Turgor pressure is what helps plants remain upright and contributes to the characteristic shape of the plant cell, and loss of turgor pressure results in wilting. The cell wall also prevents plant cells from lysing should a plant find itself in a hypotonic environment.

The aerial tissues in plants, that is, those which are visible above ground, contain immense numbers of round-to-oval-shaped chloroplasts. Like mitochondria, chloroplasts have their own genome and associated transcripts. During RNA isolation procedures the chloroplasts rupture, thereby liberating very large amounts of chloroplast mRNA. These transcripts are highly abundant and characteristically appear as very distinct bands between and mostly below the ribosomal subunits (Fig. 7.1). The sheer number of these bands is often surprising, if not unsettling, to investigators who are accustomed to seeing only the predominant rRNA species from nonplant sources upon electrophoretic examination. One should consider the appearance of the chloroplast transcripts to be an indicator of the integrity of the sample, meaning that downstream applications such as RT-PCR can be performed with confidence in the template material.

The first order of business when planning to isolate plant RNA is to determine the method for breaking through the formidable cell wall. Methods that have been used to meet this challenge include pulverizing plant tissue that has been frozen in liquid nitrogen (a predominant method), and the use of an electronic homogenizer or a BeadBeater (discussed in Chapter 4: RNA Isolation Strategies), or a laboratory blender or similar apparatus. While soft tissues can often be homogenized without first freezing them, the achievable level of disruption is often suboptimal, resulting in a lower RNA yield. After pulverizing the tissue, a short centrifugation step ($10,000 \times g$) is almost always needed to remove insoluble plant cells and tissue debris. From this point forward, the processing of the cell homogenate varies depending on the type of plant tissue starting material, the intended use of the RNA, and the preference(s) of the investigator. In some instances,

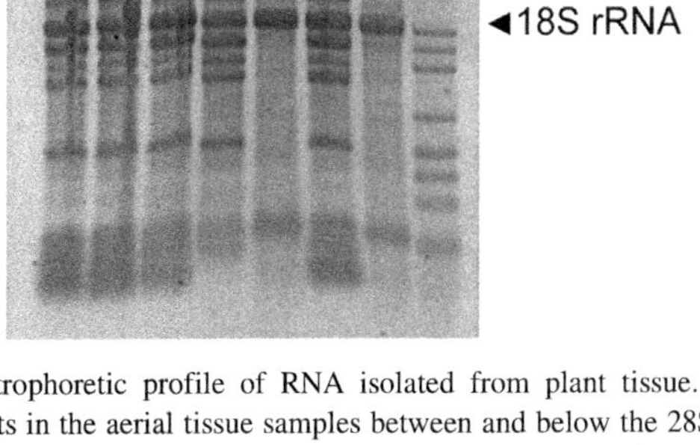

FIGURE 7.1 Electrophoretic profile of RNA isolated from plant tissue. Note the abundant chloroplast transcripts in the aerial tissue samples between and below the 28S and 18S rRNAs in lanes 1, 2, 3, 4 and 6. These transcripts, and the well-defined ribosomal species, demonstrate the integrity of the sample. Lanes 1–4, leaf RNA; lane 5, root RNA; lane 6, leaf RNA; lane 7, cotyledon RNA; lane 8, MW marker. While the stability of RNA from some plant species is legendary, it is always wise to run a gel just to be sure that the RNA is intact, especially if more than six months have elapsed since the RNA was last examined. Electrophoresis of RNA is described in detail in Chapter 14; similarly, RNA quality control procedures are described in Chapter 8.

for example, the resulting supernatant is extracted with phenol, including a heating step, to remove cellular proteins, and then chloroform-extracted to remove residual phenol. Bringing the sample to a final concentration of 2–3 M LiCl selectively drives the precipitation of RNA, allowing it to be sequestered from contaminating DNA and other impurities that would otherwise copurify with the RNA. The recovered RNA can then be precipitated again with sodium acetate and ethanol, which will allow the RNA to be cleaned up nicely because of the ease with which sodium acetate and ethanol can be removed from precipitated nucleic acids in general. This approach to RNA isolation is similar to the phenol-based methods that are used for RNA isolation from animal cells, with the exception of the initial pulverization of flash-frozen tissue and is considered by most investigators to be the "classical method" by which to approach plant RNA isolation.

A slightly different approach to RNA isolation is through the use of a guanidinium buffer, usually guanidinium thiocyanate or guanidinium hydrochloride (Jaiprakash et al., 2003), as a phenol alternative. Needless to say, great care must be taken at all levels to control RNase during plant RNA isolation procedures. The chaotropic nature of guanidinium reagents (e.g., TRIzol) characteristically inactivates nucleases on contact and tends to move proteins away from nucleic acids by solubilizing them. As always, the success of these methods depends upon using the proper ratio of lysis buffer to starting mass. A good rule of thumb is to use 500 μL lysis buffer formulation per 100 mg frozen, pulverized plant tissue unless a protocol specifies otherwise; variations are common. Keeping the sample fairly concentrated at all times is also a valuable strategy and, logistically speaking, it is much easier to work at the microcentrifuge tube scale than having to use larger (e.g., Oak Ridge and Corex or other high-strength glass) tubes and centrifuges whenever practical. In an alternative isolation method, a combination of the well-regarded chaotropic compounds ammonium thiocyanate and guanidine thiocyanate are used in the presence of acidic phenol (Bilgin et al., 2009). This method has the advantage of removing more organic compounds and rendering improved A_{260}/A_{280} ratios than previously published methods for the recovery of RNA from plants. One of the best-known and time-honored methods for RNA isolation from plant tissue features the detergent cetyltrimethylammonium bromide (CTAB), which quite efficiently promotes the separation of both DNA and RNA from polysaccharides. Other traditional methods for plant RNA recovery likewise favor aggressive disruption in the presence of high concentrations of SDS.

Regardless of the method of lysis and subsequent processing, one of the major concerns associated with the isolation of plant RNA is carbohydrate contamination, which should be suspected if the A_{260}/A_{230} ratio is low (Chapter 8: Quality Control for RNA Preparations). As with glycogen contamination in RNA isolated from animal liver tissue, carbohydrates contaminating plant RNA preps can be removed by precipitating the RNA in LiCl.

Briefly, RNA can be precipitated directly from an aqueous solution by the addition of LiCl to a final concentration of 3 M LiCl and then incubating overnight at −20°C. The RNA is then recovered by microcentrifuging the sample(s) for 10 minutes at 12,000 × *g*, preferably at 4°C. Interestingly, washing precipitated RNA with 3 M NaOAc is also effective for the removal of potentially problematic polysaccharides. Fortunately, several of the protocols (Tattersall et al., 2005) and products designed for plant RNA isolation are likewise designed to minimize contaminates in the final RNA prep. Neutralization and removal of these contaminants is of particular concern for downstream applications involving real-time PCR assays (Chapter 10: RT-PCR: A Science and an Art Form), microarray analysis (Chapter 21: Array Analysis of Gene Expression), or RNA-seq (Chapter 24: RNA-seq: The Premier Transcriptomics Tool).

Types of RNA produced in plant cells

RNA molecules in plant cells arise from one of three cellular compartments, namely the nucleus, the mitochondria, and the chloroplasts. In eukaryotic organisms, there are (at least) five known nuclear polymerases, namely RNA polymerase I, RNA polymerase II, RNA polymerase III, RNA polymerase IV, and RNA polymerase V, as reviewed in Chapter 1, RNA and the Cellular Biochemistry Revisited (see Table 1.5). The predominant mature products of nuclear transcripts are, ranked by mass, rRNA, tRNA, and mRNA (a mature, spliced form of hnRNA). Nuclear RNA is transported into the cytoplasm through the formation of ribonucleoprotein (RNP) particles by the transient association of transcripts with various RNA binding proteins. RNA polymerase IV and RNA polymerase V in plants (Onodera et al., 2005; Pontier et al., 2005; reviewed by Huang et al., 2015) do not appear to overlap in function with RNA polymerases I, II, or III. Rather, RNA polymerase IV in plants appears to have a role in promoting methylation-associated higher-order heterochromatin formation (Kanno et al., 2005) and gene silencing. Finally, some RNA molecules are transported from the cytoplasm into the mitochondria, such as certain tRNA species that are not encoded in mtDNA but are required to support protein synthesis in that organelle.

Angiosperms also exhibit a unique group of nuclear RNA polymerases that strongly resemble the very well-characterized T7 and T3 RNA polymerases from bacteriophages. Some members of this so-called RpoT gene family can be found both in monocots and dicots. In *Arabidopsis*, RpoT1 (UniProt P92969) is responsible for the transcription of mitochondrial genes while RpoT3 (UniProt O24600) transcribes chloroplast genes. The unique N-terminal amino acid composition of RpoT2 (UniProt Q8VWF8) permits the entry of this enzyme into both chloroplasts and mitochondria (Hedtke et al., 2000; reviewed by Tadini et al., 2020). This is possible because the RpoT2 mRNA contains two in-frame AUG codons near the 5′ terminus

(Richter et al., 2002). Initiation of translation at the second AUG codon produces a mitochondria-target translation product. In contrast, initiation of translation from the first AUG codon produces a translation product with chloroplast transit sequences.

With respect to mRNA, there are some interesting differences between monocot and dicot transcripts. For example, there are significant differences in the 5′ untranslated leader sequence (UTR; located between the 5′ cap and the initiation codon) with respect to GC content, length, and context of the start codon used to initiate translation. This is important because the features of the RNA that precede the initiation codon are known to affect the efficiency of translation. In general, mRNAs with longer leader sequences tend to be translated at higher rates than short-leader sequence mRNAs (Kozak, 1991; Futterer and Hohn, 1996), though this is not a hard-and-fast rule. Dicot mRNA leader sequences also tend to be AU-rich, and many ranges from 10 to 120 nts. In contrast, the leader sequences associated with monocot mRNAs are C-rich, ranging from about 40 to 120 nts long. By comparison with their counterparts in the animal kingdom, plant mRNAs share many structural features and, in many cases demonstrate remarkable stability under a variety of conditions.

Finally, it is important to keep in mind that both the mitochondria and the chloroplasts in the cell are transcriptionally active compartments. Each of these organelles has its own genome, and genes are often encoded on both strands of the organellar chromosome. Proteins produced in the mitochondria remain in the mitochondria, though several additional proteins required for mitochondrial function are imported from the cytoplasm, having been encoded by nuclear genes. Chloroplast genomes tend to be much larger than mtDNA and, as is the case with mitochondrial proteins, chloroplast-encoded proteins remain in that organelle. Also similar to mitochondrial physiology, requisite proteins are imported into the chloroplast from the cytoplasm by virtue of a transit sequence associated with their amino terminus. Once inside the chloroplast, imported proteins either fold into the tertiary (3°) structures that impart their functionality or, in other cases, participate in the assembly of quaternary (4°) protein structure. In this regard, perhaps the most noteworthy example is RUBISCO (ribulose-1,5-biphosphate carboxylase/oxygenase; also known as RuBisCo or RuBPCase), which is the first of many enzymes in the atmospheric carbon fixation pathway commonly known as photosynthesis. Widely touted as the most abundant protein in the world, RUBISCO consists of eight nuclear-encoded (small chain, S) peptides and eight chloroplast-encoded (large chain, L) peptides and is an excellent example of cooperation among two subcellular compartments in the regulation of gene expression. In yet other cases, imported proteins are further sorted and directed toward the lumen of the thylakoid or in its membrane. For an overview of the regulation of gene expression in plants, see Jones and Vandepoele (2020).

The following sections contain examples of RNA isolation procedures that have been used successfully for a number of plant tissue types. Keep in mind that many vendors sell products that often streamline the isolation procedure and, in many cases, reduce or altogether eliminate toxic compounds that are traditionally associated with RNA recovery. The amount of RNA that is to be expected is highly dependent upon the plant species, the specific tissue, the season, and the isolation procedure. For example, one gram of leaf tissue may yield 25–100 μg total RNA while three to four grams of fruit tissue may be needed to isolate an equivalent amount of RNA. Regardless of the isolation method, it is incumbent upon the investigator to assay the purified RNA for quantity and quality per standard quality control procedures (Chapter 8: Quality Control for RNA Preparations).

Protocol: RNA isolation from leaf

The following protocol is a modification of the procedures of Jaakola et al. (2001) and Chang et al. (1993) that is useful for deciduous leaf and pine needles. It is also very effective for tissues and organs with high RNase- and phenolic compound content.

1. Flash-freeze leaf tissue in liquid nitrogen and, if not ready for immediate use, store at −80°C. Leaf tissue harvested in the field should ideally be flash-frozen on site. If this is not possible, the leaf tissue should be transported back to the lab on ice and then processed immediately.
2. Grind the frozen tissue with a mortar and pestle which have been precooled with liquid nitrogen. Place the precooled mortar and pestle on a bed of dry ice; be sure to work quickly so that neither the mortar and pestle nor the tissue has an opportunity to warm.

   Note 1: After grinding, leaf tissue can be stored at −80°C until ready for use.

   Note 2: If more liquid nitrogen is required during grinding to avoid any thawing of the sample, be sure to add it very slowly to the mortar to avoid potentially dangerous splashing and sample loss. Maintaining the mortar on dry ice usually precludes the need to add more liquid nitrogen. In any event, wear gloves and protective eyewear at all times.
3. Weigh pulverized frozen tissue in a weigh boat that was precooled on dry ice to ensure a suitable mass of starting material. Transfer the frozen tissue to a prechilled tube on ice.
4. For each 100 mg pulverized tissue, add 750 μL extraction buffer (2% CTAB; 2% PVP-40 (polyvinylpyrrolidone, 40,000 MW); 10 mM Tris, pH 8.3; 10 mM EDTA; 2.0 M NaCl; 0.05% spermidine; add β-ME to 100 mM just prior to use) that has been preheated to 65°C. Incubate tubes at 65°C for 10 minutes.

Note 1: This procedure can be performed in a microfuge tube or scaled up, as needed. As always, be sure to check for buffer and solvent compatibility with all centrifugation tubes.

Note 2: In order to maximize the yield of RNA, vortex the tube 3–4 times during the incubation period.

5. Centrifuge samples at 10,000 × *g* for 10 minutes. If possible, perform this centrifugation at 4°C.
6. Transfer the supernatant to a fresh prechilled tube, taking care not to disturb the pelleted debris.
7. Add an equal volume of chloroform:isoamyl alcohol (24:1) and mix carefully.
8. Centrifuge tubes for 2 minutes at 10,000 × *g*.
9. Transfer the aqueous (upper) phase to a fresh tube.
10. Add an equal volume of chloroform:isoamyl alcohol (24:1) and mix carefully.
11. Centrifuge tubes for 2 minutes at 10,000 × *g*.
12. Transfer the aqueous (upper) phase to a fresh tube.
13. Add 0.25 vol 12 M LiCl to the aqueous material and mix thoroughly.
14. Incubate at 4°C overnight to precipitate the RNA.
15. Collect the RNA by centrifugation at 12,000 × *g* for 20 minutes. Decant the ethanol carefully.
16. Wash the RNA pellet twice with 500 μL aliquots of 70% ethanol. Decant each ethanol wash.

    Note: It is not necessary to centrifuge the tubes between washes unless the RNA dislodges from the side of the tube.
17. Wash the RNA once with 1000 μL 95%–100% ethanol. Decant the ethanol carefully and allow the tube to air-dry (less than 5 minutes). Ensure that no residual ethanol remains in the tube before proceeding to the next step.
18. Dissolve the RNA in 500 μL of a solution of 1.0 M NaCl; 0.25% SDS; 10 mM Tris, pH 8.3, and 5 mM EDTA.
19. Add an equal volume of phenol:chloroform:isoamyl alcohol (25:24:1). Mix carefully and thoroughly and then incubate for 5 minutes at 55°C.
20. Carefully mix again and then centrifuge for 5 minutes at 12,000 × *g*.
21. Transfer the aqueous (upper) phase to a fresh tube, taking care not to disturb the protein interface, if any, or the lower organic phase.
22. Add an equal volume of chloroform:isoamyl alcohol (24:1) to the aqueous material. Mix carefully.
23. Centrifuge to 2 minutes at 12,000 × *g*.
24. Transfer the aqueous (upper) phase to a fresh tube. Add 2.2 volumes of 95%–100% ethanol. Store at −80°C for at least 2 hours or at −20°C overnight.
25. Collect precipitated RNA by centrifugation at 12,000 × *g* for 15 minutes. If possible, perform this centrifugation step at 4°C.

26. Decant the ethanol carefully. Wash the RNA pellet twice with 500 μL aliquots of 70% ethanol. Decant each ethanol wash.

    Note: It is not necessary to centrifuge the tubes between washes unless the RNA dislodges from the side of the tube.
27. Wash the RNA once with 1000 μL 95%–100% ethanol. Decant the ethanol carefully and allow the tube to air-dry (less than 5 minutes). Ensure that no residual ethanol remains in the tube before proceeding to the next step.
28. Dissolve the RNA in the smallest possible volume of nuclease-free water, and assess purity, sample integrity, and concentration as described in Chapter 8, Quality Control for RNA Preparations. If the RNA is not to be used immediately, store it in suitable aliquots at −80°C.

Protocol: RNA isolation from bark

The following protocol is a modification of the procedures of Artlip et al. (1997) and Callahan et al. (1993). It has been used successfully by this lab group for a number of applications, including suppression subtractive hybridization (Chapter 22: Subtractive and Nonsubtractive Methods for the Analysis of Gene Expression) for global analysis of gene expression in peach bark (Bassett et al., 2006). The bark is a rather attractive tissue for RNA isolation, particularly for seasonal studies, because it is readily available in large quantities and does not separate from the plant in the manner of deciduous leaves, flowers, and fruit (abscission).

1. Collect bark tissue by gently scraping small pieces from twigs using a nuclease-free scalpel or other implements.

   Note: If collected in the field, place bark samples on ice for transport back to the lab.
2. Flash-freeze the bark tissue in liquid nitrogen and then grind the frozen tissue. Place the precooled mortar and pestle on a bed of dry ice; be sure to work quickly so that neither the mortar and pestle nor the tissue has an opportunity to warm.

   Note 1: After grinding, bark tissue can be stored at −80°C until ready for use.

   Note 2: If more liquid nitrogen is required during grinding to avoid any thawing of the sample, be sure to add it very slowly to the mortar to avoid potentially dangerous splashing and sample loss. Maintaining the mortar on dry ice usually precludes the need to add more liquid nitrogen. In any event, wear gloves and protective eyewear at all times.
3. Rapidly transfer the pulverized, frozen tissue to a weight boat that was precooled on dry ice to ensure a suitable mass of starting material.

4. Resuspend 1 g tissue in 20 mL of preheated (65°C) extraction buffer (0.1 M Tris-Cl, pH 9.0; 0.1 M NaCl, 0.1% SDS, 1% PVP-40 (polyvinylpyrrolidone, 40,000 MW). Just prior to use, bring to 10 mM ascorbic acid; 100 mM β-ME; 100 μg/mL proteinase K). Carefully resuspend all particulate matter and then incubate for 10 minutes at 65°C with gentle agitation.
5. Centrifuge samples for 10 minutes at 15,000 × *g* to pellet cellular debris.
6. Carefully transfer the supernatant into a fresh tube. Add an equal volume of Tris-saturated phenol that was preheated to 65°C. Mix carefully by vortexing.

   Note: As always, be sure to check for buffer and solvent compatibility with all centrifugation tubes.
7. Centrifuge at 12,000 × *g* for 10 minutes at room temperature.
8. Note the appearance of the tube at the end of the centrifugation. If the protein interface is abundant, then repeat steps 6 and 7. Otherwise, transfer the aqueous (upper) phase to a fresh tube and then add an equal volume of phenol:chloroform:isoamyl alcohol (25:24:1). Mix carefully by vortexing.
9. Centrifuge at 12,000 × *g* for 10 minutes at room temperature.
10. Transfer the upper aqueous phase to a fresh tube and then add an equal volume of chloroform:isoamyl alcohol (24:1). Mix carefully by vortexing.
11. Centrifuge at 12,000 × *g* for 3 minutes at room temperature.
12. Transfer the upper aqueous phase to a fresh tube that has been prechilled on ice.
13. Add 0.1 vol 3 M NaOAc, pH 5.2. Mix by gently vortexing the tube.
14. Add 0.1 vol 10% SDS and 0.1 vol 5 M NaCl. Mix by gently vortexing the tube.

    Note: The addition of sodium acetate, SDS, and sodium chloride in steps 12 and 13 will drive the precipitation of polysaccharides when incubated on ice in the next step.
15. Incubate tubes on ice for 2 hours.
16. Centrifuge at 15,000 × *g* for 20 minutes. If possible, perform this centrifugation at 4°C.
17. Carefully transfer the supernatant to a fresh tube. Add 0.2 vol 10 M LiCl and then mix by gentle vortexing the tube.

    Note: Be careful to avoid disturbing the pelleted material.
18. Incubate at −20°C overnight to precipitate the RNA.
19. Centrifuge at 15,000 × *g* for 20 minutes. If possible, perform this centrifugation at 4°C.
20. Carefully remove the supernatant as completely as possible, taking care not to disturb the pellet.
21. Resuspend the pellet in a minimum volume of nuclease-free water. Aggressive vortexing may be needed to dissolve the pellet completely.

Note: If possible, perform this step and all remaining steps in microcentrifuge tubes.

22. Add 0.02 vol 5 M NaCl and 2.5 vol ice-cold 95% ethanol. Mix thoroughly by gentle vortexing and then incubate at −20°C overnight or for 3 hours at −80°C.
23. Centrifuge at 12,000 × *g* for 20 minutes. If possible, perform this centrifugation at 4°C.
24. Remove the ethanol supernatant as completely as possible. Wash the pellet twice with 1 mL 70% ethanol to remove the excess salt.

    Note: It is not necessary to centrifuge the sample between washes unless the precipitate dislodges from the side of the tube.
25. Wash the precipitate once with 1 mL 100% ethanol, and then remove it as completely as possible.
26. Allow the sample to air-dry (less than 5 minutes), but do not dry it out completely.

    Note: Keeping the RNA slightly damp with ethanol will facilitate dissolving it aqueous buffer.
27. Resuspend the RNA in a small volume of nuclease-free water and allow the sample to dissolve on ice. Micropipette up-and-down thoroughly to create a homogeneous solution.
28. Measure concentration and estimate purity spectrophotometrically, as described in Chapter 8, Quality Control for RNA Preparations. Store RNA in suitable aliquots at −80°C until ready to use.

Protocol: RNA isolation from fruit

Plant fruit tissue is frozen in liquid nitrogen immediately upon harvesting and then stored at −80°C until ready for processing. Alternatively, fruit tissue can be lyophilized (Fig. 7.2) prior to commencing the isolation protocol, though the lyophilization process may take several days and the RNA quality may be compromised (A. Callahan, personal communication), not to mention the increase in carbohydrate concentration as the water content decreases. Therefore, effective methods for the isolation of RNA from fruit (Sajeevan et al., 2014) focus on expedient removal of metabolites and other potentially inhibitory contaminants. In still other cases, the juice is literally squeezed from fruit tissue with high water and sugar content, followed by RNA extraction from the resulting pulp (Davis et al., 2006). The following protocol for the isolation of RNA from fruit is a modification of the procedures of Callahan et al. (1989) and Morgens et al. (1990).

1. Grind frozen tissue in liquid nitrogen using a precooled mortar and pestle. Place the precooled mortar and pestle on a bed of dry ice; be sure to work quickly so that neither the mortar and pestle nor the tissue has an opportunity to warm.

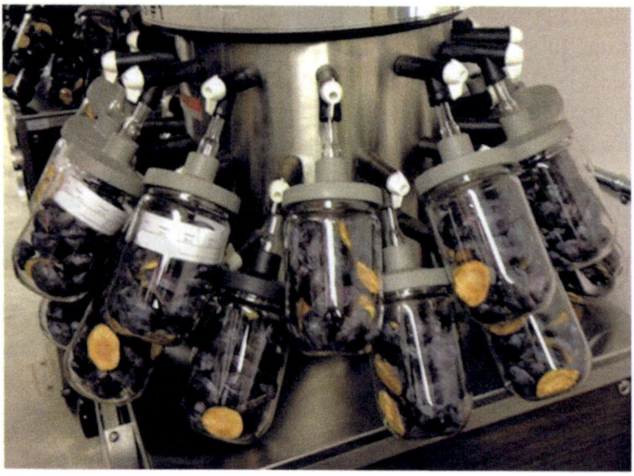

FIGURE 7.2 Lyophilization of plums prior to fruit processing. When performed properly, the removal of excess water enhances the recovery of RNA from various types of fruit.

Note: If more liquid nitrogen is required during grinding to avoid any thawing of the sample, be sure to add it very slowly to the mortar to avoid potentially dangerous splashing and sample loss. Maintaining the mortar on dry ice usually precludes the need to add more liquid nitrogen. In any event, wear gloves and protective eyewear at all times.

2. Rapidly transfer the pulverized tissue to a weight boat that was precooled on dry ice to ensure a suitable mass of starting material.
3. In a 50-mL Oak Ridge tube, resuspend up to 1 g tissue in 20 mL of extraction buffer (0.1 M Tris-Cl, pH 9.0; 0.1 M NaCl, 0.1% SDS, 1% PVP-40 (polyvinylpyrrolidone, 40,000 MW) that was preheated to 65°C. Just prior to use, bring to 10 mM ascorbic acid; 100 mM β-ME; 100 μg/mL proteinase K). Carefully resuspend all particulate matter.

   Note: The mass of starting material is variable depending on the tissue type. For example, use 0.3 g if working with leaf tissue, 0.5 g if working with early fruit, or 2 g if working with very ripe fruit.
4. Centrifuge for 10 minutes at 12,000 × *g*.

   Note: Be attentive the manufacturer's recommendations pertaining to the proper use of Oak Ridge tubes.
5. Carefully transfer the supernatant into a fresh Oak Ridge tube containing 20 mL of Tris-saturated phenol that was preheated to 65°C. Mix carefully by vortexing.

   Note: As always, be sure to check for buffer and solvent compatibility with Oak Ridge and other types of centrifuge tubes prior to use.
6. Centrifuge at 12,000 × *g* for 10 minutes at room temperature.

7. Examine the contents of the Oak Ridge tube. If the protein interface is abundant, then repeat steps 5 and 6. Otherwise, transfer the upper aqueous material to a fresh Oak Ridge tube containing 18 mL phenol:chloroform:isoamyl alcohol (25:24:1). Mix carefully by vortexing.
8. Centrifuge at 12,000 × *g* for 10 minutes at room temperature.
9. Transfer the (upper) aqueous material to a fresh Oak Ridge tube containing 18 mL chloroform:isoamyl alcohol (24:1) using a nuclease-free pipette. Mix carefully by vortexing.
10. Centrifuge at 12,000 × *g* for 3 minutes at room temperature.
11. Transfer the (upper) aqueous phase to a fresh Oak Ridge tube that has been prechilled on ice.
12. Add 0.1 vol 3 M NaOAc, pH 5.2. Mix by gently vortexing the tube.
13. Add 0.05 vol 20% SDS and 0.1 vol 5 M NaCl. Mix by gently vortexing the tube.
14. Incubate tubes on ice for 2 hours.
15. Centrifuge at 12,000 × *g* for 20 minutes. If possible, perform this centrifugation at 4°C.
16. Carefully transfer the supernatant to a fresh tube. Add 0.2 vol 10 M LiCl and then mix by gentle vortexing the tube.
17. Incubate at −20°C overnight to precipitate the RNA.
18. Centrifuge at 12,000 × *g* for 20 minutes. If possible, perform this centrifugation at 4°C.
19. Carefully remove the supernatant as completely as possible, taking care not to disturb the pellet.
20. Resuspend the pellet in a minimum volume of nuclease-free water, keeping in mind that more than 1 mL may be required. Aggressive vortexing may be needed to dissolve the pellet completely.

    Note: If possible, perform this step and all remaining steps in microcentrifuge tubes.
21. Add 0.02 vol 5 M NaCl and 2.5 vol ice-cold 95% ethanol. Mix thoroughly by gentle vortexing and then incubate at −20°C overnight.
22. Centrifuge at 12,000 × *g* for 20 minutes. If possible, perform this centrifugation at 4°C.
23. Remove the ethanol supernatant as completely as possible. Wash the pellet twice with 1 mL 70% ethanol to remove the excess salt.

    Note: It is not necessary to centrifuge the sample between washes unless the precipitate dislodges from the side of the tube.
24. Wash the precipitate once with 1 mL 100% ethanol, and then remove it as completely as possible.
25. Allow the sample to air-dry, but do not dry it out completely.

    Note: Keeping the RNA slightly damp with ethanol will facilitate dissolving its aqueous buffer.

26. Resuspend the RNA in 100–500 μL nuclease-free water and allow the sample to dissolve on ice. Micropipette up-and-down thoroughly to create a homogeneous solution.
27. Measure concentration and estimate purity spectrophotometrically, as described in Chapter 8, Quality Control for RNA Preparations. Store RNA in suitable aliquots at −80°C until ready to use.

Protocol: RNA isolation from plant tissue with hot borate

The use of hot borate as a means of purifying RNA has been a well-respect procedure for nearly three decades. It is an aggressive method, though most investigators agree that it is helpful when the recovery of RNA is marginal from recalcitrant tissue using other isolation methods. In this procedure, borate removes carbohydrates and inhibits the oxidation of the phenolics found in plant tissue. Polyvinylpyrrolidone then removes phenolics that are liberated from plant tissue upon cellular disruption. The protocol which follows is a modification of the procedures described by Wan and Wilkins (1994) and Gudenschwager et al. (2012) and is particularly well-suited for RNA isolation from fruit tissue.

1. Grind frozen tissue in liquid nitrogen using a precooled mortar and pestle. Place the precooled mortar and pestle on a bed of dry ice; be sure to work quickly so that neither the mortar and pestle nor the tissue has an opportunity to warm.

   Note: If more liquid nitrogen is required during grinding to avoid any thawing of the sample, be sure to add it very slowly to the mortar to avoid potentially dangerous splashing and sample loss. Maintaining the mortar on dry ice usually precludes the need to add more liquid nitrogen. In any event, wear gloves and protective eyewear at all times.
2. If not prepared to recover the RNA immediately after grinding, return the pulverized tissue to −80°C storage for extended storage. Otherwise, proceed to step 3.
3. Rapidly weigh out 2.5 g of pulverized tissue in a weight boat that has been precooled on dry ice, and then transfer the tissue to a nuclease-free Oak Ridge tube.
4. Carefully add 12.5 mL extraction buffer that has been prewarmed to 75°C. The extraction buffer consists of 200 mM sodium borate decahydrate, pH 9.0, 25 mM EGTA, 1% sodium deoxycholate, 0.5% IGEPAL CA-360 (Sigma Chemical Co.), 2% polyvinylpyrrolidone (360,000 MW), and 1% 2-mercaptoethanol (added just prior to use).
5. Carefully vortex the tube(s) and incubate for 1 hour at 48°C in an aerated shaker incubator at 50 rpm. If a shaker incubator is not available, then vortex the tube gently every 15 minutes throughout the incubation period.

6. At the conclusion of the incubation period, add 1000 μL 2 M KCl. Incubate on ice for a minimum of 30 minutes. Place tubes and ice bath on an orbital platform at 50 rpm. If an orbital platform is not available, gently invert the tube(s) every 10 minutes.
7. Centrifuge Oak Ridge tubes at 10,000 × *g* for 30 minutes, preferably at 4°C.
8. Carefully decant the supernatant into a prechilled (on ice) fresh Oak Ridge tube. Add an equal volume of 4 M LiCl. Gently vortex samples and then mix by inversion.
9. Incubate tubes for 12–16 hours at 4°C.
10. Centrifuge Oak Ridge Tubes at 20,000 × *g* for 30 minutes at 4°C.
11. Carefully decant and discard the supernatant, taking care not to dislodge the pellet.
12. Resuspend the pellet in 500 μL nuclease-free water and then transfer to a 1.7 mL microcentrifuge.

    Note 1: If the pellet does not dissolve completely, add more water in 100 μL increments. If the pellet was dissolved in more than 500 μL, split the volume equally between two microfuge tubes. Transfer no more than 500 μL to the microfuge tube. If the total volume is ≤ 500 μL, transfer all of it to one 1.7 mL microcentrifuge tube.

    Note 2: The use of 1.7 mL tubes is preferred because the shape of a larger (e.g., 2.0 mL) microcentrifuge tube makes access to RNA pellets and complete sample recovery something of a challenge.
13. Add 0.1 vol 3 M NaOAc, pH.5.2 to the microfuge tube(s).
14. Add an equal volume of chloroform:isoamyl alcohol (24:1) to each sample tube. Mix carefully by tube inversion or by vortexing for 1 minute.
15. Centrifuge samples for 3 minutes at 12,000 × *g*, preferably at 4°C.
16. Transfer the upper aqueous phase to a fresh tube and place it on ice.
17. Return to the tube with the residual organic phase and re-extract it with the addition of 400 μL sterile water. Mix carefully by tube inversion or vortexing for 1 minute.
18. Centrifuge samples for 3 minutes at 12,000 × *g*, preferably at 4°C.
19. Combine the upper aqueous phase with the sample resting on ice from step 16, above.
20. Add 0.1 vol 3 M sodium acetate, pH 5.2. If the total volume is ≤ 800 μL, add an equal volume of isopropanol, mix thoroughly by tube inversion or vortexing, and place tubes on ice for a minimum of 1 hour. If the total volume is more than 800 μL, divide the tube evenly between two 1.7 mL microcentrifuge tubes. Add an equal volume of isopropanol to each tube, and place it on ice for a minimum of 1 hour.

    Note: If necessary, the RNA isolation procedure can be stopped here, and the tubes stored at −20°C overnight.
21. Centrifuge tubes for 20 minutes at 12,000 × *g*, preferably at 4°C.
22. Carefully decant and discard the supernatant, taking care not to disturb the RNA pellet.

23. Wash pellets with 500 μL 70% ethanol; carefully decant and discard ethanol wash. Repeat.
24. Wash pellets with 500 μL 95% ethanol; carefully decant and discard ethanol wash.
25. Place open tubes on a Kimwipe and allow to air dry for 5–10 minutes. Placing the tubes in a 37°C incubator will facilitate air drying. Ensure that no residual ethanol remains in the tube.
26. Dissolve the RNA pellet in the smallest volume of sterile water, typically 25 μL.

    Note: As needed, add additional water in 5 μL increments to ensure that the sample dissolves completely. Mix thoroughly by micropipetting.
27. Measure concentration and sample purity spectrophotometrically, as described in Chapter 8, Quality Control for RNA Preparations. Store purified RNA is suitable aliquots at −80°C until ready to use.

Strategies for RNA isolation from other plant tissues

One of the primary complaints associated with RNA isolation from plant tissue is an unacceptably low yield. There are several reasons why this could happen. First, and most assuredly, the plant cell wall is always a challenge. While it is true that the wall can be dismantled enzymatically, the process is rather inefficient and time-consuming. It is also true that a fair amount of damage can be inflicted upon plant tissue by dropping it into a chaotropic lysis buffer and homogenizing it, though a large number of cells may remain intact, reducing the yield of RNA that might otherwise be possible. Moreover, some tissues are rather reluctant to give up their RNA. Thus, a majority of procedures in this domain begin with pulverizing the starting tissue in liquid nitrogen down to a fine powder. When in doubt, pulverize. In instances where it is not practical to maintain field-harvested tissue in a frozen state until returning to the lab, preparing a guanidinium lysate on-site is usually a satisfactory alternative.

Second, tissues from some plants, including pine, grape, and tea leaves contain high concentrations of phenolics, including tannins; polyphenols tend to be more prevalent in conifers than in deciduous trees (Schneiderbauer et al., 1991). Elimination of these compounds is essential because phenolic compounds form quinones upon oxidation which crosslink nucleic acids, rendering them useless for molecular biology applications. Moreover, endogenous RNase is also a common problem, and a surprising source of exogenous RNase contamination in the isolation of plant RNA are the mortar and pestle, both of which should be squeaky-clean when used to grind flash-frozen plant tissue. One easy way to ensure RNase neutralization is to soak the mortar and pestle in H_2O_2 for 15–20 minutes, as described in Chapter 2, Creating a Ribonuclease-Free Environment, followed by several rinses with nuclease-free water. It is possible to eliminate toxic compounds such as phenol and guanidinium buffers altogether and still be able to rapidly recover total RNA using a CTAB lysis system, though chloroform is used to remove the proteins.

Third, high concentrations of polysaccharides, waxes, and sometimes latex, can be problematic. Occasionally, plant RNA isolation protocols raise eyebrows among investigators who previously have not worked with plant tissue. Certain protocols incorporate three or more precipitations of the RNA, particularly Tris-lithium chloride procedures, as a means of eliminating the massive amount of carbohydrate that fruit, in particular, can render. Some investigators actually rinse pelleted RNA with 3 M NaOAc because of its ability to wash away polysaccharide contaminants (Logemann et al., 1987; Lal et al., 2001). It is true that minimizing the number of manipulations to which the RNA is subject is quite favorable, but it is also true that contaminants which copurify with the RNA must be dealt with assertively. Plant RNA lysis buffers also often feature the inclusion of spermidine (as high as 0.5% w/v) and/or PVP (up to 2% w/v). Both of these compounds can be quite helpful for maximizing high purity RNA recovery. While well-known for its ability to enhance RNA polymerase activity in transcription assays, spermidine in this case forms a low-solubility complex with genomic DNA in the lysate, making it easier to remove the DNA. PVP is a rather large polysaccharide that essentially takes up physical space, thereby increasing the effective concentration of other molecules and making it easier to precipitate them. Further, PVP and β-ME, another common lysis buffer component, are effective reducing reagents, which will minimize quinone formation due to the oxidation of phenolic compounds.

The good news is that plant RNA in general is more stable than RNA isolated from animal cells and, according to some investigators, is legendary in terms of capacity for long-term storage. For example, we have cloned genes and mapped transcription start sites using morning glory RNA that had been stored at $-80°C$ for years.

Troubleshooting RNA isolation from plant tissue

1. When working with plant RNA, low A_{260}/A_{230} may indicate polysaccharide contamination. Consequently, the RNA isolation protocol may need to be modified to ensure polysaccharide elimination. Further, a low A_{260}/A_{280} ratio is an indicator of protein contamination. See Chapter 8, Quality Control for RNA Preparations for a detailed discussion.
2. As certain plant tissues mature, particularly leaf, the yield of RNA tends to decrease, while certain plants, for example, *Arabidopsis*, tend to yield less RNA in general, compared to other species.
3. If the RNA yield is low, it could also be due to inefficient homogenization or incomplete dispersal of pulverized tissue. If the former is suspected, one should examine a few microliters of the lysate microscopically to assess the level of cell disruption and then adjust the procedure accordingly. If the latter

is suspected, be sure that all pulverized tissue is dispersed uniformly before continuing with the isolation procedure. Low RNA yield may also reflect low levels of RNA in the sample to begin with. In this case, it is possible to drive the precipitation of whatever RNA is present by adding 1 μL 20 mg/mL ultra-pure glycogen per mL of RNA solution just prior to the precipitation step, that is, the addition of salt and/or alcohol.

4. For plant mRNA isolation, multiple precipitations with LiCl are particularly valuable for the selective precipitation of RNA when polysaccharide contaminants are present.
5. A very nice method for removing impurities and concentrating a sample, particularly if using homemade reagents, is one of the commonly used silica-based spin columns. Briefly, dissolve the RNA in the chaotropic (guanidinium) lysis buffer that accompanies the kit, bind the RNA to the column, wash the column, and then elute it in nuclease-free water, according to the manufacturer's instructions.
6. Degraded RNA could be a sign that (a) the tissue had high levels of endogenous RNase that was not controlled; (b) the sample thawed prior to the addition of the lysis/extraction solution; or (c) RNase was accidentally introduced by the user, meaning that a modification in technique is in order.
7. As indicated above, the best strategy for RNA isolation from a particular plant tissue will need to be determined empirically. Kits for plant RNA isolation abound, and one should not by shy about asking vendors for a first-time evaluation (free) kit.

References

Artlip, T.S., Callahan, A.M., Bassett, C.L., Wisniewski, M.E., 1997. Seasonal expression of a dehydrin gene in sibling deciduous and evergreen genotypes of peach (*Prunus persica* [L]). Batsch. Plant Mol. Biol. 33, 61–70.

Bassett, C.L., Wisneiwski, M.E., Arlip, T.S., Norelli, J.L., Renaut, J., Farrell Jr., R.E., 2006. Global analysis of genes regulated by low temperature and photoperiod in peach bark. J. Am. Soc. Hort. Sci. 131, 551–563.

Bilgin, D.D., DeLucia, E.H., Clough, S.J., 2009. A robust RNA isolation method suitable for Affymetrix GeneCHip analysis and quantitative real-time RT-PCR. Nat. Protoc. 4, 333–340.

Callahan, A., Morgens, P., Walton, E., 1989. Isolation and in vitro translation of RNAs from developing peach fruit (*Prunus persica* L. Batsch 'Reliance'). HortScience 24, 356–358.

Callahan, A., Morgens, P., Cohen, R., 1993. Isolation and initial characterization of cDNAs for mRNAs regulated during peach fruit development. J. Am. Hort. Soc. 118, 531–537.

Chang, S., Puryear, J., Cairney, J., 1993. A simple and efficient method for isolating RNA from pine trees. Plant Mol. Biol. Rep. 11, 113–116.

Davis, A.R., Levi, A., Kim, S., King, S.R., Hernandez, A., 2006. RNA extraction method from fruit tissue high in water and sugar. HortScience 41 (5), 1292–1294.

Futterer, J., Hohn, T., 1996. Translation in plants – rules and exceptions. Plant Mol. Biol. 32, 159–189.

Gudenschwager, O., González-Agüero, M., Defilippi, B.G., 2012. A general method for high-quality RNA isolation from metabolite-rich fruits. S. Afr. J. Bot. 83, 186–192.

Hedtke, B., Börner, T., Weihe, A., 2000. One RNA polymerase serving two genomes. EMBO Rep. 51, 435–440.

Huang, Y., Kendall, T., Forsythe, E.S., Dorantes-Acosta, A., Li, S., Caballero-Pérez, J., et al., 2015. Ancient origin and recent innovations of RNA polymerase IV and V. Mol. Biol. Evol. 32 (7), 1788–1799.

Jaakola, L., Pirttilä, A.M., Halonen, M., Hohtola, A., 2001. Isolation of high-quality RNA from bilberry (*Vaccinium myrtillus* L.) fruit. Mol. Biotechnol. 19, 201–203.

Jaiprakash, M.R., Pillai, B., Venkatesh, P., Subramanian, N., Sinkar, V.P., Sadhale, P.P., 2003. RNA isolation from high-phenolic freeze-dried tea (*Camellia sinensis*) leaves. Plant Mol. Biol. Rep. 21, 465a–465g.

Jones, D.M., Vandepoele, K., 2020. Identification and evolution of gene regulatory networks: insights from comparative studies in plants. Curr. Opin. Plant Biol. 54, 42–48.

Kanno, T., Huettel, B., Mette, M.F., Aufsatz, W., Jaligot, E., Daxinger, L., et al., 2005. Atypical RNA polymerase subunits required for RNA-directed DNA methylation. Nat. Genet. 37, 761–765.

Kozak, M., 1991. Structural features in eukaryotic mRNAs that modulate the initiation of translation. J. Biol. Chem. 266, 19867–19870.

Lal, L., Sahoo, R., Kumar-Gupta, R., Sharma, P., Kumar, S., 2001. RNA isolation from high-phenolic tea leaves and apical buds. Plant Mol. Biol. Rep. 19, 181a–181f.

Logemann, J., Schell, J., Willmitzer, L., 1987. Improved method for the isolation of RNA from plant tissues. Anal. Biochem. 163, 16–20.

Morgens, P., Callahan, A., Dunn, L.J., Abeles, F., 1990. Isolation and sequencing of a cDNA clone encoding an ethylene-induced peroxidase from cucumber cotyledons. Plant Mol. Biol. 14, 715–725.

Onodera, Y., Haag, J.R., Ream, T., Nunes, P.C., Pontes, O., Pikaard, C.S., 2005. Plant nuclear RNA polymerase IV mediates siRNA and DNA methylation-dependent heterochromatin formation. Cell 120, 613–622.

Pontier, D., Yahubyan, G., Vega, D., Bulski, A., Saez-Vasquez, J., Hakimi, M.A., et al., 2005. Reinforcement of silencing at transposons and highly repeated sequences requires the concerted action of two distinct RNA polymerases IV in Arabidopsis. Genes Dev. 19, 2030–2040.

Richter, U., Kiessling, J., Hedtke, B., Decker, E., Reski, R., Borner, T., et al., 2002. Two RpoT genes of Physcomitrella patens encode phage-like RNA polymerases with dual targeting to mitochondria and plastids. Gene 290, 95–105.

Sajeevan, R.S., Shivanna, M.B., Nataraja, K.N., 2014. An efficient protocol for total RNA isolation from healthy and stressed tissues of mulberry (Morus sp.) and other tissues. Am. J. Plant Sci. 5, 2057–2065.

Schneiderbauer, A., Sandermann Jr., H., Ernst, D., 1991. Isolation of functional RNA from plant tissues rich in phenolic compounds. Anal. Biochem. 197, 91–95.

Tadini, L., Jeran, N., Peracchio, C., Masiero, S., Colombo, M., Pesaresi, P., 2020. The plastid transcription machinery and its coordination with the expression of nuclear genome: plastid-encoded polymerase, nuclear-encoded polymerase and the genomes uncoupled 1-mediated retrograde communication. Phil. Trans. R. Soc. B 375, 20190399. Available from: https://doi.org/10.1098/RSTB.2019.0399.

Tattersall, E.A.R., Ergul, A., AlKayal, F., DeLuc, L., Cushman, J.C., Cramer, G.R., 2005. Comparison of methods for isolating high-quality RNA from leaves of grapevine. Am. J. Enol. Vitic. 56 (4), 400–406.

Wan, C.Y., Wilkens, T.A., 1994. A modified hot borate method significantly enhances the yield of high-quality RNA from cotton. Anal. Biochem. 223, 7–12.

Further reading

Ainsworth, C., 1994. Isolation of RNA from flora tissue of *Rumex acetosa.* Plant Mol. Biol. Rep. 12, 198–203.

Bassett, C.L., Wisniewski, M.E., Artlip, T.S., Richart, G.A., Norelli, J.L., Farrell Jr., R.E., 2009. Comparative expression and transcript initiation of three peach dehydrin genes. Planta 230, 107–118.

Bassett, C.L., Nickerson, M.L., Farrell Jr., R.E., Harrison, M., 2004. Multiple transcripts of a leucine-rich repeat receptor kinase from morning glory originate from different TATA boxes in a tissue-specific manner. Mol. Gen. Genom. 271 (6), 752–760.

Bassett, C.L., Nickerson, M.L., Farrell Jr., R.E., Artlip, T.S., Ghaouth, A.E., Wilson, C.L., et al., 2005. Characterization of an S-locus receptor protein kinase-like gene from peach. Tree Physiol. 25, 403–411.

Farrell Jr., R.E., 2007. The regulation of gene expression in plants and animals. In: Bassett, C.L. (Ed.), Regulation of Gene Expression in Plants: The Role of Transcript Structure and Processing. Springer Publishers, pp. 1–38.

Farrell Jr., R.E., Bassett, C.L., 2007. Multiple transcript initiation as a mechanism for regulating gene expression. In: Bassett, C.L. (Ed.), Regulation of Gene Expression in Plants: The Role of Transcript Structure and Processing. Springer Publishers, pp. 39–66.

Gallego-Bartolomé, J., Liu, W., Kuo, P.H., et al., 2019. Co-targeting RNA polymerases IV and V promotes efficient de Novo DNA methylation in Arabidopsis. Cell 176, 1068–1082. e19.

Graham, G.C., 1993. A method for extractions of total RNA from *Pinus radiata* and other conifers. Plant Mol. Biol. Rep. 11, 32–37.

Kanazawa, A., O'Dell, M., Hellens, R.P., Hitchin, E., Metzlaff, M., 2000. Mini-scale method for nuclear run-on transcription assay in plants. Plant Mol. Biol. Rep. 18, 377–383.

Lewinsohn, E., Steele, C.L., Croteau, R., 1994. Simple isolation of functional RNA from woody stems of gymnosperms. Plant Mol. Biol. Rep. 12, 20–25.

López-Gómez, R., Gómez-Lim, M.A., 1992. A method for extracting intact RNA from fruits rich in polysaccharides using ripe mango mesocarp. Hort. Sci. 27, 440–442.

Marasco, M., Li, W., Lynch, M., et al., 2017. Catalytic properties of RNA polymerases IV and V: accuracy, nucleotide incorporation and rNTP/dNTP discrimination. Nucleic Acids Res. 45, 11315–11326.

Mitra, D., Kootstra, A., 1993. Isolation of RNA from apple skin. Plant Mol. Biol. Rep. 11, 326–332.

Norelli, J.L., Farrell Jr., R.E., Bassett, C.L., Baldo, A.M., Lalli, D.A., Aldwinckle, H.S., et al., 2009. Rapid transcriptional response of apple to fire blight disease revealed by cDNA suppression subtractive hybridization analysis. Tree Genet. Genomes 5, 27–40.

Rymen, B.L., Ferrafiat, Blevins, T., 2020. Non-coding RNA polymerases that silence transposable elements and reprogram gene expression in plants. Transcription 11 (3–4), 172–191. Available from: https://doi.org/10.1080/21541264.2020.1825906.

Soni, R., Murray, J.A.H., 1994. Isolation of intact DNA and RNA from plant tissues. Anal. Biochem. 218, 474–476.

Wisniewski, M.E., Bassett, C.L., Renaut, J., Farrell Jr., R.E., Tworkoski, T., Artlip, T.S., 2006. Differential regulation of two dehydrin genes from peach (*Prunus persica* L. Batsch) by photoperiod, low temperature, and water deficit. Tree Physiol. 26, 575–584.

Zhang, H., Ding, Y., 2021. Novel insights into the pervasive role of RNA structure in post-transcriptional regulation of gene expression in plants. Biochem. Soc. Trans. 49 (4), 1829–1839.

Chapter 8

Quality control for RNA preparations

Rationale

Every laboratory application involving purified RNA requires a quality assessment of the sample, particularly the mRNA fraction. Because of the labor-intensive and costly nature of most downstream assays, it is prudent to show that the template material is intact and of sufficient purity so as to maximize productivity, reproducibility, and assay sensitivity. One would certainly not begin with degraded or undetectable quantities of RNA and simply "hope for the best" for the same reasons that one would not attempt to build a house with wood that had rotted. In the RNA world, folks, it's all about quality control.

Checking an RNA sample for integrity and level of purity may be accomplished in a variety of ways, some of which are also very good positive-control techniques. Of the methods delineated below, spectrophotometric analysis and examination of the electrophoretic profile of each sample (quality control techniques 1 and 2, respectively) should always be performed shortly after purification and, if the RNA has been stored for a while, a second aliquot should be tested just prior to use. Never make assumptions about the quality of a sample because of the isolation method used; nor should one assume that a previously characterized sample remains intact after storage for several months, even at −80°C.

Quality control technique 1: ultraviolet spectrophotometry and absorption ratios

Nucleic acid concentration can be determined easily by spectrophotometric analysis. This is a mainstream tool in any molecular biology laboratory. The absorbance of ultraviolet (UV) light provides a direct means of determining both nucleic acid concentration and purity, as well as an indirect means of calculating the total yield from the biological source and even an estimation of the mass of RNA per cell, assuming that the approximate cell number is known at the onset of the isolation procedure. Moreover, looking for up- or downregulation of specific transcripts requires some basis for normalization

RNA Methodologies. DOI: https://doi.org/10.1016/B978-0-323-90221-2.00016-3

of samples to each other. This normalization is performed, first and foremost, at the level of RNA input mass.

The UV absorption profile of a nucleic acid sample is the easiest, most rapid method for determining yield, purity, and probable utility for the intended application(s). The importance of making this measurement cannot be overstated. Pure samples of nucleic acid produce a characteristic absorbance profile between 230 and 320 nm (Fig. 8.1). Deviations from the standard shape of the curve indicate the presence of contaminants, the nature of which is suggested by the location of the distortion of the curve. Having this information is critical importance because downstream reaction efficiency is heavily dependent on the input mass and sample purity.

Determination of nucleic acid concentration

The concentration and purity of RNA and DNA samples are determined readily by taking advantage of the ability of nucleic acids to absorb UV light, with an absorbance maximum of 260 nm. In order to accurately calculate nucleic acid concentration, consideration must be given to the extinction coefficient of the material being measured. According to the Beer-Lambert Law, originally known as the Bouguer–Beer Law, or simply Beer's Law, $A = \varepsilon bc$ where A is absorbance, ε is the molar extinction coefficient, b is the light path length (1 cm is standard), and c is the concentration of the absorbing material, such as the nucleosides adenosine, cytidine, guanosine, thymidine, and uridine. This equation can be rearranged such that $\varepsilon = A/bc$. In a mixture of polynucleotides, the average extinction coefficients (μg/mL)(cm^{-1}) for dsDNA, ssRNA, and

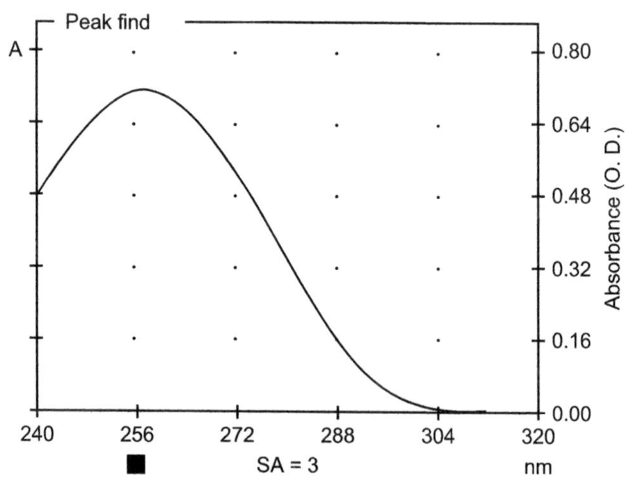

FIGURE 8.1 Typical ultraviolet absorbance profile of purified RNA. Note the positive slope of the curve below 260 nm, the negative slope above 260 nm, the absorbance maximum at 260 nm, and 0 absorbance at 320 nm. The spectrophotometric profile for purified DNA is similar.

ssDNA are 0.020, 0.025, and 0.027, respectively; the units of absorption are optical densities (ODs). Assuming a standard 1 cm light pathlength, an absorption of 1 OD_{260} corresponds to 50 μg/mL for dsDNA, 40 μg/mL for ssRNA, and 33 μg/mL for ssDNA.

With this information, absorbance measurements at 260 nm permit the direct calculation of nucleic acid concentration in a sample:

$$[\text{RNA}]\,\mu\text{g/mL} = A_{260} \times \text{dilution} \times 40$$

where A_{260} = absorbance, in optical densities, at 260 nm (OD_{260}), dilution = dilution factor, 40 = standard extinction coefficient multiplier for RNA (40 μg/OD_{260}).

The concentration of a sample of double-stranded DNA (dsDNA) can be determined in a similar manner:

$$[\text{dsDNA}]\,\mu\text{g/mL} = A_{260} \times \text{dilution} \times 50$$

where A_{260} = absorbance, in optical densities, at 260 nm (OD_{260}), dilution = dilution factor, 50 = standard extinction coefficient multiplier for dsDNA (50 μg/OD_{260}).

The concentration of single-stranded DNA (dsDNA) such as oligonucleotides can also be determined easily:

$$[\text{ssDNA}]\,\mu\text{g/mL} = A_{260} \times \text{dilution} \times 33$$

where A_{260} = absorbance, in optical densities, at 260 nm (OD_{260}), dilution = dilution factor, 33 = standard extinction coefficient multiplier for ssDNA (33 μg/OD_{260}). Note that the extinction coefficient for each oligonucleotide is unique and is always reported on the product "spec sheet" from the manufacturer that accompanies the product. If the precise extinction coefficient for a particular oligonucleotide is unknown, use a value of 33 μg/OD260 for the standard extinction multiplier.

The above equations render concentration data in units of μg/mL; for the molecular biologist, the mathematics associated with sample dilution and reaction assembly is often greatly simplified when concentration is expressed in μg/μL, so remember to divide by 1000. Also, the units of OD_{260} are mL/cm, which is useful to know if converting between mass and molarity (see Appendix B for clarification and examples).

For greater precision in determining nucleic acid concentration, the extinction coefficient for a particular (unique) RNA or DNA sample can be calculated, again using Beer's Law, if the G + C content of the organism is known. For example, the precise extinction coefficient multiplier for human RNA is 44.19. In practice, however, the average extinction coefficient multipliers of 40, 50, and 33 for RNA, dsDNA, and ssDNA, respectively, are the standards for most molecular biology applications and are preprogrammed into later model spectrophotometers.

The measurements described above reflect the total mass of nucleic acid in a sample, and not the relative contribution of any single component, such as mRNA in a mixture of total cellular RNA. Since the extinction coefficients of thymine and uracil differ, however, DNA contaminating RNA samples, or RNA contaminating DNA samples, can falsely elevate or depress spectrophotometric readings. DNA can be removed from an RNA sample by incubation in RNase-free DNase I. This technique is described in detail in Appendix C. Similarly, incubation of a DNA sample with RNase A, followed by reprecipitation, should remove virtually all contaminating RNA from the sample, as described in Appendix D. In addition, when contaminating RNA is present, DNA molecules often display increased electrophoretic mobility, a phenomenon that may well result in an underestimation of DNA fragment size or size range.

For older-style spectrophotometers, UV measurements are best taken by diluting the sample 1:200–1:500 in scrupulously cleaned quartz cuvettes. Alternatively, disposable plastic cuvettes are now manufactured that permit measurement in the UV portion of the spectrum. The reader is cautioned that not all plastic cuvettes are UV-rated, and one should double-check the manufacturer's description of the product before ordering. Ideally, UV-rated plastic cuvettes should be used only once, but certainly, not more than 3–4 times, after which the plastic is changed by exposure to UV light and the data become wholly unreliable.

When determining nucleic acid concentration, one should be certain to use a sufficient mass to obtain at least the minimum absorbance recommended by the manufacturer of the spectrophotometer. For example, spectrophotometric measurements of less than 0.1 OD are often unreliable in older instrumentation. Larger-volume cuvettes require a minimum volume to be readable and, when used, the increased volume may have the unfortunate consequence of diluting the sample below the level of detection. One solution to this type of problem is to invest in more expensive, smaller-volume quartz cuvettes.

A significant improvement in spectrophotometric instrumentation accompanied the development of the NanoDrop product line (Thermo Fisher Scientific, Waltham, MA). These fiber-optic-based instruments are ideal for processing samples in rapid succession and are capable of full-spectrum visible and UV wavelength measurements. As shown in Fig. 8.2, as little as 1ul of sample is required for rapid, precision profiling of the sample. In this laboratory we routinely load 2 μL per sample directly onto the pedestal, and there are no required consumables, except for a micropipette tip for each sample. Best of all, cuvettes are not required; the sample is retained by surface tension alone while the absorption measurements are performed. The latest generation of NanoDrop spectrophotometers will also accommodate a cuvette, if needed. In addition to nucleic acids, the NanoDrop instrument can be used to measure protein concentration and some NanoDrop instrument models are capable of microvolume fluorescence measurements.

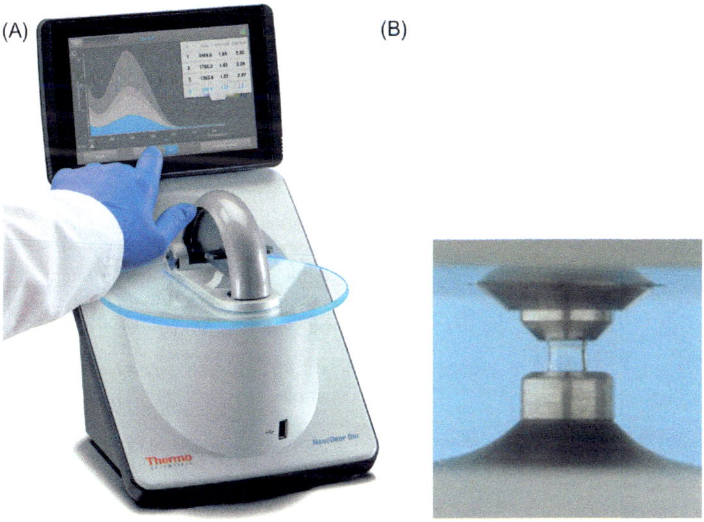

FIGURE 8.2 Thermo Scientific NanoDrop One Microvolume ultraviolet-visible spectrophotometer. (A) The small footprint is convenient for placing it almost anywhere in the lab. (B) Surface tension holds a 1–2 μL aliquot of a sample in place between the pedestal and the arm while the instrument measures concentration, displays the absorbance profile, calculates standard ratios for assessing purity, and identifies common contaminants. *Reproduced with permission of Thermo Fisher Scientific.*

Determination of nucleic acid purity

Calculations based on absorbance at 260 nm provide little information about the quality and purity of the sample; in the presence of excess salt, contaminating proteins and/or carryover organic solvents, the absorbance value will change significantly. For this reason, calculation of the so-called "260 to 280 ratio" (A_{260}/A_{280}), the "260 to 230 ratio" (A_{260}/A_{230}), and the "260 to 240 ratio" (A_{260}/A_{240}) provide a reasonable estimate of the purity of the preparation (Table 8.1). A pure sample of RNA has an A_{260}/A_{280} ratio of 2.0 ± 0.1, and a pure sample of DNA has an A_{260}/A_{280} ratio of 1.8 ± 0.1. For both RNA and DNA, the A_{260}/A_{230} should be greater than 2.0 and less than 2.4. Variations outside this range generally indicate contaminants, and investigators attuned to this detail may wish to take corrective action. A low A_{260}/A_{280} usually indicates protein contamination that carried over during the RNA isolation. A low A_{260}/A_{230} may be caused by carry-over guanidinium (from chaotropic lysis buffer) or β-mercaptoethanol, and this is much more of a problem than most people think. One method for removing excess guanidinium buffer is to reprecipitate the RNA with sodium acetate and ethanol, followed by washing the pellet 2–3 times with 70% ethanol or even soaking the RNA in 70% ethanol overnight at −20°C. This same procedure should be followed if the A_{260}/A_{230} is greater than 2.4. When working with plant

TABLE 8.1 Key indicators of nucleic acid purity.

	DNA	RNA	Problem ratios
A_{260}/ A_{280}	1.8	2.0	Ratios below 1.8 suggest protein or another organic compound contaminant.
A_{260}/ A_{230}	2.0–2.4	2.0–2.4	Ratios below 2.0 suggest organic compound contamination including phenol, guanidinium formulations, or carbohydrates. Ratios above 2.4 suggest that the instrument was blanked with a different solution than the one in which the test sample is prepared.
A_{260}/ A_{240}	1.4	1.4	Ratios below 1.4 suggest an excessive amount of salt in the sample.

Note: Absorbance ratios within the ranges shown suggest that the nucleic acid sample is suitable for subsequent experimentation. Be sure to examine the full absorption profile between 230 and 320 nm as well.

RNA, a low A_{260}/A_{230} may indicate polysaccharide contamination. Finally, use the A_{260}/A_{240} ratio, which should be around 1.4 for pure nucleic acids, to detect excess amounts of salt. When A_{260}/A_{240} is less than 1.4, one should consider reprecipitation of the sample followed by multiple washes with 70% ethanol; a final wash with 95% ethanol will accelerate drying the pellet. It is also worth noting that the pH of the diluent does, in fact influence the absorbance ratio, and is most reliable when measurements are made between pH 7.5–8.5. It is now commonplace to dilute both DNA and RNA samples in TE buffer to establish a pH within this range. The take-home lesson is that biochemical quality water, which is usually slightly acidic with pH around 6.0, may not be the best choice for accurate measurement of the concentration of nucleic acid in a sample.

Another common problem associated with the isolation of RNA is the incomplete removal of cellular protein from a cell lysate. Since amino acids with aromatic rings (tryptophan, tyrosine, phenylalanine) increase the absorbance at 280 nm, the resulting decrease in the A_{260}/A_{280} ratio is a reliable indicator of a purity problem with a nucleic acid sample. There are two ways to prevent this. In first instance in which the RNA was isolated using a silica column, the RNA should be redissolved in the high-salt bind buffer that accompanied the isolation kit, after which the RNA should be processed as if it were a cell lysate, for example, binding to a fresh silica column, washing, and then eluting. In the second instance where the use of organic compounds (i.e., phenol and chloroform) resulted in incomplete protein extraction, the sample should be dissolved 100–200 μL sterile (nuclease-free) TE buffer, extracted once with an equal volume of phenol:chloroform:isoamyl alcohol (25:24:1) and then once with an equal volume of chloroform or chloroform:

isoamyl alcohol (24:1). The appearance of opaque material, no matter how little, between the lower organic phase and the upper aqueous phase clearly indicates protein contamination. Performing one extra phenol:chloroform extraction to remove contaminating protein *during* the isolation procedure is definitely worthwhile if an additional extraction appears to be warranted and may well preclude a lot of aggravation later.

Following cleanup, RNA samples in large volumes may then be reprecipitated to increase the concentration of the sample. This type of manipulation should not be used gratuitously, however, as the stability and recovery of the sample are compromised by repeated precipitation, not to mention the potential for accidental introduction of nuclease activity into the sample. Precipitated samples of RNA should be washed with 70% ethanol and the concentration determined as described earlier.

Although the absorbance ratios obtained for a particular sample is a good indicator of the quality of the sample, it is also of paramount importance to examine the complete absorbance profile of the sample, usually between 230 and 320 nm. Pure samples of nucleic acids will produce a characteristic skewed bell curve (Fig. 8.1), which exhibits an absorbance maximum close to 260 nm. The absorbance of UV light by nucleic acids is 0 at 320 nm, and below 230 nm the absorbance profile has little bearing on the computation of the concentration and determination of the purity of a sample. High-quality nucleic acid samples manifest an absorption profile with a positive slope between 230 and 260 nm and a negative slope above 260 nm. This is not observable when measuring absorbance only at 260 nm and at 280 nm.

The benefits of these preliminary analyses to gain information about the quality of a sample should be obvious. In the absence of UV spectrophotometry, the amount of nucleic acid in a sample would be largely unknown, thereby making accurate normalization among samples all but impossible. Without computation of the various absorbance ratios, it may not be immediately obvious if contaminants have copurified with the sample. In the absence of a wavelength scan between 230 and 320 nm, there will be no indication of the shape of the curve. It is frequently noted in the laboratory that contaminants change the characteristic shape of the curve at wavelengths other than 260 nm or 280 nm. For example, excessive quantities of carryover salt (nucleic acid coprecipitate) can mask the positive slope of the curve that is usually associated with the 240–260 nm portion of the spectrum. Washing a nucleic acid coprecipitate with 70%–75% ethanol immediately following centrifugation (as suggested in several protocols) will remove most of the salt that was used to drive the precipitation of the nucleic acid molecules in the first place; sodium acetate (NaOAc) is particularly easy to remove with 70% ethanol washes. In the absence of a wavelength scan, the presence of excess salt, and probable negative impact on downstream applications, would not be evident. Finally, negative absorption readings may result when the spectrophotometer is blanked with a buffer different from that in which the sample is dissolved.

Nonspectrophotometric methods

In the absence of spectrophotometric equipment, there are other rather unsophisticated methods through which a reasonable attempt can be made to estimate the concentration of nucleic acid in a sample. First, the prevalence of image analysis software in the molecular biology laboratory affords an opportunity for quantifying nucleic acid concentration by digitizing a fluorescence image of a sample that has been run out on a gel and comparing it to a known mass standard. One may purchase mass standards from any of a number of suppliers. Dilutions of the commonly used molecular weight standards ΦX174 or λ DNA can be electrophoresed side-by-side on a gel along with RNA or DNA of unknown concentration (Fig. 8.3). Image analysis software can then count the pixels that make up the image of all bands and smears in a lane, and then compare the value for each unknown sample against a mass standard. This approach for determining nucleic acid concentration has been used successfully in this lab to assess the efficiency of cDNA synthesis by "quantifying the smear." The method is rapid and usually quite accurate, especially in view of the ready availability of sensitive nucleic acid stains such as SYBR Green, GelStar, and SYBR Gold. The subject of image analysis is presented in detail in Chapter 15, Photodocumentation and Image Analysis.

Second, several companies sell kits for the fluorescence-based determination of nucleic acid concentration. One mixes a fluorescent compound with the sample, which then fluoresces in proportion to the mass of nucleic acid with which it has bound. This procedure requires a fluorescence detector and also requires that the user prepare a standard curve against known samples in order to make precision concentration determinations.

A third, nonspectrophotometric determination of nucleic acid concentration involves making a *rough estimate* by simple visual inspection. The

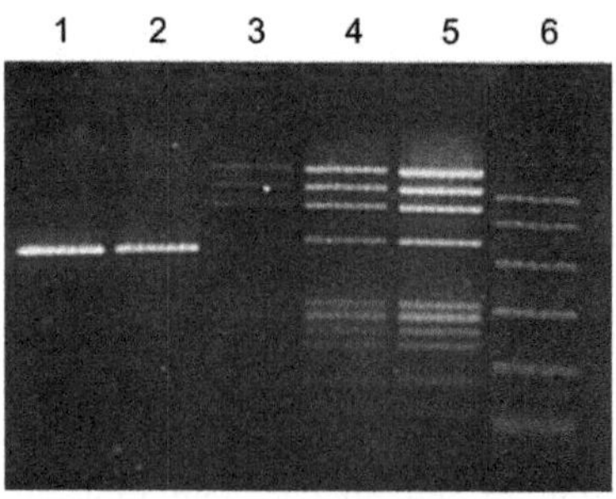

FIGURE 8.3 Determination of sample mass and molecular weight by comparison with dilutions of a DNA mass standard. Electrophoresis was performed on a 2.5% agarose gel. Lane 1, 2 μL sample of unknown concentration; lane 2, 1 μL sample of unknown concentration; lanes 3–5, 100, 200, and 400 ng, respectively, of ΦX174 DNA; lane 6, PCR Marker (Promega): molecular weights are 1000, 750, 500, 300, 150, and 50 bp. Digital image analysis is very helpful; for determining the closest fluorescence "match" among the individual bands.

method involves spotting dilutions of known amounts of nucleic acid and samples of unknown concentration onto an agarose plate containing a fluorescent dye. Upon UV irradiation, the dye will fluoresce in proportion to the amount of nucleic acid present, and then compare the fluorescence intensity between the known and unknown samples. Performing this rather imprecise assay requires the preparation of agarose plates containing either 0.5 μg/mL ethidium bromide or 1X SYBR Green. To prepare ethidium bromide/agarose plates, prepare a solution of 1% agarose in water, cool to 55°, and then add ethidium bromide stock solution (10 mg/mL) to a final concentration of 0.5 μg/mL. Carefully swirl the flask to mix, and avoid inhaling vapors. The molten agarose/ethidium bromide mixture can then be poured into 100 mm Petri dishes and allowed to solidify. SYBR Green plates are prepared in the same manner, substituting 1X SYBR Green in place of the ethidium bromide. Dilutions of the standards and the test material are best applied in aliquots no greater than 5 μL so as to avoid sample diffusion, which would make quantitation even more difficult. The plates can be observed directly on the surface of a UV transilluminator or irradiated from above with a handheld UV lamp. A word of caution: handheld UV irradiating devices can inflict injuries as serious as those from transilluminators. Be sure to wear proper eye protection and cover exposed skin when working with UV light. Be sure to allow samples to be absorbed into the agarose before assessing the fluorescence intensity of the ethidium bromide. This method only has value as a teaching tool when funds are limited.

Quality control technique 2: electrophoretic profiling of RNA

In general, running out an RNA sample on an agarose gel is the single best quality control measure available to the molecular biologist; parameters and protocols for electrophoresis are described in detail both in Chapter 14, Electrophoresis of RNA, and in Appendix E. Chemically intact, biologically competent RNA always produces a characteristic banding profile when denatured, which can be observed simply by heating the sample briefly before loading it onto an agarose gel. Along these same lines, one may incubate a high-quality RNA sample with an aliquot of a test buffer or other solution to assay for the presence of nuclease activity. If, after incubation, the RNA shows any signs of degradation, then the solution being tested is probably tainted with RNase.

There is no better indicator of the probable utility of an RNA sample than the measurement of the fluorescence ratio between the 28S rRNA and the 18S rRNA. For mammalian RNA, a minimum fluorescence ratio of 2:1 is desirable and a ratio of 2.5:1 or greater is better yet (Fig. 8.4). It has been the experience of this author and many others that the smaller the 28S:18S ratio, the less likely the sample will yield representative data or do so in a reproducible manner. One may simply "eyeball" the gel to estimate the

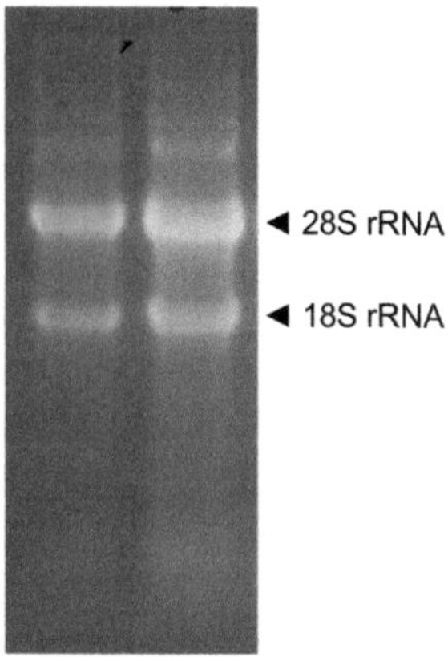

FIGURE 8.4 Electrophoretic profile of high-quality mammalian RNA. The rRNA 28S and 18S bands are well-defined while the remainder of the sample (mRNA) appears as faint smearing slightly above, between, and just below the major ribosomal species. The light "smudge" at the leading edge of the gel is a mixture 5S and 5.8S rRNA, along with tRNA. Ideally, the 28S:18S fluorescence ratio should be approximately 2:1, or greater.

28S:18S ratio or, as needed, the measurement can be performed using image analysis software. For higher eukaryotes, the larger the ratio, the better. Note that other aspects of the electrophoretic profile are also of key importance.

Excellent RNA (eRNA) samples show a minimum amount of smearing above, between, and below the 28S and 18S rRNAs (Fig. 8.5A). Transfer RNA (tRNA), and the low-molecular-weight 5S and 5.8S rRNA species all co-migrate at the leading edge of the gel and, when stained, commonly appear as an "indiscrete splotch" in the vicinity of the bromophenol blue tracking dye that is often part of standard gel loading buffers. Lack of definition to the 28S and 18S rRNAs usually means that the sample has been subjected to nuclease assault, especially if the smearing is confined to the lower portion of the gel (Fig. 8.5B), below the level of the 18S rRNA, which can be observed in intact RNA in adjacent lanes on the same gel. Heavy smearing along the length of the gel may indicate limited degradation of the sample, though it is possible that the RNA was not denatured properly prior to electrophoresis: the resultant smear may very well reflect persistent secondary RNA structure. Further, this Author has also observed situations in which previously precipitated RNA was not completely dissolved, resulting in a characteristic smear along the length of the lane, as described in Chapter 14, Electrophoresis of RNA. While it is desirable to keep nucleic acids concentrated, as described above, it is also necessary to make sure that the entire RNA or DNA sample is completely dissolved so that each aliquot removed from the stock tube is a representative aliquot. Detergents and excess salt in the sample can also cause smearing of the RNA and, in some cases, inhibit the entry of the RNA into the gel itself. If this is a chronic difficulty, the use

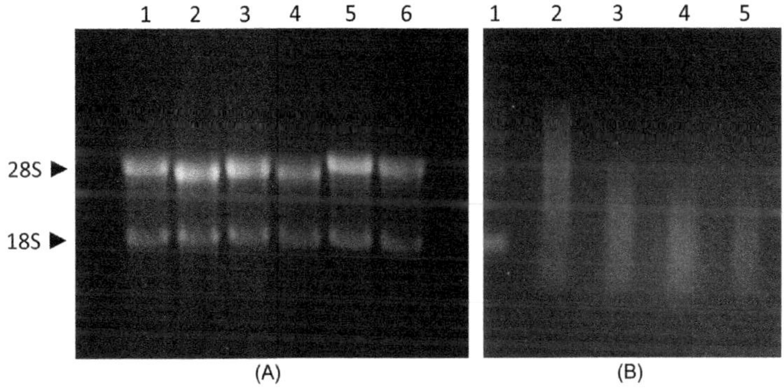

FIGURE 8.5 Assessment of the integrity of RNA samples. (A) Minigel (7 cm × 8 cm) analysis of RNA isolated from Swiss mouse 3T3 fibroblasts. Five micrograms of purified RNA was electrophoresed in a 1.2% agarose-formaldehyde gel for 1.5 h at 50 V and then stained with SYBR Green II. The clear definition of the 28S and 18 S rRNA species and the comparatively stronger fluorescence of the 28S rRNA in all six lanes demonstrate the integrity and quality of the samples. (B) Going, going, gone. Minigel (6 cm × 9 cm) electrophoresis of HT-1080 human fibroblast RNA. A representative aliquot from four different samples was electrophoresed in a 1.2% agarose-formaldehyde gel for 1.5 h at 50 V and then stained with ethidium bromide. The RNA molecular weight standard is barely visible in lane 1. Samples in lanes 2–5 show increasing degrees of degradation due to RNase contamination during the isolation procedure. This is an excellent example of what not to do.

of *N*-laurylsarcosine (sarkosyl) may alleviate detergent-related symptoms. Moreover, if care was not taken to remove DNA from an RNA prep, as described in Chapter 4, RNA Isolation Strategies, and Appendix C, contaminating DNA will coprecipitate with the RNA in the final stages of RNA purification, wreaking havoc in downstream applications. Briefly, options for the removal of DNA from RNA preparations include differential centrifugation of intact nuclei, treatment of the sample with RNase-free DNase I, preferential binding to and elution from a silica column, repeated extraction with acid–phenol, or isopycnic partitioning of the RNA. The appropriate choice depends on the volume of the sample as well as the total amount (mass) of the RNA involved.

The inclusion of a denaturant, for example, formaldehyde or urea, which is necessary when performing Northern analysis, is optional at this stage; the ribosomal RNAs (rRNAs) are quite evident when heated, quick-cooled, and electrophoresed under nondenaturing conditions. It should be noted that the inclusion of 2 M urea will facilitate sample denaturation, producing a gel that is easier to interpret, compared to the omission of a denaturant. Following a relatively brief period of nondenaturing electrophoresis, the presence of the predominant 28S and 18S rRNA species confirms that the RNA is intact.

Genomic DNA on an EtBr- or SYBR Green-stained RNA gel appears as an area of fluorescence within and just below the well into which the sample was loaded (Fig. 8.6). If this phenomenon is observed, it is wise to ascertain

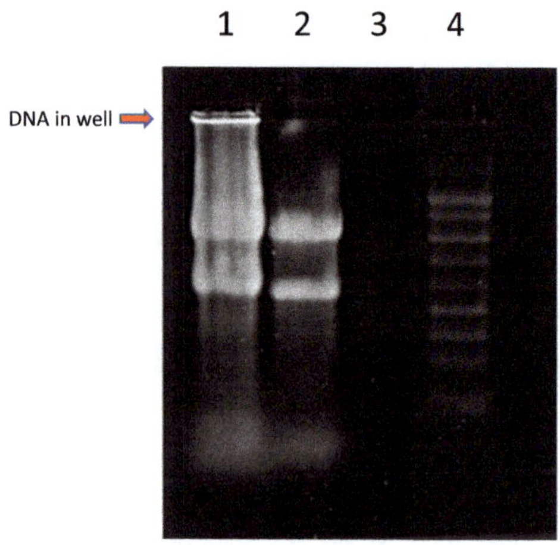

FIGURE 8.6 SYBR Green II staining of RNA samples. Fluorescence confined to a well (lane 1) into which a sample was loaded strongly suggests genomic DNA contamination. This lane was also overloaded with too much RNA. Lane 2, high-quality RNA; lane 3, no sample; lane 4, RNA MW standard.

the cause of the DNA carryover before using the RNA for any type of quantitative analysis, particularly those methods involving PCR. Needless to say, contaminating DNA has a profound negative influence on the outcome of most RT-PCR assays in its capacity as an alternative, competitive template for primers. Note that the absence of high-molecular-weight fluorescence does not mean that the RNA sample is devoid of DNA, only that if present the DNA is below the level of detection. The only reliable methods to ensure the absence of DNA from RNA samples are to incubate the sample with RNase-free DNase *and* run a negative control reaction that will produce a result only if genomic DNA is present in the sample, for example, a "no reverse transcriptase control," as described in Chapter 10, RT-PCR: A Science and an Art Form.

Protocol: nondenaturing agarose electrophoresis

The integrity of experimental RNA can be determined by minigel (gel dimensions: 7 cm × 8 cm) electrophoresis through a standard, nondenaturing 1.2% agarose gel. The RNA sample is briefly heat-denatured just prior to electrophoresis in order to disrupt secondary structure.

1. Prepare a 1.2% solution of molten agarose in 1 × TAE buffer. Cast gel when agarose has cooled to 65°.

   Note 1: 50 × TAE (per liter) = 242 g Tris base; 100 mL 0.5 M Na2-EDTA, pH 8.0; 57.1 mL glacial acetic acid; autoclave. Melt agarose in

1 × TAE buffer and also use it as the electrophoresis running buffer at a 1 × concentration.

Note 2: If desired, EtBr can be added directly to the molten agarose to a final concentration of 0.5 μg/mL. In so doing, the gel can be examined as soon as the electrophoresis has been completed. Do not add SYBR Green to gels prior to electrophoresis because it will profoundly interfere with the formation of sharp bands. Stain gels with SYBR Green after electrophoresis has been completed.

2. While the gel is solidifying, dilute 1–5 μg RNA in sterile H_2O or TE buffer in a final volume of 10 μL.
3. Heat to 65° for 5 minutes. Pulse centrifuge to collect sample at bottom of tube.

   Note: If it is discovered that the RNA proves difficult to denature by heating alone, add freshly prepared urea to a final concentration of 0.2 M prior to heating. This is the preferred method. Alternatively, add 2.5 μL formaldehyde (37% stock solution) and 5.0 μL formamide to the 10 μL aliquot of RNA. Handle formaldehyde and formamide cautiously and in a chemical fume hood, according to the manufacturers Safety Data Sheet (SDS; formerly MSDS).
4. Add 1 μL of 10 × loading buffer (50% glycerol, 1 mM EDTA, 0.25% bromophenol blue) and then load the gel.

   Note 1: Use 2 μL loading buffer if formaldehyde and formamide were added to the RNA sample.

   Note 2: A molecular weight marker is usually not needed; this is a quick and dirty assessment of the integrity of the sample.
5. Electrophorese at 75–100 V until dye front has migrated 2–3 cm into the gel. This is all that is necessary to visualize the sample; however, allowing the gel to run longer will certainly favor greater resolution.
6. Unless the dye was added directly to the agarose, stain gel for 10 minutes in a solution of EtBr (0.5 μg/mL in H_2O). Alternatively, stain the gel in 1 × SYBR Green II (diluted in 1 × TAE) for 25–30 minutes. Inspect gel as usual for indications of degradation and/or DNA contamination.

   Note 1: EtBr is a mutagen that must be handled with care and disposed of properly (Appendix G).

   Note 2: When working with a transilluminator or hand-held UV light source, be sure to protect eyes and skin from UV light.
7. Photograph the gel and archive the image. When performing quality control checks on RNA preps, photodocumentation (Chapter 15: Photodocumentation and Image Analysis) is important for future reference.

Quality control technique 3: RNA integrity number

A state-of-the art method for assessing RNA integrity and measuring sample concentration is microfluidics-based profiling on a 2100 Bioanalyzer instrument (Agilent Technologies; Santa Clara, CA). Best thought of as a miniaturized electrophoresis platform, up to twelve RNA samples can be processed on a single "chip" in about thirty minutes; the system is also configurable for DNA and protein analysis. The sample is run in the presence of a fluorescent dye and the image generated is similar to that observed on a full-size agarose gel, accompanied by an electropherogram in which peak amplitude is indicative of the abundance of each RNA species resolved by the microfluidics device. A proprietary algorithm is applied to the distribution of RNA species observed in each channel, resulting in the assignment of an RNA integrity number (RIN). The RIN for an RNA sample ranges from 1 to 10; a RIN of 10 is fully intact, and a RIN of 1 is completely degraded (Fig. 8.7). Generally speaking, a RIN of 8 or higher is indicative of high-quality RNA. It is noteworthy that the RIN assigned to a sample takes into account all of the RNA species along the entire channel, rather than just examining the fluorescence ratio associated with the 28S and 18S rRNA species. This can be quite useful for those investigators working primarily with miRNA. Moreover, as a sample begins to experience degradation, the decrease in the fluorescence signal from the rRNA species will be accompanied by a shift in the location of dye in each channel because smaller RNAs, and RNA fragments, have greater mobility. Succinctly, RIN assessment eliminates individual interpretation and provides the investigator with a nonbiased method for rating and comparing the quality of RNA samples. It has quickly become the industry standard for integrity and quality assessment.

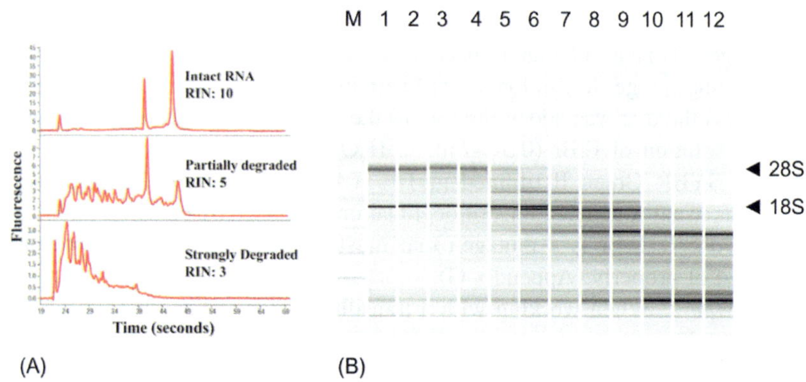

FIGURE 8.7 RNA integrity check on a Bioanalyzer microfluidics chip. Each sample is assigned a RIN (RNA integrity number) score using a proprietary algorithm. (A) RIN examples include RIN 10 (completely intact), RIN 5 (partial degradation), and RIN 3 (heavily degraded). (B) Lanes 1–12, RNA samples showing gradual loss of integrity and usefulness for downstream applications. Note the dramatic mobility shift of the sample from top to bottom, particularly in lanes 7–12. Lane M = MW standard. The automation of the Bioanalyzer approach provides a universal and unbiased comparison of RNA samples. *Images © Agilent Technologies, Inc. 2016. Reproduced with permission, courtesy Agilent Technologies, Inc.*

The method can be used to assay total RNA, poly(A)$^+$ RNA, and cRNA in order to predict the reliability of a sample for some type of transcription analysis.

Bioanalyzer instrumentation is useful for abundance and integrity checks, but less so as an indicator of sample purity. This type of information usually comes from the calculation of the various absorption ratio permutations described above in quality control technique 1. However, the RIN assigned to an RNA sample is usually a very reliable indicator of the probable utility of an RNA sample in any of a variety of gene expression assays; additional assessments or purity checks are usually not necessary but could be helpful.

While there is much to be gleaned from an RNA integrity check performed in this manner, the RIN number is not a tell-all with regard to downstream applications. For example, depending upon primer design, RT-PCR can be performed successfully when starting with partially degraded RNA when the amplicon is short. In the case of nucleic acids that have been purified from archived pathology specimens, RNA (and DNA) recovered from formalin-fixed paraffin-embedded tissues and tissue sections always exhibit some degree of degradation. In this case, it is possible to use RIN analysis as part of one's SOP for archived tissue by simply establishing a threshold RIN for FFPE samples, below which the RNA sample will not be used under any circumstances. In all cases, having RIN data is a great experimental asset.

Quality control technique 4: ultraviolet shadowing

On occasion, the amount or concentration of a nucleic sample may be too low to observe by electrophoresis, even with the very sensitive SYBR family of dyes. Regardless of the amount of the sample, it is still important to ascertain the integrity of the sample. An alternative visualization method, known as UV shadowing, is perfect for observing very low amounts of sample on a gel, particularly low molecular weight molecules run on a polyacrylamide gel. In contrast to gel staining with fluorescent dyes (e.g., ethidium bromide or SYBR Green), UV shadowing is performed by resting a gel (wrapped in plastic) over a thin layer chromatography (TLC) fluor-coated silica plate and irradiating it from above in an image analysis workstation or through the use of a hand-held UV light lamp (254 nm). RNA in the gel will absorb the fluorescence emanating from the plate, creating a shadow. The same principle applies to the analysis of DNA and proteins in this manner. As always be sure to observe all safety procedures, including the use of (UV protective glasses/face shields and gloves) when working with a UV light.

Two common applications come to mind when thinking about UV shadowing. First, the synthesis of a sense (+) or antisense (−) RNA by in vitro transcription is always followed by electrophoresis in order to demonstrate that the transcript is intact and of the predicted size. Owing to its sensitivity, UV shadowing requires only a small amount of the sample for electrophoresis; the shadowing procedure will also display contaminating truncated transcription products, even if they are of very low abundance. Be aware, however, that even short exposure of RNA to the UV shadowing method can incur substantial damage to the sample (Kladwang et al., 2012). Second, many laboratories now synthesize their own primers to support various PCR and RT-PCR applications. It is wise to electrophorese and examine an aliquot of each new primer prior to use to demonstrate successful synthesis and to look for the presence of nonfull-length oligonucleotides, commonly known as failure sequences (Fig. 8.8). Depending on the application, failure sequences are readily removed from among the full-length sequences by cartridge purification (see Chapter 10: RT-PCR: A Science and an Art Form).

Protocol: ultraviolet shadowing

1. Carefully remove the gel from the glass plates used for the electrophoresis and place it on a sheet of plastic wrap. If necessary, gently stretch the plastic wrap to remove wrinkles that may have formed under the gel.
2. Carefully position the plastic wrap (and gel) on top of a fluor-coated TLC plate.

   Suggestion: Think about designating a single TLC plate for UV shadowing and using it exclusively for that purpose. This is an economical

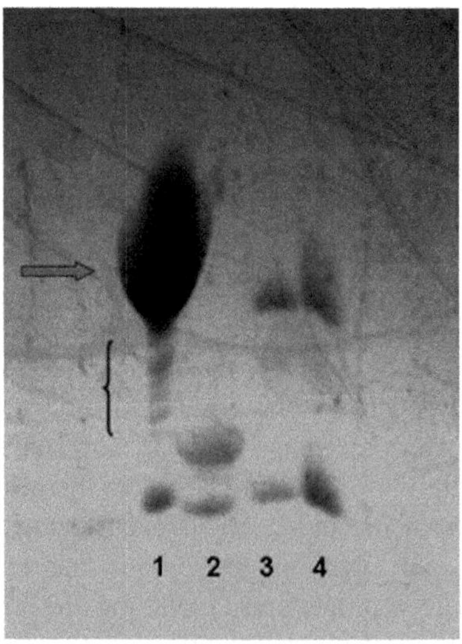

FIGURE 8.8 Assessment of oligonucleotide purity by ultraviolet shadowing. The major product (22-mer) in lane 1 is indicated by the arrow. Note the presence of the failure sequences (bracket) directly beneath the major band. This oligonucleotide was only desalted, which does nothing to remove nonfull-length molecules. None of these bands was visible upon staining with SYBR Green even though the lanes were intentionally overloaded. Material visible in lanes 2–4 is unrelated to the quality control check. *Photograph courtesy Aptagen, LLC (Jacobus, PA).*

way the perform UV shadowing routinely; keeping the plate near the image analysis station will also save time.

3. Taking care to avoid skin and eye exposure to UV light, position the hand-held UV lamp above the gel. When possible, a fully enclosed image analysis workstation should be used instead of a hand-held UV lamp for greater protection of the investigator.

   Suggestion: It is helpful to place the UV light source on a slight angle above the gel in order to see the bands. Also, rather than holding the UV light source, consider clamping it onto a ring stand so that both hands remain free.

4. Mark the location of the band(s) of interest if they are to be cut out of the gel for further purification. This is easily accomplished by placing a second sheet of plastic wrap on top of the gel and then marking the position of band(s) with a lab marker.

   Note 1: RNA and DNA bands appear as very dark blue bands against a pale green fluorescent background.

   Note 2: Be sure to note the position of the loading dyes bromophenol blue and/or xylene cyanol, which are often mistaken for nucleic acid bands. Consider running a lane with only the loading dye(s) for comparison with samples in adjacent lanes.

5. Photograph the gel with the position of the loading dyes clearly indicated. This will preclude confusion when referring back to the photograph at later date.

Quality control technique 5: sample capacity to support RT-PCR

The polymerase chain reaction (PCR) is a standard molecular biology tool which, without a doubt, revolutionized the way many experimental questions are approached. In any given laboratory, RT-PCR is likely in widespread use for transcript detection using primers and conditions that have long since been optimized. The frustration index rises when one is unable to generate a product using a freshly prepared sample of RNA, or from an RNA sample that "used to work." As such, one standard protocol in this laboratory is to use primers that are "known to work" each time RNA is isolated intact and clean from a particular biological source. Positive control primers most often fall into this category. Testing a sample in this way can be performed using very boring primers, such as those for β-actin or GAPDH, just to show that the sample in question can at least support the amplification of *something* from the RNA template material. If a PCR product is evident with control primers, then the reason for failure to generate a product from experimental primers can be investigated more intelligently. In summary: no signal from a fairly abundant transcript probably means that the RNA sample is poor and unable to support reverse transcription or amplification of anything. PCR quality control testing may be performed using either the traditional end-point approach or using a real-time platform. Specific techniques pertaining to the use of RNA template material for RT-PCR applications are detailed in Chapter 10, RT-PCR: A Science and an Art Form, and Chapter 11, Quantitative PCR Techniques.

Quality control technique 6: sample capacity to support in vitro translation

The biological integrity of mRNA is reflected by its ability to direct the translation of its encoded polypeptide. This is most often accomplished by adding the RNA to a cell-free translation system, such as a rabbit reticulocyte lysate, wheat germ extract, or even an *E. coli* cell-free system, to name just a few common examples. Biologically competent mRNAs will, when translated, generate a defined and reproducible pattern of protein bands. This technique, commonly known as in vitro translation (*recently reviewed by* Gregorio et al., 2019) is often coupled with immunoprecipitation, predicated upon the availability of a suitable antibody. Whereas this technique is valuable as a quality control method in certain specialized proteomics applications, generally speaking it is infrequently used as an RNA quality control measure.

References

Gregorio, N.E., Levine, M.Z., Oza, J.P., 2019. A user's guide to cell-free protein synthesis. Methods Protoc. 2, 24. Available from: https://doi.org/10.3390/mps2010024.

Kladwang, P., Hum, J., Das, R., 2012. Ultraviolet shadowing of RNA can cause significant chemical damage in seconds. Sci. Rep. 2, 517. Available from: https://doi.org/10.1038/srep00517.

Further reading

Cavaluzzi, M.J., Borer, P.N., 2004. Revised UV extinction coefficients for nucleoside-5′-monophosphates and unpaired DNA and RNA. Nucleic Acids Res. 32 (1), e13. Available from: https://doi.org/10.1093/nar/gnh015.

Hammerling, M.J., Krüger, A., Jewett, M.C., 2020. Strategies for in vitro engineering of the translation machinery. Nucleic Acids Res. 48 (3), 1068–1083. Available from: https://doi.org/10.1093/nar/gkz1011.

Pamudurti, N.R., Bartok, O., Jens, M., Ashwal-Fluss, R., Stottmeister, C., Ruhe, L., et al., 2017. Translation of circRNAs. Mol. Cell 66 (1), 9–21. Available from: https://doi.org/10.1016/j.molcel.2017.02.021. e7.

Wilfinger, W.W., Mackey, K., Chomczyski, P., 1997. Effect of pH and ionic strength on the spectrophotometric assessment of nucleic acid purity. BioTechniques 22, 474–481.

Chapter 9

cDNA: a permanent biochemical record of the cell

Rationale

RNA is an easily accessible molecule within the cell and therefore a commonly assayed parameter of gene expression. When adequate attention is given to the subtleties of isolation and postisolation handling, the RNA will ordinarily be of sufficient quality to support any of a number of transcription assays. It must be emphasized, however, that there is an upper limit to the sensitivity that each of these methods can offer. This ceiling is due to the labile nature of RNA and the fact that, while variable in abundance, individual RNA molecules cannot be amplified directly in order to boost the signal.

Fortunately, there is a simple solution to this sensitivity impediment. First, RNA serves as a template for the in vitro enzymatic synthesis of complementary DNA (cDNA), a straightforward and routine procedure. The immediate product of this reaction, known as first-strand cDNA, is much more stable than the corresponding RNA template material from which it is synthesized. Next, PCR and other amplification techniques are then applied to amplify the cDNA with amazing speed and robustness, thereby improving the ability to detect low abundance cDNAs that correspond to low abundance RNAs. This facilitates extensive analysis in both the short- and long-term by any of a number of techniques. Alternatively, the cDNA can be sequenced directly. The golden rule of cDNA synthesis is this: when you make cDNA, what you are really doing is making a permanent biochemical record of transcription and, by extension, a biochemical record of the cell. One makes a biochemical "snapshot" since the only cDNA that *can* be synthesized comes from the RNA products of transcriptionally active loci. Transcriptionally silent loci are not represented in a cDNA pool because there no corresponding RNA template material being produced. This chapter addresses the subtleties of making cDNA, while the next chapter addresses the subtleties of amplifying it.

cDNA synthesis—an overview

The synthesis of cDNA is a central component of gene expression research. cDNA is useful in both the single-stranded- and double-stranded forms, and

RNA Methodologies. DOI: https://doi.org/10.1016/B978-0-323-90221-2.00032-1

it is chemically equivalent to genomic DNA, though there are important structural differences. While the original methods for cDNA propagation (e.g., cloning into plasmids) have given way to PCR and next-generation sequencing (NGS), a historical perspective of the mechanics and significance of traditional cDNA synthesis remains central to the proper understanding of the technology, its flexibility, and the interpretation of data.

cDNA is a product of enzymatic in vitro synthesis using an enzyme known as reverse transcriptase (RT) (Temin and Mizutani, 1970; Baltimore, 1970). It is prepared one strand at a time and the need for high-quality RNA is the first and foremost requirement. Generally, much less than a microgram of RNA is needed for most routine applications, and more contemporary techniques require as little as a few nanograms of starting material. Although the selection of poly(A)$^+$ RNA as the starting template for the synthesis of cDNA has historical significance, it is now commonplace to synthesize cDNA directly from total cellular RNA or total cytoplasmic RNA without prior poly(A)$^+$ selection. Given the power of PCR, poly(A)$^+$ selection is often neither necessary nor recommended because the mechanics of poly(A)$^+$ selection often result in the loss and further under-representation of very low abundance transcripts. Thus, methods that are intended to enrich in favor of poly(A)$^+$ mRNA are often counterproductive. In any event, poly(A)$^+$-selected mRNA from previously purified cellular or cytoplasmic RNA sample is a perfectly acceptable starting material when it is available.

The synthesis of cDNA offers many advantages to the investigator wishing to characterize gene structure or expression, develop nucleic acid probes or primers, express proteins, or perform global transcriptome sequencing. First, because RNA is a naturally labile single-stranded molecule, the synthesis of more stable single- or double-stranded cDNA facilitates very long-term storage of these sequences. Second, by synthesizing cDNA, the investigator creates a method for propagating the cDNA as needed, either via PCR directly or, less commonly, through the construction of a cDNA library. Third, when compared with genomic DNA sequencing information corresponding to the same genetic locus, it is possible to identify exon-intron splicing patterns and characterize flanking 5′ and 3′ sequences. Fourth, since cDNA cannot be synthesized from mRNA that is absent, the emphasis is on active gene expression. Fifth, direct cDNA sequencing is useful for generating a comprehensive transcriptional profile of the cell under a defined set of experimental conditions. Sixth, since cDNA is DNA, a population of these molecules can be transcribed to produce more RNA in direct proportion to its relative abundance in the cell (Eberwine linear amplification) without distortion of sequence abundance relationships that sometimes occurs with PCR. Seventh, because cDNA is DNA, it can be expressed ectopically to determine the consequences for a cell in terms of normal biochemical processes or disease initiation or progression.

Since cDNA is synthesized directly from RNA, the cDNA population present at the end of a synthesis reaction directly mirrors the RNA complexity at the beginning of the synthesis reaction. While cDNA populations from two different cell or tissue types will share at least those sequences necessary for cellular viability, the actual members of a cDNA population are quite unique to the biological source; recall that the phenotype and physiology of a cell result from induction and repression of specific genes and gene pathways. Moreover, because the RNA makeup of a cell is subject to change upon an experimental challenge or even simply by progression through the cell cycle, one should expect that the cDNA from a particular source is also subject to change. This contrasts sharply to the genomic identity of two different tissue types taken from the same organism in which the membership of sequences should be identical (e.g., gDNA from splenocytes vs. hepatocytes from the same mouse). The take-home lesson: in a given organism the cDNA made from various tissues differs, while genomic DNA (mutations and telomeres notwithstanding) does not.

cDNA is synthesized in a stepwise fashion, always demanding the highest quality RNA possible. A representative listing of the enzymes associated with cDNA synthesis is found in Table 9.1. These days, most cDNA synthesis kits are quite good and are widely available; it is highly advisable to invest in such a kit when performing cDNA synthesis for the first time because of the rather sensitive nature of the reaction components, particularly the RT. A disproportionate number of man-hours and other laboratory resources are likely to be wasted trying to optimize several homemade reagents, with no success guaranteed. On a per-reaction cost basis, these kits are easily the most effective and economical way to make cDNA, and these systems make a lot of cDNA. The following is a description of the steps involved in the synthesis of cDNA and, for the hearty scientist, a cDNA synthesis protocol follows.

First-strand considerations

There are two absolute requirements that must be fulfilled in order to support the synthesis of nucleic acids in vivo and in vitro. First, there must be high-quality RNA template material that can direct the synthesis of a cDNA strand; working with a suboptimal RNA preparation is the leading cause of poor cDNA synthesis. Second, there must be a primer molecule that provides a free 3′-OH, a chemical group necessary for all DNA (and RNA) polymerases to support elongation by the sequential addition of nucleotides. As polynucleotides are assembled $5' \rightarrow 3'$, and because whenever double-stranded molecules form the individual strands must be base-paired in an antiparallel manner, it is clear that the primer molecule with the requisite 3′-OH must base-pair close to the 3′ end of the mRNA template if the synthesis of cDNA is to be supported.

TABLE 9.1 Enzymes associated with cDNA synthesis and cloning.[a]

Enzyme	Activities	Role in cDNA synthesis
Reverse transcriptase (wild type) (RNA-dependent DNA polymerase; RT)	$5'\rightarrow3'$ polymerase RNase H+ (minor)	Enzymatic conversion of RNA into cDNA. Substrates may include mRNA, hnRNA, tRNA, rRNA, viral RNA.
Reverse transcriptase (modified)[b] (RNA-dependent DNA polymerase; RT)	$5'\rightarrow3'$ polymerase RNase H^-	Enzymatic conversion of RNA into cDNA. Substrates may include mRNA, hnRNA, tRNA, rRNA, viral RNA. RNase H^- formulations produce longer cDNAs.
DNA polymerase I (Kornberg enzyme)	$5'\rightarrow3'$ polymerase $5'\rightarrow3'$ exonuclease $3'\rightarrow5'$ exonuclease	Synthesis of second-strand cDNA by the Gubler and Hoffman (1983) method ($5'\rightarrow3'$ exonuclease activity removes nicked mRNA).
DNA polymerase I (Klenow fragment)	$5'\rightarrow3'$ polymerase $3'\rightarrow5'$ exonuclease	Synthesis of second-strand cDNA by the classical method of cDNA synthesis.
RNase H	Nicks mRNA component of RNA: DNA hybrid molecules	Nicking activity provides numerous primers (3′-OH) for synthesis of second-strand cDNA.
S1 nuclease	Single-strand-specific nuclease (RNA and DNA)	Removal of hairpins in classical synthesis (enzyme has specificity for single-stranded domains).
T4 DNA ligase	Ligation of cohesive ends; ligation of blunt ends	Seal phosphodiester backbone (. . .3′-OH + p5′X. . .).
T4 DNA polymerase	$5'\rightarrow3'$ polymerase $3'\rightarrow5'$ exonuclease	Preparation of blunt ends on dsDNA (cDNA or gDNA); exonuclease activity is 200 × greater than DNA polymerase I.
DNA methylase (*Eco RI* methylase)	Methylase	Methylation of internal restriction sites on cDNA; protects cDNA from cleavage by *Eco* RI.
Terminal deoxynucleotidyl transferase (terminal transferase; T.T′ase)	$5'\rightarrow3'$ polymerase	Addition of nucleotides to blunt ended (Co^{++}-dependent) or protruding 3′-OH groups (Mg^{++}-dependent).

(*Continued*)

TABLE 9.1 (Continued)

Enzyme	Activities	Role in cDNA synthesis
Thermus aquaticus polymerase (*Taq* polymerase)	Reverse transcriptase 5′→3′ polymerase	Poor reverse transcriptase activity for first-strand synthesis; second-strand synthesis/ amplification by PCR; does not proofread.
Thermus thermophilus polymerase (*Tth* polymerase)	Reverse transcriptase 5′→3′ polymerase	High-temperature first-strand synthesis (Mn^{2+}-dependent); second-strand synthesis/ amplification by PCR (Mg^{2+}); does not proofread.
Other thermostable polymerases	Variable	First- and second-strand synthesis; amplification by PCR; enzymes with 3′→5′ exonuclease activity have proofreading capability.

[a]Some of the enzymes listed here are infrequently utilized now that RT-PCR is commonplace.
[b]Most modern reverse transcriptases have been genetically engineered to be RNase H-deficient. Their respective properties and reaction optima are variable by vendor.

The synthesis of representative cDNA depends upon the disruption of RNA intramolecular folding. Following heat denaturation, a short poly(dT) primer (oligodeoxythymidylic acid), commonly referred to as oligo(dT), is annealed to the poly(A) tail of the RNA as the sample cools. A major improvement in oligo(dT)-priming was the innovative use of an equimolar mixture anchored oligo(dT) primers, having the generic structure $T_{(n)}V$, where V = A, C, or G. For example, the sequences of 13-mer poly(dT)-anchored primers are

5′ TTTTTTTTTTTTA

5′ TTTTTTTTTTTTC

5′ TTTTTTTTTTTTG

The anchor nucleotide at the 3′ end of the primer (V) forces the primer to base-pair to the 5′-most end of the poly(A) tail, that is, the last base just before the beginning of the poly(A) tail. Priming in this manner tends to produce much more representative cDNAs containing greater amounts of the coding portion of each mRNA template molecule; no one really cares about copying the poly(A) tail into a "T-tract." These primers work wonderfully, though the intended template molecules will be $poly(A)^+$.

The antiparallel base pairing that will occur between the primer and the RNA template will support $5' \rightarrow 3'$ synthesis of a molecule complementary to the RNA template, hereafter known as first-strand cDNA.

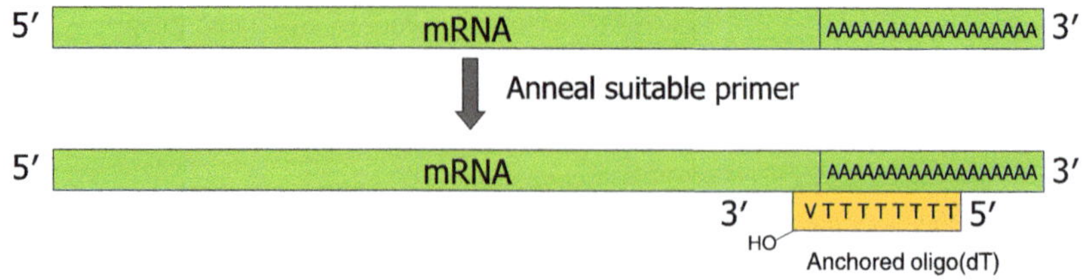

Oligo(dT) primers may also be appended at their 5′ end with overhanging sequences of all types that will support the addition of restriction enzyme sites or other useful sequences for directional cloning, that is, the forced ligation of an insert into a vector in a particular orientation. Directional cloning is necessary when the investigator plans to express (DNA → RNA → protein) a cloned sequence. In an event, primers with additional 5′-end sequences are known as adapters.

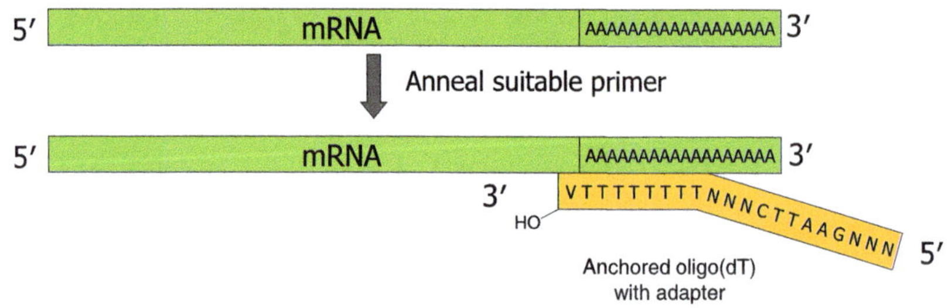

The oligo(dT) primer may also be attached to a magnetic bead, thereby supporting on-bead synthesis of the first-strand cDNA. Either of these priming methods fulfills the template and primer requirements needed to support the synthesis of first-strand cDNA.

It is important to note that, if primed as described, the resulting cDNA is really poly(A)$^+$-generated cDNA, and not total cDNA, because not all mRNAs are polyadenylated. For example, older-style cDNA libraries prepared in this fashion do not contain clones corresponding to poly(A)$^-$ mRNA. Scientists should be aware that such libraries persist and surface periodically when a −80°C freezer is (finally!) cleaned out and reorganized. When screening a library that may have been synthesized some years ago, knowledge of the method of cDNA priming may influence interpretation of the results. To circumvent the potential shortcomings of oligo(dT)-primed cDNA, an aliquot of random hexamers or random nonomers can be used instead to prime the synthesis of the first strand cDNA. Inclusion of these oligomers results in the internal priming of first-strand cDNA instead of priming only from the 3′ end of the template. In general, random-primed cDNA is usually longer and more representative than cDNA primed by oligo(dT) alone. Due to their shorter sequence, first-strand cDNA synthesis using random hexamers is usually performed at 37°C to ensure stable annealing with the template.

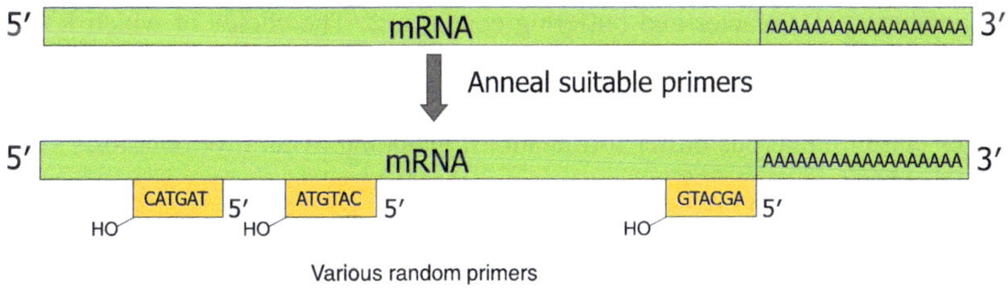

In this laboratory, the most efficient cDNA synthesis reactions occur when a combination of anchored oligo(dT) *and* random primers are used together in the same reaction tube. This approach eliminates the worry about the possible exclusion of poly(A)$^-$ transcripts and tends to produce a population of much larger cDNA molecules, many of which are close to full length. Random primers also work very well with mRNAs that exhibit strong secondary structures near the 5′ terminus. The parameters governing the synthesis of the first strand need not be changed; the only difference is the simultaneous use of one-half of the recommended mass of *each* type of primer. This is now our standard operating procedure and has been especially valuable in the assay of low abundance genes (e.g., Bassett et al., 2000, 2004). If using total RNA, some investigators prefer to use oligo(dT) alone in order to prevent reverse transcription of any non-mRNAs into cDNA; the use of oligo(dT) primers alone, however, without the inclusion of random primers, tends to produce cDNA molecules which significantly underrepresent the 5′ end of the transcript. Conversely, the use of random primers targets all RNAs for reverse transcription, including the redundant rRNA species; methods for rRNA depletion are described in Chapter 24. It is the opinion of this Author that oligo(dT) and random primers together perform in a very satisfactory manner. If the cDNA is intended for PCR, or any other downstream application, and the primers are designed correctly, then using total cellular RNA and double priming is a nonissue.

Reverse transcriptase options

The type of enzyme required for the synthesis of first-strand cDNA is properly known as an "RNA-dependent DNA polymerase," far more commonly known as RT or simply "RT." The natural function of RT is to support retroviral (RNA virus) replication. As with many molecular biology enzymes, RT may manifest as many as three different, measurable activities, namely

RNA-dependent DNA polymerase (DNA synthesis from an RNA template);
RNase H (an RNase specific for RNA:DNA hybrids);

DNA-dependent DNA polymerase (DNA synthesis from a DNA template).

RTs function in a Mg^{++} or Mn^{++}-dependent manner and the relative activity of each of these activities are controlled by the reaction conditions (e.g., salt, temperature, pH, cofactor and buffering conditions). The choice of which RT to use has become a bit more complicated in the past few years and requires some forethought. There are many varieties of RT which are sold under various trade names and in numerous buffer formulations, including avian myeloblastosis virus (AMV); Moloney murine leukemia virus (MMLV); human immunodeficiency virus (HIV) RT; various RNase H^- RT enzymes; r*Tth*, from *Thermus thermophilus* (Mn^{2+}-dependent); and other genetically modified thermostable RTs.

The classical enzymes are the AMV and MMLV RTs. Some of the properties of these enzymes are compared in Table 9.2. Investigators tend to

TABLE 9.2 Properties of wild-type avian myeloblastosis virus and Moloney murine leukemia virus reverse transcriptase.

	AMV RT	MMLV RT
5′ → 3′ polymerase (RNA-dependent DNA polymerase)	Yes	Yes
5′ → 3′ exonuclease	No	No
3′ → 5′ exonuclease (proofreading ability)	No	No
Template switching ability (adds up to three nontemplated bases, usually C, to the 3′ end of the first-strand cDNA)	Yes; template-independent DNA polymerase activity	Yes; template-independent DNA polymerase activity
Typical error rate[a]	≈ 1 per 30,000 nts	≈ 1 per 35,000 nts
Natural RNase H[b]	Yes	Yes
cDNA synthesis	Preferred for shorter templates	Preferred for longer templates
Enzyme activity	More processive than MMLV	Less processive than AMV
Natural temperature optimum	42°C–48°C	37°C–42°C
Typical enhanced temperature optimum[c]	42°C–58°C	37°C–55°C
Typical heat inactivation	85°C for 5 min	80°C for 10 min

[a]*Sebastián-Martín et al. (2018).*
[b]*Genetically engineered enzymes are often RNase H deficient and vary by vendor.*
[c]*Variable by vendor.*

prefer the very processive AMV enzyme for reverse transcribing shorter templates; because it is active at temperatures of 50°C (or more) it, is also the preferred enzyme for RNA templates with more secondary structure. In contrast, MMLV is the favored enzyme when attempting to reverse transcribe longer templates. Unfortunately, some natural preparations AMV and MMLV RT harbor endogenous RNase H activity. The endonucleolytic action of RNase H will nick the RNA strand of an RNA:DNA hybrid, precisely that which forms when a DNA primer anneals to template RNA just prior to reverse transcription. The cleavage of the template and resulting instability of the primer greatly diminish the yield of cDNA: the primer cannot be extended because it is no longer associated with an intact mRNA. When present, RNase H activity is sure to interfere with the production of full-length cDNA since the growing hybrid molecule is also a potential substrate. Natural MMLV RT generally has lower levels of RNase H activity and is therefore preferred for reverse transcribing longer templates, ordinarily between 37°C and 40°C. In view of this intrinsic difficulty, several genetically engineered AMV and MMLV RTs have been developed which are certified as being RNase H^- and/or which possess enhanced thermostable properties and which are in widespread use. These traits are highly desirable and result in a greatly improved first-strand synthesis, particularly in situations where cDNA synthesis is somewhat recalcitrant.

RT is naturally error-prone, given that it lacks the proofreading activity ($3' \rightarrow 5'$ exonuclease) commonly associated with several other enzymes in the molecular biology world. The error rate ranges from approximately 1 in 30,000 nts for AMV RT and 1 in 35,000 nts for MMLV RT, to as much as 1 in 1500 nts for the lesser-known RT from HIV (Arezi and Hogrefe, 2007; Roberts et al., 1998). A newer RT xenopolymerase (RTX) with proofreading capability has been described (Ellefson et al., 2016). The increased fidelity of this polymerase translates into a great increase in the accuracy, throughput, and reliability of data in the area of transcriptomics, including direct RNA sequencing. In addition, it is useful under thermocycling conditions due to its enhanced thermostability. This enzyme has found success in a SARS-Cov-2 detection assay (Bhadra et al., 2021).

Each enzyme has an optimal set of reaction conditions (ionic strength, pH, monovalent cation, divalent cation), all of which are provided by the 10× or 5× first-strand buffer that is specific to the RT. This buffer is diluted to a working concentration of 1× and the synthesis reaction is conducted at a temperature compatible with the specific RT being used. The investigator must further supplement the reaction with the primer(s), a ribonuclease inhibitor to protect the RNA until it is reverse transcribed, and a nucleotide cocktail consisting of an equimolar mixture of dATP, dCTP, dGTP, and dTTP, hereafter referred to as dNTP. First-strand cDNA synthesis requires anywhere from 10 minutes to 1 hour. All of the first-strand reaction cDNA kit components are compatible with the chemistries that are subsequently required for amplification of the resulting cDNAs by PCR or other applications.

If reverse transcription remains something of a challenge in your lab, here are nine ideas that should be considered for possible inclusion or modification of one's current approach.

1. The integrity of the RNA should be checked. Always have evidence that the RNA being used in a cDNA synthesis reaction is intact. An elaborate electrophoretic setup is not necessary; all that is required is a minigel separation followed by staining to show the presence of the rRNAs and the degree of smearing within each lane (Chapter 8). This is especially important if the RNA was in long-term storage.
2. Denaturation of RNA immediately prior to cDNA synthesis should take place in the absence of the reaction buffer because metal-induced RNA hydrolysis is likely to occur in the presence of Mg^{++} or Mn^{++} at elevated temperatures. The enzyme and reaction buffer are added to initiate cDNA synthesis after the heat-denatured sample is quickly cooled on ice for a minute or two. Rapid cooling impedes the re-formation of secondary structures and also gives the primers an opportunity to anneal to the RNA template.
3. Be sure that the RT formulation being used has been certified as being RNase H^-, or at least has reduced RNase H activity, in order to enhance cDNA synthesis. Examples of these enzymes include Superscript III, ArrayScript, and Maxima H Minus Reverse Transcriptase (ThermoFisher Scientific), PrimeScript™ (Takara), ProtoScript II (New England BioLabs), and EpiScript™ (Lucigen). Other vendors offer similar excellent products as well.
4. Include gelatin in the first-strand reaction mix. The function of gelatin, a component in some cDNA synthesis kits, is to stabilize the RT and improve the efficiency of the reaction. When used, gelatin is added to a final concentration of 10 μg/mL.
5. Use an enhanced enzyme that is able to reverse transcribe at temperatures greater than 50°C. A number of remarkable enzymes are now readily available that can synthesize cDNA at temperatures unheard of just a few years ago; for example, EAMV RT (Sigma-Aldrich) functions up to 65°C. Not only do the higher temperatures promote higher fidelity synthesis of cDNA, but they are also very effective at disrupting secondary structures in the RNA, well known for their role in reducing the length of first-strand cDNA.
6. Consider adopting the popular one-tube, one-reaction buffer format, if the cDNA will be used for RT-PCR (Chapter 10). With the innovative enzyme and buffering systems, the one-tube method can greatly streamline and enhance the entire process of making cDNA and then amplifying it by PCR. After sealing, the tube need not be opened again until after cDNA synthesis and amplification have been completed and the sample is ready for analysis. In the case of real-time PCR, the tube need not be re-opened at all.

7. Always prepare a cDNA master mix when multiple cDNA synthesis reactions are to be performed. This is very much the same strategy as in the preparation of a PCR master mix, in which all but one component are pipetted together in a large volume, and then aliquoted into the individual reaction tubes. For large-scale investigations in this laboratory, 1–2 mL of cDNA master mix is prepared (excluding the RT and RNasin) and stored at −20°C. Using this strategy, the investigator has access to identical chemistries for follow-up and repeat applications, thereby minimizing procedural variations and pipetting errors. The RT and RNasin (RNase inhibitor, discussed in Chapter 2) are supplied in a glycerol storage buffer and should never be subjected to repeated freeze/thaw cycles. The other reagents are quite stable at −20°C for extended periods.
8. Consider the template-switching approach to cDNA synthesis (Chenchik et al., 1998; Matz et al., 1999). Succinctly, it is the habit of MMLV RT to add a few extra nucleotides (most often up to three deoxycytidines) to the 3′ end of first-strand cDNA. This occurs in a nontemplate-dependent manner and occurs efficiently only if the RT reaches the 5′ end of the RNA template. Consequently, a template-switching oligonucleotide (rG), that is, an oligo with a guanosine tract at its 3′ end, base-pairs with the newly synthesized first-strand poly(dC) tract, causing the RT to switch templates; the primer with the oligo(G) sequence at its 3′ end is now the template for continued cDNA synthesis activity. Thus, known sequences are thereby incorporated at the 3′ end of first-strand cDNA (Fig. 9.1). This methodology strongly enriches full-length cDNAs for subsequent PCR-based assays. This technology is widely known as SMARTer cDNA synthesis (Takara) and is a patented process. Contact Takara for details.
9. You would be amazed how infrequently micropipettors are actually calibrated. Check yours.

Second-strand considerations

The mechanics of first-strand cDNA synthesis are remarkably similar to the method used at the inception of the technology. In contrast, the greatest advances in the synthesis and cloning cDNA pertain to the strategies and innovations associated with the synthesis of the second-strand cDNA. In all cases, the synthesis of second-strand cDNA is mediated at a minimum by a DNA polymerase. The mechanics and second-strand synthesis reaction conditions are such that the original RNA is separated from the first-strand cDNA and then destroyed. Due to the chemical make-up and natural stability of DNA, the first-strand cDNA is unaffected by the dismantling of the RNA.

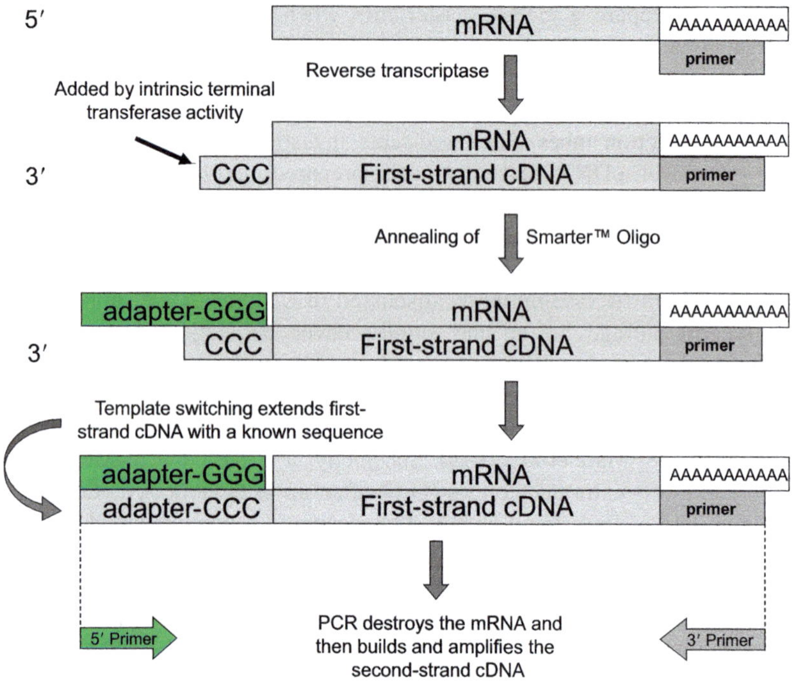

FIGURE 9.1 Template-switching cDNA synthesis using SMARTer (Takara) technology.

PCR-based methods

Most first-strand cDNA synthesis reactions are coupled with PCR for the synthesis of second-strand cDNA. This is true whether the intent is to assay the abundance of specific cDNAs, for transcript mapping purposes, or for the synthesis of a traditional cDNA library. PCR is performed in a variety of formats, often featuring well-designed individual primers, and cleverly designed primer pairs. As the permutations are numerous and far-reaching and the strategies diversified, to say the least, a comprehensive discussion of these approaches is best placed under the heading of "RT-PCR."

Legacy methods

Very early methods involved alkaline hydrolysis of the original mRNA template after synthesis of the first strand, and actually relied on the transient formation of a hairpin at the 3′ end of the first strand, which, acting as primer, supported the synthesis of the second-strand cDNA by using the first-strand as a template. The resulting double-stranded molecule was joined by a single-stranded hairpin, requiring S1 nuclease digestion to remove the hairpin for subsequent vector insertion. Primitive by today's standards, this original approach resulted in the loss of sequences corresponding to the 5′ end of the mRNA, as well as exposing the newly synthesized cDNA to the extremely aggressive and often uncontrollable degradative activity commonly associated with the S1 enzyme. No one does this anymore.

Okayama and Berg (1982) subsequently developed an elegant, novel method by which an oligo(dT)-tailed vector (e.g., a plasmid) was used to prime the synthesis of the first-strand cDNA. The resulting RNA:DNA was circularized by linker ligation. The real innovation in this technique involved the use of a mixture of the enzymes RNase H and DNA polymerase I to remove the mRNA template *enzymatically* after the synthesis of the first-strand cDNA. In this approach, the RNA template, now annealed to first-strand cDNA, is nicked by RNase H and degraded by the intrinsic 5′→3′ exonuclease activity of *Escherichia coli* (*E. coli*) DNA polymerase I while at the same time the 5′→3′ polymerase activity of this same enzyme synthesizes the second-strand cDNA, using the 3′-OH groups from the nicked RNA as primers. Finally, as DNA polymerase is unable to seal the nicks made by RNase H, DNA ligase was included in the enzyme cocktail to repair the backbone of second-strand cDNA to maximize its stability during the subsequent cloning steps. A major improvement to this approach was the procedure of Gubler and Hoffman (1983), in which oligo$(dT)_{12-18}$ rather than a (dT)-tailed vector was used to prime the first cDNA strand, followed by the synthesis of the second cDNA strand using Okayama and Berg's RNase H, DNA polymerase I, DNA ligase cocktail.

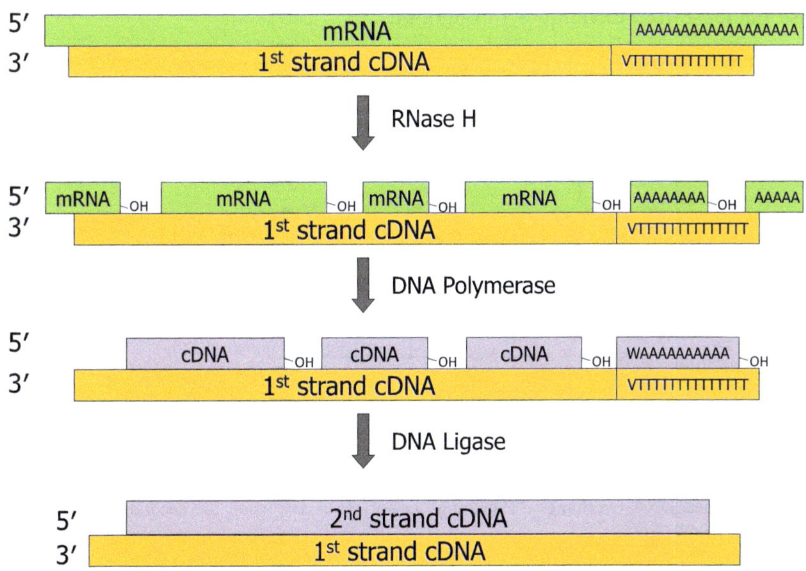

The Gubler and Hoffman method greatly simplified cDNA synthesis by eliminating the vector and concomitant circularization of the cDNA. It remains the standard platform for cDNA synthesis, PCR notwithstanding. cDNA prepared by this method is usually a bit shorter than the original mRNA template due to the unavoidable loss of sequence at each end of the molecule. Moreover, because cDNA synthesis in this manner does not prepare perfectly blunt ends, double-stranded cDNAs are generally incubated for a short time in the presence of T4 DNA polymerase to "polish,"that is, create blunt ends to facilitate ligation into a vector (Fig. 9.2).

Protocol: first-strand cDNA synthesis

This protocol is just one example of how to put together a typical first-strand reaction; as noted above, the use of an optimized first-strand cDNA synthesis kit is strongly preferred. Typical conditions for a first-strand cDNA synthesis reaction with AMV RT are as follows: 10 mM Tris, pH 8.3; 50 mM KCl; 5 mM $MgCl_2$; 1 mM dNTP; 3.0 μg primer [mixture of oligo(dT) and random hexamers]; 20U RNasin; 20U AMV RT; 50–1000 ng RNA; all in a final volume of 20 μL. Conveniently, the required first-strand reaction buffer accompanies the RT, minimizing or altogether eliminating the need to optimize the reaction conditions. Note that the reaction is put together in two parts, primarily to facilitate template mRNA denaturation without compromising the stability of the RNasin or the RT. As always, be sure to use nuclease-free techniques throughout.

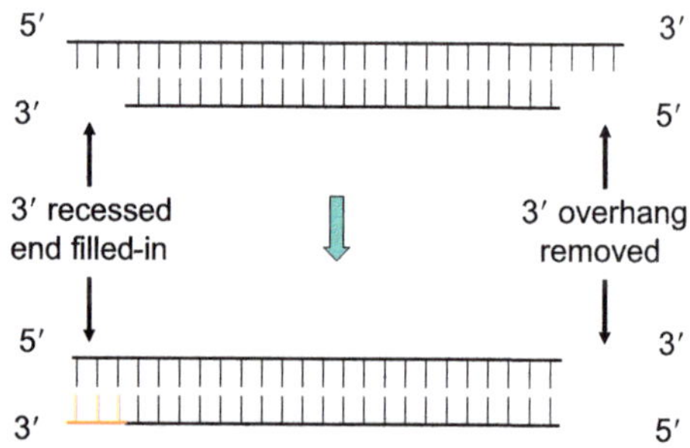

FIGURE 9.2 After synthesis of the second strand, cDNA molecules often require polishing of one or both ends to maximize cloning efficiency. The enzyme T4 polymerase adds missing nucleotides in the case of a 5′ overhang and degrades 3′ overhangs in order to produce double-stranded molecules with two blunt ends. This is also known as "repairing" or "polishing" single-stranded DNA overhangs. "Polishing" is not necessary when the second strand is synthesized by PCR.

1. Prepare master mix 1:

Component	**Per Rx**
Sterile H_2O	5 μL
dNTP (10 mM)	1 μL
Random primers (2.0 μg/μl)	1 μL
Oligo(dT)$_{12-18}$ (1.0 μg/μl)	1 μL
Total	**8 μL**

2. Distribute 8 μL of master mix 1 into each sample tube.3. Add 2 μL (50–1000 ng) of RNA. Ensure that an equal mass has been assessed for each tube.

Note: Do not add RNA directly to the master mix. Individual RNA samples are added after the master mix has been aliquoted.

4. Heat tubes to 70°C for 5 minutes.
5. Immediately place the samples on ice for at least 2 minutes.
6. While the RNA is being heat-denatured (step 4), prepare master mix 2:

Component	**Per Rx**
Sterile H_2O	6 μL
10× first-strand buffer	2 μL
RNasin (20U/μL)	1 μL
AMV RT (20U/μL)	1 μL
Total	**10 μL**

7. Distribute 10 μL of master mix 2 into each sample tube containing the denatured RNA.

8. Incubate at 25°C for 5 minutes.

Note: This short room temperature incubation (1) facilitates primer annealing to the template, and (2) gives the RT an opportunity to extend the random primers and/or oligo(dT) for greater thermostability when the incubation temperature is increased in the following step. This almost always has the effect of producing much longer cDNA molecules.

9. If using a thermostable or enhanced RT, incubate samples per the recommendations of the manufacturer. Otherwise, incubate the sample(s) at 42°C for 60 minutes.

Note: There are numerous RT formulations that support first-strand cDNA synthesis at elevated temperatures. These are widely available from any of a number of vendors.

10. Heat samples to 85°C for 5 minutes to destroy the RT. Failure to do so may be detrimental to subsequent PCR-based procedures.
    Note: The extent to which active RT can inhibit subsequent PCR amplification of the cDNA is a function of the ratio of enzymes and the amount of template material in the reaction tube.
11. The newly synthesized first-strand cDNA is now ready for any of several types of direct analysis or for long-term storage. Depending on the intended application, the cDNA can be cleaned through the use of a commercial cDNA clean-up kit, by phenol-chloroform extraction followed by salt and alcohol precipitation, or by size fractionation by running the reaction over a spun column (e.g., a homemade or purchased Sephadex G-50 column). The details about the many RT-PCR options are presented in the next chapter.

Assessing cDNA synthesis efficiency

To quantitatively determine the efficiency of reverse transcription, and as a basis of normalization among various samples, one may wish to consider spiking the first-strand reaction with an aliquot of [α-^{32}P] dCTP or [α-^{3}H] dCTP; the measured amount of label incorporation is directly proportional to the mass of the newly synthesized cDNA. The mass and size distribution of radiolabeled cDNAs can be assessed by electrophoresis of an aliquot of the first- and second-strand synthesis reactions, coupled with autoradiography directly from the gel. Ideally, the exposure should show a lot of smearing toward the top of the gel: the closer the smearing is to the wells, the better. It is also possible, albeit a bit more cumbersome, to electrophorese an aliquot of cDNA on a gel, stain it with SYBR Green and, using image analysis software, quantify and compare the resultant smears of cDNA distributed among the samples on the gel. Running an alkaline denaturing gel is really a matter of personal preference; one may also elect to examine the products of cDNA synthesis by briefly denaturing a small aliquot with 0.1 volume 0.1 N NaOH followed by electrophoresis through a neutral 1.2% agarose gel.

In this laboratory, first-strand synthesis products are cleaned up using one of the silica-based purification products (e.g., High Pure PCR Cleanup Kit; Roche) after which the concentration is determined by NanoDrop spectrophotometry (Chapter 8). Finally, the size distribution of the cDNA products is examined by running a small aliquot of the sample on a 1% agarose gel and staining it with SYBR Green or SYBR Gold. A typical result is shown in Fig. 9.3. An alternative method for efficiency assessment by real-time PCR is described in the next chapter.

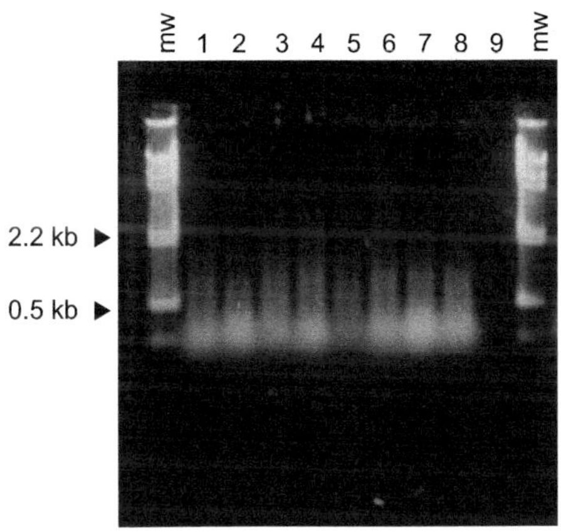

FIGURE 9.3 Assessment of first-strand cDNA synthesis by electrophoresis. The intensity and distribution of products within each lane are reliably indicative of the quality of the cDNA and, by comparison with other samples on the same gel, the relative abundance of the cDNA overall. The first-strand reaction products shown here are of good quality, with many molecules among the smear in each lane in the 1–2 kb range (and larger), which is more than adequate for many contemporary applications. After cleaning up the first-strand reaction, one may quantify the amount of cDNA produced by sacrificing another microliter for NanoDrop spectrophotometric analysis.

Cloning cDNA

A principal reason for cloning cDNA area gene expression, is the physical separation of a mixture of different cDNA sequences, or for the characterization of a novel sequence such as a retroviral genome for which primers or probes have yet to be developed. Cloning cDNA for general purpose sequencing is still performed, albeit much less common due to the availability of contemporary NGS methods. Upon completion of the first- and second-strand cDNA synthesis reactions, the ends of the resultant molecules are sometimes modified to yield cohesive ends suitable for ligation to a vector (e.g., plasmid) with a similar end structure. If the cDNA was primed with adapters containing rare restriction enzyme sites, then cleavage of these sequences creates "sticky ends" that will improve ligation efficiency. Although cDNA library synthesis is somewhat a thing of the past, kits for direct cloning, in vitro expression, and site-specific recombination (Hartley et al., 2000; Ohara and Temple, 2001; Ohara et al., 2002) remain available for purchase.

In the absence of adapter-mediated cDNA synthesis, blunt end ligation, while rather inefficient, can be performed to join linkers to the cDNA or ligate the cDNA directly to a vector.

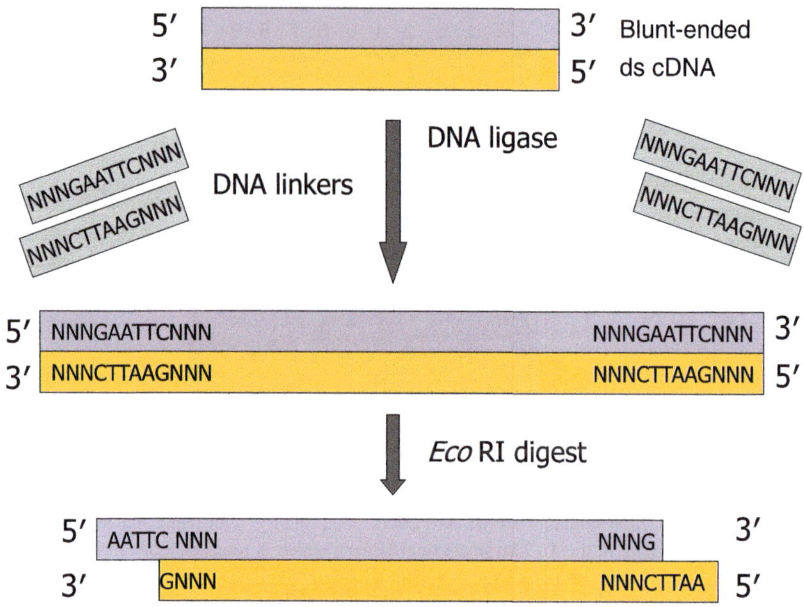

cDNA with *Eco* RI sticky ends

The new restriction enzyme sites added via linker ligation are cleaved with the correct restriction enzyme to yield cohesive ends. When directional cloning is required, particularly if the plan is to sequence or express the cDNA, it is most easily accomplished through the use of adapters with different restriction enzyme sites; these can be purchased as part of a kit or made in the lab. If the restriction enzyme used to generate cohesive ends also cuts an internal recognition site, the cDNA will be diminished in size. To preclude this possibility, one strategy is to methylate the cDNA (e.g., with Eco RI methylase) prior to linker ligation and restriction enzyme digestion. In this example, methylated Eco RI sites in the body of the cDNA would be resistant to Eco RI cutting. Alternatively, given the vast amount of sequence information available in public databases, it should be fairly easy to look up the gene and avoid using restriction enzymes that will cut it. When using PCR, the primers must be designed in advance to provide an end structure compatible with the intended method of cloning. Finally, many of the intrinsic cloning difficulties described here can be avoided altogether by TA cloning or by using the TOPO cloning approaches that are described in the next chapter.

Ligation considerations

Maximum ligation efficiency is favored when the molecules to be ligated, in this case vector and insert, are present in a 1:1 molar ratio. In the synthesis of a library, a near 1:1 ratio of vector to insert will favor completeness in cloning. In the event that the generation of recombinant clones is problematic, ligation optimization is accomplished by calculating, to the extent possible, the number of picomole (pmol) ends involved in the reaction, and then increasing or decreasing the mass of one of the "participants" in other to achieve 1:1 vector:insert ends. One must calculate, or at least estimate, pmol ends for both the insert and the vector, and then equalize them in the ligation reaction.

$$\text{pmol ends} = \frac{\mu g \text{ DNA}}{660 \times \text{number of bases}} \times 2 \times 10^6$$

where

pmol ends = number of molecules ends available to participate in the ligation reaction,
μg DNA = mass of the DNA (either the insert or the vector),
660 = average molecular weight of a nucleotide base pair,
number of bases = the length of the DNA (either insert or vector),
2×10^6 = conversion factor from μg (10^6) to pmol (10^{12}) and the fact that each molecule has two ends capable of ligation.

Thus, 575 ng of a DNA sequence that is 852 bp long contains 2.045pmol ends:

$$\text{pmol ends} = \frac{0.575\ \mu g}{660 \times 852} \times 2 \times 10^6$$

Assuming, for cloning purposes, that this DNA is intended to be an insert, the investigator would also need to determine the mass of the intended vector needed to give an equivalent number of pmol ends, thereby ensuring an equimolar concentration of each in the ligation reaction.

This calculation is very simple and straightforward when an insert of uniform length is intended to be cloned. The situation is quite a bit more complicated when a mixed population of cDNAs is to be cloned because the cDNAs are all of different lengths, and some are much more abundant than others. In this case, the most worthwhile strategy is to vary the insert:vector ratio in a series of ligations, an example of which might be:

1:0.25
1:0.5
1:0.75
1:1
1:1.25
1:1.5
1:1.75
1:2

At the conclusion of the ligation reaction, the now recombinant molecules are used to (a) transform competent *E. coli* cells, most often using a chemical transformation process and ideally with competent cells that have been purchased, or (b) undergo an in vitro packaging reaction to produce infectious virion particles for bacteriophage library construction and screening. cDNA library construction is most efficient when cloning into one of the popular λ bacteriophage vectors rather than into plasmids. Because λ bacteriophage plaques are very small, the library can be plated at a much higher density than a corresponding number of bacterial transformants. False positives are also much less frequent when screening λ libraries, thereby conferring enhanced sensitivity and resolution, compared to colony hybridization. After screening a λ library, the insert can be rescued from putative positives by PCR for subsequent characterization.

If, however, one is simply attempting to clone a single type of molecule, rather than a library, ligation ratios are much less of a concern. After bacterial transformation and plating onto selective media (e.g., 50 μg/mL ampicillin, 50 μg/mL kanamycin, or 12.5 μg/mL tetracycline), if even a single colony appears and is found to possess the desired insert, then the experiment is a success. Since false positives do occur, it is best to have at least a few candidate clones to examine for the presence of the recombinant vector. The take-home lesson: a perfect 1:1 ratio is often not a critical parameter for simple cloning applications.

Enzymes used for ligation

The master enzymes for ligation are DNA ligase and topoisomerase. For most routine cloning experiments T4 DNA ligase is in widespread use. The natural source of this enzyme is T4 (virus)-infected *E. coli*. Like all DNA ligases, this enzyme creates phosphodiester linkages between 5′ phosphate groups on one molecule and 3′ hydroxyl groups on a proximal molecule. The action of DNA ligase is best thought of as "molecular paste" which can either undo the action of "molecular scissors" (restriction enzymes) or can be used to create novel DNA sequences by cleverly planning a series of ligation reactions. DNA ligase will stably join DNA molecules with cohesive (sticky) ends or blunt ends. Another useful activity associated with ligases is the ability to repair nicks in one of the two strands of dsDNA, dsRNA, and DNA:RNA hybrid molecules. Ligation reactions often require incubations ranging from 2 to 24 hours. T4 RNA ligase is likewise available for the ligation of both ssDNA and ssRNA.

TOPO cloning (ThermoFisher Scientific) has become a widely respected alternative to traditional ligase-based approaches for cloning dsDNA. In this system, the insert of interest is combined with a linearized TOPO vector to which the enzyme topoisomerase I has already been joined. These "activated" vectors are therefore ready for immediate use and do not exhibit self-ligation. This method of cloning is rapid and efficient, often rendering >95% cloning efficiency in only a few minutes at room temperature. There are several TOPO vectors that support a variety of cloning formats. Upon ligation, the topoisomerase I enzyme is released from the now recombinant plasmid construction. Bacterial transformation proceeds as usual. This approach to cloning is described in greater detail in the next chapter.

Applications

The ability to synthesize cDNA that is representative of the starting mRNA pool is an important part of molecular biology. The emphasis in cDNA synthesis and cloning is on active gene expression; by its very nature, the synthesis of cDNA creates a permanent snapshot of gene expression at the moment the RNA was harvested from the biological source. Unlike intrinsically labile RNA, cDNA can be archived for years in the freezer and when constructed with appropriately characterized ends, the cDNA can be propagated indefinitely. Further, the creation of cDNA allows the investigator to use the same set of tools that are commonly used to study DNA in other contexts.

When a cDNA molecule is cloned, it can be expressed and the encoded protein studied in vitro, in vivo, or in silico. Alternatively, cloning cDNA into a vector facilitates sequencing. It is worth noting that a reference to cDNA or RNA library construction these days usually means the preparation of cDNA for NGS. cDNA has very important clinical implications, too. Retroviral detection, after all, involves the conversion of a viral RNA chromosome into cDNA, which may then be amplified several million-fold by PCR or other amplification procedures. Finally, appropriately designed cDNA synthesis experiments are quite useful for mapping the transcription start site (TSS) of an RNA molecule, mapping the 3′ end of a transcript, and the discovery of alternative splicing patterns associated with the RNA molecules produced from a single genetic locus. These and other applications are presented in greater detail in the following chapter.

References

Arezi, B., Hogrefe, H.H., 2007. *Escherichia coli* DNA polymerase III epsilon subunit increases Moloney murine leukemia virus reverse transcriptase fidelity and accuracy of RT-PCR procedures. Anal. Biochem. 360, 84–91.

Baltimore, D., 1970. RNA-dependent DNA polymerase in virions of RNA tumour viruses. Nature 226, 1209–1211.

Bassett, C.L., Nickerson, M.L., Cohen, R.A., Rajeevan, M.S., 2000. Alternative transcription initiation and novel post-transcriptional processing of a leucine-rich repeat receptor-like kinase gene that responds to short-day photoperiodic floral induction in morning glory (Ipomoea nil). Plant. Mol. Biol. 43, 43–58.

Bassett, C.L., Nickerson, M.L., Farrell Jr., R.E., Harrison, M., 2004. Multiple transcripts of a leucine-rich repeat receptor kinase from morning glory originate from different TATA boxes in a tissue-specific manner. Mol. Gen. Genomics 271 (6), 752–760.

Bhadra, S., Maranhao, A.C., Paik, I., Ellington, A.D., 2021. A One-enzyme RT-qPCR assay for SARS-CoV-2, and procedures for reagent production. Bio-protocol 11, e3898. Available from: doi.org/10.21769/BioProtoc.3898.

Chenchik, A., Zhu, Y.Y., Diatchenko, L., Li, R., Hill, J., Siebert, P.D., 1998. Generation and use of high-quality cDNA from small amounts of total RNA by SMART PCR. In: Siebert, P.D., Larrick, J. (Eds.), Gene Cloning and Analysis of RT-PCR. BioTechniques Books, Natick, MA, pp. 305–319.

Ellefson, J.W., Gollihar, J., Shroff, R., Shivram, H., Iyer, V.R., Ellington, A.D., 2016. Synthetic evolutionary origin of a proofreading reverse transcriptase. Science 352, 1590–1593.

Gubler, U., Hoffman, B.J., 1983. A simple and very efficient method for generating cDNA libraries. Gene 25, 263–269.

Hartley, J.L., Temple, G.F., Brasch, M.A., 2000. DNA cloning using in vitro site-specific recombination. Genome Res. 10, 1788–1795.

Matz, M., Shagin, D., Bogdanova, E., Britanova, O., Lukyanov, S., Diatchenko, L., et al., 1999. Amplification of cDNA ends based on template-switching effect and step-out PCR. Nucleic Acids Res. 27, 1558–1560.

Ohara, O., Nagase, T., Mitsui, G., Kohga, H., Kikuno, R., Hiraoka, S., et al., 2002. Characterization of size-fractionated cDNA libraries generated by the in vitro recombination-assisted method. DNA Res. 9 (2), 47–57.

Ohara, O., Temple, G., 2001. Directional cDNA library construction assisted by the in vitro recombination reaction. Nucleic Acids Res. 29, e22.

Okayama, H., Berg, P., 1982. High efficiency cloning of full-length cDNA. Mol. Cell. Biol. 2, 161–170.

Roberts, N.A., Craig, J.C., Sheldon, J., 1998. Resistance and cross-resistance with saquinavir and other HIV protease inhibitors: theory and practice. AIDS 12, 453–460.

Sebastián-Martín, A., Barrioluengo, V., Menéndez-Arias, L., 2018. Transcriptional inaccuracy threshold attenuates differences in RNA-dependent DNA synthesis fidelity between retroviral reverse transcriptases. Sci. Rep. 8, 627. Available from: https://doi.org/10.1038/s41598-017-18974-8.

Temin, H.M., Mizutani, S., 1970. RNA-dependent DNA polymerase in virions of Rous sarcoma virus. Nature 226, 1211–1213.

Further reading

Boone, M., De Koker, A., Callewaert, N., 2018. Capturing the 'ome': the expanding molecular toolbox for RNA and DNA library construction. Nucleic Acids Res. 46 (6), 2701–2721.

Sasagawa, Y., Hayashi, T., Nikaido, I., 2019. Strategies for converting RNA to amplifiable cDNA for single-cell RNA sequencing methods. Single Molecule Single Cell Sequencing 1–17.

Chapter 10

RT-PCR: a science and an art form

Rationale

In the mid-1980s, a series of papers were published describing a novel method for in vitro synthesis of DNA. The method is a form of primer extension, which was a fairly well-known mapping and probe-labeling technique at the time. The innovation associated with the method is the repetitive application of a series of heating and cooling steps, resulting in impressive levels of DNA synthesis. This method, now universally known as the polymerase chain reaction or simply PCR, revolutionized the fledgling biotechnology industry. PCR changed everything and has had a positive impact on literally all facets of research in life science, and far beyond. Forensics, basic and applied research, diagnosis of infectious disease, and maternity and paternity testing all owe their current level of accuracy, sensitivity, and sophistication to PCR. It is a mainstream tool in the world of molecular biology and is perhaps the most widely recognized technique.

There is no direct method for the exponential of RNA molecules. Instead, RNA supports the synthesis of complementary DNA (cDNA) sequences using any one of several reverse transcriptases (RTs). In turn, newly synthesized cDNA can be amplified using standard PCR procedures. Because of the role of RT in the synthesis of first-strand cDNA, this approach to studying gene expression is universally known as RT-PCR. Interestingly, this technique was originally referred to as RNA PCR, which was something of a misnomer. Because of the speed and efficiency of PCR, RT-PCR is now the preferred method for the synthesis, expansion, and analysis of cDNA.

The utility of RT-PCR is readily apparent when compared to the traditional methods of RNA analysis, which pale by comparison. The emphasis of RT-PCR, as well as all methods involving cDNA, is on transcriptionally active loci: a transcript must be present in a cell lysate or tissue homogenate to support reverse transcription and amplification by PCR. Genes that are transcriptionally silent, therefore, are not assayable if no template exists in the reaction tube. The unexcelled sensitivity of PCR has made the detection and quantification of low abundance transcripts commonplace.

RNA Methodologies. DOI: https://doi.org/10.1016/B978-0-323-90221-2.00039-4

Without a doubt, very few nonmolecular folks had ever heard of PCR, let alone RT-PCR or even mRNA, prior to the emergence of the COVID-19 pandemic. These terms have become well-known owing to the fact that the detection of viral RNA genomes via qRT-PCR is one of two major approaches for SARS-CoV-2 detection, the other being rapid antigen testing that is directed at virion surface proteins. While requiring more time to perform than rapid antigen testing, RT-PCR is a much more sensitive assay because of its ability to amplify RNA (i.e., cDNA) from as little as a single virus particle to a level at which its presence is easily observable.

PCR—an overview

Unquestionably, the widespread implementation of PCR has fundamentally influenced all aspects of basic research in the life sciences. The popularity of many classical labor-intensive methods of vintage molecular biology (c. 1985) continue to fade in favor of the faster, more discriminating method of what is precisely defined as a "primer-mediated, enzymatic amplification of specific genomic or cDNA sequences" (Saiki et al., 1985; Mullis et al., 1986; Mullis and Faloona, 1987). That minute quantities of template material can be amplified several million-fold in a matter of a few hours unleashed enormous potential in the study of infectious disease, gene mutations, forensics, ontological relationships among members of gene families, and transcriptional activity in cells and tissues, using RNA as a parameter of gene expression. One should not, however, consider abandoning the classical methods entirely, because of several intrinsic PCR shortcomings, namely:

1. In order for the reaction to occur, the investigator must have direct or indirect knowledge of the sequence to be amplified so that appropriate primer sequences can be designed. This has become less of a problem in recent years, however, given the vast amount of sequence information available at GenBank. Further, artificial known sequences can be ligated to the 5′ and 3′ ends of samples and to which the primers are complementary such that everything that lies between the primers, known or unknown, can be amplified.
2. A primer, when base-paired to the template, cannot have a mismatch at its 3′-OH terminus; if this were to occur, the amplification process would not proceed.
3. The template for the reaction (i.e., the target) must be of sufficient purity to support primer annealing and subsequent elongation.
4. The various components of the reaction cocktail must be optimized to yield unambiguous data.
5. The length of a PCR product is generally much shorter than the length of the original template.

6. PCR is a biased amplification procedure. Some sequences amplify with greater efficiency than other sequences.
7. From a relative abundance perspective, PCR tends to distort the true abundance relationships among cDNAs (and therefore the mRNAs they represent), and the magnitude of which is related to the number of cycles performed.

PCR should not be viewed as a cure-all for the difficulties that occasionally frustrate both seasoned and novice molecular biologists. One should not attempt implementation of PCR to the exclusion of other molecular methods, even if one is reasonably proficient with basic isolation, handling, and storage methods for nucleic acids. More than one student of molecular biology has mistakenly believed that the immediate implementation of PCR will solve all problems in the lab. PCR should be viewed as complementary to other molecular approaches, and not independent of them.

PCR is dependent, first and foremost, on the primers included in the reaction. Primers, in this context, are short, single-stranded DNA sequences, also known as oligonucleotides, which have been artificially synthesized to the specifications of the investigator. Their job is to base-pair to the two strands of the template to be amplified and provide the requisite 3′-OH group needed by all DNA polymerases to support the synthesis of new DNA molecules. Among the characteristics of well-designed primers:

1. Primers are short, usually less than 30 bases. A primer that is 20–25 bases long is ideal for routine applications.
2. Primers are single-stranded.
3. Primers manifest a free 3′-OH group to which nucleotides will be added by a DNA polymerase. The 5′ end of a primer, in contrast, may carry all types of exotic modifications without compromising its specificity. This can be helpful when one wishes to add a restriction endonuclease site or a transcription promoter to the PCR product for a downstream application.
4. The 3′ terminal base, and often the penultimate base, must be perfectly base-paired to the template in order to support primer extension. In contrast, the 5′ end of a primer does not need to be base-paired to the template at all, as will be the case when adding novel sequences as described in item 3, above.
5. The primers must show at least some degree of complementarity with the cDNA that the investigator wishes to amplify.
6. Primers are used in pairs. The primers must base-pair to opposite strands of the heat-denatured template in such a way that their respective 3′ ends face each other. Thus, the anticipated PCR product is that which is framed by the 5′ termini of a primer pair. If, however, the template is modified so that the ends have identical sequences, read 5′→3′ on each strand, it is possible to perform single-primer PCR.

It is imperative that the primers be present in the amplification reaction at equimolar concentrations so that each strand of the template is amplified symmetrically, hence the terminology "symmetric PCR" (Saiki et al., 1985). This is the norm for standard reactions. If the primer concentrations are different, an asymmetric reaction (Gyllensten and Erlich, 1988) may result in which one of the target strands is amplified preferentially, thereby producing a potentially large amount of single-stranded product. As a procedural aside, a single-stranded asymmetric product is usually visible upon electrophoresis as a "phantom band" that runs faster and with a size of about one-half that of the expected PCR product. Always check primer concentrations spectrophotometrically before using!

PCR is divided into cycles, with each cycle typically consisting of three components in which the sample experiences substantial temperature changes (Fig. 10.1). The orderly application of these functionally distinct components,

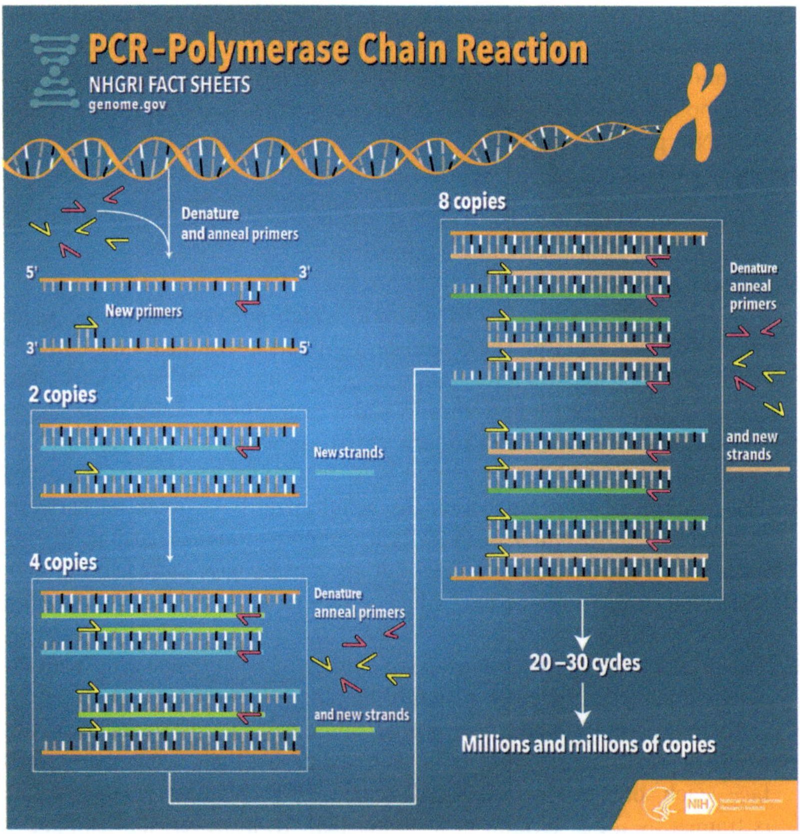

FIGURE 10.1 The polymerase chain reaction results in exponential amplication of specific target sequences. *Courtesy the National Human Genome Research Institute (http://www.genome.gov).*

typically over 25–30 cycles, results in the exponential amplification of template molecules (Table 10.1). The three distinct components of each cycle are:

1. *Denaturation*, the goal of which is high-temperature dissociation of double-stranded template molecules into their constituent single strands. This includes denaturation of any secondary structures that may have formed within or among the primers; the section on primer design (below) describes strategies on how to avoid intramolecular and intermolecular base-pairing of primers. Complete template denaturation in each cycle is required if the amplification is to proceed efficiently.

TABLE 10.1 PCR theoretical exponential amplification.

Cycle number	Relative product mass	Base 2 exponent (2^n)
1	2	2^1
2	4	2^2
3	8	2^3
4	16	2^4
5	32	2^5
6	64	2^6
7	128	2^7
8	256	2^8
9	512	2^9
10	1024	2^{10}
11	2048	2^{11}
12	4096	2^{12}
13	8192	2^{13}
14	16,384	2^{14}
15	32,768	2^{15}
16	65,536	2^{16}
17	131,072	2^{17}
18	262,144	2^{18}
19	524,288	2^{19}
20	1,048,576	2^{20}
21	2,097,152	2^{21}

(Continued)

TABLE 10.1 (Continued)

Cycle number	Relative product mass	Base 2 exponent (2^n)
22	4,194,304	2^{22}
23	8,388,608	2^{23}
24	16,777,216	2^{24}
25	33,554,432	2^{25}
26	67,108,864	2^{26}
27	134,217,728	2^{27}
28	268,435,456	2^{28}
29	536,870,912	2^{29}
30	1,073,741,824	2^{30}
31	2,147,483,648	2^{31}
32	4,294,967,296	2^{32}
33	8,589,934,592	2^{33}
34	17,179,869,184	2^{34}
35	34,359,738,368	2^{35}

Notes: The theoretical accumulation of PCR product for any number of cycles is $2n$, where n = number of cycles completed. In reality, however, the actual amplification of the template is much less, due to the depletion of available primer and dNTPs, as well as the cumulative negative impact of repeated sample denaturation (92°C–95°C) on the polymerase, and other issues. Product accumulation is exponential in the earlier cycles and subsequently demonstrates what is known as a "plateau effect" in later cycles. This means that although product accumulation may continue, it does so in a nonexponential manner and only as long as the polymerase remains active.

a. The precise temperature and time required for denaturation is a direct function of the complexity of the target as well as the G + C content. Thus, the more knowledge one has about the template, the easier any necessary optimization will be later on. Genomic DNA requires a longer denaturation time than does cDNA to ensure complete dissociation of the template; failure to produce PCR product from a genomic template may be the result of incomplete target denaturation from the start, which greatly impedes the efficiency of subsequent cycles. Typically, initial denaturation is conducted at 94°C for as little as 10–30 seconds for short, linear DNA sequences, and as long as 8–10 minutes for higher complexity, GC-rich DNA.

2. *Primer annealing*, the goal of which is the hybridization of the primers to complementary sites on the template. In so doing, the primers provide the 3′-OH group needed to support DNA synthesis in the next step. When a set of primers anneals to the template, the 5′ ends of the primer frame and define the size of the PCR product in cycle 1 (cycle 2 for RT-PCR). For annealing to occur, the temperature inside the reaction tube should not ordinarily exceed the melting temperature (T_m) of either of the oligonucleotide primers. For this reason, the annealing temperature (T_a) is usually at least 1°C–2°C below the lowest calculated T_m of the primers involved, though there are exceptions (e.g., see "touchdown PCR," below). Moreover, primer annealing will not take place unless the region of the target material-containing sequences complementary to the primers has been sufficiently denatured; see the section on optimization procedures for suggestions. Since each target sequence and therefore each primer sequence is different, this is the reason that the annealing temperature is the most variable component in a PCR amplification protocol. Typical annealing temperatures range from 50°C to 68°C, for 10–60 seconds.

3. *Primer extension*, the goal of which is a synthesis of the desired PCR products by sequential addition of nucleotides to the 3′ end of each primer that annealed to a template molecule in the previous step. This is the heart of PCR product synthesis. A thermostable DNA polymerase such as *Taq* polymerase, from the thermophilic microbe *Thermus aquaticus*, supports primer extension at elevated temperatures. These enzymes can withstand the high temperatures associated with repeated denaturation as well as the primer extension component, most often 68°C–72°C, for 10–60 seconds.

In cycles one and two, primer extension occurs beyond the region defined by the 5′ end of the primer on the opposite strand, resulting in the production of the "long" and "intermediate" products of PCR, respectively. By cycle three the reaction is producing the "short" product, that is, the desired product that accumulates exponentially in the reaction tube. The intermediate product is produced throughout the reaction, but only accumulates arithmetically (Table 10.2) and is really a nonissue with respect to the quantitative character of the assay. Nonspecific product, however, is another issue entirely. Further, thermostable polymerases make fewer mistakes at elevated temperatures, though it is important to note that not all thermostable enzymes have a proofreading activity. Because the error rate of the *Taq* polymerase can be significant, most protocols now use either a proofreading polymerase or, ideally, two different thermostable enzymes (an enzyme blend) functioning in a permissive manner, to increase the yield and fidelity of the amplification process.

In summary, PCR consists of the repeated application of the three steps well-defined steps: template denaturation, primer annealing, and primer extension. While the PCR product yield is quite unimpressive for the first several cycles, the later cycles (cycle 20 and beyond) are responsible for generating the bulk of the usable product. Regardless of the PCR variation, the

TABLE 10.2 PCR product inventory.

Relative product distribution by type				
Cycle number (n)	Long product (constant)	Intermediate product ($2n-2$)	Short product (2^n-2n-2)	Total (2^n)
1	2	0	0	2
2	2	2	0	4
3	2	4	2	8
4	2	6	8	16
5	2	8	22	32
6	2	10	52	64
7	2	12	114	128
8	2	14	240	256
9	2	16	494	512
10	2	18	1004	1024
11	2	20	2026	2048
12	2	22	4072	4096
13	2	24	8166	8192
14	2	26	16,356	16,384
15	2	28	32,738	32,768
16	2	30	65,504	65,536
17	2	32	131,038	131,072
18	2	34	262,108	262,144
19	2	36	524,250	524,288
20	2	38	1,048,536	1,048,576
21	2	40	2,097,110	2,097,152
22	2	42	4,194,260	4,194,304
23	2	44	8,388,562	8,388,608
24	2	46	16,777,168	16,777,216
25	2	48	33,554,382	33,554,432
26	2	50	67,108,812	67,108,864
27	2	52	134,217,674	134,217,728

(*Continued*)

TABLE 10.2 (Continued)

Relative product distribution by type				
Cycle number (n)	Long product (constant)	Intermediate product ($2n-2$)	Short product (2^n-2n-2)	Total (2^n)
28	2	54	268,435,400	268,435,456
29	2	56	536,870,854	536,870,912
30	2	58	1,073,741,764	1,073,741,824

Notes: The accumulation of PCR products in a typical symmetric reaction is distributed among three forms, namely, the long product, the intermediate product, and the short product. Examination of the "short product" and "total" columns reveals the rapid, exponential accumulation of product in a reaction tube, the vast majority of which is the short product. The actual amount of PCR product present at the end of the reaction is also a direct function of the amplification efficiency, which is always less than the 100% efficient model depicted here.

fundamental replication of these steps will always apply. Depending on the laboratory needs, productivity in general, available instrumentation, and throughput, examples of format options for performing PCR include 200 μL- or 500μL-microfuge tubes, 96-well plates, 384-well plates or microfluidic lab chips/cards, or specialized 3072-well plates.

RT-PCR—general approach

It is easy to see how PCR amplification is accomplished by starting with cellular or vector DNA as a template. The same principles apply with respect to the amplification of RNA, with an additional step required to first synthesize cDNA from RNA "templates." The process of cDNA synthesis from RNA and then amplifying it by PCR, known universally as RT-PCR, stands for reverse transcriptase PCR. A common point of confusion is incorrectly referencing real-time PCR as RT-PCR. For the sake of absolute clarity, using the verbiage "real-time RT-PCR" is not redundant and is strongly recommended. Alternatively, the symbols qRT-PCR and qPCR are often used as shorthand to refer to PCR procedures in which the detection method is "real-time." Given the complex nature of any assay designed to study gene expression, conditions governing the success or failure of RT-PCR require careful planning and occasional tweaking. Accompanying improvements in cDNA synthesis and amplification efficiency were dramatic improvements in detection sensitivity and methodology. For the sake of clarity, the standard conventions shown in Table 10.3 are widely recognized.

The historical focus on cDNA has centered on the synthesis of full-length or *representative* cDNA molecules, and the extent to which this can be

TABLE 10.3 Comparison of PCR platforms.

Nomenclature	Template	Analysis platform
Standard PCR	Any DNA	End-point electrophoresis
RT-PCR	RNA, then cDNA	End-point electrophoresis
qPCR	Any DNA	Real-time detection
qRT-PCR	RNA, then cDNA	Real-time detection
Direct PCR	Unpurified template directly from the biological source	End-point electrophoresis or real-time detection
Digital PCR	Any RNA or DNA	Multiple replicated, partitioned reactions in microwells or in oil emulsion droplets, coupled with fluorescence detection

Note: For the sake of clarity, the following standard conventions are used throughout this and subsequent chapters.

accomplished is directly related to the quality of the RNA starting material, the method of priming for the first-strand cDNA, and the ability of RT to plow through secondary RNA structures. The ability of RT to disrupt intramolecular base-pairing is limited and it is for this reason that RNA is heat denatured just prior the commencing with cDNA synthesis. Because of the unsurpassed sensitivity and amplification potential of PCR, the quantity of starting RNA is less important in terms of its ability to be PCR-amplified, by comparison with more traditional molecular biology methods. Moreover, because PCR primers frame the domain that will be amplified, this method is a bit more tolerant of partially degraded RNA than traditional cDNA synthesis techniques. It is the amplification of cDNA by PCR that facilitates the detection of rare or low abundance transcripts.

RT-PCR is a two-step process. First, purified RNA is reverse transcribed by a RT (e.g., MMLV, AMV, or SuperScript) via an appropriate method for priming. In this reaction, which is first-strand cDNA synthesis, one may utilize traditional downstream primers, such as oligo(dT)$_{12-18}$, random primers, or gene-specific primers. The net result is the synthesis of stable, single-stranded cDNA molecules capable of acting as a template for amplification by PCR. Second, first-strand cDNA is amplified using some variant of PCR. As with all multistep procedures, the outcome of the experiment is determined by the efficiency of each individual reaction. Along these lines, an important issue is whether the process of RT-PCR is performed in one tube or in two. The original method, which is still in widespread use today, has been to prepare cDNA in one reaction and then remove aliquots for subsequent amplification in a second tube using a completely different reaction chemistry. In contrast, single-tube or one-tube RT-PCR streamlines the assay of specific transcripts and is particularly suited for clinical and diagnostic applications and most often when a "yes" or "no" answer is sought. In the one-tube reaction format, reverse transcription of the template RNA and the amplification of the resulting cDNA occur in a single tube.

There are also several variants on the one-tube RT-PCR theme. In the original one-tube format, the thermostable enzyme *rTth* polymerase was used to create first-strand cDNA in a Mn^{2+}-dependent reaction in a bicine (KCl and $(NH_4)_2SO_4$) buffering environment. Then, chelation of the Mn^{2+} and addition of a Mg^{2+} buffer coaxes the same enzyme to support PCR amplification. While rather innovative at the time of its development, the assay did require opening the tube in order to add reagents after making cDNA but before amplifying it. In the present-day approach, all necessary cDNA synthesis *and* PCR components are premixed from the onset. One common method involves mixing RT and the other reaction components ($MgSO_4$ is often substituted for $MgCl_2$) in a single tube along with *Taq* polymerase in an inactivated state by association with an anti-*Taq* antibody. Reverse transcription proceeds as usual. Then, heating to 95°C destroys the RT and the antibody, thereby liberating the *Taq*, and PCR then proceeds as usual. Obviously, the one-tube reaction format is ideal for any of a number of applications, particularly the detection of pathogens. The key advantage of this latter method is that once the reaction is assembled and the tube closed, no additional components are required nor is there any reason to open the tube until it is time to analyze the PCR products; the ensuing development of real-time PCR eliminated the need to reopen the tube for analysis altogether. Improvements such as these, better instrumentation, the fact that PCR is better understood by its numerous practitioners, and other innovations have drastically reduced the required mass of input RNA, sometimes to subnanogram quantities and thereby opening the door to single-cell PCR.

The first-strand cDNA synthesis reaction produces an RNA:DNA hybrid. In the second reaction, the mechanics of PCR are responsible for the displacement of the original RNA template: at the onset of cycle 1 the heat denaturation step and quasialkaline environment produced by the PCR reaction buffer cause RNA:DNA strand denaturation and RNA alkaline hydrolysis, respectively. This clears the way for sequence-specific upstream primer annealing needed to support the synthesis of the second strand. The primer extension component of the first cycle will, in turn, generate the second-strand cDNA. After cycle 1 is completed, the amplification of both strands of the newly synthesized cDNA will proceed symmetrically (Fig. 10.2). The primary advantages of RT-PCR over traditional methods of cDNA synthesis are speed, amplification potential, and sensitivity.

In the days before PCR, second-strand cDNA synthesis was initiated by the addition of the enzyme DNA polymerase, often without prior

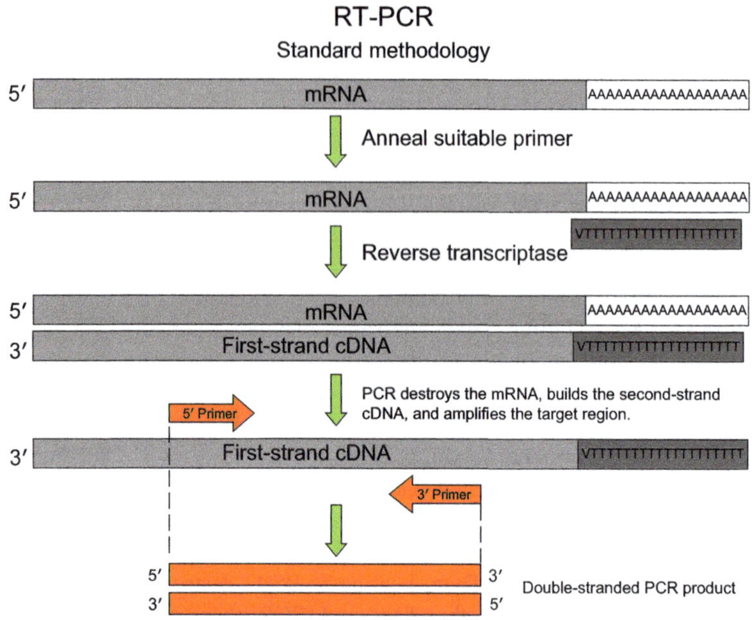

FIGURE 10.2 RT-PCR. First-strand cDNA is synthesized from mRNA by the action of reverse transcriptase. The mRNA template is then destroyed by the mechanics of the ensuing PCR component of this technique. The second-strand cDNA is synthesized in PCR cycle 1, followed by the exponential amplification of double-stranded cDNA beginning with cycle 2. Numerous alternative priming strategies are possible. See text for details.

removal or inactivation of the RT used to make first-strand cDNA. DNA polymerase and RT working together have an enhanced ability to push beyond secondary structures that can compromise the length of the resulting double-stranded cDNA. With respect to PCR, however, failure to heat-inactivate RT may interfere with subsequent applications (Kawasaki, 1990; Sellner et al., 1992).

One might think that the assay of transcripts by RT-PCR should be just as easy as amplification of genomic DNA because cells make multiple copies of the mRNA, even of the low abundance variety. Just the opposite is true, however, because the investigator is at the mercy of the RT. The potential exists for significant variability in RT efficiencies among multiple samples, even when pipetting from the same master mix. Consequently, normalization of samples and the inclusion of appropriate controls becomes all the more important in RT-PCR, and the efficiency of first-strand cDNA synthesis is one of the most important determinants of the success or failure of this method. For this reason, it is strategically better to make a large cDNA pool from which aliquots may be drawn for individual applications rather than repeating the same cDNA synthesis reaction over and over.

Moreover, it makes even more sense from a quantification point of view to actually spike an RNA sample with a known quantity of exogenous RNA and then follow the reverse transcription of that transcript and its subsequent PCR amplification to assess the overall efficiency of the reaction, knowing that the same physical parameters were acting on the experimental RNA as well as the spiked-in template. This process is described in greater detail in Chapter 11, Quantitative PCR Techniques.

PCR carryover prevention

The major advantage of PCR is also the major disadvantage. The amplifying power associated with this process occasionally causes problems when (1) *non*specific products are coamplified, often in competition with the intended template and (2) when products from a previous reaction manage to contaminate subsequent reactions. In the case of the latter, amplicon contamination can occur almost anywhere. An amplicon is an amplifiable unit, meaning a DNA or RNA molecule that can function as a PCR template. It is imperative that well-defined procedures be in place to prevent amplicon contamination, and that unambiguous negative controls be performed along with each set of experimental samples. False-positive PCR data are widely observed as a consequence of uncontrolled amplicon release into the workflow.

Invariably, failure to address the potential for amplicon contamination will, sooner or later, lead to full-blown contamination in the lab. Suddenly, negative controls will show up as positives, the reliability index of quantitative assays will approach zero, clinical samples from healthy individuals will show up as false positives, and any manufacturing process that involves a PCR product will come to a grinding halt. These are just a few of the dire, far-reaching consequences associated with contamination and they all come with a heft cost.

Amplicon contamination is known more commonly as carry-over contamination because the chief source of the contamination is a PCR product that "carried over" from a previous experiment. There are specific procedures, equipment, and laboratory policies that, if followed, should prevent carry-over contamination in the lab. These topics can be broadly classified into four categories: laboratory design and workflow, various procedural methods, the use of aerosol-resistant tips and positive-displacement micropipettors, and the appropriate use of the enzyme uracil-N-glycosylase and other chemical treatments. In most cases, amplicon contamination can be reasonably well controlled through the use of barrier micropipette tips, a laminar-flow biological hood or bench-top system, and careful monitoring of samples through the liberal use of no-template controls. The procedures and policies described below apply to low throughput and high throughput laboratories alike.

Laboratory design

The major objective in laboratory design is to segregate PCR products from all other areas involved in preamplification activities, such as template preparation, preparation of stock solutions and master mixes, and PCR reaction assembly. A typical lab layout is shown in Fig. 10.3. While it is clearly desirable to have adequate laboratory square footage available to maintain separate rooms for various pre- and post-PCR tasks, this is not always possible for smaller laboratories. Instead, to minimize the inherent dangers of amplicon contamination (PCR carryover), it is best to at least define areas within the lab where the template alone will be handled, and then identify areas exclusively for reagent preparation and PCR cycling. Further, investigators should be very cognizant of which tubes are moved into each area, and in which direction. Unidirectional movement is the key to making this amplicon control strategy successful.

In general, four areas should be designated. The movement or flow of materials should always be from the area with the cleanest or most highly purified materials (reagent preparation) to a lesser purity area (template preparation), to the final area containing the least pure or dirtiest materials (the location of the thermal cycler) and then finally an area where product analysis is performed (e.g., electrophoresis, if applicable). Each area should have a dedicated set of micropipettors, lab coats and other instrumentation, including microfuges. Nothing should be moved from one area to the next, except for the sample tubes and aliquots were taken from stock solutions that are to be used for the experiment at hand. Organizing the workflow for unidirectional movement will reduce the incidence of carryover contamination. When laboratory space is at a premium, it is most judicious to run gels or otherwise perform PCR product analysis as far away as possible from the other three areas. Keep in mind also that the additional steps involved in RT-PCR (format-dependent) afford greater opportunities for amplicon

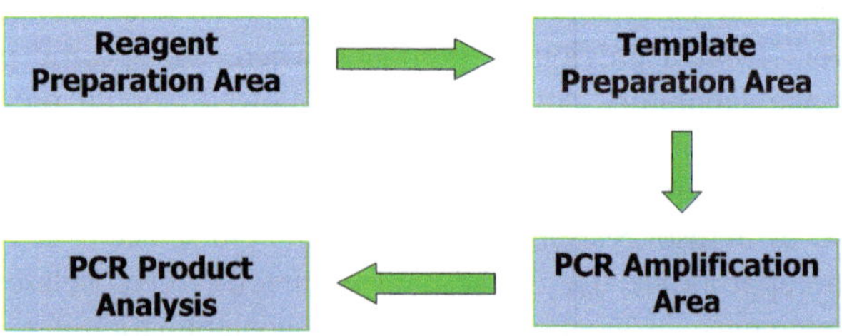

FIGURE 10.3 Preferred arrangement of a work area or laboratory utilizing PCR. Reagents, templates, and other materials move only in the direction indicated by the arrows and are *never* returned to a previous work area. Moreover, a set of micropipettors and laboratory coats should be designated for each station and should not travel between one station and the next. Unidirectional movement of stock solutions and sample tubes is critical for preventing amplicon contamination.

contamination of the stock solutions and downstream reactions, compared to the normal peril associated with non-RT-PCR. In order to address this issue, the investigator is well advised to perform the RT reaction in the sample preparation area, as opposed to the PCR amplification area in the lab.

Procedural methods

It is incumbent upon the head of the laboratory to establish, update, and enforce standard operating procedures associated with the use of PCR. While some of these procedures may be taken for granted by the more senior members of the lab, appropriate procedures and behaviors may not be immediately intuitive to new personnel. Each new lab member should be required to not only go through an orientation to familiarize that person with the proper and safe operation of equipment but should also be given a series of simple experiments to perform in order to assess their competency in performing PCR.

Beyond learning solid laboratory skills, there are other procedures that may be helpful in minimizing carry-over contamination. In some laboratories, for example, the thermal cycler is maintained in a laminar flow hood (the type used for cell culture). At some predetermined interval, perhaps at the end of each day or each week, the block into which the sample tubes are placed is briefly UV-irradiated in order to cross-link any amplicons, thereby eliminating their ability to cause experimental turmoil. Alternatively, a handheld UV device can be mounted over the block for the same purpose. As always, be sure to avoid any eye or skin exposure to UV light.

If preparing master mixes in-house, it is acceptable to UV-irradiate the water and 10 × reaction buffer as another measure to control amplicon carryover. This must be done prior to the addition of the template, primers, or enzyme. While this measure is not foolproof because enzyme, primers, and template are added postirradiation, it may help. One might also consider the use of isopsoralen which, when exposed to long-wave UV light at the end of the amplification protocol, will form covalent adducts involving pyrimidines in the DNA products, thereby preventing them from functioning as a PCR template again (Isaacs et al., 1991; Fahle et al., 1999). Isopsoralen is added to the reaction tube(s) prior to the cycling and has no apparent effect on amplification efficiency. UV irradiation occurs after cycling is complete but before the tube is opened.

Aerosol-resistant tips

Aerosol contamination is a problem of immense magnitude when it occurs in the lab and is probably the most common means by which carryover contamination occurs. Aerosols are formed when a liquid is drawn up into a micropipette tip. The vacuum that is created by the micropipettor can draw amplicons into the bottom of the micropipettor shaft, thereby making the instrument itself

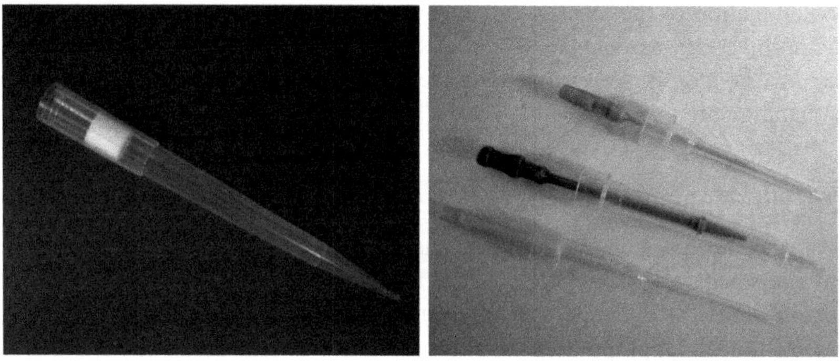

(A) (B)

FIGURE 10.4 Micropipette tips for PCR. (A) A standard barrier tip (Rainin), is designed to prevent aerosols that form within the tip from contaminating the micropipettor. (B) Positive displacement tips (Tri-Continent Scientific). The microsyringe style of these tips precludes aerosol formation altogether. Unless aerosol formation is minimized, amplicon chaos will ensue. Barrier tips are also quite useful for preventing radioactive contamination of micropipettors when working with isotopes as well for tissue culture applications in which an aseptic technique is required.

a vector for contamination. Aerosol-resistant tips, also known in some circles as barrier tips (Fig. 10.4A), contain a hydrophobic polyethylene filter positioned within the tip that not only effectively prevents aerosols from moving up past the filter and into the shaft, but also prevents anything that might already be in the shaft from coming down and contaminating the aliquot in the tip. Barrier tips are supplied sterile and are not to be autoclaved. Alternatively, the use of positive displacement micropipette tips (Fig. 10.4B) may also be quite helpful in this regard, though positive displacement tips are more expensive than barrier tips. Regardless of the means, it is important to block aerosol-borne template which can otherwise easily get into stock solutions and nonintended reactions. Always practice safe pipetting.

Uracil-*N*-glycosylase

The enzyme uracil-*N*-glycosylase (UNG) is quite useful for the inhibition of contaminating carryover template. PCR reactions are assembled with dUTP, a dTTP analog, meaning that the amplified PCR product contains the base uracil. The inclusion of dUTP in the reaction does not influence the specificity of efficiency of product amplification. Prior to the next amplification, samples are incubated with the UNG enzyme which removes uridine wherever it is located. Consequently, the DNA backbone is susceptible to alkaline hydrolysis at the abasic sites created by UNG; the breakage of these molecules from previous reactions prevents reamplification going forward. A fresh, thymine-containing template added to the reaction mix by the investigator is not a substrate for the UNG enzyme and is therefore unaffected by this incubation. The UNG itself is subsequently destroyed by heating.

Primer design

There are many excellent primer design tools available online. Some reside at vendor websites where primer orders are placed, while others are stand-alone resources, for example, Primer3Plus (primer3plus.com). A great deal of information about the process of primer design is included in this section because anyone working in this area should have a clear understanding of how all of the molecules involved are able to interact with each other. While the obvious emphasis here is on RT-PCR, the basic rules and guidelines apply equally to standard reactions as well. Successful PCR requires properly designed primers.

One common source of confusion pertains to the nomenclature used to distinguish the two primers. For the sake of clarity, the nomenclature used to describe and distinguish each of the primers is presented in Table 10.4. For example, the *downstream primer* is that which base-pairs closest to the 3′ end of the mRNA template, thereby supporting the synthesis of first-strand cDNA. This primer is also referred to as the 3′ primer, among other terms. The *upstream primer* base-pairs to newly synthesized first-strand cDNA sequences closer to the 5′ end of the original mRNA (Fig. 10.5). This primer is also referred to as the 5′ primer, among other terms. Most often, primers are designed by direct scrutiny of a published DNA sequence, be it genomic DNA or cDNA. By convention, only one strand of the DNA, the coding strand, written 5′→3′, is published. When designing primers from sequences appearing in the literature, the reader is expected to know that there is indeed another strand base-paired and antiparallel to the published strand. Although the software will do this automatically, these two rules are important to know if one is to

TABLE 10.4 Essential primer nomenclature.

5′ primer	3′ primer
Upstream primer	Downstream primer
Forward primer	Reverse primer
Sense primer	Antisense primer
(+) Primer	(−) Primer
Watson	Crick

Notes: Terms in the left-hand column are used interchangeably to describe the primer associated with the 5′ end of the mRNA. Similarly, all of the terms in the right-hand column are used interchangeably to describe the primer responsible for amplifying sequences found closer to the 3′ end of the mRNA. It is also correct to refer to primers more formally as oligonucleotides or just "oligos."

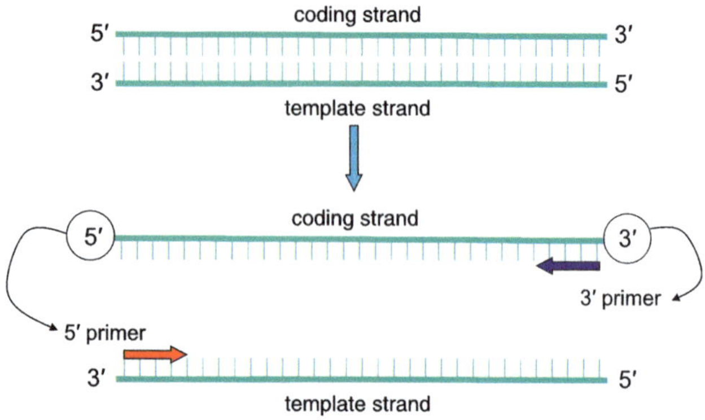

FIGURE 10.5 Essential primer nomenclature. The 5′ primer base-pairs closest to the 5′ end of the coding strand, while the 3′ primer base-pairs closer to the 3′ end of the coding strand.

understand how the primer sequences are actually picked. With respect to published sequences in journals or in databases:

1. The upstream primer will have the same sequence as the published coding strand sequence.
2. The downstream primer sequence will be the inverted complement of the published coding strand sequence.

Following these rules will ensure that the primers have the correct sequence in the correct orientation. It is often very helpful for the novice to "visualize" primer:template interaction by writing out and labeling both strands of the published sequence, ensuring that the primers base-pair in an antiparallel configuration to opposite strands and that their respective 3′-OH groups point toward each other. The 5′ ends of the primers will thereby frame the sequence that will be amplified. Any nucleic acid sequence can be amplified through the use of a properly designed primer pair under suitable reaction conditions. The most common type of primer is that which base-pairs perfectly to the template; these are commonly known as full match primers.

Similar rules apply in the design of primers for RT-PCR: if an mRNA sequence is known, recall that it is the same sequence as the coding strand of the gene from which it was transcribed, except for the inclusion of the base uracil, rather than thymine, in the transcript. Thus, the downstream gene-specific primer for RT-PCR, if oligo(dT) is not selected, should be the inverted complement of a region of the RNA upstream from the beginning of the poly(A) tail. The upstream primer will be the same sequence as the targeted 5′ region of the mRNA, except for the presence of thymine, rather than uracil, in the primer.

Another very important convention is that all primer sequences are always written 5′→3′. This applies equally to the upstream and downstream primers. If a primer sequence is entered incorrectly on an order form, the resulting primer, which won't work, will be entirely the fault of the person who filled out the form because there is an expectation that design conventions will be followed. In terms of publication format, the sequences are published 5′→3′ from left to right across the page, and in blocks of ten bases (nucleotides) making the sequence easier to read.

Practice problem: From the following mRNA sequence design a pair of primers, each 20 bases long would generate a PCR product, after first-strand cDNA synthesis, spanning nucleotides 6–190. Then, determine the expected size of the PCR product.

GAGCCGCACA	CCUGGGAUCC	GAUAAGAACA	CGGAGCGAUU	AUUUUAUCUA
GCUAUGCUUA	GAGAGAUGAC	CGAGUAGCGA	CUAUUUUACC	AGCUUUUUUG
CGCGAUCGAC	CGAUCGACGA	AACAACAACA	CCCGCAACAG	CACGCGAGCG
AGACGAGAUU	CUACUAUGGG	UAAGUAGAAC	CCCAGACGAU	CACUCGCUAA
CGCUAUCGAG	CUAGUAGGAU	GCGUGUGCGC	UCGCUGCUAG	CGCUCGAGGC
UAAGUAGAAC	CCCAGACGAU	AGAAAAAAAA	AAAAAAAAAA	AAAAAAAAAA
AAAAAAAAAA	AAAAAAAAAA	AAAAAAAAAA	AAAAAAAAAA	AAAAAAAAAA
AAAAAAAAAA	AAAAAA			

Answer:

The sequence of the 5′ primer: 5′ GCA CAC CTG GGA TCC GAT AA.
The sequence of the 3′ primer: 5′ ATC GTC TGG GGT TCT ACT TA.
The expected size of the PCR product: 185 bp.

In addition to sequencing data, oligonucleotide primers can be designed based on knowledge of the nucleotide sequence of a related gene in either the same or a different species, or knowledge of the amino acid sequence of the protein encoded by the gene of interest. The problem with the latter approach is the degeneracy of the genetic code: some amino acids have multiple codons. When designing primers based on amino acid sequence information, one strategy is to examine the peptide for clusters of amino acids with only one or two possible codons. The single-codon amino acids are methionine (Met, M) and tryptophan (TRP, W). One must realize, however, that the occurrence of five or six tandem single-codon amino acids is quite rare. The other amino acids all have two or more possible codons (Table 10.5). In the case of multiple codons, the investigator may have knowledge of preferential usage of one codon over another in a particular species, and this information can be acquired from a codon utilization table (e.g., http://www.kazusa.or.jp/codon). Without direct nucleic acid sequence information, the likelihood of designing a definitive oligonucleotide for use as a primer or as a hybridization probe is slim. As a general guiding principle, the sixfold degenerate codons (for arginine, leucine, and serine) should be avoided; if necessary, several 5′ and 3′ primers can be synthesized and then tested in PCR reactions in various combinations.

A degenerate primer is a mixture of primers having nearly identical sequences, with specific permutations at certain locations. Thus, the

TABLE 10.5 Codon utilization table for degenerate primer design.

Fold degeneracy	Amino acid	DNA codons	RNA codons	Standard abbreviation	Symbol
1	Methionine	ATG	AUG	Met	M
1	Tryptophan	TGG	UGG	Trp	W
2	Asparagine	AAT AAC	AAU AAC	Asn	N
2	Aspartic acid	GAT GAC	GAU GAC	Asp	D
2	Cysteine	TGT TGC	UGU UGC	Cys	C
2	Glutamine	CAA CAG	CAA CAG	Gln	Q
2	Glutamic acid	GAA GAG	GAA GAG	Glu	E
2	Histidine	CAT CAC	CAU CAC	His	H
2	Lysine	AAA AAG	AAA AAG	Lys	K
2	Phenylalanine	TTT TTC	UUU UUC	Phe	F
2	Tyrosine	TAT TAC	UAU UAC	Tyr	Y
3	Isoleucine	ATA ATC ATT	AUA AUC AUU	Ile	I
4	Valine	GTA GTC GTG GTT	GUA GUC GUG GUU	Val	V
4	Proline	CCA CCC CCG CCT	CCA CCC CCG CCU	Pro	P
4	Threonine	ACA ACC ACG ACT	ACA ACC ACG ACU	Thr	T

(*Continued*)

TABLE 10.5 (Continued)

Fold degeneracy	Amino acid	DNA codons	RNA codons	Standard abbreviation	Symbol
4	Alanine	GCA GCC GCG GCT	GCA GCC GCG GCU	Ala	A
4	Glycine	GGA GGC GGG GGT	GGA GGC GGG GGU	Gly	G
6	Arginine	AGA AGG CGA CGC CGG CGT	AGA AGG CGA CGC CGG CGU	Arg	R
6	Leucine	CTA CTC CTG CTT TTA TTG	CUA CUC CUG CUU UUA UUG	Leu	L
6	Serine	TCA TCC TCG TCT AGT AGC	UCA UCC UCG UCU AGU AGC	Ser	S
Nonsense codons					
		TAA	UAA (Ochre)	Stop	
		TAG	UAG (Amber)	Stop	
		TGA	UGA (Umber)	Stop	

molecules differ in one or more base positions. Degenerate primers are useful when there are uncertainties pertaining to the expected homology with the template or when attempting to characterize sequence conservation across evolutionary time. The degenerate primer strategy facilitates annealing to and amplification of a variety of related sequences, some of which may vary significantly, depending on the exact cycling profile. For example, the primer

5′ GGN CCG TCR TCR AAW GTC ARG TA

is a 23-mer that exhibits 64-fold degeneracy because of the presence of nucleotide variability at positions 3, 9, 12, 15, and 20. The overall degeneracy of the primer is determined by multiplying together the fold-degeneracy observed at each position (Table 10.6), along with the standard symbols for each nucleotide and for several nucleotide groupings. This means that the

TABLE 10.6 Standard IUB base abbreviations for primer design.

Base	Common symbol(s)	Abbreviation	Fold degeneracy
Adenine	A	A	1
Cytosine	C	C	1
Guanine	G	G	1
Thymine	T	T	1
Uracil	U	U	1
Inosine	I	I	1
Purine	A/G	R (puRine)	2
Pyrimidine	C/T	Y (pYrimidine)	2
Adenine, Cytosine	A/C	M (aMino)	2
Guanine, Thymine	G/T	K (Keto)	2
Adenine, Thymine	A/T	W (Weak bonds)	2
Cytosine, Guanine	C/G	S (Strong bonds)	2
Cytosine, Guanine, Thymine	C/G/T	B (not A; B next letter)	3
Adenine, Guanine, Thymine	A/G/T	D (not C; D next letter)	3
Adenine, Cytosine, Thymine	A/C/T	H (not G; H next letter)	3
Adenine, Cytosine, Guanine	A/C/G	V (not T; V next letter)	3
Any base	A/C/G/T	N (aNy)	4

Note: The table contains the conventions for representing the specific order of bases, and degree of primer degeneracy, for the unambiguous synthesis of specific oligonucleotides needed to support PCR. Note also that inosine is actually a nucleoside that contains the base hypoxanthine. *IUB*, International Union of Biochemistry.

"primer" described here actually consists of 64 different types of molecules, corresponding to each of the 64 possible base permutations. Because certain genes show highly conserved regions across a breadth of organisms, degenerate primers can be helpful for fishing out a specific sequence from a genome. Now for the bad news: as primer degeneracy increases, primer specificity decreases. At least, however, the sequence being sought will not get way! One way to reduce primer degeneracy is to incorporate inosine, which is a nucleoside containing the purine base hypoxanthine, into the primer. Inosine can base-pair with adenine, cytosine, and guanine and, with slightly less efficiency, thymine. As such, inosine is particularly useful for primers designed to span known single nucleotide polymorphisms (SNPs). Obviously, the inclusion of more than one inosine at positions where uncertainty exists in the design of a primer can substantially reduce the fold-degeneracy. Working with degenerate primers can also be useful when an amino acid sequence is known and one needs to work backward to design primers to isolate the corresponding gene.

General guidelines

The following are some of the characteristics of useful primers for PCR-associated applications. Be aware, however, that no matter how much brain power goes into primer design, vindication comes only when the primers actually work!

1. Ideally, primers should be 19–24 bases long in the region that base-pairs to the template. Longer primers may be used, though their thermodynamic behavior becomes less predictable as length increases. With shorter primers, one runs the great risk of promoting nonspecific primer annealing, because as a primer becomes shorter, its discriminatory abilities diminish. To an extent, the minimum primer length is species-specific. A good rule for determining minimum primer length whilst maintaining primer specificity is to solve for the exponent:

$$4^X > Y$$

where

4 = the number of possible nucleotides that make up the primer;
X = the length of the primer;
Y = the genome size of the organism for which the primers are being designed.

For example, the diploid size of the human genome is about 6.6×10^9 bp (equivalent to about 6 pg DNA per cell). Thus, $4^X > 6.6 \times 10^9$. By solving for X, it is determined that, at least theoretically, a 17-base primer (a 17-mer) should be long enough to find only the DNA sequence to which the primer is uniquely targeted and not another unrelated sequence due to the random occurrence of bases. After all, 6.6×10^9 base pairs is a lot of sequence. In practice, a 17-mer remains in what should be thought of as the gray area, where a primer may or may not impart the necessary level of sensitivity. Therefore, a bit longer is better.

2. In a PCR reaction, it is far more important to match primers by T_m than by length or by sequence. T_m differences of more than one or two degrees, especially among shorter primers, are likely to have a profound negative effect on the outcome of the reaction in terms of specificity and yield.
3. Primer mismatches at their 3′ end are fatal. Mismatches (locations where a base in the primer does not hydrogen bond with the template) that involve the last nucleotide at the primer's 3′ end, and often the penultimate nucleotide, will not support primer extension. Polymerases require that the base containing the requisite 3′-OH to be perfectly base-paired to the template to support extension. When primers are designed by trial and error, there is often no way of knowing if the 3′ end is perfectly base-paired to the template. If such ambiguity exists, it may be wise to target a different region of the gene; often moving just a few bases 5′ or 3′ from the original primer target site may be all that is necessary.
4. The 5′ end of a primer can be designed with a considerable number of bases unrelated to the template when the investigator wishes to generate a PCR product with new sequences on one or both ends. Useful sequences that one may wish to add include any combination of RNA transcription promoters, restriction enzyme sites, sequences that allow the PCR product to be recognized and amplified by a completely different primer pair for downstream applications, and even older-style DNA sequencing promoters.
5. Each primer base sequence should be examined to ensure that it does not exhibit the potential for intramolecular hybridization due to the presence of inverted repeats or for any other reason (discussed in "ΔG Considerations," below). The formation of what is best thought of as a primer hairpin occurs when a primer base-pairs to itself, thereby causing a marked reduction in free primer concentration which can drastically reduce product yield. This is more of a problem than is commonly recognized. Because of the extremely rapid hybridization kinetics associated with oligonucleotides, it should be intuitive that primer molecules will base-pair to the spatially closest sequence to which it is complementary. If the 5′ end of a primer is complementary to its own 3′ end, then hybridization is sure to occur, and the magnitude of the problem is dependent on the annealing temperature. Moreover, intramolecular base pairing can promote nonspecific base pairing to the correct template,

though in the wrong location, due to the presence of a few nonhybridized bases at the 5′- and 3′ ends of the primer, possibly resulting in a variant of an "insertion by PCR." Happily, the Internet is rife with free software tools that can be used to analyze primer sequences for potential difficulties. One of the best ways to survey the variety of gratis primer characterization tools is to go to any search engine and type in "primer analyzer" or "primer design." These websites will calculate T_m, molecular weight, propensity for hairpin formation, and the likelihood of interaction between primers, to name a few useful applications. Interestingly, analyzing the same primer sequence at different websites does not always produce identical results. The only real test of primer design is how the primer(s) perform in the thermal cycler.

6. Each primer pair should be examined to ensure that the primers are not complementary to each other near their 3′ ends. When primers base-pair to each other, the synthesis of primer-dimer is favored, detracting greatly from the synthesis of the intended PCR product. The incidence of primer-dimer is greatest when the primers are GC-rich at their 3′ ends. If primer scrutiny suggests that primer-dimer may occur, changing the length of one or both primers by as few as one or two nucleotides will usually rectify this problem.

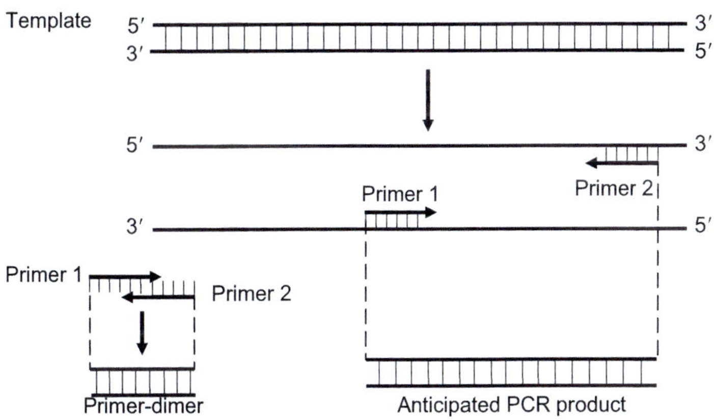

7. Primers should be designed with average GC content. In humans, the GC content of the genome is about 41%. In this laboratory, primers with 40%–55% G + C are considered average. This parameter is important because guanine and cytosine base-pair with three hydrogen bonds between them, while adenine and thymine base-pair with only two. This will preclude one primer from base-pairing to the template more tenaciously than the other primer.

8. A particular primer should be designed with an average distribution of nucleotides, as opposed to having a disproportionate density of guanine and cytosine nucleotides at one end of the primer, and with adenine and thymine at the other. This will preclude having one end of the primer base-pair with and denature from the template with a significantly different efficiency than the other end.

9. Repeated motifs should be avoided. For example 5′ ATGATGATGATG ATGATG may cause specificity problems.
10. There should be no more than two Gs or Cs among the last five bases at the 3′ end of a primer. For example, using the primers 5′ TAGCTAG CCTAATATATCTGCGC and 5′ TCATCCAATATCGATCGTAGCGC would strongly favor primer-dimer formation with themselves and with each other. This would be disastrous.
11. PCR can often be improved by designing a primer(s) with a single G or C at the 3′ end. Some authors argue in favor of the absolute necessity of using PCR primers with GC dinucleotide at or near the 3′ end. The logic behind this argument is that a GC-rich clamp (a string of guanine and cytosine bases at the 3′ end of primers) will increase the efficiency of priming by stabilizing the oligonucleotide on the template. While this approach may be helpful in cases where a primer shows relaxed specificity, it may also cause primer-dimer formation depending on the calculated annealing temperature and the base sequence in the remainder of the oligonucleotide. Most of the established high-precision primers used in this lab do not end with one of the GC dinucleotide permutations.
12. A single gene-specific primer may be all that is necessary to support PCR when one end of the template has been ligated to an adapter sequence or otherwise tailed. For example, the downstream primer imparts template specificity while the upstream primer anneals to the adapter. This approach is characteristic of RACE PCR, described below, and the template-switching strategy described in Chapter 9, cDNA: A Permanent Biochemical Record of the Cell.

T_m considerations

The design of each primer is of fundamental importance to the success of PCR; of equal importance is how the primers behave together in a reaction tube. Designing useful primers necessitates promoting a proper balance between template specificity, thermodynamic stability when base-paired to the template, and the capacity of one primer to function efficiently with the other(s) to support RT-PCR. The probable collaborative behavior of one or more pairs of oligonucleotide primers is best described in terms of the T_m of each primer involved. The T_m is that temperature at which 50% of the possible annealing events between primer and template have occurred and 50% have yet to occur; thus, the T_m is best thought of as an equilibrium temperature at which one primer base-pairs to the template for every primer that melts or dissociates from the template. As noted above, it is important to match primers closely by T_m.

Estimating T_m

T_m is a direct function of the length of the primer, the base sequence of the primer, and the concentration of salt (Na^+) in the reaction. One rapid method for approximating T_m is to assign 2°C for each adenine and thymine, and 4°C for each cytosine and guanine:

$$T_m = 2°C(A + T) + 4°C(G + C)$$

This method is reliable for 11-mer to approximately 28-mer primers. When the T_m is determined to be greater than 68°C, this calculation is no longer reliable. This method can be used to quickly *estimate* the probable T_m for each primer. Conveniently, when primers are synthesized commercially, the specification sheet that accompanies them includes the T_m calculation, most often in 50 mM NaCl. When designing primers, it is best to match them closely based on T_m rather than by length. As suggested previously, primer sequences should be analyzed in silico before they are actually synthesized, in order to predict their probable behavior.

The T_m defines an annealing equilibrium point; it should be fairly obvious though that one does not wish to saturate *only* 50% of all complementary target sites, but rather all of them. Thus, knowledge of the T_m of each primer allows the calculation of the temperature of annealing (T_a), the second step of each PCR cycle. To promote the greatest specificity of primer annealing, one may wish to use an annealing temperature of only a few degrees below the T_m, for example, $T_a = T_m - 2°C$. If annealing 2°C below the T_m fails to generate the product, one should consider repeating the experiment at T_m−4°C, and so forth. Incrementally decreasing the annealing temperature is most often required when one or more mismatches exist between primer and template. While mismatches will be tolerated at a lower temperature, be acutely aware that by lowering the annealing temperature, nonspecific priming is also favored.

Precision T_m *calculations*

The estimation of thermodynamic behavior based on inventorying the number of each base, as described above, is usually a fairly accurate predictor. For greater precision, especially in cases where multiple primers are to be synthesized and used in various combinations, it is best to use primer design software. Biotech vendors that manufacture custom oligonucleotides all have some type of primer design or analysis resource on their website. These tools are remarkably handy not only for basic primer interaction analysis and T_m determination but also for the amount of related information that is calculated instantly. For example, when a primer sequence is an input using an online oligonucleotide analyzer, the software computes optical density (OD), molecular weight (MW), the number of nanomoles per OD (nmol/OD), the number of micrograms per OD (μg/OD), extinction coefficient, GC-content, the sequence of the complementary strand to which the primer is expected to base-pair, and propensity for the primer to participate in primer–dimer formation or the exhibit intramolecular base-pairing. The other nice thing about primer analysis websites is that the value for monovalent- and divalent cation concentrations can be changed. This gives the investigator some flexibility with respect to reaction chemistry adjustments (e.g., Mg^{2+} concentration) as well as the potential to use either or both of the primers as hybridization probes for other applications. This information is convenient to have on hand for performing concentration determinations and when the transition between mass and molar concentrations (reviewed in Appendix B).

One final note about T_m. Recall that degenerate primers actually consist of two or more different primer species, depending on the degree of degeneracy. For example, the 20-mer sequence GCTARATGCTATCTACBATC actually consists of a mixture of the molecules:

GCTAAATGCTATCTACCATC
GCTAAATGCTATCTACGATC
GCTAAATGCTATCTACTATC
GCTAGATGCTATCTACCATC
GCTAGATGCTATCTACGATC
GCTAGATGCTATCTACTATC

because of the twofold degeneracy at nt 5 (*R* = A, G) and the threefold degeneracy at nt 17 (*B* = C, G, T). In instances such as this, online primer design tools will also provide the minimum, maximum, and mean T_m for the primer pool.

ΔG considerations

Primer design typically focuses on primer T_m and ensuring that primer pairs are closely matched, which is very appropriate. Consideration should also be given to ΔG (change in Gibbs free energy) as a means of ensuring complete predictability of primer behavior. The concern is the degree to which a primer is likely to acquire secondary structure through intramolecular base-pairing, rather than being completely available for base-pairing to the intended template. Briefly, $-\Delta G$ favors double-strandedness, while $+\Delta G$ favors single-standedness. As a general rule, one should avoid primers with a large $-\Delta G$ (less than -5). Fortunately, many primer design resources now calculate ΔG. UNAFold (http://www.unafold.org) is one of several software tools that can predict the likelihood of secondary structure formation and is helpful in this regard.

Multiplex primer design

"Multiplexing" is the process of using several primers simultaneously to amplify one or more target sequences. The targets may or may not be related, and the primers need not be degenerate. It is important, however, to maintain fairly tight control over the melting temperature range of the primers involved. One worthwhile strategy is to prepare separate pools of primers, the members of each of which exhibit a similar T_m. Of course, it is also possible to use only one primer, a "best-guess" oligonucleotide sequence (also known as a "guess-mer"), at a lower annealing temperature. While relaxed stringency may favor semi-nonspecific hybridization, the important point is that the desired sequence is amplified, albeit among false positives, which can be eliminated in the later stages of sequence identification. At lower stringency, not all of the bases of a primer will necessarily base-pair with the template. Mismatches, individual bases that do not hydrogen-bond to the template, are expected to abound at lower temperatures, though the primer may continue to anneal enough to support elongation. If mismatches are suspected it may be wise to perform the first 2–3 PCR cycles 5°C–10°C below the desired annealing temperature. This strategy will stabilize the primer and permit primer extension along the template in the presence of multiple mismatches (except at the 3′ end), which is critical for successful PCR. Lastly, remember that if a pool of primers is used, necessitated by target sequence ambiguity, the effective concentration of the correct primer sequence will be much lower compared to a homogeneous primer.

Multiplexing offers numerous advantages, first among which is the ability to coamplify a reference (housekeeping) gene or exogenous templates that may have been added to the sample. Thus, the *exact* same reaction conditions influence the amplification efficiency of all templates. This is particularly convenient when there is very little template material or the sample is irreplaceable. In addition, more data is generated per reaction, which also translates into more data per dollar since fewer reaction tubes are needed.

Optimization procedures

PCR is as commonplace in the molecular biology laboratory as a set of micropipettors. Having a high confidence level in a PCR-based assay comes from reproducibility which, in turn, is highly dependent on optimization of the reaction. Optimization procedures are frequently required whenever a new set of primers is utilized, when reaction conditions change, or when the same primers are used in a different model thermal cycler.

There are numerous potential difficulties, some obvious, others not so obvious, that the investigator must address. Several of these are listed below and are best taken to heart. Succinctly, the efficiency and specificity of PCR are influenced by *each* component of the reaction.

1. The standard method for sample preparation involves the preparation of a large volume of the master mix which contains all but one of the reaction components. To save time, most investigators purchase a premade

PCR master mix. Regardless of the source, the master mix is aliquoted into individual reaction tubes, followed by the one variable component (usually cDNA) that was missing from the master mix. Using a master mix minimizes tube-to-tube variations attributable to subtle errors in pipetting technique. Also be sure to prepare slightly more master mix than the anticipated need. For example, if four samples are to be tested, multiply the per-tube volumes of each component by five. This will prevent any master mix shortage as each sample tube is prepared.

A typical PCR reaction consists of 10 mM Tris-Cl, pH 8.3 (at 20°C); 1.5 mM $MgCl_2$; 50 mM KCl; 200 μM dNTP; 0.5 μM of each oligonucleotide primer; 0.5 U *Taq* polymerase; 10^2–10^5 copies of template per 20 μL reaction. Each of these components is subject to change, though the parameters listed here represent a good place to begin.

2. Pay particular attention to the final magnesium concentration when using a 10× PCR reaction buffer without $MgCl_2$, requiring its addition by the investigator. This divalent cation binds the dNTPs and the template (negatively charged phosphodiester backbone) quantitatively, thereby exerting a measurable effect on the efficiency and specificity of the reaction.
3. Be sure to ascertain template quality. PCR amplifies segments of DNA that lie between the primers. Thus, the reaction is a bit more tolerant of partially degraded DNA and RNA than some of the classical techniques in molecular biology. Templates that are severely degraded are unlikely to support the amplification of anything. Wherever possible, always check template quality by minigel electrophoresis before proceeding to PCR.
4. The final concentration of each primer needed to support PCR is typically 0.1–1.0 μM, beyond which range product yield drops off precipitously. If newly designed primers fail to generate a product when annealing close to the T_m, then reduce the annealing temperature incrementally, and do so at least twice. If the primers continue to fail to produce PCR product, don't waste time with primers that don't work; select alternative primer sequences. Primers can be ordered at any scale of synthesis (Table 10.7) and are usually shipped in lyophilized form. Just prior to first use primers should be reconstituted in sterile H_2O or modified TE buffer (10 mM Tris, pH 8.0; 0.1 mM EDTA). Next, the concentration of each primer should be determined by UV absorption. *Never* assume that the mass of the primer reported on the

TABLE 10.7 Oligonucleotide synthesis scales.

Scale	0.05 μmole	0.2 μmole	1.0 μmole	10 μmole
Yield: O.D.	5 O.D.	12 O.D.	40 O.D.	400 O.D.
Yield: μg	150 μg	360 μg	1200 μg	12,000 μg

Note: Typical results for the synthesis of a 20-mer with standard desalting.

side of the tube or on the product spec sheet is completely accurate. An empirical determination is absolutely critical to ensure a symmetric reaction. Failure to use equimolar quantities of the upstream and downstream primers is a sure way to favor asymmetric (single-stranded) product synthesis.

5. Primer purity is an important parameter of amplification efficiency. It should be intuitive that highly purified primers support high reaction efficiency and specificity. Primer purification means that, at a minimum, residual compounds associated with primer synthesis are removed. It is also important to understand that full-length primer molecules are invariably contaminated with truncated molecules (also known as failure sequences; refer back to Fig. 8.8), the presence of which could negatively impact the outcome of an experiment. There are several levels of purification to which primers can be subjected by the manufacturer. While the cost of some purification methods is included in the purchase price of the primer, the more elaborate purification methods will incur an additional charge. Typical purification options include desalting, cartridge purification, reverse-phase HPLC, PAGE, and gel filtration.

   a. Desalting is usually included in the price of the primer and is generally more than satisfactory for most mainstream PCR applications, up to 35-mers. The purpose of desalting is to remove the residual chemistries and unnecessary organic materials associated with the synthesis of the primer, removal from the column, and primer deprotection. Deprotection is the process by which the chemical groups protecting the reactive base and phosphate components of the nucleotides are removed by hydrolysis with ammonium hydroxide. For primers longer than 35 bases, or when very pure primer preparations are required, more extensive cleanup procedures will be helpful.

   b. Cartridge purification is a rapid method that produces primer purity (80%–90% pure) often approaching that which is possible with HPLC. The method is based on differences in hydrophobicity due to the presence of a 5′ DMT (dimethoxytrityl) group in full-length primers and the absence of this chemical group in nonfull-length molecules, that is, those molecules that were incomplete at the end of the synthesis process. Cartridge purification is useful for primers up to 50 bases in length. The overall efficiency of quantitative assays, cloning, and site-directed mutagenesis reactions can be improved through the use of cartridge purification or HPLC.

   c. Reverse-phase high-performance liquid chromatography (HPLC) often generates primer purity exceeding 95%. The purification principle is similar to cartridge purification. HPLC can be used to purify greater amounts of a primer, meaning more than 1μmol, but is not recommended for primers greater than 50 bases. HPLC-purified primers often require an extra day or two before they are shipped from the manufacturer.

d. Polyacrylamide gel electrophoresis (PAGE) purification is the ultimate purification tool, with primer purity routinely as high as 99%. This is the method of choice for the purification of primers between 50 and 100 bases because stiff polyacrylamide gels offer the ability to separate molecules that differ by as little as a single base. Investigators should expect a reduced yield of PAGE-purified primers due to the required recovery of the primer from the gel. This method of purification also requires an extra day or two prior to shipping.

e. Gel filtration is recommended for oligonucleotides intended for use in vivo, in conjunction with cell culture, or for antisense assays because of the potential for cytotoxicity in the presence of any residual synthesis of purification chemistries. Gel filtration can be used as a stand-alone method of primer purification or in conjunction with the other methods described above. As a stand-alone method, gel filtration will remove the inorganic salts, residual reaction chemistries, and very short truncated sequences, after which the primer can be lyophilized, resuspended in a suitable buffer, and its concentration determined.

6. Test the components of each new PCR kit before using it with the precious template. Many PCR kits contain a test template and control primers, used to show that the reagents are functional. Failure to generate a product can just as easily be due to compromised reagents as to poor primer design or degraded template.

7. Optimize PCR conditions and perform subsequent reactions in the same physical thermal cycler, and not just another instrument with the same model number. Moreover, due to enormous variation in efficiency among different models, reactions thought to be optimized in the older machines suddenly manifest new PCR products in the newer machines. These "new" bands were, in fact, probably being generated in the lesser efficiency machines, but their relatively low mass on the gel put them below the level of detection. If reactions suddenly stop working altogether, check to make sure that someone has not accidentally altered the thermal cycler programming: modification of an optimized cycling profile usually is not a good thing.

8. Each component of the cycling profile (denaturation, annealing, and extension) should be kept as short as possible with the exception of certain very specific applications such as long-range PCR. Short cycles, along with the rapid movement between temperatures (ramping) that is commonplace with later-model thermal cyclers, tend to generate highly reproducible reactions. In cases where the annealing temperature (T_a) is >65°C it may be helpful or even necessary to use a two-temperature cycling profile, rather than the usual three-cycle approach. For example, the annealing and extension components can be combined into a single step that is performed between 68°C and 72°C. In this case, the thermal cycler could be programmed as follows:

Initial melt	94°C	2 min	
Denature	94°C	15 s	30 cycles
Anneal/extend	68°C	15 s	
Soak	4°C	∞	

This is possible because longer probes with higher T_m will associate with the template more rapidly than shorter primers at higher temperatures. Moreover, all thermostable enzymes commonly used for PCR have excellent polymerase activity between 68°C and 72°C.

9. Consider investing in an "optimization kit" which often consists of combinations of chemistries known to enhance the PCR process. For example, such a kit might contain several 10× buffer formulations with varying concentrations of magnesium and separate aliquots of various adjuncts to enhance the robustness of the reaction. Known PCR adjuncts include *E. coli* single-strand binding protein (Chou, 1992), formamide (Sarkar et al., 1990), ammonium sulfate (Olive et al., 1989), bovine serum albumin (BSA) (Paabo et al., 1988), bovine thrombin (Zhang et al., 2014), dimethyl sulfoxide (Cheng et al., 1994), betaine (Rees et al., 1993; Henke et al., 1997) and glycerol (Cheng et al., 1994). The influence of each of these adjuncts is summarized in Table 10.8. One worthwhile strategy is to select the best 10× PCR buffer, based on product yield and absence of nonspecific product, and then to add the various adjuncts one at a time. This will allow the investigator to identify the proper magnesium concentration and adjunct combination needed to optimize the reaction for a particular template and primer set. This will also save time and money whenever a new set of reaction conditions is used.

10. Beware of the following known inhibitors of PCR: humic acid, polysaccharides (glycogen is not a PCR inhibitor below 8 μg/mL; best used below 5 μg/mL), hemoglobin, IgG, lactoferrin, high bacterial concentrations, urea, phenol, chloroform, guanidinium thiocyanate, excess NaCl, sodium dodecyl sulfate (SDS), *N*-laurylsarcosine (sarcosyl), ethanol, formalin, bromophenol blue, and EDTA, to name but a few.

11. Avoid mineral oil as a reaction overlay. Not only is mineral oil messy but its use also correlates nicely with diminished reaction robustness and nonreproducibility of data. If you still do not have a thermal cycler with a heated top in your lab, then make friends with someone who does.

TABLE 10.8 Effects of adjuncts that enhance PCR.

PCR adjunct	Working concentration	Effect
Ammonium sulfate	15–30 mM	Influences denaturation and annealing profiles of the template and primers by increasing overall ionic strength
Betaine (*N*,*N*,*N*-trimethylglycine)	0.8–1.0 M	Reduces secondary structure in GC-rich regions; increases yield and specificity of PCR products
Bovine serum albumin (BSA; nonacetylated)	10–100 μg/mL	Stabilizes *Taq* (and other enzymes, too); particularly useful for ancient DNA or templates containing PCR inhibitors, especially humic acid
Bovine thrombin	10–100 μg/mL	Enhances PCR robustness and greatly diminishes primer-dimer formation
Dimethyl sulfoxide (DMSO)	1.0%–10.0%	Decreases incidence of template depurination; also reduces secondary structure; good for GC-rich templates
Formamide	1.0%–5.0%	Increases stringency of primer annealing, especially in GC- rich template domains
Gelatin	0.01%–0.1%	Stabilizes *Taq* and other enzymes during the amplification cycles
Glycerol	5%–10%	Increases *Taq* stability; lowers template T_m and reduces secondary structure
$MgCl_2$	1.0–2.5 mM	Divalent cation that enhances the activity of the thermostable enzyme; generally, begin at 1.5 mM and adjust up or down as needed
Nonionic detergents (NP-40, Tween 20, Triton X-100)	0.1%–1.0%	Stabilizes *Taq* polymerase; can neutralize carryover SDS from DNA extraction procedures
SSB (various single-stranded DNA binding proteins)	0.7–1.5 μg/assay	Accelerates primer annealing; inhibits extension of mismatched primers; alleviates some reaction inhibitors

Thermostable polymerases

Amplification efficiency and overall sensitivity can be improved by using combinations of thermostable polymerases. These so-called enzyme blends were originally formulated to support long-range PCR, a technology which commonly results in the synthesis of genomic-length PCR products, that is, greater than 4 kb. In this laboratory, the long-range technology has been used to support the efficient amplification of cDNAs that correspond to exquisitely low abundance mRNAs as well as longer mRNAs in the cytoplasm, and much longer unspliced hnRNA. This amplification strategy provides greater sensitivity and, overall, is quite beneficial. A typical enzyme blend, for example, might consist of *Taq* polymerase with another thermostable polymerase, such as *Pwo* polymerase. *Taq* is an efficient (processive) enzyme with respect to its intrinsic $5' \rightarrow 3'$ polymerase activity, though the half-life of this enzyme is only about 40 minutes at 95°C, and it lacks $3' \rightarrow 5'$ exonuclease activity (proofreading). Thus, if *Taq*, or any other nonproofreading enzyme, makes an error, the error becomes a permanent part of the PCR product. Other thermostable enzymes, such as those from the genus *Pyrococcus* exhibit greatly enhanced thermostable qualities and also exhibit a proofreading capability. The combination of *Taq* and one of these other enzymes, used in conjunction with a modified buffering system, is much more efficient and amplifies with higher fidelity than is possible with either enzyme alone. Keep in mind, though, that even high-fidelity polymerases make mistakes, too, with a typical error rate of 5×10^{-7} (variable among different enzymes). In instances when amplification is extended beyond 30 cycles, PCR product can continue to accumulate, because the second enzyme more or less takes over after the *Taq* has been, quite literally, burned out. The properties of some examples of thermostable enzymes are presented in Table 10.9. Various enzyme blends are readily available from many vendors.

TABLE 10.9 Stability of selected polymerases used for PCR.

Enzyme	Organism	Half-life	Proofreads?
Taq	*Thermus aquaticus*	40 min at 95°C	No
Tfl	*Thermus flavus*	40 min at 95°C	No
Tth	*Thermus thermophilus*	20 min at 95°C	No
Tli	*Thermococcus litoralis*	400 min at 100°C	Yes
Pfu	*Pyrococcus furiosus*	12 h at 95°C	Yes
Pwo	*Pyrococcus woesei*	2 h at 100°C	Yes
Deep $Vent_R$	*Pyrococcus GB-D*	23 h at 95°C	Yes
$Vent_R$	*Thermococcus litoralis*	6.7 h at 95°C	Yes
Stoffel Fragment (*Taq*)	*Thermus aquaticus*	80 min at 95°C	No

Positive controls

The number of positive controls should be limited because they can easily contaminate stock solutions and cause unmitigated havoc in various downstream applications. Positive controls should not be the same size as the experimental amplicon. That way, if the positive control template does manage to contaminate stock solutions or other components of the reaction at any level, it will be easily distinguishable as a separate band on a gel or by melting-curve analysis in the case of real-time PCR. Avoid working with highly concentrated positive control template stock solutions and be sure to assemble the positive control reaction last.

Negative controls

Each set of reactions and the corresponding gel should contain at least one negative control, the nature of which is dependent on the experiments being performed. Examples of negative controls include a tube into which no template was added (no template control; tests for amplicon contamination), a tube which contains an RNA template that was not exposed to RT (Fig. 10.6), and two tubes that contain only one of the two primers (tests for self-priming; need to perform this only once with each new primer pair). In this laboratory, negative control reactions are assembled from the master mix after all of the tubes containing the experimental template have been assembled, but before an assembly of the positive control reaction tube.

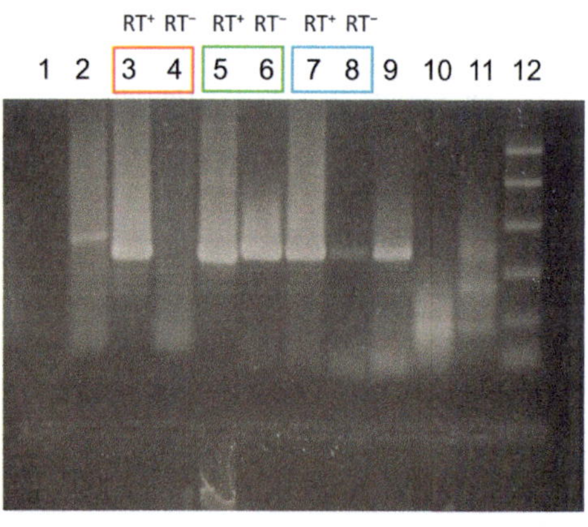

FIGURE 10.6 Negative (RT^-) control reactions for RT-PCR applications. Reverse transcriptase is intentionally omitted from a reaction tube to ensure that there is no contaminating genomic DNA in an RNA sample. PCR product in lane 3 (RT^+) and not in lane 4 (RT^-) shows that cDNA is the only template. PCR product in lanes 5 (RT^+) and 6 (RT^-) strongly suggests that both samples are contaminated with genomic DNA. Strong PCR product in lane 7 (RT^+) and a much weaker band in lane 8 (RT^-) suggest a low level of genomic DNA contamination.

Hot-start PCR

A common problem that results in spurious product formation is associated with the assembly of reactions at room temperature or on ice in a nonhot-start format: during the initial ramping (raising the reaction temperature) to 94°C needed to achieve initial melting and then to initiate the cycling, the temperature of the sample obviously must pass 72°C on the way up to 94°C (Fig. 10.7). As 72°C is the optimal temperature for primer elongation for most thermostable polymerases, nonspecific PCR product can be generated even if the primers are imperfectly base-paired to sites on the template where they would not have base-paired under more stringent conditions. This is especially true of primer-dimer formation. Any nonspecific product made at this stage is more than likely to be coamplified with the intended template throughout the reaction. To preclude this difficulty, hot-start PCR is specifically designed to keep the enzyme inactive until the temperature of the reaction is at least higher than the melting temperature of both primers. In so doing, the incidence of mispriming is drastically reduced if not eliminated altogether, which would otherwise lead to spurious PCR product formation. The application of hot-start technology is an important facet of any RT-PCR-based assay and can be the deciding success factor in assays featuring "difficult templates."

The most prevalent hot-start method involves the use of an anti-*Taq* antibody which inhibits the polymerase activity of the enzyme until the antibody is heat denatured; the development of anti-*Taq* antibodies was a significant milestone, improving the quality of PCR data significantly. Commercial preparations are widely available under various trade names. Alternatively, polymerase activity can be transiently suppressed by associating it with an aptamer, that is, a short nucleic acid molecule with a protein-binding ability. The aptamer mediates *Taq* polymerase activity by transiently binding to it in a temperature-dependent manner (e.g., OneTaq Hot Start DNA Polymerase;

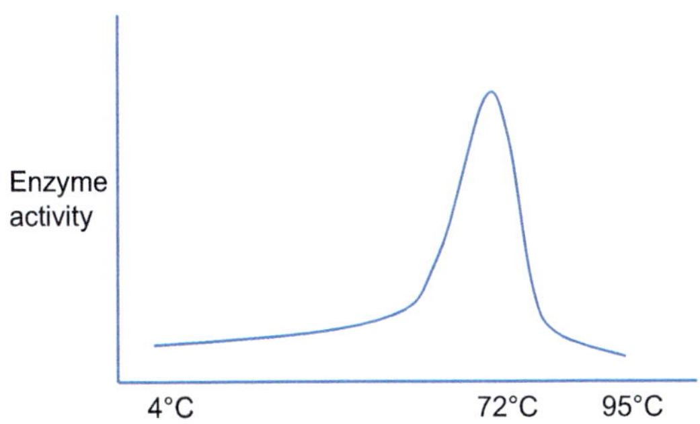

FIGURE 10.7 Spurious product formation can result prior to the initial melt in a nonhot-start PCR format because of the burst of polymerase activity that is observed as the reaction temperature passes 72°C on the way up to 94°C. Hot-start approaches inhibit polymerase activity until after the reaction components have reached denaturation temperature for the first time.

New England BioLabs) and preventing it from synthesizing nonspecific products at lower temperatures. Another hot-start innovation is nucleotides with a 3′ thermolabile protecting group. Known as the CleanAmp dNTP Mix (TriLink BioTechnologies), the initial pre-PCR melt at 95°C that characterizes most cycling profiles removes the protecting group, leaving a 3′-OH needed for sequential addition of nucleotides by primer extension (Fig. 10.8). Older methods including the use of wax beads, Mg^{2+} aliquots, and even withholding a reaction component, have fallen out of favor long ago due to the development of antipolymerase (e.g., *Taq*) antibodies and other innovations. The application of hot-start technology is an important facet of any RT-PCR-based assay and can be the deciding success factor in assays featuring "difficult templates." Be aware, however, that some polymerases show activity prior to heat activation (Stevens et al., 2016). This occurrence was noted by primer-dimer formation, which should have been minimized or prevented altogether by virtue of the nature of the hot-start method. This revelation further underscores the importance of running no-template controls to ensure the reliability of the assay. Primer–dimer formation, or for that matter the formation of any other spurious PCR product, has profound implications on the quantitative character of the assay.

Locked nucleic acid

The term "locked nucleic acid" (LNA; Koshkin et al., 1998) describes a modified RNA nucleotide in which the 2′-O and 4′-C atoms of the ribose are joined through a methylene bridge (Fig. 10.9). This additional bridge limits the flexibility normally associated with the ring, essentially locking the molecular structure into a rigid conformation. These nucleic acid analogs are also referred to in some circles as "inaccessible RNA" or "bridged RNA."

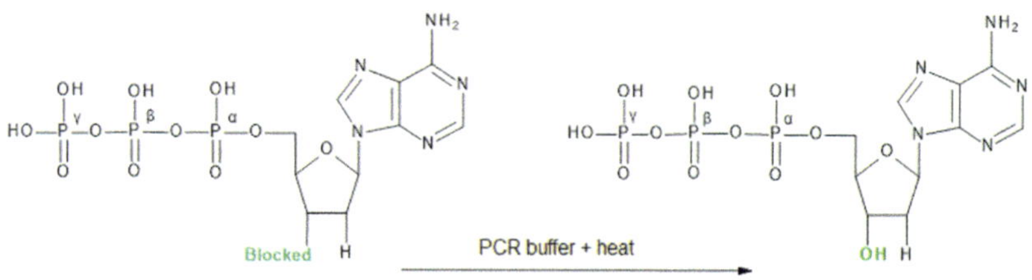

FIGURE 10.8 Hot-start nucleotides. CleanAMP dNTPs (TriLink BioTechnologies; http://www.trilinkbiotech.com) exhibit a thermolabile protecting group associated with the 3′ terminus of the nucleotides (dATP shown) which effectively blocks the sequential addition of nucleotides. The chemistry of the PCR buffer and the elevated temperature during the initial sample denaturation release the protecting group, thereby making the requisite 3′-OH available so that primer extension/amplification may proceed. All four modified dNTPs are available individually or as a hot-start nucleotide cocktail.

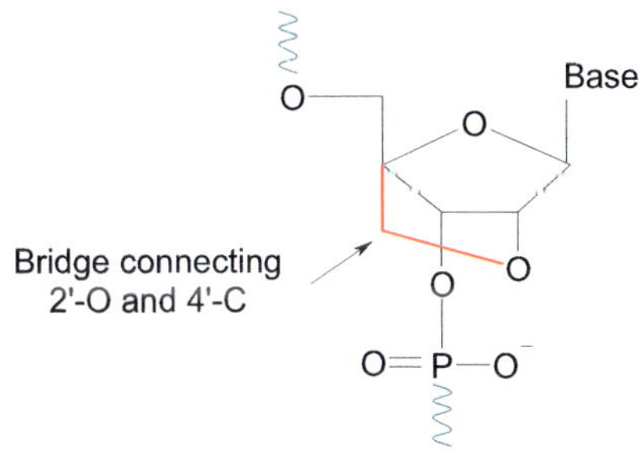

FIGURE 10.9 Locked nucleic acids (LNAs) are modified nucleotides in which the 2′ oxygen and 4′ carbon atoms are joined by an extra bridge. Oligonucleotides that include LNAs show enhanced specificity, sensitivity, and hybridization stability.

TABLE 10.10 Effect of LNA on oligonucleotides.

Oligonucleotide sequence	Number of LNAs	T_m	ΔT_m	ΔT_m per LNA
dG dT dG dA dT dA dT dG dC	0	29°C	–	–
dG + T dG dA + T dA + T dG dC	3	44°C	15°C	+5°C
+G +T +G +A +T +A +T +G +C	9	64°C	35°C	+3.9°C

Note: Data courtesy of Proligo (now part of MilliporeSigma).

LNAs can be inserted into both RNA and DNA oligonucleotides, and their presence increases the T_m by 2 to 4°C per LNA (Table 10.10). The inflexible nature of these molecules greatly enhances hybridization stability. Further, oligonucleotides containing LNAs offer tremendous discriminatory power, allowing short probes to distinguish between exact match and mismatched complementary target sequences with very little difficulty. LNAs are also widely respected for their enhanced biostability in blood and other biological fluids. LNA oligonucleotides are well-suited for antisense experiments, many real-time PCR procedures, ribozyme design, in situ hybridization, SNP (single nucleotide polymorphism) identification, microarray screening, and aptamer functionality, to name but a few possible applications. One of the nice things about LNA oligonucleotides is that the same platforms used for DNA synthesis can be used for LNA synthesis, meaning that labs with (or access to) a DNA synthesizer may begin making LNA oligonucleotides immediately.

When an oligonucleotide contains standard nucleotides and LNA nucleotides in the same molecule, the term "mixmer" is sometimes invoked to describe their nature more aptly. When describing the sequence of a mixmer, one of the conventions is to use lower case letters for standard nucleotides, and uppercase letters for LNA nucleotides, while another convention is to place a (+) symbol in front of uppercase letters to represent an LNA nucleotide. For example, the primer

gctGcacgTcatcgatcATCtcatgc

a 26-mer with LNAs in positions 4, 9, 18, 19, and 20, can also be written as

GCT + GCACG + TCATCGATC + A + T + CTCATGC

When designing LNA-containing oligonucleotides, one should follow the basic rules of primer design described above, paying particular attention to the location and number of LNAs. For example, a typical 18-mer should contain a maximum of 7–8 LNAs. Also, try to avoid stretches of more than 4–5 consecutive LNAs, which would result in very tight hybridization in that region. Stretches of LNAs are to be avoided close to the 3′ end of an oligonucleotide. Finally, be sure to match the T_m of the primers, as usual, keeping in mind that each substitution of a standard nucleotide with an LNA increases the T_m.

Touchdown PCR

Problems with RT-PCR specificity? Make sure that a hot-start formulation is being used. Also, using a touchdown PCR cycling profile can drastically improve the overall specificity of the reaction. Succinctly, touchdown PCR involves programming the thermal cycler such that the initial annealing temperature is higher than the T_m of any primer in the reaction tube. With each passing cycle, the annealing temperature is reduced by a predetermined increment, usually 0.5°C, until the annealing temperature is reduced to about 5°C less than the lowest T_m among the primers in the tube. As a result, virtually no nonspecific product is generated in the first few cycles, and when the temperature is low enough so that primer annealing does occur, the primers will only interact with exact match sequences. By the time the reaction reaches its lowest annealing temperature, the expected product is in large excess over a nonspecific product that may have formed in the last few cycles when the annealing temperature was below the T_m. This approach is renowned for imparting great specificity to reactions involving "difficult" primers (Fig. 10.10) and does a remarkable job of eliminating spurious products. Many of the newer thermal cyclers contain a generic touchdown template, and the user enters the starting and ending annealing temperatures which, of course, will be unique to the primer set.

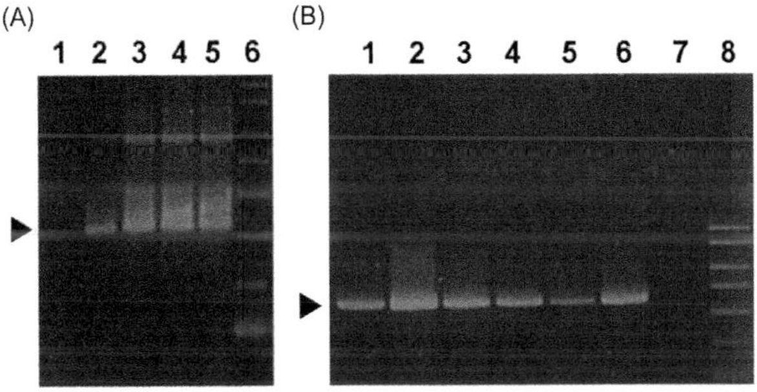

FIGURE 10.10 Touchdown PCR. Primers will occasionally interact with template in a nonspecific manner, even at temperatures close to the T_m. By gradually reducing the annealing temperature from starting temperature *above* the T_m, only perfectly matched primer:template duplexes will be tolerated and amplified. (A) Five first-strand cDNA samples (lanes 1–5) were subjected to PCR with primer annealing at 57°C. Ideally, these primers should have produced a single, 365 bp PCR product, indicated by the arrow; another nonspecific product is clearly visible. Lane 6 is the PCR molecular weight standard. (B) Six first-strand cDNA samples (lanes 1–6) to which a touch-down PCR profile was applied. The annealing temperature was reduced from 65°C to 55°C at a rate of 0.5°C per cycle. At 55°C, 15 additional cycles were performed. A single product is observed at the predicted location in lanes 1–6. Lane 7 is empty and lane 8 contains the PCR molecular weight marker. The PCR products in panels (A) and (B) are identical in size, but were not electrophoresed on the same gel.

Internal controls

Inconsistencies in technique and day-to-day variation among investigators are often responsible for the introduction of error at nearly every level of scientific inquiry, beginning with one's own laboratory skills all the way through to data analysis. To address at least one aspect of this potential problem in transcription assays, a commonly used approach is the measurement of an internal control or endogenous standard. These are often referred to as reference (formerly housekeeping) genes, the expression of which is associated with basic (basal) cellular metabolism. Reference genes, characteristically, are expressed constitutively, meaning that transcripts are produced continuously from those gene loci. The general idea is to identify at least one gene, the expression pattern of which does not change, or at least shows minimal modulation, as a function of experimental manipulation. This demonstrates that the behavior of one or a few genes changes rather than, for example, a complete shutdown in transcription.

The transcripts from the reference gene allow normalization (meaningful comparison) among two or more samples, thereby facilitating calibration of gene expression. The reference gene(s) should be assayed side-by-side with the sample series. Regardless of the method used (qRT-PCR, nuclease protection, blot analysis), an assay of at least one reference gene is essential. In addition, some investigators choose to spike an RNA sample with a known quantity of an exogenous reference template (discussed below) for precision measurement of both reverse transcription efficiency and the overall mass of PCR product at the end of the reaction. Be aware, however, that while the spiking approach has merit with respect to accurate quantification of the expression of specific transcripts, it does not help in determining the overall transcriptional health of the cell.

Investigators should be aware that there is no all-purpose internal or "reference" transcript applicable to every system and every situation. This is a fact of life. Many of the commonly assayed, constitutively expressed housekeeping genes are, alas, subject to variation by the cell. This is especially true when working with cells or tissue that exhibit a disease phenotype. It is possible to identify a reference gene or two for each model system without too much difficulty, and it is wise to assay at least two different genes so that ensuing conclusions are accurate and convincing. PCR methods have demonstrated that genes once thought to be relatively invariant can change dramatically in terms of transcript abundance in response to the experimental challenge, requiring that the suitability of each be determined empirically. Finally, it is usually a good idea to pick a reference gene with a different physiological function than the genes being assayed for variation. For example, if cells are treated with a compound that inhibits microfilament formation, one should pick as an internal control a gene that is unrelated to the cytoskeleton; in this case tubulin would not be the best choice.

Below are some of the commonly reported reference genes. While this list is by no means intended to be exhaustive, it does highlight those genes whose transcripts are most frequently reported as controls for the purpose of reliability of results. The appropriateness of a particular transcript for use as an internal control transcript is entirely dependent on the biochemistry of the system under investigation. While there is no gene that will remain invariant (constant level of expression) under all conditions, genes that show the least variation under a variety of experimental conditions are those best suited as internal control genes.

1. β-actin, a constitutively expressed transcript of medium abundance: its isoforms are found in nearly all cell types. Historically the first transcript to be used as a standard, the use of β-actin as a control transcript remains commonplace. Numerous reports, however, have demonstrated repeatedly the variation of this gene in response to a variety of external stimuli and cytoplasmic cues (e.g., Spanakis, 1993).
2. GAPDH (glyceraldehyde-3-phosphate dehydrogenase), manifested as a transcript of medium abundance encodes a key enzyme required for glycolysis. The expression of the GAPDH transcript as a reference gene or control is in widespread use, including in this lab. As is the case with β-actin, GAPDH gene expression is also subject to modulation under a variety of experimental conditions (e.g., Bhatia et al., 1994; Lau et al., 2000).

3. Cyclophilin, a protein involved in the isomerization of peptide bonds, is a highly conserved gene, the RNA from which is of medium abundance in the cytoplasm. Cyclophilin A is also known as peptidylprolyl isomerase A (PPIA). While commonly assayed as a control transcript, variation in cyclophilin mRNA has been demonstrated (e.g., Feroze-Merzoug, et al., 2002).
4. The histones are a family of proteins needed for nucleosome formation and regulation of transcription. The transcripts are abundant and are often assayed as reference genes (Farrell and Greene, 1993; Robert, et al., 2002).
5. Hypoxanthine–guanosine phosphoribosyltransferase (HPRT, *also* HGPRT), an enzyme well known for its role in the salvage pathway, whereby nucleotides are recycled for use by the cell. HPRT has been shown to be constitutively expressed at rather low levels (Pernas-Alonso et al., 1999). However, HPRT levels are known to fluctuate and are typically elevated in proliferating cells (Steen et al., 1990).
6. Dihydrofolate reductase (DHFR), an enzyme necessary for the synthesis of purines, thymidylate, and glycine. As such, the expression of this gene is absolutely necessary for cell growth. DHFR has been cited widely as a reference gene (e.g., Horikoshi et al., 1992; Balajee et al., 1999).
7. 26S proteasome, consisting of several subunits, is responsible for proteolysis in eukaryotic cells (ubiquitin-proteasome pathway). Primers and probes targeting specific subunits are often used as controls in various transcription assays (e.g., Bassett et al., 2005).
8. Other genes, including fibronectin (abundant transcript; extracellular matrix), transferrin receptor (medium-low abundance; iron uptake) ubiquitin (a small protein that targets other proteins for degradation via proteasomes), lamin (nuclear intermediate filament proteins), tubulin (microtubule formation), ribosomal protein 49 (RP49; Teng et al., 2012), and porphobilibogen deaminase (PBGD; cytosolic enzyme in heme biosynthesis), and various transcription factor and translation factor mRNAs are widely reported as control transcripts. The protein product of each of these genes has a unique role in the cell. In general, these transcripts are less likely to experience disturbing fluctuations in mRNA levels, although there are always exceptions to this generalization. Test these first!

The word on transcription controls

A common source of confusion is the relationship between transcription controls and loading controls in the discernment of the overall transcriptional response of the cell to the experimental challenge. mRNA and its unspliced precursor hnRNA are the products of the RNA polymerase II activity in eukaryotic cells. Thus, when studying changes in mRNA abundance, it makes the greatest sense to use a transcription product of RNA polymerase II as an internal control, particularly since the eukaryotic RNA polymerase shows differing sensitivities to various compounds, of which α-amanitin (Chapter 1: Ribonucleic Acid and the Cellular Biochemistry Revisited, and Chapter 19: Analysis of Nuclear RNA) is a notorious example. The products of RNA polymerases I and III, rRNA and tRNA, which are described below, are better suited as loading controls, meaning that they are used to show that an equivalent mass of RNA is present in each sample. A well-designed set of experiments should consist of transcription controls (mRNA) and loading controls (rRNA or tRNA).

1. 28S and 18S rRNA, highly abundant posttranscriptionally-regulated products of the ribosomal genes, are often cited as nearly invariant transcripts of great predictability as an internal control (de Leeuw et al., 1989; Bonini and Hofmann, 1991). These transcripts are an essential component of mature ribosomes. It is not difficult to imagine a situation where an experimental manipulation causes a large overall reduction in mRNA synthesis all the while rRNA and tRNA are being produced, or at least sustained in the cytoplasm, at near control levels. The reverse is also possible.
2. 16S mitochondrial rRNA, moderately abundant transcripts that are localized in the mitochondria, is utilized as internal control with lesser frequency than its cytoplasmic counterparts. The expression of these mitochondrial genes, purportedly, is influenced far less by experimental manipulation than genes localized in the nucleus (Tepper et al., 1992). However, the fact that these are ribosomal genes, rather than mRNA, should incite the same concerns expressed in item 1, above.
3. tRNA consists of a number of high abundance transcripts which shuttle amino acids to the ribosomes during translation. The total mass of tRNA is occasionally measured, using a tRNA-specific probe, as a loading control. tRNA is transcribed by RNA polymerase III and is, therefore, better suited as a loading control than as a gene expression control, for the same reasons cited in item 1.

Analysis of PCR products

There are two common methods by which PCR products are analyzed. The first method, end-point analysis, is performed using the time-honored method of electrophoresis at the conclusion of the amplification. The other method, known as real-time PCR, allows the investigator to collect data pertaining to PCR product accumulation at the end of each cycle. Inasmuch as real-time PCR is described in detail in Chapter 11, Quantitative PCR Techniques, the following comments pertain to routine end-point analysis.

At the conclusion of a typical reaction, usually consisting of 25–30 cycles, the PCR tube is opened and an aliquot of each sample is run out on a gel along with an appropriate MW standard. In most settings, agarose gels

(1%–3%) represent the matrix of choice for the evaluation of PCR products and are able to resolve almost any size PCR product. These gels may be prepared with or without ethidium bromide added to the gel solution after it has been melted and cooled to about 55°C. Alternatively, gels may be stained with SYBR Green or SYBR Gold *after* electrophoresis, for superior sensitivity with minimal background fluorescence.

A variety of agaroses are widely available. One may elect to use standard- or low melting temperature agarose, or an agarose grade that favors the separation of low molecular PCR products. In this lab, SeaKem GTG (Lonza) is routinely prepared as a 2.5%–3.0% solution in 1 × TAE. This formulation favors the exceptional resolution of PCR products ranging from 200 to 1000 bp. Be sure to use caution when melting high agarose concentrations such as these in a microwave, as the tendency for the agarose to boil over is quite high, both in the microwave and immediately after heating. To reduce the boil-over potential, let agarose sit in TAE buffer for about 5–10 minutes before heating in order to allow the agarose beads to swell. Then heat the agarose to boiling while continuing to exercise caution. Great care must be taken when preparing these high-concentration gels to prevent injury to the investigator and to avoid burning large clumps of agarose in the flask or beaker, especially when using a hot plate.

The intensity of a band on a gel is associated with its abundance at the end of the reaction, and a reasonable estimate of the mass on the gel can be made if it is electrophoresed along with a mass standard, which consists of well-characterized dsDNA molecules in which the concentration of each band has been precisely measured. A better method for determining reaction yield, however, is to clean up the reaction using a silica-based PCR purification column and then measure the purified PCR product spectrophotometrically.

Another concern that is often expressed when PCR products are electrophoresed pertains to the identity of the product itself. In real-time PCR, product identity, and the presence of spurious products, are determined by melting curve analysis: a dsDNA product of a particular length and base composition will denature at a predictable temperature. This analysis is performed in the real-time thermal cycler immediately after the last cycle. Of course, DNA sequencing is the ultimate identification tool, though this is not practical for every product generated in every reaction. When end-point analysis is performed in this lab, many of the products that are expected are known to contain what we refer to as a diagnostic restriction site that can be used to (putatively) identify a product (Fig. 10.11). Every so often an aliquot of a PCR product is digested with the restriction enzyme that is expected to cut it. If the appropriate banding pattern is observed, this is taken as in indicator of the identity of the product without the need for sequencing. This approach is very worthwhile.

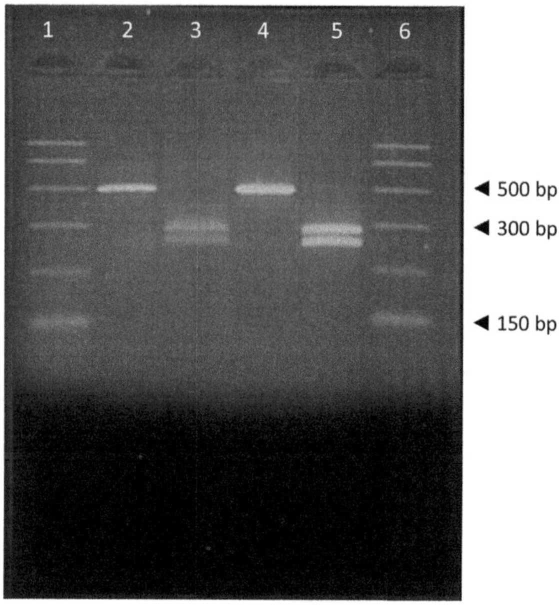

FIGURE 10.11 The presence of a known restriction enzyme site in routinely produced PCR products can be used for putative product identification. The 500 bp PCR product in lanes 2 and 4 has a single *Mbo* I site which, upon digestion, renders the expected 227 and 273 bp fragments (lanes 3 and 5). This precludes the need for constant sequencing and is a good spot check of the specificity of the assay. If the PCR product did not cut, then an immediate examination of all aspects of the amplification would be performed. Lanes 1, 6: PCR MW standard.

RT-PCR quality control points

Among the more notorious difficulties encountered when preparing a nucleic acid sample for PCR is the integrity of the template material, an aspect of the reaction that takes on an entirely new meaning when working with RNA. For those laboratories actively investigating gene expression, it is likely that some PCR variant is being used to amplify one or another transcript with primers and parameters that have long since been optimized. The frustration index rises when one is unable to generate a product from a freshly prepared sample of RNA, or from an RNA sample that "used to work." The following should be pondered in the context of the optimization items discussed previously.

1. Assess the quality of the RNA just prior to use by electrophoresis. The minigel format is more than adequate to simply show the presence of the 18S and 28S rRNAs. RNA that has been stored in aqueous buffer, rather than as an ethanol precipitate, degrades rapidly. This is exacerbated when the sample has been subjected to repeated freeze–thaw cycles. In this lab, previously purified RNA is heated to 65°C for 5 minutes and then

electrophoresed in a 1.2% agarose minigel in 1 × TAE (no denaturants needed; no blotting), the sole purpose of which is to assess the integrity of the sample. As long as there is a reasonable definition of the 28S and 18S rRNAs, the investigator may proceed confidently.

2. Test each pair of primers without a template. This is necessary to show that the stock solutions are not contaminated with an amplicon as well as to give any indication of primer-dimer. This is the so-called "no template control."
3. Test each primer alone, with template. This is the only true way to show that self-priming is not a possibility, and that the primer pair is specific. The sometimes-surprising results of precisely this type of QC experiment in an evaluation of a new set of primers are shown in Fig. 10.12.
4. Test each sample of RNA by omitting the RT from the first-strand synthesis reaction. Then use the contents of this tube as template material for PCR. In this sample, known as the "no RT control" "RT^-," or the "RT-minus control," no products should be observed. Without reverse transcription of the RNA template, the only bands that could possibly be generated will have come from a contaminating genomic DNA in the sample, assuming that there is no amplicon contamination anywhere in the pipeline. The standard protocol in most labs performing any type of RT-PCR is to DNase the samples (Appendix C) at the time of isolation, after which the RT^- control is performed.

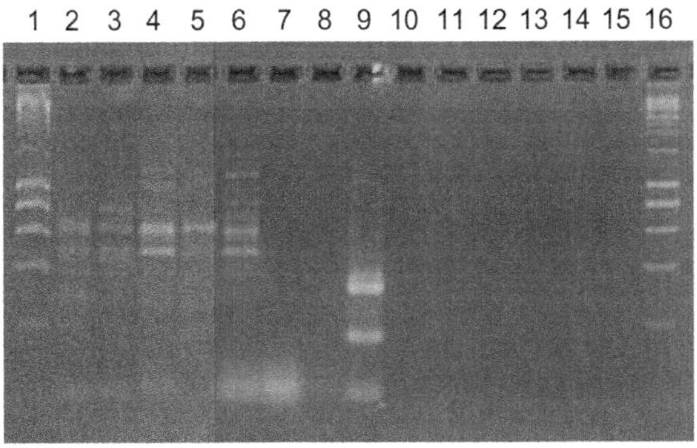

FIGURE 10.12 Testing for primer self-priming. Two primers designed for use in a transcription assay were tested independently for a propensity for self-priming. Lanes 2–6: 100 ng first-strand cDNA from five different samples was added to aliquots of PCR master mix that contained only the 5′ primer. Lanes 10–14: 100 ng first-strand cDNA from the same six samples were added to aliquots of master mix that contained only the 3′ primer. Lanes 7 and 15 were negative controls (no primer), Lane 8 was empty, lane 9 contained a positive control, and lanes 1 and 16 contain PCR molecular weight standard. The numerous, unexpected variety of PCR products generated with only the 5′ primer clearly demonstrated its capacity for self-priming. Upon observing these data, the 5′ primer for the assay was redesigned.

5. Test difficult RNA samples with primers that may not be of direct experimental interest, but that are at least known to generate a PCR product each time RNA is isolated, intact, and clean, from a particular biological source. This can be performed using very boring primers, such as those for β-actin, histone, or GAPDH just to show that the sample is able to support the amplification of something (anything). If a PCR product is evident with control primers, then the reason for failure to generate a product from experimental primers can be investigated more intelligently. Succinctly, no signal can mean no transcription *or* that the RNA sample is poor and unable to support amplification of anything. This latter possibility will have been validated, or ruled out, through the use of these control primers.
6. Phantom bands may be asymmetric products. Upon electrophoresis, an unexpected band that appears to be about one-half the size of the expected product is often attributed to the preferential amplification of one strand. This occurs when one primer is present in a molar excess over the other primer, and this is a more frequent occurrence than one might think. This can be remedied easily by checking the primer concentrations spectrophotometrically, regardless of what is printed on the label on the tube. This will ensure that subsequent dilutions of primer stock solutions produce equimolar primer concentrations in future reactions. If this was the problem, the unexpected bands will disappear.
7. Be judicious in the use of positive control templates, the products from which should be distinguishable electrophoretically, or by melting curve analysis, from the expected product. If a positive control amplicon manages to contaminant a stock solution, everything in the lab will come to a grinding halt.

Non-PCR methods for confirming PCR-derived data

PCR is an efficient amplification tool, but it is also a biased reaction. It is clear that some sequences amplify with greater efficiency than others and, as the number of cycles increases, the true abundance relationships among the starting material (mRNA → cDNA) are distorted. One of the most convincing approaches for validating changes in gene expression from RT-PCR experiments is to use a non-PCR method to show that same trend. For example, demonstrating up- or downregulation of a transcript by nuclease protection (Chapter 18: Quantification of Specific mRNAs by Nuclease Protection) or by reverse Northern analysis (Fig. 10.13) is hard to contradict. There are other methods by which to confirm PCR-derived data, too. Some of these alternative methods, however, may be difficult to employ if the transcript of interest is below the level of detection of a lesser sensitivity technique.

Many of the methods described in this book are simple to perform while others are rather labor-intensive. The intention here is not to burden the

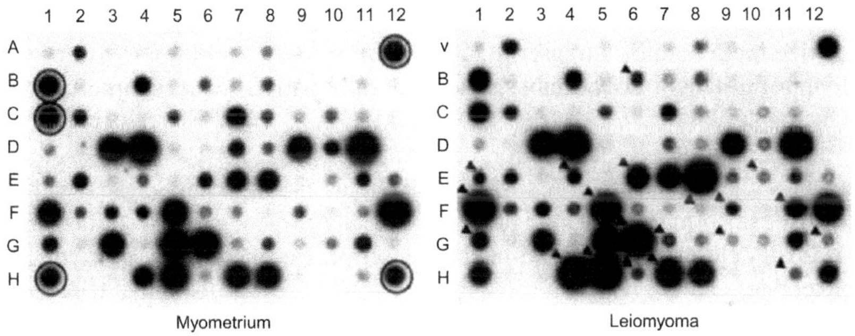

FIGURE 10.13 Reverse Northern analysis. Differentially expressed genes are detected by reverse Northern analysis using mRNAs as probe template. Two sets of blots are made with an equal quantity of PCR products. Positive controls (G3PDH) are marked with circles. Genes that have been upregulated are indicated with arrows. Leiomyoma is a benign smooth muscle tumor and myometrium is normal smooth muscle tissue from the wall of the uterus. *Reprinted with permission from Macmillan Publishers Ltd: Cell Res. 12, 215–221, copyright 2002.*

investigator with additional tasks prior to submitting data for publication. Rather, the emphasis here is the importance of confirming one's findings using other methods, to the extent possible. This becomes particularly important when only subtle changes in gene expression are detected; given the sensitivity of PCR to all aspects of reaction composition and assembly, even minor variations in pipetting can have an amazing impact on the product abundance at the end of the amplification protocol, not to mention reproducibility among replicates. Moreover, because labs often have principal investigators, students, and/or postdocs working on various aspects of the same project, tubes occasionally become mixed up (date, template ID, concentration), leading to conflicting or erroneous data. Just as one may drive to a destination via a number of different routes, one should also be able to draw the same conclusions via different methods.

This strategy goes a long way toward convincing colleagues, critics, competitors, and peer-reviewers.

Related techniques

5′ RACE PCR

One of the key drawbacks of traditional cDNA synthesis is the failure to generate full-length cDNA molecules that capture sequences corresponding to the 5′ end of the mRNA. This is due to challenges with RT in the synthesis of first-strand cDNA, the mechanics of PCR, up- and downstream primer design, and of course the quality of the input RNA. The resulting loss of sequence representing the 5′ end of transcripts makes it difficult to clone full-length sequences or map the transcription start site (TSS) at a particular locus. A method known as rapid amplification of cDNA ends (RACE;

Frohman et al., 1988) was developed mainly for cloning sequences that characterize the 5′ end of a transcript (5′ RACE) or the 3′ end (3′ RACE), or for cloning full-length cDNA molecules when only partial sequence information is available.

There are two strategies for performing 5′ RACE. The older method involves the synthesis of first-strand cDNA as usual, accomplished with a downstream primer, either oligo(dT) or, ideally, a gene-specific primer that base-pairs closer to the 5′ end of the transcript. Recall that any deoxyoligonucleotide can provide the requisite 3′-OH needed to initiate reverse transcription. Next, the newly synthesized first-strand cDNA molecules are tailed, meaning that extra nucleotides of known sequence are added enzymatically to the 3′ end of the cDNA by the somewhat unique enzyme terminal deoxynucleotidyl transferase. Known a bit more casually as "terminal transferase" this enzyme will add a few additional nucleotides in a nontemplate-dependent manner to either blunt- or protruding 3′-OH groups on either single- or double-stranded DNA, as shown in Fig. 10.14. The extra nucleotides added in this manner constitute what is known as the anchor sequence; years ago, this gene cloning method was known as anchor PCR and one-sided PCR. The addition of one species of nucleotide is often referred to as homopolymeric tailing; thus, a poly(G) anchor can anneal a poly(C) anchor (upstream) primer to support PCR. Similarly, a poly(A) anchor will be able to anneal a poly(T) anchor primer. Using this approach, amplification by PCR will only require one gene-specific (downstream) primer, not two. A major limitation of this method, however, is the fact that any first-strand cDNA molecules made by RT are candidates for the homopolymeric tailing, even truncated cDNAs that did not reach all the way to the 5′ end of the mRNA template. For this reason, false positives are to be expected and, depending on the positioning of the downstream primer, may be present in large numbers. Even worse, the co-amplification of the truncated species is in effect a competition reaction which greatly detracts from the ability of the reaction to amplify the desired product, especially when it is present in low

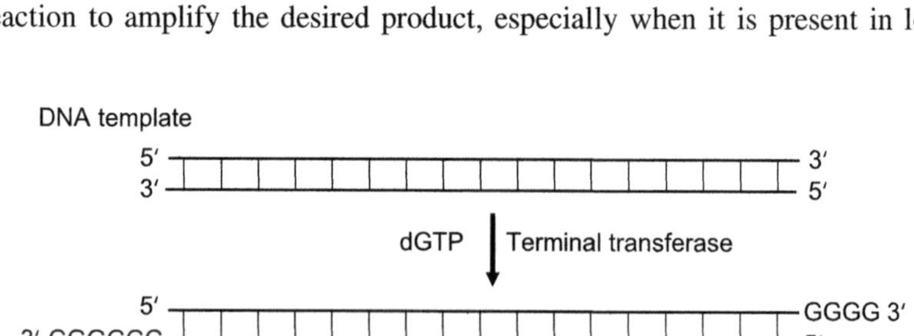

FIGURE 10.14 Action of terminal deoxynucleotidyl transferase. This unusual enzyme, also known as terminal transferase, is capable of adding nucleotides to blunt- and protruding 3′ ends of DNA in a nontemplate-dependent manner. This activity is useful for a number of PCR and cloning applications. Terminal transferase adds nucleotides to RNA, too, but does so with much lower, impractical efficiency.

abundance. Finally, the homopolymeric nature of the anchor primer creates a substantial T_m imbalance when used with to the downstream gene-specific primer needed to support amplification.

5′ RLM-RACE

The current method for cloning the 5′ end of transcripts is known as RNA ligase-mediated RACE (5′ RLM-RACE; Maruyama and Sugano, 1994; Shaefer, 1995). This technique drastically improved the ability to capture sequences corresponding to 5′ end of the mRNA, rather than a mixture of truncated cDNAs or cDNAs that were primed from partially degraded transcripts. Succinctly, the method involves using the enzyme T4 RNA ligase to join a sequence-defined RNA oligomer to the 5′ end of only full-length, undegraded mRNA. Upon reverse transcription, the ligated RNA oligomer becomes the 3′-most end of the cDNA to which the upstream primer will anneal.

The RLM-RACE process is outlined in Fig. 10.15. After running a gel to ensure that the RNA sample is very high quality, RLM-RACE begins by incubating the RNA in the presence of calf intestinal phosphatase (CIP) to remove any exposed 5′ phosphate groups. The 5′ cap structure is resistant to CIP, so the mRNAs are dephosphorylated are those from which the cap has already been removed or those molecules which are already partially degraded; full-length mRNAs are not modified by CIP. The result is that damaged, nonfull-length transcripts now exhibit a 5′-OH structure. Next, incubation with the enzyme tobacco acid pyrophosphatase (TAP) removes the ^{7m}G nucleoside along with the α and β phosphates from full-length, capped mRNA, thereby exposing the single remaining phosphate group at the 5′ end of the molecule. This phosphate group is a required substrate for T4 RNA ligase, which ligates an RNA anchor (adapter) sequence to the decapped mRNAs. Any other molecules that were partially degraded or already decapped at the onset of the experiment are not candidates for ligation because those molecules now have a 5′-OH group, and not the required 5′-PO_4. The RNA to which the anchor was ligated is now ready for reverse transcription. While it is true that all RNA molecules that are able to anneal the downstream primer will be reverse transcribed, the only first-strand cDNAs that will be amplified are those which contain the adapter sequence and can anneal the upstream anchor-specific primer. The RLM-RACE approach has the added advantage of using nested anchor primers that conform more closely to the standard rules of primer design with respect to base distribution and average GC content: anchor primers direct toward homopolymerically tailed cDNAs are either 100% GC-rich or 0%! RLM-RACE has been used in this lab with great success to identify multiple TSSs from one gene locus and in a tissue-specific manner (Bassett et al., 2004).

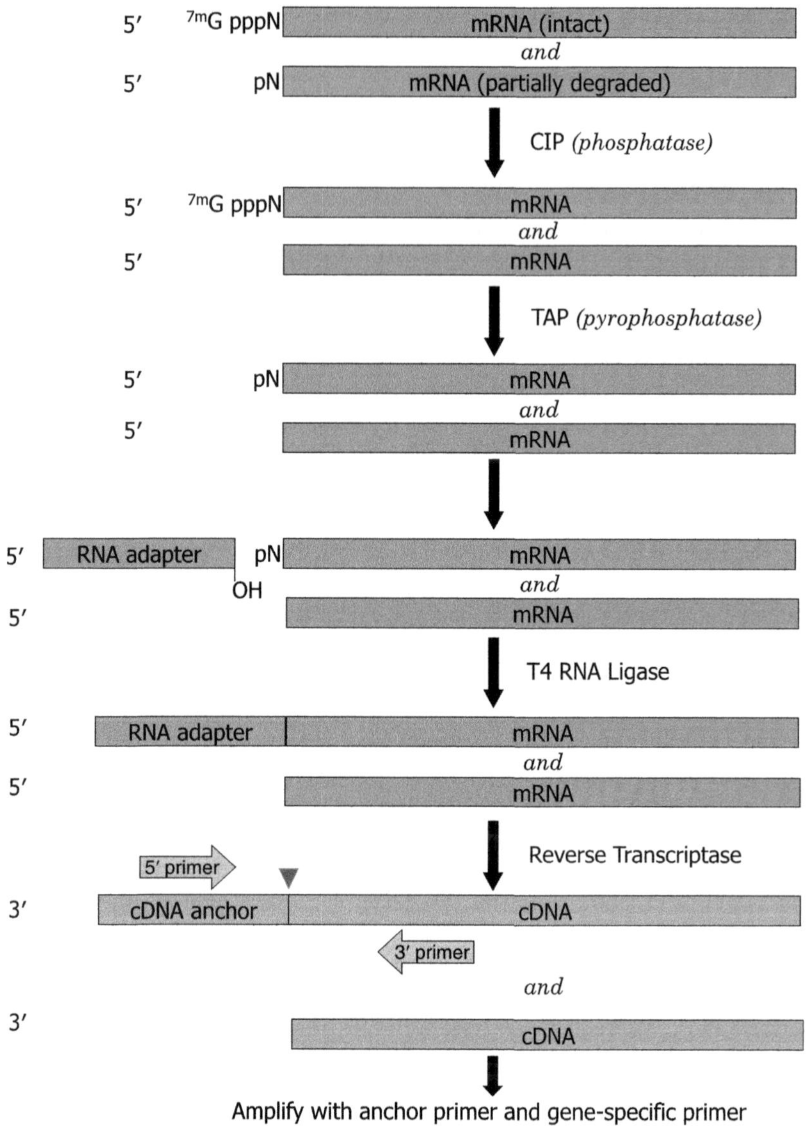

FIGURE 10.15 RLM-RACE. The mechanics of this assay favor the cloning of only intact mRNA molecules that manifest a 5′ cap. The novelty of this approach is the ligation of an RNA adapter after CIP-mediated dephosphorylation of partially degraded mRNA. Then, incubation of full-length mRNA with tobacco acid pyrophosphate (TAP) removes the cap, thereby exposing the requisite 5′ for the ligation event. Non-mRNA, corrupted mRNA, and first-strand cDNA molecules that failed to reach the RNA adapter during the ensuing reverse transcription are non-amplifiable. RLM-RACE is a very effective tool for transcription start site (TSS; green arrow) identification.

3′ RACE PCR

Cloning the sequences at the 3′ end of the transcript (3′ RACE) does not pose the numerous challenges associated with 5′ RACE. The standard 3′ RACE method involves the priming mRNAs first-strand cDNA synthesis using a downstream primer with the generic structure:

$$5' - \text{anchor sequence} - T_{11}V - OH\ 3'$$

where

1. The anchor sequence is a unique and user-defined sequence located 5′ to an oligo(dT) tract;
2. T_{11} is the minimum length of the oligo(dT) tract that should be used for stability reasons;
3. The 3′ end of the primer (V) is a single nucleotide anchor consisting of adenine, cytosine, or guanine (see Table 10.6). This ensures that the primer positions itself at the 5′-most end of the poly(A) tail so that reverse transcription proceeds as far into the coding region of the mRNA as possible. It is also possible to design the downstream primer as 5′-anchor sequence-T_{11}VN-OH 3′ with a dinucleotide anchor at the 3′ end for added stability.

Following cDNA synthesis, PCR amplification of cDNAs is supported through the use of an anchor (downstream) primer that is complementary to the anchor, and the use of a gene-specific (upstream) primer (Fig. 10.16). While one might envision 3′ RACE by simply using oligo(dT) as the downstream primer, experience strongly advises against the use of any AT-rich sequence, including oligo(dT) alone, which is 100%! If working with poly $(A)^-$ RNAs, incubation with poly(A) polymerase will add an artificial poly (A) tail to the template material, after which it can be analyzed using the strategies described above.

Nested PCR

The failure to generate a PCR product may be related to the exquisitely low abundance of the template or, perhaps more commonly, to the purity of the template and/or specificity of the primers. In one strategy, an aliquot of a completed PCR reaction, which manifests no product, is diluted 1:100 into a fresh reaction mix and then subjected to an additional 30 cycles. However, a technique involving two sets of primers, known as nested PCR, is a widely regarded alternative amplification strategy, compared to simply running through more cycles. This is especially valuable when there is a purity issue to be resolved.

In the nested PCR approach, one of the two sets of primers (the inner primer set) is designed to base pair completely within the region amplified by the other set of primers (the outer primer set). The relative positioning of

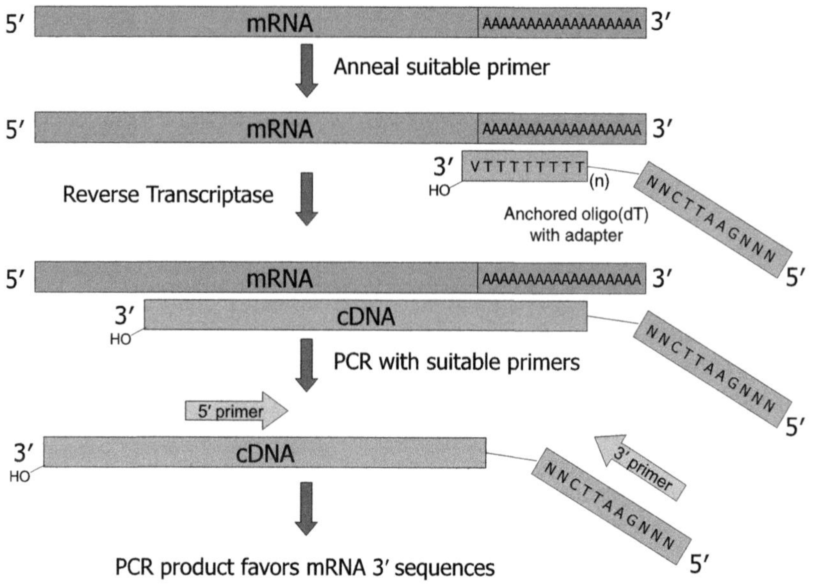

FIGURE 10.16 3′ RACE. The approach is used when the goal is to amplify sequences that correspond to the 3′ end of the original mRNA. This method is far less complicated than either traditional 5′ RACE or RLM-RACE. The use of an anchored oligo(dT) primer with an adapter at the 5′ end positions the primer at the 5′ end (the beginning) of the poly(A) tail, and provides a novel sequence at the 5′ end of the first-strand cDNA to which a 3′ primer can be annealed (following synthesis of second-strand cDNA).

these primers is shown in Fig. 10.17. Succinctly, the sample is amplified using the outer primers for 30 cycles, after which an aliquot (dilution recommended) is amplified for an additional 15–30 cycles using the inner primers. When properly designed, the outcome of the nested PCR approach is usually most gratifying. Upon electrophoresis, the product generated with the nested primers (secondary product) migrates more rapidly than the product(s) produced with the outer primers (primary product), because the secondary product is shorter by at least the combined size of the outer primers. One must be certain to examine all four of the proposed primer sequences to ensure that they conform to the standard rules for primer design, as discussed above.

Long-range PCR

The concept of long-range PCR applies to the synthesis of genomic-length PCR products which are widely accepted to be greater than 4000 bp. Almost unthinkable in the early days of PCR, improvements in component purity, modifications of the reaction chemistry, and the clever selection of thermostable polymerases now facilitate the routine amplification of very large pieces of DNA (Fig. 10.18). The key to the long-range PCR process is

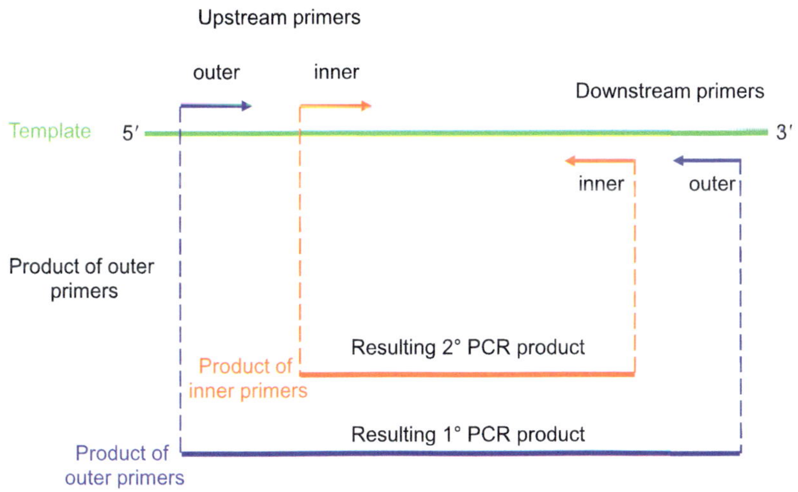

FIGURE 10.17 Nested primers. Two sets of gene-specific primers are often utilized when issues of product purity surround a PCR-based assay. A primary PCR amplification is performed using the outer the primers. This is followed by a second PCR amplification using the inner primers, and a diluted aliquot of the product(s) of the primary reaction. This approach confers tremendous specificity to the assay by minimizing the amplification of any nonspecific product that may have tainted the primary reaction.

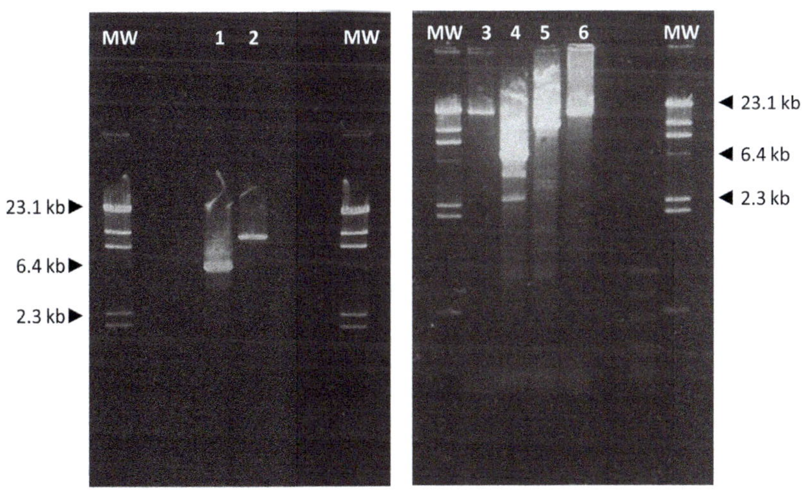

FIGURE 10.18 Long-range PCR. The products on this gel range from 2.3 to 23.0 kb. Note the location of the λ-*Hind* III MW standard in the outside lanes on the gels. This procedure is very likely to require optimization to eliminate smearing (truncated product formation) between the major bands observed upon electrophoresis. Lane 1, 4.8 kb; lane 2, 9.3 kb; lane 3, 23.0 kb; lane 4, major product is 9.4 kb; lane 5, 15.0 kb; lane 6, 23.0 kb.

the innovative use of two thermostable enzymes in the reaction tube, rather than one. These so-called enzyme blends often consist of *Taq* polymerase, which is highly processive but does not proofread, and another thermostable enzyme that usually has a longer half-life at elevated temperatures which is less processive but does proofread. Inasmuch as *Taq* polymerase makes mistakes, the two-enzyme format can produce very long PCR products with high fidelity. Given the unique amplification conditions associated with this form of PCR, long-range reactions are far more likely to require optimization than standard reactions. Consider the following:

1. Primers used for long-range applications tend to have a T_m in 60°C–68°C range and, as always, the T_m should be similar for both.
2. Long-range reaction buffers are more alkaline (pH > 8.8). This improvement minimizes the risk of depurination of the template during cycling. Since the upstream and downstream primers often anneal to the template thousands of base pairs away from each other, it is imperative that the template is of sufficient purity to support amplification.
3. Long-range PCR is far less tolerant of partially damaged or lower MW template than more traditional forms of the reaction. Template material should be checked electrophoretically (size) and spectrophotometrically (concentration and purity) before use owing to the rather unforgiving nature of the long-range reaction format.
4. Long-reaction reactions are often supplemented with one of the PCR adjuncts described in Table 10.8 which lowers the T_m of the template so that denaturation can be performed at 92°C rather than 95°C. Further, primer extension can be reduced from 72°C to 68°C. While these temperature reductions may at first seem trivial, they go a long way toward extending the life of the polymerase once cycling is underway.
5. The $MgCl_2$ concentration is more likely to require titration and is often higher (1.75–2.25 mM) than is associated with standard PCR applications.
6. The concentration of nucleotides (dNTPs) is usually increased to 300–500 μM for long-range PCR because of the very large size of the product that will be built in each PCR cycle. Some investigators favor the use of ultrapure nucleotides while others find the cocktail from a standard PCR kit to be more than satisfactory.
7. The cycling profile is usually greatly modified, especially the primer extension step. As a general rule, allow 1 minute of primer extension at 68°C for each 1000 bp to be amplified from the template.

Long-range technology can also be used to increase the sensitivity of standard PCR procedures (Fig. 10.19). When long-range reaction components are used to amplify typical length PCR products (200–900 bp), there is almost always an increase in sensitivity. When used in an RT-PCR format, the ability to detect very low abundance cDNAs improves. The day is coming when all routine PCR-based transcription assays will feature the

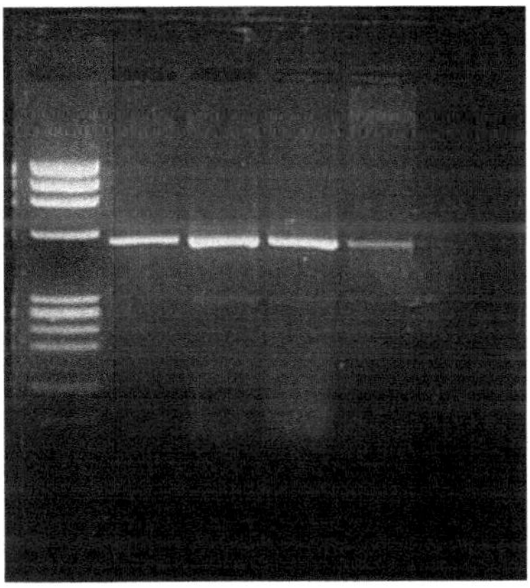

FIGURE 10.19 Enhanced PCR amplification using enzyme blends. 50 ng of RNA was reverse-transcribed, purified, and then aliquoted into four different PCR formulations. After 25 cycles, reaction products were analyzed on a 3% agarose gel containing 0.5 μg/mL ethidium bromide. Lane 1: ΦX174 mw standard. Lanes 2–4: three different enzyme blends (*Taq* with another thermostable, proof-reading polymerase). Lane 5: *Taq* polymerase alone. Digital image analysis revealed at least a twofold increase in the mass of the resulting 580 bp product when an enzyme blend was used, compared to *Taq* alone. *Reproduced with permission from Farrell, R.E., 1999. Long-range PCR: innovations in the long- and short terms. Biotech. Int. 11(2), 13.*

two-enzyme format. There are several excellent commercial enzyme blends that support both long-range and short-range applications.

Single-cell PCR

Everyone wants to be able to amplify PCR products from a single cell, though some people would be happy if they could amplify something from just a few cells. The good news is that the technology to amplify a single-copy gene or low abundance mRNAs from a single cell is available and is becoming increasingly available. Characterizing individual cells from a heterogenous population has profound implications in many facets of biology and medicine, including understanding normal developmental pathways and the progression of the disease. The most current trend utilizes microdroplet isolation of individual cells followed by one-step RT-PCR with fluorescence-labeled primers in a digital PCR format (see Fig. 11.13). A recent innovation involves the addition of the T7 bacteriophage gene 2.5 protein (gp2.5) to the PCR mix. A single-stranded DNA binding protein (SSB), gp2.5 effectively alleviates inhibitors that would otherwise be problematic in a single-cell RT-PCR reaction (Ma et al., 2021). The use of SSBs as PCR adjuncts is not new (refer to Table 10.8) and the use of the success of gp2.5 in that capacity is a contemporary addition to tools available for PCR optimization. SSBs are perhaps best known for their natural role in DNA replication.

Some points to keep in mind when attempting single-cell analysis: (1) well-characterized, PAGE- or HPLC-purified primers are very important for successful amplification of the intended product only without competing for spurious product formation; (2) individual cells can be isolated by limit dilution, microdissection, partitioning into and oil droplet emulsion, or by other methods that may be available; (3) standard PCR formulations can be used (primer concentration, dNTPs, Mg^{2+}), though the addition of 1 ng/μL prokaryotic carrier RNA (e.g., *E. coli* tRNA or rRNA) is recommended to drive forward hybridization kinetics between primer and template; and (4) a hot-start approach is almost always strongly recommended, and the inclusion of nested primers in a secondary reaction format as described above can be very helpful in terms of amplification specificity.

Single-cell PCR has numerous advantages. First, the template material is completely unique to that cell. In contrast, nucleic acids purified from thousands or millions of cells are a heterogeneous mixture and at best represent an "average." One of the more common methods for this type of assay involves transferring a single cell to a thin-wall PCR tube in 1 × PCR reaction buffer, snap freezing it on dry ice (do *not* place this tube in liquid nitrogen), after which it can be stored at −70°C for up to 6 months. When ready to perform PCR, slowly thaw the tube on ice and then immediately add sufficient reaction components to produce a 50 μL reaction. Be sure to perform an initial melt of 5 minutes at 95°C, followed by 40 amplification cycles. If no product is visible, transfer a 1 μL aliquot into 49 μL of fresh PCR reaction mix and perform an additional 40 cycles using either the same primers of a nested primer set. The only real disadvantage is to single-cell PCR is that it is not practical when large numbers of cells need to be analyzed individually.

Analyzing transcriptional activity in a single cell (single-cell RT-PCR) is a bit more of a challenge, though it really isn't that difficult to perform. Depending on one's application, there are many ways to recover live cells as well as cells collected by laser capture from a tissue section (Hodne and Weltzien, 2015; Ståhlberg and Kubista, 2018). Some investigators perform a mini-RNA isolation procedure in a PCR tube containing a single cell, while other investigators prefer to simply handle the single cell (which will have lysed upon thawing) as if it were a purified RNA prep: the sample is heat denatured in the presence of the downstream primer(s), after which RNasin, RT, and other reaction components are added. cDNA is synthesized for 30−60 minutes and, after heat inactivation of the RT, the cDNA is used as a template for PCR either in a multiplex format or by splitting the cDNA into multiple tubes for analysis with different primer sets. It is not uncommon to perform 40−50 cycles. As suggested above, we have had excellent success detecting very low abundance transcripts by performing 35−40 cycles, transferring a 1 μL aliquot into 49 μL of fresh PCR master mix, and then performing additional cycles.

Splinkerette PCR

A splinkerette is a modified adapter molecule with a hairpin structure involving the 3′ end of one of the two strands (Fig. 10.20A). One of the major applications of splinkerette PCR is older-style chromosome walking in order to obtain upstream and downstream regions flanking a known sequence (Devon et al., 1995), and for mapping transposable elements (Potter and Luo, 2010). Better known techniques that are used for the same objective are inverse PCR and vectorette PCR (Riley et al., 1990; Horn et al., 2007).

The impetus behind the development of splinkerette PCR is a phenomenon known as end-repair priming. This occurs when the adapter- and insert DNA molecules that are not ligated undergo a filling-in reaction in which the 3′ end(s) of molecules are extended in PCR cycle 1. Recessed 3′ ends are present when restriction enzyme digestion produces fragments with a 5′ overhang. These ends have a proclivity to anneal following template denaturation in PCR cycle 1, resulting in a competition reaction of sorts in which the amplification of nonspecific, unwanted products detracts from the sensitivity of the assay. In the case of nonligated splinkerette adapters, the 3′ end of stand with the hairpin primes its own extension, resulting in a looped domain that is unable to support subsequent nonspecific amplification (Fig. 10.20B). Thus, splinkerette PCR minimizes end-repair priming and, because the splinkerettes are not kinased, there is no opportunity for ligation between the lower stand of the splinkerette and the DNA template that is intended for amplification.

The splinkerette amplification strategy is attractive (1) when there is only enough sequence information to construct one gene-specific primer;

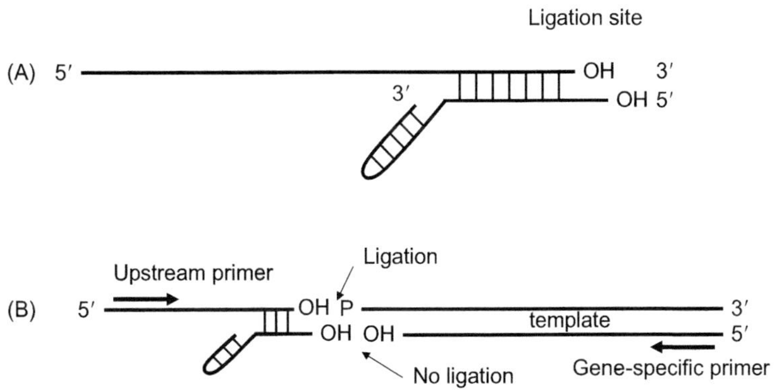

FIGURE 10.20 Splinkerette PCR. (A) Structure of a splinkerette adapter. Splinkerettes do not have a 5′ phosphate, so only one of the two strands is able to be ligated to the intended template. (B) Formation of a nonamplifiable loop domain in nonligated adapters minimizes the formation of spurious products due to 3′ end-repair priming.

(2) when there are issues with adapter self-ligation, especially when the DNA to be ligated to the adapter(s) has a 5′ or 3′ overhang; or (3) when nonspecific priming is an issue. Moreover, splinkerette PCR does not require circularization of the molecules under investigation. It is very useful for identifying uncharacterized sequences which flank a known region, particularly when working with genomic length templates. For contemporary applications (see Sato et al., 2022; Alquezar-Planas et al., 2020; Macias et al., 2017).

The hunt for alternative transcription start sites

An important question associated with gene expression is the possibility that transcripts representing more than one TSS may result from a single gene locus. This has already been observed widely in nature in both plant and animal species and appears to be an important evolutionarily conserved process. Most often, transcription from a particular genetic locus produces mRNAs representing a primary TSS and perhaps several nearby secondary TSSs corresponding to a single TATA box (Wray et al., 2003). Reports of multiple TATA boxes within the same promoter, while rare, have been described in animal models (Fra et al., 2000; Yamada et al., 2001; *reviewed by* Farrell and Bassett, 2007).

In plant models, multiple TATA boxes initiating transcripts at different TSSs are less frequent than is observed in animals, though some interesting examples have been reported. For example, 5′ RLM-RACE was used to demonstrate that a leucine-rich repeat receptor kinase gene (*inrpk1*) in morning glory (*Ipomoea* [*Pharbitis*] *nil*) produces transcripts from at least nine different TSSs using three different consensus and nonconsensus TATA boxes (Bassett et al., 2004), and does so in a tissue-specific manner. Similar types of gene regulatory events have been observed in *Arabidopsis* and tobacco (Dehesh et al., 1994; Adams et al., 1995). Thus, there are examples of multiple functional TATAA elements just as there are examples of multiple promoters controlling certain nuclear genes, at least in animal models. The nuclear-encoded rat mitochondrial glycerol-3-phosphate dehydrogenase (G3PD) locus, for example, produces transcript variants encoding three different first exons due to the action of three distinct promoters (Weitzel et al., 2000, 2001).

This type of information is important because when the TSS changes, there may be an accompanying impact on the cognate protein, should translation occur. For example, the dehydrin 2 gene (*PpDhn2*) in peach (Wisniewski et al., 2006) initiates from a consensus TATA box approximately 100 bp upstream of the translation start site in bark, while in mature fruit, initiation occurs an additional 125 bp upstream of the conventional ATG from an unidentified element which could be a nearby nonconsensus TATA box. The resulting fruit-specific transcript contains an upstream AUG in good translational context, followed by an intron, which contains the consensus TATA box used in bark tissues. The result of splicing out of this intron places the upstream AUG in frame with the conventional translation start of the protein. This protein variant would therefore acquire an additional 34 amino acids at the N-terminus, and bioinformatics software has predicted this to be a mitochondrial transit sequence (Bassett et al., 2009).

Classical RACE, as described above, has the unfortunate habit of producing many false positives as a consequence of the mechanics of the method itself, namely, that nonfull-length cDNAs, often representing nonfull-length mRNAs, are represented at the end of the experiment and must be characterized. 5′ RLM-RACE, also described above, is now the definitive method for unambiguous TSS identification.

Protocol: first-strand cDNA synthesis

The protocols which follow are examples only. Use of commercial master mix, enzyme choice, reaction volume, and other parameters are all subject to change based upon SOPS that may already be in place in the lab.

In advance: assess the quality of the RNA template by minigel electrophoresis.

1. Prepare master mix 1:

Component	Per reaction[a]
H_2O	5 μL
dNTP (10 mM)	1 μL
Random primers (2.0 μg/μL)[b]	1 μL
Oligo$(dT)_{12-18}$ (1.0 μg/μL)	1 μL
Total	8 μL

[a]If preparing a master mix for multiple cDNA syntheses, be sure to multiply the "per reaction" volumes by one more than the total number of required reactions. This is a standard procedure to ensure an adequate volume of master mix in the event of subtle micropipetting errors. This same procedure also applies to the preparation of the PCR master mix.
[b]Alternatively, use a gene specific primer at a final concentration of about 1 μM (required volume variable) instead of priming with oligo(dT) and/or random primers.

2. Distribute 8 μL of master mix 1 into each sample tube.
3. Add 2 μL of RNA (50–1000 ng) to each tube and do not vary the mRNA mass that is added to each tube.
4. Heat tubes to 70°C for 5 minutes. At the conclusion of this RNA denaturation step, place samples on ice for at least 2 minutes.
5. While the RNA is being heat-denatured (step 4), prepare master mix 2:

Component	Each
H_2O	6 μL
10× first-strand buffer[a]	2 μL
RNasin (20U/μL)	1 μL
AMV R.T. (20U/μL)	1 μL
Total	10 μL

[a]Typical 10× first-strand buffer: 100 mM Tris, pH 8.3; 500 mM KCl; 40 mM $MgCl_2$. This buffer is provided with the reverse transcriptase.

6. Distribute 10 μL of master mix 2 into each sample tube containing the denatured RNA.
7. Incubate at 25°C for 5 minutes.
   Note: This short room temperature incubation (1) facilitates primer annealing to the template, and (2) gives the RT an opportunity to extend the random primers and/or oligo(dT) for greater thermostability when the incubation temperature is increased in the following step. This almost always has the effect of producing much longer cDNA molecules.
8. If using a thermostable or enhanced RT, incubate samples at 50°C for 50 minutes. Otherwise, incubate the sample at 37°C–42°C for 60 minutes.
   Note: There are numerous RT formulations that support first-strand cDNA synthesis at elevated temperatures. These are widely available from any of a number of vendors.
9. Heat samples to 95°C for 5 minutes to destroy the RT. Failure to do so can be detrimental to subsequent PCR-based procedures.
10. The newly synthesized first strand cDNA is now ready for any one of several RT-PCR applications and should be placed on ice. Alternatively, cDNA can be stored at −20°C until ready to use.

Protocol: PCR amplification of cDNA

1. Prepare PCR master. Typically, for each 20 μL reaction to be performed, mix 2 μL 10× PCR buffer (without Mg^{2+}); 1.2 μL 25 mM $MgCl_2$ (subject to change if optimization is required); 0.1 μM upstream primer (volume variable); 0.1 μM downstream primer (volume variable); 0.5 μL *Taq* DNA polymerase (1.0–2.5U per reaction); water to 18 μL.
   Note 1: If different sequences are to be amplified in each reaction tube, then omit the primers from the master mix, adding each pair to the proper reaction tube directly. If the cDNA template is shared among all samples, it may be added directly to the PCR master mix.
   Note 2: Never add all of cDNA to the master mix. By reserving a small aliquot, it will be possible to support several rounds of PCR from the same experimental source using various primers at a later date.
2. Aliquot 18 μL of the master mix into each reaction tube.
3. Add 2 μL of cDNA first-strand products to each reaction tube.

4. Cycle the samples according to the following suggested generic profile. These parameters have been optimized for an Applied Biosystems 9700 thermal cycler using thin-wall 500 μL PCR tubes. Remember that cycling parameters are subject to optimization for each new set of primers, changes in the reaction composition, and model of the thermal cycler.

Initial melt	94°C for 2 minutes	
Cycling	94°C for 30 s	
Denaturation	55°C for 30 s }	30 cycles
Primer annealing[a]	72°C for 1 min	
Primer extension		
Final extension	72°C for 5 min	
Soak	4°C until ready to use	

[a]The correct temperature at which to anneal (T_a) is a direct function of the Tm of the oligonucleotides being used. Consult the manufacturer's specification sheet that accompanied the primers. If uncertain where to begin, use $T_a = T_m - 5°C$.

5. Remove an aliquot (2 μL from a 20 μL reaction) and analyze size and variety of products by minigel electrophoresis. PCR products are commonly analyzed in 2%–3% agarose gels prepared in 1 × TAE buffer, as described earlier in this chapter.

Cloning PCR products

The ability to amplify and reamplify DNA sequences by PCR has rendered routine vector-insert ligation something of a rarity. After all, if one requires more of the same product, it can be generated again by PCR using the same primers. Cloning of PCR products is commonly performed to (1) facilitate older-style sequencing; (2) facilitate in vitro transcription; (3) physically separate a mixture of PCR products; and (4) support the expression of recombinant proteins. Of these, sequencing and in vitro transcription can be supported without cloning: reamplification of PCR products with modified up- and downstream primers consisting of appropriate 5′ promoter sequences precludes the requirement for a vector altogether. Further, multiple PCR products can be separated and recovered from agarose gels or directly from the reaction tube. Numerous kits are available for PCR product clean-up, all of which are simple to use, expedient, and economical. For example, a gel slice containing a band of interest can be melted in a solution of guanidinium isothiocyanate (GTC) to create a chaotropic, high-salt environment. Alternatively, GTC can be added directly to the tube in which the cycling was performed. In either case, the PCR products will become bound to a silica matrix. Residual reaction components, including primers, nucleotides, enzyme, and melted agarose, are washed away. The purified DNA is eluted from the silica matrix under very low-salt conditions, using 10 mM Tris, pH 8.0 or nuclease-free water.

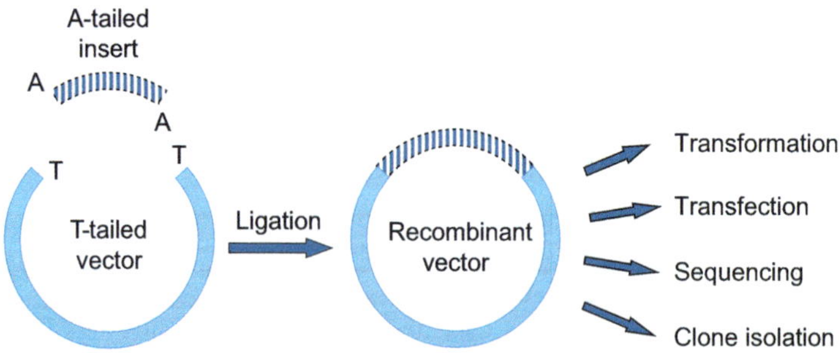

FIGURE 10.21 TA Cloning. PCR products generated with *Taq* polymerase exhibit a single base over-hang, most often an adenine, at the 3′ end of each strand. This facilitates ligation into a linearized vector that has been tailed with a thymine overhang (T-tailed). Such vectors are widely available. Cloning PCR products in general supports DNA sequencing, transformation into prokaryotes, transfection into eukaryotes for expression studies, and an overall strategy for physically separating multiple PCR products from each other.

Should cloning be deemed necessary, there are several options. *Taq* DNA polymerase exhibits a very useful habit of adding an extra nucleotide, most often an A, at the 3′ end of each strand of the product that it generated. This is the basis for a convenient method known as TA cloning, in which the investigator anneals and then ligates the *Taq* A-tailed product into a vector that exhibits a 3′ "T over-hang" on each strand (Fig. 10.21). These vectors are widely available and most have a least one selectable marker (e.g., Amp^r, Tet^r, or Kan^r) as well as a lacZ gene for blue/clear color selection.

There are several advantages associated with A-tailed products: (1) A-tailed products may be cloned directly without any knowledge of the base sequence of the PCR product, (2) A-tailed products eliminate the need for primers carrying 5′-overhanging restriction enzyme- or other sequences, (3) PCR products can usually be cloned directly, though purification can be helpful, (4) directional cloning is possible when hemiphosphorylated PCR products are generated (5′ phosphorylation of one of the primers), and (5) vector cloning sites are usually located in a polylinker region (multiple cloning site, or MCS). This will facilitate excision using any of a number of restriction enzymes.

The ligation itself is mediated by T4 DNA ligase in a reaction requiring several hours or much more expediently via TOPO cloning. Note that some thermostable enzymes used for PCR demonstrate a $3' \rightarrow 5'$ exonuclease (proofreading) activity and produce blunt-ended PCR products, which cannot be used for TA cloning unless they are tailed. This is easily accomplished by cleaning up a completed PCR reaction to remove the proofreading enzyme (PCR product cleanup kit), followed by incubation of the blunt-end PCR products with *Taq* polymerase (and other reaction components) for 15 minutes at 72°C, and followed by another clean-up procedure. Failure to perform this second clean-up procedure will reduce the efficiency of ligation in subsequent steps.

Protocol: A-tailing of blunt-end PCR products

1. Purify PCR products with a silica-based clean-up kit.
2. Prepare the tailing reaction:

Blunt-end PCR products	10.0 μL
H_2O	7.0 μL
10× PCR Reaction Buffer (Mg^{2+})	2.0 μL
10 mM dNTP mix	0.5 μL
Taq polymerase	0.5 μL

3. Incubate samples at 72°C for 15 minutes.
4. Purify tailed PCR products as in step 1.
5. Begin TA-cloning procedure immediately, as the 3′-overhang will be lost quickly. If not ready to begin, store the newly tailed molecules at −20°C.

Protocol: TA cloning ligation reaction

Two or three ligation reactions for each PCR product should be assembled in order to maximize cloning efficiency: remember that the precise ratio of vector to insert influences the ligation efficiency. Note that multiple ligations are suggested only for first-time users. It is generally necessary to perform only one ligation reaction because a large number of transformants (i.e., maximum ligation efficiency) is not required when only a single type of insert is expected.

1. Assemble the ligation reactions listed here. Be sure to add the enzyme last. As needed, the precise ratio of vector:insert can be modified to suit the investigator.

	10× Ligase buffer	T-tailed vector	PCR product	H_2O	T4 DNA ligase
Reaction 1	1 μL	10 ng	0.1 μL	*q.s.* 9 μL	1 μL
Reaction 2	1 μL	10 ng	1.0 μL	*q.s.* 9 μL	1 μL
Reaction 3	1 μL	10 ng	2.0 μL	*q.s.* 9 μL	1 μL

2. Mix the reaction(s) gently after addition of the ligase.
3. Incubate for a minimum of 3 h at 16°C.
   Note: Ligation reactions that are performed overnight generally produce the highest number of clones.
4. Thaw an aliquot of frozen (−80°C) competent cells on ice.
   Note: These cells may be purchased or prepared in advance. Be sure that the bacterial host strain is compatible with blue/clear color selection if using this selection scheme.
5. Pipette 20 μL aliquots of competent cells into prechilled microfuge tubes on ice.

6. Add 1 μL ligation reaction to each tube. Stir gently with pipette tip; avoid aspirating cells up and down.
7. Incubate tubes on ice for 20 minutes.
8. Place the tubes in a 42°C water bath for *exactly* 45 seconds. Do not shake these tubes.

   Note: This heat shock is the most critical step of all, often determining the success or failure of the transformation. Do not heat at a higher temperature or for a longer period than recommended here.
9. Immediately return the tubes to the ice for 2 minutes.
10. Add 400 μL room temperature LB medium or room temperature SOC medium.

    Note 1: LB medium (per liter): 10 g tryptone, 5 g yeast extract, 10 g NaCl; autoclave.

    Note 2: SOC medium (per liter): 20 g tryptone, 5 g yeast extract, 0.6 g NaCl, 0.19 g KCl, 2 g $MgCl_2$, 2.5 g $MgSO_4$, 3.6 g glucose; autoclave.
11. Incubate at 37°C for 1 h with shaking (200 rpm).
12. Plate an aliquot (10–100 μL) of each transformation reaction onto standard LB agar plates that have been supplemented with the appropriate antibiotic. Reserve the unused portion of the transformation by storing it at 4°C.
13. Spread the transformation reaction aliquot over the surface of the agar as completely as possible using a sterile bent glass rod or similar implement.
14. Keep the plates upright for at least 5 minutes to allow all of the liquid to absorb into the agar.
15. Invert the plates and incubate at 37°C overnight.
16. The following morning, remove the plates from the incubator as soon as possible. This will minimize the formation of satellite colonies.
17. To test whether a colony on the dish is a true transformant, pick a small portion of the colony from the plate and inoculate the live bacteria directly into a 100 μL PCR reaction mix: 10.0 μL 10 × PCR reaction buffer (containing Mg^{2+}); 2 μL 10 mM dNTP; 0.5 μM downstream primer (volume variable); 0.5 μM upstream primer (volume variable); 1 μL Taq polymerase; water *q.s.* 98 μL.

    Note 1: The heating steps associated with PCR will break open the bacterial cells and denature the DNA, thereby making the plasmid accessible to the primers.

    Note 2: Do not transfer too much of the colony. Otherwise, amplification will be inhibited. Simply touching the surface of a colony with a toothpick is all that is needed.

Note 3: There are ongoing reports of PCR inhibitors found in some brands of wooden toothpicks. It would be judicious to test toothpicks for obvious reaction inhibition before using them to pick experimental clones. A better, simple solution is to use plastic toothpicks or one end of a sterile, disposable inoculating loop.

18. Cycle with the same amplification protocol used to generate the PCR product.

19. At the conclusion of the amplification, remove a representative aliquot of the reaction (5 μL) and analyze by agarose gel electrophoresis. Colonies that contain a true insert will manifest a band identical in size to that observed for the PCR product prior to cloning.

20. Return to any of the original colonies that generated a PCR product. Use the remainder of the colony to streak a new agar plate and to prepare a glycerol stock for long-term storage.

TOPO cloning

A widely respected alternative approach for cloning PCR products, or any double-stranded DNA, is the TOPO Cloning method (ThermoFisher Scientific). The key to TOPO Cloning is the use of the *Vaccinia* virus enzyme topoisomerase I. This enzyme functions both as a sequence-specific restriction enzyme (CCCTT) as well as a ligase. The method of cloning is rapid and efficient, often rendering >95% cloning efficiency following a room-temperature incubation for as few as 5 minutes. The TOPO Cloning method is compatible with TA-cloning, blunt-end ligations, and directional cloning.

There are several TOPO vectors that support a variety of cloning formats. In each case the TOPO vector has already been linearized and the topoisomerase I enzyme is already attached via a 3′ phosphotyrosyl bond (Shuman, 1994); such vectors are said to be "activated" and will not exhibit self-ligation (Fig. 10.22). Successful ligation to a TOPO vector requires that the intended DNA insert exhibit free 5′ –OH ends, meaning that PCR products must be generated with primers that do not have 5′ phosphate groups. If the cloning of cDNA, genomic DNA, or restriction fragments is intended, the DNA must be dephosphorylated first, which is itself a rather simple manipulation. The typical ligation incubation period using a TOPO vector is remarkably short, on the order of 5 minutes. Upon ligation, the topoisomerase I enzyme is released from the now recombinant plasmid construction. Bacterial transformation proceeds as usual.

Other amplification procedures

PCR is by far the most widely used and best-known nucleic amplification procedure. It is not, however, the only amplification method. PCR has its

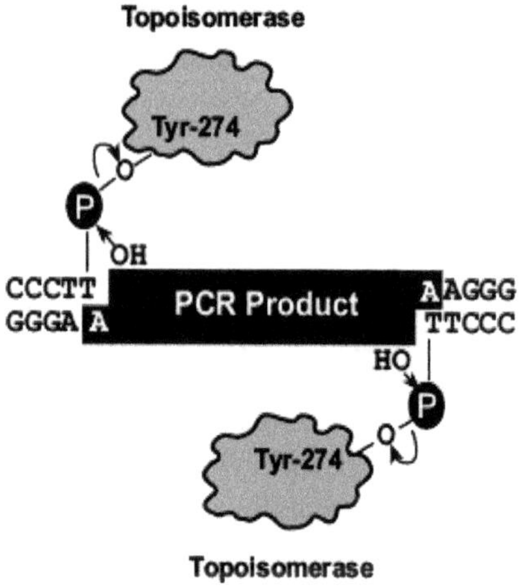

FIGURE 10.22 TOPO Cloning. TOPO vectors are linearized and already have topoisomerase I attached. This enzyme is linked to the vector via a 3′ phosphate. The PCR product, that is, the insert, contributes at 5′ -OH. *Courtesy Invitrogen.*

shortcomings, and some of the alternative amplification procedures circumvent these shortcomings and, in some cases, rival or exceed the amplification potential of PCR. Some of the amplification options include isothermal procedures, in which reaction takes place at a constant temperature. Isothermal is useful for analysis in the field or, in the case of medical intervention, at the point of service, because these methods preclude the need for a thermal cycler.

Linear RNA amplification (Eberwine process)

Procedures examining RNA abundance can be compromised when PCR is used to amplify first-strand cDNA for use in a downstream application that may involve the production of more RNA by in vitro transcription. This is a consequence of the bias that PCR has toward certain molecules, resulting in a distortion of the true abundance relationships among the various molecular species after even a few cycles. The need for any amplification at all occurs when only very small quantities of RNA are available (typically nanogram quantities) from a biological source because microarrays, as one example, and experiment replicates may require more RNA than is available.

The preferred method for generating more RNA from a paucity of template material is known as the Eberwine process or, more commonly, linear RNA amplification (Van Gelder et al., 1990). As shown in Fig. 10.23, the process involves converting RNA into double-stranded cDNA, after which the cDNA is transcribed to produce more RNA. Because the cDNA directly reflects the abundance of the RNA, subsequent transcription of the cDNA

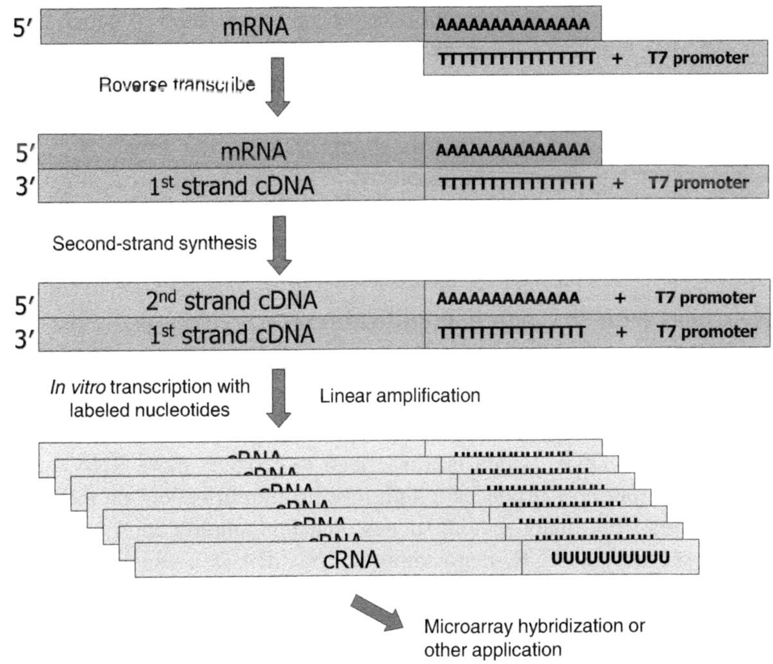

FIGURE 10.23 Eberwine linear amplification process. In order to produce more RNA while preserving the abundance relationships among transcript species, cDNA is synthesized by using a downstream primer-adapter that encodes an in vitro transcription promoter. Double-stranded cDNA is then transcribed, often in the presence of label, to produce cRNA which is then used for array hybridization.

maintains the relative abundance of the resulting RNA while providing more of it. The incorporation of a transcription promoter via a downstream primer-adapter, supports the synthesis of labeled complementary RNA (cRNA, *also* aRNA), which can then be hybridized into an array. This procedure may be used to generate sense (+) or antisense (−) transcripts, depending on the requirements of the investigator and the orientation of the promoter(s). The overall bias using this method of amplification is quite low, in comparison with PCR. Linear RNA amplification is of great value for the study of RNA binding protein, studying RNA regulatory roles, detection of low abundance RNAs, RNA-seq, and other applications.

Strand displacement amplification

SDA (strand displacement amplification) is a patented isothermal, in vitro method for sequence amplification which was developed by Becton Dickinson (Walker et al., 1992a,b). The method routinely produces at least 10^8-fold amplification, and much higher levels of amplification have been reported. SDA typically begins with a target generation step in which template molecules that are flanked by a restriction enzyme site are produced. The exponential amplification stage follows. Succinctly, the method is based on a restriction endonuclease cutting its recognition site which then allows a polymerase to extend the nick from the 3′ end. This has the effect of displacing the downstream strand. The displaced (+) strand serves as a target for (−) strand reaction. Repeated nicking, strand displacement, and annealing of primers to the displaced strands result in exponential growth.

Nucleic acid sequence-based amplification

Nucleic acid sequence-based amplification (NASBA) (Compton, 1991) is a primer-dependent isothermal process for RNA synthesis process. It is also commonly referred to as "self-sustained sequence replication" whereby RNA is converted into cDNA, which is transcribed to produce more RNA, all at 41°C. NASBA routinely produces a high level of amplification, on the order of 10^9. Reports from many sectors suggest that this amplification method is at least as sensitive as PCR, if not more so. NASBA is a widely respected molecular diagnostics technique, particularly for retroviral and microbial detection (Clancy et al., 2016). Real-time PCR, however, remains the gold standard primarily because of its widespread recognition and because so many processes have been developed and optimized using a PCR platform.

Rolling circle amplification

Rolling circle amplification (RCA) is another example of an isothermal amplification process (reviewed by Xu et al., 2021). The unique feature of this method is the formation of a circular template, mediated by a ligation reaction. A primer is then annealed to the circular DNA which can then be extended by DNA polymerase or RNA polymerase to synthesize either DNA or RNA products, respectively. The manner in which this proceeds is similar to the naturally occurring method of rolling circle replication observed in some viruses. The long synthesis product, a concatemer, is then processed into smaller linear segments which, in turn, are circularized and then able to support another cycle of amplification with a greater number of template molecules, continuing the process. Given the design flexibility with respect to the template, RCA has been used successfully for various nanotechnology applications and has great potential in the area of molecular diagnostics.

Ligase chain reaction

The ligase chain reaction (LCR) is an amplification process that differs from PCR in that it involves a thermostable ligase to join two probes or other molecules together which can then be amplified by standard PCR cycling

(Barany, 1991). Thus, LCR requires two completely different enzymes to operate properly: ligase, to join probe molecules together, and a thermostable polymerase to amplify those molecules involved in successful ligation. The probes involved in the ligation are designed such that the 5′ end of one probe is directly adjacent to the 3′ end of the other probe, thereby providing the requisite 3′-OH and 5′-PO_4 group substrates for the ligase.

LCR was originally developed to detect point mutations; a single base mismatch at the junction of the two probe molecules is all that is needed to prevent ligation. By performing the ligation at the T_m of the oligonucleotide probe, only perfectly matched primer:template duplexes will be tolerated. LCR can also be used to amplify template molecules that have been successfully ligated for the purpose of assessing ligation efficiency and producing a large amount of product with even greater specificity than PCR. Thus, LCR is not necessarily an alternative, but rather a complement, to PCR.

LAMP assay

Loop-mediated isothermal amplification (LAMP; Notomi et al., 2000) is an alternative, isothermal method designed primarily for diagnostic applications. It involves the use of four-six primer sets that recognize distinct regions of the target and which support ongoing amplification via loop-mediated (dumbbell structure) product synthesis using a strong strand displacement DNA polymerase (Fig. 10.24) This detection method has enjoyed a high-visibility profile since the onset of the COVID-19 pandemic in that has become a widely used, rapid (usually less than 20 minutes) diagnostic for the presence of SARS-CoV-2 in a clinical sample (Dao Thi et al., 2020; Park et al., 2020); particularly attractive is that this method does not require the expensive instrumentation needed to support qRT-PCR. When used to detect viral RNA genomes, the method is referred to as RT-LAMP or qLAMP, owing to the role of RT to produce an amplifiable first-strand cDNA template. It has the significant advantage of being a point-of-care assay.

The RT-LAMP procedure is a dual-reaction, single-tube process. In the first reaction, cDNA is made by template reverse transcription. In the second reaction, *Bst* DNA polymerase amplifies the cDNA by incubation at 60°C–65°C. The outcome of the assay (Fig. 10.25) is determined by visual inspection of easily discernible changes in the appearance of the sample which, depending on the precise detection chemistry, may include fluorescence, a colorimetric change, or the production of a heavy precipitate (magnesium pyrophosphate) in the reaction tube; for greater accuracy in the case of the latter, quantification can be measured with a turbidimeter. If desired, the reaction products may also be run out on an agarose gel for direct visualization. Many of the same fluorescence dyes and intercalating agents used in real-time PCR are usable with this amplification platform as well, facilitating the measurement of a fluorescence signal using real-time RNA

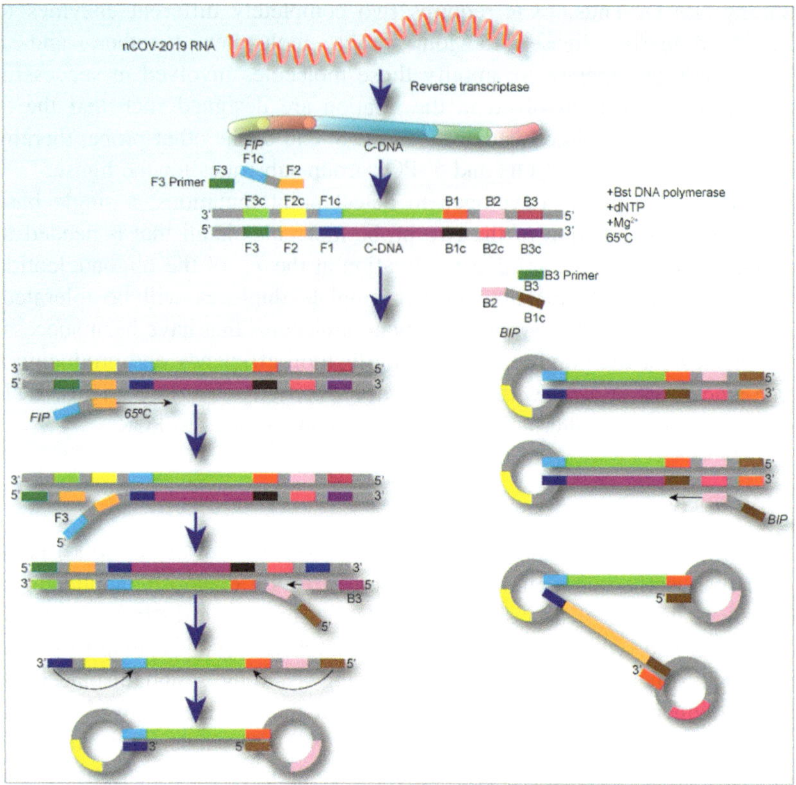

FIGURE 10.24 RT-LAMP assay. Initially, the LAMP primers bind to the complementary target cDNA sequences (e.g., RNA viral detection), and dumbbell-shaped DNAs are produced. Then, during the cycling amplification step, several copies of such dumbbell DNAs are continuously produced. The products formed during the cycling amplification step are used in the elongation phase to generate amplified DNA with various sizes, which produce a characteristic laddering appearance upon electrophoresis (Augustine et al., 2020). *Courtesy Augustine, R., Hasan, A., Das, S., Ahmed, R., Mori, Y., Notomi, T., et al., 2020. Loop-mediated isothermal amplification (LAMP): a rapid, sensitive, specific, and cost-effective point-of-care test for coronaviruses in the context of COVID-19 pandemic. Biology 9 (8), 182. https://doi.org/10.3390/biology9080182 and reproduced with permission under the terms the Creative Commons Attribution License (CC BY 4.0; https://creativecommons.org/licenses/by/4.0/).*

instrumentation. The method has a high tolerance for inhibitors, as is often the case when it is not possible or practical to isolate nucleic acids of sufficient purity to support other assays. In general, however, the assay is lacking in sensitivity compared to qRT-PCR.

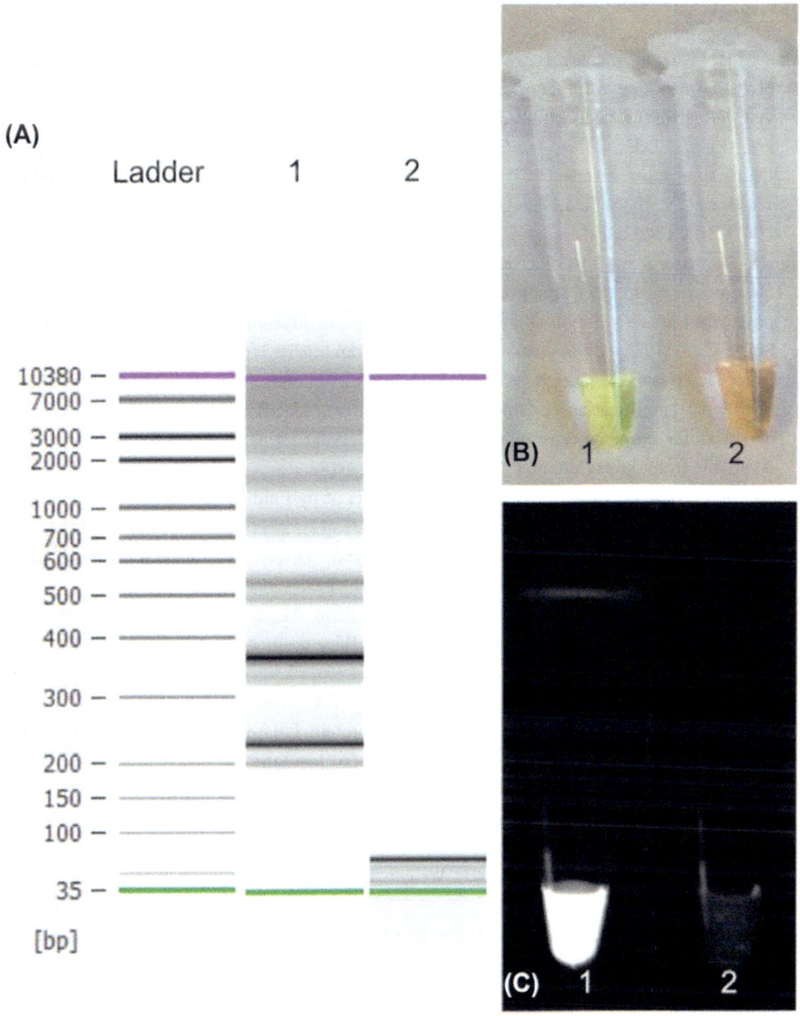

FIGURE 10.25 Visualization of loop-mediated isothermal amplification (LAMP) products. The LAMP products were separated by (A) capillary gel electrophoresis or stained with SYBR Green (Life Technologies, Carlsbad, CA) for observation under (B) visible and (C) UV light. 1 = positive control, containing 1000 copies of genomic DNA of *Mycoplasma bovis* type strain Donetta PG45 ATCC 25523; 2 = negative control. *Courtesy Appelt, S., Aly, S.S., Tonooka, K., Glenn, K., Xue, Z., Lehenbauer, T.W., et al., 2019. Development and comparison of loop-mediated isothermal amplification and quantitative polymerase chain reaction assays for the detection of* Mycoplasma bovis *in milk. J. Dairy Sci. 102 (3), 1985-1996. https://doi.org/10.3168/jds.2018-15306 and reproduced with permission from Elsevier.*

References

Adams, E., Kozma-Bognar, L., Dallmann, G., Nagy, F., 1995. Transcription of tobacco phytochrome-A genes initiates at multiple start sites and requires multiple cis-acting regulatory elements. Plant Mol. Biol. 29, 983–999.

Alquezar-Planas, D.E., Löber, U., Cui, P., Quedenau, C., Chen, W., Greenwood, A.D., 2020. DNA sonication inverse PCR for genome scale analysis of uncharacterized flanking sequences. Methods Ecol. Evol. 12 (1), 182–195. Available from: https://doi.org/10.1111/2041-210X.13497.

Appelt, S., Aly, S.S., Tonooka, K., Glenn, K., Xue, Z., Lehenbauer, T.W., et al., 2019. Development and comparison of loop-mediated isothermal amplification and quantitative polymerase chain reaction assays for the detection of *Mycoplasma bovis* in milk. J. Dairy Sci. 102 (3), 1985–1996. Available from: https://doi.org/10.3168/jds.2018-15306.

Augustine, R., Hasan, A., Das, S., Ahmed, R., Mori, Y., Notomi, T., et al., 2020. Loop-mediated isothermal amplification (LAMP): a rapid, sensitive, specific, and cost-effective point-of-care test for coronaviruses in the context of COVID-19 pandemic. Biology 9 (8), 182. Available from: https://doi.org/10.3390/biology9080182.

Balajee, A.S., May, A., Bohr, V.A., 1999. DNA repairs of pyrimidine dimers and 6–4 photoproducts in the ribosomal DNA. Nucleic Acids Res. 27 (12), 2511–2520.

Barany, F., 1991. Genetic disease detection and DNA amplification using cloned thermostable ligase. Proc. Natl. Acad. Sci. 88, 189–193.

Bassett, C.L., Nickerson, M.L., Farrell Jr., R.E., Harrison, M., 2004. Multiple transcripts of a leucine-rich repeat receptor kinase from morning glory originate from different TATA boxes in a tissue-specific manner. Mol. Gen. Genom. 271 (6), 752–760.

Bassett, C.L., Nickerson, M.L., Farrell Jr., R.E., Artlip, T.S., Ghaouth, A.E., Wilson, C.L., et al., 2005. Characterization of an S-locus receptor protein kinase-like gene from peach. Tree Physiol. 25, 403–411.

Bassett, C.L., Wisniewski, M.E., Artlip, T.S., Richart, G.A., Norelli, J.L., Farrell Jr., R.E., 2009. Comparative expression and transcript initiation of three peach dehydrin genes. Planta 230, 107–118.

Bhatia, P., Taylor, W.R., Greenberg, A.H., Wright, J.A., 1994. Comparison of glyceraldehyde-3-phosphate-dehydrogenase and 28S ribosomal RNA gene expression as RNA loading controls for Northern blot analysis of cell lines of varying malignant potential. Analyt. Biochem. 216, 223–226.

Bonini, J.A., Hofmann, C., 1991. A rapid, accurate nonradioactive method for quantitating RNA on agarose gels. BioTechniques 11, 708–710.

Cheng, S., Fockler, C., Barnes, W.M., Higuchi, R., 1994. Effective amplification of long targets from cloned inserts and human genomic DNA. Proc. Natl. Acad. Sci. 91, 5695–5699.

Chou, Q., 1992. Minimizing deletion mutagenesis artifact during Taq DNA polymerase PCR by *E. coli* SSB. Nucleic Acids Res. 20, 4371.

Clancy, E., Coughlan, H., Higgins, O., Boo, T.W., Cormican, M., Barrett, L., et al., 2016. Development of internally controlled duplex real-time NASBA diagnostics assays for the detection of microorganisms associated with bacterial meningitis. J. Microbiol. Methods 127, 197–202.

Compton, J., 1991. Nucleic acid sequence-based amplification. Nature 350, 91–92.

Dao Thi, V.L., Herbst, K., Boerner, K., Meurer, M., Kremer, L.P.M., Kirrmaier, A., et al., 2020. A colorimetric RT-LAMP assay and LAMP-sequencing for detecting SARS-CoV-2 RNA in clinical samples. Sci. Transl. Med. 12 (556) Available from: https://doi.org/10.1126/scitranslmed.abc7075.

de Leeuw, W., Slagboom, P., Vijg, J., 1989. Quantitative comparison of mRNA levels in mammalian tissues: 28S ribosomal RNA level as an accurate internal control. Nucleic Acids Res. 17, 10137–10138.

Dehesh, K., Franci, C., Sharrock, R.A., Somers, D.E., Welsch, J.A., Quail, P.H., 1994. The Arabidopsis phytochrome A gene has multiple transcription start sites and a promotor sequence motif homologous to the repressor element of monocot phytochrome A genes. Photochem. Photobiol. 59, 379–384.

Devon, R.S., Porteous, D.J., Brookes, A.J., 1995. Splinkerettes – improved vec-torettes for greater efficiency in PCR walking. Nucleic Acids Res. 23, 1644–1645.

Fahle, G.A., Gill, V.J., Fischer, S.H., 1999. Optimal activation of isopsoralen to prevent amplicon carryover. J. Clin. Microbiol. 37, 261–262.

Farrell, R.E., 1999. Long-range PCR: innovations in the long- and short terms. Biotech. Int. 11 (2), 13.

Farrell, R.E., Bassett, C.L., 2007. Multiple transcript initiation as a mechanism for regulating gene expression. In: Bassett, C.L. (Ed.), Regulation of Gene Expression in Plants. Springer Science, New York, pp. 39–66.

Farrell, R.E., Greene, J.J., 1993. Regulation of c-myc and c-Ha-ras oncogene expression by cell shape. J. Cell Physiol. 153, 429–435.

Feroze-Merzoug, F., Berquin, I.M., Dey, J., Chen, Y.Q., 2002. Peptidylprolyl isomerase A (PPIA) as a preferred internal control over GAPDH and β-actin in quantitative RNA analyses. BioTechniques 32, 776–782.

Fra, A.M., Pasqualetto, E., Mancini, M., Sitia, R., 2000. Genomic organization and transcriptional analysis of the human genes encoding caveolin-1 and caveolin-2. Gene 243, 75–83.

Frohman, M.A., Dush, M.K., Martin, G.R., 1988. Rapid production of full-length cDNAs from rare transcripts: amplification using a single gene-specific oligonucleotide primer. Proc. Natl. Acad. Sci. 85, 8998–9002.

Gyllensten, U.B., Erlich, H.A., 1988. Generation of single stranded DNA by the polymerase chain reaction and its application to direct sequencing of the HLA-DQA locus. Proc. Natl. Acad. Sci. 85, 7652–7656.

Henke, W., Herdel, K., Jung, K., Schnorr, D., Loening, S.A., 1997. Betaine improves PCR amplification of GC-rich DNA sequences. Nucleic Acids Res. 25, 3957–3958.

Hodne, K., Weltzien, F.A., 2015. Single-cell isolation and gene analysis: pitfalls and possibilities. Int. J. Mol. Sci. 16 (11), 26832–26849.

Horikoshi, T., Danenberg, K.D., Stadlbauer, T.H., Volkenandt, M., Shea, L.C., Aigner, K., et al., 1992. Quantitation of thymidylate synthase, dihydrofolate reductase, and DT-diaphorase gene expression in human tumors using the polymerase chain reaction. Cancer Res. 52, 108–116.

Horn, C., Hansen, J., Schnütgen, F., Seisenberger, C., Floss, T., Irgang, M., et al., 2007. Splinkerette PCR for more efficient characterization of gene trap events. Nat. Genet. 39, 933–934.

Isaacs, S.T., Tessman, J.W., Metchette, K.C., Hearst, J.E., Cimino, G.D., 1991. Post-PCR sterilization: development and application to an HIV-1 diagnostic assay. Nucleic Acids Res. 19, 109–116.

Kawasaki, E.S., 1990. Amplification of RNA. In: Innis, M.A., Gelfand, D.H., Sninsky, J.J., White, T.J. (Eds.), PCR Protocols: A Guide to Methods and Applications. Academic Press, San Diego, CA, pp. 21–27.

Koshkin, A., Singh, S.K., Nielsen, P., Wengel, J., 1998. LNA (locked nucleic acid): synthesis of the adenine, cytosine, guanine, 5-methylcytosine, thymine and uracil bicyclonucleoside monomers, oligomerization, and unprecedented nucleic acid recognition. Tetrahedron 54 (14), 3607–3630.

Lau, W.Y., Lai, P.B., Leung, M.F., Wong, N., Chen, G., Leung, T.W., et al., 2000. Differential gene expression of hepatocellular carcinoma using cDNA microarray analyses. Oncol. Res. 12, 59–69.

Ma, J., Tran, G., Wan, A.M.D., Young, E.W.K., Kumacheva, E., Iscove, N.N., et al., 2021. Microdroplet-based one-step RT-PCR for ultrahigh throughput single-cell multiplex gene expression analysis and rare cell detection. Sci. Rep. 11, 6777. Available from: doi.org/10.1038/s41598-021-86087-4.

Macias, V.M., Jimenez, A.J., Burini-Kojin, B., Pledger, D., Jasinskiene, N., Phong, C.H., et al., 2017. nanos-Driven expression of piggyBac transposase induces mobilization of a synthetic autonomous transposon in the malaria vector mosquito, *Anopheles stephensi*. Insect Biochem. Mol. Biol. 87, 81–89. Available from: doi.org/10.1016/j.ibmb.2017.06.014.

Maruyama, K., Sugano, S., 1994. Oligo-capping: a simple method to replace the cap structure of eukaryotic mRNAs with oligoribonucleotides. Gene 138, 171–174.

Mullis, K.B., Faloona, F.A., Scharf, S., Saiki, R., Horn, G., Erlich, H., 1986. Specific enzymatic amplification of DNA in vitro: the polymerase chain reaction. Cold Spring Harb. Symp. Quant. Biol. 51, 263–273.

Mullis, K.B., Faloona, F.A., 1987. Specific synthesis of DNA in vitro via a polymerase-catalyzed chain reaction. Methods Enzymol. 155, 335–350.

Notomi, T., Okayama, H., Masubuchi, H., Yonekawa, T., Watanabe, K., Amino, N., et al., 2000. Loop-mediated isothermal amplification of DNA. Nucleic Acids Res. 28 (12), e63. Available from: https://doi.org/10.1093/nar/28.12.e63.

Olive, D.M., Simsek, M., Al-Mufti, S., 1989. Polymerase chain reaction assay for detection of human cytomegalovirus. J. Clin. Microbiol. 27, 1238–1242.

Paabo, S., Gifford, J.A., Wilson, A.C., 1988. Mitochondrial DNA sequences from a 7000-year-old brain. Nucleic Acids Res. 16, 9775–9787.

Park, G.S., Ku, K., Baek, S.H., Kim, S.J., Kim, S.I., Kim, B.T., et al., 2020. Development of reverse transcription loop-mediated isothermal amplification assays targeting severe acute respiratory syndrome coronavirus 2 (SARS-CoV-2. J. Mol. Diagn. 22 (6), 729–735. Available from: https://doi.org/10.1016/j.jmoldx.2020.03.006.

Pernas-Alonso, R., Morelli, F., di Porzio, U., Perrone-Capano, C., 1999. Multiplex semi-quantitative reverse transcriptase-polymerase chain reaction of low abundance neuronal mRNAs. Brain Res. Protoc. 4, 395–406.

Potter, C.J., Luo, L., 2010. Splinkerette PCR for mapping transposable elements in Drosophila. PLoS One 5 (4), e10168. Available from: https://doi.org/10.1371/journal.pone.0010168.

Rees, W.A., Yager, T.D., Korte, J., Von Hippel, P.H., 1993. Betaine can eliminate the base pair composition dependence of DNA melting. Biochemistry 32, 137–144.

Riley, J., Butler, R., Ogilvie, D., Finniear, R., Jenner, D., Powell, S., 1990. A novel, rapid method for the isolation of terminal sequences from yeast artificial chromosome (YAC) clones. Nucleic Acids Res. 18, 2887–2890.

Robert, C., McGraw, S., Massicotte, L., Pravetoni, M., Gandolfi, F., Sirard, M.-A., 2002. Quantification of housekeeping transcript levels during the development of bovine preimplantation embryos. Biol. Reprod. 67, 1465–1472.

Saiki, R.K., Scharf, S., Faloona, F., Mullis, K.B., Horn, G.T., Erlich, H.A., et al., 1985. Enzymatic amplification of β-globin genomic sequences and restriction site analysis for diagnosis of sickle cell anemia. Science 230, 1350–1354.

Sarkar, G., Kapelner, S., Sommer, S.S., 1990. Formamide can dramatically improve the specificity of PCR. Nucleic Acids Res. 18, 7465.

Sato, M., Saitoh, I., Kiyokawa, Y., Akasaka, E., Nakamura, S., Watanabe, S., et al., 2022. Electroporation-based nonviral gene delivery to adipose tissue in mice. OBM Genet. 6 (2), 151. Available from: https://doi.org/10.21926/obm.genet.2202151.

Sellner, L.N., Coelen, R.J., Mackenzie, J.S., 1992. Reverse transcriptase inhibits Taq polymerase activity. Nucleic Acids Res. 20, 1487–1490.

Shaefer, B., 1995. Revolution in rapid amplification of cDNA ends: new strategies for polymerase chain reaction cloning of full-length cDNA ends. Analyt. Biochem. 227, 255–273.

Shuman, S., 1994. Novel approach to molecular cloning and polynucleotide synthesis using vaccinia DNA topoisomerase. J. Biol. Chem. 269, 32678–32684.

Spanakis, E., 1993. Problems related to the interpretation of autoradiographic data on gene expression using common constitutive transcripts as controls. Nucleic Acids Res. 16, 3809–3819.

Ståhlberg, A., Kubista, M., 2018. Technical aspects and recommendations for single-cell qPCR. Mol. Asp. Med. 59, 28–35. Available from: doi.org/10.1016/j.mam.2017.07.004.

Steen, A.M., Luthman, H., Hellgren, D., Lambert, B., 1990. Levels of hypoxanthine phosphoribosyltransferase RNA in human cells. Exp. Cell Res. 186, 236–244.

Stevens, A.J., Appleby, S., Kennedy, M.A., 2016. Many commercial hot-start polymerases demonstrate activity prior to thermal activation. BioTechniques 61 (6), 293–296.

Teng, X., Zhang, Z., He, G., Yang, L., Li, F., 2012. Validation of reference genes for quantitative expression analysis by real-time RT-PCR in four lepidopteran insects. J. Insect Sci. 12, 60. Available from: https://doi.org/10.1673/031.012.6001.

Tepper, C.G., Pater, M.M., Pater, A., Xu, H., Studzinski, G.P., 1992. Mitochondrial nucleic acids as internal standards for blot hybridization analyses. Analyt. Biochem. 203, 127–133.

Van Gelder, R.N., von Zastrow, M.E., Yool, A., Dement, W.C., Barchas, J.D., Eberwine, J.H., 1990. Amplified RNA synthesized from limited quantities of heterogeneous cDNA. Proc. Natl. Acad. Sci. 87, 1663–1667.

Walker, G.T., Little, M.C., Nadeau, J.G., Shank, D.D., 1992a. Isothermal in vitro amplification of DNA by a restriction enzyme/DNA polymerase system. Proc. Natl. Acad. Sci. 89, 392–396.

Walker, G.T., Fraiser, M.S., Schram, J.L., Little, M.C., Nadeau, J.G., Malinowski, D.P., 1992b. Strand displacement amplification – an isothermal, in vitro DNA amplification technique. Nucleic Acids Res. 20, 1691–1696.

Weitzel, J.M., Grott, S., Radtke, C., Kutz, S., Steitz, H.J., 2000. Multiple promoters direct the tissue-specific expression of rat mitochondrial glycerol-3-phosphate dehydrogenase. Biol. Chem. 381, 611–614.

Weitzel, J.M., Kutz, S., Radtke, C., Grott, S., Steitz, H.J., 2001. Hormonal regulation of multiple promoters of the rat mitochondrial glycerol-3-phosphate dehydrogenase gene. Eur. J. Biochem. 268, 4095–4103.

Wisniewski, M.E., Bassett, C.L., Renaut, J., Farrell Jr., R.E., Tworkoski, T., Artlip, T.S., 2006. Differential regulation of two dehydrin genes from peach (*Prunus persica* L. Batsch) by photoperiod, low temperature, and water deficit. Tree Physiol. 26, 575–584.

Wray, G.A., Hahn, M.W., Abouheif, E., Balhoff, J.P., Pizer, M., Rockman, M.V., et al., 2003. The evolution of transcriptional regulation in eukaryotes. Mol. Biol. Evol. 20, 1377–1419.

Xu, L., Duan, J., Chen, J., Ding, S., Cheng, W., 2021. Recent advances in rolling circle amplification-based biosensing strategies – a review. Anal. Chim. Acta 1148, 238187. Available from: https://doi.org/10.1016/j.aca.2020.12.062.

Yamada, M., Ozawa, A., Ishii, S., Shibusawa, N., Hashida, T., Ishizuka, T., et al., 2001. Isolation and characterization of the rat Prolactin-releasing peptide gene: multiple TATA boxes in the promoter region. Biochem. Biophys. Res. Comm. 281, 53–56.

Zhang, Y., Li, X., Zou, R., Xue, Y., Lou, X., He, M., 2014. Bovine thrombin enhances the efficiency and specificity of polymerase chain reaction. BioTechniques 57, 289–294.

Further reading

Bustin, S., Nolan, T., 2004. Pitfalls of quantitative real-time reverse-transcription polymerase chain reaction. J. Biomol. Tech. 15 (3), 155–166.

Corman, V.M., Landt, O., Kaiser, M., Molenkamp, R., Meijer, A., Chu, D.K., et al., 2020. Detection of 2019 novel coronavirus (2019-nCoV) by real-time RT-PCR. Eurosurveillance 25 (3), 2000045.

Mohon, A.N., Oberding, L., Hundt, J., van Marle, G., Pabbaraju, K., Berenger, B.M., et al., 2020. Optimization and clinical validation of dual-target RT-LAMP for SARS-CoV-2. J. Virol. Methods 286, 113972. Available from: doi.org/10.1016/j.jviromet.2020.113972.

Sellner, L.N., Turbett, G.R., 1998. Comparison of three RT-PCR methods. BioTechniques 25 (2), 230–234.

Chapter 11

Quantitative PCR techniques

Rationale

The polymerase chain reaction (PCR) revolutionized every aspect of biotechnology since its inception. In short, it is a primer-mediated, enzymatic amplification of specific genomic DNA or cDNA sequences. The successive application of the very familiar format of sample denaturation, primer annealing, and primer extension, routinely over as many as 25–30 cycles, results in the exponential amplification of target molecules. Readers who are not yet familiar with the mechanics of RT-PCR are strongly encouraged to read Chapter 10, RT-PCR: a science and an art form, in its entirety before moving forward.

PCR in its various iterations offers unsurpassed amplification potential. For simple "yes or no"-type experiments there are less stringent requirements for the input mass and purity of RNA compared to more traditional methods. This is true partly because PCR primers "frame" the domain that will be amplified, making PCR a bit more tolerant of lesser quality RNA than traditional cDNA synthesis and cloning procedures. For accurate quantification, however, input mass of RNA, transcript quality, overall amplification efficiency, and detection platform are key parameters.

Historically, the major difficulties associated with the synthesis of cDNA, both for cloning purposes and as an indirect measure of transcript abundance, have been (1) the inability to generate near full-length, *representative* cDNA molecules and (2) the overall low efficiency of the process of reverse transcription itself. Currently, the extent to which these difficulties can be overcome is related directly to (1) the quality of the RNA template; (2) the quantity of the RNA template; and (3) the cDNA synthesis methodology, which is described in Chapter 9, cDNA: a permanent biochemical record of the cell.

As a result of remarkable technology innovations, the achievable level of sensitivity in PCR-based transcription assays is extraordinary. Each month, improvements appearing in the literature offer new or improved approaches for maximizing sensitivity. As close as we might come, however, there is no such thing as absolute quantification, though, with the rise of digital PCR (dPCR), some might dispute that statement. Human error, no matter how small, will always be a confounding variable in the unending quest for absolute accuracy.

RNA Methodologies. DOI: https://doi.org/10.1016/B978-0-323-90221-2.00037-0

Sensitivity index

This laboratory guide presents a number of methodologies for the isolation and subsequent characterization of RNA. It should be abundantly clear that each of these methods offers distinct advantages and disadvantages which extend into the realm of the overall reliability and sensitivity of the particular assay. Failure to include appropriate controls or to adhere precisely to protocols will absolutely have a negative impact on the sensitivity of an experiment and the interpretation of the resulting data.

Most often, levels of RNA in an experimentally related set of samples are described in terms of relative abundance, meaning that one of the samples is designated as the control or reference sample and to which all other samples are compared. Typically, an observed hybridization signal or the intensity of a reference PCR product is assigned a value of 1.0 and the behavior of each of the other samples is compared to this baseline value. Be aware, however, that determination of relative abundance is directly related to the sensitivity of the assay being performed; evaluating the same set of samples using a battery of different methods or even the same method on different days will often produce startlingly different results.

The relative sensitivity of several common transcription assay methods is herein described in terms of what this author affectionately calls "Farrell's RNA sensitivity index." As shown below, RNA-based assays for transcript quantification are arranged vertically, moving down from highest sensitivity to lowest. The sensitivity index ranking is based on three factors: the intrinsic difficulty of the technique, reproducibility, and current average level of sensitivity under optimal conditions. The cost of utilizing the method is not taken into account.

It should be clear from reading earlier parts of this text that techniques which rank lower on the sensitivity index are nonetheless able to provide useful, if not unique, information. For example, the time-honored Northern analysis provides information about the native size of the mRNA produced at a specific locus, a datum not associated with methods of higher sensitivity ranking.

Highest Sensitivity
RNA-seq
Digital PCR
Real-Time PCR
End-Point PCR
Microarray Technology
Nuclear Runoff Assay
RNase Protection Assay
S1 Nuclease Protection Assay
Northern Analysis
Dot Blot Analysis
Lowest Sensitivity

Note that the sensitivity associated with RNA-seq is read depth-dependent. Further, RNA-seq and quantitative PCR (qRT-PCR) should be viewed as complementary methods, owing to the advantages and disadvantages of each. Some molecular biology practitioners might rate digital PCR as having a sensitivity edge over RNA-seq; keep in mind, however, that PCR only amplifies sequences for which there are known primers. RNA-seq identifies all products of transcription, known and unknown, including the myriad ncRNA species produced by the cell.

Quantitative approaches

Quantitative approaches of all persuasions have been developed for the purposes of quantifying transcript abundance and determining gene copy number: the former is the focus of this chapter. Historically, quantification has been something of a challenge because of sample-to-sample variations in the efficiency of cDNA synthesis often due to residual chemistries associated with the recovery of RNA from its biological source, including ethanol, isopropanol, phenol, or (commonly) excess salt, all of which are easy to resolve using contemporary isolation methods. A primary concern among investigators wishing to measure transcript abundance by RT-PCR is the realization that the final mass of the PCR product(s) is dependent on the efficiency of the reverse transcriptase *and* the efficiency of the thermostable enzyme(s) used to support PCR, even if said actions occur in the same tube and by the action of the same enzyme (one-tube PCR). Thus, any reliable quantification method must factor in the relative efficiencies of both enzyme-mediated reactions because the overall efficiency of a technique is the product of the efficiencies of the individual steps involved. Keep in mind also that the "efficiency index" is user-dependent. At present, one variant or another of real-time PCR is available to most investigators and is widely considered to be the gold standard. Looking ahead, however, as the cost of whole transcriptome RNA-seq continues to decrease, the prevalence of RNA-seq data in the literature will continue to increase. Alternative approaches are also discussed below.

In order to ensure the meaningful comparison between all samples, the process of normalization begins very early in an investigation, and there are a number of ways to approach this challenge. With respect to the reverse transcriptase reaction products, normalization of cDNA samples since the early days of molecular biology has been based on sample input mass, performed most often (1) by preparing uniform dilutions of a single aliquot of the cDNA, (2) based on A_{260} measurements, or (3) based on the amount (cpm) of incorporated radiolabel in instances where cDNA was performed in the presence of a small amount of isotopic tracer. Although these methods offer a reasonably accurate glimpse of cellular biochemistry, a more reliable normalization procedure is to measure some type of control or reference sequence, comparing its behavior to that of the experimental sequences under investigation. The inclusion of such a control is of prime importance because of the variegated nature of cDNA synthesis, with efficiencies commonly ranging from 10% to 90%. As discussed below, the reference transcript can be an endogenous mRNA or an artificial transcript of precisely known mass that is spiked into the biological sample prior to first-strand cDNA-synthesis. In so doing, reverse transcription and PCR amplification should influence both control and experimental templates equally, which is the rationale for closely monitoring a control or reference transcript.

With respect to RT-PCR, inasmuch as PCR is an exponential process, the degree to which the template is amplified is described by the equation

$$N = N_o(1+\text{eff})^n$$

where N = the number of copies of PCR product at the end of amplification, N_o = the initial number of copies of template, eff = the efficiency of the reaction, expressed as a decimal, n = the number of cycles performed.

It is important to note that this mathematical model can be used only to make an estimate of the amplification potential of PCR in a particular reaction tube because of the efficiency of *each* round of "denature–anneal–extend" changes with each passing cycle, particularly in later cycles when the amplification efficiency of each cycle declines significantly. In great measure, this is due to what is known as the plateau effect (Fig. 11.1). PCR product accumulates exponentially in the early, high-precision stages of the reaction because the reaction components are present in abundance, making the data very reliable. As PCR progresses, the enzyme(s) is becoming less efficient (they are not thermostable forever), the free nucleotide concentration is diminishing, as is the available primer concentration, and there is an increasing number of

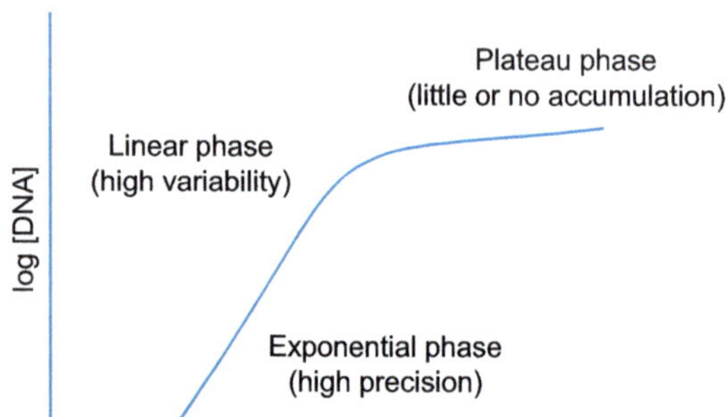

FIGURE 11.1 PCR amplification dynamics. The greatest precision in quantification results when product accumulation measurements and comparisons are made very early in the exponential phase, as is done in real-time PCR. As the reaction progresses, the true abundance relationships among samples are lost because the reaction is becoming less efficient with each passing cycle and because the reagents become limiting, due to a reduction in raw materials, namely, primers, nucleotides, and functional enzymes. Invariably side reactions will then occur.

new template molecules after each passing cycle which are competing for the reduced amounts of the primers, nucleotides, and functional polymerase. Thus, the reaction enters the linear phase in which there is great variability with respect to reaction efficiency. Finally, in the plateau phase, there is little, if any, product accumulation with each passing cycle, meaning that additional cycles within the plateau phase are essentially pointless.

Why not, therefore, just run PCR for 8–12 cycles and the run an aliquot on a gel? The problem is that there is almost never enough product after only a dozen or so cycles to see anything on a gel, even using the sensitive SYBR dyes. Plus, the opening of the tube provides yet another opportunity for amplicon contamination to occur. An older attempt to circumvent the product mass issue was to add ^{32}P to the reaction; the resulting PCR products could then be run on a gel and quantified by autoradiography. The problems with this method are twofold: first, there is the potential for a huge radioactive mess, and second, the X-ray film has a narrow linear response, meaning the film has a limited ability to record the true abundance differences among samples. For these and other reasons, real-time PCR has been widely embraced.

The MIQE guidelines

The development of multiple real-time PCR platforms has been a tremendous asset to all types of laboratory investigations, from basic research to applied diagnostics and everything in between. Productivity increased dramatically, as did the number of journal articles featuring real-time PCR data. It quickly became apparent that the materials and methods sections of these papers were seriously lacking in terms of critical experimental detail, thereby preventing the replication of results and greatly impeding critical analysis. These serious shortcomings were exacerbated by the fact there really is no such thing as a standard PCR reaction, and that "nonstandard" reactions were being performed on instrumentation with varying levels of efficiency.

Fortunately, common sense prevailed and the MIQE (minimum information for publication of quantitative real-time PCR experiments) guidelines were compiled by an international group of collaborators and published in the journal *Clinical Chemistry* (Bustin et al., 2009). The MIQE (*pronounced* MY-KEY) guidelines (Fig. 11.2) present a series of checklists in areas that include experimental design, nucleic acid extraction, qPCR target information, qPCR protocol, qPCR validation, data analysis, and other areas. Within each of the groups, specific items are ranked as being essential (E) for inclusion in a manuscript, or desirable (D) for inclusion in a manuscript if that information is available. The MIQE guidelines also include recommendations about standard terminology for specific methodologies. An important update to the MIQE guidelines was subsequently published in 2020 (Fig. 11.3) which was expanded to include digital PCR applications.

Clinical Chemistry 55:4
611–622 (2009)

Special Report

The MIQE Guidelines: *M*inimum *I*nformation for Publication of *Q*uantitative Real-Time PCR *E*xperiments

Stephen A. Bustin,[1*] Vladimir Benes,[2] Jeremy A. Garson,[3,4] Jan Hellemans,[5] Jim Huggett,[6] Mikael Kubista,[7,8] Reinhold Mueller,[9] Tania Nolan,[10] Michael W. Pfaffl,[11] Gregory L. Shipley,[12] Jo Vandesompele,[5] and Carl T. Wittwer[13,14]

FIGURE 11.2 The MIQE guidelines. This article provides a comprehensive of recommendations pertaining to details that should be included when real-time PCR-based studies are published. Also included are recommendations for standardized procedural and technical terminology. *Bustin, S.A., V. Benes, J.A. Garson, J. Hellemans, J. Huggett, M. Kubista, R. Mueller, T. Nolan, M.W. Pfaffl, G.L. Shipley, J. Vandesompele, and C.T. Wittwer. (2009). The MIQE guidelines: minimum information for publication of quantitative real-time PCR experiments.* Clinical Chemistry *55(4), 611–622. doi:10.1373/clinchem.2008.112797. Reproduced with permission from the American Association for Clinical Chemistry.*

Clinical Chemistry 66:8
1012-1029 (2020)

Special Report

The Digital MIQE Guidelines Update: Minimum Information for Publication of Quantitative Digital PCR Experiments for 2020

The dMIQE Group*

FIGURE 11.3 Updated MIQE guidelines which include important information pertaining to digital PCR applications. *Huggett, J.F. et al. (2020). The digital MIQE guidelines update: minimum information for publication of digital PCR experiments for 2020.* Clinical Chemistry *66(8), 1012–1029. https://doi.org/10.1093/clinchem/hvaa125. Figure courtesy of the American Association for Clinical Chemistry and reproduced according to the terms of the Creative Commons Attribution 4.0 License (https://creativecommons.org/licenses/by/4.0/).*

The purpose of these guidelines is to ensure greater inter- and intralaboratory consistency, which is critical for collaborators. Given the widespread use and high-visibility profile of the real-time PCR process, there is a risk of flooding the primary literature with inaccuracies and with conflicting information unless standards for the reporting of real-time PCR data are in place. In so doing, greater transparency is achieved and thereby speeding up the journal peer-review process. Investigators working in this area are strongly encouraged to become familiar with these guidelines in order to streamline one's publication pipeline or if for no other reason, to ensure that experimental records are as complete as possible. It is very important to become MIQE-compliant. All of this makes perfect sense, particularly given that there are reporting guidelines, controlled vocabulary standards, and recommended data formats already in place for many other subdisciplines within the biological sciences. Information pertaining to reporting standards in multiple areas is available from FAIRsharing, which was previously known as BioSharing until 2018. FAIRsharing ins managed by the Research Data Alliance Foundation and can be accessed at https://fairsharing.org.

Real-time PCR

A major breakthrough in gene- and transcript quantification was the development of what is commonly known as "real-time" PCR (Higuchi et al., 1992, 1993). In this fluorescence-based approach, PCR products are quantified as they accumulate, in real-time, as opposed to performing a fixed number of cycles and then assessing the amount and variety of products by electrophoresis, now commonly known as end-point PCR. Early on, many reports in the scientific literature referred to real-time PCR as "the new RT-PCR." This nomenclature led to confusion, however, because of the immediate association of RT-PCR with the use of reverse transcriptase to synthesize cDNA, and the amplification thereof, which may not involve real-time PCR at all. Real-time PCR should be referred to as "real-time PCR" so that there is no question in the mind of the reader as to precisely which molecular techniques have been performed. It is currently a widespread practice to refer to real-time PCR as simply qPCR (stands for quantitative PCR).

Real-time PCR requires a fluorescent label in the reaction and a thermal cycler equipped with fluorescence excitation and optical fluorescence detection devices, all of which are automated. Real-time PCR variants may involve the reaction vessel design, the reaction format, the amplification and detection chemistries, or the design and functionality of the thermal cycler instrumentation, an example of one of which is shown in Fig. 11.4. Succinctly, the abundance of PCR products is interpolated from the increasing level of fluorescence that is generated during PCR; these data are collected once in each PCR cycle and then related to a standard curve for high-precision quantification. The fluorescence may be the result of chemically modified oligonucleotides acting as primers or as probes, or by direct binding of a fluorescent dye to dsDNA molecules as they are synthesized.

The real-time approach proffers several advantages over traditional methods for measuring transcript abundance including greater speed because this is largely an automated process and a wide dynamic range which often spans eight to nine orders of magnitude (Table 11.1). The ability to recognize subtle variations in gene expression is possible because quantification of the original template mass is calculated early in the exponential phase of cDNA amplification. With respect to concerns regarding the identity of the amplified DNA, the closed-tube system reduces the risk of laboratory-wide amplicon contamination and data variations due to sample handling, since there is no post-PCR

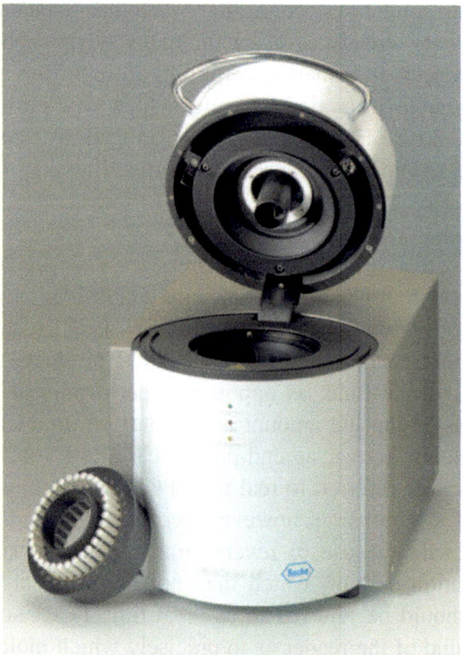

FIGURE 11.4 LightCycler 2.0 real-time PCR system shown with the top open. The instrument holds up to 32 samples in specially designed capillary tubes that travel on a rotating carrousel (shown) during the PCR process. This is but one example of real-time PCR instrumentation. *Courtesy of Roche Applied Science.*

TABLE 11.1 Real-time PCR transcript quantification: advantages and disadvantages.

Advantages
• The format is simple and versatile. • The reaction is fast. • The data are collected in real time. • Very low abundance mRNAs can be detected due to fluorescence sensitivity. • Plateau phase associated with end-point PCR is a nonissue. • Small numbers of cells or tissue mass are needed. • Poly(A)$^+$ selection is neither necessary nor recommended. • Formats include standard curve analysis, relative quantification, and digital PCR. • The reaction vessel need not be opened at the end of the experiment. • The format precludes difficulties of blot-associated analysis.
Disadvantages
• Reverse transcriptase reaction is the major source of variability among samples. • Small variations in micropipetting or technique can cause drastic changes in product yield and variety. • Sequence-specific probe, in addition to primers, is needed for the *Taq*Man assay • True abundance relationships among transcripts may be impacted by the intrinsic amplification bias of PCR. • The reaction is sensitive to contaminants, especially genomic DNA. • Specialized instrumentation is required.

handling of samples. For each type of thermal cycler, the accompanying software is capable of making sophisticated measurements; for example, melting curve analysis routinely performed which allows discrimination between true PCR products, primer–dimer artifacts, and "unexpected" amplicons.

The basic unit of measure in real-time PCR is the threshold cycle (C_T). This is the cycle at which the fluorescence emanating from a sample, due to PCR product accumulation, crosses a threshold level that is predetermined by the user (e.g., 10 times above background). The higher the amount of template at the onset of a reaction, the sooner a significant fluorescence increase will be detected (Fig. 11.5). As a general rule, each three-C_T difference between samples reflects about a ten-fold difference in template concentration at the onset. The important points to keep in mind are (1) low $C_T =$ high template abundance and (2) early detection is best, so that plateau effects (nonexponential product accumulation) do not compromise the sensitivity of the assay. Note that C_T is also referred to as C_q per the publication of the original MIQE guidelines (Bustin et al., 2009).

There are two ways to approach data analysis in real-time PCR applications. The first is the relative quantification method, whereby the amplification of an experimental template is compared to that of a control sample. This is similar to the older, well-known relative abundance concept. The relative quantification method is also known as the $\Delta\Delta C_T$ method because one compares the C_T values of unknown samples to the amplification behavior of the same gene in a control or reference sample, or to reference genes, previously referred to as housekeeping genes. The other method, known as the standard

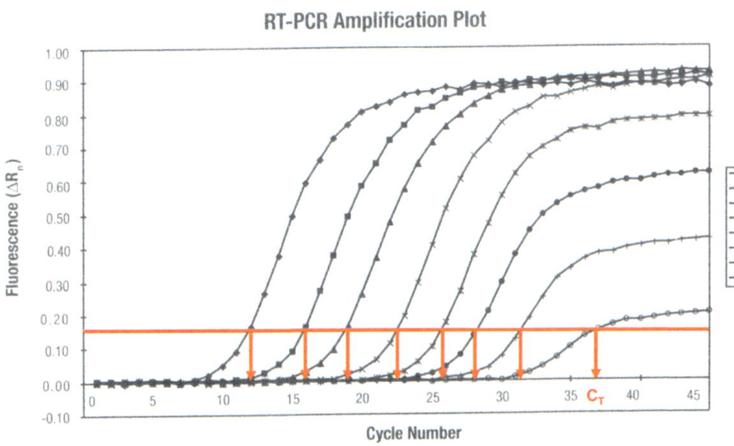

FIGURE 11.5 Real-time PCR measures the amount of product at the end of each cycle, which is then plotted as a function of cycle number. The important parameter is the cycle threshold, or C_T, where the product mass in each reaction tube accumulates to predetermined level (red line). The lower the amount of starting material (cDNA or genomic DNA) the higher will be C_T. The C_T is labeled in this figure for one sample only. *Courtesy of Dr. K.J. Livak, Applied Biosystems, Inc.*

curve method, involves comparing the amplification behavior of an experimental sample against a standard curve (Fig. 11.6) and, in so doing, is used to calculate the actual copy number of template molecules with statistical confidence. While all of the measurements and calculations are performed automatically by the real-time thermal cycler, they are included as figures here to enhance the understanding of readers who may be unfamiliar with standard curve-type analyses. One historical note: the standard curve real-time PCR method was originally referred to as the absolute quantitation method, which is something of a misnomer.

Real-time systems are attractive for a number of obvious reasons, the most obvious of which is that data collection in real time greatly expands the linear dynamic range of this assay. The downside to implementation in the laboratory is the cost of purchasing the instrumentation, often representing an up-front investment of tens of thousands of dollars. There is also the cost of the consumables needed for each experiment to be considered as well. It is important to note that real-time PCR is not a suitable technique for the simultaneous assay of more than a few genes unless the lab is set up for high-throughput analysis, including instrumentation capable of accommodating larger numbers of samples. Some investigators perform a preliminary

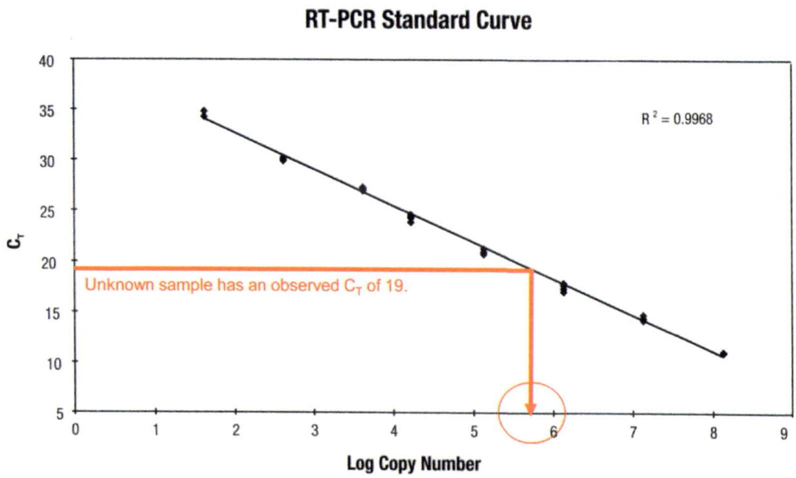

FIGURE 11.6 If absolute quantification is to be performed then a standard curve is generated by the usual method of preparing and assaying template serial dilutions of known mass, similar to the data shown in Fig. 11.5. To generate the standard curve, the starting number of copies of the template is plotted as a function of threshold cycle (C_T). The C_T of an experimental sample (red line) is then plotted against the standard curve, and the corresponding amount of the starting template is read from the x-axis; note that this is a completely automated calculation. If changes in gene expression are to be reported in terms of relative abundance then the real-time format would involve the assay of a reference gene (e.g., GAPDH or β-actin), or an entirely different control sample, to which the experimental gene product accumulation is compared directly. *Courtesy of Dr. K.J. Livak, Applied Biosystems, Inc.*

screening for differentially expressed genes using microarrays, as described in Chapter 21, Array analysis of gene expression. Following identification of purported up- or downregulated sequences, real-time PCR is then used to confirm differential expression and to determine the magnitude of the change.

The real-time PCR format has been invaluable across many disciplines, including pathogen detection, molecular diagnostics, mutation detection, miRNA profiling, assay of alternative splicing, and the analysis of tissue samples that may have been archived long ago. In the case of interlaboratory collaboration, all parties involved must standardize their protocols and identify some type of control or reference RNA to be used for quantitative analysis. Many of the leading biotech suppliers manufacture a control RNA product, sometimes known as universal RNA, which is a pool of RNA from several different cell lines representing different tissues and transcripts at different abundance levels. Because each lot of reference RNA is manufactured on an industrial scale, one may obtain RNA from the same lot for a very long time, and geographically dispersed users have ready access to the same control material.

Real-time PCR platforms

Real-time PCR combines amplification and detection, meaning that no gels are needed to analyze the product. Real-time PCR can generate fluorescence signals in a number of interesting ways, the choice of which is based on the desired level of sensitivity as well as the nature of the data that is sought. Laboratory economics is also a concern because the instrumentation and consumables are expensive. Each of the standard platforms is described here, though it is important to realize that there are also permutations within each of these real-time methodologies.

SYBR Green assay

The first published reports of the real-time PCR process actually described the use of ethidium bromide (EtBr) in the reaction. This intercalating agent associated itself with double-stranded PCR products as they accumulated and the resulting increase in fluorescence was proportional to the product mass. This was known as intercalator-based detection. While somewhat more reliable than many of the classical methods for PCR-based assay of transcript abundance available at the time, intercalating agents will associate both with specific and nonspecific products generated in the reaction. Moreover, the intrinsic background fluorescence associated with EtBr was a huge problem from the beginning, as anyone who has ever run an EtBr-stained gel knows it all too well. No one uses EtBr for real-time applications.

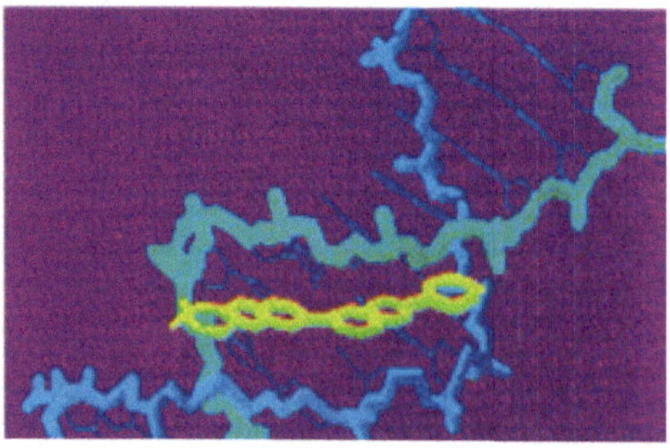

FIGURE 11.7 SYBR Green assay for real-time PCR. The dye binds to the minor groove of dsDNA, resulting in a fluorescence signal that is proportional to the accumulating product. This methodology does not discriminate between specific or nonspecific products. All double-stranded molecules will contribute to the fluorescence signal. *Courtesy of Applied Biosystems, Inc.*

The development of the fluorescent dye SYBR Green greatly fine-tuned the real-time PCR assay, mostly into the format which is used today. Unlike EtBr, SYBR Green binds to the minor groove of DNA (Fig. 11.7), meaning that it fluoresces only when bound to dsDNA (read: PCR products), thereby greatly reducing background fluorescence. The issue of a fluorescence contribution by nonspecific products (false positives) persists: any dsDNA products that are synthesized, specific or nonspecific, will result in a measurable level of fluorescence increase. Nonspecific products are revealed, however, during an ensuing postamplification melting curve analysis.

The SYBR Green assay has the advantage of being cost-effective relative to some of the other real-time PCR methods. Because the fluorescence signal is in proportion to the product as it accumulates, there is no reason to open the tube at the end of the assay. The commonly cited disadvantages to the SYBR Green assay include the fact that SYBR Green will base-pair to any double-stranded molecules at are produced which may, therefore, result in a lower sensitivity assay. Optimization of SYBR Green-based real-time PCR may also be required, though tube-to-tube variations have been reduced through the widespread use of commercial SYBR Green master mix formulations.

*Taq*Man assay

To address many of the intrinsic difficulties associated with DNA binding-based detection (SYBR Green assay), a modification known as probe-based detection was developed that provides a substantially higher degree of resolution. Originally known as the fluorogenic 5′ nuclease assay (Holland et al., 1991; Lee et al., 1993; Livak et al., 1995) the method is now known universally as the TaqMan assay. Whereas the assay was performed originally via the liberation of a 5′ ^{32}P end-label as an indicator and quantified by thin layer chromatography, the current form of this assay features a dual-labeled fluorochrome oligonucleotide probe, typically 18–22 bases, that is designed to anneal to the template between the region flanked by the 5′ primer and the 3′ primer. This means that there are three oligonucleotides involved in the process: the 5′ primer, the 3′ primer, and the *Taq*Man probe. At its 5′ end, the probe is labeled with a reporter (R) fluorochrome, commonly FAM, TET, or VIC. The 3′ end the probe is labeled with a nonfluorescent quencher (Q) or (NFQ), commonly Dabcyl or one of the BQH quenchers. Examples of common reporter–quencher combinations are shown in Table 11.2, and the

TABLE 11.2 Common reporter–quencher probe combinations.

5′ Reporter dye	3′ Quencher
FAM	MGB-NFQ
TET	
VIC	
HEX	Dabcyl
TET	
FAM	
HEX	BHQ-1
TET	
FAM	
JOE	
VIC	
TAMRA	
Yakima Yellow	
Cy3	BHQ-2
ROX	
TAMRA	
Texas Red	
Cy5	BHQ-3
ROX	
Texas Red	

Reporters: *FAM*, 6-carboxyfluoroscein; *TET*, tetrachloro-6-carboxyfluoroscein; *VIC*, "Victoria"; *HEX*, hexachloro-6-carboxy-fluorescein; *JOE*, 6-carboxy-4′,5′-dichloro-2′, 7′-dimeth-oxyfluorescein, succinimidyl ester; *Cy3*, indocarbocyanine; *Cy5*, indodicarbocyanine; *ROX*, 6-carboxy-X-rhodamine, succinimidyl ester; *TAMRA*, tetramethyl-6-carboxyrhodamine; *Texas Red*, sulforhodamine 101 acid chloride; *VIC*, 2′-chloro-phenyl-1,4-dichloro-6-carboxyfluorescein; *Yakima Yellow*, yakima yellow phosphoramidite. Quenchers: *MGB-NFQ*, minor groover binder nonfluorescent quencher; *Dabcyl*, 4-(4′-dimethylaminophenylazo)benzoic acid; *BHQ*, Black hole quencher.

various available reporter fluorochromes facilitate multiplexing. The spatial proximity of reporter and quencher in an intact probe suppresses fluorescence of the reporter by Förster resonance energy transfer, also known as fluorescence resonance energy transfer, or simply FRET (Förster, 1946, 1948).

*Taq*Man probes are designed to have a higher T_m than the primers (typical 8°C–10°C) so that the probe will anneal to the template sooner than the primers are able to anneal to their respective sites. It is also possible to enhance assay sensitivity by the incorporation of locked nucleic acid (LNA) nucleotides (Chapter 10: RT-PCR: a science and an art form), which endow the probe with a higher Tm and, therefore, greater thermal stability. A significant improvement to the *Taq*Man assay accompanied the development of minor groove binder (MGB) probes. These innovative molecules feature a standard fluorescent label at the 5′end and MGB moiety and nonfluorescent quencher at the 3′ end. This has the effect of increasing the T_m and thereby reducing the length of the probe, resulting in greater sensitivity and resolution. As such, MGB probes have proven to be particularly useful for SNP discrimination, multiplexing, allelic discrimination, or when simply working with "difficult" templates.

During the primer extension phase of PCR which occurs between 68°C and 72°C, when the *Taq* polymerase reaches the probe as the upstream primer is extended, the 5′ exonuclease activity associated with this enzyme cleaves the probe that is hybridized to the target strand in a hydrolysis reaction. In so doing, the reporter fluorochrome is released from the probe and, no longer under the influence of the quencher, begins to fluoresce upon laser excitation. The 5′→3′ polymerase activity of the *Taq* enzyme can continue through the region previously blocked by the now-displaced probe, result in the production of PCR product (Fig. 11.8). As the process is repeated, more and more fluorescence is detected as increasing numbers of 5′ reporter fluorochromes are liberated from the probe and diffuse away from the quencher. The current generation of real-time PCR thermal cyclers are equipped with several detection channels that will allow the investigator to perform confidently multiplex reactions. By partnering several primers, probes, and fluorochrome labels, real-time multiplex reactions greatly reduce the amount of time required to complete a set of reactions while increasing productivity dramatically.

Molecular beacons

Molecular beacons are probe molecules that exhibit a characteristic stem-loop structure through which the 5′ and 3′ ends are maintained in close proximity (Fig. 11.9). In a manner similar to the TaqMan probes described above, fluorescence from a fluorochrome at one end of the probe is suppressed by a nearby nonfluorescent quencher. In contrast to the *Taq*Man probes described above, molecular beacons are not degraded during the amplification process

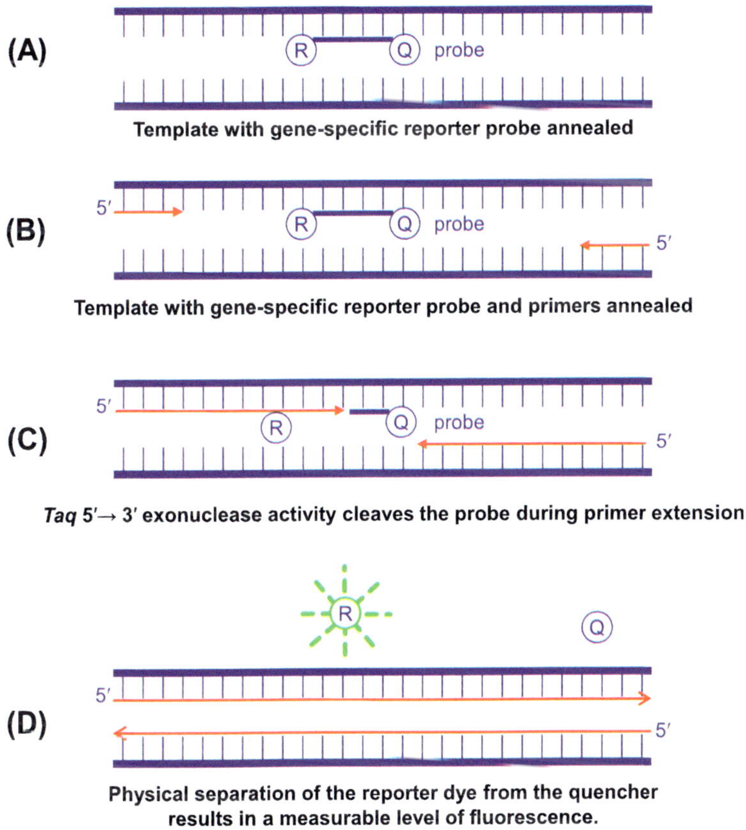

FIGURE 11.8 5′ nuclease assay for real-time PCR. (A) Fluorogenic probes, consisting of both a reporter (R) dye and quencher (Q), are annealed to the denatured template. As long as probe integrity is maintained, reporter dye fluorescence is quenched due to its proximity to the quencher dye. (B) Primers, which have a lower T_m than the probe, anneal to the template as the temperature is lowered. (C) During PCR, the 5′→3′ exonuclease activity of the *Taq* polymerase causes the cleavage of the probe by hydrolysis. (D) The fluorescence of the reporter dye, once separated from the quencher, is measured in real time (as it occurs and accumulates). Fluorescence intensity is directly related to amount of PCR product being produced. PCR primer extension then proceeds through the now-exposed region, completing the synthesis of the PCR product. This is methodology is widely known as the *Taq*Man assay.

but rather remain intact and must bind to the target in every cycle in order to produce measurable fluorescence.

Molecular beacons are added to real-time PCR reactions prior to amplification. The disruption of base-pairing in the stem, upon denaturation, allows the probe to hybridize to a complementary target. In so doing, the reporter dye and quencher are separated sufficiently to induce reporter dye fluorescence. As expected, simultaneous targets (multiplexing) can be detected through the use of various color fluorophores, and the technology is such

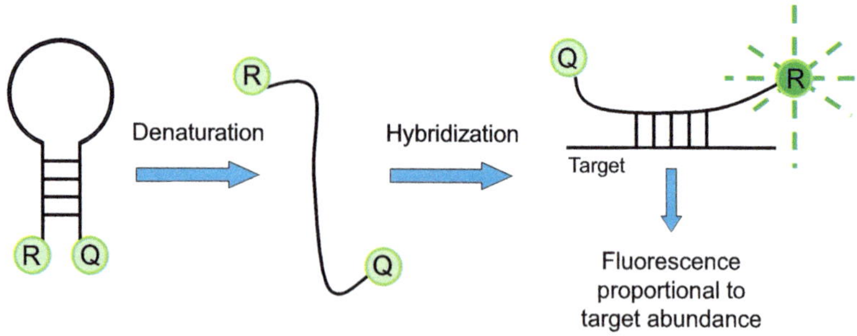

FIGURE 11.9 Molecular beacons. The characteristic stem-loop structure suppresses fluorescence until the beacon is denatured and able to hybridize to the accumulating product. See http://www.molecular-beacons.org for licensing information.

that molecular beacons are able to distinguish among targets that differ by a single base, making them ideal for single nucleotide polymorphism (SNP) profiling (Bonnet et al., 1999). Molecular beacons are also quite useful in diagnostic-type assays. Licensing and other technical information pertaining to the use of molecular beacons can be accessed at http://www.molecular-beacons.org.

Scorpions

Scorpions probes are quite unique in that the priming and product detection are accomplished through the use of a single oligonucleotide that functions both as primer and probe (Whitcombe et al., 1999). As with the TaqMan and molecular beacon probes, Scorpions probes exhibit a 5′ fluorophore reporter and, downstream, a 3′ nonfluorescent quencher. Scorpions probes (Fig. 11.10), by design, are either Uni-Probe (single molecule structure) or Bi-Probe (two molecules, base-paired). A stem-loop structure (Uni-Probe) limits fluorescence while the 3′ end anneals to the template and supports PCR product synthesis. After a second high-temperature denaturation, the stem-loop structure is disrupted, allowing the loop, which is complementary to the new strand that was synthesized by extension of the 3′ end of the scorpion, to base pair in an intramolecular manner to the PCR product which is now part of the same molecule. The opening of the loop and physical separation of the 5′ fluorophore from the quencher results in fluorescence signal emission. Similarly, fluorescence is suppressed in the case of Bi-Probe design as long as the 5′ reporter and 3′ quencher remain in proximity. This method is sometimes referred to as scorpion PCR and the resulting product might be thought of as a self-probing amplicon. Theoretically, the reaction is faster than other real-time methods because the reaction is unimolecular.

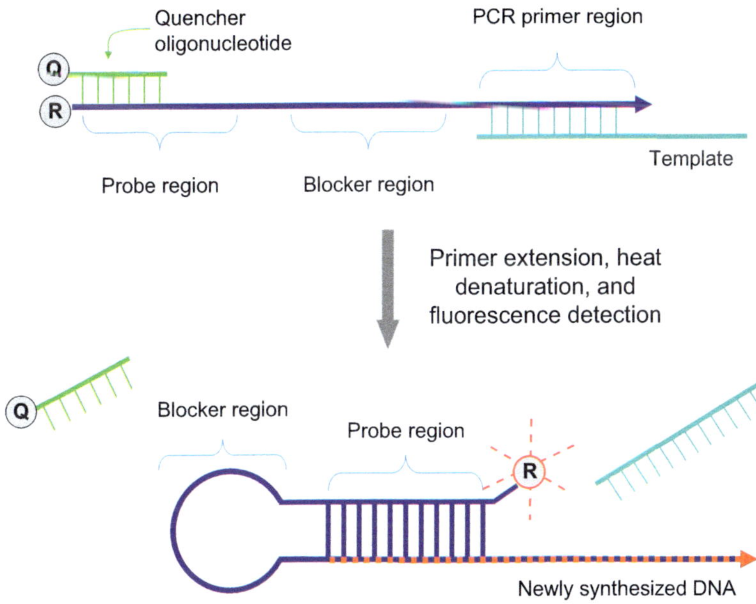

FIGURE 11.10 Scorpions Bi-probe. These molecules consist of an oligonucleotide primer linked directly to a probe and which is base-paired to another oligonucleotide with a quencher. Extension of the scorpion primer proceeds along the template to which it is base-paired. After subsequent denaturation and cooling, the scorpion probe region hybridizes in an intramolecular manner to the newly synthesized complementary DNA target. The resulting three-dimensional architecture distances the 5′ fluorophore from the quencher so that fluorescence emission begins. The blocker region is chemically modified to prevent copying of the probe region. In a related design, Scorpions Uni-probes contain both the quencher and fluorophore in the same molecule (not shown).

Melting curve analysis

The real-time PCR platform relies on an increasing level of fluorescence as an indicator of product accumulation. Those who fear amplicon contamination (should be everyone) feel very confident using real-time assays because the tube or capillary need not be opened postamplification, nor is gel analysis performed. How, then, may one determine if the correct product was synthesized and how can nonspecific products be discerned? Needless to say, nonspecific products, including primer–dimer artifacts will certainly impact sensitivity, not to mention the interpretation of the data.

These questions can be answered by performing melting curve analysis, a well-established and reliable method for product identification or for genotyping applications. The idea is that molecules of different lengths and/or base compositions will dissociate or “melt apart” into their constituent single strands at different temperatures. The standard procedure is to run the melting curve protocol immediately following the completion of the last cycle. As the temperature of a sample increases, double-stranded DNA will begin to melt as its T_m is approached and will complete the melting process after the T_m has been exceeded. Consequently, SYBR Green or oligonucleotides with a fluorophore label will no long associate with single-stranded DNA, resulting in diminished fluorescence as the product(s) melt, thereby producing the melting curves (Fig. 11.11B and E). To make the interpretation of the data easier, the melting curve profile is generally presented as the negative derivative ($-dF/dT$) so that each PCR product in the sample appears as a melting peak (Fig. 11.11C and F), from which the associated product T_m may be quickly deduced. If a unique product was produced, on should observe a single, prominent peak associated with the sample melting profile. Having a priori knowledge of the base composition of the expected product should also allow the investigator to predict exactly where the melting curve peak should occur.

Problems with the reaction products are detected when the melting profile is other than expected or when a second peak occurs at a lower temperature. In the case of the former, it is possible that a variant of the expected PCR product was synthesized or that a different product was produced altogether. In the case of the latter, low-temperature T_m peaks often result from the spurious formation of nonspecific product or, very commonly, from primer–dimer formation. In the case of a variant or otherwise unexpected product, one should examine the template, re-evaluate the primers, recalculate the expected product size, consider raising the annealing temperature, and ensure that amplicon contamination has not occurred. Primer–dimer formation can be minimized by analyzing potential primer sequences in silico before ordering them. Free software is readily available that will not only examine individual primers for the likelihood of secondary structure formation and self-primer–dimer formation but will also characterize the probable interaction of two primers that are being considered for use. These tools reside in many places on the internet, including the vendor websites where primers are ordered.

Digital PCR

There is no question that real-time PCR rose quickly to the top of the sensitivity ladder almost immediately upon its development. The method has been in widespread use for decades and enjoys many successes in research and clinical settings. Despite these successes, real-time PCR is perceived as being somewhat imprecise in the assay of rare mutations, gene copy number, and quantification of very low abundance transcripts, all of which are dependent on the construction of a standard curve and assignment of a C_T. Real-time PCR, rightly, is recognized as the gold standard owing to its widespread

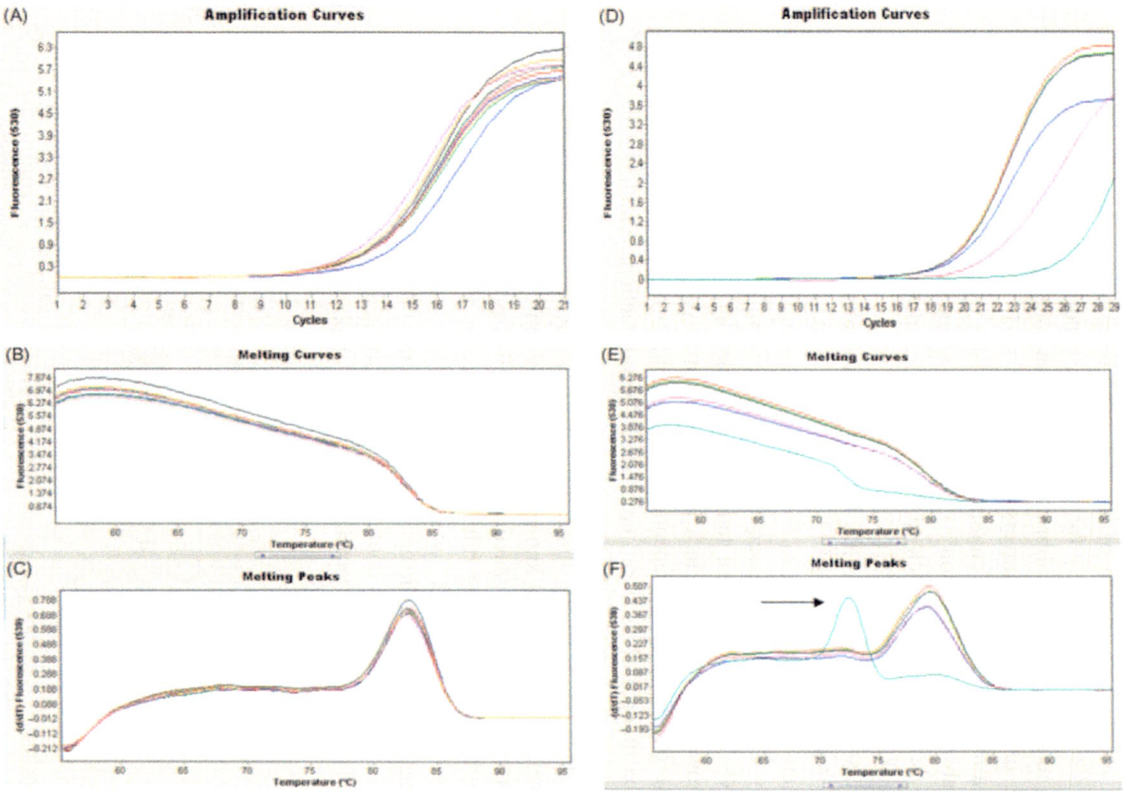

FIGURE 11.11 Real-time PCR and melting curve analysis. Strand dissociation causes a decrease in fluorescence as the fluorophore is released from its association with the PCR product. (A) Amplification profile of eight identical samples run in parallel. (B) Melting curve analysis of the PCR products generated in (A), showing the fluorescence decrease as the temperature increases and the SYBR Green moves away from the denatured PCR product. (C) The data is most often presented as the negative derivative of the fluorescence decrease so that specific melting curve peaks can be readily identified. The single peak strongly suggests a single, identical PCR product among the eight samples. (D) Amplification of five samples run in parallel. (E) Melting curve analysis of the PCR products generated in (D). (F) The appearance of a second peak (arrow) at a lower-than-expected melting temperature strongly suggests primer–dimer or the formation of another nonspecific product. Moreover, the divergence of amplification curves between cycles 17–21 (panel D) strongly suggests pipetting errors or unknown contaminants in some of the samples.

use, but it is far from perfect, and the need for higher-precision quantification and sequence discriminatory capacity are common needs.

Enter digital PCR (dPCR). A state-of-the-art method, digital PCR (Vogelstein and Kinzler, 1999; Sykes et al., 1992) was developed to altogether eliminate the construction of and dependence upon a standard curve and the need to calculate the C_T for each sample. Since dPCR eliminates many of the required real-time calibration metrics, one need not worry about running out of reference or universal RNA. As such, dPCR is widely regarded as a higher-precision technique by comparison with real-time PCR. The term "absolute quantification" is often used when describing dPCR, though many colleagues are quick to suggest that no method is capable of absolute or error-free quantification.

The chemistry which supports dPCR is very similar to the *Taq*Man assay, requiring primers, a fluorescent probe, a nucleotide cocktail, and other reaction components. It is the format of the method, however, that eliminates the need for any external reference sequences. In one form of this technique, dPCR involves the distribution of template molecules into very large numbers of microwells (Fig. 11.12) in a process known as partitioning. It is in each of these partitions that independent PCR amplifications occur in what might best be thought of as very numerous PCR subtractions. Dilutions of each template are prepared such that, on average, one template molecule is expected to end up in every third well. This is analogous to a well-established cell biology method known as cloning by limit dilution, which is designed to isolate individual, genetically unique cells from a heterogeneous mixture. Interestingly, "limiting dilution PCR" was an early descriptor of this method before "digital PCR" was coined.

Partitioning cDNA or gDNA template molecules in this manner reduces competition among targets and, in so doing, reduces background and increases sensitivity. Wells that receive a PCR template will support amplification, and will be

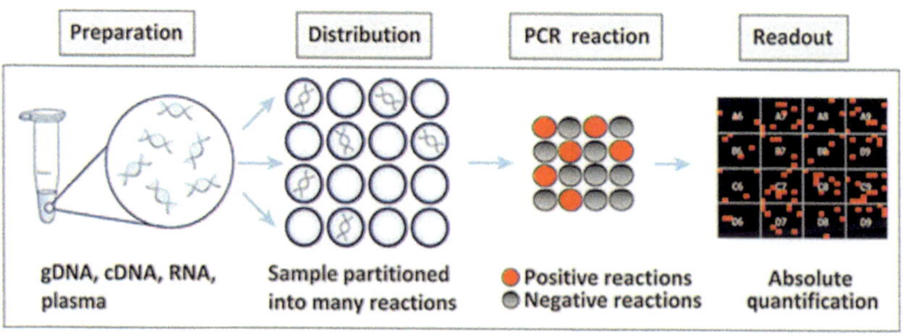

FIGURE 11.12 Digital PCR. Partitioning of the template into individual wells creates large numbers of individual clonal amplification reactions. This format for PCR supports high-precision quantification at the end of the assay. See text for details. *Image courtesy of Thermo Fisher Scientific.*

fluorescence-positive (positive reaction); as suggested above, this is sometimes referred to as clonal amplification because of the way the template molecules are distributed. In contrast, a fluorescence-negative well (negative reaction) is one in which no PCR product synthesis occurred because the well did not receive a template molecule. Therefore, the number of positive reaction wells is in direct proportion to the number of template molecules present at the onset. This binary-type quantification (positive or negative well) is the reason why dPCR is so named: it's a "yes = 1" or "no = 0" answer for each well. This method is much more straightforward than product accumulation measurements associated with traditional real-time PCR platforms. Needless to say, a situation in which more than one template molecule ends up in a single reaction must be taken into account. To this end, Poisson distribution modeling is applied to the results in order to compensate for this distinct possibility. By virtue of its direct counting format, dPCR is able to greatly reduce statistical errors associated with other forms of PCR, including the traditional real-time format. With respect to primer designer, smaller amplicons (100–200 bp) are preferred, and multiplexing is easily accomplished through the use of different color fluorescence labels.

The dPCR approach to quantification has great value in rare mutation detection because of the high number of technical replicates that are performed in this assay. The higher the number of replicates, the more likely that rare events (e.g., SNPs in gDNA) will be observed and the greater the precision of the assay. For this same reason, the method is useful for the assay of very rare transcripts that produce a late C_T value when assayed by real-time PCR, for viral load determinations, and for the analysis of complex template mixtures. The dilution process and the partitioning of template material make dPCR a bit more tolerant of PCR inhibitors than other PCR formats. Perhaps most noteworthy of all is the ability to share meaningful data with collaborators because dPCR data are independent of instrument configuration, reference sample(s), and C_T. The method shows great promise as a diagnostic tool, as having an important role in biomarker validation, and as a method for validating next-generation sequencing results.

An up-and-coming variant of dPCR is known as droplet digital PCR (dd™PCR), a method in which template molecules in a typical 20 μL reaction are partitioned into as many as 20,000 nanoliter-volume droplets in an oil emulsion, with each droplet functioning as a unique real-time PCR amplification reaction chamber (Fig. 11.13A). This is the equivalent of almost two million reactions that can be performed in a single 96-well plate. Within each nanoliter droplet, there are still hundreds of millions primer and probe molecules which are able to support robust amplification when a target sequence is present. After the amplification protocol has been completed, each droplet is examined to determine if any PCR product was synthesized in it (Fig. 11.13B). The instrumentation currently in use (e.g., Bio-Rad® QX200) is able to read more than one fluorescent channel, thereby supporting sample multiplexing even greater latitude in experimental design.

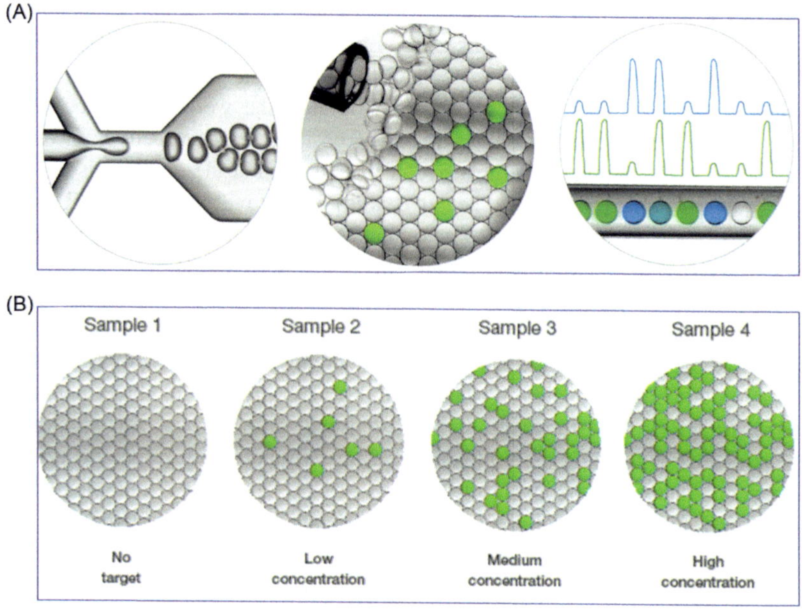

FIGURE 11.13 Droplet digital PCR. (A) DNA template molecules are partitioned into oil droplets, each of which functions as a microreactor. PCR occurs independently in each droplet that receives template. Fluorescence indicates whether amplification occurred. (B) Direct counting of positive oil drops reveals the actual number of template molecules in the original sample, thereby eliminating the need for standard curves on reference samples. *Permission to publish this image has been granted by Bio-Rad Laboratories, Inc.*

Internal controls

Observed changes in the transcriptional pattern of one gene are meaningful only when (1) they are compared to the behavior of the same gene in a reference (untreated or control) sample and (2) when the transcriptional behavior of the gene of interest is compared to the transcription of other genes, specifically reference gene mRNA transcripts such as GAPDH (glyeraldehyde-3-phosphate dehydrogenase) or HGPRT (hypoxanthine-guanine phosphoribosyltransferase) within each sample, the expression of which is (are) not expected to change as a function of experimental manipulation. Examples of such genes are presented in detail in Chapter 10, RT-PCR: a science and an art form, and the same considerations apply here as well. They are considered to be internal (endogenous) controls because they are already part of the RNA samples. The transcripts from the reference gene allow normalization (meaningful comparison) among two or more samples, and their assay is an indispensable component of any transcription assay. While there is no all-purpose control (reference) transcript applicable to every system and every circumstance, the assay of one or more internal controls is not to be taken lightly.

Exogenous controls

In addition to the assay of a reference gene, some investigators choose to spike an RNA sample with a known quantity of an exogenous reference template (discussed in the next section) for precision measurement of reverse transcription efficiency as well as the mass of PCR product at the end of the subsequent reaction. One must be aware, however, that while the spiking approach has merit with respect to accurate quantification of the expression of specific transcripts, it is not helpful in determining the overall transcriptional health of the cell.

For accurate calculation of transcript template copy number, rather than relative abundance, one technique is to spike an RNA sample with a known mass of an exogenous transcript. These RNA molecules can be made by in vitro transcription or simply purchased, examples of which include ribulose biphosphate carboxylase, commonly known as RuBisCO or rabbit globin mRNA (Smith et al., 2000, 2003). These are two examples of exogenous controls. Depending on the format of the investigation, the exogenous template can be completely unrelated to the experimental transcripts under investigation. Adding a plant transcript to animal RNA may be a particularly useful strategy in this regard: because the ribulose biphosphate carboxylase is associated with the photosynthetic pathway, no background is expected, not even basal level expression. To proceed with this type of control, add 1–5 ng of RuBisCO (0r other exogenous) mRNA to the purified RNA, the cell lysate or directly to the lysis buffer. Thus, the plant mRNA is copurified and subsequently reverse transcribed along with the RNA under investigation, thereby exposing both the exogenous and endogenous transcripts to precisely the same chemical environments and physical manipulations during the course of the isolation procedure.

Alternatively, if the exogenous transcript is associated with at least one of the experimental transcripts to be assayed, it should be a truncated version so that (1) it is recognized by the same primers and/or probe with the same thermodynamic stability as the experimental transcript and (2) it appears as a distinct band upon electrophoresis or as a distinct melting curve peak, thereby ensuring adequate resolution of control and experimental transcripts. Similar transcript templates such as these are sometimes referred to as "mimics" because they mimic the characteristics of the authentic template and, as such, are coamplified. The idea is that the exogenous mRNA will be reverse transcribed and amplified by PCR with the same relative efficiency as other transcripts normally found in the sample. When the precise amount of exogenous template mRNA spiked into each sample is known, a comparison with the amount of the corresponding PCR product can serve as a basis for normalization, and provide an indication of the efficiency of reverse transcription in each tube. This type of approach is often a useful identifier of the source(s) of data variation such as inhibitors at either the cDNA synthesis- or cDNA amplification levels, particularly if there are data reproducibility issues in the lab. Further, by spiking different amounts into various samples, one is able to generate a standard curve to correlate observed experimental band intensity with the mass of the standard(s).

Using an exogenous transcript as both a reverse-transcription and PCR amplification control was first described by Wang et al. (1989). The vector, pAW108, contained sequences recognized by both 5′ and 3′ primers for twelve different target genes. The gene sequences were tandemly joined together, flanked by a 5′ T7 RNA polymerase promoter and an encoded poly (A) tail downstream. This vector construction supported the synthesis of large quantities of complementary RNA (cRNA) which were then purified by affinity chromatography (Chapter 5: Isolation of polyadenylated RNA) and the precise concentration determined spectrophotometrically (Chapter 8: Quality control for RNA preparations). When spiked into the experimental RNA prior to reverse transcription, the cRNA is subject to the same physical and chemical environment as the RNA under investigation, thereby minimizing concerns about the relative efficiency of reverse transcription and PCR amplification between reference genes and the transcripts of interest, among various samples and compensates for "tube effects." The use of cRNA standards for enhancing the quantitative dimension of Northern analysis was described shortly thereafter (Wang et al., 1993).

cRNA molecules can be "sense" or "antisense," the former behaving almost identically to the native mRNA, and the latter being complementary to the native mRNA. Antisense transcripts can be wonderful probes with the potential to form very thermodynamically stable duplexes with mRNA in applications such as the RNase protection assay and for in situ hybridization applications. For the quantification strategy described here, however, the cRNA of interest is the sense (+RNA) in vitro transcript because it must function as a minor competitor with the native mRNA during cDNA synthesis and ensuing amplification by PCR. In order to make the product of cRNA amplification distinguishable from the native RT-PCR product it is usually best to add or delete a few bases to the in vitro construct such that the two products, while very similar with respect to base sequence are, in fact, distinguishable. In the case of end-point analysis, another strategy is to engineer a restriction enzyme site into the template so that the resulting reference template will be exactly the same size as the experimental template and yet distinguishable from the template by post-PCR digest. In either case, the same set of primers will coamplify the two templates, the mass of each of which is related to each other by standard curve analysis. The preparation and assay of cRNA standards along with experimental samples, however, can be a rather tedious process. In many applications, determining the actual mass or copy number of a transcript may not be necessary and data presentation in terms of relative abundance may be satisfactory.

Control reaction formats

Primers that are designed for both the control- and experimental transcripts can be used in a variety of formats, some of which are much better strategies than others. While these formats where established prior to the development of real-time, PCR, if desired they can be easily adapted for real-time analysis. The following list is not intended to be exhaustive, but it does put these various historical permutations into perspective.

1. Use of a second set of primers in a different tube with an aliquot of the cDNA from the same experimental source. This is an early form of transcript quantification using PCR. This approach, which is to be avoided, has several severe problems:
   a. The second set of primers probably will not have the same thermodynamic character as the first set and will support the synthesis of the second type of PCR product with a different efficiency;
   b. Quantification in this manner requires that the mRNAs for the target and control genes be present at similar levels to permit meaningful interpretation;
   c. mRNA from reference genes often does not remain constant throughout the cell cycle;
   d. The control and experimental samples are in different tubes, introducing the distinct possibility of "tube effects," particularly with respect to reverse transcription efficiency.
2. Use of a second set of primers in the same PCR tube. This is also known as multiplex PCR or simply "multiplexing." Although this is a significant improvement over the two-tube approach, the potential for serious inaccuracy persists for the same reasons cited in 1a, 1b, and 1c, above. At least both primer sets are in the same tube!
3. Competitive PCR is a format in which an exogenous transcript that can be recognized by the *same* set of primers is spiked into the sample. Succinctly, the cDNA under investigation *and* what is best thought of as a secondary target both compete for the same primers and other reaction components. This methodology was actually described prior to the emergence of real-time PCR (Li et al., 1991). In a nonreal-time format, stating that this was a labor-intensive set of procedures is generous. Samples were typically amplified over 25 cycles and the products were analyzed on agarose gels. Next, the amplification of competitor templates can be quantified in real-time along with experimental templates; the experimental and competitor PCR products are distinguishable and can be quantified by melting curve analysis and through the use of difference color fluorescence primers.

   There are two forms of competitive PCR. The first method involves the in vitro synthesis of an artificial transcript (cRNA), a precise mass of which is used to spike the experimental RNA. The rationale behind the in vitro transcription approach is that an artificial transcript mixed directly with the experimental material will be subjected to the same physical conditions; therefore, the extent of reverse transcription is determined by assaying the amount of cDNA from the artificial transcript, whose input mass is known precisely, and using the data to construct a standard curve (Fig. 11.14). In vitro transcription of an artificial RNA requires the ligation of a DNA template next to a suitable transcription promoter (Fig. 11.15). Alternatively, it is a simple matter to construct a transcription template by PCR in which a bacteriophage RNA polymerase promoter is appended 5′ to a DNA sequence intended for transcription (Fig. 11.16). In either case, the resulting control transcript ideally exhibits a deletion or an insertion, compared to the naturally occurring sequence. Thus, the control transcript is known as a homologous competitor because experimental and control transcripts essentially share the same sequence, including PCR primer recognition sequences; they are expected to be reverse transcribed and amplified by PCR with the same efficiency and will be distinguishable by their size difference upon electrophoresis or melting curve analysis.

In the second approach, what might best be thought of as a nonhomologous template, sharing only the primer recognition sequences with the endogenous sequence(s) under investigation, is constructed. This is accomplished by first using PCR to place primer recognition sites on the end of the competitor DNA. Composite primers, the 5′ end of which is sequence (gene)-specific and the 3′ end of which is competitor specific, are used to generate the competitor (Fig. 11.17). When constructed in this manner, both competitor and the corresponding cDNA will be amplified using a single set of primers. Often, nonhomologous competitor sequences are used for assaying gene-specific cDNA abundance when short PCR products (less than 300 bp) are expected: longer templates are likely to experience variations in amplification efficiency due to differences in the order of bases. Further, if spiked into the cDNA, rather than into the RNA, quantification in this manner does not factor in the efficiency of reverse transcription. Depending on the context of the experiments being performed, this issue may or may not be of importance.

4. Assay of a reference gene, the expression of which is expected to be invariant, and related changes in the expression of the gene of interest to it. This is particularly well-suited for real-time PCR platforms. As needed, control primers and/or the external cRNA templates described above may be included in the reaction for the generation of a standard curve (Fronhoffs et al., 2002). A contemporary reaction includes two or more primer sets, each with a similar T_m and a different color fluorescent label in order to distinguish and quantify the various products at the end of the reaction.

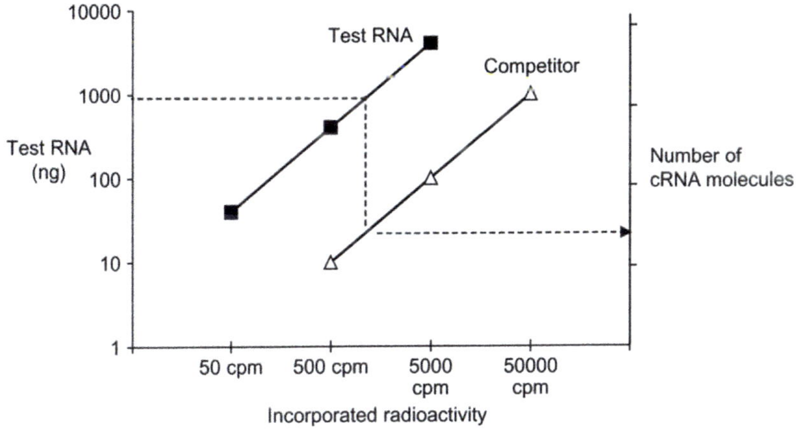

FIGURE 11.14 Interpolation of transcript abundance with a standard curve. In the original description of this approach (Wang et al., 1989), PCR was performed using ^{32}P-labeled primers. Following visualization by electrophoresis, the bands were cut out of the gel and label incorporation determined by Cerenkov counting. In this example, ■ represents the mass of cellular RNA from a sample under investigation and Δ corresponds to the number of competitor RNA molecules used to spike the reaction tube. Equal amounts of label incorporation in the test RNA and the cRNA competitor correlates with equal numbers of molecules. The resulting data might then be expressed as the number of molecules of the specific transcript (determined by comparison with the coamplified cRNA) per ng or μg of test RNA from the biological source.

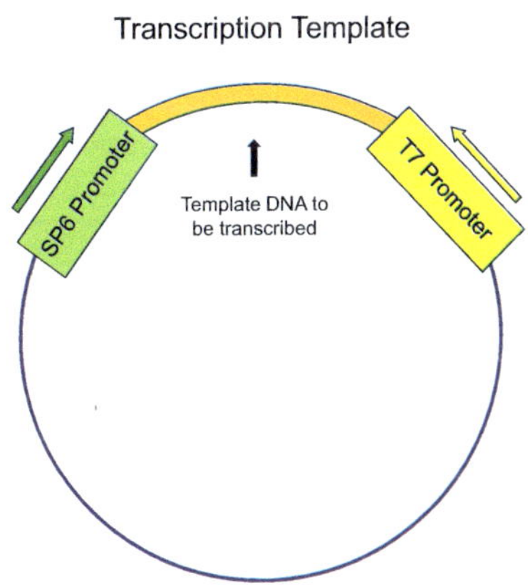

FIGURE 11.15 Template construction for in vitro transcription. The ligation of a template DNA between two different transcription promoters, such as T7 or SP6, allows the synthesis of large quantities of single-stranded sense or antisense RNA molecules from the same vector, which must be linearized prior to transcription to ensure that the resulting RNA molecules are of uniform length. The RNA can be used to spike an experimental sample as a means of assessing the efficiency of reverse transcription into cDNA and subsequent amplification by PCR. Such transcripts are useful also as nucleic acid probes, details pertaining to which are found in Chapter 17, Nucleic acid probe technology.

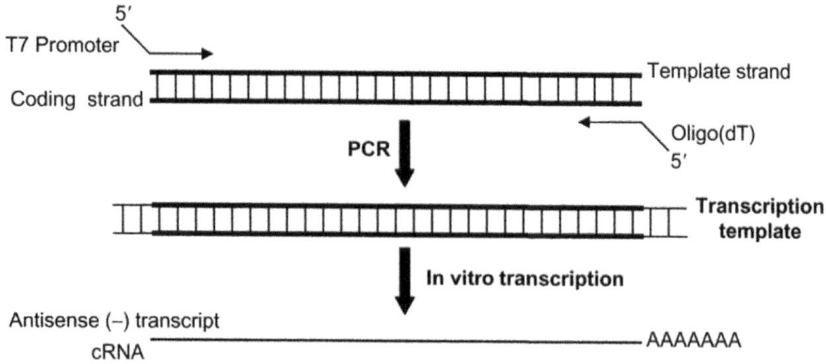

FIGURE 11.16 Synthesis of a transcription template by PCR. Antisense cRNA sequences are produced by in vitro transcription. The transcription template can be synthesized using one primer with a 5′ T7 overhang and the other primer with a 5′ oligo(dT) overhang. The thymidylate tract is necessary for producing an artificial poly(A) tail, which is required to support some forms of reverse transcription. A similar strategy can be used to synthesize antisense cRNA molecules for use a hybridization probes.

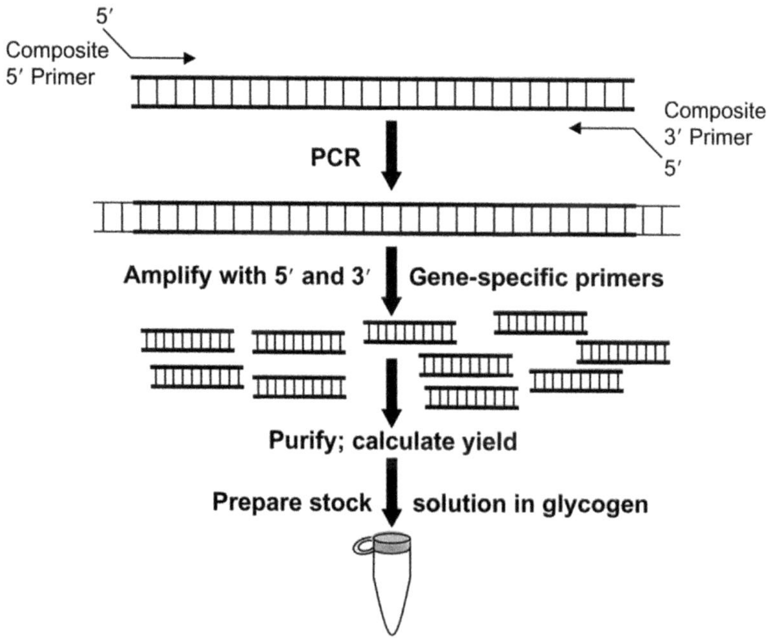

FIGURE 11.17 Construction of a nonhomologous competitor DNA template. Both the upstream and downstream composite primers contain 5′ gene-specific sequences which, following PCR, will become part of the competitor. Subsequently, in the competitive PCR reactions, shorter primers, consisting only of gene-specific sequence, are used to amplify both the competitor and the cDNA under investigation. The extent of coamplification of each of these species will be a direct function of their molar ratio in the reaction tube.

Negative control considerations

Negative control reactions are just as important as positive controls, quantification controls, loading controls, and enzyme efficiency controls. These reactions are designed to ensure that any observed PCR product is authentic and not an artifact due to sloppy micropipetting or, worse, amplicon contamination. Many laboratories now routinely use portable PCR workstations, which look like miniature, tabletop biological safety cabinets (Fig. 11.18). These workstations typically produce high-efficiency particulate filtered air to create a sterile pipetting environment as well as a built-in ultraviolet light that is used to crosslink rogue amplicons in the pipetting zone. Many of the more versatile workstations also feature thermoplastic construction that will protect the user from exposure to ^{32}P in experiments involving the use of isotopes.

Among the more important PCR control reactions that must be performed for each sample is the RT^- (RT-minus) control. This consists of a tube which contains all of the components for cDNA synthesis (RNA, nucleotides, primers) *except* for reverse transcriptase. Because cDNA cannot be synthesized without reverse transcriptase, any PCR product from the RT^- tube is the result of contaminating genomic DNA template material. Thus, it is commonplace to treat all RNA samples with RNase-free DNase I to destroy endogenous DNA template prior to any RT-PCR assay. This has been a standard practice for RT-PCR since the inception of the technique.

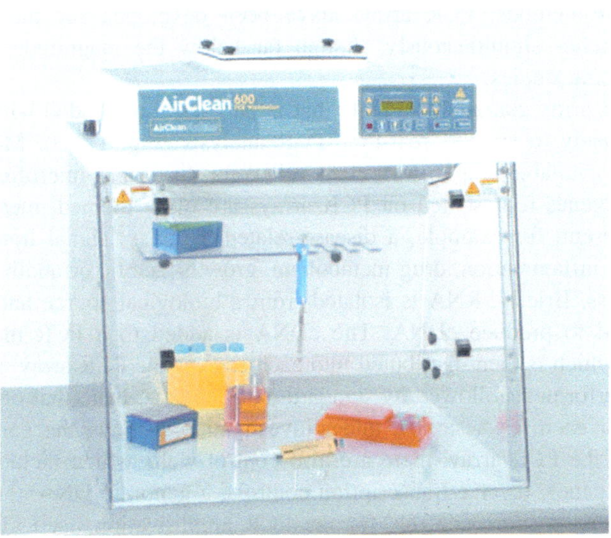

FIGURE 11.18 Portable PCR workstation. The clean air environment with the ability to perform UV sterilization provides optimal protection from sample contamination. *Courtesy of AirClean Systems (www.aircleansystems.com).*

In a variation on this theme, it has been suggested that an RT-PCR assay can be streamlined by treating a sample with DNase I *after* the first-strand synthesis (Flohr et al., 2003). The notion is that mRNA:cDNA hybrids are resistant to DNase I, so only genomic DNA, and not the newly synthesized cDNA, will be eliminated. Briefly, synthesize cDNA at 37°C for 50 minutes with MMLV; heat inactivate the enzyme for 15 minutes at 50°C. It is important not to exceed this temperature because dissociation of the mRNA:cDNA hybrid will render the first-strand cDNA susceptible to nuclease degradation should intramolecular base-pairing occur (transient formation of a double-stranded region). The cDNA:mRNA hybrids are incubated with 10U DNase I for up to 2 hours at 37°C, after which the DNase is removed by standard silica technology for cleanup. It is true that DNase I will cleave DNA:RNA and ssDNA, but does so at a much lower rate compared to dsDNA template. This approach is not for the faint-hearted.

PCR arrays

The vast improvements in laboratory throughput have greatly increased productivity. Newly developed methods for the assay of gene expression have turned the focus for many laboratories on the global analysis of gene expression, as opposed to the older strategies and protocols of examining one or a few genes at a time. To a great extent, this is possible in the era of transcriptomics, microarrays, and RNA-seq. In addition to these sensitive and high-throughput methods, PCR arrays have been developed for the assay of multiple genes simultaneously, though far below the magnitude of whole transcriptome studies.

A PCR array consists of a plate that is purchased preloaded with primers that are ready to support real-time PCR analysis (Fig. 11.19). Most often, arrays are available in a 96-well, 384-well, rotor-disk, or in microfluidics format. The genes represented on PCR arrays are often themed, meaning that they represent, for example, a disease-related pathway, signal transduction, apoptosis, inflammation, drug metabolism, growth factors, or another biological process. Briefly, RNA is isolated from a biological source and reverse-transcribed to produce cDNA. The cDNA is added to a PCR master mix cocktail, which is then distributed into each well of the PCR array. Real-time PCR is performed, followed by standard data analysis. Note that one plate is needed for each RNA sample under investigation. Among the genes represented on the PCR array, there are also control wells which include several reference genes, reverse transcription controls, a genomic DNA control, and one or more positive controls for the PCR amplification itself. The entire process can be automated and robotic-controlled in high-throughput laboratories.

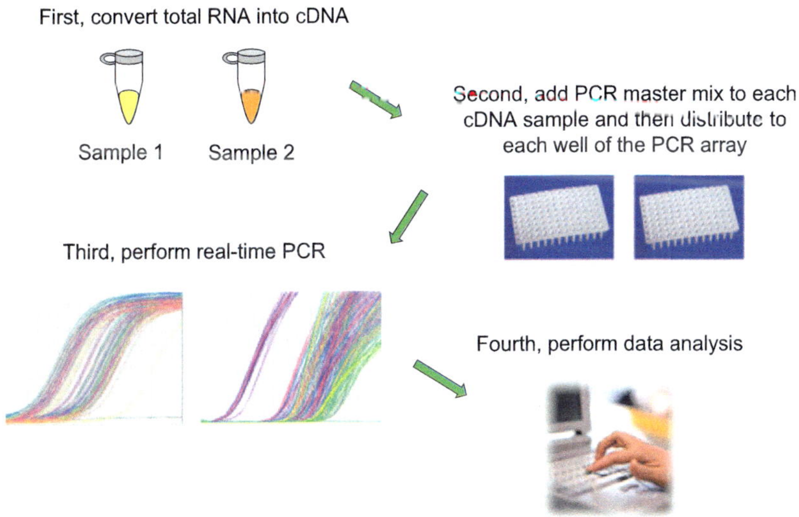

FIGURE 11.19 One example of a PCR array. Multiple primers, often themed, are predistributed into a 96-well plate or other format. Real-time amplification facilitates higher productivity by focusing on multiple genes per assay, and only those of direct experimental interest to the investigator.

Competitive PCR: key considerations

Competitive PCR is exactly what the name implies: a form of PCR reaction in which two templates compete for amplification (Diviacco et al., 1992; Siebert and Larrick, 1993). This is possible only if both the experimental cDNA and the competitor have identical primer recognition sites. The competitor template can be introduced either in the form of cRNA, described above or in the form of a double-stranded cDNA after reverse transcription. Further, competitive PCR amplification can be evaluated using a real-time PCR format or by end-point analysis. The discussion which follows describes the design and synthesis of a competitor molecule which can be used in either format. The advantages and disadvantages of competitive PCR are presented in Table 11.3.

The magnitude of amplification of experimental and competitor templates in each reaction tube is a function of their molar ratios: while it might be possible to amplify both of these templates, the template represented by the larger number of molecules will yield more product, at the expense of the other template. It follows, therefore, that if both templates are present in equimolar amounts, they should be amplified with equal efficiency. Therefore, to determine the mass of an unknown cDNA in an experimental sample, dilutions of the competitor template are made until an equimolar concentration is identified (Fig. 11.20). The following are the key considerations in the design of a competitive PCR experiment:

TABLE 11.3 Competitive PCR: advantages and disadvantages.

Advantages

- One set of primers recognizes both the competitor and experimental templates.
- cRNA control transcripts, produced by in vitro transcription, are useful indicators of the efficiency of reverse transcription.
- Physical parameters in the reaction tube influence competitor and cDNA amplification equally, including plateau effects in the case of end-point PCR.
- Quantification is based on ratios between both PCR products.
- The reaction data are highly reproducible.
- Products are easily distinguishable by melting curve analysis or by electrophoresis.
- Radioisotopes not required.

Disadvantages

- Time-consuming sample preparation procedures.
- Several dilutions required to identify the correct ratio of target with competitor.
- A new, unique competitor is required for each new target to be assayed.
- A second set of primers is needed to construct the competitor.
- Spiking a sample series with cDNA, rather than cRNA, does not address efficiency of reverse transcription.
- If used, a nonhomologous competitor must be confirmed to coamplify along with the experimental template with equivalent efficiency.
- For end-point analysis, labor-intensive post-PCR processing is required.

1. Quantity of the competitor must be known. The competitor is synthesized by reverse transcription, PCR, or both, after which it must be purified, and its concentration determined either by UV absorption (Chapter 8: Quality control for RNA preparations) or by digital image analysis (Chapter 15: Photodocumentation and image analysis). For maximum accuracy, the exact number and identity of nucleotides that constitute the competitor must be known, from which the concentration of the newly synthesized competitor can be expressed in micromoles (μmol). When working with homologous RNA competitor transcripts, typical dilutions ranging from 1 to 10,000 femtograms (fg) of a competitor are spiked into 100 ng of cellular RNA. Nonhomologous competitor templates are generally diluted down to a new stock concentration in the attomole (amol) range and are often expressed in terms of the actual number of molecules spiked into the tube (Table 11.4). For quantitative analyses, the competitor is further diluted along with identical aliquots of the sample. When the amount of the experimental and competitor templates is identical, the mass of each PCR product should likewise be the same.
2. Competitor and the target template amplification must be identical. One set of primers must recognize both types of the template and amplify them with the same efficiency. To accomplish this, the investigator first adds sequences to the ends of a DNA molecule which are complementary

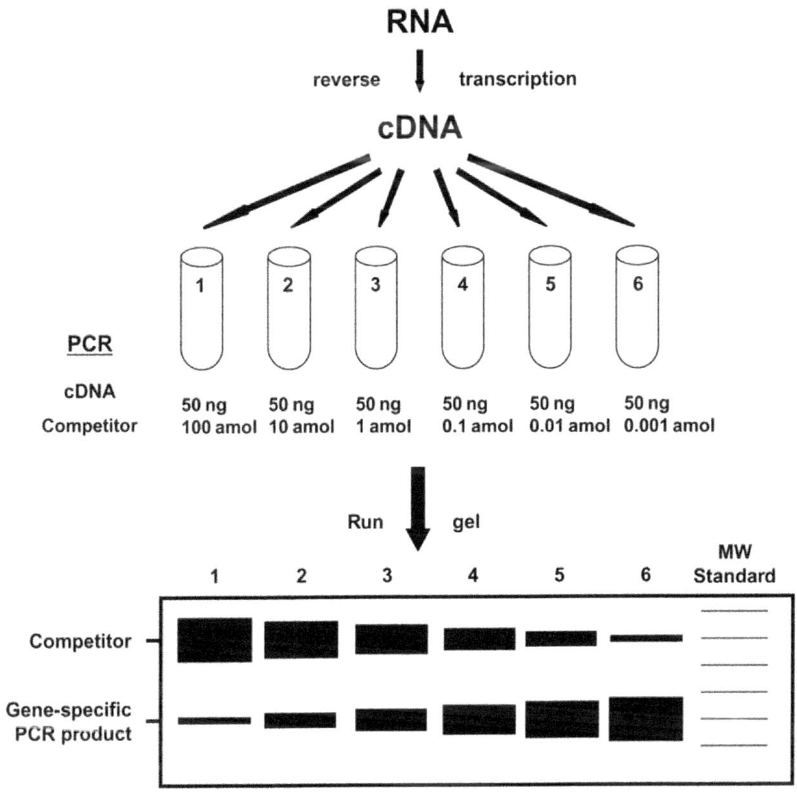

FIGURE 11.20 Transcript quantification by competitive PCR. Coamplification of both test cDNA and competitor, when properly designed, produces two distinct PCR products which can be discerned by electrophoresis (shown) or by melting curve analysis. In each reaction tube, both products are generated using the same set of primers and are subject to the influence of the same physical parameters. The objective is to identify the reaction in which both bands are of equal intensity, meaning that the molar quantity of each was equivalent at the onset of amplification. In cases where image analysis software is used, the abundance of the test template under investigation can be quickly determined by relating the lane where $IOD_{sample} \div IOD_{competitor} = 1.0$ to the mass of competitor that was spiked into that reaction tube along with the experimental template.

to the primers. This modified DNA molecule will become a competitor. The intended DNA competitor should show as much homology with the experimental template as possible in order to render the most accurate comparison at the end of the experiment. Thus, no one can argue that observed differences are the result of preferential amplification of one sequence over another. Very useful homologous competitor templates can be prepared by simply making an insertion or a deletion in the cDNA that is normally produced with a given set of primers. In so doing, the competitor will be either a bit smaller or a bit larger than the

TABLE 11.4 Unit analysis.

Unit	Abbreviation	Mass	No. of molecules
1 mole	mol	10^0 mole	6.02×10^{23}
m, millimole	mmol	10^{-3} mole	6.02×10^{20}
μ, micromole	μmol	10^{-6} mole	6.02×10^{17}
n, nanomole	nmol	10^{-9} mole	6.02×10^{14}
p, picomole	pmol	10^{-12} mole	6.02×10^{11}
f, femtomole	fmol	10^{-15} mole	6.02×10^{8}
a, attomole	amol	10^{-18} mole	6.02×10^{5}
z, zeptomole	zmol	10^{-21} mole	6.02×10^{2}
y, yoctomole	ymol	10^{-24} mole	0.6

experimental template and will be easily resolvable electrophoretically or by melting curve analysis. In instances when a homologous competitor sequence is not available, one may still confidently perform a competitive PCR experiment using a nonhomologous competitor template, that is, a sequence which is not related to the experimental cDNA under investigation, as long as:

a. The sequence of the intended competitor is known.
b. The intended competitor has average GC-content.
c. The intended competitor does not have AT- or GC-rich domains.
d. The size of the intended competitor is ±200 bp of the size of the experimental cDNA.

3. Reverse transcription efficiency of the template and the competitor must be identical. To correlate data from any RT-PCR assay across a range of samples one must factor in the efficiencies of both the reverse transcriptase reaction and PCR. The coamplification of the experimental cDNA and the competitor addresses the issues surrounding the efficiency of PCR: whatever influences are exerted on the cDNA are also exerted on the competitor. Adding a DNA competitor to newly synthesized cDNA, however, does not address the issue of the relative efficiency of reverse transcription, which is always the most problematic issue surrounding cDNA synthesis. By spiking the original RNA sample(s) with a transcript produced in vitro prior to reverse transcription, a standard curve can be generated to address the efficiency of the reverse transcriptase in each tube. To do this, the control transcript must be designed so that it will be

recognized by whatever primer is used to support reverse transcription, for example, a poly(A) tract that can be recognized by oligo(dT).

4. Heteroduplex formation must be minimized. Heteroduplexes are PCR products that form as a result of single-stranded cross hybridization (i.e., recombination), in this case between the competitor sequence and the experimental cDNA, resulting in novel dsDNA molecules. This is common when the competitor is a deletion or insertion form of the wild-type cDNA, and these are readily identifiable as higher molecular weight products, compared to the competitor or the target cDNA alone. The greater the number of PCR cycles performed, the more likely heteroduplex formation. Further, heteroduplex formation in the last PCR cycle is especially troublesome because these molecules will not be subsequently denatured. Obviously, heteroduplex formation compromises accurate quantification of authentic PCR products. Reducing the number of PCR cycles reduces the likelihood/severity of heteroduplex-associated difficulties.
5. Detection method must be optimized. In the case of real-time competitive PCR, the detection methodology will be melting curve analysis. However, there are many laboratories that may want to perform competitive PCR analysis using alternative detection procedures. Most importantly, radiolabeling the products is neither required nor recommended. Routinely, 25–30 cycles are performed, thereby favoring the synthesis of an adequate mass of product by simple gel staining. In this regard, SYBR Green and SYBR Gold work well because, compared to EtBr, these dyes offer a significant enhancement in sensitivity with the minimal background when viewing, photodocumenting, or digitally capturing the image directly from the gel. Although a reduction in the number of cycles is in order to minimize heteroduplex formation, the concomitant decrease in product accumulation may not necessarily be a detection problem because SYBR Gold in particular allows the visualization of low abundance bands to an extent not possible with EtBr. Moreover, in the instance of adequate band mass, visualization with EtBr is not recommended because the high background commonly associated with this dye detracts greatly from the sensitivity of this assay. If real-time PCR is being performed then the detection is automated, as is the determination as to which cycles constitute the period of linear accumulation of product.

In summary, competitive PCR is a sensitive method for template quantification which can be performed in a number of different formats. As with all PCR-based assays, the sensitivity of optimized reactions is a major strength. The inclusion of an internal competitor that mimics the experimental template precludes the requirement of having to determine in which cycles the exponential phase occurs: if the cDNA is no longer experiencing exponential amplification, neither is the competitor. Among the disadvantages associated with the method are the requirement for serial dilutions of each sample which are needed to titrate the competitor template; and the fact that two sets of primers must be synthesized, first a composite primer pair to synthesize the competitor, and then the gene-specific primer pair for the actual competitive PCR reactions.

Competitive PCR: major steps involved

1. *Prepare PCR competitor*
   a. Nonhomologous competitor
      i. Design gene-specific and composite primers (both sets of primers required). Perform the primary PCR amplification with the composite primers and then perform the secondary PCR amplification with the gene-specific primers. Try 25-mers for use as gene-specific primers, with T_m of at least 60°C. This will favor higher yields and greater fidelity of cDNA amplification. As always, use standard guidelines for PCR primer design (Chapter 10: RT-PCR: a science and an art form).
      ii. Remember that competitor templates are usually 200–700 base pairs.
      iii. Recall that competitor templates should be a maximum of 200 bp larger (or smaller) than the experimental target.
      iv. Purify competitor products and determine the concentration
      v. Prepare titration stock solutions, usually 50–100amol/μL, just prior to use, ideally in silanized (Appendix H: Silanizing centrifuge tubes and glassware) microfuge tubes.
   b. Homologous competitor prepared by in vitro transcription
      i. Ligate cDNA adjacent to an RNA polymerase promoter (e.g., T7, SP6, or T3).
      ii. Perform in vitro transcription, purify transcript, and determine concentration.
      iii. Store at −80°C until ready to use.
2. *Isolate high-quality RNA*
   a. Use good RNase-free technique.
   b. Perform expedient recovery from cells or tissues.
   c. Silica-based purification or a guanidinium–acid–phenol isolation techniques are preferred.
   d. Treat sample(s) with RNase-free DNase.

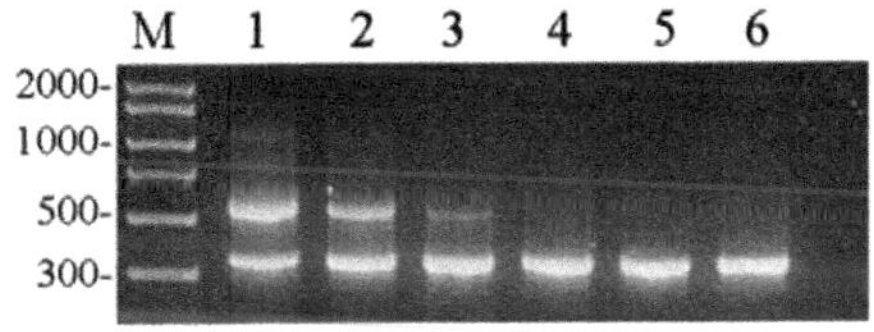

FIGURE 11.21 Primary 10-fold dilution series for mass determination. Coamplification of mouse β-actin cDNA and a nonhomologous competitor. Competitor template (540 bp) was generated using composite primers, as described. β-actin (365 bp) cDNA synthesis was supported by oligo(dT)- and AMV-mediated reverse transcription. PCR was performed for 25 cycles, after which the products of all six reactions where electrophoresed on a 2.5% agarose gel prepared in 1 × TAE, along with a molecular weight standard (Sigma). At the conclusion of the run, the gel was stained in 1 × SYBR Green diluted in 1 × TAE, and then photographed. The 10-fold dilution series represents a primary characterization of the sample in order to estimate the approximate mass of actin cDNA in the sample. Mass of competitor in each sample is as follows: lane 1, 10.0 amol; lane 2, 1.0 amol; lane 3, 0.1 amol; lane 4, 0.01 amol; lane 5, 0.001 amol; lane 6, 0.0001 amol. Best match ($IOD_{sample}/IOD_{competitor} = 1.0$) is in lane 2.

   - **e.** If precipitated, wash RNA pellets with 70% ethanol prepared in nuclease-free H_2O (cationic salts influence PCR reactions by quantitatively binding of dNTPs).
   - **f.** Use proper long-term storage of purified RNA: ethanol precipitate at −80°C or, if stored in aqueous solution, store in suitable aliquots at −80°C.

3. *Synthesize cDNA*
   - **a.** Use an RNase H^- reverse transcriptase.
   - **b.** Synthesize first-strand cDNA and purify products.
   - **c.** Store cDNA at −20°C until ready to use.

4. *Perform primary competitive PCR amplification* (Fig. 11.21)
   - **a.** Prepare serial 10-fold dilutions of the competitor.
   - **b.** Use an aliquot of each PCR competitor dilution with first-strand cDNA samples.
   - **c.** Perform PCR amplification; analyze products.
   - **d.** Determine competitor concentration that most closely approximates target cDNA concentration.

5. *Perform secondary competitive PCR amplification* (Fig. 11.22)
   - **a.** Prepare serial twofold dilutions of the competitor, beginning with a competitor stock solution that is 10-fold more concentrated than the dilution that generated competitor and experimental bands of near-equal intensity in the primary amplification.
   - **b.** Use an aliquot of the first-strand cDNA with each twofold competitor dilution strand.
   - **c.** Perform PCR amplification and then analyze products.
   - **d.** Determine competitor concentration that most closely approximates target cDNA concentration.
   - **e.** Expect detection and resolution of as little as a twofold change in gene expression (transcript abundance).

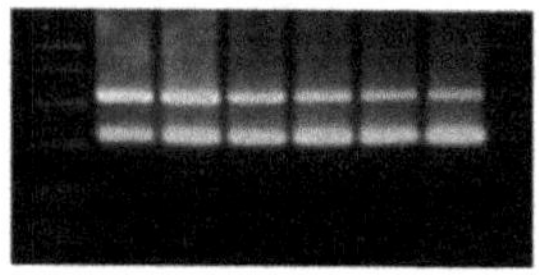

FIGURE 11.22 Secondary twofold dilution series for mass determination. Reaction series and analysis was identical to that described for the serial 10-fold dilutions prepared for the primary PCR samples, with the exception of the competitor mass added to each tube. Primary amplification showed the best match between competitor and experimental PCR products at 1.0 amol; thus, a twofold dilution series was utilized in the secondary amplification beginning with the previous dilution: lane 1, 10 amol; lane 2, 5 amol; lane 3, 2.5 amol; lane 4, 1.25 amol; lane 5, 0.625 amol, lane 6, 0.3125 amol. By chance, the best match is again in lane 2 (refer back to Fig. 11.21), meaning that the mass of actin cDNA in the sample under investigation is about 5amol, or approximately 120,000 molecules!

Alternative approach: nonreal-time competitive PCR

Real-time PCR is in widespread use, though it is still beyond the means of smaller laboratories. The protocol that follows is an older end-point analysis form of competitive PCR. It begins with the synthesis of a nonhomologous competitor mediated by composite primers. The product of this reaction is then used to spike cDNA samples, thereby promoting competitive coamplification of templated by PCR. If using an artificial control transcript generated by in vitro transcription, both the experimental RNA and the control RNA are placed in the same tube and then reverse-transcribed. Although this is no longer a mainstream method, it is an option for laboratories without access to a real-time platform; it is also included in this chapter because it is a useful teaching tool.

Protocol: competitive PCR

Synthesis of nonhomologous competitor

Each PCR competitor is synthesized in a separate reaction tube using a unique set of composite primers. Once synthesized, an aliquot of the product is used as a template for reamplification of the competitor. Note that the PCR cycling profile will vary depending on the type of thermal cycler. The reaction volumes may also be scaled up or scaled back, as needed.

1. Mix:

10× PCR buffer	5 μL
25 mM $MgCl_2$	3 μL
10 mM dNTPs	1 μL
Competitor template	X μL (1 ng suggested)
10 μM 5′ composite primer	2 μL
10 μM 3′ composite primer	2 μL
Sterile H_2O	36.5−X μL
Taq polymerase	0.5 μL

2. Perform cycling protocol:

Initial melt	94°C	3 min	
Amplification	94°C	1 min	20 cycles
	X°C	1 min	
	72°C	1 min	
Final extension	72°C	10 min	
Soak	4°C		

Note: Because of the extensive 5′ overhang structure exhibited by each of the composite primers, it may be necessary to reduce the annealing temperature several degrees below the calculated Tm to facilitate annealing to the competitor template.

3. Prepare a 1:500 dilution (in water) of the PCR reaction from step 3, an aliquot of which is used for reamplification. Be sure to reserve an undiluted aliquot of this same reaction as a diagnostic for electrophoresis.

4. Mix:

10× PCR buffer	5 μL
25 mM $MgCl_2$	3 μL
10 mM dNTPs	1 μL
Dilution from step 4	2.5 μL
10 μM 5′ cDNA primer	2 μL
10 μM 3′ cDNA primer	2 μL
Sterile H_2O	34 μL
Taq polymerase	0.5 μL

5. Perform cycling protocol:

Initial melt	94°C	3 min	
Amplification	94°C	1 min	20 cycles
	X°C	1 min	
	72°C	1 min	
Final extension	72°C	10 min	
Soak	4°C		

Note: If amplification is weak after 20 cycles, return the tube to the thermal cycler and perform five more cycles.

6. Remove an aliquot of the PCR products from step 5 and electrophorese along with the undiluted aliquot reserved at the beginning of step 3. Stain and photograph.
7. The product generated in step 6 is the competitor DNA, and it must be purified to remove the reaction components. The most efficient purification method is the use of one of the silica column-based rapid PCR clean-up kits, such as the High Pure PCR kit (Roche).
8. Determine competitor concentration and yield, either by A_{260} or by image analysis, as described in Chapter 15, Photodocumentation and image analysis. Store the competitor at −20°C.
9. Just prior to use, prepare a stock solution of competitor template by diluting a portion of the competitor to a concentration of 100 amol/μL. The dilution should be in 10 μg/mL ultrapure glycogen or in PCR TE buffer (10 mM Tris, pH 7.5; 0.1 mM EDTA), ideally in a silanized microfuge tube. Dilutions should be used shortly after preparation, and the remainder should be discarded.
   Note: The use of silanized tubes in which to prepare the dilutions is strongly recommended to prevent loss of any of the templates by sticking to the inside wall of the tube. At such low concentrations, the loss of any template material can strongly impact the outcome of the experiment.

Synthesis of first-strand cDNA

1. Ensure that high-quality RNA is available to support the synthesis of first-strand cDNA. This is ascertained by minigel electrophoresis and

visualization of the ribosomal 28S and 18S RNA, with minimal smearing below the 18S rRNA. See Chapter 8, Quality control for RNA preparations, for RNA quality control options

2. Prepare the cDNA mix:

		(Final)
Nuclease-free H_2O	3 μL	–
25 mM $MgCl_2$	4 μL	5 mM
10 × first-strand buffer	2 μL	1 ×
10 mM dNTPs	2 μL	1 mM
Oligo(dT)[a]	2 μL	2.5 μM
RNasin	1 μL	2.5U/μL

[a]Alternatively, random primers (1 μg) can be used with or in place of oligo(dT) to support cDNA synthesis.
Set aside on ice.

3. Transfer 500 ng RNA (2 μL; 250 ng/μL) to a sterile microfuge tube and then add 3 μL of nuclease-free H_2O. Heat to 65°C for 2 minutes, quickly pulse-centrifuge, and then cool on ice for 1 minute.
4. Add the entire cDNA mix (from step 2) to the tube containing the heat-denatured RNA.
5. Incubate at room temperature for 3 minutes to facilitate primer annealing.
6. Add 1 μL reverse transcriptase; gently pipette up and down to mix. Do not vortex, ever.
7. Incubate at 42°C for 30–60 minutes (AMV) or at 37°C for 30–60 minutes (MMLV).
8. Heat to 95°C for 5 minutes to destroy the reverse transcriptase, and then store on ice until ready to use.

Primary amplification

The purpose of the primary reactions is to determine the approximate mass of the cDNA under investigation. Although the mass is expected to vary drastically among different cDNAs reflecting the abundance of the original transcript in vivo, it is likely that the primary reactions will be required only once. After identifying the competitor dilution that most closely approximates the mass of the cDNA, secondary twofold dilutions of the competitor are run in order to zero in on the mass of the experimental template. Of course, profound changes in mRNA abundance, in response to an experimental challenge, will mandate repetition of the 10-fold competitor dilution series.

1. For each sample to be tested, prepare PCR master mix for 13 tubes (12 sample tubes +1 extra) as shown in Table 11.5.
2. Label six PCR tubes P1–P6.

TABLE 11.5 Master mix preparation for competitive PCR.

PCR master mix	Per 50 μL Px	[Final]	6 samples (×7)	(×13) 12 samples
10× PCR buffer	5.0 μL	1 mM	35.0 μL	65.0 μL
25 mM $MgCl_2$ stock	3.0 μL	1.5 mM	21.0 μL	39.0 μL
Sterile H_2O	34.5 μL	–	241.5 μL	448.5 μL
10 mM dNTP mix	1.0 μL	0.2 mM	7.0 μL	13.0 μL
5′ primer (20 μM)	1.0 μL	0.4 μM	7.0 μL	13.0 μL
3′ primer (20 μM)	1.0 μL	0.4 μM	7.0 μL	13.0 μL
cDNA	2.0 μL	Variable	14.0 μL	26.0 μL
Taq polymerase[a] (5U/μL)	0.5 μL	2.5U/ reaction	3.5 μL	6.5 μL
Total volume	46.0 μL	–	322.0 μL	598.0 μL

[a]*Other thermostable enzymes may yield superior results compared to the use of* Taq *alone. Regardless of enzyme choice, be sure to use a hot-start format, which may be antibody- or aptamer-based, or by the inclusion of modified hot-start nucleotides.*

3. To prepare the primary 10-fold serial dilution series, pipette 9 μL of 10 μg/mL ultrapure glycogen in TE buffer (10 mM Tris, pH 7.5; 0.1 mM EDTA) into each of the six tubes.
4. Designate an aliquot of the competitor stock (100 amol/μL) as Tube P. Ten-fold dilution of the competitor is performed as follows:

Tube	Dilution method	(Final)
P_1	Add 1 μL P, mix	10 amol/μL
P_2	Add 1 μL P_1, mix	1.0 amol/μL
P_3	Add 1 μL P_2, mix	10^{-1} amol/μL
P_4	Add 1 μL P_3, mix	10^{-2} amol/μL
P_5	Add 1 μL P_4, mix	10^{-3} amol/μL
P_6	Add 1 μL P_5, mix	10^{-4} amol/μL

5. Label six new PCR tubes.
6. To each tube add:
   - **a.** 2 μL one of the dilutions just prepared (P1–P6);
   - **b.** 48 μL PCR master mix (step 1);
   - **c.** 50 μL total volume.
7. Gently pulse centrifuge to collect the entire reaction volume at the bottom of the tube.

8. Perform cycling protocol.

Initial melt	94°C	3 min	
Amplification	94°C	1 min	25 cycles
	X°C	1 min	
	72°C	2 min	
Final extension	72°C	5 min	
Soak	4°C		

9. Remove a 10 μL aliquot from each tube. Add 1.0 μL 10× loading buffer to each sample.

10. Electrophorese samples in a 2.5% agarose gel along with an appropriate range molecular weight standard.

Note: The slower the gel is run, the better will be the resolution. The concentration of the agarose can also be increased or decreased as needed.

11. At the conclusion of the run, stain gel with 1× SYBR Green I diluted in 1× TBE or 1× TAE buffer and then photograph the gel.

12. Determine the dilution of the competitor that generated a PCR product most similar to the mass of the experimental product. Visual inspection is usually adequate for this measurement involving serial 10-fold dilutions. However, the integrated optical density (IOD) of each band and ratio for each pair of bands can be determined using image analysis software for a more objective measurement.

Secondary amplification

Having identified the competitor dilution in which the PCR products of the competitor and the cDNA are nearly identical, the competitor dilution that is 10-fold *more* concentrated will be used as the starting point for making the secondary set of dilutions. The purpose of making twofold dilutions for use in the second set of reactions is to fine-tune the mass determination of the gene-specific cDNA in the sample(s) under investigation. For the sake of consistency and reproducibility, use aliquots of the same PCR master mix which was prepared to support the primary reactions. The master mix should be maintained on ice; if prepared well in advance it may be frozen, as long as the *Taq* (or other) polymerase has not been added.

1. Label six PCR tubes S1–S6.
2. To prepare the twofold serial dilution series, place 5 μL of 10 μg/mL glycogen in PCR TE buffer (10 mM Tris, pH 8.0; 0.1 mM EDTA) into each of the six tubes.
3. Beginning with the competitor concentration that is 10-fold greater (10×, designated as tube P×) than that which matched the mass of the cDNA in the primary amplification, prepare the twofold dilution series as follows:

	Tube	Dilution method	(Final)
Dilutions:	S_1	Add 5 μL P_x, mix	$5 \times 10^{x-1}$ amol/μL
	S_2	Add 5 μL S_1, mix	$2.5 \times 10^{x-1}$ amol/μL
	S_3	Add 5 μL S_2, mix	$1.25 \times 10^{x-1}$ amol/μL
	S_4	Add 5 μL S_3, mix	$6.25 \times 10^{x-2}$ amol/μL
	S_5	Add 5 μL S_4, mix	$3.125 \times 10^{x-2}$ amol/μL
	S_6	Add 5 μL S_5, mix	$1.56 \times 10^{x-2}$ amol/μL

4. Label six new PCR tubes.
5. To each tube add:
   a. 2 μL: one of the dilutions just prepared (S1–S6);
   b. 48 μL: PCR master mix (refer to Table 11.5);
   c. 50 μL: total volume.
6. Gently pulse centrifuge to collect all reaction volume at the bottom of the tube.
7. Perform cycling protocol:

Initial melt	94°C	3 min	
Amplification	94°C	1 min	25 cycles
	X°C	1 min	
	72°C	2 min	
Final extension	72°C	5 min	
Soak	4°C		

Note: Be sure to use precisely the same cycling parameters as with the primary 10-fold dilution series, including the selection of annealing temperature.

8. Remove a 10 μL aliquot from each tube. Add 1.0 μL 10× loading buffer to each sample.
9. Electrophorese samples in a 2.5% agarose gel, along with an appropriate range molecular weight standard.
10. At the conclusion of the run, stain gel with 1× SYBR Green I diluted in 1× TBE or 1× TAE buffer and then photograph the gel.
11. Determine the dilution of the competitor that generated a PCR product most similar to the mass of the experimental product. Because the samples represent only twofold dilutions of the competitor, it may be helpful to take advantage of image analysis software to determine the closest match between competitor and cDNA, that is, the sample in which $IOD_{sample} \div IOD_{competitor} = 1$.

Image analysis considerations

The ability to make sense of the gels used to analyze competitive PCR reactions is highly dependent on the quality of the appearance and image of the gel from the start. Note that the degree to which gel analysis is accurate and reproducible is a direct function of what the gel and/or photograph look like from the onset; to a very great extent, image analysis optimization is just as important as optimizing PCR or other reactions. No image analysis system, regardless of the cost, will compensate for poorly run gels and low-quality images. The reader is encouraged to review the "Digital Image Analysis" section in Chapter 15, Photodocumentation and image analysis, for suggestions pertaining to the optimization of image analysis.

With respect to competitive PCR, EtBr-stained gels are often difficult to extract data from because of the intrinsically high levels of background associated with the use of this dye. Gels stained with SYBR Green or SYBR Gold are clearly superior. Better quality gels facilitate greater accuracy both in visual inspection and in digital image analysis. If desired, labeled dNTPs may be included in the PCR reaction, thereby supporting detection by autoradiography and mass determinations via scintillation counting. Radiolabeling for this type of assay is not required and is often counterproductive; as such, it is not recommended.

Troubleshooting quantitative PCR techniques

The exquisite sensitivity of PCR to the precise concentrations of all reaction components is indisputable. For assays designed to generate quantitative data, accurate micropipetting and the use of master mixes for both cDNA synthesis and PCR are of critical importance. The items below outline what may be done to minimize aggravation and maximize reproducibility and productivity.

1. *Nonnegotiables*
   a. High-quality RNA:
      i. Use good RNase-free technique.
      ii. Silica-based purification or guanidinium-acid-phenol techniques are preferred.
      iii. Perform double precipitation with guanidinium lysis buffer, especially when starting with a tissue sample.
      iv. Treat samples with RNase-free DNase.
      v. If precipitated, wash RNA pellets thoroughly and extensively with 70% ethanol prepared in nuclease-free H_2O. Recall that cationic salts bind dNTPs quantitatively and can cause havoc in the cDNA synthesis and PCR reactions.
      vi. Properly store purified RNA.
   b. Reverse transcription into cDNA and subsequent PCR:
      i. Use equal mass of RNA per cDNA reaction (200–500 ng).
      ii. Use an RNase H − reverse transcriptase.
      iii. Always work with a cDNA master mix.
      iv. Always work with a PCR master mix.
2. *Common problems*
   a. Heavy smearing in one or more lanes:
      i. Optimize Mg^{2+} concentration by decreasing its concentration incrementally by 0.1 mM.
      ii. Reduce the number of PCR cycles.
      iii. Decrease the amount of enzyme in the reaction.
      iv. Use a hot-start PCR format.
   b. No product:
      i. Check the RNA!
      ii. Add an RNase inhibitor (e.g., RNasin).
      iii. Test the components of the RT and PCR reactions, especially the enzymes, using control templates that generally accompany the purchase of a kit.
      iv. Check off each component as the reactions are assembled.
      v. Check PCR cycle parameters.
   c. Low yield:
      i. Check for incomplete homogenization or lysis of samples.
      ii. Check for a final RNA pellet that is incompletely redissolved.
      iii. Determine if A260/A280 <1.65 (suggests protein or other contaminants).
      iv. Check for RNA degradation:
      v. Tissue was not immediately processed/frozen after harvesting.
      vi. Purified RNA was stored at −20°C, instead of −80°C.
      vii. Aqueous solutions or tubes were not RNase-free.

d. DNA contamination:
   i. Often occurs when the sample was not DNase-treated.
   ii. Always validate with RT − controls.
e. Nonreproducibility of data:
   i. Use thin-walled PCR tubes.
   ii. Extend cycle parameters.
   iii. Use a hot-start PCR format.
   iv. Prepare a single, large-volume master mix per cell type and aliquot for each reaction.
   v. Match Tm of upstream and downstream primer pairs more closely.
   vi. Consider enzyme mixtures (e.g., Taq + proofreading enzyme) designed for long-range PCR. Working together, these enzymes generate more product and do so with high fidelity.
   vii. Occurs when various RNA samples are not isolated by the same method.
   viii. Occurs when RNA samples are not isolated on the same day.
   ix. Occurs when RNA is not isolated by the same investigator.
   x. Ascertain the growth state of the starting cells or tissue. If cells are in different parts of the cell cycle or the tissues are isolated from biochemically different individuals, data may be nonreproducible.
   xi. Check if a different master mix formulation was used.
   xii. Verify that micropipettors were calibrated recently. This is a bigger problem than one might imagine.
   xiii. Optimize and run reactions in the same thermal cycler (not the same model number).
   xiv. Consider the purchase of precast gels; both agarose and polyacrylamide are available.
   xv. If applicable, consider elimination of mineral oil: use a thermal cycler with a heated lid.

3. General suggestions:
   a. Keep denaturation steps as short as possible.
   b. Keep denaturation temperature as low as possible (90°C−92°C).
   c. Reduce elongation temperature to 68°C.
   d. Extend elongation time for each successive cycle (5−20 seconds/cycle).
   e. Ensure that the input mass of RNA is exactly the same in each sample. Competitive PCR, for example, is exquisitely sensitive to the input mass of RNA (for cDNA synthesis) and cDNA (for PCR amplification). As with most RT-PCR reactions "less is more," meaning that it is generally counterproductive to overload the reaction with too much template. When putting the PCR reactions together, make sure that unintentional droplets on the *outside* of the tip do not contribute to the outcome of the reaction.

f. Observe standard methods for the prevention of carryover contamination, including the use of barrier tips or positive displacement micropipettors.
g. Consider the use of silanized microfuge tubes. These tubes may be helpful in minimizing losses during the various manipulations because cDNA/PCR products are present in picogram quantities (or less) prior to amplification (see Appendix H: Silanizing centrifuge tubes and glassware, for protocol). Making dilutions of minute quantities of template in ultrapure glycogen is helpful, too.
h. Hone micropipetting skills, which are critical for reproducibility. Have micropipettors recalibrated at regular intervals.

References

Bonnet, G., Tyagi, S., Libchaber, A., Kramer, F.R., 1999. Thermodynamic basis of the enhanced specificity of structured DNA probes. Proc. Natl. Acad. Sci. USA 96, 6171–6176.

Bustin, S.A., Benes, V., Garson, J.A., Hellemans, J., Huggett, J., Kubista, M., et al., 2009. The MIQE guidelines: minimum information for publication of quantitative real-time PCR experiments. Clin. Chem. 55 (4), 611–622.

Diviacco, S., Norio, P., Sentilin, L., Menzo, S., Clementi, M., Biamonti, G., et al., 1992. A novel procedure for quantitative polymerase chain reaction by coamplification of competitive templates. Gene 122, 313–320.

Flohr, A.M., Hackenbeck, T., Schlueter, C., Rogalla, P., Bullerdiek, J., 2003. DNase I treatment of cDNA first strands prevents RT-PCR amplification of contaminating DNA sequences. BioTechniques 35 (5), 920–926.

Förster, T., 1946. Energiewanderung und Fluoreszenz. Naturwissenschaften 6, 166–175.

Förster, T., 1948. Zwischenmolekulare Energiewanderung und Fluoreszenz. Ann. Phys. (Leipz.) 2, 55–75.

Fronhoffs, S., Totzke, G., Stier, S., Wernert, N., Rothe, M., Brüning, T., et al., 2002. A method for the rapid construction of cRNA standard curves in quantitative real-time reverse transcription polymerase chain reaction. Mol. Cell. Probes 16, 99–110.

Higuchi, R., Dollinger, G., Walsh, P.S., Griffith, R., 1992. Simultaneous amplification and detection of specific DNA sequences. Biotechnology 10, 413–417.

Higuchi, R., Fockler, C., Dollinger, G., Watson, R., 1993. Kinetic PCR: real-time monitoring of DNA amplification reactions. Biotechnology 11, 1026–1030.

Holland, P.M., Abramson, R.D., Watson, R., Gelfand, D.H., 1991. Detection of specific polymerase chain reaction product by utilizing the 5′-3′ exonuclease activity of *Thermus aquaticus* DNA polymerase. Proc. Natl. Acad. Sci. 88, 7276–7280.

Lee, L.G., Connell, C.R., Bloch, W., 1993. Allelic discrimination by nick-translation PCR with fluorogenic probes. Nucleic Acids Res. 21, 3761–3766.

Li, B., Sehajpal, P.K., Khanna, A., Vlassara, H., Cerami, A., Stenzel, K.H., et al., 1991. Differential regulation of transforming growth factor beta and interleukin 2 genes in human T cells: demonstration by usage of novel competitor DNA constructs in the quantitative polymerase chain reaction. J. Exp. Med. 174, 1259–1262.

Livak, K.J., Flood, S.J., Marmaro, J., Guisti, W., Deetz, K., 1995. Oligonucleotides with fluorescent dyes at opposite ends provide a quenched probe system useful for detecting PCR product and nucleic acid hybridization. PCR Meth. Appl 4, 357–362.

Siebert, P.D., Larrick, J.W., 1993. PCR MIMICS: Competitive DNA fragments for use as internal standards in quantitative PCR. BioTechniques 14 (2), 244.

Smith, R.D., Malley, J.D., Schechter, A.N., 2000. Quantitative analysis of globin gene induction in single human erythroleukemic cells. Nucl. Acids Res. 28, 4998–5001.

Smith, R.D., Brown, B., Ikonomi, P., Schechter, A.N., 2003. Exogenous reference RNA for normalization of real-time quantitative PCR. BioTechniques 34 (1), 88–91.

Sykes, P.J., Neoh, S.H., Brisco, M.J., Hughes, E., Condon, J., Morley, A.A., 1992. Quantitation of targets for PCR by use of limiting dilution. BioTechniques 13 (3), 444–449.

Vogelstein, B., Kinzler, K.W., 1999. Digital PCR. Proc. Natl. Acad. Sci. 96, 9236–9241.

Wang, A.M., Doyle, M.V., Mark, D.F., 1989. Quantitation of mRNA by the polymerase chain reaction. Proc. Natl. Acad. Sci. 86, 9717–9721.

Wang, S., Murtagh Jr., J.J., Luo, C., Martinez-Maldonado, M., 1993. Internal cRNA standards for quantitative Northern analysis. BioTechniques 14 (6), 935–941.

Whitcombe, D., Theaker, J., Guy, S.P., Brown, T., Little, S., 1999. Detection of PCR products using self-probing amplicons and fluorescence. Nat. Biotechnol. 17, 804–807.

Further reading

Li, D., Zhang, J., Li, J., 2020. Primer design for quantitative real-time PCR for the emerging Coronavirus SARS-CoV-2. Theranostics 10 (16), 7150. Available from: https://doi.org/10.7150/thno.47649.

Roden, A.C., Vrana, J.A., Koepplin, J.W., Hudson, A.E., Norgan, A.P., Jenkinson, G., et al., 2021. Comparison of in situ hybridization, immunohistochemistry, and reverse transcription–droplet digital polymerase chain reaction for severe acute respiratory syndrome coronavirus 2 (SARS-CoV-2) testing in tissue. Arch Pathol Lab Med 145 (7), 785–796. Available from: https://doi.org/10.5858/arpa.2021-0008-SA.

Tan, C., Fan, D., Wang, N., Wang, F., Wang, B., Zhu, L., et al., 2021. Applications of digital PCR in COVID-19 pandemic. View 2 (2), 20200082. Available from: https://doi.org/10.1002/VIW.20200082.

Vanden Heuvel, J.P., Tyson, F.L., Bell, D.A., 1993. Construction of recombinant RNA templates for use as internal standards in quantitative RT-PCR. BioTechniques 14 (3), 395–398.

Chapter 12

miRNA and other noncoding RNAs

Rationale

Accompanying the completion of the 13-year human genome sequencing project in 2003 was the then shocking revelation that there are relatively few protein-encoding genes distributed across 24 distinct chromosomes. The 20,000+ genes in this category, which encode approximately 50,000 different mRNAs, occupy roughly 4% of the genome; this is far fewer than expected, given that there are more than 1 million known functional proteins. In sharp contrast to what was commonly held only a few years ago, comprehensive transcriptome deep-sequencing has revealed that most of the typical eukaryotic genome is transcriptionally active, producing an astonishing array of noncoding RNA (ncRNA). The majority of the transcriptome consists of staggering numbers of ncRNA, the major categories of which are small noncoding RNA (sncRNA; <200 nts), long noncoding RNA (lncRNA; 200+nts), long intergenic RNA (lincRNA; 200+nts), circular RNA (circRNA; variable size), catalytic RNA (Fig. 12.1). There are undoubtedly other categories in which to sort noncoding RNAs, most likely by function; those functions, however, have yet to be discerned. It is worth noting that when referring to ncRNA, one does not commonly think about rRNA or tRNA because of their well-defined roles in supporting protein synthesis and because of their limited varieties in the cell.

ncRNA genes are found in intergenic regions, completely within introns, and even overlapping protein-encoding genes in both the sense (+) and antisense (−) orientations. Given the variegated sequence, length, and capacity for secondary structure formation, it comes as no surprise that these molecules have both known and suspected regulatory roles in such diverse areas as cell proliferation, growth and development, telomere maintenance, genome maintenance, cellular immunity and, assuredly, other yet to be determined aspects of cellular physiology. As one might naturally suspect, dysregulation of these important transcripts is well known to contribute to the initiation and progression of the disease state. With respect to occasional reports that some lncRNAs do, in fact, have the capacity to support translation, there is considerable ongoing debate regarding the possible misannotation of these transcripts.

RNA Methodologies. DOI: https://doi.org/10.1016/B978-0-323-90221-2.00035-7

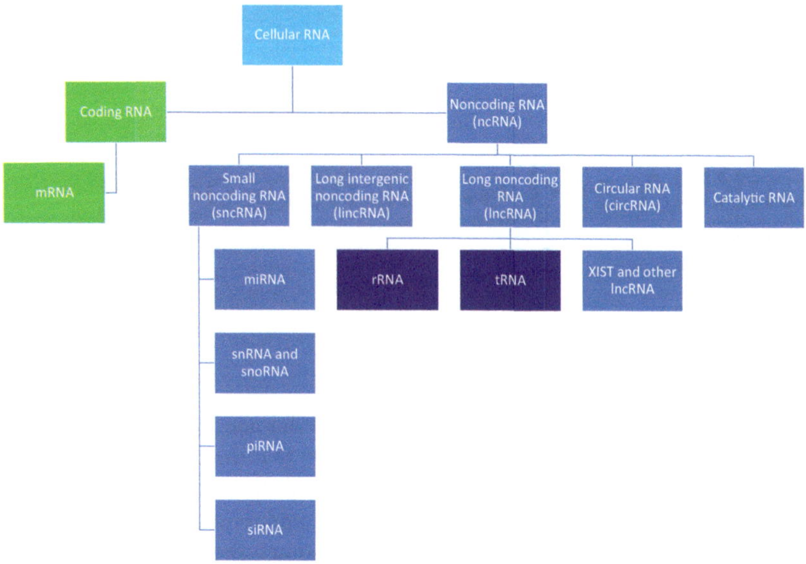

FIGURE 12.1 Classification of noncoding RNAs in a typical mammalian cell. While rRNA (28S, 18S, 5.8S, 5S) and tRNA (codon-specific) are technically noncoding, they are often categorized separately owing to their abundance and unique roles in protein synthesis.

One prevalent type of small noncoding transcript, microRNA (miRNA), has been documented to have important roles in the modulation of gene expression by profoundly influencing mRNA stability and translatability. Originally believed only to suppress protein synthesis, miRNAs are now known to have the ability to promote protein synthesis (Vasudevan, Tong, and Steitz, 2007), sometimes doing so epigenetically via histone modification (Xiao, et al., 2017). While the functions of some miRNAs have yet to be determined, the fact that many of these transcripts are strongly conserved suggests a central regulatory role across evolutionary time.

Without a doubt, the transcriptome is far more complex than anyone could have ever imagined a few short years ago.

Overview of noncoding RNAs

Transcripts known widely as ncRNA are diverse and abundant in eukaryotic cells. By definition, ncRNAs do not encode proteins, a job that is uniquely the responsibility of mRNA. Once thought of as "junk RNA," nothing could be more of a misnomer when referring to ncRNA. Almost unknown a decade ago, it is now evident that at least 90% of the human genome is transcribed, resulting in copious quantities of ncRNA, examples of which are presented in Table 12.1. The identification of transcripts, in several species, suggests a central regulatory role for these molecules.

TABLE 12.1 Examples of noncoding RNA.

RNA type	Name	Symbol	Basic function	Prokaryotic	Eukaryotic
Long noncoding RNA (lncRNA)	Ribosomal RNA	rRNA	Forms scaffolding of the ribosomal subunits.	Yes	Yes
	Transfer RNA	tRNA	Transports amino acids to the ribosome to support translation.	Yes	Yes
	Long intergenic (or intervening) noncoding RNA	lincRNA	Transcripts produced within intergenic regions.	No	Yes
	XIST	Xist	Sex-linked lncRNA involved in X-chromosome inactivation and Barr body formation.	No	Yes
Small noncoding RNA (sncRNA)	Small nuclear RNA	snRNA	Facilitates splicing of hnRNA into mature mRNA as well as rRNA processing. snRNA molecules exist as an RNA-protein complex, referred to as a snRNP, or snurp.	No	Yes
	Small nucleolar RNA	snoRNA	Processing of immature rRNA transcripts in the nucleolus; some have a role in gene silencing.	No	Yes
	Small cytoplasmic RNA	scRNA	Facilitates protein trafficking and secretion, and possibly mRNA degradation. scRNA molecules exist as an RNA-protein complex, referred to as a scRNP, or scyrp.	Yes	Yes
	microRNA	miRNA	Short antisense RNAs that participate in the regulation of gene expression by blocking mRNA and inhibiting translation.	No	Yes
	Guide RNA	gRNA	RNAs that lead enzymes to another RNA or a DNA target to alter or destroy it.	Yes	Yes

(Continued)

TABLE 12.1 (Continued)

RNA type	Name	Symbol	Basic function	Prokaryotic	Eukaryotic
	Piwi-interacting RNA	piRNA	Involved in epigenetics and post-transcriptional silencing via Argonaute protein interactions. Largest group of sncRNA in animals; usually 23–30 nts.	No	Yes
	Exosomal RNA	exRNA	miRNA and often other RNA types that travel via the circulatory system within vesicles.	Yes	No
Catalytic RNA	Ribozyme	Varies by type	An RNA molecule with a catalytic function.	Yes	Yes
	Telomerase RNA	TERC	Noncoding RNA portion of the enzyme/RNA complex that repairs chromosome telomeres (TERC: telomerase RNA component).	No	Yes

A particular prevalent type of ncRNAs is the family of small miRNA transcripts. These powerful regulators of gene expression suppress mRNA translatability via transient binding to an mRNA or by mediating the precision, directed cleavage of the mRNA itself. With apparent ubiquity, these highly specialized molecules elicit profound changes in gene expression patterns despite their remarkably small size range of 19–26 nts, depending on their origin and function in the cell. Much of the early work in the realm of molecular biology missed these small but powerful molecules primarily because the original RNA isolation protocols, and later the RNA isolation kits available at the time, were not designed to capture such small transcripts. It is now evident that miRNAs are of supreme importance in terms of modulation of gene expression, and commercially available kits and devices specifically designed to isolate small RNAs abound. The molecular biology and applications of miRNA and the relationship between miRNA, siRNA, and RNA interference (RNAi) are discussed in detail in Chapter 13, RNA Interference and Gene Editing.

sncRNA

Transcripts classified as sncRNAs consist of fewer than 200 nts. As with most other types of ncRNA, their membership is variable both in form and function. There are currently 7500+ known sncRNA genes in humans including miRNA, snRNA, snoRNA, scRNA, piRNA. Examples of their diversified functions include post-transcriptional processing of precursor RNA, facilitating protein trafficking, regulating transcription factors, mRNA destabilization, translational repression, gene silencing, and telomere maintenance.

lncRNA

Transcripts classified as lncRNAs are high variable in length (200–22,000+nts) and range from a few to more than a thousand copies per cell (for a recent review, see Wu et al., 2021). Although the typical abundance level of lncRNAs is about one log less than what is typically observed for mRNA, the amount and variety of lncRNA transcripts in the cell are impressive; the human genome encodes 18,000+ lncRNAs that directly or indirectly influence virtually all cellular processes. They are distributed throughout the nucleus and the cytoplasm and interact in a variety of ways with proteins and other molecules (Fig. 12.2). In so doing, lncRNAs exert influence over a variety of cell functions including gene expression at the transcriptional and translational levels. In addition, lncRNAs are known to function as tumor suppressors or oncogenes, roles which portend excellent utility are RNA biomarkers.

Some types of lncRNAs are immediately recognizable to even the most casual reader of cell biology literature, wit, ribosomal RNA (rRNA), and

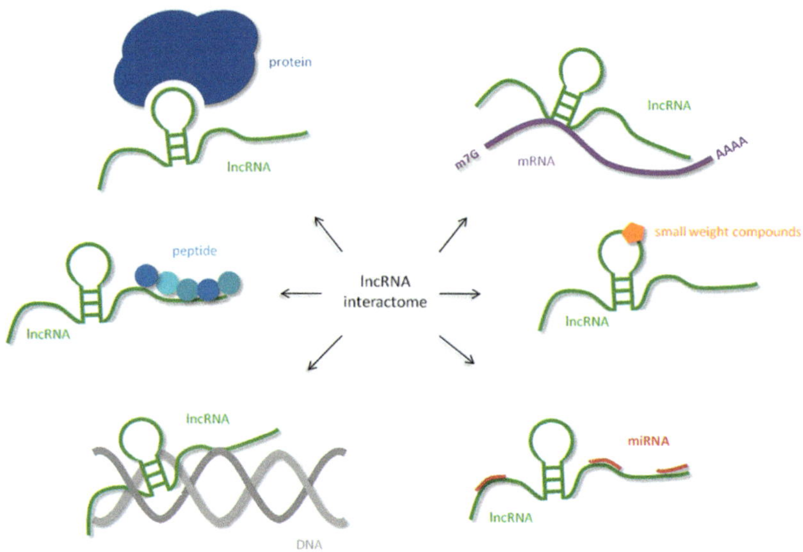

FIGURE 12.2 The lncRNA interactome. Long noncoding RNA has the ability to interact with other macromolecules in a variety of ways, thereby influencing multiple cell functions (Kazimierczyk et al., 2020). *Figure courtesy Kazimierczyk, M., Kasprowicz, M.K., Kasprzyk, M.E., Wrzesinski, J., 2020. Human long noncoding rna interactome: detection, characterization and function. Int. J. Mol. Sci. 21 (3), 1027. https://doi.org/10.3390/ijms21031027 and reproduced according to the terms of the Creative Commons Attribution 4.0 License (https://creativecommons.org/licenses/by/4.0/).*

transfer RNA (tRNA). With respect to translation, rRNAs serve as the scaffolding for the assembly of the large and small ribosomal subunits, formed when these predominant rRNA species bind 80 + proteins to become functional protein synthesis factories. The rRNA species are not particularly diverse, though variations in base composition allow identification of the source organism. Supporting the role of the ribosomes are tRNA molecules which, when charged, carry a unique amino acid to the ribosome to support protein elongation.

lincRNA

Another important category of ncRNAs is the class representing the transcription of intergenic regions [long intergenic noncoding RNA (lincRNA)]. In general, lincRNAs are longer than lncRNAs, are autonomously transcribed, and they tend to reside in the nucleus while lncRNA is mainly cytoplasmic. It is believed that at least half of the lncRNAs in humans are, in fact, lincRNAs. While clearly an abundant class of ncRNA, very little is actually known about the biological role of many of these transcripts. What is known about lincRNA is that they are directly involved in regulating the structural features of chromatin and, in so doing, indirectly impact both transcription and translation. Some lincRNA transcripts have also been identified as useful biomarkers (Chapter 25, RNA Biomarker Discovery and Validation). By definition, lincRNA transcripts do not have open reading frames (ORFs) and, being the product of RNA polymerase II activity, lincRNA transcripts are frequently capped and exhibit poly(A) tails. This means that any experiments involving cDNA synthesis for cloning, sequencing, older-style library construction, or RNA-seq experiments more than likely have lincRNAs among their "membership."

A somewhat unusual example of well-known and very important lincRNA, Xist, is an approximately 17,000 nts transcription product from a sex-linked locus and that is essential for X-chromosome inactivation, that is, dosage compensation, and the concomitant formation of a cytologically significant Barr body observed in the nuclei of female mammals (Brown et al., 1992).

Y RNA

Y RNAs are a class of highly conserved sncRNAs. They have been observed across all vertebrates, and were once thought to be random degradation products of larger transcripts. Although some related sequences are observed among the invertebrates, they have not been observed in insects, plants, or fungi. Although they were discovered more than 40 years ago (Lerner et al., 1981), Y RNAs were, until recently, the least studied among their ncRNA cousins. The nomenclature for these molecules came from their discovery in the cYtoplasm and they were so named to avoid confusing them with previously identified U ncRNA in the nUcleus in the course of studying autoimmune disorders at the time. In humans, Y RNAs appear to be expressed in all cells, although at varying abundance levels in a tissue-specific manner; elevated levels are found in brain and heart tissue, while the lowest abundance levels appear in hepatocytes. In humans there are four known Y RNA genes, producing transcripts ranging from 84 to 113 bases; there are also hundreds of pseudogenes distributed across the human genome. Each of the canonical genes has its own promoter, and the genes lie in proximity to each other on the same chromosome, in this case 7q36 (Maraia et al., 1994) in the vicinity of the telomere on the long arm of chromosome 7. The genes are transcribed by RNA polymerase III. The transcripts from the four functional Y genes are characterized by extensive intramolecular base-pairing involving the 5′ and 3′ ends of the transcript as well as a 3′ polyuridine tract (Fig. 12.3) that acts as a binding site for La, which is a ubiquitous eukaryotic stabilizing protein commonly associated with transcripts produced by RNA polymerase III. Interestingly, both human and mouse Y RNAs contain few or no modified nucleotides. As with all transcripts, mature Y RNAs fold into sequence-dependent loop and stem structures and associate with various proteins, particularly Ro60, a conserved ring-shaped protein found in most animal cells and which accompanies at least some YRNA-La complexes into

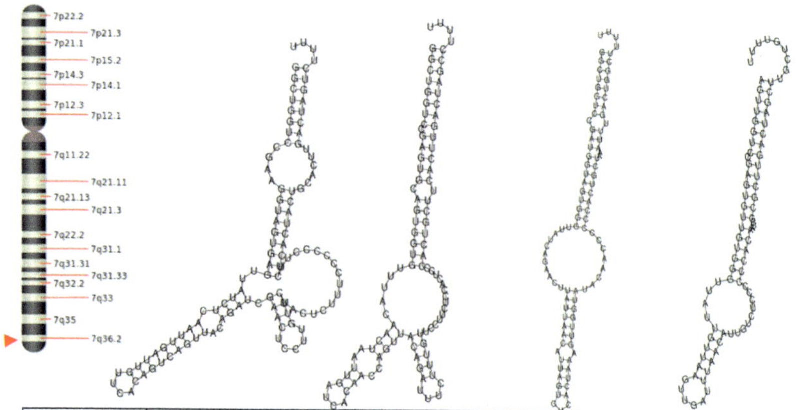

Transcript name	hY1	hY3	hY4	hY5
Gene name	*RNY1*	*RNY3*	*RNY4*	*RNY5*
Size	113 nt	102 nt	96 nt	84 nt
Chromosomal location	7	7	7	7
Oligo U 3′ end structure	U-U-U-U 3′	U-U-U-U 3′	U-U-U-U 3′	U-U-G-C-U-G-U-U-U-U 3′

FIGURE 12.3 Characteristics of the four canonical human Y RNA. *Red* arrow indicates approximate location of the Y RNA genes near the end of the long (q) arm of chromosome 7. Folded structures predicted using RNAfold (http://rna.tbi.univie.ac.at).

the cytoplasm. The normal functions of Y RNAs are associated with cellular stress responses, apoptosis, ncRNA quality control, initiation and replication of DNA and concomitant progression through the cell cycle, and differentiation (for a recent review, see Valkov and Das, 2020). Y RNA dysregulation appears to have a role in the disease state, including cancer, cardiovascular disorders, and some autoimmune disorders (reviewed by Boccitto and Wolin, 2019), thereby suggesting a possible role for these unique molecules as biomarkers.

circRNA

Once believed to be a rare consequence of a spliceosome machinery error, circRNAs (Hsu and Coca-Prados, 1979; Cocquerelle et al., 1993) are abundant, though mysterious, single-stranded molecules that are a widely observed transcriptome feature across of breadth of species from viruses to mammals. Found in both the nucleus and the cytoplasm, circRNA is produced by an alternative splicing phenomenon known as back-splicing, a process by which the 3′ end of exon "n" is joined covalently to the 5′ end of the same exon, that is, exon "n," rather than to the 5′ end of exon "n + 1." It appears that the positioning of repetitive *Alu* inverted sequences adjacent to intronic sequences or within them, in concert with *trans*-acting splicing factors, directs either the formation of excised circularized exon(s) or linear splicing, respectively (Zhang et al., 2014; Ivanov et al., 2015). Almost certainly, there are other contributing variables that orchestrate the formation of circRNA. Since these are covalently closed structures, circRNAs lack the 5′ cap and 3′ poly(A) tail structures commonly associated with vertebrate mRNAs and which were present in the primary transcript prior to processing.

Usually consisting of one or a few exons, the number of known eukaryotic circRNAs is in excess of 400,000. Although the first circRNAs identified were exclusively exon-encoded sequences, many subsequently identified circRNAs also contain intron sequences. It seems that intron-containing circRNAs are primarily confined to the nucleus, while exon-only circRNAs are capable of nucleocytoplasmic transport in a size-dependent manner. Structurally, circRNAs are classified based on the "type" of material from which they were transcribed, namely ciRNA (intronic circRNA), EIciRNA (exon-intronic circRNA), or EcircRNA (exonic circRNA). The nonlinear shape of circRNA imparts exonuclease resistance and great stability to these molecules; the extended half-lives of these molecules, estimated to be anywhere between 3 and 10 × as long as their linear counterparts, presumably allow them enough time to perform their intended function(s).

While the expression of circRNAs is widespread, the highest incidence of circRNAs is found in the mammalian brain, with the greatest density of these molecules observed in the synaptic region of nervous tissue cells (Rybak-Wolf et al., 2015). circRNAs are believed to play a role in neuronal differentiation, as their expression is upregulated during brain development (You et al., 2015). In a surprise development, it was shown that the number of circRNAs from some genetic loci exceeds the linear counterpart by a factor of as much as 10 (Salzman et al., 2012). One clear function is that at least some circRNAs are able to absorb miRNAs much like a molecular sponge (Hansen et al., 2013), thereby sequestering them as a means of controlling the expression of specific genes associated with any of a variety of pathways (Fig. 12.4). Similarly, the formation of circRNAs may regulate the concentration of certain types of the more than 1000 known RNA binding proteins by transiently binding them. Not unexpectedly, certain circRNA signatures are associated with several human diseases (reviewed by Lu, 2020).

Since circRNAs are the result of splicing events intended to bring exons together, it is reasonable to assume that many gene loci are capable of producing an even greater array of processes transcripts, further diversifying the transcriptome. The bases that constitute circRNAs are also subject to epigenetic modification, for example, the formation of m^6A (N^6-methyladenosine). Originally believed to be exclusively noncoding molecules, as they are lacking free 3′ and 5′ ends, some circRNAs have been demonstrated to be translatable (Pamudurti et al., 2017). This is further supported by the

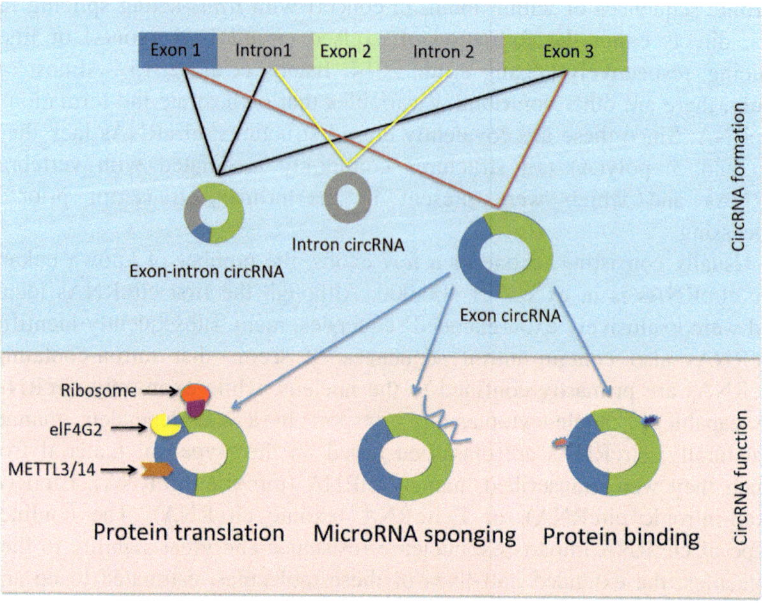

FIGURE 12.4 Form and function of circRNA. circRNAs may consist of exonic sequences, intronic sequences, or both. In some cases, translation is supported by recruitment of initiation factor elF4G2, METTL3 (methyltransferase-like 3), METTL14 (methyltransferase-like 14), and ribosomes. circRNAs may also function as microRNA sponges or sites from the binding of other proteins (Hadad and Lorenzen, 2019). *Figure courtesy Hadad, G., and J.M. Lorenzen. (2019). Biogenesis and function of circular RNAs is health and disease. Front. Pharmacol. 26. https://doi.org/10.3389/fphar.2019.00428. Reproduced according to the terms of the Creative Commons Attribution 4.0 License (https://creativecommons.org/licenses/by/4.0/).*

presence in some circRNAs of an AUG initiation codon as well as an ORF. Taken together, these observations make sense when one considers that 5′ cap-independent initiation of translation from two perspectives. First, translation initiation via an internal ribosome entry site (IRES) typically located close to the 5′ end is known to occur with some mRNAs under cellular stress conditions. This is the same strategy that occurs with the translation of polycistronic mRNAs in prokaryotic cells. Second, the presence of m^6A may also recruit ribosomes by first attracting translation initiation factor 3 to certain circRNAs, followed by the preinitiation complex and other requisite molecules needed to support translation (reviewed by Lei et al., 2020).

Owing to their great stability, circRNA has the potential to be a powerful platform for the therapeutic delivery whole host of proteins. The Endless RNA (eRNA) technology (http://www.laronde.bio), for example, has already engineered a proprietary, nonimmunogenic platform toward this goal (Fig. 12.5). Briefly, this technology involves the engineering of an IRES into a circRNA which thereby promotes ribosome attachment and the initiation of

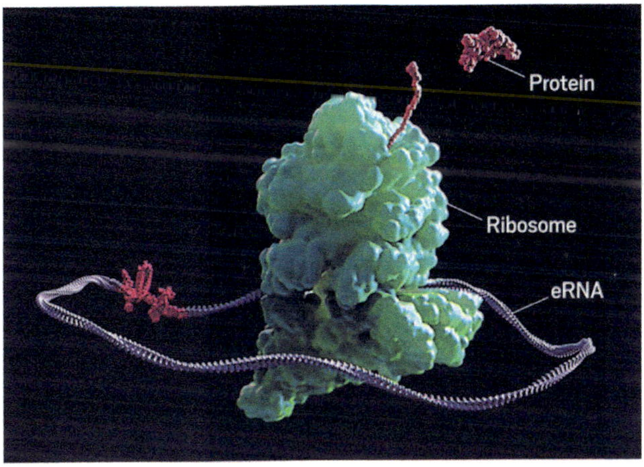

FIGURE 12.5 Endless RNA. An IRES (internal ribosome entry site) engineered into a circRNA promotes ribosome attachment, initiation of translation, and sustained production of the encoded polypeptide. *Image courtesy Laronde (http://www.laronde.bio).*

translation. This strategy would produce sustained production of the encoded polypeptide. As noted above, in the absence of a 5′ cap ordinarily used to identify mRNAs in eukaryotes, the circRNA IRES motif mimics the translation method observed in the prokaryotic world, including some RNA viruses.

miRNA structural and functional characteristics

miRNAs are endogenous, noncoding regulatory molecules with a typical length of approximately 22 nts when mature. They are known to govern cellular and organism physiology at the levels of cell proliferation, differentiation, signal transduction, cellular senescence, apoptosis, and the initiation and progression of the disease state, to name a few examples. In general, they base-pair close to the 3′ end of their target mRNA(s) and either repress translation or induce cleavage of the mRNA. Because of their short length miRNAs have been referred to as "micromanagers of gene expression" (Bartel and Chen, 2004) and, characteristically, a single miRNA can target dozens if not hundreds of different mRNAs.

The ability to silence a gene is highly desirable, offering the possibility to discover what a gene does via intentional loss of function. Inasmuch as miRNAs have this ability naturally, they are recognized as powerful regulators of gene expression. The thousands of known miRNAs, it would seem, have global regulatory responsibility for cellular functions. For example, the use of small RNAs in plants, including miRNAs, is part of a natural antiviral defense pathway (for a recent review, see Jin et al., 2022). Small RNA activity may also embellish the innate immune system observed in higher animals. In addition to targeting specific mRNA, miRNAs also have a ligand-binding function in the cell (Fabbri et al., 2012), demonstrating that this type of molecule is also able to influence cell signaling pathways.

The first miRNA was reported in 1993 (Lee et al., 1993). Discovered in the nematode *Caenorhabditis elegans*, the miRNA lin-4 was demonstrated to have powerful regulatory control over the production of the LIN-14 protein by interfering with the stability of the cognate lin-14 mRNA. Given the technology that was available at the time, it would be another 7 years before the next miRNA to be discovered, let-7, was reported, and in the same organism (Reinhart et al., 2000). It was later demonstrated that let-7 has direct regulatory control over RAS oncogene expression in human cells (Johnson et al., 2005); this miRNA is highly conserved and observed across many species. By 2001, abundant and diverse populations of miRNA were widely demonstrated in animals, and they are now observed in most eukaryotes.

Tens of thousands of miRNAs have already been identified in many organisms and have been added to the rapidly expanding miRNA database known as miRbase, located at http://www.miRBase.org and which is maintained by the University of Manchester. The increased availability of next-generation sequencing and improvements in deep-sequencing have been the impetus behind the exponential increase of newly discovered miRNA sequences. Presently, there are over 38,500 miRNAs, from nearly 300 different species, which may be accessed from miRbase. Presented in Table 12.2 are several online tools and smaller subset databases that are indispensable for investigators working with miRNA.

As noted above, lncRNAs are an abundant and diverse group spread across the genome. It is from lncRNAs that the much shorter miRNAs arise. Collectively, the miRNA content of the cell is known as the miRNome and is subject to constant change. This miRNA flux is not at all unexpected, given the influence of these small transcripts on gene expression. It is likely that miRNAs have an important role in the establishment and maintenance of steady-state levels of gene expression.

miRNA biogenesis

miRNA genes are generally found in intergenic areas, though they are transcribed from coding regions of the genome as well. Some miRNAs are organized into tandem repeat-like clusters (estimated 15%), some are expressed independently (estimated 60%), and yet other miRNAs are intron-associated (estimated 25%). They tend to be concentrated in the vicinity of the centromere of each chromosome, an area once considered rather static and more structural than functional. The miRNA genes are transcribed mostly by RNA polymerase II (Lee et al., 2004), though some miRNA genes are transcribed by RNA polymerase III (Borchert et al., 2006; Ozsolak et al., 2008). The primary product of transcription is a long, single-stranded RNA which folds

TABLE 12.2 Select miRNA and ncRNA resources.

Name	Location	Description
miRBase	http://www.miRBase.org	Centralized repository for miRNAs and miRNA annotation.
Tools4miRs	https://tools4mirs.org	Various miRNA collections and tools for miRNA analysis.
miRDB	http://mirdb.org/miRDB/	miRNA target prediction.
miRNA Cancer Association Database	mircancer.ecu.edu	Comprehensive collection of miRNA expression profiles in human cancers.
circRNADb	http://reprod.njmu.edu.cn/circrnadb	Comprehensive database of human circRNAs.
circBase	http://www.circbase.org	Repository of known circular RNAs (circRNAs).
circAtlas	circAtlas.biols.ac.cn	Integrated resource of circRNAs in vertebrates.
MiOncoCirc	https://mioncocirc.github.io/	Compendium of circRNAs compiled from cancer clinical samples.
Exobase 2.0	http://www.exoRBase	Repository of extracellular vesicle RNA-seq data analyses in different human body fluids (lncRNA, circRNA, and mRNA).
Other useful resources		
ViennaRNA Web Services	http://rna.tbi.univie.ac.at/	Various tools for the study of RNA secondary structures.
NCBI	http://www.ncbi.nlm.nih.gov	Portal to multiple bioinformatics resources.
GENCODE	http://www.gencodegenes.org/	Classification of all gene features in human and mouse genomes.

over on itself to produce a single-stranded loop, a stem consisting of about 20 base pairs, and unpaired sequences dangling from the 5′ and 3′ ends (Fig. 12.6). In a manner not unlike the synthesis of mRNA, transcription of a miRNA DNA sequence produces a large 5′-capped and 3′-polyadenylated nuclear molecule known as pri-miRNA, that is, a primary transcript. The pri-miRNA may contain one or several regions of intramolecular base-pairing referred to as hairpins or stem-loops, and may also contain more than one

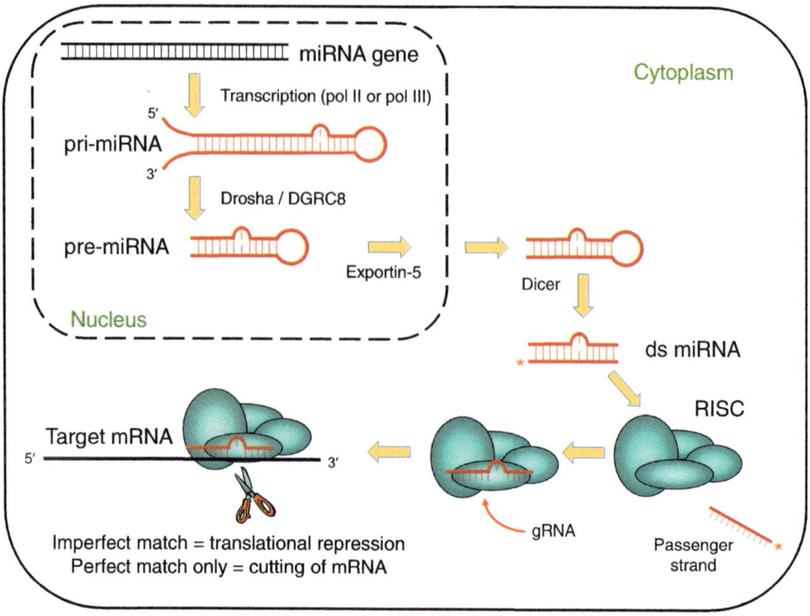

FIGURE 12.6 miRNA biogenesis. Transcription of a miRNA gene produces a large pri-miRNA which is spliced to produce a smaller pre-miRNA, which is then exported from the nucleus and spliced again in the cytoplasm to produce miRNA. The type of interaction that a RISC-associated miRNA has with its cognate target mRNA, including the degree of mismatching, determines the fate of the mRNA.

miRNA. While still considered to be noncoding, it has been demonstrated that in some plants, specifically *Medicago* and *Arabidopsis*, several pri-miRNAs are actually translated, producing short peptides on the order of 9–18 amino acids. These peptides seem to be involved in a positive feedback loop in which they upregulate transcription of their own pri-miRNA gene (Lauressergues et al., 2015) and thereby causing accumulation of the respective miRNAs in the cell.

The processing of each pri-miRNA begins in the nucleus. It is trimmed by a protein complex consisting of the enzyme Drosha, a type III RNase, and its partner, Pasha (a contraction of "Partner of Drosha" and typically associated with invertebrates), the action of which removes the nonbase-paired 5′ and 3′ ends; in humans, Drosha partners with the RNA-binding protein DGRC8. The resulting trimmed transcript is known as a pre-miRNA transcript, that is, precursor miRNA, and is on the order of 80–85 nts and in the form of a stem-loop-type structure which assumes a mini-helical configuration. Exportin-5 is a double-stranded RNA binding protein which, in a RanGTP-dependent manner, is able to orchestrate the egress of the pre-miRNA across the nuclear membrane and into the cytoplasm (Ran, a GTP-binding protein, is a contraction of Ras-related nuclear protein). In contrast, mirtrons, that is, miRNAs produced from intron sequences (Westholm and Lai, 2011) and which bypass Drosha processing, exit the nucleus through coordination with the RNA-binding protein Exportin-1. Once in the cytoplasm, the enzyme Dicer, another type III RNase, acts to remove the single-stranded loop from the pre-miRNA. The result of this latter cleavage event is a short double-stranded miRNA of approximately 20–22 nts (Lee et al., 2002).

Within the double-stranded miRNA, each strand has a name and a function. The passenger RNA strand has the same sequence as the cognate mRNA and is pretty much just along for the ride (sorry). The guide RNA (gRNA) strand is complementary to the intended mRNA target and will therefore be able to base-pair to it once the gRNA has become associated with a ribonucleoprotein complex formally known as an RNA-induced silencing complex (RISC). The dsRNA is sometimes designated miRNA-miRNA*, with the asterisk representing the passenger strand that will not become RISC-associated and which will, in fact, be degraded by it. The unwinding and ensuing strand separation of the passenger RNA from the gRNA is helicase-mediated and is an intrinsic component of the RISC. Thus, miRNA biogenesis may be described as

$$\text{miRNA gene} \xrightarrow{\text{transcription}} \text{pri-miRNA} \rightarrow \text{pre-miRNA} \rightarrow \text{ds miRNA} \rightarrow \text{miRNA}$$

Upon strand separation, the gRNA then becomes associated with the RISC, whereby direct interaction with the target mRNA becomes possible. This interaction tends to occur close to the 3′ end of the mRNA, often with the 3′ UTR trailer region, beyond the coding regions. In order for that interaction to occur, there is an essential requirement for what is known as the miRNA seed sequence, a conserved heptametrical motif that must base-pair perfectly (no mismatches) with the target mRNA, even though mismatches may be tolerated elsewhere in the miRNA-mRNA duplex that forms. The seed sequence is typically situated at nucleotides 2–8, counting from the miRNA 5′-end. As shown in Fig. 12.7, perfect base-pairing between the miRNA and the target mRNA results in mRNA scission, mediated by the Argonaute (Ago) protein component of the RISC; this is essentially RNAi (detailed in Chapter 13, RNA Interference and Gene Editing). In contrast, mismatching results in translational repression of mRNA (Vasudevan et al., 2007). Translational repression is understood to mean impediments to translation and/or accelerated RNA degradation, mediated in part by blocking downstream migration of ribosomes by the physical presence of the gRNA and the RISC components, and by recruiting molecules that speed up mRNA deadenylation and decapping at the 3′ end and the 5′ end of the mRNA, respectively. There is also evidence that translation repression may, in some cases, be transient (Mazumder et al., 2013).

With respect to nomenclature, miRNAs are represented in the literature according to established conventions. A complete description of miRNA

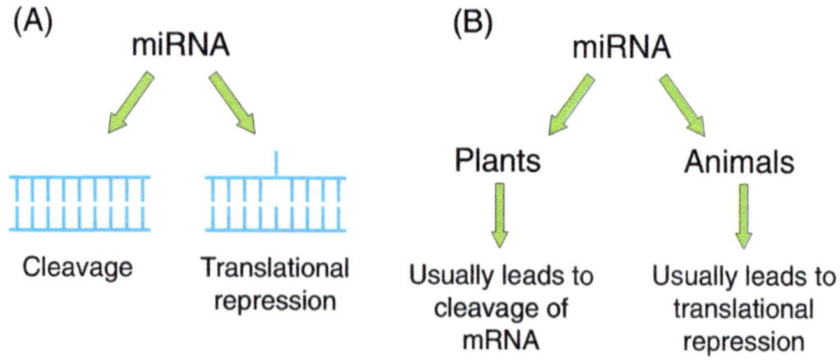

FIGURE 12.7 Regulatory processes for miRNA. (A) Perfectly base-paired miRNAs quickly direct the cleavage of the mRNA while mismatched miRNA molecules induce translational repression. (B) miRNAs almost always base-pair (imperfectly) with the target mRNA in animal cells, inducing translational repression; in plant cells, however, full-match base-pairing directs the miRNA-mediated scission of the mRNA.

naming standards is accessible at http://www.mirbase.org/help/nomenclature.shtml. While the exceptions to these guidelines pertain to miRs that were discovered and named before the guidelines were established, there is an expectation that the naming of all newly discovered miRs will conform to these conventions to minimize confusion. When represented in lowercase font, reference is being made to the either the gene locus or the pre-miRNA (stem-loop sequence). miRNA names begin with a three-letter prefix that is determined by taking the first letter of the genus and then the first two letters of the species: all in the lower-case font. Some examples include

Organism prefix	Organism	Common name
hsa	*Homo sapiens*	Humans
gga	*Gallus gallus*	Chicken
dre	*Danio rerio*	Zebrafish
mmu	*Mus musculus*	Mouse
ptr	*Pan troglodytes*	Chimpanzee
ath	*Arabidopsis thaliana*	Thale cress, that is, Arabidopsis
ppe	*Prunus persica*	Peach

After processing is completed, the organism prefix is followed by a hyphen and then "mir," indicating that reference is being made to a functional miRNA. Another hyphen is then followed by a number which indicates the order of discovery in that organism. So, the standard format is

organism prefix-mir-number

Each entry, the immature pre-miRNA as well as the mature miR, has its own accession number which also cross-references its entry in GenBank

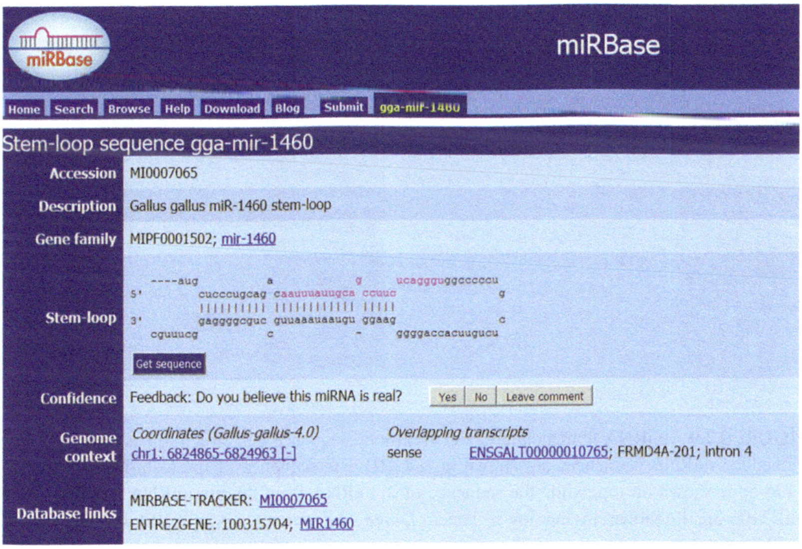

FIGURE 12.8 Example miRNA entry in miRBase. Each miRNA, in this case gga-mir-1460 from *Gallus*, has an accession number which cross-references it to GenBank. Note the presence of the mature miR sequence (in *red* font) embedded in the stem-loop (pre-miRNA) structure. *Image and data courtesy miRBase (http://www.miRBase.org).*

(Fig. 12.8). When two different loci produce miRs with the same sequence, postprocessing, those miRs are distinguished by expanding the suffix with a hyphen followed by another digit. An example of this, in humans, would be the loci

hsa-mir-124-1 (accession number MI0000443)

hsa-mir-124-2 (accession number MI0000444)

hsa-mir-124-3 (accession number MI0000445)

In the event that two or more miRNAs are members of the same family, the members are distinguished from each other by expanding the suffix with a hyphen followed by a letter. An example of this, in *Arabidopsis*, would be the loci:

ath-mir-156a (accession number MI0000178)

ath-mir-156b (accession number MI0000179)

It is also possible for a single pre-miRNA to produce two distinct miRs, one originating from each arm (Fig. 12.9). In such an event, the miRs carry an addition suffix, either 5p or 3p, which designates from which side or "arm" the miR was liberated; hsa-miR-7106–5p came from the arm closest

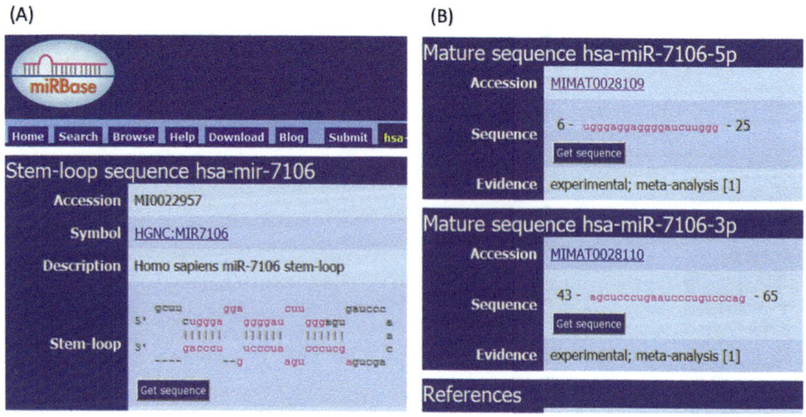

FIGURE 12.9 miRNA nomenclature. (A) Stem-loop sequence and structure has-mir-7106. The embedded miRNA sequences are shown in *red*. (B) The sequence of the 5′ miRNA, hsa-miR-7106-5p is shown on top, while the sequence of 3′ miRNA from the pre-miRNA transcript, hsa-miR7106-3p, is shown in the lower panel. *Image and data courtesy miRBase (http://www.miRBase.org).*

to the 5′ end of has-mir-7106 while has-miR-7106−3p was positioned closest to the 3′ end of the same pre-miRNA.

miRNA profiling

Purified miRNAs can be profiled in any of a number of ways, including array analysis, qPCR, RNA-seq, and even modified Northern analysis using a stiff polyacrylamide gel. In terms of assay, qPCR is the most sensitive technique for profiling miRNA. One popular method for rapidly profiling miRNA expression is the use of PCR plates that are preloaded with primers for the assay of specific miRNAs. A PCR reaction mix is added to the wells of a plate followed by real-time PCR. Various formats are available that will accommodate different thermocycler platforms. A major advantage of this approach is that primers have been optimized and validated for this precise application. As with traditional preloaded arrays, miRNA-prepared plates are themed as well, examples of which include various types of cancer, stem cells, diabetes, apoptosis, cell differentiation, tumor suppressor miRNAs, and many others. Microarrays are also useful for producing a miRNA profile from a biological source as is miRNA sequencing (Hafner et al., 2008; Liu, 2018).

Two additional suggestions. First, a worthwhile strategy to profile miRNA is to perform an Argonaute complex pull-down to discover what types of miRNAs are associated with them (Fields et al., 2021; Li et al., 2020). Second, since Drosha-processing is a prerequisite for pre-miRNA export, knocking out Drosha will cause the accumulation of large amounts of unprocessed pri-miRNAs in the nucleus (Yang and Lai, 2011; Kim et al., 2016); it has also lead to the discovery of noncanonical, Drosha-independent miRNA pathways. For laboratories that would like to jump on the miRNA bandwagon, several companies offer outsourcing services for miRNA profiling: you send in your sample, and they report back which miRNAs are being expressed. Outsourcing high-throughput RNA sequencing or other miRNA profiling service is a viable option when a laboratory does not have that capability. It is a great way to get a lot of data quickly and add another dimension to one's research enterprise.

A logical next step in the characterization of miRNA expression is the implementation of functional analysis studies pertaining to one or more miRNA species, the expression of which are found to be modulated. These types of studies do not differ markedly from the strategies that have been used to characterize other genes. They include gene inhibition or downregulation to discover loss-of-function consequences, gene induction or upregulation to discover gain-of-function consequences or influencing other elements in an expression pathway to discover the implications for the natural expression of the miRNAs of interest.

miRNA as key regulator of gene expression

The understanding of gene regulation, at least in the traditional sense, was turned upside-down by the discovery of miRNAs. miRNAs are tightly coupled to the post-transcriptional control of nearly all cellular processes, particularly differentiation, growth and development, and the maintenance of homeostasis. In the cell, the abundance of each type of transcript, including miRNAs, is a well-orchestrated balance between the overall rate of gene transcription, nuclear processing and export, association with the RISC, and the biological life of the molecule. This is not unexpected, given the powerful influence that these molecules exhibit over mRNA stability and all associated downstream cellular processes.

A great deal of evidence points to the dysregulation of miRNA expression, at any of a number of levels, as being tightly coupled with the onset and progression of various pathologies in humans. As with mRNA and its role in the disease state, dysregulation of miRNAs may be a result of mutations or gross chromosomal structural changes such as deletions or gene duplications, epigenetic modifications, abrogation of normal transcriptional controls, or defective processing and maturation of miRNA. With this in mind, miRNAs have great potential as biomarkers for a wide range of pathologies as well as patient prognosis.

It is easy to envision how a gene duplication event could impact miRNA production: more copies of the same gene are likely to result in more transcriptional products. Envision, for example, too much suppression of a tumor suppressor gene. Conversely, a gene deletion may reduce or altogether eliminate miRNA modulation of an oncogene; the same type of effect might result from an unexpected translocation event, placing a gene nearer to a promoter exhibiting a different transcription initiation efficiency.

It is also easy to understand how the loss of transcription factor control could likewise lead to the disease state. For example, activation of *c-myc*, a well-known transcription factor gene associated with the capacity for cellular proliferation, causes enhanced transcription of certain clusters of miRNAs that not only enhance the expression of certain oncogenes but also result in the downregulation of p53 and some of the other all-important tumor suppressor genes (Chang et al., 2008; Wang et al., 2014a,b). In other cases, miRNA dysregulation has been shown to induce progression through the cell cycle at the G_1 to S phase interface (Han et al., 2013).

Epigenetic changes associated with genomic DNA are a well-established feature of many types of malignancies. Recall that an epigenetic modification is not a change in the base sequence of a gene but rather the addition or removal of a functional group from a nucleotide or an amino acid, most often to induce a transient change in gene expression. For example, CpG base methylation is associated with gene silencing, which includes the silencing of miRNA genes (Fazi et al., 2007). Demethylation, by contrast, often signals an imminent change in the transcriptional status of a gene. Coupled with methylation patterns as a prognosticator of gene expression is the acetylation pattern of the associated histone proteins packaged in nucleosomes. Histone deacetylation relaxes the tight association between the histone protein bundle and the DNA wrapped around it, providing easier access to the transcription factors needed to initiate RNA synthesis, including miRNA genes. In the context of cancer, miRNAs that dysregulate oncogenes or tumor suppressor genes in any way are therefore prime candidates as biomarkers and have become known as oncomirs.

Finally, and as noted above, miRNA levels may be controlled by circRNA, a most unusual class of transcripts. Thousands of circRNAs (http://www.exoRBase.org) have been identified in animals, plants, archaea, and even viruses, and all evidence points to these molecules as powerful post-transcriptional regulators of gene expression. For example, the circRNA CiRS-7, also known as Cdr1as, is an inhibitor of miR-7. The circular CiRS-7 acts as a sponge of sorts, by soaking up mir-7 in cases when it is over-expressed; this has been demonstrated to be the case in diabetic murine models (Xu et al., 2015). This same circRNA has been recently recommended as a prognostic biomarker and potential gene therapy target for certain forms of kidney cancer (Mao, et al., 2021). Thus, it is clear from multiple lines of evidence that miRNAs are not only gene regulators, but also regulatees. So, it would seem that mimics, molecular decoys of a sort which could soak up dysregulated miRNAs, could potentially be used to restore homeostasis. Owing to their powerful regulatory abilities, miRNAs are prime candidates as molecular biomarkers and are discussed in this context in Chapter 25, RNA Biomarker Discovery and Validation.

References

Bartel, D.P., Chen, C.-Z., 2004. Micromanagers of gene expression: the potentially widespread influence of metazoan microRNAs. Nat. Rev. Genet. 5, 396–400.

Boccitto, M., Wolin, S.L., 2019. Ro60 and Y RNAs: structure, functions, and roles in autoimmunity. Crit. Rev. Biochem. Mol. Biol. 54 (2), 133–152. Available from: https://doi.org/10.1080/10409238.2019.1608902.

Borchert, G.M., Lanier, W., Davidson, B.L., 2006. RNA polymerase III transcribes human microRNAs. Nat. Struct. Mol. Biol. 13, 1097–1101.

Brown, C.J., Hendrich, B.D., Rupert, J.L., Lafrenière, R.G., Xing, Y., Lawrence, J., et al., 1992. The human XIST gene: analysis of a 17kb inactive X-specific RNA that contains conserved repeats and is highly localized within the nucleus. Cell 71, 527–542.

Chang, T.C., Yu, D., Lee, Y.S., Wentzel, E.A., Arking, D.E., West, K.M., et al., 2008. Widespread microRNA repression by Myc contributes to tumorigenesis. Nat. Genet. 40, 43–50.

Cocquerelle, C., Mascrez, B., Hetuin, D., Bailleul, B., 1993. Mis-splicing yields circular RNA molecules. FASEB J. 7, 155–160.

Fabbri, M., Paone, A., Calore, F., Galli, R., Gaudio, E., Santhanam, R., et al., 2012. MicroRNAs bind to toll-like receptors to induce prometastatic inflammatory response. Proc. Natl. Acad. Sci. USA 109, E2110–E2116.

Fazi, F., Racanicchi, S., Zardo, G., Starnes, L.M., Mancini, M., Travaglini, L., et al., 2007. Epigenetic silencing of the myelopoiesis regulator microRNA-223 by the AML1/ETO oncoprotein. Cancer Cell 12, 457–466.

Fields, C.J., Li, L., Hiers, N.M., Li, T., Sheng, P., Huda, T., et al., 2021. Sequencing of argonaute-bound microRNA/mRNA hybrids reveals regulation of the unfolded protein response by microRNA-320a. PLOS Genet. 17 (12). Available from: https://doi.org/10.1371/journal.pgen.1009934.

Hadad, G., Lorenzen, J.M., 2019. Biogenesis and function of circular RNAs is health and disease. Front. Pharmacol. 26. Available from: https://doi.org/10.3389/fphar.2019.00428.

Hafner, M., Landgraf, P., Ludwig, J., Rice, A., Ojo, T., Lin, C., et al., 2008. Identification of microRNAs and other small regulatory RNAs using cDNA library sequencing. Methods 44, 3–12.

Han, H., Sun, D., Li, W., Shen, H., Zhu, Y., Li, C., et al., 2013. c-myc-MicroRNA functional feedback loop affects hepatocarcinogenesis. Hepatology 57, 2378–2389.

Hansen, T., Jensen, T., Clausen, B.H., Bramsen, J.B., Finsen, B., Damgaard, C.K., et al., 2013. Natural RNA circles function as efficient microRNA sponges. Nature 495, 384–388. Available from: https://doi.org/10.1038/nature11993.

Hsu, M.T., Coca-Prados, M., 1979. Electron microscopic evidence for the circular form of RNA in the cytoplasm of eukaryotic cells. Nature 280, 339–340.

Ivanov, A., Memczak, S., Wyler, E., Torti, F., Porath, H.T., Orejuela, M.R., et al., 2015. Analysis of intron sequences reveals hallmarks of circular RNA biogenesis in animals. Cell Rep. 10 (2), 170–177. Available from: https://doi.org/10.1016/j.celrep.2014.12.019. **13**.

Jin, L., Chen, M., Xiang, M., Guo, Z., 2022. RNAi-based antiviral innate immunity in plants. Viruses 12 (2), 432. Available from: https://doi.org/10.3390/v14020432.

Johnson, S.M., et al., 2005. RAS is regulated by the let-7 microRNA family. Cell 20, 635–647.

Kazimierczyk, M., Kasprowicz, M.K., Kasprzyk, M.E., Wrzesinski, J., 2020. Human long noncoding rna interactome: detection, characterization and function. Int. J. Mol. Sci. 21 (3), 1027. Available from: https://doi.org/10.3390/ijms21031027.

Kim, Y.-K., Kim, B., Kim, V.N., 2016. Re-evaluation of the roles of DROSHA, exportin 5, and DICER in microRNA biogenesis. Proc. Natl. Acad. Sci. 113 (3), e1881–e1889.

Lauressergues, D., Couzigou, J.M., Clemente, H., Martinez, Y., Dunand, C., Bécard, G., et al., 2015. Primary transcripts of microRNAs encode regulatory peptides. Nature 520, 90–93. Available from: https://doi.org/10.1038/nature14346.

Lee, R.C., Feinbaum, R.L., Ambros, V., 1993. The *C. elegans* heterochronic gene lin-4 encodes small RNAs with antisense complementarity to lin-14. Cell 75, 843–854.

Lee, Y., Jeon, K., Lee, J.T., Kim, S., Kim, V.N., 2002. microRNA maturation: stepwise processing and subcellular localization. EMBO J. 21, 4663–4670.

Lee, Y., Kim, M., Han, J., Yeom, K.H., Lee, S., et al., 2004. MicroRNA genes are transcribed by RNA polymerase II. EMBO J. 23, 4051–4060.

Lei, M., Zheng, G., Ning, Q., Zheng, J., Dong, D., 2020. Translation and functional roles of circular RNAs in human cancer. Mol. Cancer 19, 30. Available from: https://doi.org/10.1186/s12943-020-1135-7.

Lerner, M.R., Boyle, J.A., Hardin, J.A., Steitz, J.A., 1981. Two novel classes of small ribonucleoproteins detected by antibodies associated with lupus erythematosus. Science 211 (4480), 400–402.

Li, X., Pritykin, Y., Concepcion, C.P., Lu, Y., La Rocca, G., Zhang, M., et al., 2020. High-resolution in vivo identification of miRNA targets by halo-enhanced Ago2 pull-down. Mol. Cell 79 (1), 167–1179. Available from: https://doi.org/10.1016/j.molcel.2020.05.009.

Liu, P., 2018. MicroRNA expression analysis: next-generation sequencing. In: Raghavachari, N., Garcia-Reyero, N. (Eds.), Gene Expression Analysis. Methods in Molecular Biology, vol 1783. Humana Press, New York, NY. Available from: https://doi.org/10.1007/978-1-4939-7834-2_8.

Lu, M., 2020. Circular RNA: functions, applications and prospects. ExRNA 2, 1. Available from: https://doi.org/10.1186/s41544-019-0046-5.

Mao, W., Wang, K., Xu, B., Zhang, H., Sun, S., Hu, Q., et al., 2021. ciRS-7 is a prognostic biomarker and potential gene therapy target for renal cell carcinoma. Mol. Cancer 20, 142. Available from: https://doi.org/10.1186/s12943-021-01443-2.

Maraia, R.J., Sasaki-Tozawa, N., Driscoll, C.T., Green, E.D., Darlington, G.J., 1994. The human Y4 small cytoplasmic RNA gene is controlled by upstream elements and resides on chromosome 7 with all other hY scRNA genes. Nucleic Acids Res. 22, 3045–3052.

Mazumder, A., Bose, M., Chakraborty, A., Chakrabarti, S., Bhattacharyya, S.N., 2013. A transient reversal of miRNA-mediated repression controls macrophage activation. EMBO Rep. 14, 1008–1016.

Ozsolak, F., Poling, L.L., Wang, Z., Liu, H., Liu, X.S., Roeder, R.G., et al., 2008. Chromatin structure analyses identify miRNA promoters. Genes. Dev. 22, 3172–3183.

Pamudurti, N.R., O. Bartok, M. Jens, R. Ashwal-Fluss, C. Stottmeister, L. Ruhe, et al. 2017. Translation of circRNAs. DOI: 10.1016/j.molcel.2017.02.021

Reinhart, B.J., Slack, F.J., Basson, M., Pasquinelli, A.E., Bettinger, J.C., Rougvie, A.E., et al., 2000. The 21-nucleotide let-7 RNA regulates developmental timing in *Caenorhabditis elegans*. Nature 403, 901–906.

Rybak-Wolf, A., Stottmeister, C., Glažar, P., Jens, M., Pino, N., Giusti, S., et al., 2015. Circular RNAs in the mammalian brain are highly abundant, conserved, and dynamically expressed. Mol. Cell 58 (5), 870–885.

Salzman, J., Gawad, C., Wang, P.L., Lacayo, N., Brown, P.O., 2012. Circular RNAs are the predominant transcript isoform from hundreds of human genes in diverse cell types. PLOS ONE 7 (2), e30733. Available from: https://doi.org/10/1371/journal.pone.0030733.

Valkov, N., Das, S., 2020. Y RNAs: biogenesis, function and implications for the cardiovascular system. Ad Exp. Med. Biol. 1229, 327–342. Available from: https://doi.org/10.1007/978-981 15-1671-9 20.

Vasudevan, S., Tong, Y., Steitz, J.A., 2007. Switching from repression to activation: microRNAs can up-regulate translation. Science 318, 1931–1934.

Wang, B., Hsu, S.H., Wang, X., Kutay, H., Bid, H.K., Yu, J., et al., 2014a. Reciprocal regulation of microRNA-122 and c-Myc in hepatocellular cancer: role of E2F1 and transcription factor dimerization partner 2. Hepatology 59, 555–566.

Wang, P.L., Bao, Y., Yee, M.-C., Barrett, S.P., Hogan, G.J., Olsen, M.N., et al., 2014b. Circular RNA is expressed across the eukaryotic tree of life. PLOS ONE 9 (4), e95116. Available from: https://doi.org/10.1371/journal.pone.0095116.

Westholm, J.O., Lai, E.C., 2011. Mirtrons: microRNA biogenesis via splicing. Biochimie 93 (11), 1897–1904.

Wu, M., Yang, L.-Z., Chen, L.-L., 2021. Long noncoding RNA and protein abundance in lncRNPs. RNA 27, 1427–1440.

Xiao, M., Li, J., Li, W., Wang, Y., Wu, F., Xi, Y., et al., 2017. MicroRNAs activate gene transcription epigenetically as an enhancer trigger. RNA Biol. 14 (10), 1326–1334.

Xu, H., Guo, S., Li, W., Yu, P., 2015. The circular RNA Cdrias, via miR-7 and its targets, regulates insulin transcription and secretion in islet cells. Sci. Rep. 5, 12453. Available from: https://doi.org/10.1038/srep12453.

Yang, J.-S., Lai, E.C., 2011. Alternative miRNA biogenesis pathways and the interpretation of core miRNA pathway mutants. Mol. Cell 6 (16), 892–903. Available from: https://doi.org/10.1016/j.molcel.2011.07.024.

You, X., Vlatkovic, I., Babic, A., Will, T., Epstein, I., Tushev, G., et al., 2015. Neural circular RNAs are derived from synaptic genes and regulated by development and plasticity. Nat. Neurosci. 18 (4), 603–610. Available from: https://doi.org/10.1038/nn.3975.

Zhang, X.O., Wang, H.B., Zhang, Y., Lu, X., Chen, L.L., Yang, L., 2014. Complementary sequence-mediated exon circularization. Cell 159 (1), 134–147.

Further reading

Ambros, V., Bartel, B., Bartel, D.P., Burge, C.B., Carrington, J.C., Chen, X., et al., 2003. A uniform system for microRNA annotation. RNA 9, 277–279.

Chen, L., Wang, C., Sun, H., Wang, J., Liang, Y., Wang, Y., et al., 2021. The bioinformatics toolbox for circRNA discovery and analysis. Brief. Bioinforma 22 (2), 1706–1728.

Luo, Y., Na, Z., Slavoff, S.A., 2018. P-bodies: composition, properties, and functions. Biochemistry 57, 2424–2431.

Memczak, S., Jens, M., Elefsinioti, A., Torti, F., Krueger, J., Rybak, A., et al., 2013. Circular RNAs are a large class of animal RNAs with regulatory potency. Nature 495 (7441), 333–338. Available from: https://doi.org/10.1038/nature11928.

Meyers, B.C., Axtell, M.J., Bartel, B., Bartel, D.P., Baulcombe, D., Bowman, J.L., et al., 2008. Criteria for annotation of plant microRNAs. Plant Cell 20 (12), 3186–3190.

O'Brien, K., Breyne, K., Ughetto, S., Laurent, L.S., Breakefield, X.O., 2020. RNA delivery by extracellular vesicles in mammalian cells and its applications. Nat. Rev. Mol. Cell Biol. 21, 585–606. Available from: https://doi.org/10.1038/s41580-020-0251-y.

Patop, I.L., Wüst, S., Kadener, S., 2019. Past, present, and future of circRNAs. EMBO J. 38 (16), e100836.

Ransohoff, J., Wei, Y., Khavari, P., 2018. The functions and unique features of long intergenic non-coding RNA. Nat. Rev. Mol. Cell Biol. 19, 143–157. Available from: https://doi.org/10.1038/nrm.2017.104.

Zhou, W.Y., Cai, Z.R., Liu, J., Wang, D.S., Ju, H.Q., Xu, R.H., 2020. Circular RNA: metabolism, functions and interactions with proteins. Mol. Cancer 19, 172. Available from: https://doi.org/10.1186/s12943-020-01286-3.

Chapter 13

RNA interference and gene editing

Rationale

RNA interference (RNAi), succinctly, is the use of double-stranded RNA (dsRNA) to mediate the dismantling of a targeted mRNA and, in so doing, hinder protein synthesis. As the mechanism is very efficient, RNAi has emerged as a powerful regulator of gene expression, functioning in a sequence-specific manner and guided by small noncoding RNAs. RNAi is a natural phenomenon that was considered an oddity when it was first observed in petunias (Napoli et al., 1990; van der Krol et al., 1990). It exists in many organisms as an endogenous means of protecting against viruses and transposons, molecular invaders that would otherwise plague a host genome and wreak molecular havoc. In eukaryotes, this method of protecting the integrity of the genome is highly conserved: dsRNA entering the cell is targeted for immediate destruction. Cellular processes mediated by RNAi include the natural turnover of both wild-type and mutant mRNAs, and translational regulation during the development of many other cellular activities. The RNAi process is also referred to as gene knockdown, posttranscriptional gene silencing (PTGS, in plants), cosuppression (older term, in plants), and quelling (in fungi). Without a doubt, RNAi represents a vast network of gene regulatory pathways and, surely, other control mechanisms in the cell that have not yet come to light. Components of the RNAi process also have a role in regulating DNA methylation (reviewed by Martienssen and Moazed, 2015) through what might be considered to be pretranscriptional gene silencing. There are also indicators that components of the RNAi pathway may have a role in the upregulation of gene expression (Li et al., 2006; Jankowski et al., 2007). Further, RNAi mechanisms have been implicated in cellular quiescence, also by promoting epigenetic chromatin methylation directed at histone 3 lysine 9 (H3K9), at least in yeast (Roche et al., 2016).

Since the first demonstrations of the power of this technique in *Caenorhabditis elegans* (Fire et al., 1998) and subsequently in human and other mammalian cells (Hammond et al., 2000; Elbashir et al., 2001a) and in plants, refinements in the methodology have had a substantial multidisciplinary impact, particularly given the potential to halt the production of illness-causing

RNA Methodologies. DOI: https://doi.org/10.1016/B978-0-323-90221-2.00043-6

detrimental proteins by disabling the cognate mRNA. The use of RNAi as a means of studying the effects of gene expression in a cell or in an organism is known as reverse genetics, the goal of which is to determine the consequences for a cell when a protein is *not* produced. In other words, the function of a gene can often be discovered by silencing it. This has profound ramifications in the realm of developmental biology, the study of the progression of the disease state, and the treatment of infectious and other diseases that result from inappropriate protein expression, for example, through gain-of-function mutations as occur in oncogenes (contrast with loss-of-function mutations that affect tumor suppressor genes). The major strength of RNAi is that it permits the study of the function of one gene at a time and, if properly designed, allows one to do so over an extended period. Comparatively speaking, RNAi is much faster and more economical than creating knockout animals in order to study the function of specific genes.

Whereas microarrays are used to correlate an overexpressed or underexpressed gene and a particular disorder, RNAi is used to elucidate a causal link. This important technology was voted the "Breakthrough of the Year" by the prestigious journal *Science* in 2002 (Science 298, 2296–2297). Not unexpectedly, services and reagents, including premade sets of highly efficient RNA oligonucleotides are widely available to support the market created by this technology. In addition to gene silencing, short dsRNA has a role in epigenetics as well, especially with regard to protein and DNA methylation modifications that influence gene expression. It is also worth noting that another posttranscriptional phenomenon known as RNA enhancement (RNAe) has been discovered (do not confuse with RNA-induced epigenetic silencing). RNAe, perhaps the "opposite" of RNAi, drives translation rather than repressing it (Carrieri et al., 2012; Yao et al., 2015). The methodology relies on an lncRNA and other elements to improve translational efficiency by improving the recruitment of ribosomes to a specific mRNA. As a counter-RNAi-type pathway, RNAe may well be the next big gene expression improvement in the area of functional proteomics. Without a doubt, RNAi is as much an art as it is a science.

Essential RNAi terminology

The following are common terms associated with RNAi methods and with which one should be familiar in order to be conversant in this specialty area of molecular biology.

Antisense RNA: Single-stranded RNA that is complementary to mRNA (+ RNA) and thus has the ability to base pair with it. Like other single-stranded RNAs, antisense RNA (- RNA) molecules are unstable unless they have been chemically modified.

Argonaute: A highly conserved family of proteins associated with the RNAi process by interaction with small, single-stranded, noncoding guide

RNA, leading to the formation of the RNA-induced silencing complex (RISC). It has been suggested that different combinations of argonaute proteins may facilitate RNAi via different pathways and have other roles in the cell. The Argonaute family proteins are often referred to as the "AGO" proteins (AGO1–4 in humans).

Cpf1: A dual endonuclease capable of gene editing and used as an alternative to the CRISPR-Cas9 system. The nomenclature comes from **C**RISPR from ***P****revotella* and ***F****rancisella***1**.

DGCR8: The human counterpart of *Pasha*, which is found in lower eukaryotes.

Dicer: A type III ds-specific RNase that processes long (endogenous or foreign) dsRNA into 21- to 25-bp fragments known as siRNAs through endonucleolytic cleavage. This cytoplasmic enzyme is also able to remove the loop from short hairpin RNA (shRNA), thereby producing short interfering RNA (siRNA) which has a dinucleotide 3′ overhang. Cleavage of precursor microRNA molecules (pre-miRNA) into functional miRNA molecules is also a Dicer enzyme function.

Drosha: A nuclear type III RNase, similar to Dicer, which is involved in the initial maturation step of newly transcribed pri-miRNA (not to be confused with cytoplasmic pre-miRNA).

dsRNA (double-stranded RNA): A duplex consisting of two complementary RNA molecules (sense and antisense) that can be cut by Dicer into siRNAs. dsRNA may exhibit mismatches or may be perfectly base-paired.

ddRNAi (DNA-directed RNA interference): An in vivo method for producing siRNA sequences, often under the direction of RNA polymerase III-promoters in mammalian cells. The process involves cell transfection with an expression vector which first produces an shRNA intermediate which then matures into siRNA. siRNAs produced in vivo are structurally and functionally equivalent to the siRNAs produced in vitro.

Euchromatin: Regions of the genome which may become transcriptionally active, as needed. The DNA exhibits lower levels of methylation than is observed in heterochromatin and is less condensed.

Heterochromatin: Tightly condensed and heavily methylated regions of the genome which are transcriptionally silent. Heterochromatin tends to segregate itself from euchromatin and accumulates along the inner periphery of the nuclear membrane in interphase cells.

miRNA (microRNA): Single-stranded, approximately 21 nts RNAs that either (1) block translation of specific mRNAs (translational repression) through mismatched base-pairing; or (2) cause the direct cleavage of specific mRNAs through perfectly matched base-pairing. miRNAs are important for the temporal regulation of gene expression during development, cellular differentiation, and myriad housekeeping gene functions. A single miRNA often targets multiple mRNA species. miRNAs also appear to have nuclear functions associated with the maintenance of the genome.

Pasha: A partner protein of Drosha, found in lower eukaryotes. The human counterpart of Pasha is the nuclear protein DGCR8.

piRNA (piwi-interacting RNA): A short RNA that is involved in protecting germline cells from transposon and other foreign gene attacks by silencing them. piRNA are a bit longer (25–33 nts) than miRNA and siRNA and are produced in the cell in a DICER-independent manner. PIWI proteins are RNA-binding proteins and members of the AGO protein family.

PTGS (posttranscriptional gene silencing): A term describing the inhibition of gene expression, especially in plants, due to any of a number of causes after a transcript has been produced by a cell. In contemporary literature, the term RNAi has all but replaced PTGS, with which it is synonymous.

RISC (RNA-induced silencing complex): A multiprotein complex, including some of the Argonaute family proteins, which associates with the short siRNA or with miRNA. The antisense or "guide" siRNA or miRNA is used to base pair with the target RNA (e.g., mRNA, viral RNA), thereby marking it for destruction.

RNAi (RNA interference): A naturally occurring mechanism for gene silencing induced by the presence of siRNA. Technically, RNAi is PTGS induced by dsRNA!

Seed sequence: Nucleotides 2–8 in a siRNA or miRNA molecule, which determines RNAi efficiency.

shRNA (short hairpin RNA): Single-stranded RNA molecules possessing sense and antisense domains which facilitate a moderate degree of intramolecular base pairing. The resulting quasi-double-stranded molecule resembles one arm of a typical transfer RNA (tRNA) molecule, with an 8-base single-stranded loop connecting the sense sequence with the antisense sequence. shRNA is a substrate for Dicer, which removes the loop, after which shRNAs effectively become siRNA molecules.

siRNA (short interfering RNA): Short, dsRNA molecules, usually 21 bp long, with 2-nucleotide 3′ overhangs (usually a UU dinucleotide). These molecules are at the heart of RNAi functionality and are the most commonly used tool to promote RNAi. siRNAs are designed to base-pair perfectly with the intended mRNA target.

RNA interference—how it works

At the crux of RNAi is a family of strategies which can be used to reduce protein synthesis by transiently or permanently compromising the ability of mRNAs to support translation. This is mediated through the formation of dsRNA in the cell. In the world of RNAi, three types of RNA molecules play this type of regulatory role, namely microRNA (miRNA), short interfering RNA (siRNA), and piwi-interacting RNA (piRNA). The functions of miRNA and siRNA are well-understood, allowing the scientific community to harness them for targeted gene silencing. In contrast, piRNAs have highly specialized functions in maintaining heterochromatin and protecting germline cells from transposons, retrotransposons, or other foreign genes. As mainstream work in RNAi revolves around siRNA and miRNA only, the discussion here is focused on these two types of specialized molecules.

There are multiple approaches by which RNAi can be induced, each of which has several mechanistic permutations. RNAi is, fundamentally, a two-step process (Fig. 13.1). The first step involves a key enzyme in the RNAi process, a type III endoribonuclease aptly named Dicer; the human-specific family member is known as DICER-1. This enzyme, a ubiquitous member of the eukaryotic proteome, is involved in the ATP-dependent digestion of longer dsRNA into 21- to 23 bp dsRNA molecules with characteristic 3′ dinucleotide overhangs on both strands (Zamore et al., 2000), 3′ hydroxyl termini, recessed 5′ phosphate termini, and suitable for presentation to the RISC. Dicer consists of an amino-terminal helicase domain, two tandem RNase III domains, a PAZ domain (from the names of the proteins PIWI, Argonaute, and Zwille) which specifically recognizes 3′ overhangs, and a carboxy-terminal dsRNA-binding domain (dsRBD). Introducing long dsRNAs into a cell for endogenous Dicer processing can be problematic, especially for mammalian cells. A better, standard approach is to use recombinant Dicer to generate siRNA in vitro after which these short molecules are introduced into the cell by transfection. Dicer also assists in miRNA biogenesis.

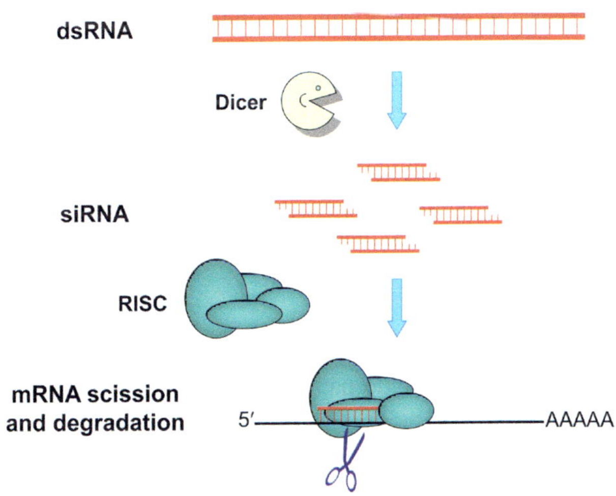

FIGURE 13.1 Key steps in the RNAi process. Long dsRNA from any of a number of sources and in any of a number of configurations is cut by the enzyme Dicer into shorter double-stranded siRNA or miRNA molecules that, in turn, become part of the multicomponent RISC. This ultimately leads either to the destruction of the target mRNA and concomitant downregulation of the associated gene, or to translational repression of the synthesis of the cognate protein and in which case the mRNA is dismantled or, in rare cases, may be left intact.

In the second step, the RISC acquires siRNA or miRNA, regardless of the source, and it becomes part of a multicomponent complex. RISC, at a minimum, consists of Dicer, an endonucleolytic argonaute protein, a dsRNA-binding domain protein (dsRBP), and the RNA itself. The argonaute proteins, the number of which is species-specific, share four conserved domains: amino-terminal domain, the PAZ domain (nucleic acid-binding), the MID domain (5′ phosphate- and cap-binding), and the carboxy-terminal PIWI domain (endoribonuclease activity). It is worth noting that the 5′ cap-binding ability of some of the Argonaute proteins suggests an important regulatory role at the translational level because one function of the cap is the identification of the molecule as an mRNA, which promotes subsequent assembly of the translation apparatus. Part of RISC is a helicase that unwinds the ds-siRNA or miRNA. Upon unwinding of the dsRNA helix, the antisense RNA is retained in RISC by an association of its 5′ end with the MID domain and association of its 3′ end to the PAZ domain. The passenger strand, as it is known, is the sense strand and it does not associate with the RISC but rather is degraded by it. The antisense RNA component, however, known as the guide strand or "guide RNA," brings the RISC to, and base-pairs with, the mRNA to be silenced. The formation of a perfectly-matched double-stranded region between the antisense component of the silencing RNA and the cognate mRNA dooms it. RISC, specifically AGO2 (Elbashir et al., 2001b) or sometimes AGO3 (Park et al., 2017) cuts the mRNA between 10 and 11 nts; this activity has been referred to a "slicing." The mRNA is then further degraded, preventing any level of interaction with the cellular translation machinery. The antisense guide RNA remains stably associated with the RISC, thereby sustaining its potent ability to direct the cleavage of other identical target mRNAs. As expected, transfection of cells with only the sense siRNA strand elicits no silencing, and transfection with the antisense siRNA sometimes produces a very low level of silencing. In contrast, transfection with ds-siRNA produces a silencing response which is much greater, to the extent that synthesis of the encoded protein can be suppressed altogether.

Endogenous silencing pathways

RNAi is an endogenous catalytic pathway that is triggered by dsRNA. The trigger can occur naturally, as in the case of cellular infection by a dsRNA virus, as mediated by the presence of a miRNA produced by the cell, or by the intentional introduction of dsRNA to induce degradation of the cognate transcript(s). The net result of RNAi is the downregulation of specific genes in one of three ways:

1. by direct cleavage (known as slicing) of the targeted mRNA,
2. by translational repression which effectively blocks ribosome access to the targeted mRNA and generally leads to mRNA degradation, or

3. induction of heterochromatin formation through the formation of an RNAi -induced transcriptional silencing complex.

Prokaryotes do not appear to exhibit RNAi, at least as understood in eukaryotic organisms. Prokaryotes do, however, possess other nonhomologous defense pathways, such as CRISPR (below).

miRNA

The understanding of gene regulation, at least in the traditional sense, was turned upside-down by the discovery of miRNAs, which are endogenous 19- to 22 nts regulatory molecules that are transcribed by RNA polymerase II or RNA polymerase III from miRNA genes. In some cases, miRNAs are known to reside in introns. miRNAs are abundant and diverse; the more than 2000 known miRNAs regulate at least 30% of all human genes (Jinek and Doudna, 2009; MacFarlane and Murphy, 2010). The primary product of transcription is a long (1000 + nts), single-stranded RNA which folds over on itself to produce a single-stranded loop of approximately 10nts, a stem consisting of about 20 base-pairs, and unpaired sequences at the 5′ and 3′ ends (Fig. 13.2). This precursor, known as pri-miRNA is recognized as such and trimmed by the nuclear RNase III enzyme Drosha and its accomplice DGCR8, the action of which removes the nonbase-paired 5′ and 3′ ends; the invertebrate counterpart of the human protein DGCR8 is known as Pasha. Together, Drosha and the RNA-binding protein DGCR8 constitute what is sometimes known as the microprocessor complex. The action of this complex produces an approximately 70 nts pre-miRNA which is then exported into the cytoplasm (exportin-5; RanGTP) where the enzyme Dicer acts to remove the single-stranded loop, an action by which the former pre-miRNA becomes known as double-stranded miRNA. While in double-stranded form, the molecule is sometimes designated miRNA-miRNA*, with the asterisk representing the passenger strand that does not become associated with the RISC and which will eventually be degraded by it. The other strand, the guide strand, will direct the RISC to the target mRNA. miRNAs almost always base-pair imperfectly with target mRNAs in animal cells and manage to suppress translation; in plant cells, however, full-match base-pairing directs the miRNA-mediated destruction of the mRNA. miRNAs are known to govern cellular- and organism physiology at many levels, including cell proliferation, differentiation, cellular senescence, apoptosis, and the progression of the disease state. Thousands of miRNAs have already been identified in many organisms and have been added to the rapidly expanding miRNA database. Originally believed only to suppress protein synthesis by base-pairing to the cognate RNA, miRNAs are now known to have the ability to promote protein synthesis (Vasudevan et al., 2007; Orang et al., 2014).

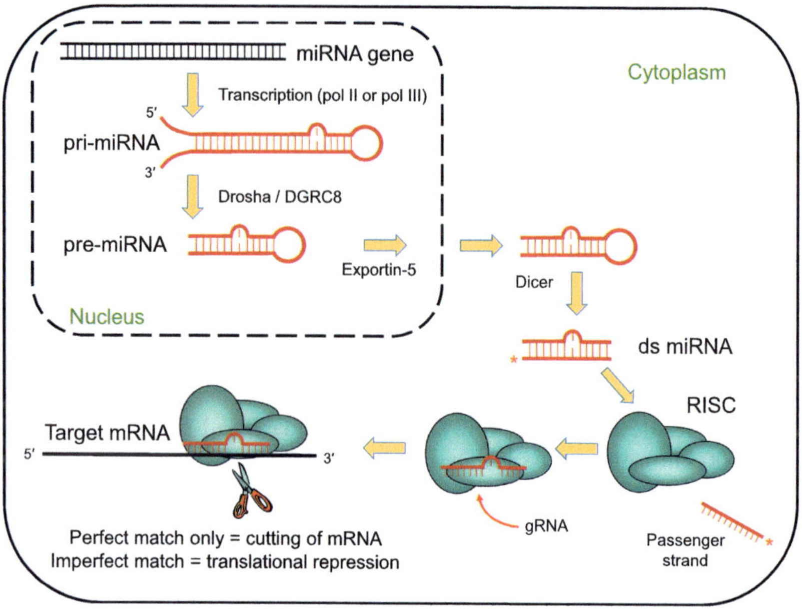

FIGURE 13.2 miRNA biogenesis. Transcription of a miRNA gene produces a large pri-miRNA, which is spliced to produce a smaller pre-miRNA, which is exported from the nucleus to the cytoplasm and spliced again to produce miRNA. The type of interaction that a RISC-associated miRNA has with its cognate target mRNA, including the degree of mismatching, determines the fate of the mRNA.

Finally, it is worth noting that there is often confusion as to the nomenclature, sources, functional similarities, and functional differences between siRNA and miRNA (Table 13.1). In terms of the net result, miRNA and siRNA actions are often indistinguishable. Thus, the major difference is not what these molecules do but rather where they come from. miRNAs are transcribed from genes resident in a cell's genomic DNA while siRNAs enter from the outside (one way or another). siRNAs can be introduced into the cell via transfection, via the processing/degradation of long dsRNA, or via the processing of shRNAs that are produced in vivo or in vitro. It is the case of the latter, that is, the processing of shRNA made within the cell to produce siRNA, which causes confusion because, technically, the shRNA is made in the cell but only as a result of an earlier introduction of an exogenous gene construction encoding it. The best way to think about the process overall and the way that miRNA and siRNA are able to affect gene silencing is to understand that they really operate on parallel tracks and, in the cytoplasm, the siRNA and miRNA pathways converge (Fig. 13.3). Both siRNAs and miRNAs mature through the action of an enzyme known as Dicer, though different Dicer-type enzymes support the miRNA- and siRNA pathways in many organisms. In either case, both types of small RNAs are loaded onto a RISC

TABLE 13.1 Comparison of siRNA and miRNA.

	siRNA	miRNA
Source	Exogenous: viruses, transposons, long dsRNA that are chemically synthesized	Endogenous: genome-encoded
How made	Degradation of viral dsRNA; vector-based foreign gene expression in cell; experimentally introduced into the cell	Transcription of endogenous genes; pri-miRNA characteristically exhibit a 5′cap and poly(A) tail prior to maturation
Synthesis enzyme	RNA polymerase III in conjunction with U6 promoter, or other transcription apparatus, via in vitro transcription	Mostly RNA polymerase II, some by RNA polymerase III; natural transcription
dsRNA	No mismatches	One or more mismatches commonly observed
Size	21–23 nts	19–22 nts
End structure	3′ dinucleotide overhang	3′ dinucleotide overhang
Target	mRNA (one)	mRNA (as many as 100 or more) by the same miRNA species, often serially attaching to and detaching from multiple mRNA targets
Interaction with target	Perfectly base-paired, by design; no mismatches	Perfect or imperfect base-pairing; one or more mismatches common
Cellular action	mRNA cleavage	mRNA cleavage in plants; translational repression common in animals, leading to degradation of the mRNA
Influence on cognate protein	Protein knockdown	Protein knockdown

(miRISC or siRISC). The major functions of miRNA are translational repression (animals) and mRNA cleavage (plants), while siRNAs tend to be associated with mRNA cleavage alone. It is worth noting that siRNA-mediated RNAi is also useful as a tool for compromising the integrity of noncoding RNAs in order to uncover previously unknown functions. Succinctly, the functions of RISC are to interfere with translation or to accelerate mRNA degradation, either of which will result in reduced output of the cognate protein.

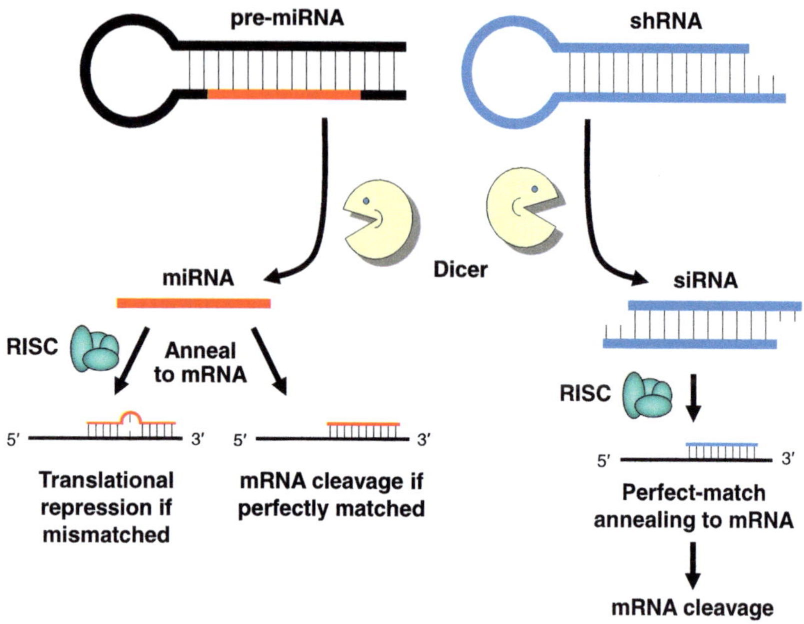

FIGURE 13.3 Comparative biogenesis of miRNA and siRNA. Regardless of the source and when mature, the net results of these small, noncoding RNAs are virtually indistinguishable.

Exogenous silencing strategies

The notion of inhibiting translation by the creation of double-stranded structures along the length of an mRNA or near its 5′ end is not new. As far back as the late 1970s small antisense DNA oligonucleotides were being used to suppress gene expression (Zamecnik and Stephensen, 1978). Antisense RNA was also shown long ago to regulate gene expression in mammalian systems in vitro (Nishikura and Murray, 1987), in animals (Knecht and Loomis, 1987), and in plants, phages, and bacteria (Green et al., 1986). Previous attempts to use antisense RNA as a therapeutic agent were thwarted by the fact that antisense RNA is not effective at low concentrations and the fear that using high concentrations of antisense RNA could have cytotoxic effects. After years of using antisense RNA as an in vitro inhibitor of gene expression, it was demonstrated (Guo and Kemphues, 1995) that the introduction of *sense* RNA was just as good as antisense RNA for targeted inhibition of gene expression. However, it was subsequently shown (Fire et al., 1998) that using dsRNA was far more efficient than using either antisense RNA or sense RNA. It was later shown that sense RNA caused gene silencing because the sense transcripts were contaminated with antisense transcripts, resulting in the formation of minute amounts of dsRNA, which is now well known for its potent ability to downregulate the associated gene.

There are advantages and disadvantages to the siRNA and shRNA approaches (Table 13.2), though it is difficult to definitively select one method as clearly superior because the technology is changing rapidly. It can be helpful to attempt the same experiment more than one way, and then decide which approach is best suited for the particulars of the investigation. Each of these approaches has many permutations, though they all have the ability to target specific mRNAs.

siRNA approach

siRNAs can be produced in a number of ways, namely, by chemical synthesis, by in vitro transcription of long dsRNA that will be processed by Dicer, or by in vivo transcription of a plasmid-borne gene construction that will produce shRNAs which will mature into siRNA through Dicer processing. Perhaps the most straightforward approach to having the individual 21 nts sense and antisense RNA oligonucleotides can be chemically synthesized, purified, and annealed in vitro. For example, a fairly standard annealing protocol is to prepare a 20 μM solution of the sense and antisense RNA (21 nts) in 100 μM KOAc, 30 mM HEPES-KOH (pH 7.4) and 2 mM $Mg(C_2H_3O_2)_2$ and then heat the sample to 90°C for 1 minute, followed by a 1 hour incubation at 37°C. At the conclusion of the incubation period, the annealed siRNAs, which are resistant to RNase degradation if synthesized with modified nucleotides, are ready for transfection. Alternatively, they may be stored for extended periods at −20°C. This siRNA stock is stable through multiple freeze-thaw cycles, as needed. It should be noted that single-stranded siRNAs, that is, prior to annealing, are very sensitive to multiple freeze-thaw cycles, which should be avoided.

An alternative strategy is to synthesize a 500–750 bp dsRNA and then digest it with Dicer to create a pool of siRNA molecules (Fig. 13.4). The entire siRNA

TABLE 13.2 Comparison of siRNA and shRNA for RNAi.

siRNA	shRNA
• Expensive • Easier to design than shRNA • Produced by chemical synthesis or by in vitro transcription and processing • Toxicity sometimes an issue, especially in primary cells and neurons • Suppression effect lasts for 5–6 days; may be less in rapidly growing cells • Less time from start to finish • Focuses on short-term effects	• Relatively economical • More difficult to design than siRNA • Primarily expression-based in vivo or in vitro, but can be produced by chemical synthesis • Toxicity usually not a problem • Sustained suppression of gene expression; can be used to generate stable cell lines • More time start to finish • Focuses on long-term effects

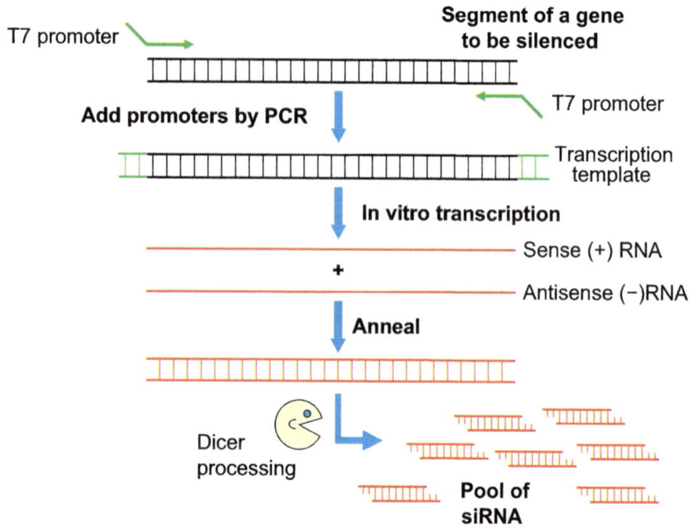

FIGURE 13.4 In vitro synthesis of a heterogeneous pool of siRNA. T7 transcription promoters are easily added to the template by PCR (Chapter 10, RT-PCR: A Science and an Art Form). The PCR product is then used to support in vitro transcription, thereby producing both sense and antisense RNA, because identical promoters are found at each end of the transcription template. After annealing, a pool of siRNA is generated by Dicer. Following transfection, the siRNA molecules are then able to interact with RISC, as usual, for RNAi induction.

pool can then be transfected into the cell to induce RNAi. This approach may be particularly beneficial when the precise siRNA sequence needed to silence a gene is unclear. By providing extensive coverage of a target mRNA with a complex siRNA pool, it is likely that at least some of the siRNA members will be able to direct the destruction of that particular target. This is essentially the brute-force approach to attacking the problem. siRNAs typically base pair perfectly with their target because they are custom synthesized to do so. In contrast, miRNAs may be or may not base-pair perfectly to their target, as noted above, and typically interact with an assortment of naturally-occurring mRNAs.

Long dsRNA can also be introduced into the cell through transfection. The length of dsRNA is of critical importance in experiments involving mammalian cells because dsRNA longer than 30 bp will induce the interferon response (Martinand et al., 1998) that is likely to result in apoptosis. For this reason, investigators prefer to transfect siRNA produced in vitro into mammalian cells. In general, the siRNA approach described here is best-suited for studying the short-term effects of gene silencing.

As noted above, there is considerable overlap between siRNA- and miRNA-induced gene silencing. The net result of RNAi is the downregulation of specific mRNAs by cleaving them or, in the case of translational repression, typically by directing them into a P-body (processing body) where they are subject to RNA decay machinery. Prokaryotes do not appear to exhibit RNAi, at least as is understood in eukaryotic organisms. Prokaryotes do, however, possess other nonhomologous defense pathways.

Unlike miRNA, siRNAs are prepared in vitro and then introduced into the cell in any number of ways, including transfection or transduction approaches. Upon entering the cell, siRNAs are available for immediate engagement with the RISC for presentation to their respective target mRNAs. If the siRNA is produced in the cell by transcription of foreign genes to produce shRNAs, these molecules must first be processed by Dicer in order to mature into function miRNAs. Once they are engaged by the RISC, the siRNA and miRNA pathways converge, and the net result regarding the fate of a targeted mRNA is identical and indistinguishable.

shRNA approach

Another strategy involves siRNA production in the cell via transcription of the (+) and (−) strands of a previously transfected gene in a suitably configured expression vector. This development was a defining moment in RNAi because it represented a method for the long-term silencing of specific genes. shRNAs (Paddison et al., 2002; Yu et al., 2002) are single-stranded RNA molecules that, by virtue of inverted repeats, exhibit intramolecular base-pairing. They are an alternative to the use of dsRNA molecules or siRNAs to suppress gene expression. Once formed, the shRNA molecules are processed by Dicer and are subsequently able to cause gene suppression, much as occurs through the use of siRNA.

shRNAs may be produced in vitro (Fig. 13.5) but are most often produced in vivo through the use of transcription vectors and under the direction of very efficient promoters, an example of which is the human U6 promoter recognized by RNA polymerase III. shRNA molecules are characterized by a sense (+) component consisting of 21−29 nts that are exactly the same sequence as the mRNA to be silenced, an 8-base single-stranded loop, and then an antisense (−) sequence that is precisely complementary to the sense sequence. The siRNA is also characterized by a terminal UU dinucleotide at the 3′ end (Fig. 13.6). If produced in vitro, shRNAs may be introduced into mammalian cells via liposome transfer though, like many other RNAi applications, delivery sometimes presents challenges.

In the case of in vivo synthesis, the transcription vector may be introduced into the cell by any standard transfection technique, though lentivirus (retroviral) and adeno-associated virus (DNA) vectors are known for high-efficiency transduction of both dividing and quiescent cells, highly differentiated cells such as neurons, and primary cells. It is for this reason that a majority of clinical trials involving gene therapy utilize these types of vectors. This approach is often referred to as ddRNAi (http://www.benitec.com). This method occasionally induces the interferon response in mammalian

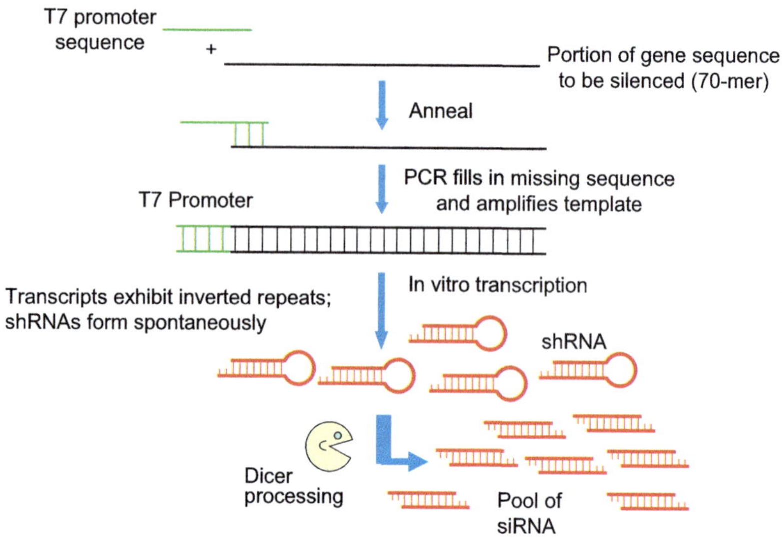

FIGURE 13.5 In vitro method for production of shRNA. PCR is used to construct a transcription template. In contrast to the method shown in Fig. 13.4, a T7 promoter is added only to one side of the template. The template is constructed in such a way that it contains inverted repeats consisting of both sense (+) and antisense (−) domains. The structure of the resulting transcripts favors the rapid formation of shRNA molecules. Subsequent cleavage by Dicer prepares a pool of siRNA that can induce RNAi.

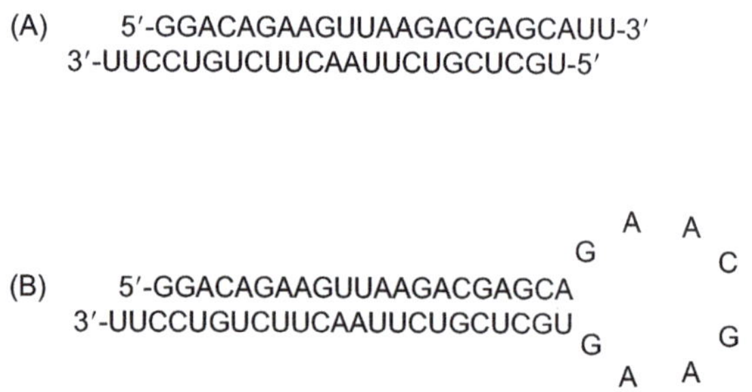

FIGURE 13.6 Comparative anatomy of siRNA and shRNA. (A) siRNA molecules consist of two complementary RNA molecules, characterized by a 3′-terminal UU dinucleotide overhang on each strand. The exact sequence of the overhang is far less important than the other bases, which will guide RISC to the proper target. (B) shRNA molecules begin as one, single-stranded molecule. Intramolecular sense and antisense domains favor the rapid formation of a hairpin-like structure. The folded molecule has an 8-base, single-stranded loop and, like siRNA molecules, exhibits a 3′ UU overhang. The loop joining the sense and antisense domains is subsequently removed by Dicer, at which point the shRNA functions as siRNA.

cells, similar to that which occurs when long dsRNA enters from the outside and which the cell perceives as a viral assault. Some scientists use this approach, however, because it represents significant advantages such as cost savings, capacity for long-term interference, the ability to silence multiple genes from the same vector, and the ability to turn the RNAi mechanism on and off, as required, via inducible promoters.

Transcription vectors used for this purpose contain inverted repeat sequences such that newly transcribed RNA will form a hairpin structure that, as a substrate for Dicer, will be processed into siRNA (Fig. 13.7). It is now commonplace to combine several shRNA cassettes into a single vector. This technology emerged as a consequence of the need to knock down the expression of more than one gene associated with a particular pathway. At some point, it may be possible to produce transgenic animals which contain these types of constructions, but only insofar as the encoded shRNA does not target genes that are required for fertility or viability.

siRNA delivery methods into mammalian cells

While the science behind RNA silencing is fairly well understood, a major impediment in moving forward with the development of RNAi-based drugs has been the delivery of the molecules into the intended target cell. Fortunately, much progress has been made in this area, and recent innovations portend therapeutic successes very soon (reviewed by Lam et al., 2016). The following are the strategies that have been used to deliver siRNAs to cells:

1. Transfection with cationic liposomes involves encapsulating nucleic acids, or anything else, in an artificial phospholipid bilayer which is able

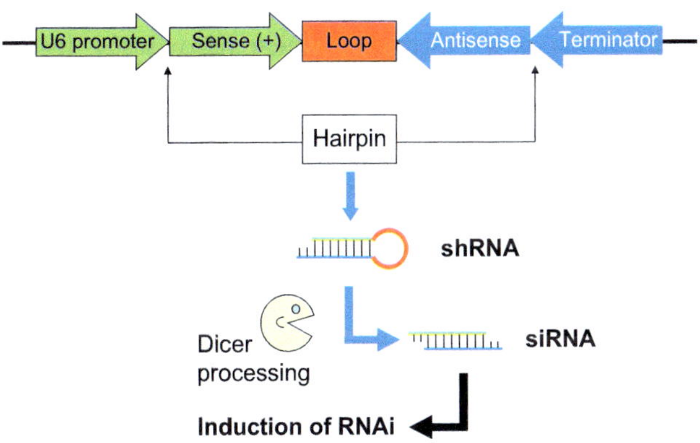

FIGURE 13.7 In vivo transcription template for the production of shRNA. The construction is under the influence of a U6 promoter, which is recognized by RNA polymerase III. Notice that the sense and antisense domains are oriented facing each other, separated by a short region encoding what will become the single-stranded loop. The resultant shRNA transcripts are substrates for Dicer which, through loop removal, will “promote” them to siRNA and induce the RNA interference pathway. In plants, the efficient constitutive cauliflower mosaic virus 35S promoter is often used.

to fuse with cells with which it comes into contact and deliver its contents. This method is also known as lipofection. For in vitro applications, the cells should be dividing, the culture should be in exponential growth, and the cell culture density should be subconfluent. Each cell type requires optimization even at the level of determining the proper lipid-to-siRNA ratio when preparing the liposomes. Not all cells respond well to transfection procedures.

2. Nanotech-type devices are an important new tool in the RNAi arsenal. They are designed to seek out and interact with only a predetermined target cell, often based on a ligand that is part of the nanodevice and which is able to bind to a membrane receptor on the target cell. Receptor-mediated entry to the cell is accompanied by the release of the siRNA molecule that the nanodevice carries, thereby inducing RNAi. Lipid nanoparticles are also very useful in this regard in that they interact directly with lipids and pass with ease across a cell membrane via nonendosomal means.
3. Electroporation is a method by which nucleic acids are able to cross a cell membrane through transient pores that are produced by momentarily disrupting the membrane with a measured, controlled jolt of electricity. Following electroporation, the cells are allowed to recover for a few minutes, replated, and then returned to the incubator. This method for siRNA is most commonly used with cells growing suspension cultures. Optimal electroporation efficiency occurs when cells are actively dividing and at a density of 10^6–10^7 cells/mL. The procedure requires specialized equipment, and some cell types do not tolerate electroporation well, though it is an often-used alternative for cells that do not tolerate transfection well.
4. Viral transduction is a process by which retroviruses, lentiviruses, or adeno-associated viruses efficiently shuttle nucleic acids into cells; after all, they've been doing it for millions of years. By altering its genetic material, the virus becomes a vehicle for delivering that material to the cell by using its membrane recognition and binding properties. As with most viruses, immunogenicity is always a concern.
5. Transfection by calcium phosphate coprecipitation is a well-established method for introducing nucleic acids into cells growing in vitro. The method is easy enough and is based upon the uptake of a coprecipitate of naked nucleic acid and calcium phosphate into the cell by endocytosis. There are some cell types, usually established cell lines, with which this type of transfection works quite well, while there are other cell types, often diploid cells, which have little-to-no tolerance for it. This method is used most often for delivering plasmids into which the shRNA sequences have been ligated. siRNAs will be produced, via the formation of shRNA, as long as the plasmid is maintained within the nucleus.
6. Infusion of siRNA refers to direct entry into the bloodstream, often intravenously. In this case, everything passes through the liver, making it the organ

about which most siRNA effects are known. The critical issues are determining the proper concentration, dosage, and cell homing.

7. Microinjection is also a well-established method, long used to deliver nucleic acids and proteins to cells. The method requires specialized equipment, is very time-consuming, and requires a fair amount of skill. Microinjection is impractical for attempting to introduce siRNAs into large numbers of cells.

Effective design of siRNAs

Without a doubt, the field of functional genomics has become supercharged by RNAi technology. As such RNAi has become a widely acclaimed method for examining gene function in vivo and in vitro. Now that the human and many other genomes have been sequenced, bioinformatics is indispensable in the process of RNA oligonucleotide design, upon which the success of RNAi is completely dependent. Conveniently, companies that sell RNAi-related products generally provide free web-accessible design tools (Table 13.3). Most of the software utilize specific algorithms for selecting the most potent siRNA sequences for each unique target and at the same time minimizing off-target effects by, for example, excluding frequently occurring seed sequences corresponding to known miRNAs.

At its most fundamental level, dsRNA induces the inactivation of the intended mRNA. It is the sequence of the dsRNA that dictates precisely where the mRNA will be cleaved. To be useful, it is important that the

TABLE 13.3 Selected siRNA and shRNA design tools.

Company	Design tool	Location
InvivoGen	siRNA Wizard Software	https://www.invivogen.com/sirnawizard/design.php
Snowformatics	Si-Fi	http://www.snowformatics.com/si-fi.html
Eurofins Genomics	siRNA Design Tool	https://eurofinsgenomics.eu/en/ecom/tools/sirna-design/
GenScript	siRNA Design Center	http://www.genscript.com/design_center.html
Horizon	siDESIGN Center	https://horizondiscovery.com/en/ordering-and-calculation-tools/sidesign-center
GeneLink	RNAi Explorer	https://www.genelink.com/sirna/RNAicustomorder.asp
Biosettia	shRNA Designer	https://biosettia.com/support/shrna-designer/

dsRNA does not perturb the expression of any other gene(s). Only great specificity and the absence of "side-effects" will help to reveal the function of specific genes and provide additional insight into the physiology of the cell. It is widely believed that all genes can be silenced by RNAi, although it is abundantly clear that not all siRNAs are capable of doing that job.

An easy way to conceptualize the design of an effective siRNA for transcript targeting is to examine the sequence of the mRNA itself. Proceeding from the 5′ end of the transcript, look for the first AA or NA dinucleotide after the translation initiation codon AUG; typically, one tries to stay from the first 100–200 bases because of potential interference from mRNA binding proteins in that region of the transcript. After finding the first AA dinucleotide, record the next nineteen bases that follow. This will constitute a 21-base silencing construct. Then, BLAST the sequence to ensure that there is no significant homology with any other mRNA targets, as this would obviously compromise the experiment; when the siRNA directs the destruction of other than its intended target, it is known as an "off-target" effect. In some cases, off-target effects may result from the mere presence of dsRNA in the cell or even the siRNA delivery process. As with the design of primers for PCR, ΔG is also a concern.

The most important aspect of proper RNAi design pertains to the order of bases that constitute the seed sequence, defined as nucleotides 2–8, beginning from the 5′ end of the siRNA. The bases in this seven-nucleotide region are of the utmost importance in terms of mRNA target recognition. In addition to the base identity of the seed sequence, modified nucleotide chemistry, as noted below, is a means of discouraging unintended siRNA/mRNA interactions. The most common mRNA region targeted for interaction with miRNAs is in or near the 3′ UTR (untranslated trailer region) which lies downstream from the coding region's stop codon; less frequently, miRNA seed sequence recognition may occur in the 5′ UTR or even within the coding region itself (AUG $\rightarrow$ UAA, UAG, or UGA).

It is wise to avoid using genomic DNA or protein sequence information for siRNA design. Genomic DNA contains intron sequences, which are (usually) not present in the mature mRNA. If chosen inadvertently, the siRNA will not be able to silence the mRNA that is produced at the gene locus. One should not use protein sequence information, either, because of the degeneracy of the genetic code: recall that many amino acids are represented by more than one possible codon. Given the tremendous amount of bioinformatics data, it is an easy task to simply use the annotated sequence of the intended mRNA target and then build the siRNA from there.

There are basic principles to which one should adhere in rational siRNA design, and most companies involved in RNAi technology have proprietary algorithms that will maximize the probability of effective gene silencing. Note that highly efficient siRNAs are those which also exhibit minimal off-target effects. Some of these design ideas, at one time known informally as

Tuschl's Rules (Elbashir et al., 2001a,b), are intuitive and reminiscent of the standard guidelines for designing oligonucleotides for PCR applications. Briefly, these rules and other practical experience suggest that one should limit the GC content to 50% or less (or at least keep it within the 30%–50% range) and endeavor to identify targets that exhibit at least a three- or four-base difference from any other similar-sized sequence in the genome (usually a 21-base sequence). The 5′ end of the antisense strand, that is, the seed sequence, should be AU-rich and exhibit a T_m of less than 22°C; this will minimize off-target effects tide to the seed sequence. As with PCR primers, one should avoid areas that are likely to form undesired secondary structures so as to maintain full access to the target sequence, so care should be taken to avoid palindromes or other internal repeat sequences. Canonical base-pairing between 2 and 10 nts of the miRNA and the target is also important, and siRNAs should be designed with a favorable (low) free energy associated with siRNA-mRNA duplex formation. Effective siRNAs are 19–21 nts duplexes exhibiting symmetrical UU 3′ overhangs, the mRNA target site should begin with AA, and targeting a conserved region of the mRNA away from the 5′ end of the molecule seems to work best. Other suggestions include avoiding G in position 13 (sense strand), including U in position 10 (sense strand), including A at position 3 (sense strand), and having a strong G + C content at 3′ end of the antisense strand. A summary of some of these general principles for siRNA design is shown in Fig. 13.8.

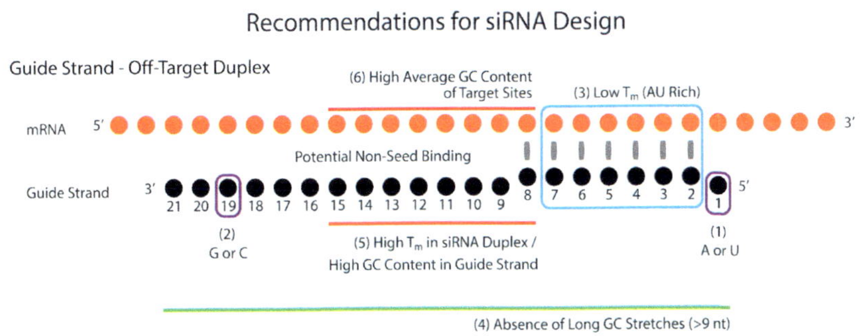

FIGURE 13.8 Example of design guidelines for effective siRNA with reduced potential for off-target effect. With respect to the guides strand, highly recommend are (1) A or U in position 1, (2) G or C in position 19, (3) AU-rich, low Tm base identity in the seed region (position 2–7), (4) no more than 9 consecutive G + C bases, (5) high T_m and G + C content in positions 8–15, and (6) elevated average G + C content all off-target sequences corresponding to positions 8–15 in the siRNA nonseed region (Kamola et al., 2015). *Image courtesy Kamola, P.J., Nakano, Y., Takahashi, T., Wilson, P.A., Ui-Tei, K. (2015). The siRNA non-seed region and its target sequences are auxiliary determinants of off-target effects. PLoS Comput. Biol. 11(12), e1004656. https://doi.org/10.1371/journal.pcbi.1004656 and reproduced with permission under the Creative Commons Attribution 4.0 International License (CC BY 4.0; https;//creativecommons.org/licenses/by/4.0/).*

Often, siRNAs are designed that target regions of the mRNA at least 50−75 nts downstream from the initiation (AUG) codon and at least 50−75 nts upstream from the 5′ end of the poly(A) tail or at least that distance upstream from the 3′ end of nonpoly(A) transcripts. That RNA binding proteins already interacting with target mRNA may reduce the binding potential for RISC is the reason why most investigators tend to stay away from the ends of these molecules. As noted above, and perhaps most importantly, one should perform a BLAST search to determine the extent to which a siRNA will interact with mRNAs other than that which is intended. Finally, the use of locked nucleic acid (LNA) and unlocked nucleic acid (UNA) nucleotides may be helpful in terms of specificity and minimizing off-target effects, particularly when positioned in the seed region of the guide strand; reviewed by Lam et al. (2016). UNAs are acyclic RNA nucleotides that lack the normal C2′-C3′ bond.

Rates of RNAi suppression vary among transcripts, so it is neither unusual nor unreasonable to test three or more siRNAs to find at least one sequence that will produce a profound knockdown in target gene expression (>90%) by targeting different areas of the mRNA to ensure its silencing. It is usually a good idea to demonstrate silencing through the use of two separate siRNAs to demonstrate specificity. Postvalidation, one may wish to try using both siRNAs simultaneously in an attempt to maximize the silencing effect. Alternatively, synthesizing large numbers of siRNAs representing the entire coding region of a transcript is often an effective means for ensuring silencing. This can be performed beginning with a very long dsRNA which will then be processed by Dicer to produce a heterogeneous pool of siRNAs capable of binding along most of the length of the transcript. In the event that an siRNA does not function as expected, especially using a cell culture model, it is worth taking the time to verify identity of the cells. As bizarre as it may seem, there have been many situations in which cell cultures have unwittingly become contaminated with and then overrun by other cells (e.g., HeLa cells), often from a different species.

The traditional chemical method for RNA oligonucleotide synthesis suffers from an overall lack of coupling efficiency with respect to the length of the molecules produced and achievable level of purity. The RNAi industry benefitted greatly from the development of 5′-silyl-2′-acetoxy ethyl orthoester chemistry (Scaringe et al., 1998), also known simply as 2′-ACE. Succinctly, this method is based on a novel protecting group scheme and which is superior in terms of speed, efficiency, and product purity and stability during shipping and are widely available. Upon arrival the RNAs can be easily deprotected in aqueous solutions and, if desired, the individual RNAs can be deprotected and annealed to form siRNA simultaneously. This latter approach imparts an added level of stability. On a related note, and in an attempt to improve the efficiency of RNAi, the use of 2′-F-CTP and 2′-F-UTP (fluorine in the pentose 2′ position normally occupied by oxygen) in the synthesis of dsRNA greatly stabilizes the RNA, making it resistant to many of the commonly observed nucleases found in human cells and on human hands (Capodici et al., 2002: Kariko et al., 2004).

Finally, be sure to perform all of the controls recommended for the system being used for the gene knockdown. These include, but are not limited, to a transfection control (determines the effect of the transfection process), one or more nontarget siRNA controls (shows the effect of having dsRNA, in general, inside the cell), and one or more positive silencing controls (directed at a housekeeping/reference gene that is known to be expressed at moderate to high levels; commonly includes lamin, cyclophilin B, and/or GAPDH).

RNAi and alternative transcript splicing

The impressive size of the transcriptome is a result of the alternative splicing of transcripts from a single locus as well as the potential for alternative transcription start sites in a temporal- or tissue-specific manner. Even the most rationally designed siRNAs can be duds, perhaps due to the plethora of alternative splicing events that characterize the mRNAs produced from many, if not all, higher eukaryotic loci. For example, if a siRNA is directed toward an exon that is spliced out during mRNA maturation (Fig. 13.9), then that particular siRNA will not be able to cause gene silencing. As it is not always possible to predict the transcriptional behavior of a genetic locus or the post-transcriptional behavior of the resulting transcript, designing several different siRNAs per target is a good idea, not to mention the overall unpredictability of any newly designed siRNA.

For the very same reasons, one should avoid targeting introns unless, for example, it is known that a particular tissue maintains a "cryptic intron" as a consequence of more than one functional TATAA-box element in the

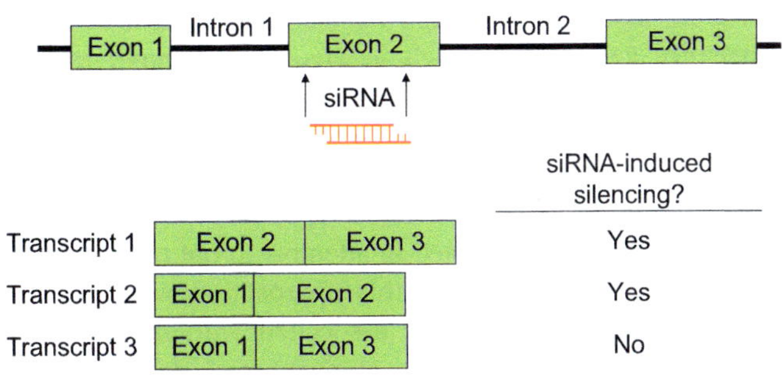

FIGURE 13.9 Differential splicing can affect the success of gene silencing by RNAi. Alternative splicing of hnRNA is a normal part of mRNA biogenesis for many genes. If siRNA is target toward an exon that has been spliced out of the mRNA precursor, RNAi will not be induced because of an inability of the siRNA to associate with the template. A similar consequence may result from the use of an alternative transcription start site. Splicing patterns and the precise location of transcription initiation may well be principal reasons why many well-designed siRNAs fail to suppress the expression of some genes.

associated promoter (Bassett et al., 2009). Similarly, one should avoid targeting known introns and single nucleotide polymorphisms because a single mismatch between mRNA and siRNA is all that is necessary to change the outcome of an experiment drastically.

In vitro and in vivo issues

RNAi has become a mainstream tool for both basic- and applied research, though there have been significant stumbling blocks along the way. The burst of excitement immediately following the first in vitro demonstrations of the power of RNAi quickly subsided with the realization that delivery to the correct target cells, both in vitro and in vivo, represented technical challenges. RNAi is an efficient process, for sure, but the trick is to get the siRNA therapeutic *to* the right cell and then *into* that cell. Many of the initial obstacles in this regard have been overcome in the last 10 years by coupling RNAi with nanotechnology applications. The resulting lipid nanoparticle "devices" are efficient at finding the right target cell by carrying the proper ligand, binding to the membrane surface receptor for which it has specificity, and then unloading its siRNA cargo once inside the cell.

Among the biggest problems associated with RNAi functionality is the resistance of cells grown in suspension culture and primary cells to siRNA uptake, compared to anchorage-dependent cell lines. Keep in mind that the degree of confluency in a tissue culture vessel as well as the passage number (associated with the number of cell population doublings), may both impact transfection efficiency. For example, cell populations with low passage numbers and which are about 50% confluent when exposed to the transfection complex usually produce superior results.

Some of the toxicity difficulties associated with classical $Ca_3(PO_4)_2$ transfection were assuaged years ago by the development of a plethora of liposome delivery vehicles, although in vivo inflammatory responses, and in vitro cytotoxicity, are sometimes noted in response to the use of lipid transfection products. Even when these liposome preparations don't kill cells outright, it has been noted that liposome transfection may alter the transcription file of the cell or even activate the apoptotic pathway (Sledz et al., 2003). Much effort has been invested in attempts to optimize liposome delivery of RNAi molecules and while this approach now works fairly well in vitro, there remain toxicity issues associated with in vivo applications. Overcoming these obstacles has been of paramount importance in moving the application of RNAi technology to the next clinical level.

Methods will also need to be improved upon to allow the efficient introduction of RNAi tools into primary cells since these most closely mirror the normal biochemistry of the cell. Let's face it...cell lines have something wrong with them and tend to experience significant changes in the architecture and composition of their genome as they are continually passaged in culture. Gene redundancy may also hamper discovering the true function of a gene. Gene duplication or the need the knock down the expression of more than one gene associated with a biochemical pathway may put a damper on productivity. In order to address this potentially serious difficulty, one should consider expression vector constructions encoding more than one shRNA, and there are vectors available specifically for that purpose. There are some mammalian cells, neurons and primary cells in particular, that are naturally resilient to transfection, often mandating nearly heroic efforts to achieve even low transfection efficiency. Novel methods for improving transfection efficiency are constantly being explored for the introduction of plasmids, dsRNA, siRNA, and a plethora of other potentially useful compounds.

Another issue associated with the use of siRNAs for in vivo applications is the presence of cytotoxic compounds that are associated with their in vitro synthesis. In particular, the removal of trace heavy metals and the maintenance of an endotoxin-free environment are of paramount importance. Often, investigators who are unfamiliar with the aseptic measures needed to maintain cells in culture or when working with animals, are surprised by the magnitude of contamination problems that can result and the ensuing inflammatory response in animals, respectively. In the case of cells maintained in culture, lipopolysaccharide (LPS; classic endotoxin) may trigger the up- or downregulation of genes that might not otherwise be associated with the investigation at hand. Surprising, even aliquots of allegedly sterile water can be a source of LPS if the sterility of the aliquot has been compromised. Several laboratories and reagent suppliers test each product lot, often using the standard *Limulus* amoebocyte lysate assay, to ensure that endotoxin levels are exquisitely low levels or absent altogether. Succinctly, one should seek guidance from someone who is familiar with standard aseptic techniques associated with the propagation of eukaryotic cells before moving forward.

It is ironic that even though RNAi appears to be widespread throughout the plant, animal, and fungi kingdoms, much still remains to be learned about the regulatory mechanisms dsRNA evokes. While there appear to be several common denominators in the RNAi process from one eukaryotic organism to the next, it is becoming increasingly clear that there are subtle differences. Many investigators still consider RNAi to be in its early years, and each new development in the technology requires careful evaluation of the model system for which its use is intended. Some of the advantages and disadvantages associated with RNAi technology are presented in Table 13.4. RNAi as a tool is currently a long way from perfection, but it is attractive for the study of gene function. If gene silencing fails completely, it is probably time to design new siRNAs. Another appropriate action is to check sequencing data and *consider its source*.

It is clear that there are a number of strategies which can result in the suppression of gene expression. Often, the choice of one approach over

TABLE 13.4 RNAi advantages and disadvantages.

Advantages	Disadvantages
Stable, protracted silencing is possible via shRNA-expressing cassettes	Incomplete gene expression knockdown is common, as are off-target effects
Examine function(s) of one gene at a time.	Cost: RNA oligos are more expensive than DNA oligos
Highly specific to the gene(s) of interest	Extensive training of technicians is required
Systematic approach to investigate gene function and which genes in involved in which pathways	dsRNA > 30 bp upregulates interferon pathway genes in mammalian cells, leading to apoptosis
Highly evolutionarily conserved process more efficient at mRNA destruction than ribozymes	Some cells, especially neurons and primary cells, tend to be resistant to RNAi
Much less toxic than phosphorthioate-based oligonucleotides	Some siRNAs, no matter how rationally designed, simply do not work, mandating the trial-and-error approach
Easy to deliver to mammalian cells	Reproducibility is sometimes a problem
Well suited for drug target identification	Interlaboratory comparisons often difficult because of variations in protocols
Potent: dsRNA is effective at much lower concentrations than antisense oligos	Synthetic RNAi oligos need to be extremely pure (e.g., desalting, deprotection, HPLC purification, mass spec)

another is based on the delivery method. Low transfection efficiency is the primary cause of poor- or altogether unsuccessful RNA silencing. For example, if a particular cell population does not transfect efficiently, then shRNA produced in vivo may be the answer. For each new cell type, it will be necessary to optimize the transfection procedure and the reagents selected in order to promote successful silencing. The efficiency of gene silencing must ultimately be validated using one or more of the methods described below.

RNAi validation

In order to claim a "silencing victory" in a given experiment, it is necessary to demonstrate that the flow of genetic information was disrupted. There are three common methods for demonstrating that posttranscriptional silencing has occurred. These methods involve the direct assay of the abundance of the mRNA, the protein, or both.

RT-PCR approaches

Without a doubt, direct mRNA assay by RT-PCR is a sure way to determine if the abundance of a transcript has increased, decreased, or remained unchanged. As described extensively in Chapter 10, RT-PCR: A Science and an Art Form, the RNA is purified from the cells under investigation, reverse-transcribed in cDNA, and then amplified by PCR with a suitable set of primers. If a particular transcript is being targeted for RISC-mediated demolition, its abundance in the cell is expected to decrease in a manner directly associated with the efficiency of silencing. A logical way by which to approach RT-PCR for this specific application is to design at least two sets of primers for each transcript under investigation. The first and most obvious set of primers would span the region where mRNA cleavage would be expected; cDNA synthesis cannot extend beyond the cleavage point and, therefore, the 5′ primer would be unable to participate in amplification by PCR because of the truncated nature of the 3′ end of the first-strand cDNA. The second set of primers should anneal to the cDNA in a region corresponding to the 3′ end of the cDNA in order to demonstrate that the cDNA had, in fact, been synthesized and its integrity was compromised by the mechanics of RNAi.

Northern analysis

RT-PCR is a fine quantitative tool for the rapid assay of transcriptional activity associated with a particular gene locus. However, the PCR product represents only a small section of the mRNA, rather than the full-length transcript. Some investigators prefer to demonstrate successful RNAi through the assay of total cellular or total cytoplasmic RNA by Northern analysis so as to be confident that the size of the transcript species targeted for destruction is consistent with the known or predicted size of that transcript. Of course, successful Northern analysis is predicated upon having a sufficiently abundant transcript to assay, at least in the control samples, because Northern analysis is a rather low-sensitivity assay (Chapter 16: Northern Analysis). If using mRNA abundance to monitor gene knockdown efficiency, be sure to measure the abundance of a nontarget mRNA, too, as a suitable control.

Western analysis and other protein methods

The most straightforward means by which to demonstrate a change in the abundance of a protein is to assay it directly. Through Western analysis, specific antibodies are an effective means by which to monitor the reduction of a protein when the cognate mRNA is targeted for destruction via RNAi. While the mechanics of RNAi are undoubtedly initiated immediately after the appearance of siRNA in the cell, the ability to measure gene knockdown, particularly at the level of protein synthesis may not be apparent for a day or two; if the protein of interest has a long half-life, even longer periods may be necessary. If protein knockdown is not observed, one may wish to assay the abundance of the corresponding mRNA either by RT-PCR or, in the case of abundant transcripts, by Northern analysis. Enzyme-linked immunosorbent assay is another protein detection method that can be performed easily and without the fanfare of running a gel. Protein immunoprecipitation pull-down methods are also used to assess cellular protein levels in response to RNAi knockdown; in this method, proteins of interest are sequestered from a cell lysate with antibodies, and the recovered protein is quantified by polyacrylamide gel electrophoresis. As called for by the experimental design, it is also possible to monitor changes in the quantity and subcellular location of a protein by in situ immunofluorescence.

RNAi applications

The immense potential of RNAi has yet to be tapped, primarily because the scientific community is just now beginning to understand the natural silencing role of miRNA as well as other regulatory roles of the superabundant noncoding RNAs in the cell. Many of the challenges regarding siRNA entry into the cell are being resolved. Recent developments in plants and in animals involving siRNA and miRNA have suggested methods by which these molecules might be controlled in order to provide some type of health or economic benefit. For example,

1. RNAi is a valuable technique for drug target discovery. The precision silencing of specific genes also makes RNAi an attractive platform for the discovery of potential biomarkers and the development of life-saving pharmaceuticals. To date, four siRNA-based drugs have been approved for use by the FDA. The first drug, patrisiran (ONPATTRO, used to treat polyneuropathy associated with hereditary transthyretin-mediated amyloidosis), was approved by the FDA in 2018, followed shortly thereafter by approval of givosiran (GIVLAARI, used to treat acute hepatic porphyria) in 2019 and lumasiran (OXLUMO, used to treat primary hyperoxaluria type 1) in 2020. A fourth siRNA drug, vutrisiran (AMVUTTRA, a follow-on to ONPATTRO) received FDA approval in 2022. Several other drugs are in the pipeline and are likely to receive FDA approval soon.
2. The control of cellulose production in barley is controlled by a siRNA (Held et al., 2008). By interfering with this siRNA-mediated downregulation of the cellulose synthesis pathway, it should be possible to induce enhanced cellulose production in certain plant species for biofuel production.
3. Also in plants, *trans*-acting small interfering RNAs (tasiRNA; Dunoyer et al., 2007) are of great interest. Briefly, miRNA mediates the cleavage of a single-stranded pre-siRNA transcript in the vicinity of the 3′ end.

The resulting cleavage product is converted into dsRNA by the action of an RNA-dependent RNA polymerase which, in turn, becomes a Dicer substrate and from which siRNA(s) are produced. Finally, the siRNA becomes RISC-associated, thereby directing the cleavage of the target mRNA. *Arabidopsis*, for example, has at least eight known loci capable of producing tasiRNA transcripts.

4. Overexpression and underexpression of various miRNAs have been linked to many diseases, affecting many tissues. Some miRNAs, when dysregulated, act like oncogenes, while others have properties that resemble tumor suppressor genes. Regaining control of these molecules, either at the level of their synthesis or by targeting them for destruction, could prove to be a valuable weapon against many deadly diseases.

The power of RNAi has profound ramifications in many facets of cell biology. Through the ability to target for study one gene at a time, RNAi has become another popular mainstream tool of the molecular biologist. Its ubiquity, functionality, and compatibility with basic- and applied research make RNAi an attractive platform for addressing many problems in the life sciences and in the developing area of personalized medicine.

CRISPR-Cas9 and gene editing

Another biotech revolution is upon us! Clustered regular interspersed short palindromic repeats (CRISPR) is a natural defense mechanism that evolved in prokaryotes to protect against viral DNA. So, CRISPR is one of several types of microbial immune systems. What is remarkable about this system, which is found in about 50% of bacteria and about 90% of archaea (Makarova et al., 2015), is that it has a heritable memory, similar to the adaptive immunity observed in vertebrates. Thus, when exposed to the same type of virus for the second time, a prokaryotic cell (or its descendent) is able to mount a quick and aggressive response designed to disable the virus by cutting its genome. A bacterial cell, upon infection, archives a piece of DNA from the invading virus within its own genome in a region known as the CRISPR array locus. The viral DNA segment stored in the CRISPR region is known as a spacer; one spacer is separated from the next by a short (20–40 nts) palindromic repeat, from which the CRISPR region of the genome gets its name (Fig. 13.10). Succinctly, the spacers in the CRISPR region are an infection history. Upon re-exposure, rapid transcription of the CRISPR locus in the bacterial cell produces an antisense transcript (pre-crRNA →→→crRNA). This defense system works collaboratively with an array of CRISPR-associated (Cas) proteins, perhaps the best known of which is Cas9. Each mature crRNA associates with a Cas family protein, in which DNA endonuclease ability resides, and it is this complex which then compromises the integrity of the viral genetic material in a crRNA-guided manner.

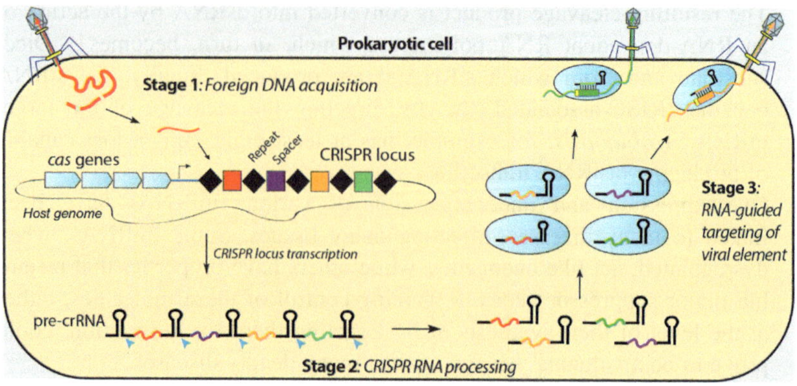

FIGURE 13.10 Expression of the prokaryotic CRISPR array locus. Spacer segments in the CRISPR locus represent previous interactions with viruses (stage 1). Transcription of the CRISPR locus produces a large pre-crRNA, representing multiple spacers which, during the maturation process, liberate individual crRNA molecules (stage 2). Mature crRNAs bind to Cas proteins and, in a crRNA-guided manner, direct the silencing of foreign genes by disabling incoming genomes (stage 3). *Image courtesy Dr. Jennifer Doudna, University of California, Berkeley.*

This is known as a CRISPR type II interference system (Makarova et al., 2011). Owing to the role of RNA in the functioning of the CRISPR-Cas9 system, the topic is presented briefly here. It is interesting to note the commonality among RNAi, CRISPR, and natural miRNA function: all three are based on the formation of dsRNA or RNA:DNA hybrid molecules.

The best way to understand a gene is to be able to control it. From a contemporary molecular biology perspective, there are two useful things that the harnessing of the CRISPR-Cas9 system as facilitated: the ability to edit a gene or altogether silence it. In the case of the former, the guide RNA leads the CRISPR-Cas9 complex to the target gene and then mediates cutting the gene on both strands. Although cells attempt to repair the gene, including re-ligation, the process is error-prone (nonhomologous end joining) and invariably introduces mutations. The net result is that the gene becomes nonfunctional and is effectively silenced. In contrast, when a repair template is made available during the gene editing process, the new information that the repair template carries becomes incorporated into the DNA such that the target gene is either repaired (if it was defective) or acquires a new function.

Changing genes in living cells has never been easy. The CRISPR approach is a very big step toward accomplishing that goal; in reality, it is gene editing in vivo rather than using the in vitro methods associated with gene cloning in the early years of molecular biology. Moreover, the system can be used to edit multiple genes at the same time, owing to its high level of specificity. The key nucleic acid components of the gene editing system are:

1. Guide RNA (gRNA), which specifies the DNA target. The gRNA is a single molecule consisting of two parts (Fig. 13.11):
   a. crRNA (CRISPR RNA), which base-pairs directly to the target genomic DNA.
   b. tracrRNA (*Trans*-activating crRNA), which enjoins the guide RNA to the Cas9 protein.
2. Protospacer adjacent motif (PAM), a conserved three-base motif 5′...NGG...3′ (or less commonly 5′...NGA...0.3′), which is located immediately downstream from the target site on the strand that shares the

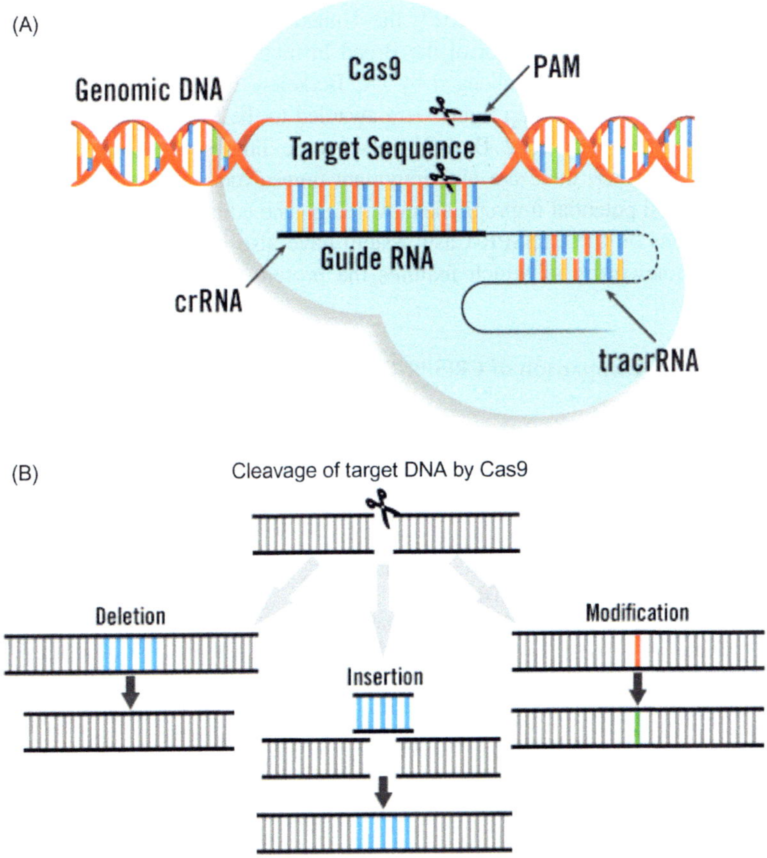

FIGURE 13.11 CRISPR-Cas9 mechanism of gene editing or gene inactivation. (A) The guide RNA, a single molecule consisting of the crRNA and tracrRNA, interacts with the Cas9 protein, thereby leading the CRISPR RNA-Cas9 complex to its intended target. (B) Cleavage of the target DNA, mediated in part by the location of the PAM element, results in gene inactivation or, if properly directly, in the editing of the gene, including the replacement of a defective or mutated region. *Images courtesy Mirus Bio (Madison, WI).*

same coding sense as the crRNA. This is a molecular recognition element which facilitates the alignment of the guide RNA with the target DNA and the ensuing cutting of the target DNA.

The CRISPR revolution is just beginning. As suggested in Table 13.5, CRISPR-Cas9 is a complementary technique to RNAi, rather than a competitor. The current technology of precision genome editing is based on the biochemistry of well-characterized microbial species. It is highly likely that alternative approaches exist among the "unculturables," referring to the bacteria that exist in the metagenome and have not yet been characterized.

There have been several controversies surrounding the CRISPR-Cas9 technology, the most noteworthy of which pertained to ownership of the use of the technology in eukaryotic cells. In 2017, the United States Patent and Trademark Office (USPTO) ruled in favor of the Broad Institute of Harvard and MIT; that claim of ownership was challenged by UC Berkeley. Following a protracted 7-year patent battle, ownership rights were awarded to the Broad Institute/Harvard/MIT group in February 2022. Beyond the scientific dimension of DNA modification in this manner, there are also important unresolved issues surrounding the ethical use and potential improprieties involving gene editing technology.

Variations on the CRISPR-Cas9 system have also been identified, perhaps the most noteworthy of which features the protein Cpf1, a class 2 CRISPR

TABLE 13.5 Comparison of CRISPR-Cas9 with RNAi.

	RNAi	CRISPR-Cas9
Target	RNA	DNA
Natural source	Eukaryotic	Prokaryotic
Action	• Cleavage of RNA via formation of fully-matched dsRNA • Translational repression via the formation of mismatched dsRNA followed by mRNA dismantling	• Gene inactivation via cleavage and inefficient repair mechanisms • Corrective or new gene editing in a site-directed manner
"Active" site	RNA-induced silencing complex (RISC)	CRISPR-Cas9 complex
Sustained effect	Yes, with shRNA	Yes, with permanent gene edit
Reversible	Yes	No, changes require an entirely new crRNA
Off-target effects	High, but can be controlled through rationale siRNA design	Lower

endonuclease, used in place of Cas9, which is a class 1 CRISPR endonuclease. Among the differences is that the class 1 endonucleases require multiple effector proteins whereas the class 2 endonuclease requires only one. Succinctly, Cpf1 functions as both an endoribonuclease (substrate − crRNA) and as an endodeoxyribonuclease (substrate = dsDNA gene target) and, compared to CRISPR-Cas9, produces fewer off-target effects. As it is smaller than Cas9, CRISPR/Cfp1 produces staggered cuts in the target DNA as opposed to the blunt ends produced by CRISPR-Cas9, among other subtle differences. This technology has enjoyed success in the genetic modification of several agriculturally important crops (reviewed by Alok et al., 2020). As one might imagine, numerous patent filings claiming ownership of these alternative processes have likewise been delivered to the USPTO as well as overseas patent offices.

CRISPR has great untapped potential in many areas, including drug development, agriculture, basic research and, eventually, treating people with genetic diseases. A few success stories so far:

1. In mid-2021, six transthyretin (TTR) amyloidosis patients were successfully treated using CRISPR technology to disable the mutated TTR gene, shortly after which the abnormally high levels of TTR plummeted and remained low for many months.
2. In early 2020, CRISPR was used to attempt to correct the genetic disorder behind Leber's congenital amaurosis, which is a type of visual impairment that often leads to total blindness. The CRISPR components were injected directly into the volunteer patients' eyes. By the following year, at least some of the patients in the study were experiencing improved vision including the perception of more vivid colors as well as more peripheral vision.
3. In early 2017, the CRISPR-Cas9 technology was used to create artificial restriction endonucleases (Enghiad and Zhao, 2017). These enzymes are valuable molecular biology tools because of their precision, predictable, and reproducible cutting of double-stranded DNA. This recent success opens the door for generating restriction endonucleases that will recognize any desired sequence.
4. Work is also underway to control the spread of malaria through the use of CRISPR technology to modify populations of mosquitoes as well as to control rodent populations using a strategy known as gene drive, in which CRISPR-modified genes are perpetuated and move rapidly through a population (reviewed by Bier, 2022).
5. In mid-2016, the first CRISPR clinical trials were approved by NIH. The protocol involves removing T-cells from cancer patients and performing three CRISPR edits on those cells before returning them to their respective donors. The gene edits are intended to facilitate the T-cells finding and then target cancer cells.

And the list goes on.

References

Alok, A., Sandhya, D., Jogam, P., Rodrigues, V., Bhati, K.K., Sharma, H., et al., 2020. The rise of the CRISPR/Cpf1 system for efficient genome editing in plants. Front. Plant. Sci. 11, 264. Available from: https://doi.org/10.3389/fpls.2020.00264.

Bassett, C.L., Wisniewski, M.E., Artlip, T.S., Richart, G.A., Norelli, J.L., Farrell Jr., R.E., 2009. Comparative expression and transcript initiation of three peach dehydrin genes. Planta 20, 107–118.

Bier, E. 2022. Gene drives gaining speed. Nat. Rev. Genet. 23, 5–22. Available from: https://doi.org/10.1038/s41576-021-00386-0.

Capodici, J., Kariko, K., Weissman, D., 2002. Inhibition of HIV-1 infection by small interfering RNA-mediated RNA interference. J. Immunol. 169, 5196–5201.

Carrieri, C., Cimatti, L., Biagioli, M., Beugnet, A., Zucchelli, S., Fedele, S., et al., 2012. Long non-coding antisense RNA controls Uchl1 translation through an embedded SINEB2 repeat. Nature 491, 454–457.

Dunoyer, P., Himber, C., Ruiz-Ferrer, V., Alioua, A., Voinnet, O., 2007. Intra- and intercellular RNA interference in *Arabidopsis thaliana* requires components of the microRNA and heterochromatic silencing pathways. Nat. Genet. 39, 848–856.

Elbashir, S.M., Harborth, J., Lendeckel, W., Yalcin, A., Weber, K., Tuschl, T., 2001a. Duplexes of 21-nucleotide RNAs mediate RNA interference in cultured mammalian cells. Nature 411, 494–498.

Elbashir, S.M., Lendeckel, W., Tuschl, T., 2001b. RNA interference is mediated by 21- and 22-nucleotide RNAs. Genes. Dev. 15, 188–200.

Enghiad, B., Zhao, H., 2017. Programmable DNA-guided artificial restriction enzymes. ACS Synth. Biol. Available from: https://doi.org/10.1021/acssynbio.6b00324.

Fire, A., Xu, S., Montgomery, M.K., Kostas, S.A., Driver, S.E., Mello, C.C., 1998. Potent and specific interference by double-stranded RNA in *Caenorhabditis elegans*. Nature 391, 806–811.

Green, P., Pines, O., Inouye, M., 1986. The role of antisense RNA in gene regulation. Ann. Rev. Biochem. 55, 569–597.

Guo, S., Kemphues, K.J., 1995. Par-1, a gene required for establishing polarity in C. elegans, encodes a putative Ser/Thr kinase that is asymmetrically distributed. Cell 81 (4), 611–620.

Hammond, S.M., Bernstein, E., Beach, D., Hannon, G.J., 2000. An RNA-directed nuclease mediates post-transcriptional gene silencing in *Drosophila* cells. Nature 404, 293–296.

Held, M.A., Penning, B., Brandt, A.S., Kessans, S.A., Yong, W., Scofield, S.R., et al., 2008. Small-interfering RNAs from natural antisense transcripts derived from a cellulose synthase gene modulate cell wall biosynthesis in barley. Proc. Natl. Acad. Sci. 105, 20534–20539.

Jankowski, B.A., Younger, S.T., Hardy, D.B., Ram, R., Huffman, K.E., Corey, D.R., 2007. Activating gene expression in mammalian cells with promoter-targeted duplex RNAs. Nat. Chem. Biol. 3, 166–173.

Jinek, M., Doudna, J.A., 2009. A three-dimensional view of the molecular machinery of RNA interference. Nature 457 (7228), 405–412.

Kamola, P.J., Nakano, Y., Takahashi, T., Wilson, P.A., Ui-Tei, K., 2015. The siRNA non-seed region and its target sequences are auxiliary determinants of off-target effects. PLoS Comput. Biol. 11 (12), e1004656. Available from: https://doi.org/10.1371/journal.pcbi.1004656.

Kariko, K., Bhuyan, P., Capodici, J., Weissman, D., 2004. Small interfering RNAs mediate sequence-independent gene suppression and induce immune activation by signaling through toll-like receptor 3. J. Immunol. 172, 6545–6549.

Knecht, D.A., Loomis, W.F., 1987. Antisense RNA inactivation of myosin heavy chain gene expression in *Dictyostelium discoideum*. Science 236, 1081–1086.

Lam, J.K.W., Chow, M.Y.T., Zhang, Y., Leung, S.W.S., 2016. siRNA vs miRNA as therapeutics for gene silencing. Mol. Ther. - Nucleic Acids 4, e2552. Available from: https://doi.org/10.1038/mtna.2015.23.

Li, L.C., Okino, S.T., Zhao, H., Pokot, D., Place, R.F., Dahiya, R., 2006. Small dsRNAs induce transcriptional activation in human cells. Proc. Natl. Acad. Sci. 103, 17337–17342.

MacFarlane, L.-A., Murphy, P.R., 2010. MicroRNA: biogenesis, function and role in cancer. Curr. Genomics 11 (7), 537–561.

Makarova, K.S., Haft, D.H., Barrangou, R., Brouns, S.J., Charpentier, E., Horvath, P., et al., 2011. Evolution and classification of the CRISPR–Cas systems. Nat. Rev. Microbiol. 9 (6), 467–477.

Makarova, K.S., Wolf, Y.I., Alkhnbashi, O.S., Costa, F., Shah, S.A., Saunders, S.J., et al., 2015. An updated evolutionary classification of CRISPR–Cas systems. Nat. Rev. Microbiol. 13, 722–736. Available from: https://doi.org/10.1038/nrmicro3569.

Martienssen, R., Moazed, D., 2015. RNAi and heterochromatin assembly. Cold Spring Harb. Perspect. Biol. 7 (8), a019323. Available from: https://doi.org/10.1101/cshperspect.a019323.

Martinand, C., Salehzada, T., Silhol, M., Lebleu, B., Bisbol, C., 1998. The RNase L inhibitor (RLI) is induced by double-stranded RNA. J. Interf. Cytok. Res. 18, 1031–1038.

Napoli, C., Lemieux, C., Jorgensen, R., 1990. Introduction of a chalcone synthase gene into Petunia results in reversible co-suppression of homologous genes in trans. Plant Cell 2, 279–289.

Nishikura, K., Murray, J.M., 1987. Antisense RNA of proto-oncogene c-fos blocks renewed growth of quiescent 3T3 cells. Mol. Cell Biol. 7, 639–649.

Orang, V.A., Safaralizadeh, R., Kazemzadeh-Bavili, M., 2014. Mechanisms of miRNA-mediated gene regulation from common downregulation to mRNA-specific upregulation. Int. J. Genomics 970607. Available from: https://doi.org/10.1155/2014/970607.

Paddison, P.J., Caudy, A.A., Bernstein, E., Hannon, G.J., Conklin, D.S., 2002. Short hairpin RNAs (shRNAs) induce sequence-specific silencing in mammalian cells. Genes. Dev. 16, 948–958.

Park, M.S., Phan, H.D., Busch, F., Hinckley, S.H., Brackbill, J.A., Wysocki, V.H., et al., 2017. Human Argonaute3 has slicer activity. Nucleic Acids Res. 45 (20), 11867–11877.

Roche, B., Arcangioli, B., Martienssen, R.A., 2016. RNA interference is essential for cellular quiescence. Science 354 (6313). Available from: https://doi.org/10.1126/science.aah5651.

Scaringe, S.A., Wincott, F.E., Caruthers, M.H., 1998. Novel RNA synthesis method using 5′-silyl-2′-orthoester protecting groups. J. Am. Chem. Soc. 120, 11820–11821.

Sledz, C.A., Holko, M., de Veer, M.J., Silverman, R.H., Williams, B.R., 2003. Activation of the interferon system by short-interfering RNAs. Nat. Cell Biol. 5, 834–839.

van der Krol, A.R., Mur, L.A., Beld, M., Mol, J.N., Stuitje, A.R., 1990. Flavonoid genes in Petunia: addition of a limited number of gene copies may lead to a suppression of gene expression. Plant Cell 2, 291–299.

Vasudevan, S., Tong, Y., Steitz, J.A., 2007. Switching from repression to activation: microRNAs can upregulate translation. Science 318, 1931–1934.

Yao, Y., Shouhong, J., Long, H., Yu, Y., Zhang, Z., Cheng, G., et al., 2015. RNAe: an effective method for targeted protein translation enhancement by artificial non-coding RNA was SINEB2 repeat. Nucleic Acids Res. 43 (9), e58. Available from: https://doi.org/10.1093/nar/gkv125.

Yu, J.-Y., DeRuiter, S.L., Turner, D.L., 2002. RNA interference by expression of short-interfering RNAs and hairpin RNAs in mammalian cells. Proc. Natl. Acad. Sci. 99 (9), 6047–6052.

Zamecnik, P.C., Stephensen, M.L., 1978. Inhibition of Rous sarcoma virus replication and cell transformation by a specific oligodeoxynucleotide. Proc. Natl. Acad. Sci. 75, 280–284.

Zamore, P.D., Tuschl, T., Sharp, P.A., Bartel, D.P., 2000. RNAi: double-stranded RNA directs the ATP-dependent cleavage of mRNA at 21 to 23 nucleotide intervals. Cell 101, 25–33.

Further reading

Brummelkamp, T.R., Bernards, R., Agami, R., 2002. A system for stable expression of short interfering RNAs in mammalian cells. Science 296, 550–553.

Gantz, V.M., Jasinskiene, N., Tatarenkova, O., Fazekas, A., Macias, V.M., Bier, E., et al., 2015. Highly efficient Cas9-mediated gene drive for population modification of the malaria vector mosquito Anopheles stephensi. Proc. Natl. Acad. Sci. 112 (49), E6736–E6743. Available from: https://doi.org/10.1073/pnas.1521077112.

Jiang, F., Doudna, J.A., 2017. CRISPR-Cas9 structures and mechanisms. Annu. Rev. Biophys. 46, 505–529. Available from: https://doi.org/10.1146/annurev-biophys-062215-010822.

Parrish, S., Fleenor, J., Xu, S., Mello, C., Fire, A., 2000. Functional anatomy of a dsRNA trigger: differential requirement of the two trigger strands in RNA interference. Mol. Cell 6, 1077–1087.

Steuber-Buchberger, P., Wurst, W., Kühn, R., 2008. Simultaneous Cre-mediated conditional knockdown of two genes in mice. Genesis 46, 144–151.

Wright, A.V., Nunez, J.K., Doudna, J.A., 2016. Biology and applications of CRISPR systems: harnessing nature's toolbox for genome editing. Cell 164, 29–44.

Xie, M., Yu, B., 2015. siRNA-directed DNA methylation in plants. Curr. Genomics 16 (1), 23–31.

Chapter 14

Electrophoresis of RNA

Rationale

Electrophoresis is a chromatography technique by which a mixture of charged molecules is separated according to size when placed in an electric field. It is a standard laboratory technique in widespread use for resolving complex mixtures of DNA molecules, RNA molecules, or protein molecules. Electrophoresis is also used routinely to assess the purity of a sample, whether a molecule has been successfully modified, for example, restriction enzyme cutting of a DNA fragment, ligation of RNA or DNA molecules, or to simply profile or fingerprint a sample. The electrophoresis technique is performed using high-voltage power supplies that have the potential to cause life-threatening injuries. It is therefore imperative that one always exercise great caution when working with such equipment; the novice should seek assistance from someone who is familiar with safe operating procedures.

In the RNA world, there are three major reasons why electrophoresis is performed. First, and perhaps the most common reason, is an integrity check of an RNA sample before using it for any of a number of applications. As discussed in Chapter 1, Ribonucleic Acid and the Cellular Biochemistry Revisited, and Chapter 2, Creating a Ribonuclease-Free Environment, RNA is a labile molecule; one would not proceed with an RNA-based investigation if the starting material had been degraded. Not only is it important to assess RNA integrity immediately after isolation, but it is also strongly advised to do so again after an extended period of storage, even in an ultracold freezer. Second, RNA electrophoresis may be performed as the first step for Northern analysis (Chapter 16: Northern Analysis), or as the penultimate step in a nuclease protection analysis (Chapter 18: Quantification of Specific mRNAs by Nuclease Protection). Third, it is possible to size select RNA molecules for cloning experiments or, concomitantly, to ensure that a previously performed size selection procedure, for example, column chromatography, produced the desired size fractionation outcome. RNA samples may be purified by any of a variety of RNA extraction procedures described in Chapters 4–7, or elsewhere. Electrophoresis is an extremely versatile, established technique that is compatible with an array of contemporary and time-honored molecular biology methods. Running a gel, even a minigel, is an excellent evaluative tool for assessing the integrity and probable utility of a sample.

RNA Methodologies. DOI: https://doi.org/10.1016/B978-0-323-90221-2.00018-7

Normalization of samples by nucleic acid concentration

In the preparation of RNA samples for electrophoresis or any type of side-by-side analysis, it is incumbent upon the investigator to ensure that the samples are "normalized," meaning that equal quantities of RNA are loaded in each lane of the gel. But what, exactly, constitutes a battery of normalized samples? Does the normalization of samples of total cellular RNA, total cytoplasmic RNA, and poly $(A)^+$ mRNA have the same meaning? Is it appropriate to normalize samples by mass (equal μg quantities per lane)? What is the difference between a loading control and a transcription control? Normalization must constitute a logical and proper basis for the comparison and interpretation of signal intensity observed in each experimental sample; it must also give meaning to signals, if any, generated from positive and negative controls. The ramifications of normalization are far-reaching indeed.

There are three issues to consider with respect to the normalization of nucleic acids for electrophoresis:

1. Normalization is based on the total mass of nucleic acid in each sample, as determined spectrophotometrically. In general, the sample with the lowest concentration is designated the "baseline" sample, meaning all of the other samples are diluted to match the concentration of the baseline sample. This is the most common starting point in most instances for normalization among samples.
2. Visual inspection of RNA after electrophoresis and staining. This is to ensure that each sample is intact and to ensure uniformity in the abundance of the rRNA species from one sample to the next. Some investigators rely on digital image analysis software to normalize the samples based on fluorescence intensity in each lane of the gel. It is important to note that rRNA is a satisfactory loading control, but not an appropriate transcription control.
3. Validation of normalization by examination of the transcriptional behavior of a "housekeeping" gene, that is, a gene whose expression pattern is not expected to change as a function of an experimental stimulus or manipulation. Popular housekeeping genes include β-actin, fibronectin, histone, GAPDH, and transferrin receptor mRNA. Some investigators even use the total mass of tRNA as a method by which to verify normalization. The reader is cautioned, however, that there is no single transcript that is optimal for normalization because different cells respond differently to various experimental stimuli: subtle transcriptional and/or post-transcriptional variations may accompany experimentally induced permutations in the cellular biochemistry. This aspect of the assessment of gene expression is described in detail in Chapter 11, Quantitative PCR Techniques, in the section on internal controls.

The most widely used procedure for normalization is to measure the concentration of each RNA sample spectrophotometrically, and then load equal

(A_{260}) microgram quantities of RNA in each lane on a gel, and this is generally acceptable practice. When purified poly $(A)^+$ mRNA is available, loading 2–5 μg per lane, that is, the same amount in each lane, is sufficient to detect many low abundance mRNAs, even by low sensitivity Northern analysis. In other cases, however, it may be desirable or even necessary to electrophorese 15–20 μg of total cellular RNA or total cytoplasmic RNA without prior selection of the poly$(A)^+$ RNA. For example, the yield of poly$(A)^+$ RNA that can be generated in a particular model system may be so small as to preclude its isolation. Moreover, because no isolation procedure is 100% efficient, steps taken to purify poly$(A)^+$ transcripts may actually serve to further underrepresent low and very-low abundance mRNAs and may be contaminated with varying amounts of nonpoly(A) material. If Northern analysis is being performed to assess the native size of a transcript, electrophoresing total cellular RNA may be more than adequate if the transcript under investigation is of sufficient abundance for easy detection, without the need for poly(A) enrichment.

Direct measurement of poly(A) content

The most prevalent component of a purified sample of total cellular or total cytoplasmic RNA is ribosomal RNA (rRNA), which constitutes 80%–85% of the sample by mass. This is clearly an enormous excess compared to the mass of the polyadenylated transcripts which, collectively, usually amount to no more than 2%–3% of the sample and comprised of thousands of different mRNA species. Thus, changes in the transcription of discrete mRNAs, even significant changes, are virtually undetectable by standard UV spectrophotometry. Yet, accurate normalization of experimental samples mandates a rationale that will ensure that any observed variations in the abundance of specific mRNAs truly indicate differential transcription or posttranscriptional regulation of these species (as opposed to an overall change in the transcription of all RNA species) and that these observed changes are meaningful. These considerations have across-the-board applicability to all assays with any quantitative component.

In order to address these concerns at least partially, one might consider the assay of poly$(A)^+$ material only. Although the shortcomings of this approach should be clear from the arguments presented in "The Poly(A) Caveat" (Chapter 5: Isolation of Polyadenylated RNA), loading equal amounts of poly $(A)^+$ RNA (both hnRNA and mRNA) may be considered at least one acceptable attempt to normalize samples, though this may not always be possible or practical. When the availability of very small quantities of RNA, cells, or tissue precludes poly$(A)^+$ selection, another method for normalization (while not mainstream) involves a synthesis of a labeled poly(T) probe (Fornace and Mitchell, 1986; Hollander and Fornace, 1990; Farrell and Greene, 1992). The probe, prepared by performing what amounts to the synthesis of first-strand cDNA, can be used to estimate the overall poly(A) content of various samples. This is not a commonly used approach owing to the types of analyses offered by PCR and other methods, though an adaptation of this technique is included here in order to present the reader with as many options as possible. With this normalization approach, a small amount of cellular RNA (500 ng or less) is sacrificed for slot blot or dot blot analysis. The observed hybridization of the poly(T) probe is a barometer of the poly(A) content of the sample (Fig. 14.1). When compared with the hybridization signal intensity from other samples, this information may be used to normalize samples of total RNA such that equivalent quantities of poly(A)$^+$ material are loaded into each lane of the gel or used for any other analysis requiring precise normalization. This technique accomplishes the equivalent of UV spectrophotometric normalization of samples of purified poly(A)$^+$ mRNA; failure to normalize with respect to poly(A)$^+$ content may result in observed variations that result directly from loading different quantities of poly(A)$^+$ RNA into each lane, rather than from actual changes in gene expression. In practice, the outcome of this type of analysis often correlates well with A_{260} normalization, which is the mainstream method.

The most important point to remember with respect to the normalization of several samples is that Northern analysis, while historically a venerable technique, is very close to the bottom of the sensitivity index (Chapter 11: Quantitative PCR Techniques) with respect to the assay of RNA as a parameter of gene expression. Northern analysis, in which RNA electrophoresis is an initial step, could be used to roughly gauge the behavior of transcripts in the cell while assessing the overall length of a transcript; the demand for fine-tuned analysis of gene expression should lead the investigator to supersensitive RT-PCR-based assays.

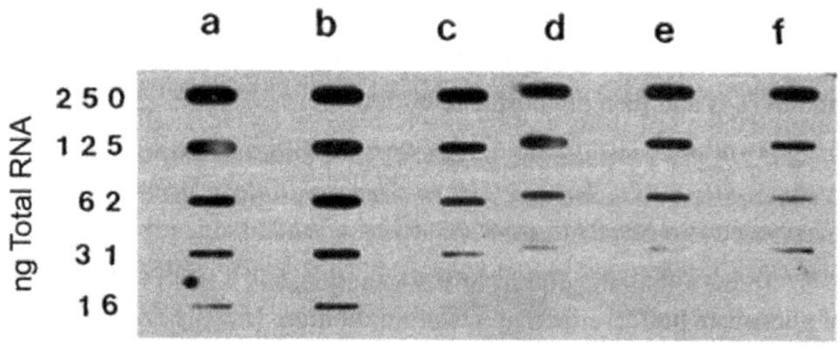

FIGURE 14.1 Slot blot normalization of poly(A)$^+$ content in samples of total cytoplasmic RNA. Serial twofold dilutions of 500 ng (A_{260}) were made in phosphate buffer and 100 μL was applied to each slot. Samples were then hybridized to a ^{32}P-labeled poly(T) probe to determine if samples of total RNA normalized by mass contained equal mass quantities of poly(A)$^+$ mRNA in the absence of oligo(dT) affinity selection. Autoradiography revealed that although an equal mass of total RNA was applied to each slot, the poly(A)$^+$ component among these samples varied from two- to fourfold.

Protocol: poly(A) normalization with a poly(T) probe

Sample preparation

1. Remove 500 ng (A_{260}) of RNA from each sample to be normalized and dilute each in 200 μL ice-cold 50 mM phosphate buffer (50 mM $Na_xH_xPO_4$), pH 6.8 prepared from the monobasic and dibasic salts (see Table 14.1 for formulation).
2. Using a Dot-Blot (Minifold I) or Slot-Blot (Minifold II) apparatus (GE Healthcare Life Sciences), apply half of each sample (100 μL = 250 ng) to a nylon filter that was prewet first in nuclease-free H_2O and then in phosphate buffer.

   Note: The slot configuration is preferred to the dot configuration because of the smaller area into which the sample is concentrated results in more quantitative data.
3. To the remaining 100 μL of RNA sample, add a fresh 100 μL aliquot of phosphate buffer, effecting a twofold dilution. Usually five twofold dilutions blotted onto a filter membrane are sufficient to accurately normalize the sample.
4. Wash each well twice with 300 μL phosphate buffer.
5. Positive controls may be generated with the dilution of commercially available pure poly(A), which, at greater dilutions, can serve as a standard for estimating the absolute poly(A) + content of the sample. For an excellent negative control, prepare a similar dilution series using tRNA.
6. Immobilize the RNA on the filter paper as described by the manufacturer, ideally by UV crosslinking. Begin prehybridization or store filter in a cool, dry place until ready to use.

Prehybridization—option 1

7. Church blots (Church and Gilbert, 1984): Equilibrate filters for a minimum of 20 minutes in prehybridization solution (1% bovine serum albumin; 1 mM EDTA; 0.5 M $Na_xH_xPO_4$, pH 7.2, 7% SDS) at 65°C. Use a minimum of 100 μL/cm^2 prehybridization solution and maintain gentle rocking of the filter on an orbital shaker platform. Proceed to step 8.

Prehybridization—option 2

8. Soak filter(s) in 5 × SSPE for at least 15 minutes and then prehybridize in a volume of at least 100 μL/cm^2 prehybridization buffer (5 × SSPE, 5 × Denhardt's solution, 0.1% SDS, 100 μg/mL denatured salmon sperm DNA, 50% formamide). Allow prehybridization to proceed for 2–3 hours at 42°C with gentle rocking on an orbital shaker incubator.

TABLE 14.1 Preparation of $Na_xH_xPO_4$ buffer.

NaH_2PO_4 (mL)	Na_2HPO_4 (mL)	Resulting pH
93.5	6.5	5.7
92.0	8.0	5.8
90.0	10.0	5.9
87.7	12.3	6.0
85.0	15.0	6.1
81.5	18.5	6.2
77.5	22.5	6.3
73.5	26.5	6.4
68.5	31.5	6.5
62.5	37.5	6.6
56.5	43.5	6.7
51.0	49.0	6.8
45.0	55.0	6.9
39.0	61.0	7.0
33.0	67.0	7.1
28.0	72.0	7.2
23.0	77.0	7.3
19.0	81.0	7.4
16.0	84.0	7.5
13.0	87.0	7.6
10.5	90.5	7.7
8.5	91.5	7.8
7.0	93.0	7.9
5.3	94.7	8.0

Theoretical proportions of 200 mM stock solutions of NaH_2PO_4 and Na_2HPO_4 were used to prepare a pH-specific sodium phosphate buffer (Dawson et al., 1959). Be sure to check pH after mixing NaH_2PO_4 and Na_2HPO_4 and subsequently diluting it. Minor adjustments in pH are frequently required. See text for details.

Synthesis of poly(T) probe

9. The poly(T) probe necessary for this technique is synthesized by the reverse transcription of a poly(A) template (polyadenylic acid). A typical reaction (100 μL volume) consists of 250 ng/μL poly(A); 2 mM DTT; 50 mM Tris-Cl, pH 8.3; 40 mM KCl; 10 mM $MgCl_2$; 0.5U/μL RNasin (Promega); 1 mM dTTP; 0.05 μg/μL Oligo dT_{12-18}; 100μCi α-^{32}P TTP (3000 Ci/mmol); 0.5U/μL AMV reverse transcriptase. Any first-strand cDNA synthesis kit is satisfactory, though a higher specific activity probe will result if the cDNA synthesis reaction is performed in the presence of dTTP alone, rather than a dNTP cocktail. The probe length will be about 200–300 bases. Alternatively, a nonisotopic poly (T) probe can be generated using any of the methods described in Chapter 17, Nucleic Acid Probe Technology.

Note: If place of ^{32}P, 250μCi of ^{35}S-thymidine 5′-(α-thio)triphosphate can be used to label the probe, though autoradiographic detection will not be as rapid.

10. Incubate at 42°C for 1.5 hours.

11. Terminate the reaction with the addition of 10 μL 1 M NaOH and 2 μL 500 mM EDTA. Incubate at 65°C for 15 minutes to hydrolyze the poly (A) template.

12. Neutralize the alkaline reaction mixture with 10 μL 1 M Tris-Cl, pH 7.5 for 15 minutes at room temperature.

13. Radiolabeled probe must be separated from unincorporated radionucleotides and other reaction components. This is easily accomplished by spun-column chromatography (Sambrook and Russell, 2001) through Sephadex G-25 (GE Healthcare) or by using another commercially available concentrating/purification device. Unused probe (and previously used hybridization buffer containing probe) should be stored with proper shielding at −20°C, at which temperature it will be stable for 4–6 weeks.

Hybridization

14. Replace the prehybridization buffer with a fresh aliquot of the same prehybridization buffer formulation (7a or 7b, above), using no more than 100 μL/cm^2 filter. Add the poly(T) probe at a concentration of 100 ng/mL or 1×10^6 cpm/mL.

15. Hybridize blots with gentle agitation for 3 to 4 hours at 44°C.

Posthybridization washes

16. At the conclusion of the hybridization period remove the probe solution. The probe can be saved (frozen) for subsequent hybridization analyses. If filters were prehybridized and hybridized in Church blot buffer (Option 1, step 7a), then wash blots twice for 10 minutes each in

200 mM NaCl at room temperature, once for 30 minutes at 44°C, and finally once in 75 mM NaCl at room temperature for 5 minutes. If filters were hybridized in SSPE-containing buffer (Option 2, step 7b), then wash the filters twice in 2 × SSPE, 0.1% SDS for 10 minutes at room temperature and then wash twice in 0.1 × SSPE, 0.1% SDS for 10 minutes at 37°C. Use a volume of at least 100 mL of buffer for each wash.

Note: In all cases, washes should be performed on an orbital shaker platform at the appropriate temperature. This will ensure efficient and effective washing of the filter(s) to remove nonhybridized probes that could contribute to the background.

17. Briefly blot the filters to remove excess was buffer. This is readily accomplished by placing the filter, probe-side up, on a piece of Whatman 3MM paper; usually, 20–30 seconds is adequate. Do not allow the filter to dry out.
18. Wrap the filter(s), while still damp, in plastic wrap and then set up autoradiography. With this type of hybridization a single overnight exposure at −70°C with an intensifying screen is quite sufficient. Depending on the specific activity of the probe, a film exposure of as little as 4 hours may produce a quantifiable image. At the conclusion of the autoradiographic exposure, develop the film (Chapter 17: Nucleic Acid Probe Technology).

    Note 1: The resulting autoradiogram can be analyzed using image analysis software. Alternatively, each dilution can be quantified by scintillation counting; individual dilutions (i.e., the dots or slots) can be cut from the membrane and dropped directly into a scintillation cocktail for measurement if radiolabelled probes were used. Refer to Chapter 17, Nucleic Acid Probe Technology, for a comparison of the differences between autoradiography and chemiluminescence detection.
19. The poly(A) content of each sample, as determined by the degree of poly (T) probe hybridization, permits samples of total cellular or total cytoplasmic RNA to be normalized to that of a reference sample; samples with the same amount of poly(A) material generate similar exposure patterns on the film. The concentration of each RNA sample is adjusted as needed, and the samples are then prepared for electrophoresis.

Intramolecular base-pairing mandates RNA denaturation

mRNA molecules in vivo are single-stranded polyribonucleotides complex with cytoplasmic proteins, which together constitute messenger ribonucleoprotein (mRNP) particles. Within an mRNP, the mRNA exhibits formidable intramolecular base-pairing which results in considerable higher-level folding. Transfer RNA (tRNA) is the premier example of a molecule that manifests so much intramolecular base-pairing that the molecule assumes a very characteristic, easily recognized, and amazingly stable tertiary structure. Nearly all RNA molecules are believed to contain short double-helical regions consisting not only of the standard A::U and G:::C base pairs but also weaker G::U base pairs. The extensive intramolecular, and intermolecular, interactions regulate the translatability and other activities of transcripts in the cytoplasm.

As one of the parameters that govern the migration of nucleic acids in an electric field is the molecular conformation, RNA secondary structures of all types impede electrophoretic separation based solely on molecular weight. Consequently, identical species of RNA molecules exhibiting varying degrees of intramolecular base pairing will migrate toward the anode (the positive electrode) at different rates, resulting in smearing rather than distinct band formation. To circumvent this problem and to ensure that like species of RNA comigrate, the electrophoresis of RNA is routinely carried out under what is known as denaturing conditions, primarily because heating alone does not sufficiently disrupt the secondary structure. Successful electrophoresis of RNA is accomplished in two parts: (1) RNA is denatured prior to loading the gel and (2) conditions are established that maintain the denatured state during electrophoresis. Moreover, the best possible resolution is achieved when gels are poured as thin as is practical and run at low voltage.

The choice of denaturing system and gel matrix are determined primarily by the size range of the RNAs to be separated and also whether the characterization of RNA is to be preparative or strictly analytical in nature. For the optimal resolution of mixtures of very small nucleic acids (fewer than 400 bases or base-pairs) the matrix of choice is polyacrylamide (3%–20%). Applications include the S1 nuclease assay (Berk and Sharp, 1977; Favaloro et al., 1980; Sharp et al., 1980), the RNase protection assay (Zinn et al., 1983; Melton et al., 1984), and old-style DNA sequencing. However, when a sample contains thousands of different-size transcripts, the optimal balance between electrophoretic resolution and efficiency of transfer (blotting) from the gel to a membrane for hybridization is achieved with a 1.0%–1.2% denaturing agarose gel.

The most commonly used RNA denaturant for agarose gel electrophoresis is formaldehyde (Boedtker, 1971; Lehrach et al., 1977; Rave et al., 1979), which is a known carcinogen and teratogen. Older RNA denaturation methods including the glyoxal/dimethyl sulfoxide (DMSO) denaturing system (McMaster and Carmichael, 1977) and the methylmercuric hydroxide system (Bailey and Davidson, 1976; Alwine et al., 1977, 1979) have fallen out of favor and are no longer in use. As a historical note, methylmercuric hydroxide (methyl mercury) was the first, and most efficient, RNA denaturant for agarose electrophoresis (Gruenwedel and Davidson, 1966). However, due to the toxicity of this compound (Junghans, 1983), it is to be avoided altogether. For polyacrylamide gels, formamide (Spohr et al., 1976) and urea (Reijnders et al., 1973) have been used successfully as RNA denaturants. Each of these denaturing systems has unique characteristics and safety requirements (e.g., formamide is a strongly suspected carcinogen and a known teratogen) as well as distinct advantages and limitations. Methodologies pertaining to the use of formaldehyde, glyoxal/DMSO, and urea as RNA denaturants in agarose gels are presented in this chapter. Formaldehyde gels for Northern analysis usually render greater detection sensitivity than the glyoxal denaturing system, due in part to difficulties associated with the complete removal of glyoxal from the RNA prior to hybridization. Most laboratories performing RNA electrophoresis use the formaldehyde denaturing system.

In this laboratory, the urea denaturing system has been used very successfully for denaturing electrophoresis, minimizing the need to use formamide or formaldehyde. The molecular structures of these common RNA denaturants are shown in Fig. 14.2.

Formaldehyde denaturation

Formaldehyde and formamide are commonly used denaturants of RNA. Both are toxic and, as teratogens, should be handled in a chemical fume hood and avoided altogether by expectant mothers. To ensure that the RNA migrates only with respect to molecular weight, samples of RNA are denatured with both formaldehyde and formamide prior to electrophoresis, and formaldehyde is added to the gel itself to maintain the denatured state and inhibit RNase during electrophoresis. Formaldehyde, which is light sensitive, is routinely supplied in a dark brown bottle as a 37% stock solution (12.3 M) that also contains 10%–15% methanol as a preservative. It should be stored at room temperature and out of direct sunlight. Formaldehyde oxidizes when it comes into contact with the air. As such, the pH of formaldehyde should be checked prior to each use and must be at a pH greater than 4.0 for RNA work. At pH below 4.0 RNA will experience severe degradation. To raise the pH, deionize formaldehyde in the same manner as formamide (Appendix G). Gels containing formaldehyde are less rigid than other agarose gels and are considerably more slippery. It is always wise to support such gels from beneath, as with a spatula or within the casting tray when moving them from place to place.

The running buffer of choice for formaldehyde gels is 1× MOPS buffer (MOPS = 3-[N-morpholino]propanesulfonic acid). This buffer is conveniently prepared as a 10× stock (400 mM MOPS pH 7.0; 100 mM sodium acetate; 10 mM EDTA, pH 8.0) and, if made properly, assumes a characteristic straw color when autoclaved. Contrary to popular belief, autoclaving MOPS buffer *is* acceptable and will not interfere with its buffering capacity as long as the solution remains a light straw color, rather than deep yellow. Instead of autoclaving,

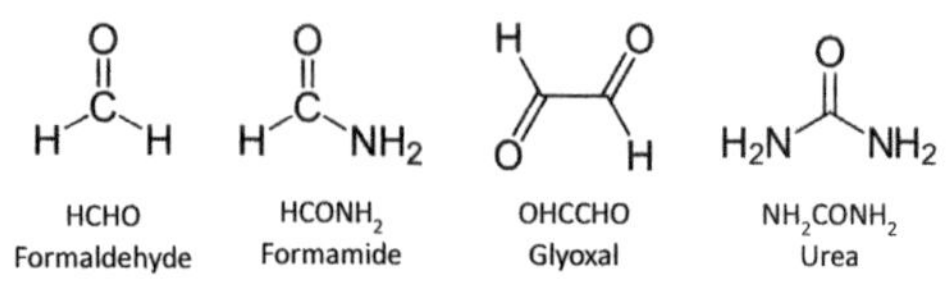

FIGURE 14.2 Molecular structure of common RNA denaturants.

MOPS buffer can be filtered twice through a nitrocellulose membrane to eliminate nuclease activity and extend its shelf life. Better yet, nuclease-free MOPS buffer can be purchased ready for immediate use. If running an RNA gel for the first time, the information at the end of this chapter may be very useful.

Protocol: formaldehyde denaturing gels

1. Wear gloves!
2. To prepare 200 mL of molten agarose solution (more than enough for two 12 cm × 14 cm gels), melt 2.4 g of agarose in 170 mL sterile H_2O in an Erlenmeyer flask on a hot plate or in a microwave oven. Be certain that the agarose has melted *completely* by gently swirling the flask and looking for undissolved beads of agarose, commonly known as "fish-eyes." When the gel has cooled to 55°C–60°C, add a prewarmed mixture of 20 mL 10× MOPS buffer and 10 mL 37% formaldehyde. This will produce a solution of 1.2% agarose in 1× MOPS buffer and 0.6 M formaldehyde. Keep this molten agarose solution in a 55°C–60°C water bath until ready to cast the gel and be sure to cover the flask to prevent the escape of formaldehyde fumes.

   Note 1: Certain older protocols recommend a final formaldehyde concentration of 2.2 M, which is much too high. Most investigators use a maximum final concentration of 0.6 M which, if desired, may be further reduced to 0.22 M.

   Note 2: In this lab, a collection of Erlenmeyer flasks are autoclaved in advance and set aside specifically for the preparation of RNA gels.
3. Cast a 0.5–0.75 cm thick gel using the solution prepared in step 2. Be sure to cast the gel in a chemical fume hood to prevent exposure to formaldehyde fumes in the room. The gel will solidify and be ready for use in about 10 minutes. Remember that the thinner the gel, the better will be the resolution of the RNA.
4. Prepare each sample for electrophoresis by mixing the following in a sterile microfuge tube:

   4.7 μL RNA (up to 20 μg);
   2.0 μL 5× MOPS buffer;
   3.3 μL formaldehyde;
   10.0 μL formamide.

   Be certain to mix thoroughly by pipetting and then pulse-centrifuge to collect the reaction components at the bottom of the microfuge tube.

   Note 1: Do not exceed 20 μg RNA per lane, beyond which one runs the risk of overloading the lanes of the gel and losing resolution. If rare or very low abundance mRNAs are of interest it may be worthwhile to load as much as 4–5 μg of poly(A)$^+$RNA per lane instead.

   Note 2: It is imperative to use freshly deionized formaldehyde and formamide.

5. Prepare 2–3 μg of RNA molecular weight standard for electrophoresis, as described above in step 4.

   Note: RNA molecular weight standards are single-stranded and need to be denatured, too.
6. Denature all samples and the molecular weight marker by heating to 55°C for 15 minutes or at 65°C for 10 minutes.
7. During the incubation of the RNA and once the gel has solidified, immerse the gel in a running buffer (1 × MOPS).

   Note: Be sure to use the smallest amount of buffer necessary to completely cover the gel, including the edges of the gel which tend to "smile" by clinging to the side of the casting tray. Carefully remove the comb by slowly pulling it upward with uniform force. Covering the gel with buffer prior to removal of the comb will lubricate the wells and reduce the vacuum effect created when the comb is pulled. If the gel is not immersed in buffer, comb removal may damage the bottom of the wells.
8. Following denaturation, add 2 μL of 10 × loading buffer (50% glycerol); 50 mM EDTA, pH 8.0; 0.20% bromophenol blue to the samples and mix thoroughly by micropipetting. Load samples onto the gel immediately.

   Note: Condensation may form on the inside of the tube during the denaturation incubation period so be sure to briefly pulse-centrifuge the tube just prior to the addition of the loading buffer.
9. Electrophorese the samples at a maximum of 5 V/cm distance between the electrodes. Continue electrophoresis until the bromophenol dye front has migrated to within 1–2 cm from the distal edge of the gel.

   Note: It is especially important to stop the electrophoresis at this point when one is evaluating RNA species of unknown size. In subsequent experiments, RNA gels may be electrophoresed for shorter or longer periods to accommodate specific needs.
10. At the conclusion of the electrophoresis, the gel may be stained in a solution of 0.5 μg/mL ethidium bromide (EtBr; in sterile water or in a fresh aliquot of 1 × MOPS buffer) or, preferably, 1 × SYBR Green II or 1 × SYBR Gold (in 1 × TAE buffer) in order to visualize the RNA within the gel. Staining techniques and options are discussed later in this chapter.

    Note: Staining formaldehyde gels with EtBr often produce unacceptable high background fluorescence. Such gels take on a strong orange/pink hue and quite often any nucleic acid in the gel is almost completely obscured. Reducing background fluorescence and, concomitantly, formaldehyde removal is easily accomplished by destaining the gel in two or three changes of nuclease-free H_2O or 5 × SSC for 10–15 minutes each, ideally with gentle agitation.
11. Regardless of how the gel was stained, it should be soaked in 5 × SSC or nuclease-free H_2O to remove formaldehyde from the gel. Typically, a 20–30 minutes soaking in two or more changes of buffer or sterile

water is adequate for formaldehyde removal from the gel. Alternatively, the gel may be stored in 300 mL sterile water overnight at 4°C.

12. The gel is now ready to be photographed.

Urea denaturation

The traditional compounds used to ensure RNA denaturation are toxic, cumbersome, or both. Urea is widely used in polyacrylamide gels for DNA and RNA analysis and can also be used easily in conjunction with agarose gels. Rather than preparing an agarose gel with urea in it (Locker, 1979), one may treat RNA with freshly prepared urea instead, briefly heating it to denature it, and then running the RNA on a nondenaturing agarose gel. This approach is used routinely in this laboratory to assess the integrity of an RNA sample prior to use in RT-PCR, nuclease protection, or other transcription assays. This method also denatures RNA completely enough to support Northern analysis. As shown in Fig. 14.3, a side-by-side comparison of formamide-denatured RNA and urea-denatured RNA shows similar levels of sensitivity, even though the urea-denatured sample shows greater mobility overall by comparison with the samples that are denatured with the formamide. While it is true that heating RNA briefly without *any* denaturants may be adequate for simply ensuring sample integrity, the extra minute required to add urea to the sample(s) is easily justified by the greatly improved appearance of RNA on the gel. Urea denaturation is an effective method for rapid assessment of RNA sample integrity without the need for toxic compounds.

Protocol: urea denaturation

1. Prepare a 1.2% agarose gel with 1× TAE.

2. Prepare a 4 M urea stock solution (2.4 g in 10 mL H_2O). This should be prepared shortly before use.

3. Prepare loading buffer by mixing 100 μL 10× standard gel loading buffer and 100 μL fresh prepared 4 M urea. The resulting 200 μL mixture is a 10× urea loading buffer stock.

Note: Typical 10× loading buffer = 50% glycerol, 50 mM EDTA (dilute 500 mM stock), 0.2% bromophenol blue.

4. For each 10 μL of RNA in an aqueous solution, add 1 μL of urea loading buffer.

5. Heat to 55°C for 10 minutes.

6. Load gel and electrophorese as usual.

7. At the conclusion of the run, stain the gel in 2× SYBR Green in 1× TAE buffer for 30 minutes and then photograph.

8. Depending on the application the gel may then be used for Northern transfer or simply discarded after confirming sample integrity.

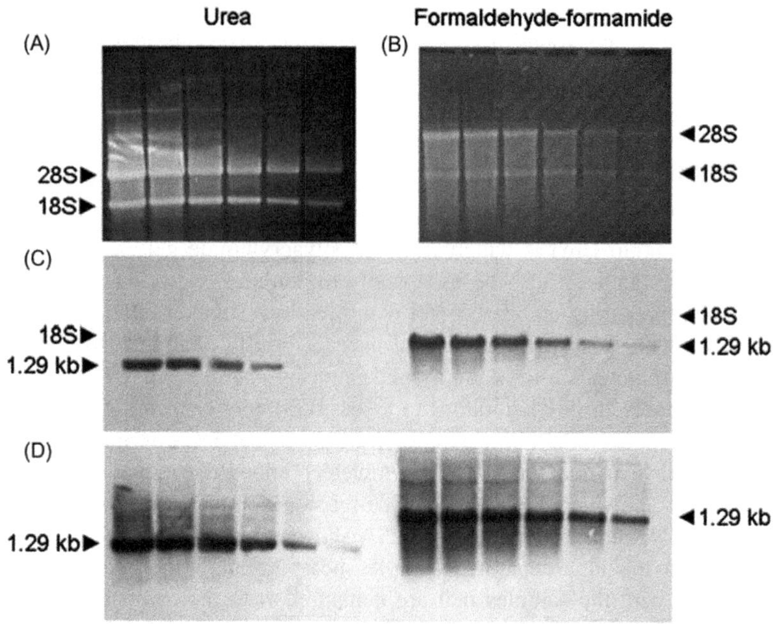

FIGURE 14.3 Comparative denaturing electrophoresis. Identical amounts of RNA were loaded onto each gel: 7.7 μg (lane 1), 5.8 μg (lane 2), 3.9 μg (lane 3), 1.9 μg (lane 4), 1.0 μg (lane 5), and 0.5 μg (lane 6). (A): RNA was denatured with urea, electrophoresed in a nondenaturing gel, and then stained with SYBR Green. (B): RNA was denatured with formaldehyde and formamide, electrophoresed on an agarose gel containing formaldehyde, and then stained with SYBR Green. The RNA has greater mobility and less background fluorescence (greater contrast) when denatured with urea. Both gels were then (Northern) blotted on the same filter paper and hybridized with a GAPDH probed. (C): Detection of GAPDH mRNA via chemiluminescence detection after a five-minute exposure. (D): Same filter as in (C) after a 20 min exposure. In both cases, the bands are sharp clear and strong and the GAPDH is in the expected position relative to the ribosomal 28S and 18S species. See Chapter 16, for additional details pertaining to Northern analysis.

Glyoxal/dimethyl sulfoxide denaturation

Another common method of denaturing RNA in preparation for electrophoresis is to treat the sample with a mixture of 1 M glyoxal and 50% (v/v) DMSO at 50°C for 1 hour (Bantle et al., 1976). Glyoxalation introduces an additional ring into guanine and, in so doing, sterically interferes with G:::C intramolecular base pairing (Hutton and Wetmur, 1973). At high concentrations, glyoxal also reacts with adenine and cytosine in both RNA and DNA, though the guanosine-glyoxal adduct is the most stable. The preparation of RNA in this fashion precludes some of the potential health hazards associated with formaldehyde. RNA samples are denatured with glyoxal prior to electrophoresis although, unlike the formaldehyde system, glyoxal is not added to the gel. While the glyoxalation process has advantages over formaldehyde denaturation, the advance preparation of glyoxal and required subsequent handling of denatured samples and Northern filters are a major disincentive and the reason why many investigators avoid this denaturing method.

Glyoxal oxidizes very rapidly; therefore, glyoxal stock solutions (40% glyoxal = 6 M) must be deionized to neutral pH before use (see Appendix G). If not deionized, the oxidation product, glyoxylic acid, will cause fragmentation of the RNA sample. As a matter of convenience, small aliquots of deionized glyoxal can be stored at −20°C in tightly capped tubes. Thawed aliquots of deionized glyoxal should be used only once, and the unused portion discarded.

Glyoxalated nucleic acids exhibit decreased electrophoretic mobility compared to an identical sample denatured with formaldehyde. In return, the banding of the glyoxalated RNA tends to be sharper than is observed in formaldehyde-denaturing systems. Further, glyoxalated DNA and RNA molecules are electrophoretically equivalent (McMaster and Carmichael, 1977); therefore, DNA molecular weight standards can be used alongside RNA samples if both have been glyoxalated.

The major drawback of the typical glyoxal gel is that the buffering capacity of the standard running buffer, 10 mM sodium phosphate, pH 7.0, is quite poor. Phosphate buffer has been used in the past because low ionic strength electrophoresis buffers at pH 7.0 maintain the maximum glyoxal denaturation effect and preclude the need for adding a denaturant to the gel itself. The inability of the sodium phosphate buffer to resist radical changes in pH, however, necessitates constant recirculation of the running buffer throughout the electrophoresis period to maintain pH 7.0. This is imperative because glyoxal will dissociate from RNA (and DNA) when the pH is greater than 8.0 (Thomas, 1980) and at which point the RNA sample is susceptible to alkaline hydrolysis. Conveniently, most electrophoresis chambers can be purchased with optional buffer ports needed to support buffer recirculation. If recirculation is not possible, the power supply should be disconnected and the buffer changed every 30 minutes during the run, though this approach is bothersome and usually quite unsatisfactory. Alternatively, glyoxal gels may be electrophoresed in 1 × MOPS buffer, rather than phosphate buffer, thereby eliminating the need to recirculate the buffer. If running an RNA gel for the first time, the information at the end of this chapter may be very useful.

Protocol: glyoxalation and electrophoresis of RNA

Sodium phosphate electrophoresis buffer should be prepared in advance by combining stock solutions of NaH_2PO_4 (monobasic) and Na_2HPO_4 (dibasic) in the appropriate ratio to yield a solution of the desired pH. Begin with 0.2 M stock solutions of Na_2HPO_4 and NaH_2PO_4 that have been autoclaved or filtered sterilized. Next, combine Na_2HPO_4 and NaH_2PO_4 in the proportions indicated in Table 14.1 to produce the desired pH (7.0) of the final mixture. Then add 90% of the remaining water required to dilute the mixture. Check the pH of the diluted phosphate buffer. It is almost always necessary to make minor pH adjustments: add NaH_2PO_4 to lower the pH or Na_2HPO_4 to raise the pH. Important: fine-tuning the pH must be performed with phosphate buffer that has been diluted to the desired final concentration, rather than using the 0.2 M stock solutions. Finally, add sterile H_2O to produce the final volume. The resulting sodium phosphate buffer is described chemically as $Na_xH_xPO_4$, indicating the inclusion of both the monobasic and dibasic salts in the formulation of this buffer.

Gel and sample preparation

1. Wear gloves!
2. Melt agarose (final concentration of 1.2%) in an appropriate volume of 10 mM $Na_xH_xPO_4$, pH 7.0.
3. When molten agarose has cooled to 55°C–60°C, cast a 12 cm × 14 cm gel and allow to solidify.

   Note: Do not add ethidium bromide to glyoxal gels, as this dye will react with the glyoxal. The inclusion of SYBR Green and SYBR Gold in the gel should also be avoided.
4. Denature the RNA in an autoclaved microfuge tube by adding:

   3.7 μL RNA (up to 10 μg per lane);
   2.7 μL 6 M glyoxal (freshly deionized);
   8.0 μL DMSO (molecular biology grade);
   1.6 μL 100 mM $Na_xH_xPO_4$, pH 7.0.

   These proportions will render final concentrations of 1 M glyoxal, 50% DMSO, and 10 mM $Na_xH_xPO_4$.
5. Close the tube tightly and incubate the mixture at 50°C for 1 hour.
6. After the gel has solidified, immerse it in running buffer (10 mM $Na_xH_xPO_4$, pH 7.0). This should be done while the RNA is being glyoxalated.

   Note: Be sure to use the smallest amount of buffer necessary to completely cover the gel, including the edges which tend to "smile" by clinging to the side of the casting tray. Carefully remove the comb by slowly pulling it upward with uniform force. Covering the gel with buffer prior to removal of the comb will lubricate the wells and reduce the vacuum effect created when the comb is pulled. If the gel is not immersed in buffer, comb removal may damage the bottom of the wells.
7. At the conclusion of the incubation period, cool the mixture from step 5°C–20°C and then add 2 μL of loading buffer (50% glycerol; 10 mM NaxHxPO4, pH 7.0; 0.20% bromophenol blue). If necessary, pulse-centrifuge the tubes for 2–3 seconds to collect the entire sample at the bottom of the tube.

8. Immediately load each sample onto the gel.
9. Electrophorese at no more than 5 V/cm distance between the electrodes, with constant buffer recirculation. Electrophoresis should continue until the bromophenol blue tracking dye has migrated about 80% along the length of the gel.

   Note: It is especially important to stop the electrophoresis at this point when one is evaluating RNA species of unknown size. In subsequent experiments, RNA gels may be electrophoresed for shorter or longer periods, as needed.
10. At the conclusion of the electrophoresis, the gel can be used for Northern transfer without prior staining, though this is not recommended. The gel can be stained in a solution of 0.5 μg/mL ethidium bromide made up in 100 mM ammonium acetate or in a solution of 2 × SYBR Green II made up in 1 × TBE buffer.

    Note 1: The investigator should be aware of potential interaction between glyoxal and ethidium bromide, which may change its spectral properties (Johnson, 1975), and reduce transfer efficiency.

    Note 2: As an alternative to EtBr or the SYBR Green dyes, the gel may be stained in a 30 μg/mL solution of acridine orange (made up of nuclease-free 10 mM sodium phosphate, pH 7.0). This is not recommended, however, because acridine orange results in notoriously high levels of background in the gel. The required destaining time for acridine orange-stained gels is considerable and is ideally performed for about 1 hour at room temperature in an enameled metal pan because enamel adsorbs acridine orange. To remove acridine orange from the enamel, rinse with running hot tap water for 10–15 minutes. Polyacrylamide gels are destained in the dark for 2 hours at room temperature or at 4°C overnight. Handle acridine orange with care because it is a DNA binding dye and, as such, a mutagen.
11. For subsequent Northern analysis, glyoxalated RNA may be transferred from the agarose gel to a filter membrane, preferably nylon, without any pretreatment to remove glyoxal. Glyoxal is removed from the RNA during a posttransfer, prehybridization wash of the filter paper. See Chapter 16, Northern Analysis, for details on Northern analysis.

Running RNA on nondenaturing gels

Many attempts have been made to streamline the process of RNA denaturation and to reduce or altogether eliminate the use of toxic denaturants. For example, an RNA sample may simply be heat-denatured for a few minutes in a modified loading buffer (e.g., containing SDS; Kevil et al., 1997) and then immediately electrophoresed on a nondenaturation gel. In this regard, some investigators have a preference for the TAE or TBE buffering systems commonly associated with DNA electrophoresis. In some laboratories, investigators report simply heating RNA samples to 65°C for 15 minutes in loading buffer without any SDS or other denaturants, and simply running the RNA on a nondenaturing agarose gel. While this may be a satisfactory method for some RNA preps, experiences in this lab have shown that not all RNA samples denature as well as others when heated only, and the observed variability seems to be directly related to the method used to isolate the RNA from the biological source (Farrell, unpublished observation). This is something that must be determined empirically in one's own laboratory. If heating alone does not produce a satisfactory electrophoretic profile, the addition of urea to the loading buffer, as described above, usually resolves denaturation issues and the RNA may still be run on a nondenaturing gel. A bit of advice: when working with RNA, keep the agarose concentration between 1.0% and 1.2% in order to maximize the resolving power of an agarose gel. Finally, because denaturants are usually powerful inhibitors of RNase and may help to protect the sample from degradation, running nondenaturing gels requires particular attention to steps for RNase control.

Proper use of molecular weight standards

The accurate determination of the size of RNA species is an important component of transcript identification, an indicator of sample integrity, and an indicator of denaturing efficiency. Transcript size determination is accomplished by comparing the electrophoretic behavior of an RNA sample against that of molecular weight standards (also known as size standards or molecular weight markers). RNA molecules migrate through gels more rapidly than double-stranded DNA size standards of comparable length, especially in agarose gels that contain formaldehyde (Wicks, 1986). Therefore, one should electrophorese RNA molecular weight standards in a gel with RNA samples. As noted above, DNA and RNA are not electrophoretically equivalent unless both have been glyoxalated; even then DNAs are useful as size standards only if they fall within the range of the majority of cellular RNAs.

Useful markers for RNA electrophoresis are easy to come by even in laboratories with minimal funding. There are two basic approaches for the size determination of RNAs of interest:

1. *External size standards* are nucleic acid molecules of known size that are electrophoresed on the same gel as experimental samples in an unoccupied lane. These standards are commercially available from most molecular biology vendors. For RNA gels, size standards typically consist of a mixture of five to 10 single-stranded RNAs that were produced by in vitro transcription. Because these markers are, in fact, single-stranded, they must be denatured by the same method used to denature the samples. Care must be taken to protect RNA molecular weight markers from degradation, which are just as susceptible to RNase degradation as the

experimental RNA. Markers that cover various size ranges are widely available. The range of RNA size standards differs from one source to the next, and each laboratory will need to decide which products are most aptly suited for the business at hand, markers that span the range of most cytoplasmic RNAs in the sample (500–10,000 bases is a good place to start). After some definition has been given to "typical results" from an experimental system, one may opt to work with very large or very small standards for greater precision.

2. *Internal size standards* are RNA species of known size that are actually part of the RNA experimental sample, most often the large and small ribosomal RNAs. As these markers are part of the sample, they will be denatured during the preparation of the sample for electrophoresis. For greater precision in molecular size determination, a strategic move is to utilize both premade size standards and abundant ribosomal RNAs. In some laboratories, investigators actually spike RNA samples with an artificial transcript. A sufficiently abundant amount of this transcript can be used either to allow its direct visualization on a gel or to permit easy detection upon hybridization (Chapter 17: Nucleic Acid Probe Technology) or by PCR (Chapter 10: RT-PCR: A Science and an Art Form).

Either internal or external size standards can be used with confidence on an RNA gel. Relying on one type or the other, or both, has several advantages as well as drawbacks. With respect to external standards, the key decisions that must be made are:

1. The amount of the standard to use.
2. The positioning of the standard in the gel.
3. The size range of the standard.
4. The method of visualizing the markers (staining vs detection following hybridization).

Several brands of premade RNA size standards often contain minute traces of the DNA transcription template upon which the standards were synthesized in the in vitro transcription reaction. Further, many of these templates exhibit limited sequence homology with plasmid vectors that were historically important for molecular cloning. A possible consequence of this type of construction is manifested when a vector-borne nucleic acid probe is labeled and used for hybridization without cutting the probe itself out of the vector, or a probe that has been removed from a vector is still flanked by limited vector sequences that may hybridize to the residual template in the marker lane. Because this hybridization is probably an exact match (nucleotide for nucleotide), even subnanogram quantities of the template are more than sufficient to generate a strong signal upon hybridization. This has the unhappy consequences of (1) generating very strong detection signals that may bleed over and obscure signals from hybridization events that may have occurred in adjoining lanes and (2) having previously unseen bands show up in the molecular weight standard lane, causing confusion with respect to size determination of hybridized experimental transcripts. If there is reason to believe that a strong signal might occur, it is judicious to leave a blank lane in the gel between the size standards and the experimental samples, and to use nanogram quantities of the RNA marker. Depending on the dye, it will be necessary to load at least 0.5–1 μg/lane in order to have sufficient mass to photograph the molecular weight standards after staining the gel.

If no vector sequences are involved or if it is known that the probe is (was) cloned into a vector that does not hybridize to components of the molecular weight standard, then one may confidently load as much as 2–3 μg/lane, a more than sufficient mass for photodocumentation. If vector sequences are involved, as little as 15–20 ng of marker will hybridize very efficiently to complementary vector sequences and render sharp bands at the detection stage.

Just as the mass of marker is an important parameter, so is the placement of the marker. *Always* load the gel asymmetrically. Very little is more frustrating than obtaining a significant banding pattern and not knowing which lane is which. The confusion can be further compounded by the fact that some investigators invert the gel during the transfer, and then fail to maintain the orientation of the filter paper after it is removed from the surface of the gel following the transfer. Were this not bad enough, inversion of the exposed X-ray film during the developing steps adds yet another level of confusion. When the markers are loaded in the center of the gel, the distinction between left and right becomes further obscured. Consistently loading the gel in an asymmetrical fashion will eliminate at least one source of ambiguity. In this lab, the molecular weight standard is always loaded in the last lane, meaning the last well in the upper right-hand corner of the gel (Fig. 14.4).

Ribosomal RNA

Eukaryotic ribosomes consist of two main components, namely the large and small subunits. The larger of the two is known as the 60S subunit and the smaller is known as the 40S subunit; "S" stands for Svedberg unit, a biochemical term used to describe the sedimentation rate of a macromolecule. Each subunit consists of a characteristic rRNA and an assortment of proteins. In the course of an RNA extraction, the 80 + ribosomal proteins are stripped away from the rRNAs which, in turn, co-purify with the other cellular RNA species. Removal of the proteins from the 60S and 40S subunits yields the 28S and 18S rRNA, respectively. The much smaller 5S and 5.8S rRNAs, also integral components of eukaryotic ribosomes, are likewise liberated from the intact ribosomes upon removal of protein. In prokaryotes, the intact ribosomal subunits are known as 50S and 30S which, upon purification,

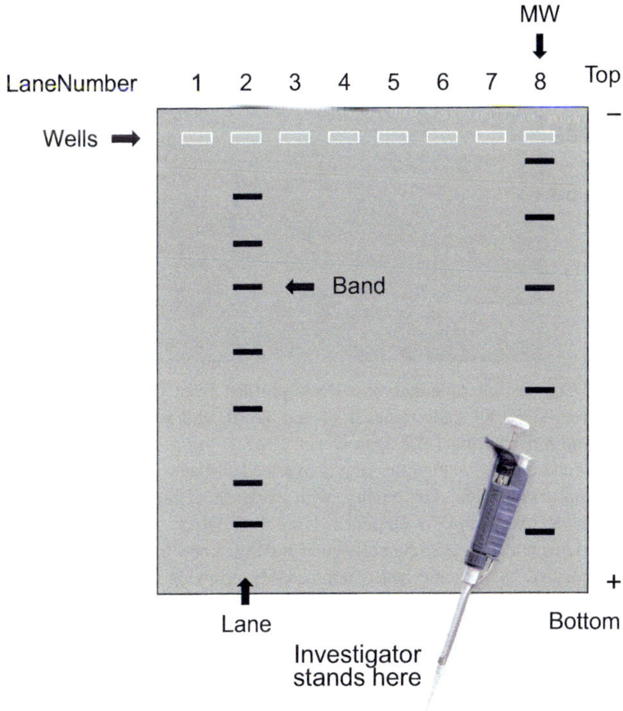

FIGURE 14.4 Standard nomenclature for referencing wells on a gel. The side of the gel with the wells is furthest away from the investigator who is standing in front of the gel, ready to load the samples. The wells are numbered consecutively, from left to right. *Pipetman micropipettor courtesy Gilson, Inc.*

render the 23S and 16S rRNA, respectively. The highly abundant rRNAs can be very useful as internal controls by:

1. acting as general molecular weight markers;
2. giving an indication of the integrity of the sample by visual inspection;
3. providing evidence that an equivalent mass of total RNA has been loaded into each experimental lane upon visual inspection or through the use of image analysis software.

Because rRNA is by far the most prevalent product of transcription (80%–85% of cellular RNA), an aliquot of total cellular or total cytoplasmic eukaryotic RNA should electrophoretically resolve into two very distinct, easily observable 28S and 18S rRNA bands. Depending on the method of lysis, a properly denatured RNA sample may also show a third, higher molecular weight band upon postelectrophoresis staining. This less obvious transcript is the 45S rRNA, a nuclear precursor to the 28S and 18S rRNAs (Fig. 14.5) that accumulate in the cytoplasm, and which form the scaffolding of functional

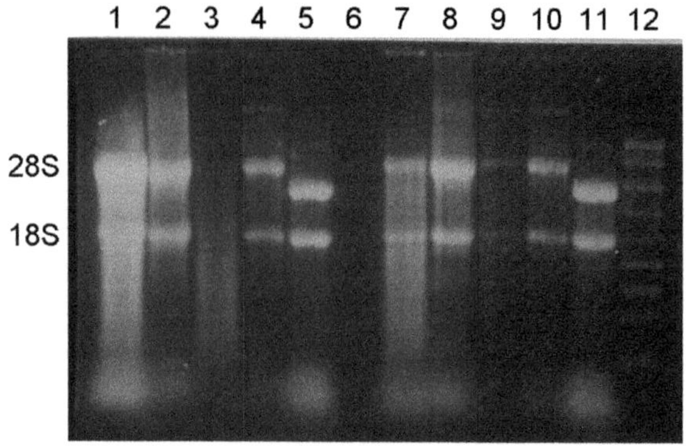

FIGURE 14.5 Comparison of eukaryotic RNA profiles from different species. Samples of RNA purified from CHO-K1 cells (lanes 1–4 and 7–10) and yeast (*Saccharomyces uvarum*; lanes 5 and 11) on a denaturing 1.2% agarose gel. Lane 1: Total cytoplasmic RNA, purified by nonionic lysis—good quality, though the lane is overloaded. Lane 2: Total cellular RNA purified using guanidinium-acid-phenol—fair quality, with evidence of high molecular weight hnRNA. Lane 3: Poly(A)$^+$ RNA—sample is degraded. Lane 4: Poly(A)$^-$ RNA—good quality. Lane 5: Total yeast RNA, isolated by acid-phenol—good quality. Lane 6: Empty. Lane 7: Total cytoplasmic RNA, purified by nonionic lysis—poor quality, showing signs of degradation. Lane 8: Total cellular RNA purified using guanidinium-acid-phenol—good quality, with evidence of high molecular weight hnRNA. Lane 9: Poly(A)$^+$ RNA—good quality. Notice the trace amount of carryover 28S and 18S rRNA that demonstrate the integrity of the sample. Lane 10: Poly(A)$^-$ RNA—good quality. Lane 11: Total yeast RNA, isolated by acid-phenol—good quality. Lane 12: Single-stranded RNA molecular weight standard from Promega. Notice the shift (greater mobility) in the 28S rRNA in the yeast sample in lanes 5 and 11, compared to the CHO-K1 samples. Notice also the presence of hnRNA and the precursor 45S rRNA in lanes 2, 4, 8, and 10.

ribosomes. The appearance of these bands is convincing evidence of the integrity of the sample, that is, that the sample has not been degraded. In high-quality mammalian RNA, the fluorescence intensity of the 28S rRNA is about twice that of the 18S rRNA. It is interesting to note, however, that this is not the case for all eukaryotes; in lower eukaryotes, with yeast as an example, the ratio approaches 1.0. The ratio also decreases when higher eukaryotic RNA begins to degrade. In addition to the 28S and 18S rRNAs, an intact RNA sample manifests its mRNA component as a subtle smear that appears slightly above, between, and just below the rRNAs. This is the normal appearance of cellular mRNA because of its extremely heterogeneous nature and low abundance overall, which is usually less than 3% of the total RNA mass in the cell. A more contemporary method for assessing RNA quality is the calculation of an RNA integrity number, or RIN, which is discussed in detail in Chapter 8, Quality Control for RNA Preparations.

To achieve maximum separation of larger molecular weight RNAs, electrophoresis is sometimes allowed to continue to the extent that the small 5S rRNA, 5.8S rRNA, and tRNA species actually run off the distal edge of the gel and into the running buffer. This is acceptable because the 5S and 5.8S rRNA species are too small to be useful as molecular weight markers for cellular RNA, as they generally migrate through the gel along with the tRNA at the leading edge of the electrophoretic separation, in the 300- to 400-base range. Although this approach is certainly not recommended when evaluating a new experimental model, after a while, it will become clear to the investigator whether the resolution can be improved by extending the electrophoresis time.

The electrophoretic profile of plant RNA from green tissue is often quite different from that observed upon electrophoresis of animal cells. In addition to the large and small rRNAs a number of rather abundant other bands are visualized upon staining. These are chloroplast transcripts (Fig. 14.6), the presence of which is taken as an indicator of the integrity of the sample. As long as there is little or no smearing in the lower half of the lane, the RNA will probably support any of a number of downstream applications. While the stability of certain plant transcripts is legendary, it is wise to run a gel just to be sure that the RNA is intact, especially if more than six months have elapsed since the RNA was last examined.

Samples of oligo(dT)-purified poly(A)$^{+}$ RNA commonly show trace amounts of carryover rRNA, the banding intensity of which may or may not support size calibration. The very fact that the rRNAs are visible at all is strong evidence of the integrity of the sample. One common mistake made

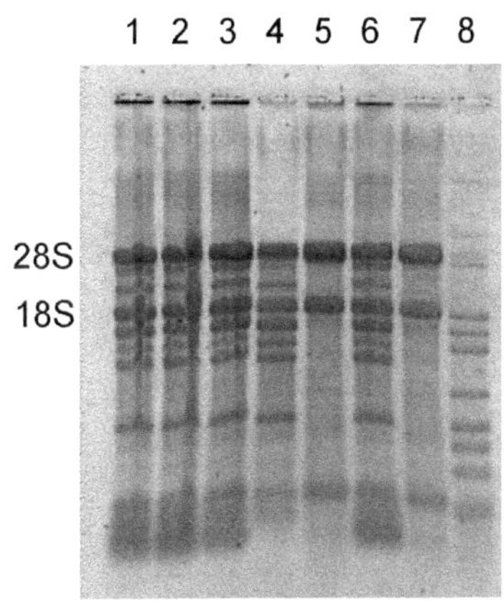

FIGURE 14.6 Electrophoretic profile of total cellular RNA from Morning Glory, *Ipomoea* sp. Ten micrograms of RNA from each tissue were subjected to standard denaturing electrophoresis at 50 V for 3 h. The gel was stained for 30 min in SYBR Gold. Lane 1: First leaf (induced). Lane 2: First leaf (vegetative). Lane 3: First leaf (induced). Lane 4: Cotyledon (induced). Lane 5: Root (induced). Lane 6: First leaf (vegetative). Lane 7: Root (vegetative). Lane 8: RNA molecular weight standard (Promega). In addition to the 28S and 18S rRNA, lanes 1–4, and 6 (RNA from aerial tissue) manifest a number of very abundant chloroplast-specific transcripts, a normal feature of RNA isolated from green tissue in plants. In contrast, the RNA samples from root (lanes 5, 7) exhibit only the 28S and 18S rRNAs. Remarkably, the RNA seen here had been stored at −80°C for more than 16 years at the time of electrophoresis (isolated in 1986 by Debra Mohnen). This high-quality RNA was subsequently used successfully in a transcript start site mapping study.

during the selection of poly(A)$^+$ mRNA is the discarding of the poly(A)$^-$ material, which consists primarily of rRNA and tRNA. If this fraction were reprecipitated and electrophoresed along with experimental samples, the poly (A)$^-$ fraction would provide suitable size standards (presence of rRNA) and also act as an excellent negative control because the depletion of poly(A)$^+$ mRNA should allow the investigator to demonstrate that hybridization signals observed in other lanes are not artifactual or otherwise indicative of nonspecific hybridization. This is especially important when the RNA species of interest are located very close to either the 28S or 18S rRNA. An excellent gel-loading strategy is to run, in order, total RNA, poly(A)$^+$, and poly(A)$^-$ RNA, as is frequently performed in this laboratory (Fig. 14.7).

Samples in which there has been almost any degree of degradation usually fail to manifest the characteristic formation of the 28S and 18S rRNAs and light smearing of the mRNA. Obliterated samples appear as heavy smears which are localized almost entirely below the level at which the 18S rRNA bands in intact samples (Chapter 8: Quality Control for RNA Preparations, Fig. 8.5B). Occasionally, the investigator will observe fairly uniform, heavy smearing along most of the length of a lane on a gel. Within this smear the 28S rRNA and 18S rRNA are usually apparent, seeming to be in the correct positions in the lane, albeit "behind" the smear. There are three possible

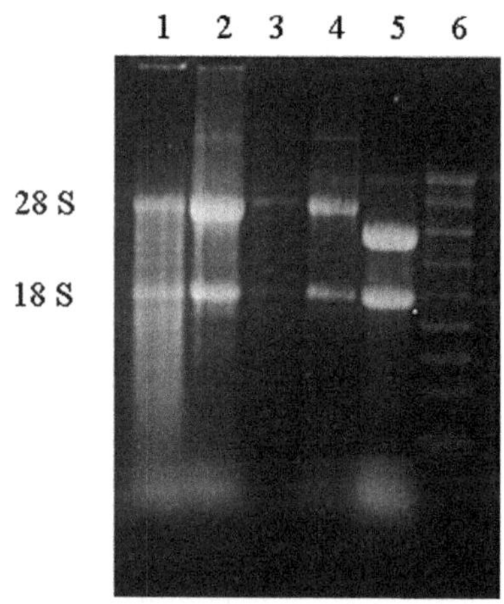

FIGURE 14.7 Comparison of total cellular RNA, poly(A)$^+$ RNA, and poly(A)$^-$ RNA from different species. Samples of RNA from CHO-K1 cells (lanes 1–4) and yeast (*Saccharomyces uvarum*; lane 5) were electrophoresed on a denaturing 1.2% agarose gel. Lane 1: Total cytoplasmic RNA, purified by nonionic lysis—poor quality. Lane 2: Total cellular RNA purified using guanidinium-acid-phenol—good quality, with evidence of high molecular weight hnRNA. Lane 3: Poly (A)$^+$ RNA—good quality, with traces of the major rRNA species. Lane 4: Poly(A)$^-$ RNA—good quality. Lane 5: Total yeast RNA—good quality. Lane 6: RNA molecular weight standard from Promega. The 28S rRNA in yeast is smaller than in mammals, and thus has greater mobility, compared to the electrophoretic migration of the 28S transcript in the CHO-K1 samples.

explanations for an RNA sample with such an appearance: (1) the RNA was not denatured properly; (2) the aliquot is not completed dissolved in buffer; or (3) the RNA is dissolved but the lane was overloaded with too much RNA. Failure to completely denature an RNA sample is the most common occurrence which may involve incompletely dissolved RNA. This occurs (1) when large amounts of very pure RNA are precipitated and *not* dissolved in a sufficient quantity of sterile H_2O or buffer to produce a concentration of 5 μg/μL or less; or (2) when a small-to-moderate amount of RNA was air-dried or dried in a SpeedVac and has not/will not go back into solution. Obstinate RNA samples can be "helped" back into solution by heating for 15 minutes at 65°C, by adding more water, or both. In order to maintain the integrity of the sample and its utility in downstream applications, be sure to avoid heating the sample repeatedly. Whatever process is decided upon, it is wise to test it first on a small aliquot of the RNA, rather than risking the entire sample.

Another bit of information that can be conveyed by examining an aliquot of RNA comes from the appearance of the well into which the sample was loaded. Fluorescence coming from within the well suggests that genomic DNA is present in the sample. Enormous fragments of chromosomal DNA, generated by shearing forces often associated with the mechanics of RNA isolation, are unable to enter the gel during the course of the electrophoresis (Fig. 14.8). Further, because UV spectrophotometry does not distinguish between pure samples of RNA and DNA-tainted samples, normalization based on A_{260} may be compromised due to the contribution to the total mass of the sample made by genomic DNA. When other than optimal conditions are suggested by the appearance of a sample on a gel, it is wise to determine the full extent of the problem as well as the quality of the remaining RNA before going on to any subsequent experiments involving a dubious preparation. If DNA tainting is a chronic problem, the symptoms are alleviated by routinely incorporating an incubation step with DNase I (RNase-free, of course), as described in Appendix C, which is standard procedure in many laboratories, including this one.

There is no reason why commercially available size standards cannot be used along with internal rRNA size standards. Having knowledge of the sizes of the 28S and 18S rRNAs, one need only to measure the distance that each of these species has migrated; recall that only two points are needed to draw a straight line and construct a reasonable size calibration curve and, these days, image analysis software instantly performs these measurements by video capture of a stained wet gel. If image analysis software is not available, it is quite useful to place a fluorescent millimeter ruler somewhere on top of the gel to facilitate measurements, and one should ensure that the placement of the ruler is consistent from one gel to the next. Thus, the UV light is used to generate an image (e.g., SYBR Green or Gold, ethidium bromide), which will also cause the fluorescent ruler to appear in the picture as well (Chapter 4: RNA Isolation Strategies, Fig. 4.5). Of course, older style Polaroid photographs can also be scanned and analyzed if no other options

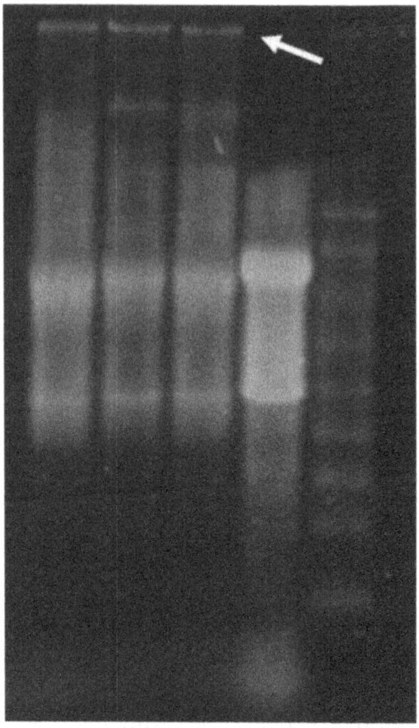

FIGURE 14.8 Presence of contaminating genomic DNA. The fluorescence apparent in the wells of this RNA gel (white arrow) are indicative of high molecular weight DNA that has carried over from the RNA isolation protocol. DNA can be removed from an RNA sample by incubation with DNase I, as described in Appendix C.

exist. Regardless of the method used, it is essential to have photographic documentation of even the most preliminary data.

In the absence of image analysis software, a reasonable size calibration curve can be constructed manually (circa 1980) with a couple of data points, a ruler, and a sheet of semilog paper. Briefly, a size calibration curve can be constructed by plotting the $\log_{10}$ of the sizes of the RNA standards against the distance migrated from the origin (the well into which the sample was loaded). A size calibration curve generated in this fashion can then be used to ascertain the sizes of RNA species observed directly in the gel or that may subsequently be detected by nucleic acid hybridization.

The utility of the ribosomal species as markers is completely dependent on knowledge of the sizes of these RNAs either from the literature or from prior empirical determination. It is unwise to proceed in this manner unless the molecular weight of the ribosomal RNAs is known for the species under investigation. There is considerable variability in the sizes of the rRNAs even among mammals and, in particular, with respect to the 28S rRNA. There is much less variation observed among 18S rRNAs. In general, 18S rRNAs range from 1.7 to 1.9 kb, while the range of 28S rRNAs is about

4.6–5.2 kb (Table 14.2). As a supplementary tool, the location of the tracking dye, most commonly bromophenol blue, may be used as an indicator of the progress of the migration of the samples. While performing electrophoresis, bromophenol blue generally runs with the leading edge of the migrating sample. Another dye, xylene cyanol, runs just behind the 18S ribosomal RNA subunit in 1.2% agarose gels. Xylene cyanol is not used routinely in this lab because of its inconvenient co-migration with the 18S rRNA, and 500–600 bp PCR products, often obscuring the fluorescence coming from the bands of interest. Succinctly, the use of xylene cyanol confers no practical advantages and only makes quantification more difficult. Get rid of it.

When characterizing a sample by Northern analysis for the first time it is judicious to continue electrophoresis just until the bromophenol blue has traveled three-fourths along the length of the gel. This will produce a very reasonable separation of RNA in medium size gels (12 cm × 14 cm), which will simplify accurate size determination of hybridized species while ensuring that small RNAs of possible interest are not lost over the distal edge of the gel into the running buffer.

The reliability of molecular weight data is only enhanced when measurements are derived from both internal and external standards, especially if the RNAs of interest are very large or very small. This is because the logarithmic separation of molecules in a gel is not linear over the entire length of the gel. Very large or very small RNAs that end up outside the linear range of the gel and whose sizes are derived from markers that are within the linear portion of the gel may be sized inaccurately. When external markers are used, it is likely that at least one of the multiple species will electrophorese close to the RNA of interest; if that RNA is outside the linear range of the gel, then some of the bands that collectively make up the marker will run outside the linear range as well. This translates into fewer errors in size determination.

TABLE 14.2 Common rRNA molecular size standards.

Size marker	Bases	Organism
28S rRNA	5025	*H. sapiens*
18S rRNA	1868	*H. sapiens*
28S rRNA	4712	*M. musculus*
18S rRNA	1869	*M. musculus*
28S rRNA	3898	*D. melanogaster*
18S rRNA	1976	*D. melanogaster*
23S rRNA	2904	*E. coli*
16S rRNA	1541	*E. coli*

Gel staining options

Visualizing nucleic acids postelectrophoresis requires staining with RNA- or DNA-specific stains, respectively. Nucleic acid stains vary widely with respect to sensitivity, toxicity, cost, and required method of disposal. A number of the stains and staining techniques described here provide a very suitable balance among these issues and remain in widespread use in the lab. Be sure to read all safety information first in order to prevent potentially harmful exposure.

Ethidium bromide

Ethidium bromide (2,7-diamino-10-ethyl-9-phenyl-phenanthridinium bromide; EtBr) remains a widely used dye for the visualization of nucleic acids in agarose gels. It is a phenanthridinium intercalator, structurally similar to propidium iodide. EtBr binds DNA with no apparent sequence preference once every 4–5 base pairs. SYBR Green and SYBR Gold are excellent alternative dyes for staining (discussed below) owing to greater sensitivity and reduced background when compared with EtBr. The issue of *when* to apply the dye is an occasional source of intralaboratory debate. Not only can EtBr be added to the gel or used to stain the gel at the conclusion of the electrophoresis, but EtBr can also be added directly to the RNA sample prior to loading the well(s), and otherwise omitted from the gel matrix itself. This latter approach cuts back on the amount of EtBr waste that is generated and will limit the fluorescence to the lanes in which a sample is loaded. Briefly, prepare a 1 μg/mL stock of EtBr in nuclease-free water. Add 2 μL of EtBr to 18 μL heat-denatured RNA and mix by pipetting up and down, and then load the gel as usual.

Before electing to use EtBr as a staining tool, it is incumbent upon the investigator to weigh four key disadvantages associated with the use of EtBr into this decision:

1. Ethidium bromide is a well-known, powerful mutagen (MacGregor and Johnson, 1977). Great care must be exercised when preparing and manipulating ethidium bromide stock solutions and dilutions thereof. Contaminated buffers must be treated as toxic waste and equipment likewise treated as such. Methods for EtBr inactivation and disposal are described in Appendix F.
2. The presence of ethidium bromide in a gel or added to samples just prior to loading the gel will retard the electrophoretic migration rate of nucleic acids by about 15%. Although this may not seriously delay the progress of an investigation, it should by no means be used as an excuse to increase the applied voltage to compensate for this phenomenon. Remember that slower-running gels demonstrate enhanced resolution.
3. Continuous exposure to ambient light or, especially, to UV light will nick the RNA, making it difficult to work with. Further, EtBr will

quickly experience photobleaching on a transilluminator, making visualization and quantitative judgments difficult.

4. The presence of ethidium bromide in a gel will reduce its transfer efficiency. If a traditional method of transfer is utilized, such as passive capillary diffusion (Chapter 16: Northern Analysis), the transfer period may need to be extended for unacceptably long periods. Destaining the gel usually improves the transfer process, even though minute amounts of ethidium bromide remaining in the gel can slow the transfer.

The only advantage of running samples in the presence of ethidium bromide is that the progress of electrophoresis can be monitored by disconnecting the power and briefly irradiating the casting tray upon which the gel is resting by placing it directly on a transilluminator or irradiating it from above with a low-intensity UV light source (remember to use eye and skin protection). If greater separation is required, the gel can be placed back in the gel box and the power supply reconnected for a suitable interval.

If the decision is made to use EtBr then the following may be helpful:

1. Always wear gloves when handling ethidium bromide contaminated solutions, pipette tips, and so forth. In this laboratory, EtBr-contaminated micropipette tips are discarded into conical 15 mL tubes before being placed in the toxic waste container.
2. EtBr is most conveniently prepared as a 10 mg/mL stock solution in water and should be stored foil-wrapped at 4°C. Microfuge tubes containing stock EtBr should be placed in small storage boxes and also maintained at 4°C.
3. Do not exceed a final EtBr concentration of 0.5 μg/mL in the gel, the running buffer, or the staining buffer (in instances when the dye is to be used after electrophoresis).
4. If adding EtBr directly to the gel, the molten agarose temperature should first be cooled to 60°C because EtBr is heat labile. In this instance EtBr should also be added to the running buffer; otherwise, the EtBr will be drawn out of the lower portion of the gel as the electrophoresis progresses (Fig. 14.9). Do not forget to dispose of the residual running buffer properly at the end of the electrophoresis period.
5. Optional, but not recommended: Consider staining only the lane(s) containing the size markers and do so after electrophoresis has been completed. To do this, cut the marker lane(s) directly from the gel with a razor blade and stain the isolated lanes as usual. After photographing, the severed molecular weight standard lane and the rest of the gel can be placed next to each other during Northern transfer. The drawback is that it will not be possible to assess the integrity of the experimental material unless the blotting membrane is reversibly stained (described below).
6. Do not add EtBr to glyoxal gels because this dye will react with the glyoxal.

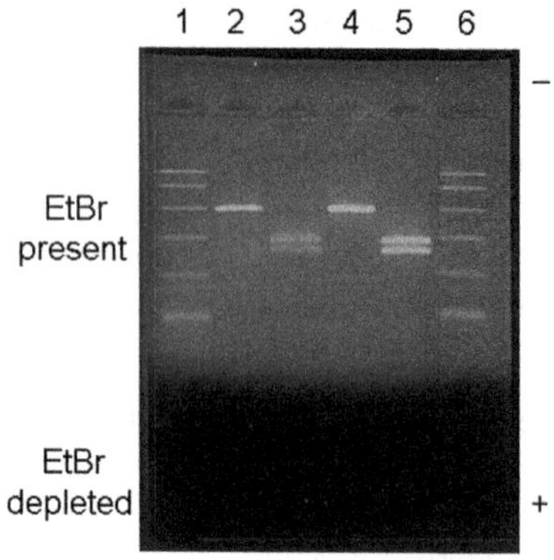

FIGURE 14.9 Effect of adding EtBr directly to an agarose gel. As the sample runs out of the gel toward the anode, the EtBr migrates in the opposite direction, toward the cathode. This has the effect of removing the ethidium bromide from the lower portion of the gel. The longer the electrophoresis continues, the more EtBr will migrate out of the gel. If the dye front happens to pass the sample, some of the fluorescence intensity will be lost, resulting in underestimation of the sample mass. This gel contains a 500 bp PCR product (lanes 2, 4) that has been cut into 273 bp and 227 bp fragments (lanes 3, 5) owing to the presence of a single *Mbo* I restriction enzyme site. The samples are flanked (lanes 1, 6) by a PCR molecular weight standard (Promega).

7. Adding EtBr to the samples alone, rather than to molten agarose, will certainly reduce the amount of the dye required; however, EtBr will still be a transfer obstacle unless it is drawn out of the gel by destaining.
8. Adding EtBr to gels containing formaldehyde can be especially troublesome because formaldehyde gels exhibit very high background fluorescence when stained with EtBr. Such gels often require extensive destaining before even the abundant rRNA species can be observed.
9. Destain gels in the same buffer that was used to stain the gel such as running buffer or nuclease-free H_2O. Alternatively, destain the gel in the buffer that will be used for Northern transfer (5 × SSPE, for example).
10. Make certain that the staining and destaining buffers were at least autoclaved to remove residual RNase. Otherwise, the sample may become partially degraded.

SYBR Green

The nucleic acid cyanine binding dyes SYBR Green I and SYBR Green II are popular stains for in-gel visualization of DNA and RNA, respectively. These dyes are usually provided as a 10,000 × stock solution in DMSO. Just prior to use, SYBR Green is diluted to 1 × in a Tris-containing buffer, usually 1 × TBE, 1 × TAE, or TE buffer. Dilution of SYBR Green in H_2O is not recommended because optimal fluorescence occurs in Tris-based buffer around pH 8. Unlike EtBr, the presence of formaldehyde in a gel does not have a negative influence on the background: SYBR Green will fluoresce only when bound to the sample.

There are many significant advantages conferred through the use of SYBR Green (Table 14.3); however, it should not be added to the gel.

TABLE 14.3 Comparison of ethidium bromide and SYBR stains.

	Ethidium bromide	SYBR stains
Stock solution	10 mg/mL in H_2O	10,000 × in DMSO
Working concentration	0.5–1.0 μg/mL in H_2O	1 × in Tris buffer (pH 7–8) (e.g., 1 × TAE, 1 × TBE or TE buffer)
Binding affinity for nucleic acids	Moderate to high	Moderate to high
Add directly to gel?	Optional	Avoid; always stain gel *after* electrophoresis
Gel background	High	Low to none
Excitation	≈302 nm transilluminator	≈302 nm transilluminator
Color	Pink/Orange	Fluorescent Green (SYBR Green) Fluorescent Gold (SYBR Gold)
Excitation-induced bleaching of the dye	Yes	No
Filtration	No. 15 deep yellow	No. 15 deep yellow
Limit of detection[a]	1.0–5.0 ng/band	60 pg/lane (DNA; SYBR Green I) 0.5–1.0 ng/band (RNA; SYBR Green II) 25 pg (DNA; SYBR Gold) 1 ng (RNA; SYBR Gold)
Mutagenicity[b]	High	Less than EtBr; must be handled with the same degree of caution.
Disposal[c]	Filtration through activated charcoal	Filtration through activated charcoal
Cost	Low to moderate	Moderate to above-average
Staining time	5–10 min	25–30 min
Destaining helpful?	Yes	Limited benefit

[a]*Limit of detection for SYBR Green is improved with epi-illumination (Haugland, 1996) and image capture with a CCD camera or a laser imaging system is best. Denaturants in the gel, particularly formaldehyde, can band detection difficult due to high background.*
[b]*Although the SYBR Green I has been shown to be less mutagenic than ethidium bromide (Singer et al., 1999), it is wise to handle all DNA binding dyes with care.*
[c]*See Appendix F for details.*

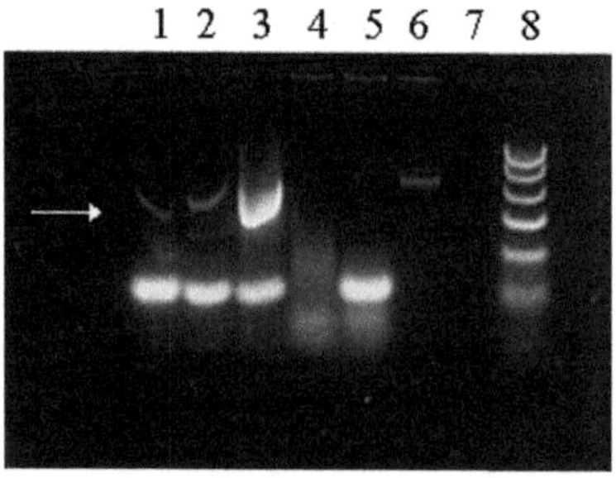

FIGURE 14.10 Effect of adding SYBR Green to agarose before casting the gel. Both DNA and RNA, at nearly all molecular weights, experience profound band distortion, confounding both mass and molecular weight determinations. In contrast, staining gels with SYBR Green or SYBR Gold *after* electrophoresis result in superior staining.

Samples electrophoresed in the presence of SYBR Green, as opposed to staining the gel afterward, routinely show wave-like, irregular-shaped bands in all lanes and at most molecular weights (Fig. 14.10). This makes mass and molecular weight determinations nearly impossible. In this laboratory, various concentrations of SYBR Green, both above and below the recommended working concentration, have been tested by inclusion in the gel. None have provided satisfactory results. While SYBR Green has been used here extensively with remarkable success, gels are always stained after the electrophoresis has been completed. SYBR Green is also compatible with a variety of in-gel manipulations and, as needed, can be removed by standard ethanol precipitation.

SYBR Gold

SYBR Gold is currently the most sensitive nucleic acid gel stain. It is a cyanine dye that can be used to visualize single-stranded RNA, double-stranded RNA, single-stranded DNA, or double-stranded DNA in either agarose or polyacrylamide gels and is comparable in sensitivity to silver staining. In this lab, SYBR Gold has been used successfully in conjunction with a standard 302 nm transilluminator as well as a laser scanner. Like SYBR Green, there is virtually no background fluorescence associated with gel staining, and we routinely stain gels consisting of various agarose concentrations for 30 minutes under subdued light: we actually cover the gel with an old cardboard box while gently agitating the gel in the staining solution at 50 rpm on an orbital platform. Upon transillumination, DNA and RNA appear as gold fluorescent bands. Like SYBR Green, SYBR Gold is provided as a 10,000 × stock solution in DMSO and should be diluted down to a 1 × concentration in either 1 × TAE or 1 × TBE. SYBR Gold is not to be added to the gel prior to electrophoresis for the same reasons for omitting SYBR Green from the gel. Conveniently, the same type of photodocumentation system that has been optimized for SYBR Green-stained gels can be used to photograph gels that have been stained with SYBR Gold.

SYBR Safe

SYBR Safe is a nontoxic alternative to ethidium bromide and SYBR Green I. In fact, SYBR Safe is not classified as hazardous waste under US federal guidelines (ThermoFisher website) and some institutions allow SYBR Safe to be disposed of by pouring it down the drain (check with your Safety Officer first). Of course, common sense dictates that all nucleic acid stains should be handled with a high level of caution. Many investigators use SYBR Safe because the annual cost reduction in hazardous material disposal can be substantial.

SYBR Safe is provided as a 10,000 × concentrate and is diluted in TBE or TAE buffer to a working concentration 1 ×. It is suitable for staining DNA or RNA. Agarose gels may be prepared with SYBR Safe in it or may be stayed after the gel has been run. It should be noted that SYBR Safe added to gels behaves like ethidium bromide, meaning that it migrates in the opposite direction of DNA migration. Thus, there may be clearing or loss of fluorescence near the bottom of the gel. Gels stained with SYBR Safe may be viewed with a standard laboratory transilluminator.

GelStar

Like the SYBR stains described above, GelStar Stain (Lonza Rockland, Inc.) is a fluorescent nucleic acid stain that offers advantages over traditional staining by ethidium bromide. While the technical data for this product indicate that it can be added directly to the gel prior to electrophoresis or stained afterward, in this laboratory the gels are always stained at the end of the run. GelStar is suitable for staining double- or single-stranded RNA or DNA on either agarose or polyacrylamide gels and offers an increase in sensitivity compared to identical gels stained with ethidium bromide. GelStar is stable for at least 12 months at −20°C when stored desiccated and protected from light. It is provided as a 10,000 × concentrate in DMSO and functions best when diluted in a Tris-buffered solution between pH 7.0–8.5. GelStar should be handled with the same care as any of the other DNA binding dyes and disposed of through an activated charcoal filter (Appendix F).

Silver staining

Silver staining is a very sensitive method for detecting small amounts of proteins and low-molecular-weight nucleic acids in polyacrylamide gels. In general, the method is about 100 times more sensitive than Coomassie blue for many proteins and several times more sensitive than EtBr for visualization of polynucleotides. It has the advantages of allowing visualization of the sample without any specialized equipment and rendering gel backgrounds that are virtually colorless. Briefly, after acid fixation, polyacrylamide gels are impregnated with soluble silver ions (Ag^+), followed by reduction to metallic silver. The initial deposit of the insoluble metallic silver initiates the autocatalytic deposit of more silver, manifested ultimately by visualization of the sample. All of the reagents needed for silver staining are readily available in kits from most major suppliers. Important: silver staining of agarose gels is *not* recommended mainly because of the drastically reduced sensitivity in this matrix.

Acridine orange

An older, alternative technique for visualizing nucleic acids molecules in a gel is to stain it with the cationic dye acridine orange (McMaster and Carmichael, 1977), historically used in conjunction with glyoxalated RNA; acridine orange does not exhibit qualities that inhibit nucleic acid transfer from gels to the extent that inhibition is observed with ethidium bromide.

Acridine orange will either (1) bind electrostatically to the phosphate groups of single-stranded molecules and fluoresce red at about 650 nm; or (2) intercalate into double-helical molecules and fluoresce green at about 525 nm. Thus, in the same gel, single-stranded nucleic acid molecules appear red/orange, while double-stranded molecules such as DNA and nondenatured RNA appear green. Given the versatility of this dye, the investigator may also detect incomplete denaturation or partial renaturation of an RNA sample before moving on to subsequent experiments.

Stain agarose gels by soaking them in a solution of 30 μg/mL acridine orange in 10 mM sodium phosphate buffer, pH 7.0 for 30 minutes. Polyacrylamide gels may also be stained, and in as few as 15 minutes. To reduce the very high background produced by staining with acridine orange, gels are destained by soaking them in 10 mM sodium phosphate buffer for 1–2 hours or by running hot water over a glyoxal-stained gel for 20 minutes (be careful not to warp the gel). It should be easy to detect as little as 0.05 μg of double-stranded nucleic acid and 0.1 μg of single-stranded nucleic acid upon UV illumination. The main reason that acridine orange has not found widespread use is because of the immensity of the background in gels so stained, and the requirement for unrealistic, heroic efforts to destain the gel.

Methylene blue

Methylene blue stains both RNA and DNA, though the background in the gel is very high, and the sensitivity of the dye is very poor. So, even after extensive destaining, methylene blue has little use as a DNA or RNA stain beyond simply showing that a band is present in the gel. As such, methylene blue staining is good for high school labs where toxic compounds such as ethidium bromide and SYBR dyes are not allowed or are too expensive. Methylene blue staining kits that may be purchased usually consist of proprietary formulations that produce less ambiguous staining than methylene blue alone. Methylene blue staining is not in the mainstream; superior sensitivity is achievable with SYBR Green and SYBR Gold.

As another alternative to the approach of staining nucleic acids in the gel with ethidium bromide or with one of the SYBR dyes, nylon or polyvinylidene difluoride filter membranes can be reversibly stained with a solution of methylene blue prior to hybridization (Herrin and Schmidt, 1988). Consider the following comparison:

Methylene blue advantages:

1. The RNA can be visualized directly on the filter membrane and the locations of the ribosomal RNA species and other size standards are identified and marked appropriately.
2. The gel can be stained *after* the transfer has been completed, rather than before, in order to verify that an efficient transfer has occurred.
3. The absence of stain in the gel (especially ethidium bromide) prior to blotting will facilitate a much more efficient transfer of RNA from the gel.
4. Methylene blue does not pose the potential health hazards and disposal difficulties associated with ethidium bromide and other nucleic acid binding dyes.

Methylene blue disadvantages:

1. Methylene blue staining does not offer nearly the same level of sensitivity as some of the other dyes, chiefly because the background on the filter tends to be unacceptably high.
2. If the electrophoresed samples are degraded or of suboptimal purity, this may not be known until after the time and expense associated with setting up the transfer.
3. It is difficult to remove completely all of the methylene blue from the filter after it has been stained.

Briefly, RNA that has been transferred to nylon should be UV cross-linked with a calibrated light source to the filter paper prior to staining; alternatively, the RNA can be baked onto the filter paper after staining. RNA may be observed directly on a filter paper within 3 minutes when placed in a solution of 0.02% methylene blue and 0.3 M sodium acetate, pH 5.5. If RNA was cross-linked to the filter, the stain is easily removed by washing the filter paper in 1 × SSPE for 15 minutes or in a solution of 20% ethanol/80% H_2O for 3–5 minutes. If the filter was stained before baking the RNA onto it, the stain may be removed from the RNA immediately prior to hybridization by washing the filter in a solution of 0.2 × SSPE; 0.1% SDS for 15 minutes at room temperature, with gentle shaking. Succinctly, methylene blue staining of filters is an interesting, though not very practical, the method in the contemporary research lab.

Safety considerations and equipment maintenance

Safety is the prime consideration when working with electrophoresis equipment and a power supply. Never attempt to manipulate the gel, gel box, or power supply while conducting electrophoresis or while the power supply is plugged into an electrical outlet. Nor should any attempt ever be made to modify the design of either a gel box or power supply; any alterations could present a serious risk to both the operator and unsuspecting colleagues in the lab. Do not operate electrophoresis equipment near sinks or other open water sources.

Well-designed systems make it impossible to open or remove the lid while maintaining the flow of electrical current. In some designs the lid slides off, breaking electrical contact before the interior of the tank is accessible. This permits the safe inspection of the progress of the electrophoresis. The lid must then be replaced to restore electrical connections and resume electrophoresis. Moreover, it is wise to disconnect the leads from the power supply as well as the gel box when accessing the tank and do so only with dry hands or gloves.

Many power supplies are equipped to accommodate two sets of leads from two different gel boxes. Be sure to leave the power supply off until both of the leads have been connected to the power supply and the gel box. Similarly, be sure to turn the power supply off before connecting the second set of leads (another gel box) to a power supply that is already operational; after all, connections have been made properly, the power supply may be turned on again. Be sure to turn the power supply off before attempting to disconnect any of the leads and do so using one hand only. Finally, never take hold of both leads simultaneously; there is a possibility that doing so may "complete the circuit," resulting in electrocution.

Electrophoresis is literally a daily ritual in most molecular biology laboratories. While good quality electrophoresis gel boxes and power supplies can represent an investment of thousands of dollars, it is an unfortunate, if not potentially dangerous, practice to ignore the upkeep of the instrumentation. Basic preventative maintenance of electrophoresis apparatuses is cost-effective and ethically compelling. All equipment for performing electrophoresis, including gel boxes and power supplies, come with suggested maintenance procedures that should be followed diligently and the use of this equipment should be guided by common sense.

The proper maintenance and use of electrophoresis gel boxes and power supplies are important, though often overlooked part of lab safety. It is clearly the responsibility of the operator to inspect wires, connections, and electrodes on a regular schedule *and* immediately prior to use to prevent potentially life-threatening injuries. All power supplies are to be plugged into properly grounded electrical outlets, as per the manufacturer's safety warnings that accompany the purchase of a new power supply, and *never* use an extension cord. Most power supplies are designed for indoor use only and should never be allowed to get wet. If the instrument does get wet, unplug it immediately and allow it to dry completely, and then contact the manufacturer prior to re-use.

Perhaps the most common problem is the exposure of bare copper wire at the junction of the power cords, electrode leads, and plug ends. This situation poses a severe shock hazard. Replacement cords are usually available at a nominal cost from the manufacturer or distributor, and it is well worth ordering a few extras when the equipment is initially purchased. Be attentive also for any indication of corrosion and also for cracks in the electrophoresis chamber or the lid itself. To be safe, all power cords, insulation, electrodes, connection nuts, and gaskets should be scheduled for inspection on a weekly basis. For a detailed regimen for electrophoresis apparatus maintenance, see Landers (1990). **All servicing should be entrusted to qualified personnel only**.

A few tips for running agarose gels for the first time

The following terms are commonly associated with the preparation of agarose gels and the ensuing electrophoretic process:

- **Band**—The visible accumulation of molecules that have identical molecular weight, as observed upon staining of the gel. The greater the intensity of the band, the greater is the number of molecules that have co-located to that particular position.
- **Casting tray**—The portion of the gel box into which the molten agarose is poured. Gel casting trays are either stationary or removable, depending on the model. The ends of the tray must be sealed, either by taping or through the use of rubber gaskets until the gel has solidified completely.
- **Gel box**—The apparatus in which the electrodes are mounted and in which the actual electrophoretic separation is carried out.
- **Lane**—The linear path that a molecule follows out of the well when an electric field is applied.
- **Loading buffer**—A dense glycerol- or sucrose-based reagent that is added to a sample prior to electrophoresis in order to increase the density of the sample. In addition, loading buffers contain a dye such as bromophenol blue which will allow the investigator to load the samples while the gel is completely submerged in the running buffer. The position of the dye can also be used to track the progress of the electrophoresis. Colloquialisms for loading buffer include "sample buffer" and "blue juice."

- **Power supply**—The electronic device that will generate constant voltage to conduct the electrophoretic separation. The power supply is connected to the gel box after all of the samples have been loaded and the safety cover closed. Only then is the power supply plugged into an electrical outlet and then switched on. Keep in mind that power supplies are potentially very dangerous; investigators using such instrumentation for the first time should seek assistance from qualified, experienced personnel.
- **Running buffer**—The electrolyte solution used to conduct the electric current and in which the gel is submerged in order to dissipate heat. A running buffer is useful only to the extent that it resists pH changes and ion depletion during electrophoresis.
- **Well**—A depression in a gel into which the sample is placed just prior to the onset of electrophoresis. The well is formed by the insertion of the comb into the gel matrix before it has polymerized (polyacrylamide) or solidified (agarose).

A few tips to keep in mind

1. There is a potential for RNase contaminating the electrophoresis apparatus, especially the teeth of the comb while preparing the gel. While nucleases may be inactive in the presence of some denaturants, they can be transferred to and interfere with the latter components of an experiment. If nothing else, the comb should be soaked in a solution of 3% H_2O_2, or in a solution of 0.1% SDS, 50 mM EDTA, and then rinsed with copious amounts of autoclaved water prior to use. Do not use DEPC or 30% H_2O_2, which will severely damage the gel box and may also harm the person using these compounds.
2. A variety of gel boxes are commercially available. Many models are outfitted with rubber gaskets on both sides of the casting tray; this precludes the necessity of taping the tray when pouring the gel (see Fig. E.1 in Appendix). In either case, the ends of the tray should be securely sealed so that molten agarose does not leak out.
3. Watch out for agarose boil-over, due to superheating, especially when melting agarose in a microwave. Boiling agarose can cause serious burns. Allowing the agarose to swell in electrophoresis buffer for 5–10 minutes prior to microwaving will reduce the likelihood of a boil-over. Continue to use caution, however, as boil-over and spillage may still occur.
4. Molten agarose is clear; upon cooling in the casting tray, the gel will take on an opaque appearance. It is important that the casting tray into which molten agarose is poured not be disturbed until it is obvious that the gel has solidified completely.
5. Gels should be cast to a thickness of 0.5–0.75 cm. Gels that are more than 0.75 cm thick will impede transfer efficiency and require exceptionally long blotting periods to elute the entire sample. Staining and destaining thick gels are also rather difficult. To accommodate samples

in large volumes, use combs with wider teeth to increase the capacity of the wells. Standard tooth sizes are 1, 1.5, and 2 mm.

6. Make certain that the casting tray is resting on a level surface before pouring the gel.
7. The best way to cast a gel is to pour the molten agarose solution slowly and evenly into the center of the casting tray before putting the comb in place. Stop pouring as soon as the entire surface of the casting tray has been covered. Alternatively, make a permanent mark on the side of the casting tray as an indicator of how much agarose to pour. In so doing, everyone in the lab will make gels the same way. These methods will generate a gel with an appropriate thickness each time. Then insert the comb without delay. Following these steps in this order will yield a gel of uniform consistency and no air bubbles. If air bubbles do form while pouring the gel, they should immediately be popped or pushed to the side of the tray using a sterile micropipette tip.
8. Agarose solutions containing denaturants (e.g., formaldehyde) should be prepared and cast in a fume hood to minimize exposure and possible health hazards.
9. Be certain to allow enough time for the gel to solidify, and then completely cover it with running buffer *before* attempting to remove the comb. The vacuum forces which are created when the comb is pulled upward are minimized if the gel is submerged in the running buffer. Otherwise, one increases the chances of damaging the bottom of the well and risking the loss of the RNA sample as soon as it is pipetted into the well.
10. If the bottoms of the wells of the gel are repeatedly damaged upon removal of the comb, let the gel "soak" in the running buffer for several minutes to lubricate and loosen the comb.
11. Submerge the gel in only the minimum volume necessary to completely cover the gel. This is because the current will travel the path of least resistance, which is around the gel, rather than through it. This will greatly reduce the electrophoresis rate. However, be sure to look at the edges of the gel that were formed against the sides of the casting tray because agarose has a tendency to "smile" by clinging to the sides of the tray. Failure to completely submerge these portions of the gel can result in overheating and melting of the agarose because there is no running buffer covering it to dissipate the heat. Finally, avoid loading samples into the outside lanes; the nonuniform thickness of those lanes will result in nonconforming sample migration (edge effects).
12. When possible, use small sample volumes for electrophoresis. In the context of this chapter, a small volume will cause the RNA to enter the gel in the narrowest cross-section, thereby maximizing resolution.
13. All gel loading buffers contain a dense material such as glycerol or sucrose which imparts an added density to the sample allowing it to be

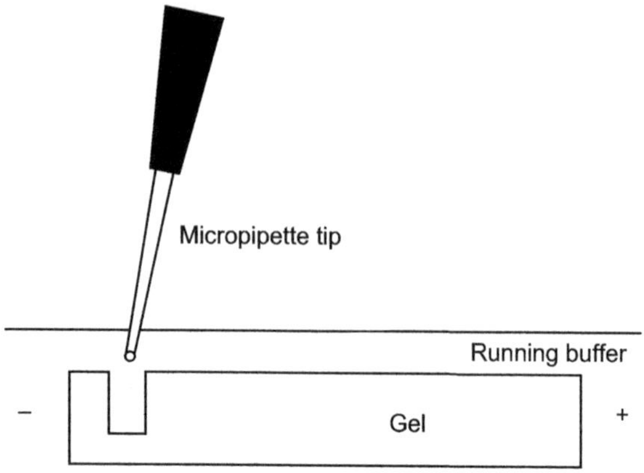

FIGURE 14.11 Proper positioning of micropipette tip for loading samples into the wells of an agarose gel.

more easily loaded into a submerged well. The investigator merely breaks the surface of the running buffer above the well and carefully extrudes the sample from the micropipette tip (Fig. 14.11). If there is difficulty seeing the well because of reflections from the surface of the buffer, it is useful to place a piece of dark tape or a piece of dark paper under the gel box just beneath the wells.

14. All wells should contain a loading buffer of the same composition, that is, don't mix and match loading buffers because high salt concentrations will cause RNA to move slower through the gel, compared to other samples in the gel. This is also the reason why RNA pellets should be washed extensively with 70% ethanol, as described in Chapter 4, RNA Isolation Strategies. Salt can drastically change the electrophoretic mobility of nucleic acids (Fig. 14.12).

15. Be certain to eliminate any air space from the end of the micropipette tip containing a sample, before attempting to load it into the well of a gel. Injecting air into a well ahead of a sample may result in the complete loss of the sample.

16. If, after loading a sample into a well, the sample floats back out, then alcohol from previous precipitation or wash step was not completely removed before resuspending the RNA in an aqueous buffer. The alcohol will need to be removed by precipitating the sample again followed by 70% ethanol washes and complete drying. As long as the RNA remains in residual alcohol, it will be very difficult to load it on a gel.

17. Do not move the gel box after the samples have been loaded into the wells unless the objective is to create a tidal wave that will displace the samples from the wells.

18. Don't even think about touching the power supply or the leads unless your hands, or gloves, are dry.

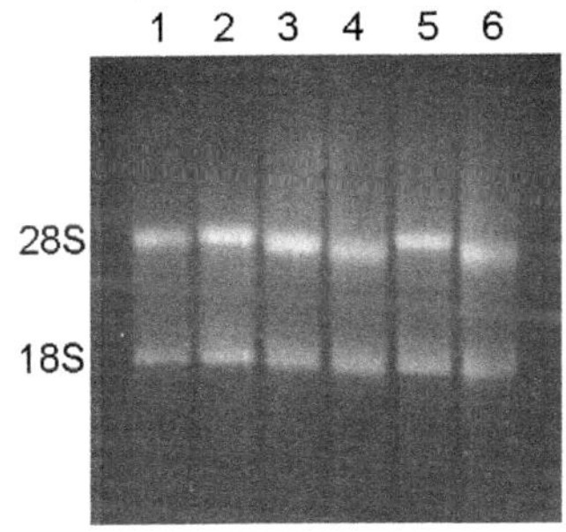

FIGURE 14.12 Effect of salt on electrophoresis of RNA. Several RNA samples from the same cell culture were isolated simultaneously. At the end of the procedure, the RNA pellets were washed with varying quantities of 70% ethanol. As it is expected that the 28S rRNA and 18S rRNA comigrate (line up) from one sample to the next, the obvious differences in electrophoretic mobility are directly related to the fact that the samples in lanes 3, 4, and 6 are in a different ionic-strength environment compared to the samples in lanes 1, 2, and 5. Previously purified RNA can be reprecipitated and washed extensively with 70% ethanol to remove excess salt, as needed.

19. When all samples have been loaded into the gel, make sure that the cover of the gel box is secured, to prevent accidental electrocution. Connect the leads to the power supply, one at a time, and be sure to use gel boxes and power supplies that are appropriately matched. Only then should the power be turned on. Electrophorese at a maximum of 5 V per centimeter of gel length. Because the gel can rest in the gel box in either orientation in many models, make certain that the RNA (with its net negative charge) runs from negative to positive (i.e., toward the anode).

20. When inspecting the gel or terminating electrophoresis, be sure to turn the power off before handling either the gel box or the power supply. Disconnect power leads one at a time and never handles both at once.

21. Never, ever, attempt to load a gel while the power supply is on, even at minimal voltages. This Author painfully recalls a scientist in a workshop who admitted that "loading the gel at low voltage" improved her data. This is a very dangerous technique that is to be frowned upon and discouraged in every instance.

References

Alwine, J.C., Kemp, D.J., Stark, G.R., 1977. Method for detection of specific RNAs in agarose gels by transfer to diazobenzyloxymethyl-paper and hybridization with DNA probes. Proc. Natl. Acad. Sci. 74, 5350–5354.

Alwine, J.C., Kemp, D.J., Parker, B.A., Reiser, J., Renart, J., Stark, G.R., et al., 1979. Detection of specific RNAs or specific fragments of DNA by fractionation in gels and transfer to diazobenzyloxymethyl paper. Meth. Enzymol. 68, 220–242.

Bailey, J.M., Davidson, N., 1976. Methylmercury as a reversible denaturing agent for agarose gel electrophoresis. Analyt. Biochem. 70, 75–85.

Bantle, J.A., Maxwell, I.H., Hahn, W.E., 1976. Specificity of oligo(dT)-cellulose chromatography in the isolation of polyadenylated RNA. Analyt. Biochem. 72, 413–427.

Berk, A.J., Sharp, P.A., 1977. Sizing and mapping of early adenovirus mRNAs by gel electrophoresis of S1 endonuclease-digested hybrids. Cell 12, 721–732.

Boedtker, H., 1971. Conformation independent molecular weight determinations of RNA by gel electrophoresis. Biochim. Biophys. Acta 240, 448–453.

Church, G.M., Gilbert, W., 1984. Genomic sequencing. Proc. Natl. Acad. Sci. 81, 1991–1995.

Dawson, R.M.C., Elliot, D.C., Elliot, W.H., Jones, K.M., 1959. Data for Biochemical Research. Oxford University Press.

Farrell Jr., R.E., Greene, J.J., 1992. Regulation of c-myc and c-Ha-ras oncogene expression by cell shape. J. Cell. Physiol. 153, 429–435.

Favaloro, J.R., Triesman, R., Kamen, R., 1980. Transcriptional maps of polyoma virus-specific RNA: analysis by two-dimensional nuclease S1 gel mapping. Meth. Enzymol. 65, 718–849.

Fornace, A.J., Mitchell, J., 1986. Induction of B2 RNA polymerase III transcription by heat shock: enrichment for heat shock induced sequences in rodent cells by hybridization subtraction. Nucleic Acids Res. 14, 5793–5811.

Gruenwedel, D.W., Davidson, N., 1966. Complexing and denaturation of DNA by methylmercuric hydroxide. I. Spectrophotometric studies. J. Mol. Biol. 21, 129.

Haugland, R.P., 1996. Handbook of Fluorescent Probes and Research Chemicals, sixth ed. Molecular Probes, Inc, Eugene, OR.

Herrin, D.L., Schmidt, G.W., 1988. Rapid, reversible staining of Northern blots prior to hybridization. Biotechniques 6 (3), 196–200.

Hollander, M.C., Fornace Jr., A.J., 1990. Estimation of relative mRNA content by filter hybridization to a polythymidylate probe. Biotechniques 9, 174–179.

Hutton, J.R., Wetmur, J.G., 1973. Effect of chemical modification on the rate of renaturation of deoxyribonucleic acid. Deaminated and glyoxalated deoxyribonucleic acid. Biochemistry 12 (3), 558–563.

Johnson, D., 1975. A new method of DNA denaturation mapping. Nucleic Acids Res. 2, 2049–2054.

Junghans, R.P., 1983. A review of the toxicity of methyl mercury compounds with application to occupational exposures associated with laboratory uses. Environ. Res. 31, 1–31.

Kevil, C.G., Walsh, L., Laroux, S., Kalogeris, T., Grisham, M.B., Alexander, J.S., 1997. An improved, rapid Northern protocol. Biochem. Biophys. Res. Commun. 238, 277–279.

Landers, T., 1990. Electrophoresis apparatus maintenance. Focus 12 (2), 54.

Lehrach, H., Diamond, D., Wozney, J.M., Boedtker, H., 1977. RNA molecular weight determinations by gel electrophoresis under denaturing conditions: a critical reexamination. Biochemistry 16, 4743–4751.

Locker, J., 1979. Analytical and preparative electrophoresis of RNA in agarose-urea. Analyt. Biohem. 98, 358–367.

MacGregor, J.R., Johnson, I.J., 1977. In vitro metabolic activation of ethidium bromide and other phenanthridinium compounds: mutagenic activity in Salmonella typhimurium. Mutation Res. 48, 103–108.

McMaster, G.K., Carmichael, G.C., 1977. Analysis of single- and double-stranded nucleic acids on polyacrylamide and agarose gels by using glyoxal and acridine orange. Proc. Natl. Acad. Sci. 74, 4835–4838.

Melton, D.A., Krieg, P.A., Rebagliati, M.R., Maniatis, T., Zinn, K., Green, M.R., 1984. Efficient in vitro synthesis of biologically active RNA and RNA hybridization probes from plasmids containing a bacteriophage SP6 promoter. Nucleic Acids Res. 7, 1175–1193.

Rave, N., Crkvenjakov, R., Boedtker, H., 1979. Identification of procollagen mRNAs transferred to diazobenzyloxymethyl-paper from formaldehyde gels. Nucleic Acids Res. 6, 3559–3567.

Reijnders, L., Sloof, P., Sival, J., Borst, P., 1973. Gel electrophoresis of RNA under denaturing conditions. Biochim. Biophys. Acta 324, 320–333.

Sambrook, J., Russell, D.W. (Eds.), 2001. Molecular Cloning: A Laboratory Manual. third ed. Cold Spring Harbor Laboratory Press, New York.

Sharp, P.A., Berk, A.J., Berget, S.M., 1980. Transcriptional maps of adenovirus. Meth. Enzymol. 65, 750–768.

Singer, V.L., Lawlor, T.E., Yue, S., 1999. Comparison of SYBR Green I nucleic acid gel stain mutagenicity and ethidium bromide mutagenicity in the Salmonella/mammalian microsome reverse mutation assay (Ames test). Mutation Res. 439, 37–47.

Spohr, G., Mirault, M.E., Imaizumi, T., Scherrer, K., 1976. Molecular-weight determination of animal-cell RNA by electrophoresis in formamide under fully denaturing conditions on exponential polyacrylamide gels. Eur. J. Biochem. 62, 313–322.

Thomas, P.S., 1980. Hybridization of denatured RNA and small DNA fragments transferred to nitrocellulose. Proc. Natl. Acad. Sci. 77, 5201–5205.

Wicks, R.J., 1986. RNA molecular weight determination by agarose gel electrophoresis using formaldehyde as denaturant: comparison of RNA and DNA molecular weight markers. Int. J. Biochem. 18, 277–278.

Zinn, K., DeMaio, D., Maniatis, T., 1983. Identification of two distinct regulatory regions adjacent to the human β-interferon gene. Cell 34, 865–879.

Further reading

Bonini, J.A., Hofmann, C., 1991. A rapid, accurate, nonradioactive method for quantitating RNA on agarose gels. Biotechniques 11 (6), 708–710.

Correa-Rotter, R., Mariash, C.N., Rosenberg, M.E., 1992. Loading and transfer control for Northern hybridization. Biotechniques 12 (2), 154–158.

Masek, T., Vopalensky, V., Suchomelova, P., Pospisek, M., 2005. Denaturing RNA electrophoresis in TAE agarose gels. Anal. Biochem. 336 (1), 46–50. Available from: https://doi.org/10.1016/j.ab.2004.09.010.

Chapter 15

Photodocumentation and image analysis

Rationale

The absolute necessity for maintaining accurate records of experiments performed and the resulting data is indisputable. Electrophoresis gels contain information critical for accurate interpretation of the outcome of every experiment performed; these invaluable data often suggest subsequent experiments that may give further definition to a scientific investigation. Gels deteriorate in a relatively short time, however, and the dyes used to stain nucleic acids in gels deteriorate even more rapidly. This mandates archiving an image of the gel electronically and/or producing a tangible photographic record (an electrophoretogram). This also applies to the archiving of images on X-ray film produced by autoradiography or by chemiluminescence detection. A stable, long-term record of results is a must for future review, conference presentations, lab meetings, and publication.

Gel documentation devices and image analysis software are commonplace in the contemporary molecular biology laboratory. At a minimum, most laboratories have access to at least one digital image analysis workstation. Higher-end image analysis systems may come with a steep price tag, though even the most modest of laboratories are usually able to procure very basic units; there is a gel documentation system to fit almost any budget. The traditional photograph generated on Polaroid film has become mostly a historical footnote.

Safety first

UV transilluminators and handheld UV light sources are both sources of intense, dangerous ultraviolet radiation. Exposure to UV light from these standard laboratory instruments, which commonly emit UV radiation at 254 nm (short wave), 302–312 nm (medium wave), or 365 nm (long wave) can quickly cause serious damage to the cornea, retina, and other structures of the eye, not to mention damage to exposed skin. Fortunately, most of the image analysis stations sold today are designed such that the transilluminator is contained entirely within the workstation, and the UV light will not

RNA Methodologies. DOI: https://doi.org/10.1016/B978-0-323-90221-2.00005-9

function unless the door is closed and sealed. This intelligent design prevents the accidental exposure of the user and bystanders to UV light unless someone foolishly bypasses the safety mechanism.

With older-style gel documentation methods, one is generally at much greater risk for exposure to UV light because of the basic design of the instrumentation: one must raise the cover of the transilluminator in order to access and photograph the gel or even simply point to a specific band for the benefit of a colleague. Because the damage is cumulative, minute amounts of exposure can be catastrophic in both the short- and the long term. Be sure to wear gloves, protective eyewear, and a face shield at all times, even when using transilluminators with a UV protective cover. Although these covers usually block most of the UV light, there may still be a small amount of transmission. Moreover, cracks, improper placement, and open areas permitting direct exposure to UV light are all serious potential health and safety risks.

In the case of older-style manual gel photography, a gel rests on the surface of a UV transilluminator on the lab bench. Although the gel is usually covered entirely by the hood of a handheld camera, the entire transilluminator surface often is not covered. The UV light should be switched on immediately prior to physically taking the picture and should be turned off immediately thereafter, even before removing the exposed film from the camera, to prevent you and other personnel in the room from being exposed to dangerous UV light. Forgetting to turn off the UV light source sometimes occurs when the room lights are on while pictures of the gel are being taken. Exposed sections of the transilluminator surface, no matter how small, pose a significant danger to the operator and should be covered at all times, Moreover, the transilluminator surface becomes hot very quickly. One should also be aware that not all safety glasses are UV rated; it is very important to utilize proper safety equipment when a UV transilluminator is used for any purpose and for any amount of time. Finally, it is important to avoid direct contact with the dyes with which the gels are impregnated, as many DNA and RNA binding dyes carry significant mutagenic potential. Although a commonplace procedure in the molecular biology laboratory, the potential dangers associated with photodocumentation are not to be misjudged or underestimated.

Digital image analysis

Traditional methods for the interpretation of photographs of gels (electrophoretograms), including visual inspection and densitometry, have been abandoned in favor of a plethora of computer hardware and software for rapid, reproducible image analysis. Useful measurements that are now automated include the determination of molecular weight, concentration, relative abundance, integrated optical density (IOD), and image saturation. Electronic archiving also provides a convenient means of cataloging and sharing stored images. An image analysis solution most often consists of a completely self-contained system (Fig. 15.1). Alternatively, image analysis software can be purchased separately and used in conjunction with currently existing instrumentation (Fig. 15.2). Many of the functions, such as identifying lanes, finding bands, determination of mass and molecular weight, calculation of relative abundance among two or more bands, and just about any other user-defined macro can be automated. Although limited input is required for use, the capacity for complete override of all automated functions allows maximum flexibility. Inasmuch as meaningful image analysis requires an understanding of the basic terminology and the benefits and deficiencies associated with this approach for analyzing gels and archiving data, a short overview of this technology and practical concerns are presented here.

Image analysis may be defined as the presentation and quantification of images. An image is a visual representation of an object, and image processing manipulates information within an image to make it more useful. Digital image processing is a specific type of image processing performed with a computer. The digitization process divides an image into a horizontal grid, or array, of very small regions, each of which is called a "picture element" or pixel (Fig. 15.3). In silico, an image is represented by this digital grid or bitmap. A pixel usually represents a very small region within an image, often 1/300th of a square inch, or less. Each pixel in the bitmap is identified by its position in the grid, referenced by its row (x coordinate) and column (y coordinate) number.

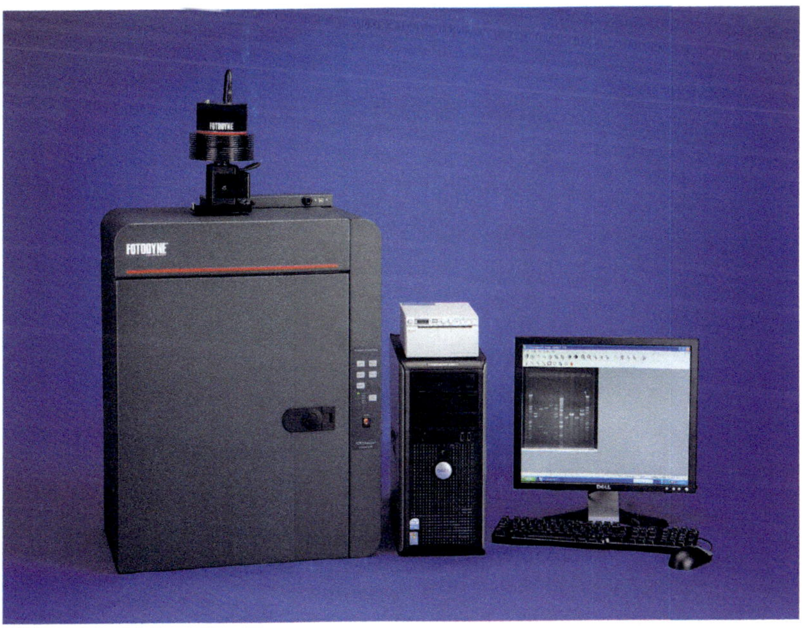

FIGURE 15.1 FOTO/Analyst Luminary/FX workstation. *Courtesy of FOTODYNE.*

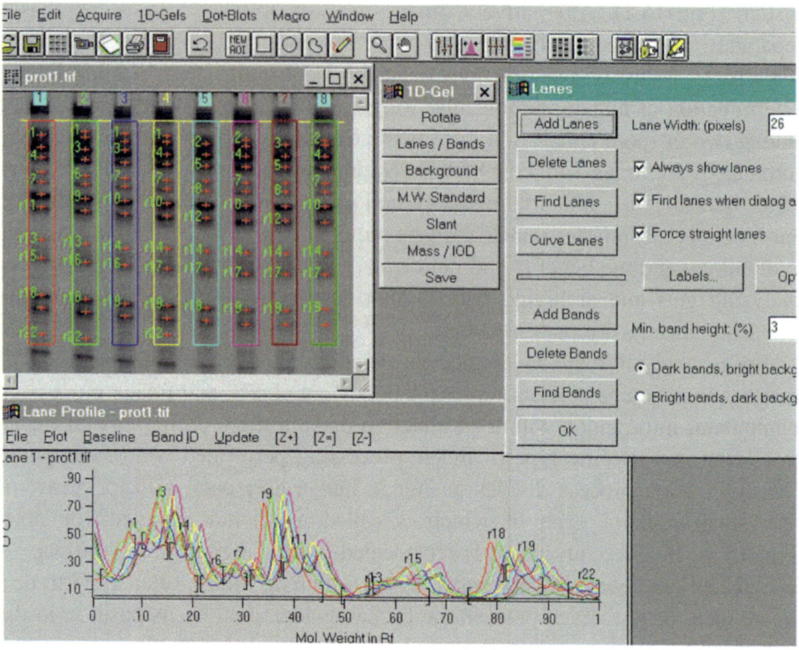

FIGURE 15.2 Example of image analysis software. *Gel analysis image courtesy of Media Cybernetics (www.mediacy.com).*

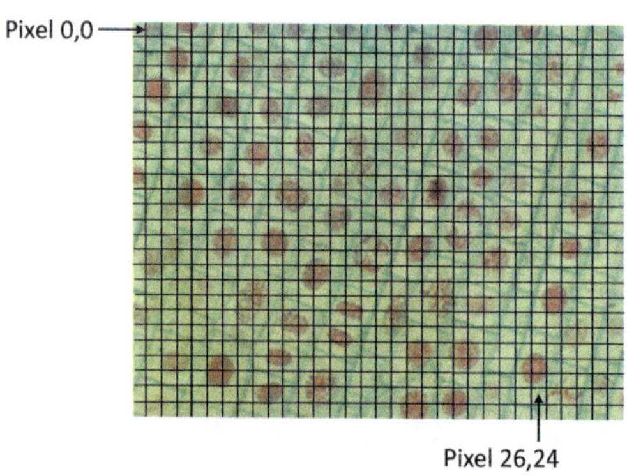

FIGURE 15.3 Pixel bitmap. *Courtesy of Media Cybernetics.*

By convention, pixels are referenced from the upper-left position of the bitmap, which is considered position 0, 0 (row 0, column 0).

When an image, such as a photograph of a gel, is digitized, it is examined in a grid fashion. This means that each pixel in the image is individually sampled, and its brightness is measured and quantified. This measurement results in a value for the pixel, usually an integer, which represents the brightness or darkness of the image at that point. This value is stored in the corresponding pixel of the computer's image bitmap. When the image is digitized, the width and height of the array are chosen and fixed. Together, the bitmap's pixel width and height are known as its spatial resolution.

Depending on the capability of the measuring hardware and the complexity of the image, anywhere from 1 to 32 bits might be used to store each pixel value. Pixel values for simple images, such as line drawings, containing only black and white information, are represented by a single bit: 0 = black, 1 = white. In contrast, photographic-like images contain much more information; 24 bits are needed to represent all the possible colors that might occur in a true color image. The number of bits used to present the pixel values in an image is referred to as its pixel depth, or bits-per-pixel (BPP). The number of BPP used to represent each pixel value determines the image's class.

Although the bit depth (BPP) indicates how many unique colors an image can possess, it does not indicate which colors are actually contained within the image. Color interpretation is determined by bit depth or image class, which includes grayscale 8, grayscale 12, and grayscale 16. Grayscale pixel values represent a level of grayness or brightness, ranging from completely black to completely white. This class is sometimes referred to as "monochrome." In an 8-bit grayscale image, a pixel with a value of 0 is completely black, and a pixel with a value of 255 is completely white. A value of 127, for example, represents a gray color exactly halfway between black and white (medium gray), and a pixel value of 64 has a gray color halfway between medium gray and black. Although grayscale images with bit depths of 2, 4, 6, 12, 16, and 32 exist, 8-BPP grayscale images are the most common because (1) its 1-byte-per pixel size makes it easy to manipulate; and (2) it can faithfully represent any grayscale image because it provides 256 distinct levels of gray. By comparison, the human eye can distinguish fewer than 200 gray levels. It is important to keep in mind that emphasis is given to monochrome-type images because electrophoresis images are typically presented as white bands against a dark background or dark bands against a white/gray background. However, many investigators prefer gel images which feature dark bands against a light background (Fig. 15.4) because of perceived improvement in sensitivity and overall image quality.

Image enhancement

Often it becomes necessary to enhance an image of a gel in order to maximize the amount of useful information that can be derived from it. Although there are numerous methods for enhancing images, the three most commonly used in digital image analysis are brightness, contrast, and gamma correction.

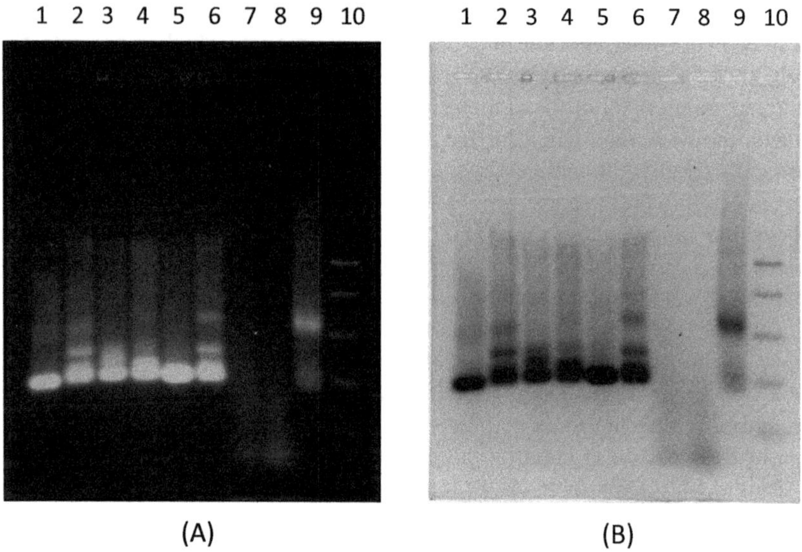

FIGURE 15.4 Image inversion. While the most common style of gel image presents white bands against a dark background (A), some investigators prefer to record dark bands against a light background (B) because of a perceived improvement in image quality. Lanes 1–6, various PCR products associated with a transcription start site mapping investigation; lanes 7–8, negative controls; lane 9, positive control; lane 10, size standards. Most digital image analysis workstations allow the user to toggle easily between these two formats.

- **Brightness** is a term used to describe the overall amount of light in an image. When the brightness is increased, the value of every pixel in the image is increased, moving each pixel closer to 255, or white. When the brightness is decreased, the value of each pixel is likewise decreased, moving it closer to 0, or black.
- **Contrast** is a term used to denote the degree of difference between the brightest and darkest components in an image. An image has poor contrast if it contains only harsh black and white transitions or contains pixel values within a narrow range. For example, an image with values ranging from 100 to 140 would have poor contrast. An image has good contrast if it is composed of a wide range of brightness values from black to white. The amount of the intensity scale used by an image is called its "dynamic range." An image with good contrast will have a broad dynamic range.
- **Gamma correction** is a specialized form of contrast enhancement that is designed to enhance contrast in the very dark or very light areas of an image. It does this by changing mid-tone values, particularly those at the low end, without affecting the highlight (255) and shadow (0) points. Gamma correction can be used to improve the appearance of an image or to compensate for differences in the way different input and output devices respond to an image. Gamma correction involves applying

standard, nonlinear γ curves to the intensity scale (Fig. 15.5). A γ value of 1 is equivalent to the identity curve, which has no effect on the image. An increase in the γ value ($\gamma > 1$) will generally lighten an image and increase the contrast in its dark areas. A decrease in the γ value ($\gamma < 1$) will generally darken the image and emphasize contrast in the lighter areas.

Another broad category of operations frequently used to collect data from gel images is known as density analysis. Optical density (OD) analysis is a common image processing application used to determine the amount of matter in a material by measuring the amount of light it transmits (passes through it). Because OD analysis measures the amount of light passing through a material, OD measurements are meaningful only in the analysis of images that have been captured with the light source radiating from *behind* them; for example, gels UV-irradiated from beneath on a transilluminator rather than by UV bulbs mounted above the gel. OD measurements are not useful in the analysis of images captured under reflected light. Molecular biologists are perhaps more familiar with OD in the context of UV spectrophotometric determination of nucleic acid concentration, as described in Chapter 8, Quality control for RNA preparations.

In general, OD of a band or other element of a digitized image assumes an exponential decay of light inside the transmitting material and is calculated as

$$\text{Optical density } (x,\ y) = \frac{-\log\ (\text{intensity } (x,\ y) - \text{black})}{\text{incident} - \text{black}}$$

where "intensity (x, y)" is the intensity at pixel (x, y), "black" is the intensity generated when no light goes through the material, and "incident" is the intensity of the incident light (the light falling on a object).

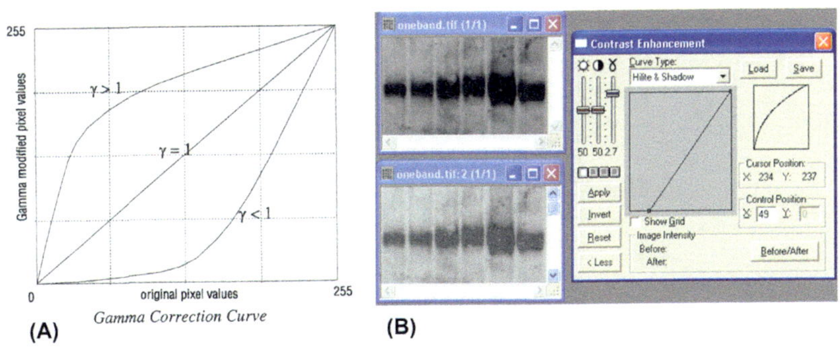

FIGURE 15.5 Gamma correction curve. (A) Effect of γ curves on pixel values 0 through 255. (B) Original Northern blot image (upper panel) and the same image γ correction (lower panel). Note how the applied γ correction of 2.7 has lightened the image while increasing the contrast of the bands.

A material through which no light is transmitted has an infinite OD, and one through which all light is transmitted has zero OD. Thus, the OD scale is inversely related to the intensity scale, that is, on an OD scale, dark pixels produce high values, and light pixels produce low values.

IOD is the sum of pixel values minus a background value for each pixel within a user-defined area of interest. In a given area, the increased size of a band or the increased brightness of a spot relates to higher IOD measurements. IODs are very useful when comparing band intensities on images, within or among lanes, in which mass standards have not been included.

Filtration

Unless the bands stand out in an image of a gel, that image will not convey a useful level of detail. Since most electrophoresis photography involves black and white film or grayscale image analysis, adequate contrast is essential. It should be noted, however, that color photographs of SYBR Green- and SYBR Gold-stained gels can be stunning. Regardless of the camera or film used, appropriate filters enhance detail and make electrophoresis bands appear more sharply defined. Without filtration, not only will contrast be lacking, but the UV bulbs from the transilluminator will be visible in the resulting image (Fig. 15.6).

A filter absorbs its own complementary colors and transmits its own color, allowing the photographer to manipulate the relative brightness of colors that would otherwise appear similar in black and white. For example, a red filter makes green and blue appear darker, and red appears relatively

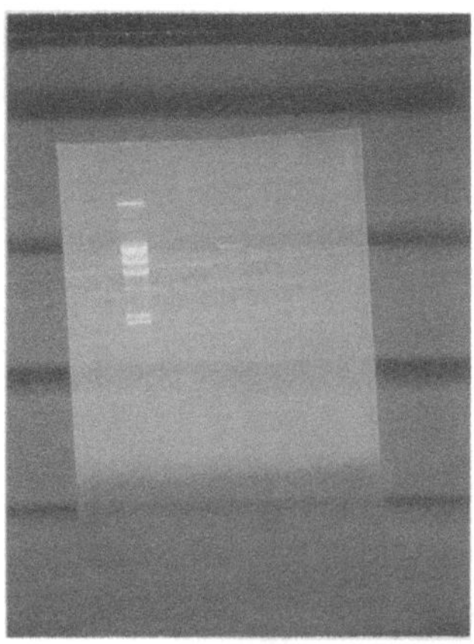

FIGURE 15.6 Image of an agarose gel without the use of an appropriate filter. The UV bulbs from the transilluminator are plainly visible, detracting greatly from the quality of this electrophoretogram.

lighter. A useful mnemonic by which to remember pairs of complementary colors and, thereby, the correct filter to use, is as follows:

"Red Cadillac by General Motors"

that is

Red Cadillac: red and cyan

by: blue and yellow

General Motors: green and magenta

Table 15.1 lists some of the common gel electrophoresis stains for nucleic acid and protein gels, the corresponding filters that offer the best results and, if needed, approximate aperture setting and shutter speeds. Because filters absorb light, it is necessary to ensure that the gel is properly illuminated; this means uniform irradiation from beneath. Burned-out UV bulbs in transilluminators should be replaced by qualified personnel as soon as possible. To an extent, one may compensate for the effect of a filter by increasing the exposure time, using a wider lens aperture, or both.

TABLE 15.1 Common electrophoretic stains and the corresponding color filters.

Light source (wavelength)	Stain	Filter	Aperture	Shutter speed[a]
UV (302 nm)	Ethidium bromide	#22 Deep orange or #15 Deep yellow	f/8	1/4 s
UV (302 nm)	SYBR Green or SYBR Gold	#15 Deep yellow	f/5.6	1/2 s
Epi-illumination (254 nm)	SYBR Green or SYBR Gold	#15 Deep yellow	f/5.6	1/2 s
White (400–700 nm)	Coomassie blue	#8 Medium yellow or #9 Deep yellow	f/16	1/30 s
White (400–700 nm)	ELISA	#8 Medium yellow	f/16	1/30 s
White (400–700 nm)	Silver	#58 Green	f/16	1/30 s

[a]*Shutter speed pertains only to the use of traditional instant films. In the case of electronic video capture, the effectiveness of filtration and image integration must be determined empirically. See text for details.*

Image formats

Digitized images are usually stored on a shared hard drive so that they can always be recalled for further analysis. It is strongly recommended that (1) each user has a personal folder primarily for organizational purposes and (2) the entire hard drive should be backed up regularly. Digital images may occupy a large amount of disk space, so it is usually best to crop the image as much as possible, though obviously not to the detriment of the interpretation of the data. When images are stored on a disk, they are stored in a particular format. These formats vary mostly in the types of image data that they support and the types of compression that they make available to reduce the storage size of the image. Digitized images of gels are most commonly stored as "TIFF" files (*.tif), an acronym for "tagged image file format" (also, "tagged information field format"). Some of the other common image file formats include BMP (bitmap), JPEG (joint photographic experts group), GIF (graphics interchange format), and several others. Once stored, these images can be retrieved and printed as any other file would be, placed into Word documents or PowerPoint slides, posted on websites, and transmitted electronically via email.

Practical considerations

The primary concern of most scientists considering the acquisition of image analysis hardware and software is "How much?" Not surprisingly, there is enormous variation in the format of equipment, the variety of functions that can be performed, the system requirements, the amount of supporting hardware required for the system to run and, of course, the cost. As suggested above, many systems are completely self-contained, are elegant to behold and have a great many functions preprogrammed. Because some of these systems are expensive, this type of instrumentation is often purchased at the department level and available to everyone in a common-use area. Other digital image analysis products are designed to be loaded directly onto a computer already in the laboratory. Beyond the purchase price of the image analysis software, the only other investment required is the purchase of some type of image acquisition devices, such as a scanner or a digital (charge-coupled device, CCD) camera. One point that is abundantly clear is that, at all levels of sophistication, the capabilities of image analysis software such as superior speed and efficient handling of data easily justify the cost. On several occasions, at the request of the manufacturer, this lab has been happy to host short visits by researchers from neighboring institutions who would like to see the instrumentation before purchasing it. This is a common courtesy to extend to colleagues in the scientific community.

Anyone who has ever used image analysis software appreciates the automation of the otherwise tedious and time-consuming determinations of the location, mass distribution, and molecular size of molecules distributed in a gel. Digital image analysis favors the reproducibility of data and reduces intra-laboratory error because subjectivity is removed; this aspect of image analysis is especially important because, upon visual inspection, most people tend to overestimate the IOD of bands on gels and autoradiograms. Further, image analysis software is often fully compatible not only with images of gels, but also with images on X-ray films, chromogenic detection methods, or other signals in either random patterns or in any type of geometric format.

A common misunderstanding about image analysis software pertains to the perceived ability of image analysis software to improve upon the image itself. One must realize first and foremost that if the investigator scans or otherwise produces a poor image of a gel, then the image analysis software has a poor image with which to work. For example, if a gel was not electrophoresed for a sufficiently long period, the software cannot further resolve the bands. A good deal of successful image analysis depends on the quality of the image provided by the investigator.

Background correction is another area of concern, particularly when high-precision quantitation of band mass and total lane mass is required. For example, an insufficiently destained, high-background gel image is likely to present significant challenges with respect to band analysis. This important aspect of image analysis is not without intrinsic difficulties, the most challenging of which is the fact that the background is never uniform in a gel or on an image. Instead, a gradient of background exists in most gel images from left to right and top to bottom. As such, there are three basic approaches to dealing with background issues:

1. Examine numerous small areas in the image, to deal with "local background" only, that is, the so-called "area of interest."
2. Position the cursor at some neutral location on the gel, sample the background at that point, and then perform a global background subtraction.
3. Position numerous background sampling symbols at various points in the image and subtract the measured background from the nearest lane in a gel. This method offers the greatest degree of accuracy and versatility for the user.

Finally, the potential exists for obscuring the true quantitative differences between bands under investigation on the same gel. Recall that 255 is the largest possible pixel value, and areas within an image containing these values are said to be saturated. While pixel values below 255 will continue to increase with longer exposures, these will also become saturated eventually. Because longer exposure times cannot increase pixel values beyond 255, calculations on bands containing saturated pixels are likely to generate data that are lower than expected, compared to nonpixel-saturated bands. Quantitative measurements are therefore compromised when saturated pixels are involved. Digital image analysis products typically alert the user, usually by turning a portion of the image into a different color (often red or blue), when some or all of the image falls into the saturation range. This allows the user to adjust the image intensity so as to avoid what this Author likes to describe as "saturation-induced misinterpretation."

Most image analysis systems also allow image integration, a feature that will enhance low-intensity bands. Integration of an image occurs by simply capturing the image and then overlaying the same image as many times as the investigator wishes. To better understand integration, imagine taking a photograph in a low light setting: as a result, many of the details of the subject matter are difficult to see. Now, imagine making an infinite number of transparent copies (referred to as frames) of this same photograph and then overlaying the copies onto the original image one by one. This is integration. As a result, the obscured details of an image become apparent, and the investigator may stop the integration process when satisfied with the enhanced appearance of the image. While this is a wonderful feature to have available for image enhancement, it is important to understand that background also increases with the number of applied frames, as do those components of the image that were easy to see in the original image. Integration is a useful tool but must be used judiciously.

One common mistake that this Author has observed repeatedly is that two different images with different degrees of integration are compared directly in order to determine the relative abundance of multiple PCR products distributed among the two gels. If one of the gel images was produced with twenty frames of integration and the other gel was produced with five frames of integration, then a direct gel-to-gel comparison is completely inappropriate. Bands that are to be compared directly for quantification or normalization are to be electrophoresed on the same gel. Though still suboptimal, an alternative would be the normalization of the bands on one of the gels to a mass standard on the same gel, followed by the same type of correction on the other gel(s) to allow for some degree of gel-to-gel comparison.

Many laboratories use electronic lab notebooks/recordkeeping exclusively, although some investigators are comfortable printing out a hard copy of each gel image for inclusion in a physical laboratory notebook. Computers are convenient, but they also crash from time to time; many people consider having a hard copy of a gel image to protect against a catastrophic loss of data to be an absolute necessity. The take-home lesson: image analysis software is a highly efficient and timesaving tool but having a photographic hard copy of a gel is just as important. In this laboratory, digital images are printed out on the very day they are generated so that a hard copy can be placed in a laboratory notebook, in addition to archiving the image on a disk or a hard drive. Old Polaroid photographs can also be scanned, digitized, recalled, analyzed, distributed, and likewise archived electronically.

Biomolecular imagers

Digital images are acquired by direct video capture from wet gels using a CCD camera, by scanning a previously generated photograph, or by using a biomolecular imager platform. Biomolecular imagers are laser-based detection devices which are considered to be next-generation imaging instrumentation. They have a wide linear dynamic range, far exceeding that of traditional gel imaging platforms. This allows the observation and quantification of very weak bands and very strong bands, even when both are present on the same gel. Biomolecular imagers may exhibit multiple functionalities, including OD measurements, red, green, and blue (RGB) fluorescence detection, near-infrared (NIR) fluorescence detection, and phosphorimaging. With respect to this latter application, phosphorimaging is a well-established alternative to autoradiography and is discussed in that context in Chapter 17. Images of gels produced by a biomolecular imager or any type of gel documentation system should be stored in the investigator's personal folder on a shared hard drive.

Traditional methods of photodocumentation

In the traditional sense, photodocumentation refers to the investigator physically taking a picture of a gel soon after electrophoresis and staining. In the past, this was most commonly performed using Polaroid instant imaging products. It was announced in late 2013 that the instant films commonly used for generating electrophoretograms were being permanently discontinued. This was due to a steady decline in demand for this type of film as a result of the explosive growth of digital imaging worldwide. However, Polaroid-compatible instant films are still available through various distributors; as such, a short overview of instant film photodocumentation is presented here for the benefit of molecular biology novices who may still be using this type of equipment, at least for now.

The traditional systems for photodocumentation included the handheld Polaroid DS-34 camera (Fig. 15.7), which was prevalent for years, and the much older-style Polaroid MP-4 + Multipurpose System. The DS-34 system was an extremely versatile, low-cost instrument that was a staple in many early molecular biology labs: the investigator would take a picture of a gel, tape it into a laboratory notebook, and then annotate the picture appropriately. Today, digital photodocumentation imaging systems usually have an attached printer, thereby allowing the investigator to print a hard copy of the gel on heat-sensitive thermal paper. Alternatively, the image is simply sent to an online laser printer. Early on, thermal prints lacked definition, though improvements have been made. The high-gloss thermal paper produces particularly attractive gel images when hard copies are needed. Traditional photographs of gels are prepared in black and white.

FIGURE 15.7 Polaroid DS-34 handheld camera. Snap-on hoods allow photodocumentation of a variety of gel sizes and they also position the camera for sharp focus. A built-in lens ensures proper magnification and focal length. *Courtesy of Polaroid.*

Camera settings

Nucleic acids are commonly visualized by staining them with SYBR Green, SYBR Gold, ethidium bromide, or other dye. Gels are then UV-irradiated from below by placing them on the surface of a standard laboratory transilluminator and photographed or scanned. Digital imaging systems often have the option of epi-illumination, that is, UV irradiation from above the gel, which may enhance sensitivity. The gel, showing the distribution of the DNA or RNA, is then photographed. Digital systems are generally focused manually and include a zoom feature to allow the investigator to photograph the entire gel or focus on a region of interest. Polaroid systems featured snap-on hoods to accommodate gels of several dimensions; the hoods produced sharp electrophoretograms by maintaining the camera at the correct focal distance from the gel and precluding the need to focus.

All image analysis systems allow the user to set the amount of time that the shutter in the lens will be open (the shutter speed) as well as the aperture setting (i.e., the *f*/stop, how wide the lens will open). If one is uncertain as to the correct parameters, try photographing ethidium bromide-stained gels for 0.5 seconds at an aperture of *f*/8; corrections can be made as needed. For SYBR Green- or SYBR Gold-stained gels, one should begin with 0.5 seconds at *f*/5.6. Of course, the exact settings will be empirically determined for each gel and are directly related to the mass of nucleic acid in the gel, the staining time, the destaining time (if any), the magnitude of background fluorescence, and whether epi-illumination or direct illumination from beneath the gel is used. Finally, for the nonphotographically inclined, the following should be noted. First, the shutter speed scale found on the lens is labeled "B, 1, 2, 4, 8, 16, 32, and 64." The "B" setting means that the lens will be open as long as the shutter release is depressed by the user. The numbers that follow are the reciprocal of the actual shutter speed [i.e., 2 is actually 1/2 (0.5 seconds); 4 is actually 1/4 (0.25 seconds), and so forth]. Second, the terms "aperture setting" and "*f*/stop" are used interchangeably and refer to how wide the lens shutter will open. Obviously, this setting influences the amount of light that will enter the camera and expose the film; the smaller the number, the larger the opening of the shutter.

Inherent limitations of photographic and X-ray films

In photographic systems, the method of image recording is an energy conversion process whereby the sensitivity of silver halide crystals is used to record the level of photons emitted by or reflected from a source or object. The photons are absorbed by the silver crystal and when the level of energy is sufficient, the crystal forms a deformation known as a latent image site. At this site, the silver and halogen bond of the crystal is broken, resulting in free halogen and metallic silver. If the crystal is exposed to too much energy in too short a time, it will be unable to react fast enough to accurately record the event; likewise, if a low level of energy strikes the crystal for a prolonged time, then the crystal will react differently than it would under normal conditions.

At some point, the crystal will be exposed to sufficient energy to fully saturate it with metallic silver. Beyond this point, it can no longer record further information, becoming uniform in response at whatever level of response that system can produce. In the case of Polaroid film, a system referred to as direct positive response, this level of saturation relates to the lowest density, known as D-Min. When the crystal reaches such a level of low absorption of photon energy that it no longer is capable of producing a latent image, it acts as if no energy has been absorbed. This produces a uniform response of maximum density, known as D-Max. Between the D-Max and D-Min are regions where the silver crystal is recording the highlight and shadow details, in a nonlinear manner, as the crystal begins to approach the linear portion of its ability to record the energy absorption. The region of useful recording between the highlight and shadow region of the film is known as the dynamic range of the film. While ideally the film and the photon source would have the same range, thus allowing the system to record in a true-to-life manner, it is often the case that the photon source exceeds the range of the film. Thus, the user of the imaging system must choose which areas are of greatest importance. By adjusting the exposure of the system, the investigator defines the range of the image that will be recorded, while losing highlights and shadows outside the dynamic range of the film. In the molecular biology laboratory, where both strong and weak bands are often observed in the same image, it is wise to make two or three exposures of the same image, using different aperture settings, to derive as much useful data from a single image as possible. The take-home lesson: traditional silver-impregnated photographic films have a very narrow linear response. This may result in an inability to record the true abundance differences among bands in an image. Fortunately, this problem can be addressed using a contemporary digital imaging platform as described above.

Tips for optimizing electrophoretograms

Many of the following suggestions were developed to optimize the appearance of gel images produced on traditional instant film. The suggestions also have applicability for optimizing the appearance and presentation of digital images and, as such, are included here.

1. *Maximize the size of the image*. An image in which the gel fills the photograph will always offer the greatest level of detail, in terms of both sensitivity and resolution. For the greatest level of detail, zoom in on the bands of interest for maximum clarity.
2. *Minimize the exposure of the gel to UV light*. In general, shorter exposures will provide the sharpest images. If working with ethidium bromide, minimizing exposure of the gel to UV light will also facilitate taking multiple photographs because ethidium bromide is rapidly bleached by UV light to the extent that it will no longer fluoresce. Shorter exposure times will also minimize the nicking of the phosphodiester backbone of nucleic acids because of overirradiation of the gel. This is a very important consideration if a band is to be eluted from the gel for cloning or sequencing. Although *minimal* exposure to UV light randomly nicks the backbone of RNA (and DNA), thereby enabling efficient Northern (or Southern) transfer, excessive nicking will render the sample incapable of probe or primer hybridization in subsequent steps.
3. *Keep the transilluminator surface clean*. Gels should be removed from the surface of the transilluminator immediately after image capture. Gels that are allowed to remain on the transilluminator are often forgotten, only to be discovered days later dried out and stuck to the surface of the transilluminator. Also, the drying of ethidium bromide or other fluorescent dyes contributes to background spots in subsequent photos/image captures. It is wise to wipe the surface of the transilluminator with dH_2O or 70% ethanol and wipe it dry with a Kimwipe immediately after each use. Nothing takes the place of photographing a gel against a clean background.

4. *Standardize digital camera settings.* For the sake of consistency, pay careful attention to the aperture setting, integration setting, and shutter speed. It is important that these settings be standardized in a laboratory to support meaningful comparison from one experiment to the next. Of course, it is understood that occasional adjustments to fine-tune an image may be necessary.
5. *Use correct filtration.* Consider filtration a standard component of photodocumentation. Color filters can improve the contrast of electrophoretograms, as previously described. Ideally, one should use screw-in glass filters whenever possible. Older-style acetate filters can sometimes distort the optical path, and, for this reason, they should be avoided.
6. *Avoid fogging the camera lens.* Pay attention to the amount of time that the transilluminator has been turned on. Because the surface of the transilluminator becomes hot in a short time, some of the liquid from the gel will vaporize and then condense on the camera lens inside the image analysis workstation. This effectively fogs the camera lens and compromises greatly the quality of the photograph (Fig. 15.8). This is easily remedied by wiping the lens dry with a Kimwipe.

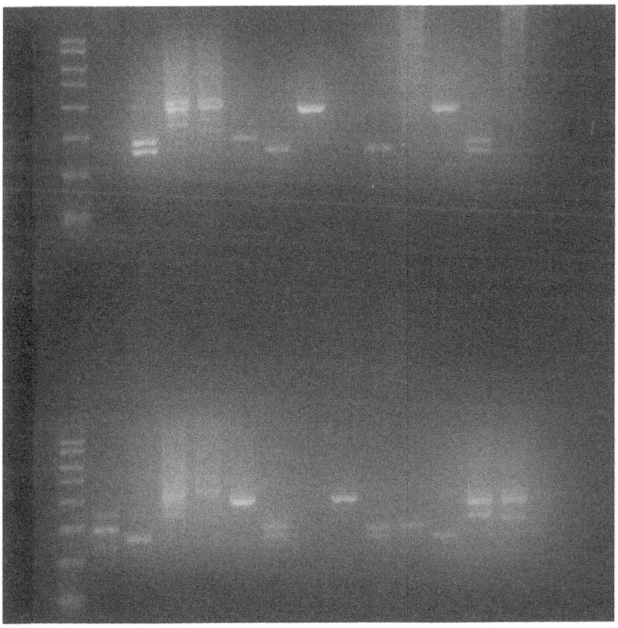

FIGURE 15.8 Effect of condensation on a camera lens. The heat generated by a transilluminator that was turned on too long fogged the resulting image, making interpretation very difficult. The fogging as completely obscured the location of the upper and the lower sets of wells and produced a halo around the bands. Wiping the lens dry with a Kimwipe and reducing transilluminator exposure time will resolve the problem.

7. *Eliminate ambient light.* If using an image analysis workstation, in most cases the transilluminator will not switch on unless the safety door is closed, and the lock engaged. This also serves to protect the user from the dangers of UV light at the same time as preventing ambient light from interfering with the production of the image. If using a handheld camera, be sure to rest the camera hood directly on the lightbox; this will prevent ambient light from reaching the film. It is usually not necessary to turn off the lights in the room with this type of camera. Be certain, though, to turn off the transilluminator immediately after making the exposure.

   In the case of older-style Polaroid films, also pay attention to the following:
8. Use small apertures. The smaller the aperture (the higher the f/stop), the greater the depth of field and the more sharply defined the resulting images. Ideally, one should begin with *f*/16 and optimize by incrementally decreasing the aperture, as much as practical. In this laboratory, ethidium bromide-stained gels are routinely photographed for 1/4 second at *f*/8. SYBR-stained gels are photographed at *f*/5.6 for 1/4 or 1/2 second, adjusted as needed.
9. Follow film development instructions, if applicable. Pay close attention to the development time and temperature recommended for each type of instant film (e.g., Type 57 or Type 667 film). Underdeveloping or overdeveloping film, a common error, can adversely affect the characteristics of the image. In general, the most commonly used black and white films require a 30-second development after the film is pulled out of the camera. Overdevelopment will enhance contrast, and this can be good or bad, depending on the extent of variation in mass among bands on the gel. By increasing the contrast, bands with the greatest mass will stand out even more (Fig. 15.9). However, bands in which the mass of a sample is minimal may begin to blend into the background of the image, leading the viewer to believe that no such bands exist.

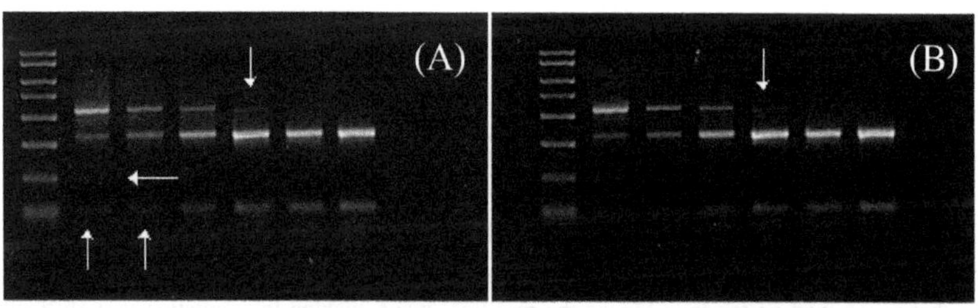

FIGURE 15.9 Effect of overdeveloping Polaroid film. (A) Correct developing time. (B) Overdevelopment of film causes loss of low-intensity bands.

10. Avoid temperature extremes. The ideal development temperature for Polaroid films is room temperature, generally 21°C–24°C (70°F–75 °F). If the lab is much warmer or cooler, or if the film was stored refrigerated until just prior to use, an adjustment in development time or exposure time may be necessary. Refer to the film package insert for guidance.

11. Minimize camera movement. Gels photographed with slower films require long exposure times when stained with ethidium bromide and, in particular, with SYBR Green or SYBR Gold. It is, therefore, more practical to use faster films. To minimize camera movement in this lab, a cable release has replaced the pistol-grip shutter release, which easily unscrews from the handheld DS-34 camera. For image analysis workstations, camera movement is usually not a concern.

12. Keep camera rollers clean. To prevent spotting and other processing defects, clean DS-34 camera rollers periodically with warm water and a cotton swab or Kimwipe. The need to clean the rollers becomes obvious when three or four spots, evenly spaced, appear on the film (Fig. 15.10). The spots are the result of debris on the rollers, which places uneven pressure on the film as it is pulled from the camera.

13. Recover the film properly. A film that has been exposed is to be pulled straight out of the camera, slowly and evenly, and perpendicular to the user. The film should not be pulled upward or downward as it is removed from the camera. It has been suggested that the user intone "Pol-ar-oid,"

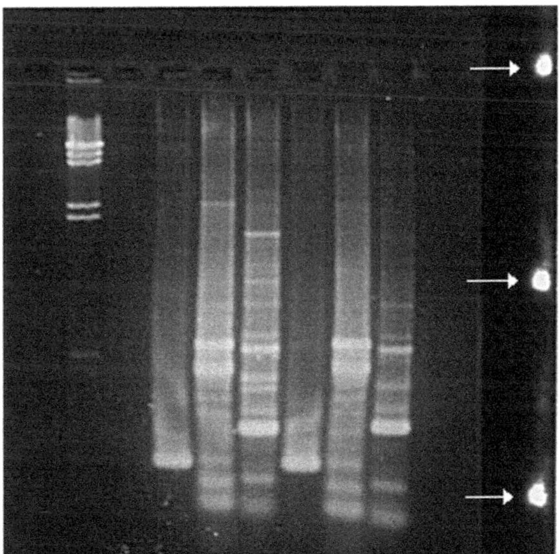

FIGURE 15.10 Effect of dirty rollers in a Polaroid camera. Notice the three heavy white spots, evenly spaced, along the right-hand edge of the image. When the rollers are cleaned, the spots will go away.

the cadence of which is used as a vocal guide to be followed as the film is pulled. When the recommended developing time has elapsed, turn the film-receiver sheet complex over, take hold of the corner of the film, where the number is printed, and peel it away from the receiver sheet, which is impregnated with the chemistries required for image development. Some films do not require any type of coating; the film may be slightly arched at first but will flatten out within 4–5 minutes. Do not touch the negative itself or the sheet to which it is attached: both contain an alkali that should not come into contact with skin.

Further reading

Baxes, G.A., 1994. Digital Image Processing – Principles and Applications. John Wiley & Sons, New York, NY.

Castleman, K.R., 1996. Digital Image Processing. Prentice-Hall, Inc, Englewood Cliffs, NJ.

Chellappa, R., 1992. Digital Image Processing. IEEE Computer Society Press, Los Alamitos, CA.

Green, M.R., Sambrook, J., 2020. Autoradiography and phosphorimaging. Cold Spring Harbor Protocols . Available from: https://doi.org/10.1101/pdb.top100446.

Medberry, S., Gallagher, S., Moomaw, B., 2004. Overview of digital electrophoresis analysis. Current Protocols in Molecular Biology . Available from: https://doi.org/10.1002/0471142727.mb1005s66Unit 10.5.

Pratt, W.K., 1991. Digital Image Processing, 2nd ed. John Wiley & Sons, New York, NY.

Russ, J.C., 1995. The Image Processing Handbook. CRC Press, Boca Raton, FL.

Shorte, S.L., Frischknecht, F. (Eds.), 2007. Imaging Cellular and Molecular Biological Functions. Springer-Verlag, Berlin, Germany.

Trussell, H.J., Vrhel, M.J., 2008. Fundamentals of Digital Imaging. Cambridge University Press., Cambridge, MA.

Chapter 16

Northern analysis

Rationale

Many variations on the theme of Northern analysis have emerged since the original descriptions of the technique (Alwine et al., 1977, 1979). Succinctly, Northern analysis is the transfer, or blotting, of electrophoretically separated RNAs from a gel to a filter membrane, also known as "the membrane" or the "transfer membrane," for subsequent fixation and hybridization to gene-specific probes. Each RNA species is transferred from the gel to the membrane exactly as it was size distributed in the gel by the electrophoresis process, thereby keeping the RNA concentrated in a relatively small area. Hybridization signals generated by Northern analysis produce a qualitative and semiquantitative profile of the sample: one gains a snapshot of mRNA size *and*, within the limits of the sensitivity of the assay, approximate abundance. Northern analysis is compatible with RNA from any source and continues to be a useful technique for assessing mRNA length and the expression patterns of moderate- to high-abundance transcripts. While Northern analysis is a comparatively low-sensitivity assay, it actually has the advantage of being a non-PCR-based method and, sensitivity limitations notwithstanding, can be used as an independent method for validating PCR-generated data.

There are various methods for isolating RNA, preparing and running denaturing agarose gels, and transferring the nucleic acid from the gel. Choices must be made in advance by the investigator regarding the type of blotting membrane, method of immobilization, type of probe, method of probe labeling, hybridization recipe and parameters, and method of detection. It is wise to outline the process in advance so as to avoid surprises partway through the assay. Optimization of Northern analysis comes from experience, and often involves "tweaking" several points throughout the process. When optimized in one's own laboratory, many of the technical difficulties and issues pertaining to reproducibility resolve themselves. Words to live by: SOPs are important.

The terminology "Northern" analysis is something of a pun, since the original method for electrophoresis and blotting of DNA was coined "Southern" blotting in honor of the person who developed it (Southern, 1975). Unlike the structure, gene organization, and gene copy number data

RNA Methodologies. DOI: https://doi.org/10.1016/B978-0-323-90221-2.00022-9

associated with Southern analysis, Northern analysis is all about gene expression, that is, the transcriptional activity of one or more genes. Previous chapters have described methodologies for RNA isolation, quantification, and electrophoresis. This chapter examines the options regarding the choice of blotting membranes, transfer techniques, and methods for nucleic acid fixation on the filter. Hybridization and detection methods are described in Chapter 17, Nucleic acid probe technology.

Choice of blotting membrane

One important decision associated with the study of RNA by Northern analysis is the selection of the membrane to which electrophoretically resolved RNA is transferred (blotted) prior to hybridization. This selection is guided by the knowledge that different types of membranes exhibit wide variations in nucleic acid binding capacity and support different transfer and immobilization methodologies. Consideration must also be given to the method of probe detection (autoradiography or chemiluminescence) because some membranes are incompatible with nonisotopic detection methods. Further, because the RNA sample on a filter paper may well represent labor-intensive efforts, many investigators perform sequential hybridizations using a battery of probe sequences. This requires, after the detection process, the removal of each hybridized probe with a high-stringency wash; some types of nucleic acid membranes withstand the removal of probes much better than other types. Suggestions for the proper handling of nucleic acid blotting membranes are described in detail by the manufacturer of the blotting membrane and some general guidelines are described here.

Membranes for nucleic acid blotting can be purchased precut to match the size of gels that are used in the lab. Alternatively, membranes are often available as three-meter rolls, allowing the user the flexibility to custom cut a membrane, as needed. In all cases, the membranes are sandwiched between two sheets of paper that protect them from scratching and from particulate matter. The protective paper is often printed with the name of the manufacturer and the trade name of the product while the membrane itself is the unmarked white sheet in the middle. It is very important to avoid touching the membrane with ungloved fingers, as finger greases will prevent the proper wetting of the membrane (described later), reduce the binding capacity of the membrane, and very likely introduce RNase into the system. As needed, the membrane can be picked up by the corner with nuclease-free forceps or gloved fingers. Laboratories utilizing nucleic acid blotting methods should invest in powder-free gloves because glove powder can contribute to background smudges following hybridization detection. Membranes should be stored in a cool dry location, out of direct sunlight and should not be refrigerated, frozen, or heated above room temperature prior to first use. If stored correctly, membranes are stable for many months.

Nylon

Nylon filters were developed to circumvent many of the difficulties inherent in nitrocellulose and have largely replaced nitrocellulose for most routine nucleic acid blotting applications. Perhaps the most significant advantage associated with their use is the great mechanical strength nylon membranes exhibit. This enables very assertive handling of the filter and repeated probing of the same filter. Nylon blotting membranes, once known as nylon 66, are widely used in the molecular biology laboratory and are sold under a variety of trade names. They fall into two categories: those with a neutral surface charge and those with a net positive charge imparted by surface amines. The latter are frequently referred to as charged, charge-modified, or simply nylon(+) membranes, and are available with varying charge densities, depending on the manufacturer. Like nitrocellulose, nylon filters used for blotting are available in 0.2 and 0.45 μm pore sizes. Nylon filters have also proven to be very compatible with nonisotopic detection technologies. Nylon membranes exhibit an enhanced nucleic acid binding capacity compared to nitrocellulose, in some cases as much as 400–500 $\mu g/cm^2$, depending on the manufacturer and the exact buffers used by the investigator. Since nylon membranes also show a particular affinity for smaller nucleic acid molecules (<500 bases), routine blotting of small molecules such as PCR products can be performed readily. Compared to nitrocellulose, the transfer buffers used in conjunction with nylon filters are of lesser ionic strength, often as low as 5 × SSC or 5 × SSPE when the nylon membrane has a net positive surface charge. While nylon(+) membranes exhibit an increased electrostatic attraction between the RNA sample (recall that RNA has a net negative charge), adequate blocking of the membrane during prehybridization and, if applicable, during the detection process is very important for producing blots with an exceptionally clean background. Failure to prehybridize or block properly, in contrast, will most certainly produce excessive background due to nonspecific interaction between the positively charged membrane and the negatively charged phosphodiester backbone of the probe.

The investigator is afforded much greater latitude when blotting onto nylon membranes. Immobilization of nucleic acids can be accomplished either by baking or by irradiation with UV light (see below). The baking of nylon filters does not require a vacuum oven and can be performed in any instrument capable of generating 65°C–70°C for 1.0–1.5 hours. Best of all, because nylon filters do not become brittle when baked, they withstand repeated probe striping much better than nitrocellulose. Finally, it has been reported that torn nylon filters can be repaired by spot-welding them with a hot metal implement such as a bacterial inoculating loop or soldering pen (Pitas, 1989). In this laboratory, partially torn filters generally do not contribute to background; by plastic wrapping the filter, the adjoining sides of the tear in the filter are usually not noticeable on the X-ray film following detection.

Nitrocellulose

Nitrocellulose is the so-called "classical membrane" of molecular biology, emerging shortly after the original RNA blotting technique described the use of diazobenzyloxymethyl paper for nucleic acid immobilization (Alwine et al., 1977). The underlying chemical basis by which nucleic acids become fixed onto nitrocellulose is hydrophobic in nature. Perhaps the most critical parameter is the need for a very high ionic strength transfer buffer to promote the binding of single-stranded molecules (denatured RNA or denatured DNA). The most common pore size for nitrocellulose used in blotting is 0.2 μm because of its enhanced ability to retain smaller nucleic acid molecules, compared to the alternative 0.45 μm pore size nitrocellulose. Although nitrocellulose is widely used for protein blotting (Western analysis), some of its properties make it less than optimal for the blotting of nucleic acids, namely

1. Nitrocellulose has a relatively low binding capacity for nucleic acids (about 80–100 $\mu g/cm^2$).
2. Nitrocellulose has a relatively poor binding capacity for smaller nucleic acid molecules (typically less than 500 bases). This precludes efficient blotting of most PCR products.
3. Nitrocellulose has a slight net negative charge that mandates the use of a relatively high ionic strength transfer buffer (Appendix K) such as 20 × SSC (saline-sodium citrate) or 20 × SSPE (saline-sodium phosphate-EDTA).
4. Nitrocellulose requires baking in a vacuum oven at 80°C to immobilize nucleic acids; a vacuum oven is a nonnegotiable requirement because humidity will interfere with the immobilization process.
5. Nitrocellulose becomes very brittle after the baking step that is required for nucleic acid immobilization. This makes subsequent manipulations difficult and requires handling of the membrane with great care.
6. The frailty of nitrocellulose after the baking step makes repeated probing impractical.
7. The background associated with the use of nitrocellulose for nonisotopic detection methods can be excessive.

The shortcomings of nitrocellulose for Northern analysis were the impetus for the development of supported nitrocellulose membranes, that is, nitrocellulose cast on an inert support. These membranes retain all of the properties of nitrocellulose, because they are much stronger than ordinary nitrocellulose, they are better able to withstand more aggressive manipulation.

Polyvinylidene difluoride

Polyvinylidene difluoride (PVDF) is a fluorocarbon polymer that historically has been a favorite membrane for Western blotting. A less common use for this membrane is Northern and Southern analysis owing to sensitivity issues; the RNA (and DNA) binding capacity of PVDF is generally less than the binding capacity of nylon and is therefore not commonly used for Northern blots. This is problematic for detecting low-abundance transcripts. In this lab, PVDF has been used previously for nucleic acid blotting in conjunction with nonisotopic probes, by handling the PVDF filters as if they were nylon membranes (when used for Northern analysis, PVDF membranes are prewet in methanol for 15 seconds, per the directions of the manufacturer), wet for a minimum of 2 minutes in water, and then for 30 seconds in 5 × SSC. The membrane is then placed in contact with the gel followed by the assembly of the transfer apparatus. PVDF is a high durability, chemically stable membrane that is sold under a number of different trade names. Like nitrocellulose and nylon, PVDF is also available with pore sizes of 0.45 and 0.2 μm. Nucleic acids can be permanently immobilized by crosslinking with UV light or by baking (vacuum oven not required).

Handling and membrane preparation

Irrespective of the type of blotting membrane is selected, it should be handled only with gloves, as described above, and cut with sharp scissors or a scalpel. When cutting the membrane to the appropriate size, which usually will be the size of the gel, the two sheets of protective wrapping in which the membrane is provided should be left in place and used to handle the membrane. Unused membrane material should be returned to storage right away in order to prevent any type of surface damage, accidental spillage of anything onto the membrane, or accidental nuclease contamination.

Blotting membranes generally require prewetting in a clean pan or tray of sterile water for 2 minutes, followed by brief equilibration in the high-salt transfer buffer selected for the particular application at hand. In this laboratory, a special supply of trays is reserved exclusively for use in the preparation of filters for Northern transfer. In addition to handling only with gloves, these trays are also soaked in 3% H_2O_2 for several minutes and then rinsed with nuclease-free H_2O prior to use. The proper way to wet a filter is to float it on the surface of nuclease-free water. An immediate change in the appearance of the membrane, from white to gray, should be observed as the filter comes into contact with the water. This method of wetting the membrane is strongly recommended because it will allow immediate identification of any areas that have become mechanically damaged, touched with bare fingers, stored incorrectly, or otherwise compromised; such areas appear as smudges or obvious uneven wetting. Once it is established that the membrane has not been damaged, it can be completely submerged by rocking the tray. Allow the filter to prewet for at least 1–2 minutes (longer periods are fine, too).

After wetting, the membrane is transferred to a tray containing the same transfer buffer that will be used for the blot. A brief equilibration of about 30 seconds is all that is required, though the membrane can rest in transfer buffer until ready to use. Note that the order of wetting is much more important than the time: water first, followed by equilibration in high-salt transfer buffer. The membrane is then ready for immediate use in the transfer process itself. One may mark the membrane asymmetrically now or, preferably, wait until the transfer is complete, as described below.

Northern transfer techniques

The transfer of electrophoretically separated RNA species from the gel to a solid support can be accomplished using any of a variety of blotting formats which differ primarily in the method by which the RNA is drawn out of the gel matrix. No matter which transfer technique is chosen, *intact* RNA must have been electrophoresed in a denaturing gel (for maximum resolution) under conditions that maintain RNA integrity (for maximum sensitivity). Otherwise, the resultant data are likely to be suboptimal. In general, the transfer techniques described here are applicable to both Northern and Southern analyses.

Capillary transfer

The method requiring the fewest accessories is the traditional passive capillary transfer method, which has also been used in the classical Southern analysis (Southern, 1975). It is also the least expensive and the most time-consuming technique. In short, a stack of absorbent material, such as paper towels or blotting paper, is used to draw (wick) the transfer buffer from a reservoir, through the gel. In so doing, the RNA sample is essentially eluted from the gel and is trapped on a juxtaposed membrane, as the buffer continues to move through the membrane and into the absorbent stack. A typical capillary transfer setup involves placing the blotting paper on the top of the stack, as shown in Fig. 16.1. Two common complaints associated with this transfer method, however, are (1) poor transfer efficiency, especially of larger RNA molecules and (2) the transfer takes too long. Regarding the former, inefficient transfer is commonly associated with the following:

1. The gel was too thick.
2. The gel consisted of more than 1.2% agarose.
3. EtBr was used to stain the gel, either during electrophoresis or afterward, without proper destaining (recall that there are excellent alternatives to staining with EtBr).

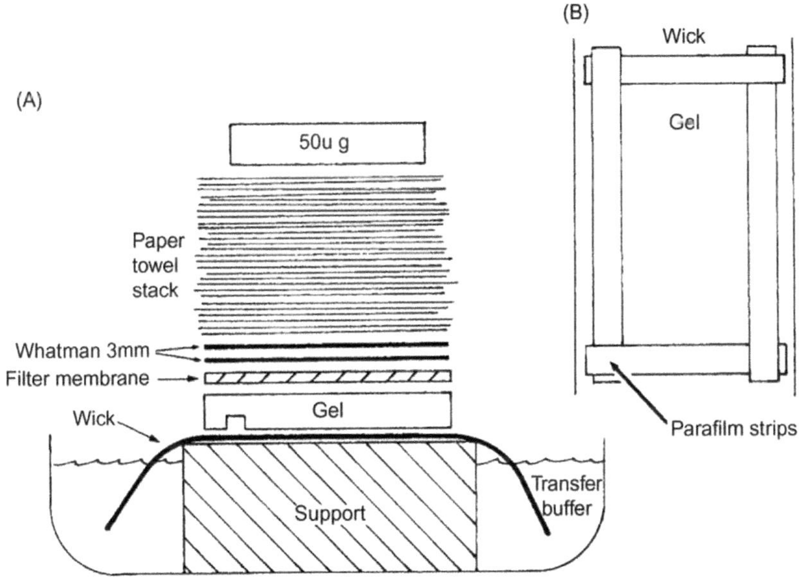

FIGURE 16.1 A typical setup for capillary transfer of RNA from denaturing agarose gels to a blotting membrane. (A) Side view. Invert the gel to reduce the blotting time and achieve greater resolution, since the RNA does not have to travel as far through the gel matrix to reach the membrane. (B) Top view. Place overlapping strips of Parafilm directly on the wick and up against the gel (not on top of the gel). Set Parafilm in place around the sides of the gel first, and then overlay membrane and other blotting materials.

4. The gel was not soaked long enough to remove the formaldehyde prior to setting up the transfer.
5. The gel was not UV irradiated (briefly) to nick the backbone and promote expedient transfer.
6. The transfer period was too brief. Most capillary transfers require 12–16 hours (overnight).
7. The RNA was degraded to begin with, so no hybridization signal could possibly be generated, no matter how efficient the transfer.
8. Any combination of the previously listed items.

One should never rule out the possibility of uneven transfer of sample material during the Northern transfer process. For example, an air bubble trapped between the gel and the membrane or between the gel and the wicking material will drastically reduce the local transfer of the sample. In one unfortunate scenario, several samples purportedly exhibit differential modulation of a gene of interest following posthybridization detection. In reality, "signal modulation" among the samples may be the result of uneven transfer: potential target sequence molecules may have been left behind in the gel! In order to assess whether an efficient and complete transfer has occurred, one or more of the following may be useful:

1. At the conclusion of the blotting period, the gel can be stained again (EtBr; SYBR Green) and inspected for residual RNA, especially rRNA (there should be none). The abundant 28S and 18S rRNAs serve as excellent transfer controls as well as gel loading controls (easy and quick).
2. The membrane itself can be reversibly stained with methylene blue to directly identify the positioning of the ribosomal 18 S and 28S rRNAs and, in so doing, serve to monitor transfer efficiency (messy; not recommended).
3. If the gel was stained prior to transfer and small amounts of EtBr or one of the SYBR stains remained in the gel and associated with the sample, then visualization of the rRNAs directly on the blotting membrane may be possible if the filter paper is briefly placed on a transilluminator or observed by overhead irradiation with a low-intensity, handheld UV light source (easy and quick; shows what's on the filter but not what remains in the gel).

In instances where incomplete transfer simply cannot be remedied and, regarding the latter complaint of time-consuming transfer, alternative transfer techniques that accelerate and improve transfer efficiency have been developed and are in widespread use, some of which are described below.

TurboBlotter

The TurboBlotter downward transfer system (Cytiva) takes advantage of gravity and, in so doing, offers enhanced transfer efficiency in a much shorter time, usually less than 3 hours, than is required by the older upward capillary transfer method (Fig. 16.2). It has also been shown empirically in this lab that, when using the TurboBlotter apparatus, most of the nucleic acid transfers out of the gel in the first hour. Precut filters and blotting papers are available to accommodate all standard-sized gels and the TurboBlotter device has a very small footprint on the lab bench. This method likewise eliminates the need for heavy weights on top of the capillary stack, thereby minimizing compression and concomitant collapsing of the pores of the gel. The technique does not require the use of any expensive equipment, such as the vacuum blot devices or electroblotting apparatuses described later and it is very easy to assemble. The record for assembling a TurboBlot transfer in this lab is 58 seconds!

Vacuum blotting

In a vacuum blotting system, negative pressure is applied to accelerate the transfer process. This technique is also known as vacuum-assisted capillary

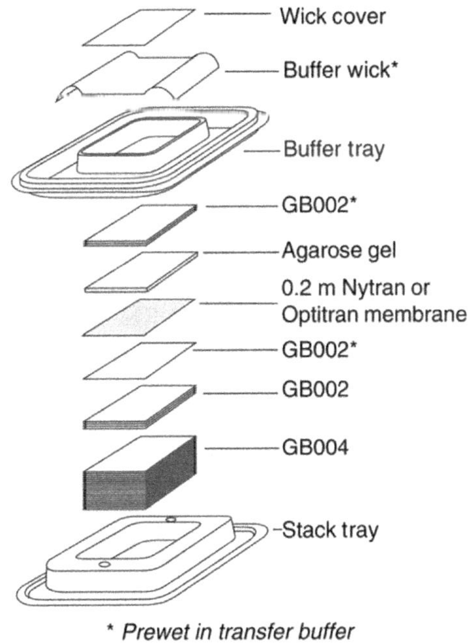

FIGURE 16.2 TurboBlotter transfer method. The principle is that of an inverted capillary transfer. By taking advantage of gravity, rapid downward movement of the high-salt transfer buffer accelerates the transfer and dramatically improves transfer efficiency. The absorbent blotting paper is at the bottom of the stack and the buffer reservoir on top, precisely the opposite of the classical capillary blot setup shown in Fig. 16.1. *Courtesy of GE Healthcare Life Sciences (now part of Cytiva).*

transfer, in which the nucleic acid sample is drawn out of an agarose gel along with the transfer buffer. The advantage of vacuum blotting is that the transfer occurs far more rapidly than in the traditional capillary transfer (typically 60 minutes vs overnight). The accelerated process reduces molecular diffusion of nucleic acid molecules on the way out of the gel, minimizing the outward (lateral) spreading of the otherwise very sharp bands observed in the gel. One must use great care when vacuum-blotting, however, not exceed the hydrostatic pressure of the gel (vacuums between 30 and 50 mbar are optimal), because gel pore collapse will trap the remaining sample within the matrix of the gel. To preclude a potential catastrophe, this type of transfer system often includes a calibrated vacuum device intended for this specific application. While no longer widely available, vacuum blot devices may often be found in established molecular biology research labs; they may also be procured from specialty suppliers. When used, nucleic acid vacuum transfer devices improve productivity in the lab.

Electroblotting

Electroblotting is widely used to transfer proteins from polyacrylamide gels. Following electrophoresis, the gel is placed adjacent to the blotting membrane in a special cassette which is then placed in a tank of electrolyte buffer (tank electroblotting). Upon application of a voltage gradient perpendicular to the direction of the initial electrophoresis of the gel, the sample migrates out of the gel and onto the membrane; this is a standard transfer methodology for Western blots.

With respect to the nucleic acid transfer, the application is known as dry or semidry electroblotting and is useful when RNA has been electrophoresed in a polyacrylamide gel. This variant method requires only a minimal amount of buffer to wet the membrane and blotting papers, thereby establishing electrical contact between the gel and apparatus itself. The advantage here is that lower voltage and current are needed for transfer. The extent to which electroblotting accelerates the transfer procedure is directly related to the current. Because some forms of electroblotting can also produce a significant amount of heat, it is, therefore, to be used judiciously to transfer nucleic acids from agarose gels. If the electroblotting method is appealing, there are several commercially available devices that have optimized the procedure. With respect to blotting membranes, nylon or PVDF are the most commonly used for electroblotting and the investigator should follow all recommendations of the membrane manufacturer. The investigator should also follow the safety instructions and exercise appropriate caution when working with power supplies and other electrical equipment.

Alkaline blotting

Alkaline blotting of nucleic acids is essentially a variation on the theme of capillary transfer. The key difference between these two approaches, however, is the method of denaturation and nucleic acid transfer. Traditional approaches to capillary blotting include the use of a high-salt buffer (15–20 × SSC or SSPE) to blot denatured RNA or DNA from the gel to the membrane. In alkaline blotting of DNA, the denaturation and transfer steps are accomplished simultaneously by the use of 0.4 N NaOH with subsequent fixation onto nylon(+) membranes. Subsequently, it was shown that RNA can be transferred in 8 mM NaOH (Chomczynski, 1992; Nature Methods, 2005) for no more than 1.5 hours; the limited alkaline hydrolysis which ensues can facilitate the transfer of RNA from the gel to a blotting membrane. Great care must be taken when performing RNA alkaline blotting to prevent excessive hydrolysis from occurring. In addition, alkaline blotting onto certain membranes precludes the requirement for posttransfer immobilization and it is compatible with nonisotopic probes. Owing to the inherent sensitivity of RNA to elevated pH environments, alkaline blotting is less frequently used compared to other methods.

Protocol: RNA transfer by passive capillary diffusion

1. Wear gloves throughout the procedure and refer to Fig. 16.1 as needed. It is imperative to prevent nuclease contamination at every level. Remember that RNAs are not safe from nuclease degradation until they have been immobilized by baking or crosslinking onto a membrane. *Suggestion: Use a wide spatula for transferring gels from one solution to another because formaldehyde gels are rather slippery and should be handled with care.*
2. **Formaldehyde gels:** Soak the gel in an excess of 5 × SSC or nuclease-free H_2O to remove the formaldehyde. Formaldehyde is typically drawn out of unstained gels by soaking them three times for 10 minutes each. Failure to do so will result in the highly inefficient transfer, loss of sample, and impaired hybridization efficiency. Ethidium bromide, if present, will likewise be drawn out of the gel. This is required to eliminate the high background fluorescence associated with EtBr-stained formaldehyde gels so they can be photographed. Thorough destaining, while not essential, may require two or more hours. In this laboratory, formaldehyde gels are occasionally destained overnight in an excess of buffer at 4°C; refrigeration temperature will minimize dye diffusion thereby keeping the bands sharp. A gel stained in SYBR Green or SYBR Gold does not require destaining, though soaking the gel in H_2O or 5 × SSC will facilitate the removal of formaldehyde. *Optional, although not recommended: Soak the gel in 10 mM NaOH, 10 mM NaCl for 30 minutes; in 100 mM Tris, pH 7.5 for 30 minutes, and then in 10 × SSC for 30 minutes. This will result in partial alkaline hydrolysis along the length of the RNA backbone. While this may facilitate the transfer of very large RNA molecules out of the gel, it is usually not necessary. Moreover, overexposure to NaOH can destroy the ability of the RNA to hybridize to any probe. Instead, random nicking of the RNA backbone and improved transfer efficiency may be accomplished by irradiating the gel on a UV transilluminator or, for greater precision, by using a calibrated UV light source.*

   Glyoxal gels: Transfer is set up immediately after staining (destain as needed) and photographing the gel. Glyoxal will be removed from the RNA in a posttransfer, prehybridization wash.
3. Trim the left and right edges of the gel (but not the top or bottom) such that there is a 1 cm margin on each side bordering the lanes of interest.
4. Cut a piece of blotting membrane (e.g., nylon) to the size of the gel. Always wear gloves when handling the blotting membrane because finger greases will compromise the membrane binding capacity. Cut the

transfer membrane and other sheets of blotting paper to the size of the gel by placing a clean, dry, gel casting tray directly on the wrapping paper that protects the membrane and then tracing the outline of the casting tray with a pencil. The gel casting tray acts as a template so that all materials (membrane, blotting papers, paper towels) are cut to the same size as the gel. This can be done while the gel is destaining or soaking in the buffer. For laboratories with enhanced budgets, all required blotting materials may be purchased precut to the proper dimensions.

5. Prepare the membrane for transfer according to the manufacturer's instructions. Typically, nylon and nitrocellulose membranes are first floated on the surface of RNase-free water and then submerged for about 2 minutes. If the membrane shows irregular patches of wetting when floated on the surface of the water, it is either old or soiled and should be discarded.
6. After wetting for 2 minutes, the membrane is next equilibrated in transfer buffer (5 × SSC or 5 × SSPE for nylon) for 30 seconds or longer. The membrane may remain in the transfer buffer until it is used.
7. Cut one sheet of Whatman 3MM paper (or the equivalent) approximately 2 cm wider than the gel and about 15 cm longer than the gel. This is the wick, and it will act as the contact between the transfer buffer reservoir and the gel itself; a wick that is cut too long can be trimmed later.
8. Saturate the wick by soaking it in transfer buffer for 10 seconds. Next, drape the wick over a plexiglass sheet resting across the shorter dimension of a baking dish or similar container.
9. Fill the baking dish with enough transfer buffer such that a minimum of 3–4 cm of each end of the wick is submerged in the transfer buffer.
10. Any air bubbles that are trapped under the wick are easily removed by gentle rolling with a sterile pipette. Failure to remove all air bubbles will compromise transfer efficiency at that location.
11. Place the gel in the middle of the wick with the wells facing down. Remove any air bubbles trapped between the gel and the wick. Mask the area surrounding the gel with strips of Parafilm to prevent short-circuiting the system during transfer (Fig. 16.1B). Occasionally, the paper towels that draw the buffer upward from the reservoir and through the gel will sag down around the sides of the gel as they become wet. This may occur when the paper towels are merely slightly larger than the gel. If the paper towels touch the wick along the side of the gel, they will draw the transfer buffer up around the gel rather than through it. Less buffer moving through the gel slows the transfer process and reduces transfer efficiency. To preclude this potential difficulty, some investigators have been known to cut away the wells from the rest of

the gel (not necessary). Parafilm along the periphery of the gel will prevent direct contact between wick and paper towels.

12. Carefully place the blotting membrane on top of the gel and remove any air bubbles trapped beneath it.

13. Cut two or three sheets of Whatman 3MM paper (or the equivalent) to the same size as the gel. Prewet at least one of these sheets in the transfer buffer.

Note: Prewet at least one of these sheets in transfer buffer just prior to use. In this lab, prewetting the sheet that will rest directly on the transfer membrane primes the wicking action efficiently.

14. Place the wet sheets of Whatman 3MM paper on top of the transfer membrane and eliminate air bubbles by gently rolling a sterile 5 mL pipette across the surface. Then place the dry piece of Whatman 3MM paper on top.

15. Cut a stack of paper towels (5–6 cm when compressed) to the size of the gel. This can be done well ahead of time. Position the paper towels on top of the sheets of Whatman 3MM paper. Finally, cover the entire stack with a 300–400 g weight or an obscure textbook to compress the entire setup. Note: Avoid paper towels with creases such as the C-fold type because inefficient transfer from the gel over the area of the crease is almost guaranteed. If no other paper towels are available, alternate the paper towels in the stack so that the creases are not aligned over any particular part of the gel. Uniformity of capillary action is an absolute requirement. A better alternative to paper towels is precut blotting paper.

16. Allow the transfer to proceed for several hours or overnight, if possible.

17. At the conclusion of the transfer period, remove and discard the saturated paper towels. Then, carefully peel the blotting membrane away from the gel with forceps or a gloved finger. Clip one corner of the filter to mark it asymmetrically and indicate orientation. This will be essential during the detection steps and ensure accurate interpretation of data.

18. Follow the posttransfer washes suggested by the manufacturer. Typically, wash the membrane for 30 seconds in a fresh aliquot of the same buffer that was used for the transfer (e.g., 5 × SSC or 5 × SSPE). Note: The brief, posttransfer washing of the filter removes any pieces of agarose that may be clinging to the filter as well as any other residual materials that could interfere with immobilization of the sample, hybridization specificity, or contribute to the background.

19. Nylon membranes should be UV-irradiated while still damp but not sopping wet. Nitrocellulose membranes should be air-dried and then baked as soon as possible.

Protocol: TurboBlotter downward transfer of RNA

1. Wear gloves throughout the procedure and refer to Fig. 16.2 as needed. It is imperative to prevent nuclease contamination at every level. Remember that RNAs are not safe from nuclease degradation until they have been immobilized, by baking or crosslinking, on a blotting membrane. *Suggestion: Use a wide spatula for transferring gels from one solution to another because formaldehyde gels are rather slippery and should be handled with care.*
2. **Formaldehyde gels:** Soak the gel in an excess of 5 × SSC or nuclease-free H_2O to remove the formaldehyde. Formaldehyde is typically drawn out of unstained gels by soaking them three times for 10 minutes each. Failure to do so will result in the highly inefficient transfer, loss of sample, and impaired hybridization efficiency. Ethidium bromide, if present, will likewise be drawn out of the gel. This is required to eliminate the high background fluorescence associated with EtBr-stained formaldehyde gels so they can be photographed. Thorough destaining, while not essential, may require two or more hours. In this laboratory, formaldehyde gels are occasionally destained overnight in an excess of buffer at 4°C; refrigeration temperature will minimize dye diffusion thereby keeping the bands sharp. A gel stained in SYBR Green or SYBR Gold does not require destaining, though soaking the gel in H_2O or 5 × SSC will facilitate the removal of formaldehyde. *Optional, although not recommended: Soak the gel in 10 mM NaOH, 10 mM NaCl for 30 minutes; in 100 mM Tris, pH 7.5 for 30 minutes, and then in 10 × SSC for 30 minutes. This will result in partial alkaline hydrolysis along the length of the RNA backbone. While this may facilitate the transfer of very large RNA molecules out of the gel, it is usually not necessary. Moreover, overexposure to NaOH can destroy the ability of the RNA to hybridize to any probe. Instead, random nicking of the RNA backbone and improved transfer efficiency may be accomplished by irradiating the gel on a UV transilluminator or, for greater precision, by using a calibrated UV light source.*

   Glyoxal gels: Transfer is set up immediately after staining (destain as needed) and photographing the gel. Glyoxal will be removed from the RNA in a posttransfer, prehybridization wash.
3. Select precut blotting papers that are as close to the size of the gel as possible. If the gel is slightly larger, the excess will be trimmed away during the assembly of the TurboBlotter. If the dimensions of the gel are less than those of the blotting paper, the exposed portion of the blotting paper will be insulated later with Parafilm. Note: The blotting membrane and all required blotting papers (steps 4, 8, and 9) are included in each TurboBlotter refill pack.

4. Cut a piece of nylon filter to the size of the gel. Always wear gloves when handling the blotting membrane because finger greases will compromise the membrane binding capacity.
5. Prepare the membrane for transfer according to the manufacturer's instructions. Typically, nylon membranes are first floated on the surface of RNase-free water and then submerged for about 2 minutes. If the membrane shows irregular patches of wetting when floated on the surface of the water, it is either old or soiled and should be discarded.
6. After wetting for at least 2 minutes, the membrane is next equilibrated in transfer buffer (5 × SSC or 5 × SSPE for nylon) for 30 seconds or longer. The filter may remain in the transfer buffer until it is used.
7. Place the stack tray of the transfer device on the bench, making sure that it is level.
8. Place 20 sheets of dry GB004 blotting paper in the stack tray.
9. Place four sheets of dry 3MM blotting paper on top of the GB004 stack.
10. Place one sheet of 3MM blotting paper, prewet in transfer buffer, onto the stack. Use a sterile 5 mL pipette to gently smooth the membrane, ensuring that no air bubbles are trapped beneath. Do not apply too much pressure with the pipette.
11. Place the equilibrated blotting membrane on the stack.
12. Carefully place the agarose gel on top of the membrane and then trim the gel to the size of the membrane by cutting away any overhanging, unused lanes. Do not move the gel once it has made contact with the filter. If any portion of the stack is exposed because the gel was smaller, mask the area surrounding the gel with strips of Parafilm to prevent short-circuiting the system during transfer. This is usually not necessary with the TurboBlotter setup unless the gel is significantly smaller than the precut blotting paper.
13. Use a sterile 5 mL pipette to gently smooth the gel, ensuring that no air bubbles are trapped between the gel and the blotting membrane.
14. Wet three sheets of 3 MM blotting paper in transfer buffer, and carefully place these on top of the stack.
15. Use a sterile 5 mL pipette to gently smooth the stack, ensuring that no air bubbles are trapped between the sheets of blotting paper.
16. Attach the TurboBlotter buffer tray to the stack tray, guided by the alignment notches to ensure proper placement.
17. Fill the upper buffer tray with transfer buffer. Do not pour any transfer buffer into the center of the stack tray where the blotting papers and gel are located.
18. Wet the wick in the transfer buffer. Initiate the transfer by connecting the gel stack with the buffer tray: Place the wick across the stack so that the shorter dimension of the wick completely covers the blotting stack, and both ends of the long dimension extend into the buffer tray.

19. Cover the stack with the wick cover (provided) or simply with Parafilm. This is strongly recommended to prevent evaporation and salt precipitation. No weight is necessary on the stack.
20. Allow the transfer to proceed until the buffer has been drawn out of the buffer tray. While this usually requires 3–4 hours to complete, it has been determined in this laboratory that most of the samples in the gel transfer within the first hour (data not shown).
21. At the conclusion of the transfer period, remove and discard the saturated wick and blotting papers above the gel. Then, carefully peel the transfer membrane away from the gel with forceps or a gloved finger. Clip one corner of the filter in order to mark it asymmetrically to indicate orientation. This will be essential during the detection steps for ensuring accurate interpretation of data.
22. Follow the posttransfer washes suggested by the manufacturer. Typically, wash the membrane for 30 seconds in a fresh aliquot of the same buffer that was used for the transfer (e.g., 5 × SSC or 5 × SSPE). Note: The brief, posttransfer washing of the filter removes any pieces of agarose that may be clinging to the filter as well as any other residual materials that could interfere with immobilization of the sample, hybridization specificity, or contribute to the background.
23. Nylon membranes should be UV-irradiated while still damp but not sopping wet. Nitrocellulose membranes should be air-dried and then baked as soon as possible.

Immobilization techniques

Although filters can be stored confidently before hybridization, it is necessary to immobilize the sample on the filter as soon as possible after transfer. Only then is the RNA sample stabilized and rendered resistant to degradation by RNase. Following immobilization, filters can be used immediately for hybridization or stored for months. Place air-dried filters between sheets of Whatman 3 MM paper and store flat, dry (in vacuo is ideal), and in the dark until ready for use.

Formaldehyde will have already been removed from the gel during the destaining process prior to gel blotting. In the case of glyoxal electrophoresis, the glyoxal that denatured the RNA prior to electrophoresis must now be removed from the filter. This may be accomplished in one of two ways. **Option 1:** Air-dry and then bake the membrane according to the recommendations of the manufacturer. This will destroy some of the glyoxal and immobilize the RNA at the same time. The remaining glyoxal can be removed by immersing the filter in 200 mL of 20 mM Tris, pH 8.0, preheated to 90°C, and then immediately cooling to room temperature. **Option 2:** After immobilization, remove glyoxal from RNA by washing the filter for 15 minutes at 65°C in 20 mM Tris, pH 8.0.

Immobilization of target RNA (or DNA) on a blotting membrane after the transfer has been completed is just as important as ensuring that the transfer process itself is complete. Nucleic acids, especially RNA molecules, are not safe from nuclease assault or natural degradation until the sample has been fixed in place on a solid support. Inefficient immobilization will invariably result in the loss of the sample during the prehybridization, hybridization, and/or posthybridization washing and detection steps. A variety of methods for nucleic acid immobilization have been described, each of which has defined advantages, limitations, and applicability. In this laboratory, nylon blotting membranes are used almost exclusively because of the great experimental flexibility that they offer. In every instance, it is strongly advised to follow the recommendations of the manufacturer of the membrane. A few general guidelines are presented here.

Baking

The original technique for nucleic acid immobilization involves baking the sample onto the surface of nitrocellulose in a vacuum oven. Following the post-Northern transfer wash, the membrane is set aside to air-dry completely, after which it is baked for 2 hours at 80°C in vacuo. The interaction between the matrix and nucleic acids is hydrophobic in nature. When working with nitrocellulose, the use of a vacuum oven is a nonnegotiable requirement because humidity interferes with the immobilization process. It is also important not to exceed 80°C during the immobilization procedure due to the flammable nature of nitrocellulose. It is also important to understand that baking nucleic acids onto a membrane is not a completely permanent process: after striping and reprobing the membrane a few times, some of the samples are lost from the surface of the membrane, thereby compromising the accuracy of all subsequent assays. This method, in spite of its numerous shortcomings, persists today.

In the case of nylon membranes, nucleic acid samples can likewise be immobilized by baking. The baking process usually requires 1.0–1.5 hours, with temperatures only as high as 65°C. Moreover, baking nylon does not require a vacuum oven. Some protocols suggest that nylon filters may be baked while still damp, although in this case, the application of a vacuum may accelerate the drying process and improve nucleic acid retention on the membrane. Rather than baking nylon membranes, however, much greater retention of nucleic acids on the membrane is promoted by UV irradiation.

Crosslinking by UV irradiation

Crosslinking of nucleic acids to filters with UV light is a reliable immobilization method. Short-wave and medium-wave UV light (254 and 302 nm, respectively) activate thymine and uracil, and other bases to a lesser degree, which become highly reactive and form covalent bonds with the surface amines that characterize many nylon matrices (Saito et al., 1981). UV crosslinking is usually more efficient with damp nylon membranes, which generally require a total exposure of 120,000 μJ/cm^2. Alternatively, air-dried membranes require approximately 160,000 μJ/cm^2, ideally delivered using a calibrated UV light source. In contrast, the practice of UV irradiation of nitrocellulose is to be avoided because it does not result in the comparable retention of nucleic acids achievable by crosslinking to nylon and, more importantly, over-irradiation of nitrocellulose poses a serious risk of fire. Don't do it.

It is worth noting that the brief irradiation of agarose gels during photodocumentation and prior to blotting causes the random nicking of the phosphodiester backbone of RNA and DNA. This takes the place of and has a similar net effect as limited hydrolysis by immersion in a dilute solution of NaOH as described above. Random nicking with UV light is highly reproducible and will facilitate nucleic acid transfer from the gel onto the membrane.

Increased stability of nucleic acid on the matrix, compared to baking procedures, is an important dividend associated with UV-mediated fixation. Laboratories routinely note an increase in sensitivity with UV-crosslinked filters compared to baked filters. Moreover, fixation by crosslinking appears to be a far more permanent immobilization technique than baking. This is of considerable importance if the investigator plans to screen a filter more than once; repeated high-stringency removal of hybridized probes usually results in the loss of target sequence from filters that were baked.

Several methods can be used to crosslink nucleic samples to the blotting membrane. The most precise energy output is, of course, delivered by a calibrated UV light source; as noted above, a UV dose of about 120,000 μJ/cm^2 is ideal for a damp nylon membrane. Calibrated UV light devices designed for this purpose are commonly found in molecular biology laboratories and are readily available for purchase from biotech vendors. Alternatively, a standard laboratory transilluminator may suffice, though this will involve some trial and error. Most transilluminators in the molecular biology laboratory emit at one of two wavelengths, either 302 nm (medium-wave energy) or 254 nm (short-wave energy). Crosslinking with a 302 nm emitter is strongly recommended over 254 nm-emitting instruments because overirradiation with the shorter wave light will reduce sensitivity due to damage to the blotted nucleic acid molecules. Damp filters are placed directly on a cleaned transilluminator surface and irradiated. For greater precision, many scientific suppliers sell UV light meters that can be used to identify the appropriate exposure time using a particular transilluminator. Rapid UV crosslinking of nucleic acids onto nylon has superseded membrane baking for most applications. It has also been reported that nucleic acid fixation onto nylon can be performed using a microwave oven: 2 minutes at 850 W

(Angeletti, et al., 1995), though this method was less efficient than baking in an oven when nitrocellulose was tested. It is the opinion of this Author that one should not attempt to microwave nitrocellulose due to its flammable nature.

Protocol: UV crosslinking RNA to nylon filters

1. Wash filters post-Northern transfer as described above and blot excess liquid by placing the blotting membrane with the RNA on it on top of one or two sheets of Whatman 3MM paper. Do not allow filters to dry out completely; it is best to irradiate filters that are still slightly damp.
2. Place filter(s) face down directly on the surface of the transilluminator, the surface of which has been wiped clean just prior to use. *Caution: Be sure to wear proper eye and skin protection because serious permanent damage can occur to the eyes and skin if not properly shielded.*
3. The appropriate exposure time is a direct function of the wavelength and age of the transilluminator. With 302 nm-emitting instruments, a total of 45 seconds is usually more than adequate. For 254 nm-emitting instruments, crosslinking is complete in about 30 seconds.

   Note 1: These are general guidelines only. The exact parameters must be empirically determined for each instrument.

   Note 2: In this laboratory, filters are irradiated for one-half of the estimated time required, rotated 90°, and then irradiated again for the remaining interval. Be certain to turn off the transilluminator before attempting to move the filter. Rotation is suggested because of the parallel orientation of the bulbs within the transilluminator and because older bulbs may have diminished output.
4. Store filters flat and sandwiched between Whatman 3MM paper in a dry location out of direct light until further use.

Postfixation handling of blotting membranes

The timeline for Northern analysis changes dramatically after transfer and immobilization are completed. It should be abundantly clear by now that up to this point all manipulations involving labile RNA should revolve around the timing of RNA isolation. This pertains to poly(A)$^+$ purification (when necessary), determination of concentration and purity, electrophoresis, blotting, nuclease protection analysis (Chapter 18: Quantification of specific mRNAs by nuclease protection), conversion into complementary DNA (cDNA; Chapter 9: cDNA: a permanent biochemical record of the cell), and many other applications. After a sample is immobilized (and stabilized) on a solid membrane support, the sense of urgency is greatly diminished and the membrane can be stored for months until probing, or reprobing is convenient.

An important point to keep in mind is that membranes should be stored dry or in a vacuum oven to prevent moisture and opportunist microbial growth from encroaching. Filters that have already been subjected to hybridization and which will be rescreened should be washed at very high stringency to remove the old probe as soon as possible after completing hybridization detection. Do not allow filters to dry out until the previously hybridized probe has been removed. If probe removal will be unavoidably delayed, keep the filter damp, and wrapped in plastic because it is very difficult to completely remove hybridized probes from filters that have dried.

Reverse Northern analysis

The traditional format of Northern analysis is the measurement of the abundance of one transcript at a time, albeit with limited sensitivity compared with other techniques. The rise of assays designed to monitor the abundance levels of dozens, hundreds, or even thousands of genes simultaneously, primarily by PCR-based approaches, microarray analysis (Chapter 21: Array analysis of gene expression), or RNA-seq (Chapter 24: RNA-seq: the premier transcriptomics tool) was the driving force behind the development of newer methods to *confirm* purported changes in gene expression. Data from this laboratory has been viewed quite favorably by journal manuscript reviewers when reported changes in gene expression from quantitative and semiquantitative PCR experiments are confirmed using a non-PCR-based approach; this can be a most worthwhile publication strategy! The reverse Northern analysis, or simply the "reverse Northern," is one such approach (Zhao et al., 1996; Zhang et al., 1997; Kang et al., 1998). In one variant of this technique, unique DNA (usually cDNA) sequences representing genes that are believed to be expressed differentially are spotted directly onto duplicate nylon filters in a dot-blot-like format (described in Appendix J: Dot blot analysis). Then, each filter is hybridized to a radiolabeled pool of total cDNA (the probe) from the control and experimental samples, respectively. The cDNA probes are synthesized as needed by performing first-strand cDNA synthesis on the original RNA samples, thereby precluding potential RNase-associated difficulties when working with an RNA probe. In a variant of this approach, the two cDNA populations are each labeled with a different fluorescent tag such as Cy3 (green) and Cy5 (red) are co-hybridized to a common membrane, and then the fluorescence pattern and intensity is examined for both fluorescence tags. In either approach, differences in hybridization intensity reflect transcript abundance differences among the genes being assayed (Fig. 16.3) because cDNA directly mirrors mRNA abundance. Confirmation of differential gene expression occurs when the hybridization signal from a specific cDNA is stronger on one filter compared to the other filter. For even greater accuracy, some investigators elect to blot each cDNA sequence in two

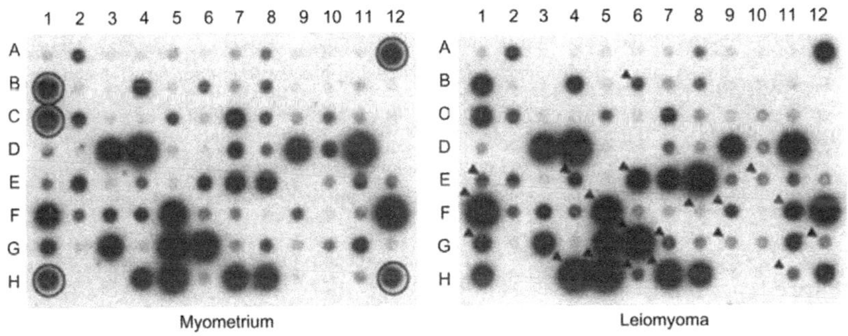

FIGURE 16.3 Reverse Northern analysis. Two sets of blots are prepared with equal quantities of PCR products. Positive controls (G3PDH, also known as GAPDH) are marked with circles. Notice the asymmetry with respect to the placement of the controls, which ensures accurate orientation of the membrane. Genes that have been upregulated are indicated with arrows. Leiomyoma is a benign smooth muscle tumor and myometrium is normal smooth muscle tissue from the wall of the uterus. *Reprinted by permission from Macmillan Publishers Ltd: Cell Res. 12, 215–221, copyright 2002.*

(usually adjacent) locations on the filter to ensure the authenticity of hybridization signal strength.

The advantage and convenience of the reverse Northern procedure are that one may choose to analyze only a few sequences at a time. This method is much less expensive than other approaches (e.g., RNA-seq, microarrays) and is ideal for laboratories with limited budgets that are interested only in genes associated with a specific biochemical pathway or pathology, such as growth and development, apoptosis, or cancer. While potentially attractive as a gene expression assay, it is important to note that reverse Northern blots suffer from the same general lack of sensitivity as other membrane-bound assays.

References

Alwine, J.C., Kemp, D.J., Stark, G.R., 1977. Method for detection of specific RNAs in agarose gels by transfer to diazobenzyloxymethyl-paper and hybridization with DNA probes. Proc. Natl. Acad. Sci. 74, 5350–5354.

Alwine, J.C., Kemp, D.J., Parker, B.A., Reiser, J., Renart, J., Stark, G.R., et al., 1979. Detection of specific RNAs or specific fragments of DNA by fractionation in gels and transfer to diazobenzyloxymethyl paper. Meth. Enzymol. 68, 220–242.

Angeletti, B., Battiloro, E., Pascale, E., D'Ambrosio, E., 1995. Southern and Northern blot fixation by microwave oven. Nucl. Acids Res. 23, 879–880.

Chomczynski, P., 1992. One-hour downward alkaline capillary transfer for blotting DNA and RNA. Analyt. Biochem. 201, 134–139.

Kang, D.C., LaFrance, R., Su, Z.Z., Fisher, P.B., 1998. Reciprocal subtraction differential RNA display: an efficient and rapid procedure for isolating differentially expressed gene sequences. Proc. Natl. Acad. Sci. 95, 13788–13793.

Nature Methods, 2005. Northern blotting: transfer of denatured RNA to membranes. Nat. Meth. 2, 997–998. Available from: https://doi.org/10.1038/nmeth1205-997.

Pitas, J.W., 1989. A simple technique for repair of nylon blotting membranes. Biotechniques 7 (10), 1084.

Saito, I., Sugiyama, H., Furukawa, N., Matsuura, T., 1981. Photochemical ring opening of thymidine and thymine in the presence of primary amines. Tetrahedron Lett. 22, 3265–3268.

Southern, E.M., 1975. Detection of specific sequences among DNA fragments separated by gel electrophoresis. J. Mol. Biol. 98, 503–517.

Zhang, H., Zhang, R., Liang, P., 1997. Differential screening of differential display cDNA products by reverse Northern. Meth. Mol. Biol. 85, 87–93.

Zhao, S., Ooi, S.L., Yang, F.C., Pardee, A.B., 1996. Three methods for identification of true positive cloned cDNA fragments in differential display. Biotechniques 20, 400–404.

Further reading

Mishima, E., Jinno, D., Akiyama, Y., Itoh, K., Nankumo, S., Shima, H., et al., 2015. Immuno-Northern blotting: detection of RNA modifications by using antibodies against modified nucleosides. PLoS One . Available from: https://doi.org/10.1371/journal.pone.0143756.

Chapter 17

Nucleic acid probe technology

Rationale

At the heart of molecular biology is the controlled and directed base-pairing of two complementary nucleic acid molecules to form an antiparallel double-stranded structure. This process is known commonly and interchangeably as hybridization, annealing, duplex formation, base-pairing, and renaturation. In a hybridization event, each of the strands has a different name. In some methods, one partner of the hybridization pair is known as the *target*, while the other partner is known as the *probe*, meaning that it carries some type of label that allows it to be followed throughout the experiment. The terminology forward hybridization kinetics refers to this process of double-stranded molecule formation and is functionally synonymous with the terms annealing, duplex formation, and probe renaturation. This chapter begins with an overview of the factors that promote or prevent hybridization, followed by specifics pertaining to DNA and RNA probe labeling and detection.

The two goals at the end of every experiment involving hybridization are to (1) localize the probe, because it is necessary to determine where on the filter paper, in a gel, or elsewhere in the assay the probe managed to base-pair stably to its target and (2) quantify the amount of the probe that has been retained at that location (e.g. in a band or within a particular region of an in situ tissue section). Nucleic acid probes can be either DNA or RNA, and either homogeneous, that is, when all probe molecules are the same, or heterogeneous in nature, as is often the case when the behavior of multiple genes is under investigation simultaneously.

The term "mixed-phase hybridization" describes the situation in which the target strand is immobilized on a solid support, in the manner of the traditional Northern or Southern blot, while the probe strand circulates in a warmed hybridization solution in search of complementary molecules with which to base-pair. Solution hybridization, in contrast, occurs when both nucleic acid strands are free-floating such that base-pairing between probe and target is driven by random collision kinetics. In the case of PCR, which is a fine example of solution hybridization, the template and the primers are the PCR cousins of the target and probe, respectively. The selection of hybridization, labeling, and detection techniques is governed primarily by the required degree of sensitivity and resolution.

RNA Methodologies. DOI: https://doi.org/10.1016/B978-0-323-90221-2.00019-9

Broadly, there are two ways by which to classify nucleic acid probes, namely (1) the type of label and associated detection strategies (radioisotopes or nonisotopic techniques) and (2) the distribution of the label in each probe molecule (end-labeling or continuous labeling). These labeling strategies are summarized in Table 17.1. Isotopic labeling methods are coupled with traditional autoradiography or with phosphorimaging, an established method which may be thought of as an electronic solid-phase form of liquid scintillation counting. While to use of radiolabeled probes was once commonplace, the nonisotopic methods have become sophisticated and reliable, accommodating detection on filter membranes, in microtiter plate format, or directly from gels using digital detection or by capturing the image on X-ray film, to list but a few of the choices. In fact, the refinements in nonisotopic systems allow the investigator to realistically question the use of radioactivity altogether.

Selecting the type of nucleic acid probe best suited for a particular application is just as important as the methodology by which hybridization events will be localized and quantified. The role of a probe, ideally, is to hybridize to every available complementary sequence; for this to occur, it must be present in a molar excess and its thermodynamic behavior predictable. Nucleic acid probes have numerous applications, owing in no small measure to their powerful discriminatory abilities. Among other things, they can be used to detect quantitative or qualitative changes in gene expression, gene amplifications and deletions, gene rearrangements, chromosomal translocations, point mutations, and the presence of new genetic sequences in cells (pathogens). Interestingly, oligonucleotides, universally used as primers to support PCR were originally used as nucleic acid probes in various applications.

TABLE 17.1 Probe labeling strategies.

	Continuous labeling	End labeling
RNA probes	In vitro transcription	T4 RNA kinase (5′ end) T4 RNA ligase (3′ end) Poly(A) polymerase (3′ end)
DNA probes	PCR Random priming Nick translation	T4 polynucleotide kinase (5′ end) *Taq* polymerase (3′ end) Terminal transferase[a] (3′ end) ssDNA: RNA ligase (3′ end)

Options for enzymatic label incorporation. In the case of synthetic oligonucleotides, nonisotopic labels such as biotin or digoxigenin may be added to the 5′ end of the molecule during synthesis.
[a]*Terminal deoxynucleotidyl transferase.*

Factors influencing hybridization kinetics and duplex stability

The rate, specificity, fidelity, and probable utility of hybridization probes are influenced by, but are not limited to: temperature, ionic strength (primarily Na^+), pH, the presence of formamide or urea, guanine and cytosine (G + C) content of probe, probe length, probe concentration, probe complexity (the total length of different probe sequences present during the hybridization), degree of complementarity between probe and target sequences, the degree of mismatching, and the viscosity of the system. So, basically everything.

The influence of each of the above-referenced variables is also dependent on the mobility of the probe and target molecules. In the case of solution hybridization, both probe and target molecules move randomly in an application-specific hybridization cocktail (e.g., the nuclease protection assays and PCR). In the case of mixed-phase hybridization, target sequences are immobilized on solid support (e.g., Northern analysis) and the probe usually circulates in a warmed, well-defined hybridization solution. The probe will base-pair with any accessible complementary sequences that have been immobilized on the membrane. Clearly, the stringency at which the hybridization and posthybridization washes are performed must be attuned to the nucleic acid identity of the type of duplexes that form (dsDNA vs DNA:DNA vs dsRNA).

When both strands are free-floating, hybridization is driven to completion more rapidly than is observed in the mixed-phase format (target immobilized on filter). The absence of the filter increases the sensitivity of these assays, usually by a factor of at least ten. Examples of solution hybridization techniques include the nuclease protection (Chapter 18: Quantification of specific mRNAs by nuclease protection) assays as well as PCR (Chapter 10: RT-PCR: A Science and an Art Form). The reason for the sensitivity differential is because the act of fixing nucleic acid molecules on a filter renders some of them incapable of hybridization, perhaps due to either their orientation on the filter or the mechanics of filter fixation.

Most probe molecules are significantly shorter than the target molecules to which they hybridize. In fact, probe molecules can be as short as twenty bases and still maintain an ability to discriminate between closely related sequences from organisms with complex genomes. Short probes afford the investigator great latitude in probe design, especially given the prevalence of and easy access to genome and transcriptome data at GenBank and other data repositories. Shorter probes tend to hybridize to complementary target sequences faster than longer probes and an enormous advantage of using short probes is that high-stringency hybridization can be performed in aqueous (formamide-free) hybridization buffers. Finally, short probes can be custom designed based on available amino acid sequence information, which is sometimes the case in the realm of new gene discovery. In this approach, very small amounts of protein sequence information are necessary: a 20-mer corresponds to about 6 amino acid codons. Probe design using this strategy is described in detail in Chapter 10, RT-PCR: A Science and an Art Form.

T_m considerations

The most frequently and easily manipulated variable that will either promote or prevent hybridization is the temperature of the system. The extent to which stable hybrids will form can be predicted by first calculating their melting temperature (T_m), which is best thought of as an equilibrium point. The T_m is that temperature at which 50% of all possible hybrids can form, while 50% remain dissociated into their constituent single strands: for each new duplex that forms at the T_m, another duplex is expected to dissociate. The current method of predicting the T_m is usually through mathematical modeling related to base composition, as shown below.

Historical perspective: in the past, the T_m of a double-stranded molecule was computed experimentally by monitoring UV light absorption at 260 nm (A_{260}) while the temperature of a DNA solution was gradually raised. As a double-stranded DNA begins to melt (denature) into its constituent single strands, there is an accompanying increase in UV light absorption known as a hyperchromic shift. This was a useful method by which to measure the thermodynamic stability of double-stranded molecules before the development of DNA binding dyes such as SYBR Green which is used commonly to characterize products synthesized by real-time PCR (Chapter 11: Quantitative PCR techniques).

The maximum rate of hybridization for long probes is typically observed 15°C–20°C below the calculated T_m in a system. In aqueous salt solutions, this corresponds to a hybridization temperature of about 60°C–65°C. The inclusion of 45%–50% formamide, or a urea equivalent (2–4 M), permits a reduction of hybridization temperature to as low as 42°C without a relaxation of stringency. However, since only 50% of all possible hybrids form at the T_m of the probe, and it is unlikely that the investigator wants only 50% of all target molecules hybridized, the actual hybridization temperature (T_{HYB}) may be 3°C–5°C below the T_m, and perhaps quite a bit more. This is especially true if there is reason to believe that the probe is not exactly matched with the target. The idea is to coax the probe to base-pair; the reality of the situation is that the investigator has far more control over the stringency of the assay in the washing steps, conducted at the conclusion of the hybridization period, before initiating the detection procedures. So powerful are the posthybridization washing steps that in most cases mixed phase hybridization can be performed at room temperature; as long as the filters are washed correctly prior to probe detection, relatively low-temperature hybridization would be unlikely to have a negative impact on the outcome of the experiment.

The thermodynamic characteristics of a probe are related to the type of hybrid that will form upon base-pairing and are influenced by the factors discussed previously, to wit, probe length, salt concentration, GC-content, and the presence of denaturants.

For double-stranded DNA (DNA:DNA) molecules (Bolton and McCarthy, 1962):

$$T_m = 81.5^\circ + 16.6\log\left[\text{Na}^+\right] + 41(\%G + C) - 0.63(\%\text{formamide}) - \left(500/L\right)$$

For DNA:RNA hybrid molecules (Casey and Davidson, 1977):

$$T_m = 79.8^\circ + 18.5\log\left[\text{Na}^+\right] + 58.4(\%G + C) + 11.8(\%G + C)^2$$
$$- 0.5(\%\text{formamide}) - \left(820/L\right)$$

For double-stranded RNA (RNA:RNA) molecules (Bodkin and Knudson, 1985):

$$T_m = 79.8^\circ + 18.5\log\left[\text{Na}^+\right] + 58.4(\%G + C) + 11.8(\%G + C)^2$$
$$- 0.35(\%\text{formamide}) - \left(820/L\right)$$

where T_m = melting temperature of duplex molecules, (%G + C) = percentage of guanine + cytosine content, expressed as a mole fraction, $\log[\text{Na}^+]$ = log of the sodium concentration, expressed in molarity, L = number of bases that participate in the actual hybridization.

For shorter probes such as oligonucleotides, the T_m of the oligomer can be quickly *estimated* by assigning 2°C for each adenine and thymine and 4°C for each guanine and cytosine (Itakura et al., 1984):

$$T_m = 4^\circ(\text{G} + \text{C}) + 2^\circ(\text{A} + \text{T})$$

It is important to realize that this relationship holds only for short oligonucleotides (11–27 bases long) and is based on hybridization in 1 M Na^+ (6 × SSC), in the absence of any organic solvents. As noted above, at the T_m only 50% of all duplexes are stable; therefore, hybridization is often performed 3°C–5°C below the calculated T_m for short oligonucleotides to promote complete target strand saturation. At T_m −5°C, all exact match duplexes will form. If the oligonucleotide probe is not exactly complementary to the target, then it will be necessary to lower the stringency of hybridization to tolerate mismatches. If more than one mismatch is involved, it will be necessary to lower the temperature of hybridization even more.

When longer oligonucleotides are involved (14–70 bases), the T_m can be calculated (Sambrook and Russell, 2001) by

$$T_m = 81.5 - 16.6\log\left[\text{Na}^+\right] + 41(\%\text{G} + \text{C}) - \left(600/N\right)$$

where T_m = melting temperature of duplex molecules, (%G + C) = percentage of guanine + cytosine content, expressed as a mole fraction, $\log[\text{Na}^+]$ =

log of the sodium concentration, expressed in molarity, N = the length of the oligonucleotide (assumes perfect match between probe and target).

These same basic principles apply to the design of oligonucleotides that will act as primers for PCR (Chapter 10: RT-PCR: A Science and an Art Form). In fact, investigators very often first become familiar with the concept of hybridization in the context of PCR. It is worth noting that there are many websites which allow investigators to determine the exact T_m of oligonucleotides prior to ordering them so as to match the probe as close as possible to one's experimental parameters.

Ionic strength

Stringency is dramatically influenced by the amount of salt in a hybridization buffer. In general, the rate of hybridization increases with salt concentration up to about 1.2 M NaCl, at which point the rate becomes constant; the T_m of a nucleic acid duplex changes approximately 16°C with each factor of 10 in salt concentration (Britten and Davidson, 1985). Further, because divalent cations have a greater influence than monovalent cations at low concentrations, restrained use of chelating agents such as EGTA and EDTA is strongly recommended. Typical hybridization buffers usually range between 0.1–1.0 M Na^+. The importance of salt concentration in determining thermodynamic behavior of nucleic acids is plainly evident from the T_m calculation equations in the previous section.

pH

Another factor that profoundly influences the stability of double-stranded molecules is the pH of the environment; most hybridization reactions are conducted at near neutral pH. Alkaline buffers promote duplex dissociation when the depletion of hydrogen ions increases the negative charge density along the backbone of each of the complementary strands, thereby causing them to repel each other. Highly acidic buffers, in contrast, will cause depurination of both probe and target molecules and may well result in strand breakage. In general, pH is among the least frequently manipulated variables in nucleic acid hybridization.

Probe length

The length of the probe is factored into the calculation of the T_m of the probe, according to the expression $D = 500/L$ where D is the reduction in T_m (in degrees Celsius) and L is the number of base pairs actually participating in duplex formation (Britten and Davidson, 1985), which is usually the length of the probe itself. The shorter the probe, the more rapidly hybridization occurs and the more discriminating the probe becomes. For example, short oligonucleotides can discriminate between closely related target sequences that differ by as little as one base. As a general rule, the shorter the probe, the more influential become many of the other variables, and temperature is the easiest variable to manipulate.

Probe concentration

The investigator is afforded considerable leeway with respect to the amount of probe added to a hybridization solution. In short, as the concentration of probe increases, forward hybridization kinetics also increase. Thus, higher probe concentrations accelerate hybridization, reducing the amount of time needed to anneal with all of the target sequences. This is especially true when oligonucleotides are used as probes. When performed properly, the probe is always present in large molar excess. Be aware, however, that the use of excessive quantities of the probe can also favor nonspecific hybridization. Think about what happens in PCR, for example, when excessive primer concentrations are used—the outcome can be disastrous.

To identify the optimal probe concentration, two variables must be considered: (1) the method of detection, that is, isotopic or nonisotopic and (2) the probable abundance of the target, for example, single-copy genes or low abundance transcripts. The amount of probe required should be based on the guidelines shown in Table 17.2 and is based either on the mass of the probe or, in the case of radiolabeling, on the specific activity of the probe. Having a large molar excess of probe drives hybridization to completion while the posthybridization washing steps have the greatest leverage over the stringency of the assay.

G + C content

The stability of a hybrid molecule is profoundly influenced by the base composition of the molecules involved. Because three hydrogen bonds form between guanine (G) and cytosine (C), and only two hydrogen bonds form between adenine (A) and thymine (T) or uracil (U), GC-rich duplexes are thermodynamically more stable than AT-rich domains. The importance of G + C content is more pronounced with shorter probes. For example, in the design of oligonucleotide probes, both for hybridization and for PCR, the paramount importance of the G + C content in duplex stability is underscored by Marmur and Doty equation (Marmur and Doty, 1961, 1962):

$$T_m = 0.41(\%G + C) + 69.3$$

where (%G + C) is the mole percentage of the duplex consisting of guanine and cytosine residues in 1 × SSC buffer (150 mM NaCl; 15 mM Na_3-citrate, pH 7.0).

TABLE 17.2 Recommended nucleic acid probe concentrations.

	Probe concentration (ng/mL)	Probe activity (cpm/mL)
Isotopic labels		
Single-copy genes or low abundance transcripts	20–50	10^6
Multiple-copy genes or high abundance transcripts	5–20	10^5
Oligonucleotide probes; all targets	1–5	$>10^4$
Nonisotopic labels		
Single-copy genes or low abundance transcripts	50–75	Not applicable
Multiple-copy genes or high abundance transcripts	20–50	Not applicable
Oligonucleotide probes; all targets	5–10	Not applicable

Note: Values indicated for probe concentration are per mL hybridization solution. Probe concentration can be expressed either by actual mass added or by label incorporation. While detection sensitivity with nonisotopic probes is often comparable to that with radiolabeling, the amount of probe added per milliliter of hybridization solution is usually greater. One may compensate for lower probe concentrations by simply extending the hybridization period.

Mismatching

Many of the hybridization recommendations that accompany the purchase of filter membranes and molecular biology kits assume a perfect or nearly perfect match between probe and target sequences. Probes that manifest this quality are known as exact-match probes which, under stringent conditions, rapidly form stable duplexes with the target. A mismatch, in contrast, is a position in a duplex where the base on one strand is not complementary to the base in the corresponding location on the other strand; this usually causes some level of helical distortion. The implication is that most or all of the other bases along the length of a probe are able to base-pair properly to the target. A mismatch could be the result of a point mutation in the organism under investigation, a naturally occurring polymorphism, or even an error in probe design. Mismatching retards the hybridization rate and at sufficiently high temperatures, mismatches may inhibit hybridization altogether. With longer probes (greater than 1 kb) the effect of a mismatch may not be noticeable under standard hybridization conditions which often include an overnight incubation. "Standard" implies that the effect of probe length is not a major variable, and abatement of the rate of hybridization due to mismatches is probably masked by the relatively long period of hybridization. As the length of the probe decreases, hybridization conditions become less "typical." Thus, the mismatch effect is more pronounced with shorter probes, especially when using oligonucleotides as probes or as PCR primers. To promote hybridization between mismatched probes and target sequences, the stringency of the system can be relaxed, usually by decreasing the hybridization temperature. The more the hybridization temperature is reduced below the T_m of the probe, the more mismatch between probe and target will be tolerated. Below a certain sequence-dependent temperature, however, random hybridization will occur.

Probe complexity

The complexity of the probe is the length of *different* probe sequences involved in the hybridization that could potentially base pair to complementary target molecules. Traditional methods, such as blot analysis, commonly involve only one probe sequence at a time (e.g., a *c-myc* probe used to assess the prevalence of cytoplasmic *c-myc* mRNA by Northern analysis). In the case of degenerate oligonucleotide probes, however, in which more than one probe sequence is present (perhaps due to target sequence ambiguity), the complexity is greater, the concentration of the correct oligonucleotide probe is reduced, and hybridization kinetics are slowed. When the complexity of the probe is very large, meaning that there is very little sequence redundancy, there is often a significant T_m range among the population of probe molecules. In such a case, it is helpful to divide the various probe sequences into pools, based on T_m similarities, and perform multiple hybridizations and stringency washes at temperatures that are compatible with the probes in a particular pool or batch. This approach is described in greater detail in the context of PCR in Chapter 10, RT-PCR: A Science and an Art Form, and is useful when significant sequence divergence across two or more species is being investigated.

Viscosity

The viscosity of the milieu is inversely proportional to forward hybridization kinetics. The anionic polymer dextran sulfate, a common ingredient in many older hybridization buffer formulations, contributes to the viscosity of the hybridization cocktail but actually increases the rate of hybridization because the dextran sulfate occupies physical space and thereby forcing the probe into a smaller volume (Wahl et al., 1979). This increases the effective concentration of the probe. A similar phenomenon occurs with the inclusion of Denhardt's solution (Denhardt, 1966; Appendix K: Useful stock solutions for the molecular biologist) in various hybridization recipes. It is an excellent alternative to dextran sulfate and alleviates the severe background problems frequently associated with the inclusion of dextran sulfate in many hybridization recipes.

Formamide

Formamide ($HCONH_2$) destabilizes double-stranded molecules by interfering with hydrogen bond formation. Thus, the inclusion of freshly deionized formamide in hybridization recipes reduces T_m (and required hybridization temperature) in a linear manner for each 1% of added formamide. This roughly corresponds to a T_m reduction of roughly 2.4°C–2.9°C per mole of formamide, depending on the G + C content and other variables (Blake and Delcourt, 1996).

As a general rule, 1% formamide reduces the T_m of DNA probes by about 0.5°C, while an equivalent amount of formamide reduces the T_m of RNA probes by about 0.3°C. As it is commonplace to perform nucleic acid hybridization in buffers consisting of 50% formamide at 42°C, there is a corresponding decrease in T_m of about 25°C for a "typical" DNA probe, and about 15°C for a "typical" RNA probe. Thus, maximum hybridization between complementary nucleic acid molecules is favored while maintaining high stringency conditions at lower temperatures. Conducting RNA analysis in formamide-containing buffers also extends the useful life of heat-labile probe molecules. One should be aware, however, that formamide is volatile at the temperatures commonly used to perform hybridization, so care should be taken to avoid exposure to its toxic fumes.

A variety of nonformamide-based hybridization buffers which drastically accelerate forward hybridization kinetics are likewise available. Many of these preparations contain a surfactant that prevents the probe from binding to the filter, thereby minimizing background hybridization. These hybridization reagents of the proprietary formulation are convenient and work remarkably well. Plus, the elimination of formamide from the hybridization recipe is an added bonus.

Urea

The toxic properties of formamide, a common component of traditional prehybridization and hybridization buffers, present intrinsic exposure dangers in the lab. Alternative formulations have been developed containing urea; hybridization solutions containing 6 M urea are equivalent to those containing 50% formamide with respect to lowering T_m. In particular, the ECL nucleic acid labeling and detection system (Cytiva) is designed in this manner. This system is compatible with the common nucleic acid blotting applications and tends to produce posthybridization detection images with very low levels of background.

Urea-based hybridization formulations are compatible with both DNA probes and RNA probes (Simard et al., 2001). The concentration of urea needed to be effectively destabilizing ranges from 2 to 6 M. Urea stock solutions should be freshly prepared prior to use.

DNA probe synthesis

DNA probes are a diverse lot. They include double-stranded cDNA or genomic DNA sequences, first-strand cDNA synthesized directly from one or a mixture of mRNAs, oligonucleotides, and PCR products. One disadvantage of DNA probes is that, when bound to a complementary molecule, they are less thermodynamically stable than RNA probes, meaning that stringency must be carefully monitored (Chapter 3: Stringency: Conditions that Influence Nucleic Acid Structure). Regarding DNA probes, note that:

1. Double-stranded probes require denaturation into their constituent single strands in order to be used as probes; several labeling techniques are described below. Double-stranded DNA probes, often consisting of restriction fragments cloned into plasmids, may or may not be excised from the plasmid prior to use. This is an example of a homogenous probe, in which all probe molecules are identical.
2. First-strand cDNA is synthesized from mRNA (or from an enriched fraction thereof) in the presence of labeled dNTP precursors. This generates a heterogeneous probe that can be used directly for hybridization. Heterogeneous probes are particularly useful for the identification of sequences that are uniquely present in one of two or more RNA populations (e.g., subtractive hybridization methods, Chapter 22: Subtractive and Nonsubtractive Methods for the Analysis of Gene Expression), for microarray screening (Chapter 21: Array Analysis of Gene Expression), and for reverse Northern blots (Chapter 16: Northern Analysis).
3. Oligonucleotides are artificially synthesized single-stranded DNA. "Oligos" were used as standard hybridization probes for years before the development of PCR. Thus, the primers that an investigator might now be used to support sequence amplification by PCR could easily function as stand-alone hybridization probes. Oligonucleotides are classified as "long" oligomers, consisting of as many as 100 + bases, or "short" oligomers, usually consisting of fewer than 30 bases. The advantage of working with oligonucleotides is that the investigator has complete control over the sequence, which is made to order. The main disadvantage of using oligonucleotides is the same as the main advantage: because the investigator has control over the sequence of the oligonucleotide, one must know *which* sequence to have synthesized. The short oligonucleotide probes offer the greatest flexibility and experimental latitude in terms of discriminating power and predictable thermodynamic behavior. In the

event that a definitive oligonucleotide probe sequence is not discernible, perhaps due to codon degeneracy, it is possible to use several oligonucleotides simultaneously in what might be thought of as a hybridization cocktail; alternatively, one may elect to use only one oligonucleotide, though at lower stringency. While this strategy reduces the concentration of the correct complementary sequence and may favor semi-nonspecific hybridization, the important point is that authentic hybridization events do not go undetected. See Chapter 10, RT-PCR: A Science and an Art Form, for tips on the design of degenerate oligonucleotide probes and primers.

4. Any PCR product can be used as a probe. Label incorporation occurs during the PCR process by incorporation of hapten-labeled nucleotides into the PCR products. Alternatively, primers can be synthesized with a hapten at the 5′ end; because the primers become part of the product, label incorporation is efficient and automatic during amplification of the template. Radiolabeling PCR products during their synthesis is also possible, though the subsequent clean-up steps tend to create a bit of a radioactive mess.

A variety of methodologies for labeling DNA are available and are classified depending on the distribution of labels within the probe molecule itself. End-labeled probes are those to which the label has been added at the 5′ end, as with the kinasing reaction (Fig. 17.1A), or at the 3′ end by the action of terminal deoxynucleotidyl transferase or simply "terminal transferase" (Fig. 17.1B). Continuously labeled probes are those in which the label has been added along the length of the backbone of the molecule (Fig. 17.2). Examples of continuously labeled probes include, but are not limited to, those generated by random priming, PCR, in vitro transcription, or

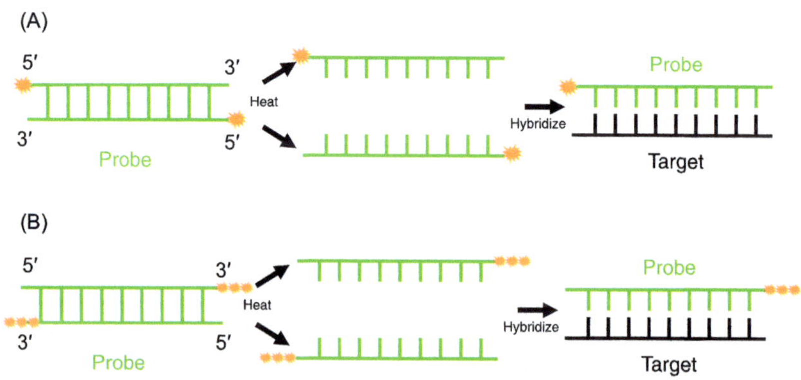

FIGURE 17.1 DNA end-labeling procedures for probe synthesis. (A) The kinasing reaction transfers a radiolabel to the 5′ end. (B) Terminal transferase is used to label the 3′ end of DNA probe molecules.

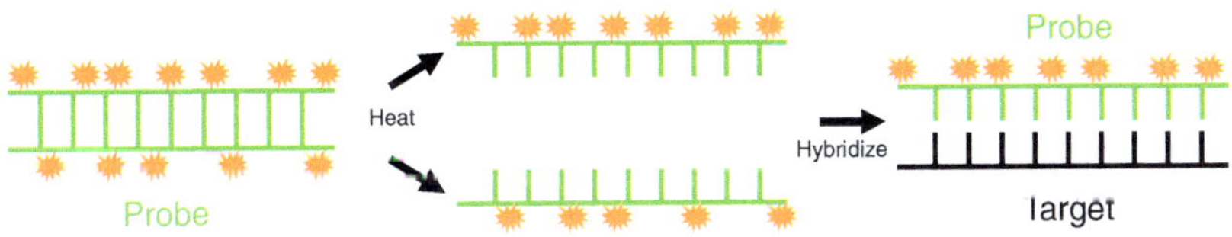

FIGURE 17.2 Continuous probe labeling. The incorporation of the label along the length of the DNA probe can be accomplished by any of several methods, including random priming, nick translation, and PCR.

cross-linking of a hapten or enzyme directly to the backbone of a molecule. The choice between end-labeling and continuously-labeling techniques depends on (1) whether the probe is DNA or RNA, (2) whether the probe is single-stranded or double-stranded, (3) whether the probe, if double-stranded, has a recessed or protruding 5′ and 3′ ends, (4) the required level of label incorporation and detection sensitivity, (5) the length of the probe to be labeled, and (6) whether the probe, if double-stranded DNA, is linear or a covalently closed circle (e.g., a plasmid).

A brief description of the more common techniques follows. All of these labeling reactions are commercially available in kit form, containing all necessary reagents, with the exception of isotopes for radiolabeling techniques. In general, most labeling systems used to generate radioactive probes can also be used to generate a variety of nonisotopic probes. When radiolabeling, be sure that the position of the isotope in the radiolabel donor (i.e., α or γ label) supports the intended labeling method.

Polymerase chain reaction

PCR is an excellent method for probe synthesis, requiring exquisitely small quantities of template material. In the presence of the radiolabeled or hapten-labeled nucleotide precursor, PCR products are labeled as they are being synthesized. Alternatively, the primers themselves may be labeled nonisotopically during their own synthesis, negating the requirement for the inclusion of labeled nucleotide precursors as part of the reaction mix; this is the preferred method.

PCR as a labeling method offers several advantages: speed, versatility, the efficiency of the reaction, and widespread acceptance. Minute quantities of starting material are required, and the resulting molecules are of uniform length. When performed in the presence of labeled nucleotides, PCR-generated probes are continuously labeled and generally show a high degree of label incorporation. Further, the elevated temperatures at which thermostable DNA polymerases function reduces the incidence of enzyme-mediated point mutations during probe synthesis. Moreover, the use of proofreading enzymes and enzyme blends, described in Chapter 10, RT-PCR: A Science and an Art Form, may be helpful in this regard. For comprehensive reviews of reaction parameters and other PCR-related strategies, see Maddocks and Jenkins (2017).

Random priming

Random priming is a type of primer extension in which a mixture of small oligonucleotide sequences, acting as primers, anneal to a heat-denatured double-stranded template (Feinberg and Vogelstein, 1983, 1984). The annealed primers ultimately become part of the probe itself because the Klenow enzyme (a DNA polymerase) incorporates the label as the random primers are extended in the 3′ direction. Random priming is significantly more efficient using linearized DNA molecules; attempts to label covalently closed, supercoiled DNA typically result in probe-specific activities 20- to 30-fold less than those of the corresponding linear DNA. DNA molecules between 200 and 2000 bp are the best candidates for random priming, although template length is not a critical parameter in this labeling reaction. When labeling with ^{32}P, random priming typically produces probes with specific activities of about $1-3 \times 10^9$ cpm/μg when 2.5–10 ng of starting template is used. Labeling usually requires 10–30 minutes.

Nick translation

Nick translation (Rigby et al., 1977) is one of the original probe labeling techniques. It involves randomly nicking the backbone of a double-stranded DNA with DNase I. At very low concentrations, this enzyme nicks a template at four or five sites, producing a free 3′-OH group that can act as a primer at each nicking location. Next, the enzyme DNA polymerase I remove the native nucleotides from the probe molecules in the 5′→3′ direction (exonuclease activity) while replacing them with labeled dNTP precursors by virtue of its 5′→3′ polymerase activity. Nick translation is efficient for both linear and covalently closed DNA molecules, and labeling requires about 1 hour. When labeling with ^{32}P, nick translation produces probes with specific activities of approximately $3-5 \times 10^8$ cpm/μg. The method of nick translation supports efficient nonisotopic labeling as well.

5′ End-labeling of DNA

An alternative method to generating a probe by continuously labeling it is to label the 5′ end of the molecule with the addition of a radiolabeled phosphate. This method of 5′ end-labeling is colloquially known as the kinasing reaction; it specifically involves the transfer of the γ phosphate of ATP (not dATP) to a 5′-OH substrate group on dephosphorylated DNA molecules. This labeling reaction is catalyzed by the enzyme T4 polynucleotide kinase. Known as the forward reaction, kinasing is an excellent method for labeling short oligonucleotides though it is much less efficient at labeling blunt-end dsDNA or dsDNA with recessed 5′ ends. The forward kinasing reaction is far more efficient than the exchange reaction, which involves the substitution of 5′ phosphates, so it is prudent to remove a 5′ phosphate group, if present, using a phosphatase. DNA kinasing reactions typically require 30 minutes, and this type of labeling is renowned for generating very high specific activity probes. Be sure to use γ-labeled ATP as the phosphate donor; a common error among novices is the incorrect use of α-labeled nucleotides that support other labeling strategies. Finally, a variety of nonisotopic labeling kits are available for 5′-end modification.

3′ End-labeling of DNA

Probe synthesis by 3′ end-labeling involves the addition of nucleotides to the 3′ end of either DNA or RNA. DNA 3′ end-labeling is often catalyzed by terminal transferase or the well-known PCR enzyme *Taq* DNA polymerase. Single- and double-stranded DNA molecules are labeled by the addition of dNTPs to 3′-OH termini; double-stranded, blunt-ended fragments and fragments with 3′ overhang structures are labeled efficiently. A most unusual feature of terminal transferase is that there is no template strand requirement, that is, this enzyme is a nontemplate-dependent polymerase. Consequently, any one dNTP (or dNTP mixture) can be selected for addition to the 3′ end of a polynucleotide. This type of labeling usually produces a 3′ overhang of three to five extra nucleotides. Should the investigator wish to add only one labeled nucleotide in this manner, then a chain-terminating nucleotide (e.g., ddATP) should be used.

The *Taq* DNA polymerase also has terminal transferase activity, as do other nonproofreading polymerases. The well-known *Taq* polymerase, for example, adds an extra nucleotide, usually an "A," to the 3′ end of each polynucleotide strand that it synthesizes. Double-stranded PCR products made by *Taq*, therefore, are expected to have an extra "A" at the 3′ end of each strand. This is the basis of the popular TA cloning process.

The disadvantage of this labeling technique can be pronounced when the labeling of short oligonucleotides is involved. The addition of nucleotides changes the length of the molecule, which may or may not modify the specificity of the oligonucleotide, as well as its thermodynamic characteristics. If an oligonucleotide is labeled by nucleotide addition to the 3′ end, it is strongly suggested that dATP be used, because of the lesser thermodynamic stability of A::T base-pairing compared to G:::C base pairs.

Sense and antisense RNA probe synthesis

RNA hybridization probes are popular because they offer some important advantages. These probes are commonly synthesized by in vitro transcription. Compared to the various methods for DNA probe synthesis there are fewer RNA labeling methods, and RNA probes can be substituted for DNA probes in nearly all applications. In the construction of the transcription template, the cDNA or gDNA to be transcribed is usually ligated between two different transcription promoters which flank a multiple cloning site and are positioned in opposite orientations. One common construction features the highly efficient SP6, T7, and/or T3 RNA polymerase promoters (Fig. 17.3). Linearizing the transcription vector with an appropriate restriction enzyme prior to initiating the transcription reaction favors the efficient synthesis of large amounts of uniform length, and continuously labeled probe molecules (Fig. 17.4). An important part of the preparation of RNA probes is the transcription of antisense and sense transcripts from the same template construction in separate reaction tubes.

Antisense RNA is, by definition, complementary to mRNA and is therefore able to base pair to it. Antisense probes are widely used for in situ hybridization studies (Chapter 20: RNA In Situ Hybridization) in which the natural tissue geometry and therefore mRNA distribution is preserved. Consider the following example. An investigator wishes to characterize the cell-specific expression of the *N-myc* gene in rat cerebellum. To make this determination, a tissue section is prepared for in situ hybridization. Because the target is mRNA, antisense RNA (cRNA) should be used as the probe. The natural thermodynamic stability of cRNA hybridized to mRNA permits very stringent assay conditions. Sense RNA probe, that is, a probe with the same sequence as the mRNA, is also generated separately and used as a negative control to

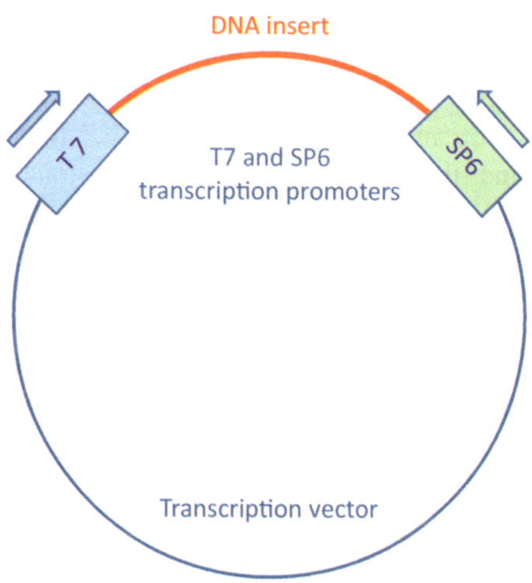

FIGURE 17.3 Typical orientation of dual RNA polymerase promoters in a plasmid construction that supports in vitro transcription. The DNA template for transcription is cloned between the two promoters. The plasmid construction is then linearized prior to initiating the in vitro transcription reaction.

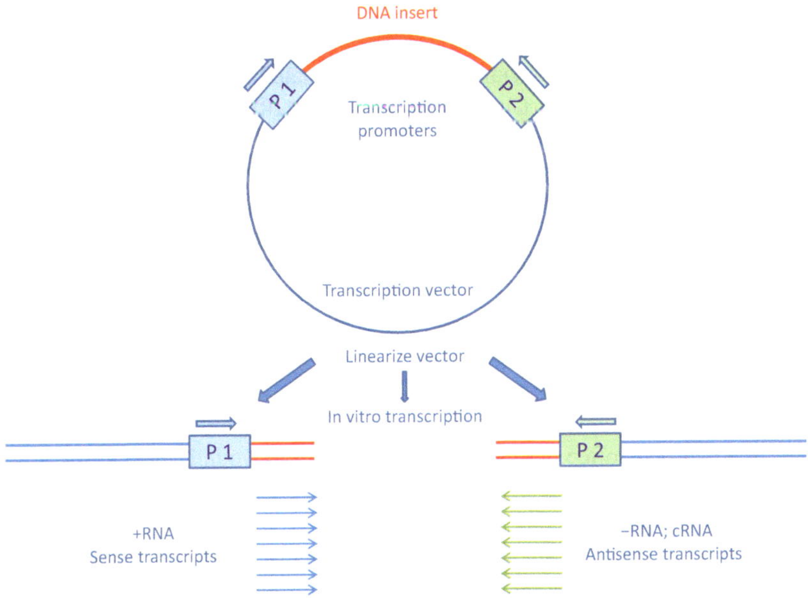

FIGURE 17.4 Linearized in vitro transcription templates favor the efficient production of large quantities of a continuously labeled probe of uniform length. Note that the positioning of the promoters influences whether the resulting transcript will be sense or antisense RNA. If the orientation of the insert cDNA is unknown, then an empirical determination will be required to discern which promoter produces which type of transcript (+RNA or − RNA).

assess the degree of nonspecific hybridization, most often using the next sequential tissue slice. Chromogenic detection is particularly well-suited for this type of assay and the distribution of color precipitate is assessed by light microscopy. As needed more quantitative information pertaining to the expression of *N-myc* can be obtained using PCR-based methods.

The following are characteristics of RNA probes.

1. RNA probes are single-stranded. While such probes do not require boiling prior to use, heating them briefly will help to disrupt any intramolecular base-pairing that may have occurred.
2. RNA probes are synthesized by in vitro transcription from a linearized template, resulting in probe molecules of uniform length.
3. RNA probes are continuously labeled as they are being transcribed, thereby generating probes with a high degree of label incorporation.
4. Large quantities of the probe can be synthesized in a single in vitro transcription reaction.
5. All RNA probe molecules are available for hybridization. Single-stranded, denatured RNA molecules cannot renature, as occurs with dsDNA. Intramolecular base-pairing, however, when not discouraged by the assay conditions, will reduce the effective concentration of the probe.

6. RNA probes show greater thermodynamic stability when base-paired to either RNA (RNA:RNA) or DNA (RNA:DNA) target molecules, compared to the thermodynamic stability associated with DNA (DNA:DNA) probes. As noted in Chapter 3, Stringency: Conditions that Influence Nucleic Acid Structure, double-stranded molecules, in order of decreasing thermostability, are RNA:RNA > RNA:DNA > DNA:DNA.
7. The SP6, T7, and T3 bacteriophage RNA polymerase promoters demonstrate virtually no cross-reactivity; therefore, transcription reactions initiated from either promoter are virtually free of transcripts of the opposite sense. Consequently, both sense and antisense RNA probes can be synthesized as needed.
8. RNA transcribed in vitro must be handled with the level of attention used any other RNA sample to control RNase activity; failure to do so will result in rapid degradation of the probe.
9. RNA probes often produce high levels of background; thus, at the conclusion of the hybridization period it is a common practice to digest all probe molecules with a cocktail of RNase A and RNase T1. This treatment will result in the degradation of all nonbase-paired molecules, that is, those that did not participate in duplex formation.
10. Radiolabeled RNA probes are often of such high specific activity that they may experience radiolysis (autodegradation resulting from radioactive decay) if stored for extended periods.

A brief description of RNA labeling techniques follows. RNA oligonucleotide probes may also be synthesized though it is more expensive compared with the cost of synthesizing a DNA oligonucleotide of the same length.

In vitro transcription

In vitro transcription is the only reliable, economical method for generating RNA probes. Large amounts of efficiently labeled probes of uniform length can be generated by transcription of a DNA sequence ligated next to an RNA promoter. As noted above, one excellent strategy is to clone the DNA to be transcribed between two promoters in opposite orientations. This allows either strand of the cloned DNA sequence to be transcribed in order to generate sense and antisense RNA for hybridization studies. This can be accomplished via the use of universally available cloning vectors. Alternatively, a template sequence can be positioned between one or two transcription promoters via PCR. RNA probes synthesized by in vitro transcription are referred to generically as riboprobes, even though Riboprobe is a trademark-protected name (Promega).

RNA probes synthesized by transcription in vitro are assembled from NTP precursors, much as occurs in vivo. Transcripts are elongated by the addition of nucleotide monophosphates into the nascent backbone of the probe. Isotopic labeling with ^{32}P-UTP requires a radiolabel in the α position, as with all continuously labeled probes. Since the probe is labeled throughout the transcription process, the degree of label incorporation is very high. In vitro transcription kits contain all necessary reagents (except isotopes). These products likewise support nonisotopic labeling nicely.

5′ End-labeling of RNA

One alternative method to generating continuously labeled RNA probes by in vitro transcription is to label the 5′ end of the molecule with the addition of a radiolabeled phosphate. Colloquially known as the kinasing reaction, it specifically involves the transfer of the γ phosphate of ATP to a 5′-OH substrate of RNA in the same manner as kinasing and is catalyzed by the enzyme T4 polynucleotide kinase (PNK).

3′ End-labeling of RNA

RNA can be 3′ end-labeled using either the enzyme T4 RNA ligase or poly (A) polymerase. In the case of the former, a labeled nucleotide can be added efficiently to the 3′ end of a single-stranded RNA. In the case of the latter, the enzyme that is responsible for natural polyadenylation can be used in vitro to catalyze the sequential addition of AMPs to the 3′ end of the molecule, using α-labeled ATP precursors. In addition to RNA probe synthesis reactions, poly(A) polymerase can be used to polyadenylate naturally poly $(A)^-$ mRNA and other RNAs in order to support oligo(dT) primer-mediated synthesis of cDNA. Although not a difficult labeling method, it is not a preferred way to make an RNA probe.

Selection of labeling system

Important considerations in the selection of a probe labeling system are (1) required sensitivity and resolution, (2) whether the probe is DNA or RNA, (3) method of label incorporation, (4) probe stability after labeling, (5) hybridization conditions, and (6) desired method of detection.

There are four basic methods for labeling nucleic acids: (1) fluorescent dye-labeled nucleotides for direct fluorescence detection (e.g., CY3, CY5, Texas Red and others), (2) direct enzyme labeling, (3) hapten incorporation [e.g., biotin or digoxigenin (DIG)] that will support nonisotopic detection by chemiluminescence, chemifluorescence, or chromogenic techniques, and (4) radiolabeling. These techniques are compatible with the labeling of DNA, RNA, and (usually) oligonucleotides for most applications, and each of which has merits and drawbacks. Numerous systems for the synthesis of nucleic acid probes are available for purchase and each has a recommended detection methodology for optimal sensitivity and resolution.

The ubiquitous dyes Cy3 and Cy5

In recent years, a new generation of fluorescent labels, and strategies for incorporating these labels, are in widespread use and have all but completely replaced fluorescein for nucleic acid labeling. Among the current favorites are the synthetic, water-soluble fluorescent dyes indocarbocyanine, more commonly known as Cy3, and indodicarbocyanine, commonly known as Cy5 (Fig. 17.5). These dyes are useful for labeling a variety of biomolecules, including RNA, DNA, and proteins. Incorporation of one or more of these labels in the synthesis of DNA probes may be performed by direct fluorochrome incorporation with labeled nucleotides, by indirect labeling via aminoallyl dUTP or aminoallyl dCTP followed by a secondary reaction to link Cy3 or Cy5 to the probe, or, in the case of oligonucleotides, during synthesis of molecule (5′ end, 3′ end, or internal base modification).

A very common combination of these dyes is Cy3 (green) and Cy5 (red). Cy2 (cyanine) may also be used with Cy3 and Cy5 when three-color fluorescence is needed. In addition to their usefulness for microarray work, these dyes are now mainstream for molecular biology applications because of the intensity and stability of their respective fluorescent signals. Fluorescence detection is a safe, cost-effective, validated, and widely recognized methodology for the detection of nucleic acids. In addition to the cyanine fluorescent dye family, dozens of other fluorochromes are available across the visible spectrum for various applications, including the Alexa Fluor and

Cy2	Cy3	Cy5
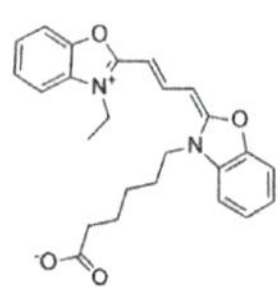	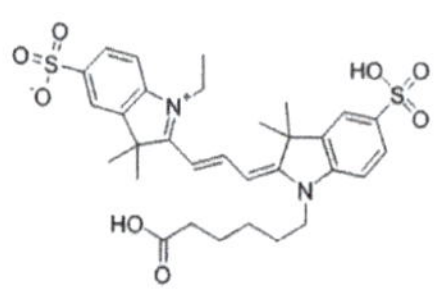	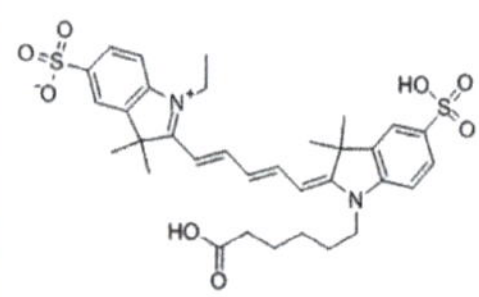
Cyanine	Indocarbocyanine	Indodicarbocyanine
MW 418.49	MW 630.77	MW 656.81
$C_{25}H_{26}N_2O_4$	$C_{31}H_{38}N_2O_8S_2$	$C_{33}H_{40}N_2O_8S_2$
Excitation 489nm	Excitation 548nm	Excitation 647nm
Emission 506nm	Emission 561nm	Emission 665nm
Approx. Color	Approx. Color	Approx. Color

FIGURE 17.5 Properties of the popular cyanine fluorochromes. These dyes are in widespread use and can be used alone or as a combination of two or more family members. They offer enhanced stability and a clearer signal with respect to fluorescence detection of labeled nucleic acids.

DyLight families (Thermo Fisher Scientific), FluoProbes (Interchim), and the Atto dyes (ATTO-TEC GmbH), to list but a few examples.

Popular nonisotopic platforms

As noted above, a nucleic acid probe is a long or short polynucleotide that carries some type of label or tag. The label allows the investigator to follow the probe throughout the experiment and provides a means for at least semi-quantitative detection. Sensitive chemiluminescence and fluorescent platforms for probe labeling and detection are now in the mainstream, often precluding the use of radioactive labeling materials altogether.

Nonisotopic approaches are, for the most part, quite compatible with nearly all molecular biology applications. These techniques include, but are not limited to, Northern analysis, Southern analysis, Western analysis, dot and slot blot analyses, colony hybridization, plaque screening, and nonfilter-based in situ hybridization. The rule of thumb that this Author frequently proffers is: "If you can see bands by autoradiography after the third day under film, then it is likely that you can switch to chemiluminescence without loss of sensitivity or resolution. Autoradiographic analysis requiring more than three days may or may not be fully compatible with nonisotopic methods." In many cases, PCR-based innovations have superseded the need to perform some of these time-honored methods altogether.

An ever-increasing variety of nonisotopic labeling and detection methodologies have been developed to minimize the inherent dangers and handling difficulties associated with the presence of radioisotopes in the laboratory. In many cases, the sensitivity achievable is comparable to that of ^{32}P, though much depends on the patience and skill of the user. Comparatively speaking, while the various nonisotopic labeling and detection systems offer similar levels of sensitivity and resolution, they differ markedly in the number of different steps which must be performed by the user, especially pertaining to the detection component of the system.

Options for nonisotopic labeling methods are biotinylation, labeling with DIG, fluorescein (FL) labeling, direct enzyme labeling, or labeling with fluorescent dyes, especially Cy2, Cy3, and Cy5. Biotin and DIG labeling support detection by chromogenic (colorimetric) methods as well as by chemiluminescence. Because of the intrinsic technical difficulties in many filter-based chromogenic techniques, this type of detection chemistry never gained widespread popularity. Improvements in nonisotopic labeling and detection have made this technology standard fare for some applications. Although the notion of eliminating radioactive probes is enticing, to say the least, the investigator must also be aware that the mechanics of nonisotopic labeling, probe purification, and detection may be drastically modified, compared to methods associated with the use of "hot" probes. The most significant change that experienced ^{32}P users will notice is the amount of dedicated time that must be allocated to perform the detection procedure. For example, the posthybridization stringency washes are usually followed by a membrane blocking step. Also, once the chemiluminescence substrates are applied, the membrane cannot be washed at higher stringency unless one repeats the entire series of posthybridization washes and blocking steps. The safety and economic benefits reaped by the elimination of isotopes far outweigh the investment of additional hands-on that this type of detection requires.

There are several ways to perform labeling reactions that support nonisotopic detection. Some of these methods involve hapten-labeling the probe, another popular format is known as direct enzyme labeling, and yet other methods involve the direct linkage of fluorescent dyes. Each approach, as described below, has advantages and disadvantages.

Direct enzyme labeling

Traditional methods for generating DNA probes involve the substitution or addition of labeled nucleotides to the probe. In contrast, direct enzyme labeling (ECL System; Cytiva) involves crosslinking a modified horseradish peroxidase (HRP) capable of supporting chemiluminescence directly to the backbone of heat-denatured DNA or RNA using glutaraldehyde as a bridge or linker arm. The method is extremely efficient and requires only a total of 20 minutes to label the probe. Succinctly, the probe is boiled for 5 minutes and then rapidly cooled on ice to minimize renaturation. Then, glutaraldehyde links the HRP to the now denatured probe. Because the probe cannot be boiled again after the labeling has been completed, and because the probe will, of course, reanneal in a relatively short time, one should label only the required mass of probe and then add it to the hybridization mix as quickly as possible after labeling. Conveniently, there are no subsequent steps required for the removal of unincorporated label. Because the enzyme is present throughout the hybridization and subsequent washes, stringency must be regulated by modifying ionic strength and by the inclusion of destabilizing agents such as SDS and urea, rather than by raising the temperature. In this laboratory the resulting X-ray films are almost always of outstanding quality with a very little background.

Biotin

Biotin is a small water-soluble vitamin that can be readily incorporated into a number of biological molecules. For probe synthesis, biotinylation can be accomplished in any one of several ways:

1. Biotinylated nucleotides. The nucleotide dUTP, functioning as a dTTP analog, is widely used for nonisotopic labeling. Biotin-labeled dUTP consistently generates the highest levels of label incorporation, by comparison with any of the other four deoxynucleotides in biotinylated form. Virtually all of the classical methods for enzymatic probe synthesis,

including random priming, nick translation, and PCR are compatible with biotinylated nucleotides. Moreover, both AMV and MMLV reverse transcriptase (Chapter 10: RT-PCR: A Science and an Art Form) will likewise recognize modified nucleotides for the synthesis of biotin-labeled, first-strand cDNA. RNA probes can also be synthesized in this manner via in vitro transcription.

The structure of the most commonly used biotinylated nucleotide, Bio-11-dUTP (biotin-11−2′-deoxyuridine-5′-triphosphate, tetralithium salt), is shown in Fig. 17.6. The prefix "Bio-11" means that the nucleotide is biotinylated and that there is an 11-atom spacer arm separating the biotin itself from the nucleoside component, the purpose of which is to prevent any steric interference. Bio-21-dUTP is also available for labeling DNA, as are Bio-11-UTP and Bio-16-UTP for labeling RNA (biotin-11−2′-uridine-5′-triphosphate and biotin-16−2′-uridine-5′-triphosphate, respectively). In general, the longer the linker arm, the greater the ease with which the biotin can interact with the required streptavidin during the detection process, as described below. In contrast, the shorter the linker arm, the more efficient is the enzymatic incorporation of the labeled. Bio-11-dUTP seems to offer the best balance between efficiency of label incorporation and detection sensitivity and is thus widely used for nonisotopic DNA probe synthesis.

2. Biotinylated primers. The easiest method of biotinylation is during the actual synthesis of oligonucleotides, which can be used directly as probes or to support the polymerase chain reaction. Primers can be biotinylated at their 5′ end during synthesis at minimal additional cost, thereby precluding the need to biotinylate later on; the biotin label will not interfere with primer specificity or amplification efficiency.
3. Psoralen biotin (Cimino et al., 1985; Wassarman, 1993). Psoralens and their derivatives are planar tricyclic compounds that intercalate into double-stranded nucleic acids upon irradiation, ideally around 350 nm. These molecules become covalently attached to nucleotide bases,

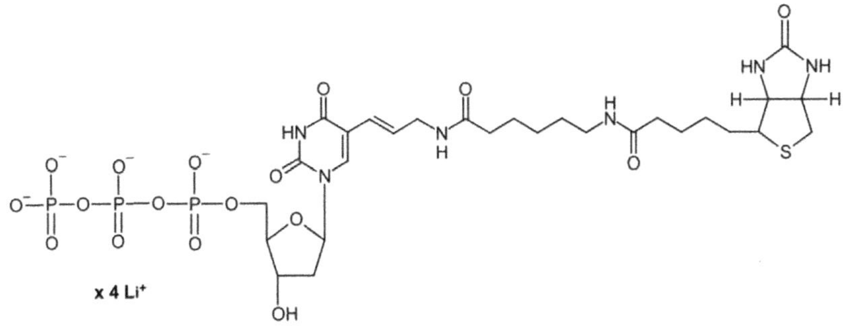

FIGURE 17.6 Biotin-11-dUTP.

preferentially to thymidine, with an average of one biotinylation event per each 10–20 nucleotides. To a lesser extent, single-stranded molecules, such as oligonucleotides, may also become biotinylated, though this strategy is not recommended; oligomers should be synthesized with biotin at the 5′ end. The most recent generation of psoralen biotin derivatives feature a three-unit, polyethylene glycol (PEG) group acting as a spacer arm, hence the nomenclature psoralen-PEG3-biotin (ThermoFisher Scientific). Psoralen biotin derivatives are very useful for the synthesis of biotinylated probes for nucleic acid hybridization. Unincorporated psoralen biotin conjugates are removed by simple extraction with *n*-butanol. Note that psoralen biotin, as well as photoreactive TFPA-PEG_3-Biotin (ThermoFisher Scientific), a trifluorophenyl azide (Soundararajan et al., 1993), have largely replaced photoactivable biotin acetate, once a staple in the early molecular biology laboratory. After labeling, biotinylated probes are stable for at least 1 year when stored at −20°C, and fewer the number of freeze-thaw cycles, the better. An aliquot of probe is thawed just prior to use and, if double-stranded, is boiled, rapidly cooled on ice, and then added immediately to the hybridization buffer. Biotinylated probes support both chromogenic detection and chemiluminescence; from a detection point of view, it really does not matter how the biotin becomes part of the probe.

Digoxigenin

DIG is a steroid hapten derived from plants of the genus *Digitalis* (*D. lanata*, aka the Grecian foxglove, and the purple foxglove *D. purpurea*), the latter of which is same source as the cardiac medication digitalis. It is an excellent nucleic acid label which supports both chromogenic detection and chemiluminescence (Martin et al., 1990). DNA probes are enzymatically labeled, usually by random priming with digoxigenin-dUTP (DIG-11-dUTP; Fig. 17.7), and RNA probes are synthesized by in vitro transcription with DIG-11-UTP. These nucleotides are linked via a spacer arm to DIG, much the same as the biotin-containing nucleotides described above. The resulting DIG-labeled molecules then function as hybridization probes in much the same manner as any other type of probe. Oligonucleotides can also be DIG-labeled during their synthesis, thereby precluding cumbersome, inefficient labeling reactions at some later date. As with biotinylation, DIG labeling does not interfere with biological activity, and probes so labeled are stable at −20°C for at least 1 year after labeling. DIG labeling supports posthybridization detection by chemiluminescence or by the formation of an insoluble color precipitate directly on the filter membrane. This method is also well respected for in situ hybridization applications (Matturro and Rossetti, 2015; Schröder et al., 2000; Crabb et al., 1992; Robbins et al., 1991).

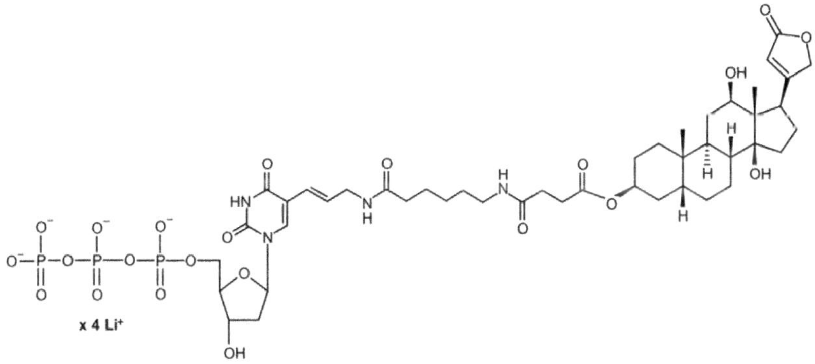

FIGURE 17.7 DIG-11-dUTP.

Fluorescein

As recently as a few years ago, the only fluorescent nucleotides readily available for either DNA or RNA probe synthesis were fluorescein (FL)-labeled: FL-12-dUTP for the synthesis of DNA probes and FL-12-UTP for the synthesis of RNA probes (Fig. 17.8). These nucleotides were used to generate probes in much the same way as biotin- and DIG-labeled nucleotides or by direct incorporation during oligonucleotide (primer) synthesis. Fluorescein as a nucleic acid label has fallen out of favor, having been replaced by newer fluorophores that have enhanced properties for this application. When used, direct fluorescence detection is strongly preferred; the use of antifluorescein antibodies is typically reserved for immunofluorescence microscopy and ELISA assays.

Isotope labeling

Isotopes offer excellent sensitivity and are compatible with many applications. Care and common sense should always be used when working with or storing radioactive materials. A lab coat, disposable gloves, eye protection, appropriate shielding, film badges to monitor exposure, and proper training are always essential, regardless of the isotope or its specific activity. In every case, the radiation safety officer should be consulted to ensure compliance with departmental or institution-wide policies.

Handling, storage, and containment issues aside, an important concern when performing radiolabeling is the half-life of the isotope. The most frequently used isotopes for nucleic acid probe synthesis are ^{32}P, ^{3}H, and ^{35}S and, less frequently, ^{33}P. Nucleotide triphosphate precursors must be labeled in the appropriate position (e.g., α or γ phosphate) to support the enzyme-mediated transfer of the portion of the nucleotide containing the isotope to each of the molecules being labeled. The intended application and the

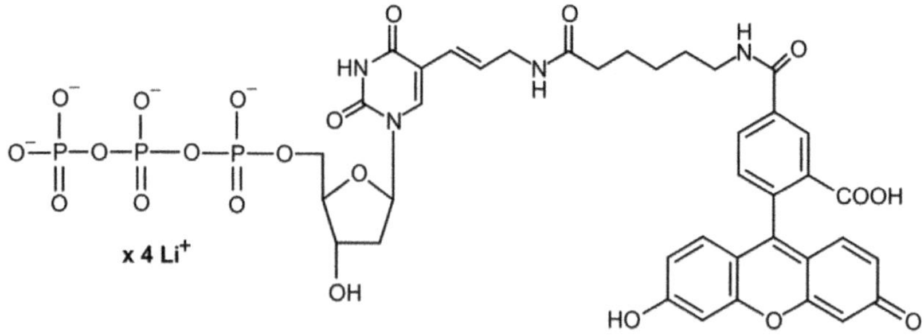

FIGURE 17.8 FL-12-dUTP.

precise method of detection dictate the selection of isotope. For example, the long path length associated with β-emission from ^{32}P-labeled probes makes this label useful for recording hybridization events on X-ray film, and for precisely the same reason, an anatomical resolution is inadequate for RNA target localization in situ. This necessitates the use of an alternative labeling and detection system.

The ready availability of ^{33}P has assuaged some of the difficulties inherent in and fears associated with the use of ^{32}P. As an alternative, ^{33}P-labeled nucleotides have been used for general molecular biology applications including DNA sequencing, at least prior to the development of next-generation sequencing platforms. The half-life of ^{33}P is 25.4 days with a β E_{max} = 0.25 MeV, which is about sevenfold less than that of ^{32}P (E_{max} = 1.71 MeV). ^{33}P can be handled on the benchtop using routine radiation safety practices but with less bulky shielding (Zagursky et al., 1991) than is required when handling ^{32}P, though one should always exercise caution nonetheless and follow guidelines established by the company or campus Radiation Safety Office. To put this into perspective and for the convenience of the reader, the half-life data for commonly used isotopes are presented in Table 17.3.

The shelf-life of a nucleotide depends on two factors: its radioactive half-life and its biological half-life. Whereas the radioactive half-life is constant, the rate of biological degradation varies depending on the handling of the material and the temperature. Storing vials of radioactive nucleotides in a refrigerator or worse, at room temperature, can greatly increase the rate of radiolysis and biological degradation. Decomposition products consist primarily of nucleotide monophosphates and inorganic phosphate, although nucleotide diphosphates can also be present. Nucleotides are shipped to arrive on dry ice and should be stored frozen below −20°C (frost-free freezers not recommended), although formulations such as EasyTides (PerkinElmer) are shipped at ambient temperature and may be confidently stored at 4°C in order to avoid multiple freeze-thaw cycles.

TABLE 17.3 Isotope decay data.

Isotope	β *E*max (MeV)	Physical half-life	Amount remaining after 30 days	Amount remaining after 60 days	Amount remaining after 1 year
^{32}P	1.71	14.29 days	0.233	0.054	0.0
^{33}P	0.249	25.34 days	0.441	0.195	0.00005
^{35}S	0.167	87.4 days	0.788	0.621	0.05541
^{3}H	0.019	12.28 years	0.995	0.991	0.94534

The following is a partial list of recommendations to help radiochemical users minimize decomposition problems. More complete details regarding the safe handling and usage of radiochemicals can be found in PerkinElmer's excellent "Guide to the Safe Handling of Radioactive Materials in Research" link which can be accessed online at http://www.perkinelmer.com/safehandleguide, and from which several of the following points have been abstracted and reproduced with permission.

1. Use the radiochemical as soon as possible after receiving it. Prolonged storage allows time for additional nuclear decay, causing increased radical generation and an accompanying greater degree of decomposition. Decomposition will occur even if the radiochemical container is left unopened and stored under the most ideal conditions.
2. Store the sample properly. Pay close attention to the recommended storage conditions, as outlined on the technical data sheet that accompanies each product.
3. Follow the technical data sheet instructions when opening an ampoule containing a radiochemical. Refer to these instructions for storage information.
4. If the radiochemicals are frozen, then they should be thawed at room temperature until melting. This includes ^{3}H-labeled radiochemicals stored in aqueous solvents, and ^{35}S-, ^{125}I- and ^{32}P-labeled biochemical. The exceptions are ^{35}S- and ^{32}P-labeled nucleotides, which should be quick-thawed. For radiochemicals stored at very low temperatures, slowly thawing the sample in the refrigerator (or on ice) is recommended.
5. Do not allow radiochemicals to sit at room temperature for prolonged periods of time, because this will significantly increase the rate of radiochemical decomposition.

6. Minimize the number of times the primary container is opened and reseal the container immediately after each use. If a compound will be used several times, aliquot the required amounts into separate vials for storage. Each time a solution is handled, impurities, especially oxygen and water, may be introduced.

Use clean syringes and pipette tips when withdrawing aliquots from the primary container, since many buffers and salts used in biological systems are harmful to the radiochemical.

Probe purification and storage

As with most labeling techniques, unincorporated nucleotides must be separated from the labeled probe. Failure to do so, particularly when working with isotopes, usually results in unacceptably high levels of background. There are three basic methods for cleaning up the probe. For optimal results, follow the manufacturer's recommendations that accompany the labeling system.

The first method, which is preferred, for probe purification involves the use of one of the numerous concentrating or absorbent devices for probe purification. These devices include the High Pure PCR system (Roche), Ultrafree filtration units, and Amicon/Centricon devices, to name but a few options. All represent rapid, convenient alternatives to traditional column chromatography or alcohol precipitation, for separating radiolabeled nucleic acids from unincorporated nucleotides. The disposable nature of these devices minimizes the spread of radiolabeled waste products as well.

The second method for probe purification is gel filtration, or a permutation thereof. One popular approach is to make or purchase a so-called spun column (Sambrook and Russell, 2001), which consists of little more than a 1 mL syringe packed with Sephadex (Cytiva); Sephadex G-50 is used for larger probes, while Sephadex G-25 is reserved for the purification of labeled oligonucleotides. Spun-column chromatography is a variant of gel filtration, involving centrifugation of the column at $1000-1200 \times g$ for approximately 4 minutes; the probe is rapidly eluted while the unincorporated nucleotides lag behind in the column. In this lab, more than 95% of the unincorporated label is routinely removed after the labeling reaction. Gel filtration probe purification is compatible with a wide variety of labeling methods, both isotopic and nonisotopic. Premade spin- and gravity-fed columns also available for purchase.

In the third approach, the probe can be ethanol precipitated, in which case the unincorporated label remains in the supernatant. Whereas this is an efficient approach for purifying larger probes, attempted precipitation of oligonucleotides can be rather difficult, especially if small masses of nucleic acids are involved. Precipitation of probes is accomplished with a combination of salt and alcohol (see Chapter 4: RNA Isolation Strategies, Table 4.2), most often 0.1 volume of 3 M NaOAc and 2.5 volumes of 95% ethanol. Note that probes synthesized by direct enzyme labeling do not require any type of clean-up or purification and should never be exposed to ethanol or excess salt.

Following probe clean-up, isotopically labeled probes are useful only as long as there is sufficient activity remaining and the probe molecules themselves remain intact. In general, radiolabeled probes should be used as soon after labeling as possible. Radiolysis of probes, especially those labeled to an extremely high specific activity, becomes increasingly problematic as the probe ages. Store probes at −20°C or −80°C until just prior to use. Avoid repeated freezing and thawing of probes.

Nonisotopic labeling systems vary with respect to the recommended post-labeling storage and handling of probes. Biotinylated and IG-labeled probes can be stored at −20°C for up to 1 year. In some labs, an entire day is invested generating isotopic probes for the experiments that are planned for the next several weeks or months. Probes are then available for hybridization reactions as needed. Probes should never be denatured until just before use.

Mixed phase hybridization: Northern and Southern blots

Experiments involving hybridization between the nucleic acid probe and target sequences consist of three major components: prehybridization (also known as blocking), hybridization, and posthybridization washing (also known as the stringency washes). The detection of hybridization events following the stringency washes, the subject of the next section, is a different art entirely.

The exact hybridization conditions for many of the universally available probes have been defined in the literature. Various generic hybridization buffers and conditions have been recommended, combinations that purportedly yield the "perfect blot every time." Although it is true that the kinetic behavior of some families of nucleic acid molecules can be predicted with a fair degree of accuracy, understanding the parameters that directly influence hybrid formation is invaluable in fine-tuning hybridization conditions. This becomes a paramount issue when using probes that are mismatched to the target(s) or perhaps when using a nucleic acid probe to screen blots or libraries in phylogenetic studies. It should be noted that temperature and salt concentration are the two variables most often adjusted, in that order, to modify hybridization conditions.

Prehybridization: filter preparation

As described in Chapter 16, Northern Analysis, nucleic acid samples should be fixed to filter membranes immediately after the transfer. The filters can then be stored for extended periods. Before hybridization, filters are subjected to a prehybridization incubation (1) to equilibrate the filter in a buffer identical or very similar to the one to be used for the actual hybridization, and (2) to block the filter paper completely. Blocking means covering the filter entirely with sheared, denatured (heterologous) DNA except for those areas that are already occupied by experimental RNA samples; salmon sperm DNA or calf thymus DNA is often selected to ensure dissimilarity with the origin of the experimental RNA. Heterologous DNA, used as a carrier and a filter membrane block is prepared as a 10 mg/mL stock solution in water. The DNA is boiled to facilitate dissolving and then sheared by drawing it repeatedly through an 18-gage needle. Stock solutions of heterologous DNA are maintained at 4°C for short-term storage or at −20°C for long-term storage. Just prior to use, an appropriate aliquot must be denatured again by boiling for ten minutes. Newer methods involve blocking filters, especially nylon, with casein or other protein, rather than with heterologous DNA. In general, protein blocking requires less time than blocking a filter with heterologous DNA. Further, blocking with protein is ideal for isotopic and nonisotopic detection systems alike, yielding an outstanding signal-to-noise ratio with a very low background. Thorough prehybridization ensures that, when added, the probe molecules do not stick in a nonspecific fashion over the surface of the filter paper, which always causes unacceptable high background levels. Numerous recipes have been concocted for prehybridization, their exact composition being a function of the type of filter and probe used in a specific experimental application. The recommendations contained herein have been used with great success in this and other laboratories.

Protocol: probe hybridization

1. Pre-wet the filter in 5 × SSC or 5 × SSPE for 2–3 minutes.
   Note 1: SSPE is a better buffer than SSC, especially in the presence of formamide because the phosphate in SSPE actually mimics the nucleic acid phosphodiester backbone and helps to block the filter and improve the background at the detection stage. This strategy is especially desirable when experimental RNA is immobilized on nylon (+) filters, due to the natural affinity of the membrane for the negatively charged phosphodiester backbone of the nucleic probe.
   Note 2: 20 × SSPE = 3 M NaCl, 200 mM NaH_2PO_4, 20 mM EDTA, adjust pH to 7.4.
   Note 3: 20 × SSC = 3 M NaCl, 0.3 M Na_3-citrate, adjust pH to 7.0.
2. Prehybridize the filter(s) for 3–4 hours at 42°C in an excess volume (250–1000 μL/cm^2) of prehybridization buffer consisting of 5 × SSPE, 5 × Denhardt's solution, 0.1% SDS, 100 μg/mL freshly denatured salmon sperm DNA, 50% deionized formamide. The following is the

proper method for assembling the components of the prehybridization buffer:

a. Prewarm freshly deionized formamide to 45°C.

b. In a separate bottle or tube, mix SSPE, Denhardt's solution, and SDS and place at 45°C. The SDS will go into solution when the remaining ingredients are added.

c. In a fresh polypropylene or glass test tube, mix the required salmon sperm DNA in the volume of sterile water necessary to bring the prehybridization buffer up to the final volume. Boil the mixture for 5–7 minutes.

d. Carefully add the boiled salmon sperm DNA to the SSPE, Denhardt's, SDS mixture, and then add the entire mixture directly to the prewarmed formamide. Swirl carefully to mix. Store this hybridization solution at 42°C until ready to use.

The reason for assembling the buffer in this fashion is to minimize the renaturation of denatured salmon sperm DNA that could occur by using the more common strategy of plunging a small volume of concentrated stock DNA on ice. The immediate addition of formamide effectively reduces the rate of renaturation, especially when all reagents are prewarmed. After mixing, the buffer is ready to be used immediately. Typically, prepare 100 mL of this buffer to prehybridize two blots and reserve 25 mL, which can be used for hybridization.

Note 1: Routine prehybridization and hybridization are most expediently carried out in a plastic container with a perfectly flat bottom, the surface area of which is only slightly greater than the area of the filter. Several configurations of Rubbermaid food storage boxes and Nalgene utility boxes work well. Important: cover the container with a securely fitting lid to prevent evaporation. This approach precludes introducing air bubbles into heat-sealable plastic bags and avoids spillage of high specific activity hybridization buffer at the conclusion of the hybridization period.

Note 2: Placing the hybridization vessel on a gently moving (30–50 rpm) orbital shaker incubator will circulate the buffer evenly over the entire surface of the filter and yield superior data. Only shaker incubators that can maintain the required prehybridization temperature should be used.

Note 3: Prehybridization with a large (excess) volume of buffer usually favors very low levels of nonspecific interaction between the probe and the filter. After blocking, the volume of the hybridization buffer can be reduced before adding the probe so as to maintain the probe at a concentration that will drive forward hybridization kinetics to completion in a few hours or overnight.

3. At the conclusion of the prehybridization period pour off the prehybridization buffer and add a fresh aliquot (optimally 100 $\mu L/cm^2$) of hybridization buffer. In most cases, the prehybridization and hybridization buffers will be identical (see step 2 for preparation).

Double-stranded probes must be denatured prior to use. Denaturation is often accomplished by boiling or by raising the pH. Because of the enormous variation in labeling approaches in nonisotopic systems, it is best to follow the instructions of the manufacturer for probe denaturation. In the case of direct enzyme labeling, the probe will already be single-stranded and can be used directly without any further manipulation. Important: Radiolabeled probes should not be denatured by boiling because of the dangers associated with radioactive steam. Instead, add 0.1 volume 1 N NaOH to the probe (which is usually in a volume of 50–100 μL) and incubate at 37°C for 10 min or at room temperature for 30 minutes. The probe can then be added directly to the hybridization solution, without prior neutralization, as long as the total probe volume is less than 150 μL (assumes a minimum of 5 mL hybridization buffer).

4. Add the denatured probe to the buffer and hybridize for 12–16 hours (overnight) at 42°C. Make certain that the cover of the hybridization chamber is securely in place.

Note 1: For the detection of rare mRNAs or single copy genes, use 5–20 ng of probe per milliliter of hybridization buffer or $1-5 \times 10^6$ cpm/mL.

Note 2: If the mass of the probe is low, increasing the concentration of the salmon sperm DNA will improve hybridization kinetics.

Note 3: A total of 25 ng probe that will base-pair with the molecular weight standards may also be added to produce a visual record of the standards along with experimental samples on the same film.

5. At the conclusion of the hybridization period, pour off the hybridization buffer and dispose of it in a manner consistent with the nature of the label (e.g., radioactive waste). Wash the filters to remove all of the probe molecules that did not participate in the formation of hybrids. Spent wash solutions should also be disposed of appropriately.

a. Wash filters for 30 seconds, in a room temperature solution of 2× SSPE, 0.1% SDS to remove most of the probe in the box and remaining on the filter membrane.

b. Wash filters twice for 15 minutes, in 200–300 mL of a room temperature solution of 2× SSPE, 0.1% SDS.

c. Wash filters twice for 15 minutes, in 200–300 mL of a solution of 0.1× SSPE, 0.1% SDS at 37°C.

Note 1: If the background is too high, a final wash in 0.1× SSPE, 0.1% SDS at 42°C–50°C may be useful. It is likely that the optimal wash conditions will need to be determined empirically for each new probe.

Note 2: Depending on the nature of the probe and the stability of the hybrid, the investigator may need to adjust the exact posthybridization washing conditions for each probe. It is very important to keep in mind that in the context of posthybridization washes, stringency is a function of ionic strength, temperature, and time; thus, it may not be suitable to cut down on the washing time by increasing the temperature and/or lowering the salt.

6. Rinse the filter very briefly in 2× SSPE and place it on a piece of Whatman paper just long enough to blot excess buffer from the surface of the filter.
   Important: do not allow the filter to dry out to any extent, especially when performing chemiluminescence detection.
7. If using chemiluminescence or chromogenic detection technology, proceed from this step directly to the detection buffers, according to the directions of the manufacturer of the system.
8. If using isotope-labeled probes, wrap the filter in plastic wrap and set up autoradiography (next section).
   Important: Be sure to keep filters damp if planning to remove the probe after autoradiography in order to hybridize to another probe(s).

Principles of detection

When hybridization is completed, the filter is washed stringently to remove the unbound, left-over probe. The magnitude of hybridization is then assessed in a manner supported by the labeling method used to generate the probe, either by autoradiography, chemiluminescence, or fluorescence detection. If the investigator labeled a small amount of probe for the size markers, the location of the markers will likewise be permanently recorded. Having the size markers visible on the film or in a digital image eliminates the guesswork regarding the size determinations of other bands related to the experimental samples.

Phosphorimaging and digital detection systems

Digital image analysis has added a new dimension to both the achievable level of assay sensitivity, unbiased measurements, and the accuracy of data interpretation; this type of technology is standard fare in most molecular biology laboratories. While digital detection systems often represent a significant investment, the purchase of such devices is a one-time expense, upgrades and service contracts notwithstanding, that can be shared by multiple labs or departments. Digital detection systems are able to capture images from gels and blots directly for completely filmless electronic analysis and archiving. Some of these systems are mentioned in Chapter 15, Photodocumentation and Image Analysis, and a few salient points are reiterated here with emphasis given to blot analysis and the elimination of X-ray film altogether.

Phosphorimagers are sophisticated laser imaging devices, the primary function of which is to produce and quantify digital images of gels and blots. The first models were developed in the late 1980s to replace traditional autoradiography for the quantification of radioactivity in gels and on filter membranes, including ^{32}P, ^{33}P, ^{35}S, and other isotopes. Known today as biomolecular imagers or laser scanners, these digital detection platforms offer sensitivity up to 100× greater than X-ray film. They also offer a much greater array of detection capabilities across several orders of magnitude, allowing the observation and quantification of very weak bands and very strong bands, even when both are present on the same gel. The resulting images are rich in quantitative data that highlight the true variations among samples. In addition to phosphorimaging, bioimager technology also supports imaging wet gels stained with, for example, SYBR Green, SYBR Gold, Cy3, Cy5, or ethidium bromide and much more. These instruments feature very powerful image analysis software and the ability to optimize, archive, and analyze data (for review, see Green and Sambrook, 2020a,b).

There are numerous advantages associated with phosphorimagers, and the mode of image capture is unique. First and foremost, there is no X-ray film involved. In the case of blot analysis, images are generated on special screens that can be used over and over and, with proper care, will last for an extended period. The exposure time can be as little as one-tenth the time required for X-ray films. The phosphorimager detection technology is completely compatible with the qualitative and quantitative assay of nucleic acids, proteins, and any other macromolecule that carries a fluorescent or radioactive tag. Depending on the model, multiple gels, filters, or microtiter plates can be scanned at the same time. Best of all, most investigators agree that, by comparison with X-ray film, the sensitivity associated with phosphorimaging is superior. Because of the popularity of phosphorimagers and the fact that a department or institution may have only one such instrument available because of the cost, it is very common for users to sign up well in advance for a time to use the machine. In high-throughput laboratories and/or laboratories with several graduate students, the time slots available to use the machine may not be until the wee hours of the morning!

Nonisotopic procedures

The methods used for nonradioactive nucleic acid probe synthesis are remarkably similar to isotopic labeling approaches. The mechanics of hybridization detection, however, are quite different. The nonradioactive methodologies are (1) chemiluminescence, a process in which light is generated by enzymatic modification of suitable substrates; (2) chromogenic detection, in which an enzymatic activity causes a color precipitate to be deposited directly on the filter at the site of the reaction; (3) fluorescence, in which molecular excitation of certain compounds at one wavelength results in the emission of light at a different, characteristic wavelength; and (4) chemifluorescence, in which a fluorogenic substrate is converted enzymatically into a fluorescent product which, upon laser excitation, may be captured using a fluorescence imaging system. Interestingly, the fluorescent product is relatively stable, meaning that blots can be scanned, or rescanned, over an extended timeline.

Detection by chemiluminescence

Chemiluminescence is the conversion of chemical energy into the emission of visible light (luminescence) as the result of an oxidation or hydrolysis reaction. This technology provides a sensitive, cost-effective detection alternative to many radioisotopes and most chromogenic detection processes (Table 17.4). There are many adaptations of this technology used in conventional filter-based techniques. While the underlying principles at work in each system are essentially the same, kits for hybridization detection by chemiluminescence differ mostly in the number of reagents and manipulations required of the investigator in order to generate a signal and, to a lesser extent, in the achievable level of sensitivity.

Labeling and detection of probe molecules by chemiluminescence generally do not influence the parameters that govern nucleic acid hybridization, which is essentially the same for radiolabeled probes. Further, many of the

TABLE 17.4 Comparison of chemiluminescence and autoradiography for hybridization detection.

Detection parameter	Chemiluminescence	Autoradiography
Sensitivity	Very good to excellent	Excellent
Resolution	Very sharp banding	Doublets difficult to resolve
Exposure time	Minutes to hours	Hours to days
Detection	X-ray film or a digital imaging platform	X-ray film, phosphorimager, or a digital imaging platform
Potential health hazards	Relatively low	Very high
Probe stability after labeling	Most stable for one year	Limited by half-life
Posthybridization handling of filters	Various detection solutions required prior to film exposure	No detection solutions required
Signal generation	HRP: immediate, lasting for 6–24 h AP: 30–60 min, lasting for 1–3 d	Immediate

HRP = horseradish peroxidase; *AP* = alkaline phosphatase.

labeling technologies that support detection by chemiluminescence are identical to those used for chromogenic detection. Differences in the detection systems are related to various chemiluminescence substrate formulations. In the case of film-based detection, the mechanics of measuring chemiluminescence are similar to those for measuring isotope decay by autoradiography. The relatively intense, albeit brief, emission of light generates a quantifiable image on X-ray film, often in a fraction of the time needed for detection by autoradiography. Alternatively, biomolecular imaging systems and other digital detection platforms often have functionality for fluorescence-based image capture. When processed correctly, samples generally do not exhibit any type of noise or background signal which would otherwise compromise the sensitivity of this type of assay.

The numerous chemiluminescence detection systems fall into what this author considers the two schools of chemiluminescence, namely the alkaline phosphatase school and the peroxidase school; note that β-galactosidase and luciferase also support chemiluminescence, though their use is commonly reserved for reporter gene assays. The classical substrates used in alkaline phosphatase systems are 1,2 dioxetane derivatives, including the Lumigen family product line (Lumigen is 4-methoxy-4-(3-phosphatephenyl)spiro(1,2-dioxetane-3,2′-adamantane), disodium salt; Lumigen, Inc.), CSPD, and CDP Star (Thermo Fisher Scientific), and others. These patented substrate formulations undergo rapid conversion into luminescent form in the presence of calf intestine alkaline phosphatase (CIP *or* CIAP). The resulting emission of blue light (≈480 nm) is in direct proportion to the amount of alkaline phosphatase and therefore the amount of probe present at a given location (Fig. 17.9). Dual chemiluminescence/fluorescence substrates are also available (e.g., Lumigen ECL Plus).

Traditional substrate formulations that contain Lumigen PPD and that are compatible with alkaline phosphatase systems include Lumi-Phos 530 (light emission maximum at 530 nm), and Lumi-Phos Plus (light emission

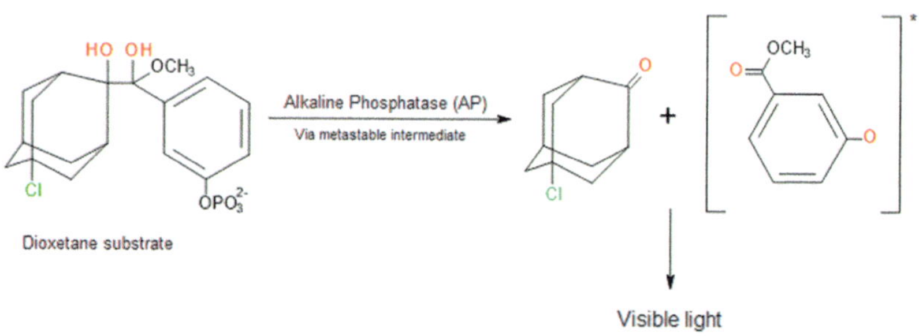

FIGURE 17.9 Enzymatic dephosphorylation of dioxetane substrate by alkaline phosphatase leads to the metastable phenolate anion which, upon decomposition, emits light at ≈480 nm. The light emission increases with time until a constant intensity is attained.

maximum at 470 nm). In addition, Lumigen APS-5, which is an acridan substrate, offers excellent sensitivity down to the low picogram to femtogram range as well as sustained luminescence, making the timing of the detection assay less demanding. Several companies also produce chemiluminescence enhancers, and these formulations are compatible with many molecular biology applications. There are other chemiluminescence substrates available to support plate-based assays, though these are beyond the scope of this chapter.

The substrate used in the peroxidase systems is hydrogen peroxide (H_2O_2). By virtue of the chemistry involved, the reduction of H_2O_2 produces oxygen radicals (O_2^-) and H_2. This, in turn, promotes the luminescent oxidation of the cyclic hydrazide luminol (5-Amino-2,3-dihydro-1,4-phthalazinedione), or any other of a variety of related substrates. The oxidation products are 3-aminophthalate and N_2, with the associated emission of light (Fig. 17.10). The location and magnitude of hybridization is most often recorded by X-ray film exposure, much as in autoradiography. Other specialized reagents in use for peroxidase-based chemiluminescence detection are likewise available.

Chemiluminescence detection represents the end of a series of steps, including posthybridization washes, hapten localization, and the application of suitable substrates. As described previously, the detection mechanics for chemiluminescence involve creating a molecular sandwich. Biotinylated probe molecules, for example, are localized by reaction with a streptavidin that is conjugated to either AP or HRP. Alternatively, DIG-labeled probes are localized by anti-DIG antibodies, which exist as an enzyme conjugate. Upon incubation with the appropriate substrate, enzyme activation results in the emission of visible light that is recorded via digital detection or on X-ray film in the manner of autoradiography. As there is no isotope involved, the resulting image might best be thought of as a chemilumigram, rather than an autoradiogram, though most people simply refer to it as "the film."

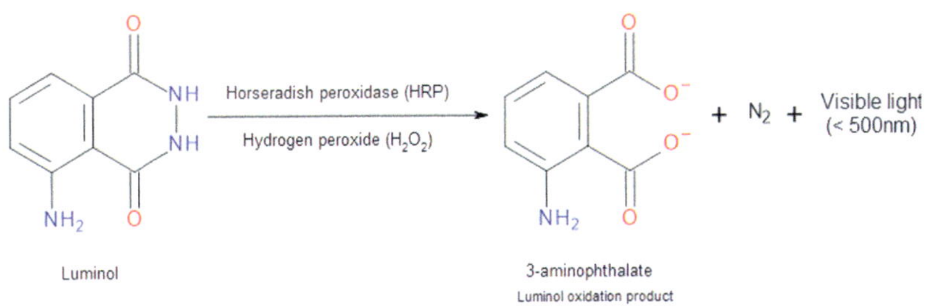

FIGURE 17.10 The reaction of luminol and H_2O_2 in the presence of the enzyme HRP produces visible light. The emission of light peaks within in few minutes after application of the substrates and then falls off precipitously. *HRP*, Horseradish peroxidase.

The most common difficulties with detection by chemiluminescence are either the incidence of unacceptably high levels of background or no signal at all. In the case of the former, this is most often due to (1) incomplete washing of filters posthybridization and during the detection steps, (2) incomplete blocking of the filter prior to addition of the enzyme conjugate, (3) allowing the filter to dry out after the detection procedure has been initiated, and (4) the use of powdered latex gloves (problematic during the processing of X-ray films, too). In this lab, stringent posthybridization washes, strict attention to the manufacturer's recommended detection protocol, and the use of powder-free gloves have consistently produced remarkably good quality exposures, with reproducible high levels of sensitivity. In the case of the latter, failure to detect a signal (including the positive control) is often the result of confusion and unfamiliarity with the unique requirements of chemiluminescence, compared to autoradiography. Finally, when using chemiluminescence, a follow-up stringency wash will mandate blocking the filter again, along with all of the subsequent steps for hapten recognition, binding of the enzyme conjugate, and delivery of the appropriate substrate. This constitutes a somewhat labor-intensive process, so it is best to wash the filter stringently the first time to minimize the likelihood that additional washes will be required.

The advent of nonisotopic labeling and detection, first by chromogenic techniques and later by chemiluminescence, was heralded as an alternative to the widespread use of radiolabeled nucleic acid probes. Many laboratories have reduced or altogether eliminated isotopes in favor of newer methods of labeling and detection, and there are many options. Chemiluminescence in particular has emerged as an excellent method for bypassing the traditional difficulties associated with chromogenic detection, not to mention circumvention of the handling and containment of radioactive materials, as well as the mounting disposal problems associated with their use. In chemiluminescence, light emission is recorded on a piece of X-ray film or through the use of a digital detection device. The exposure period associated with chemiluminescence is significantly shorter than is necessary for probe detection by autoradiography and is usually on the order of minutes, rather than hours or days. Chemiluminescence has now achieved a fair level of acceptance as a standard molecular biology tool. Generally, the mechanics of nonisotopic detection methodologies involve (1) assembly of an immunohistochemical "molecular sandwich" by the ordered application of a series of reagents directed against a hapten label or (2) activation of an enzyme that was linked directly to the probe molecules, such as in the Enhanced Chemiluminescence (ECL) system (Cytiva).

As described above, there are several methods by which probe biotinylation can be achieved. Biotinylated probes are localized posthybridization via the unusually high natural affinity observed between biotin and streptavidin ($K_d = 10^{-15} M^{-1}$). Streptavidin is a tetrameric protein (MW 60,000), isolated from the bacterium *Streptomyces avidinii* and has four biotin-binding sites. Unlike avidin (from egg), streptavidin has a neutral isoelectric point at physiological pH, with few charged groups, and contains no carbohydrate. These properties reduce nonspecific binding and background problems, thereby enhancing the sensitivity of many forms of this assay; biotin-streptavidin applications in nucleic acid hybridization detection and related techniques are widespread. In the most common detection format, streptavidin is conjugated to either alkaline phosphatase (AP) or HRP. Subsequently, the choice of substrate for the enzyme will determine the mode of detection. In the case of AP, dioxetane substrates result in the emission of light while, for example, BCIP (5-bromo-4-chloro-3-indolyl phosphate) and NBT (nitro blue tetrazolium) together support the formation of a color precipitate which is deposited directly on the filter. These reaction sites correspond to hybridization events. In contrast, HRP initiates a set of chemical reactions in the presence of H_2O_2, resulting in the emission of light (luminol-mediated) or the formation of a color precipitate, mediated by the substrate 4-chloro-1-napthal. There is no film exposure involved because this is neither autoradiography nor chemiluminescence involved. If utilized, chemiluminescence may be quantified with digital detection equipment or by simply recording light emissions on X-ray film.

DIG labeling is a well-accepted method for generating nonradioactive probes and is an alternative to biotinylation. Like biotin, DIG-labeled probes support both chromogenic and chemiluminescence detection formats. Following the posthybridization stringency washes, DIG-labeled probes are detected by enzyme-linked immunoassay, using an antibody conjugate (e.g., anti-DIG-AP conjugate or HRP conjugate). As with systems involving biotin, the emission of light or the formation of a precipitate is mediated by enzymatic cleavage of a substrate compatible with the method of detection, either chromogenic or chemiluminescence.

Reliable direct fluorescence measurements are obtainable through the use of the cyanine dyes Cy2, Cy3, Cy5, and/or other fluorescent compounds, which have all but replaced the use of fluorescein-labeled nucleotides. Fluorochromes are detected by devices capable of laser excitation and measuring the resulting fluorescence emission. For example, Cy2, Cy3, and Cy5 have unique properties that allow them to be distinguished, even when used simultaneously (refer back to Fig. 17.5).

The ECL system (Cytiva) features a modified HRP linked directly to probe molecules through a glutaraldehyde linker arm. Thus, the enzyme is present throughout the hybridization period as well as the posthybridization stringency washes. This labeling and detection method precludes the need for any type of hapten recognition, thereby supporting the immediate application of chemiluminescence substrates. It remains one of the most rapid methods for labeling and detection and usually produces low-background images. The ECL system is highly respected for Western analysis; the nucleic acid and protein detection formats are in widespread use.

Chromogenic detection procedures

Many of the chromogenic detection systems that appeared in the mid-1980s failed to gain widespread acceptance for one reason or another and then lost popularity altogether with the debut of many refined chemiluminescence systems. One noteworthy exception is the DIG Labeling System which has become well respected as a method for nonisotopic in situ hybridization detection.

Chromogenic detection and procedures for chemiluminescence are often similar up to the addition of the very last reagent, the enzyme substrate. The chromogenic detection process may often require up to several hours to reach completion. The low sensitivity observed in some chromogenic systems is due to the fact that the activity of alkaline phosphatase is inhibited as the color precipitate forms and is deposited on the filter. Enzyme inhibition in this manner quickly compromises the quantitative aspect of the assay, resulting in unreliable, nonrepresentative data. The final step of an HRP-mediated, chromogenic detection involves the application of H_2O_2 and either 4-chloro-1-naphthol or TMB (3,3′,5,5′-tetramethylbenzidine).

A bit of an anecdote: On one occasion in this lab, plaque lift filters were being processed with the intention of performing chromogenic detection, though a decision made at the last minute resulted in the substitution of chemiluminescence reagents (hydrogen peroxide and luminol) rather than hydrogen peroxide and 4-chloro-1-naphthol. This impromptu procedural change had no negative impact on the outcome of the screening and allowed the confirmation of a pure strain of positive clones pulled from a library (Fig. 17.11). While not a recommended standard procedure, this example is included here to underscore the similarity of these methods.

In order to reprobe filters that were previously subjected to chromogenic detection methods, very harsh washes of the filter are required, to remove the precipitate (and probe) from the filter. Each type of filter membrane and nonisotopic detection system has a set of recommended guidelines for stripping previously hybridized filters, and preparations for reprobing.

Autoradiography considerations

Autoradiography is a photochemical technique used to measure and visually record the spatial distribution of radiolabeled molecules within an object. The mechanics of the autoradiographic process are dependent on the energy emission spectrum of the isotope, the image quality requirements, and the amount of radioactivity in the system. There are no hard and fast rules about exposure time and exposure strategies; these parameters must be empirically determined with each experiment. There are multiple options for each step of the autoradiography process, the choice of each of which may change depending on the application and equipment on hand.

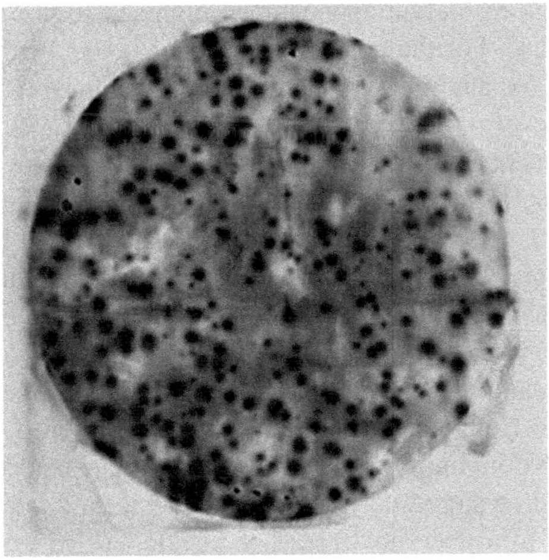

FIGURE 17.11 Plaque screening assay by chemiluminescence using Luminol as a horseradish peroxidase substrate. This image represents a 5-minute exposure. As expected, nearly every plaque shows up as positive because this was the third round of clone replating and clone enrichment after the initial recovery of a putative positive clone from a cDNA library screening. All of the clones on the filter are now genetically identical, any one of which could then be retrieved from the associated agar plate for subsequent analysis.

Handling of filter membranes

Filter membranes used for Northern analysis or other applications should not be allowed to dry out completely after the last stringency wash. Instead, these filters should be blotted briefly by resting them on a piece of Whatman 3MM paper, radioactive side up, and then wrapped in plastic. Maintaining filter membranes in a damp state will greatly facilitate the postautoradiography removal of the hybridized probe with a very high-stringency wash. Thus, a filter membrane can be screened several times with other probes of interest. Removal of the probe from membranes that are allowed to dry completely is significantly more difficult than removal from damp membranes, if not altogether impossible. *Never* place a damp or wet filter membrane in direct contact with a sheet of X-ray film. The two will form a permanent symbiosis of sorts, resulting in the ruination of both filter and film.

X-ray film

The choice of X-ray film is dependent on whether high-energy or low-energy emitters were used as a label and the required level of resolution. The BioMax film (Carestream) product line and Hyperfilm (Cytiva) are examples of X-ray films well-suited for autoradiography and chemiluminescence. Different formats of these high-performance films are available to accommodate high-resolution applications, high sensitivity applications, and general-purpose use. For the convenience of the reader, Table 17.5 includes film suggestions for autoradiography and chemiluminescence in the context of the required level of sensitivity and resolution. The maximum resolution films are recommended for use with the lower energy β-emitters ^{35}S, ^{33}P, and ^{14}C, while the high-performance and multipurpose formats can be used to detect ^{32}P, ^{33}P, ^{125}I, and ^{35}S as well as applications involving chemiluminescence detection. Films with a clear base, rather than having the traditional gray tint associated with older film formulations, allow the visualization of even the faintest bands on the film.

Unopened packages of X-ray film should be stored between 50°F and 70°F and shielded from penetrating radiation. Once opened, storage between 30%–50% relative humidity is recommended. For extended periods, boxes of the film should be stored upright, rather than stacked in a horizontal position, so that the film does not crease or buckle, which could result in the background or uneven processing.

It is very important to protect the undeveloped film from any stray light. While almost any room can be turned into a photographic darkroom very economically, it is of great importance to ensure that no light enters the darkroom from under the door(s), around the edges of windows, through openings in the walls or ceiling, or even from indicator lights on equipment that might be in the same room, including computer monitors.

TABLE 17.5 Select X-ray films for autoradiography and chemiluminescence.

	Autoradiography	Chemiluminescence
Maximum sensitivity	BioMax MS	BioMax Light
	Hyperfilm MP	Hyperfilm ECL
All-purpose usage	BioMax MS	BioMax Light
	BioMax XAR	BioMax XAR
	Hyperfilm MP	Hyperfilm ECL
Maximum resolution	Hyperfilm MP	BioMax Light
	BioMax MR	Hyperfilm ECL

Notes: Film selection for hybridization detection is based on the requirement for sensitivity and resolution. Biomax MS, BioMax Light, and Hyperfilm ECL are double-emulsion film, while BioMax MR is a single-emulsion film. Hyperfilm MP is a double-emulsion, all-purpose film, as is BioMax XAR. X-Omat LS is a double-emulsion, lower sensitivity blue tint film that requires longer (2×) exposure times. Most X-ray films are available in two configurations, either in alternate interleaved packing that protects each sheet of film during shipment and prevents static electricity from exposing the film as it is withdrawn from the box or packaged such that each sheet of film is individually wrapped in a light-tight envelop (e.g., Carestream Kodak XAR-2). *MS* = maximum sensitivity; *MR* = maximum resolution; *MP* = multipurpose; *ECL* = enhanced chemiluminescence.

Safelight

Undeveloped X-ray films remain light sensitive until processed through the photographic fixer solution, though limited low-intensity illumination is permissible to allow movement and efficient handling in the darkroom. A safelight typically consists of an ordinary 15-W frosted light bulb that shines through a filter, the color of which is compatible with the film being used. The most commonly used films for chemiluminescence and autoradiography are blue-green sensitive, so the recommended darkroom setup features a red GBX-2 safe light filter. An excellent economical safelight filter for quickly setting up a darkroom is the GBX-2 Safelight (Kodak No. 141 6627) and Model B safelight lamp, which is small, portable, and durable. The term safelight is something of a misnomer, though, because prolonged exposure to safelight illumination will eventually fog the film. Ideally, the film should never come any closer to the safelight than four feet. Be aware also that safelight filters fade over time with extended use, so testing the filter periodically is recommended. This can be performed easily by first developing an unexposed sheet of X-ray film with the safelight on and then developing another unexposed sheet of film with the safelight off. Any fogging of the first film suggests that the safelight filter needs to be replaced.

Exposure time

There is no limit to the number of exposures that can be made from a single radioactive source other than the half-life of the material itself. In the case of high-energy emitters, a Geiger counter probe placed in the vicinity of the filter is a good way to assess the activity on a filter membrane. Obviously, a source manifesting 3000 cpm will require a much shorter exposure period than a source registering but a few unimpressive "clicks" above the background. The lack of an obvious signal should not be an immediate discouragement; with autoradiography, it is not unusual to have to wait three days or more days to observe hybridization signals corresponding to rare transcripts. It may be helpful to do an overnight exposure at room temperature to assess the extent of background and nonspecific filter hybridization; subsequent exposure tactics are based on the results of this test exposure.

Intensifying screens

In the case of low-activity filters, autoradiographic detection of ionizing radiation can be improved through the use of intensifying screens, consisting of a flexible polyester base coated with luminescent phosphor crystals. X-rays striking the phosphor coating generate visible light which can further expose a sheet of X-ray film. Thus, intensifying screens are commonly used to lessen the time needed to produce an image on the X-ray film within an exposure cassette.

The latest generation of intensifying screens is known as the BioMax TranScreen product line (Carestream). When assembling the X-ray film exposure cassette, the TranScreen is placed *between* the radioactive source (that is, the membrane) and film, rather than behind the film, which was the preferred orientation for the earlier calcium tungstate ($CaWO_4$) and rare earth screens. In so doing, emitted β particles strike the phosphor first, producing green light which immediately becomes visible to the film emulsion. The proximity of the screen to the radioisotope helps to maximize spatial resolution. Higher energy emitted particles continue on, through the phosphor layer, striking the film and enhancing the exposure. Moreover, the placement of a reflective backing behind the film enhances the exposure process to an even greater extent by enabling higher-energy particles to bounce back and interact with the film emulsion again. This design greatly enhances autoradiographic detection of weak β particle emitters and, in the case of high-energy emitters, drastically reduces the required exposure time.

Prior to the development of the TranScreen methodology, there were two types of intensifying screens once in widespread use for autoradiography. First, $CaWO_4$ phosphor-coated screens, which produce blue light upon excitation, were in use in the earliest days of the molecular biology revolution. Subsequently, rare earth screens intensifying screens coated with gadolinium oxysulfide (green light), lanthanum oxybromide (blue light), or yttrium-oxysulfide were developed. Depending on the phosphor, rare earth screens produce a bright green or blue glow when excited and are much faster than $CaWO_4$-coated screens. The rare earth screens exhibit very low-level noise (background signal) and given their more efficient conversion of X-rays into light; they are nicely compatible with standard autoradiography procedures.

Autoradiography with intensifying screens is performed at $-70°C$ because film response to low light intensities is improved dramatically at low temperatures and because intensifying screens perform best when cooled to $-70°C$. The net result, then, is a greater degree of exposure per decay event than in the absence of an intensifying screen. Note that low-temperature exposure without an intensifying screen, however, will not improve the image formation process. The $CaWO_4$ intensifying screens improve the detection of the high-energy β emitters ^{45}Ca and ^{32}P and γ ray emitters such as ^{125}I, though they are not effective with low energy emitters ^{3}H, ^{35}S, or ^{14}C. In the case of ^{33}P, used mostly for old-style DNA sequencing, $CaWO_4$ screens were ineffective because the autoradiographic characteristics of ^{33}P are similar to those of ^{35}S. This shortcoming has been eliminated with the development of Carestream's TranScreen design.

Finally, it is important that the sensitivity of the film used matches the light emission spectrum of the intensifying screen. The luminescent phosphor composition of a particular rare-earth intensifying screen will result in the emission of light at a characteristic wavelength, so it is important to ensure that the film (blue- or green-sensitive) is compatible with the screen. This information may be found on the product insert that accompanies the screen or by calling the manufacturer's technical services department prior to ordering.

Preflashing film

To increase the sensitivity of film to low intensities of light and thus increase the linear response of the film, an investigator may elect to preflash (hypersensitize) a sheet of the film just prior to setting up an autographic or fluorographic exposure. By providing a short, controlled burst of light on the order of 1 ms, the film essentially becomes charged or primed, meaning that the latent image formation process is initiated. The latent image formation process is then completed by light photons from intensifying screens. Priming the film in this manner expands the rather narrow linear response of the film, thereby enhancing its sensitivity. Calibrated preflash units can be purchased from a number of suppliers, and the exact conditions for use are dependent on local darkroom conditions. In the absence of an intensifying screen, preflashing is not an effective means of enhancing latent image formation by ionizing radiation. Preflashing has been reported as useful for increasing the sensitivity of X-ray film when used in conjunction with chemiluminescence, though it is not performed in this lab.

Type of cassette

The selection of an appropriate light-tight apparatus to accommodate autoradiographic exposures is probably the least involved decision of all. There is an exposure apparatus to fit every budget, all of which can be used both for autoradiography and chemiluminescence. The older style heavy cardboard cassettes, lined with a light-tight inner envelop, may still be found in many darkrooms. This type of exposure holder is best sandwiched between two 1cm-thick sheets of Plexiglas to block high-energy β particles which would otherwise pose a potential health hazard to the investigator and others in the vicinity. The more commonly used models are constructed from lightweight, vinyl-covered aluminum and include permanently mounted intensifying screens, as do stainless-steel X-ray film holders. Within these more deluxe cassettes, polyurethane pads help to maintain continuous uniform contact between the film and the filter. When assembling the exposure apparatus, it is of paramount importance to immobilize a filter membrane; even the slightest movement of the filter at any time during the exposure period will produce a fuzzy image. Immobilization by mounting the plastic-wrapped filter on a piece of Whatman paper cut to the size of the cassette is probably the simplest way. Never tape anything directly to an intensifying screen.

Processing X-ray films

(Some material is this section adapted, in part and with permission, from Eastman Kodak Company.)

Autoradiography occurs when ionizing radiation and photons are captured by an emulsion placed in direct contact with a radioactive source. It is a simple yet sensitive technique that creates a permanent record of the decay events associated with unstable radionuclides. In the context of the nucleic acid-related study, radiolabeled probes are hybridized to target sequences that are immobilized on a solid support, in solution, or in situ. Autoradiography is subdivided into two broad groups, commonly referred to as macroautoradiography and microautoradiography, indicative of the type of specimen containing the radioactivity, the type of emulsion necessary for image formation, and the method of examining the results.

The most common applications of autoradiography in molecular biology include the quantitative analysis of nucleic acid hybridization events. If used in conjunction with electrophoresis, the resultant imagery will provide a qualitative aspect as well. For example, in Northern analysis (macroautoradiography), the extent of hybridization and probe specificity to RNA target molecules is assessed by the magnitude and location of radioisotope emissions from the filter membrane. Microautoradiography, on the other hand, is performed by coating a thin section of a larger specimen with a light-sensitive emulsion. This technique, also known as in situ autoradiography, is usually performed on a microscope slide. Because this technique has the advantage of maintaining natural cellular geometry, the intracellular distribution of radiolabel can be easily determined by light or electron microscopy (that is, by counting the grains).

In macroautoradiography, a piece of high-speed X-ray film is placed in direct contact with a plastic-wrapped filter membrane. This is commonly known as direct exposure; in some cases, such as a nuclease protection assay, isotopes can be detected directly in polyacrylamide gels without the need for blotting onto a filter. The film base is a flexible piece of polyester–films used for autoradiography are typically 0.007-inch (7 mil) thick–that has been coated on one or both sides with radio- and light-sensitive emulsion. The emulsion consists partly of sensitive silver halide (bromides, chlorides, or mixed halide) crystals, to which energy is released by ionizing radiation. During film exposure, energy is absorbed in the silver-halide grains within the film emulsion, with the release of electrons. The resulting negatively charged "sensitivity specks," that is, faults in the crystal lattice, attract positively charged silver ions, thereby forming an atom of metallic silver.

Exposure of X-ray film generates the so-called latent image, an invisible precursor to the visible image. The latent image consists only of chemically developable silver grains, and it is not visible in the darkroom (under safelight conditions). Once formed, the latent image is not stable for long exposure periods; potential latent image instability is exacerbated by the presence of oxygen and moisture on the film. As outlined in Fig. 17.12, the distinct stages in the processing of the film are: development, fixation, washing, and finally drying; these stages can be performed manually or with an automatic

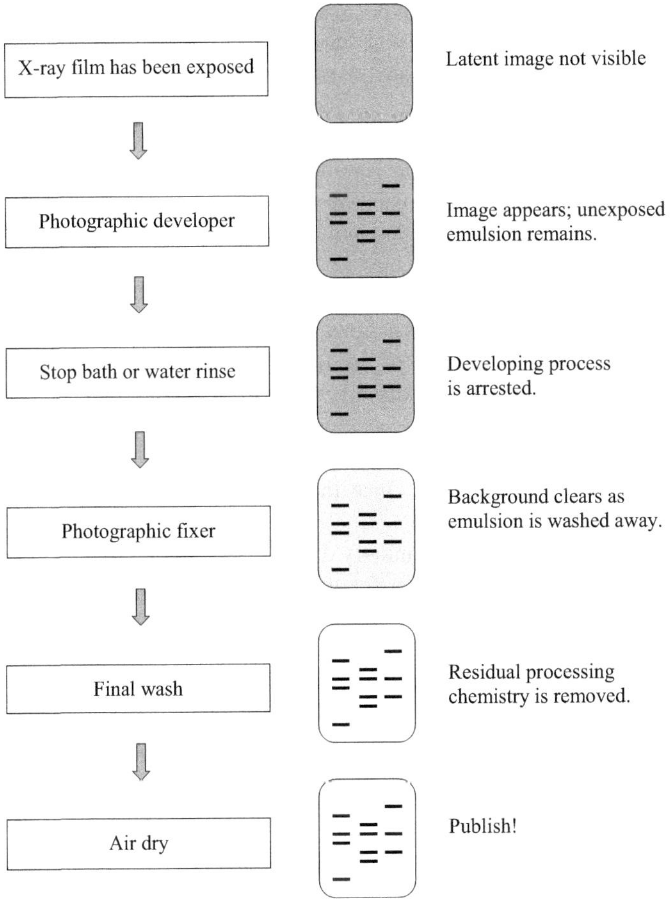

FIGURE 17.12 Developing scheme for X-ray film. Although the choice of film is dependent on the application (autoradiography vs chemiluminescence), the mechanics of X-ray film processing are identical.

film processor. In either case, it is wise to standardize the developing parameters, particularly when to perform these steps manually, so as to allow meaningful comparison from one film to the next.

Transformation of the invisible latent image into a visible autoradiographic image occurs when the film is immersed in a photographic developer, thereby reducing the exposed silver halide grains in the film's emulsion to metallic silver. The amount of time required to complete the developing process is variable, depending on the degree of exposure of the film, the temperature of the developer, the age of the developer, and the tolerance of the investigator for background. In this lab Kodak D-19, D-76, and GBX developer have been used routinely.

The action of the developer is arrested by immersion of the film in a chemical “stop bath.” A variety of stop bath solutions may be purchased, though a homemade version can be concocted by preparing a solution of 0.5% acetic acid. Arresting the activity of the developer at precisely the right time is very important because over-development can obscure any image that may have been captured on film. Keep in mind that a piece of X-ray film withdrawn from the developer tray is still coated with the developer and that the developing process will continue until the action of the developer has been neutralized. In the absence of a suitable stop bath formulation (not recommended), the film should be immersed quickly in water to dilute and wash off the developer. These actions will not only arrest the developing process but also extend the life of the fixer. Next, the film is immersed in a chemical fixer solution which dissolves away the unexposed silver halide, producing the characteristic X-ray film appearance. Thorough, room-temperature washing of the film in water, after the fixation step, is needed to remove processing byproducts. Once the film has been completely dried, a stable image is in place.

When developing films manually in the darkroom, it is necessary to develop, wash, and fix each film for the same length of time. Without this type of quality control, it may be difficult to determine the extent to which observed data variations are due to experimental manipulation or to variation in the mechanics of developing the film. One of the advantages of manual film processing is that the investigator can extend the developing time in order to accentuate faint bands or reduce the developing time in order to minimize background or the overdevelopment of extremely strong signals from one or more bands. Moreover, the equipment requirements are minimal. If autoradiographic exposure was carried out at −80°C, be sure to warm the film to room temperature before opening the cassette, otherwise, the quality of the image and the reproducibility of the assay may be compromised. Films that are developed manually should be washed in running water (16°C–24°C) for at least 15 minutes prior to air-drying to render the highest possible quality autoradiogram.

Alternatively, films can be developed automatically by machine, if such an apparatus is available. Machine-developed autoradiograms have the advantage of being developed according to a standardized process; however, one is at the mercy of the machine.

Autoradiography: suggested protocol

It is important to standardize both the exposure time and the developing procedure if one is to generate meaningful data. Evaluation of short- and long-term exposures or different developing protocols involving the same blot may suggest very different conclusions. Needless to say, be sure to follow all departmental guidelines for experiments involving isotopes; this includes proper handling and containment of radioactive material and attending to all safety guidelines to minimize potential health hazards. It is wise to manipulate radioactive filters behind Plexiglas shields.

1. At the conclusion of all posthybridization stringency washes, place the filter on a piece of Whatman 3 MM paper just long enough to blot excessive buffer from the filter.
2. Wrap the filter in plastic wrap, taking care to avoid wrinkling on the side of the filter to which the probe has hybridized. Fold excess plastic wrap behind the filter so there is no possibility that the X-ray film will get wet.
3. Hold a Geiger counter probe 10−15 cm above the surface of the plastic-wrapped filter to estimate the required time for autoradiography. It is not unusual to have very few "clicks" registering just above the background when assaying rare mRNAs. Significantly greater activity may be observed in the region of the filter corresponding to the molecular weight markers if the hybridization cocktail contained a probe capable of base-pairing to these sequences, too. High activity distributed over the surface of the filter may be indicative of unacceptable background levels. The exposure must be made to know this for sure, and a short test exposure may be helpful in this regard.
4. Place the plastic-wrapped filter, a sheet of X-ray film, and an intensifying screen together in a cassette, under safelight conditions. Store at −70°C overnight.
   Note: Be sure to close the box containing the unused film before turning on the room lights. Forgetting to do this is a common error.
5. At the conclusion of the exposure period allow the cassette to warm to room temperature. If using an automatic X-ray film developing machine, open the cassette under safelight conditions and process the film according to the protocol for the developing machine, and then proceed to step 15. If manually developing the film, follow steps 6−14.
6. Photographic developer (e.g., Kodak D-19, D-76, or GBX) should be prepared in advance and stored in a tightly sealed dark bottle because it is light-sensitive and oxidizes rapidly. Pour enough developers into a photographic developing tray so that the film is submerged when placed into a developer. The developer should be clear and colorless when poured from its storage bottle. The developer oxidizes as it ages, acquiring a yellow tint. While not ideal, a developer that is *slightly* yellowed will continue to function as long as it is clear, though the quality of the film may be suboptimal. If the developer appears dark yellow or brown, it is completely useless and should be discarded.
   Note: Care should be taken to avoid skin contact with photographic developer, stop bath, and fixer, and be certain to wear eye protection when working with these and all chemicals.

7. Fill a second photographic tray with some type of stop bath solution, or simply water.
8. Fill a third photographic developing tray with photographic fixer which, like the developer, was prepared in advance and stored in a dark, tightly sealed bottle. The volume of the fixer in the third tray should permit the film(s) to be submerged with ease.
9. Take note of the placement of everything in the darkroom before turning off the lights. Fumbling around in the dark is counterproductive.
10. Turn the safelight on and then turn the room lights off. Wearing gloves, open the X-ray exposure cassette and, before removing the film, bend one corner of the film so that the resulting asymmetry can be used to keep track of the film orientation within the cassette. Do not remove the filter from the cassette or otherwise change its relative position and be certain that the plastic-wrapped filter is not sticking to the film (which occurs frequently).
11. Slide the film into the developer and submerge it all at once; do not float the film on the surface of the developer. Begin timing immediately, and gently agitate the developer tray. An exposed film will generally begin to manifest bands within 30–60 seconds, though the developing time may be extended to as much as 2.5 minutes. When the bands begin to appear, note the elapsed time, using it as a guideline for subsequent films. Standardizing developing time gives greater meaning to data generated in different experiments.
12. At the conclusion of the developing period, quickly remove the film from the developer and place it in the stop bath. Rinse the film for 20–30 seconds.
13. Transfer the film to the third tray containing the photographic fixer and submerge. Gently agitate this tray, and notice how, after a few moments, the film itself begins to clear, rendering the characteristic X-ray film appearance. Fixation is complete when the film has *completely* cleared, usually in less than 5 minutes, though the fixation time should be extended to a full 5–10 minutes.
14. At the conclusion of the fixation step, it is safe to turn on the room lights. Carefully transfer the developed film into a sink and allow it to rinse in running warm water for at least 10 minutes. Washing away all of the residual chemicals at this postdeveloping stage will produce superior quality films. Hang the films to dry.
15. If very little is evident in the film after this initial exposure, then the background is likely to be low. Set up another exposure, store at −70°C for three days, and then repeat the steps for developing the film. If the background on the film is high upon initial exposure, usually indicated by the appearance of an outline of the filter paper on the film and with no apparent bands, consider unwrapping the filter paper and repeating the most stringent posthybridization wash; for greater stringency, wash

the filters at 55°C–60°C in 0.1 × SSPE, 0.1% SDS for 10–15 minutes. Repeat steps pertaining to wrapping the filter and setting up exposure.

16. Follow instructions of the filter manufacturer for the removal of hybridized probe and/or storage of the filter. See below for probe removal suggestions.

Protocol: generic method for probe removal

Filters with immobilized RNA or DNA sequences can be screened repeatedly with different probes if the filter is handled and stored properly. In the case of Northern analysis, one strategy is the sequential hybridization of a filter containing experimental RNA in order to assay more than one gene. After the first round of hybridization and detection, the probe is removed by very high stringency washes and then the filter is hybridized with a different probe that will assay for the presence of a different type of mRNA. Nylon membranes are very well-suited for this approach. Keep in mind, however, that the filter may begin to lose some of the bound target material as this process is repeated.

Rehybridization is often desirable because (1) the preparation of the blot may represent a lot of effort, (2) there may be limited or no sample remaining with which to prepare another blot, (3) it is important to probe a single filter for the expression of housekeeping (control) genes in addition to screening for the modulation of genes of interest, (4) in the case of dubious results or high background, re-screening the same filter with the same probe may clear up ambiguity, and (5) one may decide later on to rescreen a blot based on new information or the availability of new probes. Consequently, proper handling of the filter is important.

Prior to the first hybridization, the filter should be stored dry and out of direct sunlight. In this lab, filters that have been UV crosslinked are stored dry between two sheets of Whatman 3MM paper in the drawer set aside exclusively for filter storage. Placing a notebook or other lightweight on top of the filter(s) helps to keep them flat. Filters can be stored this way for months.

When hybridization detection has been completed, filters are subject to high stringency washes without delay to remove the old probe. In the event that probe removal is not convenient right after probe detection, the filter should remain wrapped in plastic to keep it damp. If the filter does dry out prior to probe removal, it will be all but impossible to remove the probe later on. Some investigators don't bother to remove the previously hybridized probe, claiming that it does not interfere with subsequent characterizations of the blot. The truth is that an old probe may or may not interfere with subsequent hybridization and analysis of the data—Table 17.6 is structured to help the investigator decide if probe removal is necessary prior to rehybridization. As a general rule, it is worth the investment of time to remove the old probe

TABLE 17.6 Probe removal matrix.

		Advantages	Disadvantages
Radioactive probe	Remove probe	1. No signal expected from previously hybridized probes 2. Can use the same probe again for confirmation of previous hybridization data, as needed	1. Investment of time and reagents 2. Generates liquid radioactive waste
	Retain probe	No additional processing necessary	1. Previous probe likely to produce bands unless it has gone through several half-lives 2. Unable to hybridize to the same target sequences again
Nonisotopic probe	Remove probe	1. Can use the same probe again for confirmation of previous hybridization data, as needed 2. Easy disposal of liquid waste	Investment of time and reagents
	Retain probe	No additional processing necessary	1. Residual detection chemistries may contribute to high levels of background in subsequent detections 2. Unable to hybridize to the same target sequences again

prior to rehybridization or long-term storage. Deciding, later on, to try to remove a previously hybridized probe is often a decision made too late.

Each type of filter has a recommended probe removal protocol. The protocol presented here is generic enough to work well with most nylon filters. Filters should be kept moist following the detection procedure until the following steps can be performed to remove the old probe.

1. Wearing gloves, carefully remove damp filters from protective plastic wrap. If the previously hybridized probe was radioactive, be sure to observe precautions for the safe handling of the filter.

2. Immediately immerse the filter in an excess volume of 50% formamide, 0.5 × SSPE which has been preheated to 65°C. *CAUTION: Be sure to tightly seal the container in which this wash is being performed to prevent exposure to formamide vapors.*
3. Incubate for 45–60 minutes on an orbital platform at 50 rpm and maintain the temperature throughout. If an orbital platform is not available, then the container in which the wash is being performed should be agitated manually every 10 minutes.
4. Carefully transfer the filter to a clean container and rinse in two changes of 0.1 × SSPE for 2 minutes each with gentle agitation.
5. Place the nylon filter, face up, on a sheet of Whatman 3 MM paper and blot the excess 0.1 × SSPE.
6. If a radioactive probe was used previously, it is a good idea to keep the newly washed filter damp, place it in plastic wrap and expose it to a sheet of X-ray film to ensure that all previously hybridized probe was removed. The likelihood that the probe has been removed can be estimated by passing a Geiger counter over the surface of the film to assess the amount of residual radioactivity.
   Note: If bands do appear on the X-ray film, repeat steps 1–5. If the probe fails to strip off the filter after a second attempt, the probe will probably remain on the filter permanently. In such an event, one could store the filter until the probe goes through several half-lives, after which the filter can be hybridized to a different probe sequence.

References

Blake, R.D., Delcourt, S.G., 1996. Thermodynamic effects of formamide on DNA stability. Nucleic Acids Res. 24 (11), 2095–2103.

Bodkin, D.K., Knudson, D.L., 1985. Assessment of sequence relatedness of double-stranded RNA genes by RNA-RNA blot hybridization. J. Virol. Meth. 10, 45–52.

Bolton, E.T., McCarthy, B.J., 1962. A general method for the isolation of RNA complementary to DNA. Proc. Natl. Acad. Sci. USA 48, 1390–1397.

Britten, R.J., Davidson, E.H., 1985. Hybridization strategy. In: Hames, B.D., Higgins, S.J. (Eds.), Nucleic Acid Hybridization: A Practical Approach. IRL Press, Washington, DC, pp. 3–16.

Casey, J., Davidson, N., 1977. Rates of formation and thermal stabilities of RNA: DNA and DNA: DNA duplexes at high concentrations of formamide. Nucleic Acids Res. 4, 1539–1552.

Cimino, G.D., Gamper, H.B., Isaacs, S.T., Hearst, J.E., 1985. Psoralens as photoreactive probes of nucleic acid structure and function: organic chemistry, photochemistry, and biochemistry. Ann. Rev. Biochem. 54, 1151–1193.

Crabb, I.D., Hughes, S.S., Hicks, D.G., Puzas, J.E., Tsao, G.J., Rosier, R.N., 1992. Nonradioactive in situ hybridization using digoxigenin-labeled oligonucleotides. Applications to musculoskeletal tissues. Am. J. Pathol. 141, 579–589.

Denhardt, D.T., 1966. A membrane-filter technique for the detection of complementary DNA. Biochem. Biophys. Res. Commun. 23, 641–646.

Feinberg, A.P., Vogelstein, B., 1983. A technique for radiolabeling DNA restriction endonuclease fragments to high specific activity. Analyt. Biochem. 132, 6–12.

Feinberg, A.P., Vogelstein, B., 1984. Addendum: a technique for radiolabeling DNA restriction endonuclease fragments to high specific activity. Analyt. Biochem. 137, 266–267.

Green, M.R., Sambrook, J., 2020a. Autoradiography and phosphorimaging. Cold Spring Harb. Protoc. Available from: https://doi.org/10.1101/pdb.top100446.

Green, M.R., Sambrook, J., 2020b. Detection of radioactive DNA in polyacrylamide gels. Cold Spring Harb. Protoc. Available from: https://doi.org/10.1101/pdb.prot100420.

Itakura, K., Rossi, J.J., Wallace, R., 1984. Synthesis and use of synthetic oligonucleotides. Annu. Rev. Biochem. 53, 323–356.

Maddocks, S., Jenkins, R., 2017. Understanding PCR: A Practical Bench-Top Guide. Academic Press, San Diego, CA.

Marmur, J., Doty, P., 1961. Thermal renaturation of deoxyribonucleic acids. J. Mol. Biol. 3, 585–594.

Marmur, J., Doty, P., 1962. Determination of the base composition of deoxyribonucleic acid from its thermal denaturation temperature. J. Mol. Biol. 5, 109–118.

Martin, R., Hoover, C., Grimme, S., Grogan, C., Holtke, J., Kessler, C., 1990. A highly sensitive, nonradioactive DNA labeling and detection system. BioTechn. 9, 762–768.

Matturro, B., Rossetti, S., 2015. GeneCARD-FISH: detection of tceA and vcrA reductive dehalogenase genes in Dehalococcoides mccartyi by fluorescence in situ hybridization. J. Microbiol. Meth. 110, 27–32.

Rigby, P.W.J., Dieckmann, M., Rhodes, C., Paul, P., 1977. Labeling deoxyribonucleic acid to high specific activity in vitro by nick translation with DNA polymerase I. J. Mol. Biol. 113, 237–251.

Robbins, E., Baldino, F., Roberts-Lewis, J.M., Meyer, S.L., Grega, D., Lewis, M.E., 1991. Quantitative non-radioactive in situ hybridization of preproenkephalin mRNA with digoxigenin-labeled cRNA probes. Anat. Rec. 231, 559–562.

Sambrook, J., Russell, D.W., 2001. Molecular Cloning: A Laboratory Manual, third edition Cold Spring Harbor Laboratory Press, Cold Spring Harbor, NY.

Schröder, S., Hain, M., Sterflinger, K., 2000. Colorimetric in situ hybridization (CISH) with digoxigenin-labeled oligonucleotide probes in autofluorescent hyphomycetes. Internat. Microbiol. 3, 183–186.

Simard, C., Lemieux, R., Côté, S., 2001. Urea substitutes toxic formamide as destabilizing agent in nucleic acid hybridizations with RNA probes. Electrophoresis 22, 2670–2683.

Soundararajan, N., Liu, S.H., Soundararajan, S., Platz, M.S., 1993. Synthesis and binding of new polyfluorinated aryl azides to alpha-chymotrypsin. New reagents for photoaffinity labeling. Bioconj. Chem. 4, 256–261.

Wahl, G.M., Stern, M., Stark, G.M., 1979. Efficient transfer of large DNA fragments from agarose gels to diazobenyloxymethyl-paper and rapid hybridization by using dextran sulfate. Proc. Natl. Acad. Sci. 76, 3683–3687.

Wassarman, D.A., 1993. Psoralen crosslinking of small RNAs in vitro. Mol. Biol. Rep. 17, 143–151.

Zagursky, R.J., Conway, P.S., Kashdan, M.A., 1991. Use of ^{33}P for Sanger DNA sequencing. BioTechn. 11, 36–38.

Further reading

Hanspach, G., Sven Trucks, S., Hengesbach, M., 2019. Strategic labeling approaches for RNA single-molecule spectroscopy. RNA Biol. 16 (9), 1119–1132. Available from: https://doi.org/10.1080/15476286.2019.1593093.

Chapter 18

Quantification of specific mRNAs by nuclease protection

Rationale

The study of rare or low abundance transcripts is a frequently reiterated theme in the molecular biology laboratory. The various permutations of PCR, without a doubt, offer the most compelling quantitative expression profile of a sample, though PCR is not without limitations. A solidly designed investigation should be formatted in such a way that data generated by PCR can be validated using a non-PCR-based technique. One approach that can be used to generate or validate quantitative data is generically known as a nuclease protection assay (NPA).

The principle of a NPA is the degradation resistance of double-stranded molecules by single-strand-specific nucleases. The formation of probe:RNA target hybrids is promoted by hybridization in solution under very stringent conditions. Then, the remaining nonhybridized molecules are eliminated by nuclease digestion. This assay was developed years before the advent of PCR to alleviate the lack of quantitativeness associated with Northern analysis, caused by the inaccessibility of some of the membrane-bound RNA (mixed phase hybridization) to probe sequences. Whenever a filter membrane is involved, complete hybridization is difficult, if not impossible. Further, the sensitivity and resolution of Northern analysis are often compromised by the inability to load large amounts of RNA on a gel, by the inefficient transfer from the gel to a blotting membrane, and by nonspecific hybridization between probe and transcripts exhibiting limited homology. Eliminating the membrane eliminates a lot of the problems. To that end, an ELISA-based variant of this assay has been reported recently (Filer et al., 2019).

Basic approach

Transcript assay by nuclease protection is a sensitive and precise method for quantifying specific transcripts. It is widely accepted as a reliable, independent method that can confirm results generated by RT-PCR, and it is equally effective as a standalone technique for the assay of gene expression. The investigator has the choice of using total cellular, total cytoplasmic, poly(A)$^+$ mRNA,

RNA Methodologies. DOI: https://doi.org/10.1016/B978-0-323-90221-2.00025-4

or any other RNA subset as starting target material; the enhanced sensitivity of this approach in comparison with Northern analysis usually does not mandate poly(A)$^+$ selection. Hybridization is followed immediately by exposure to a single-strand-specific nuclease that ordinarily has no affinity for double-stranded molecules, that is, either DNA:RNA hybrids or double-stranded RNA:RNA molecules. Thus, any probe sequences or target RNA molecules that do not participate in duplex formation are digested, and the resultant "protected fragments" are recovered by ethanol precipitation. The size and mass of these double-stranded molecules are then deduced by denaturing polyacrylamide gel electrophoresis, and the signal intensity is quantified by in-gel autoradiography or phosphorimaging. In one variant of this technique, the abundance of the undigested molecules is assessed by liquid scintillation counting instead. A shortcoming of an NPA is the fact that the protected fragments are not indicative of the size of the native (full-length) transcript. This information can only be ascertained by Northern analysis or by direct sequencing. Some of the advantages and disadvantages of the NPA methods are given in Table 18.1.

By virtue of the solution hybridization between the probe and the target, followed by digestion of all other sequences that do not participate in the duplex formation, the signal-to-noise ratio is boosted, often dramatically. The investigator can expect at least a ten-fold increase in sensitivity over conventional blot analysis as well as the ability to reproducibly detect as little as several femtograms (fg) of target RNA after a 3-day X-ray film exposure using high-specific activity probes. Any transcript that can be studied quantitatively by standard Northern analysis is easily detectable by a more sensitive NPA, which is especially convenient when only small amounts of RNA are available. Although autoradiography is considered to be the preferred method for maximum sensitivity, successful nonisotopic variations of both assays have been reported (Rosenau et al., 2002; Kris et al., 2007). More recently, variants of the NPA using fluorescent probes and labels have been described (Xu et al., 2016; Li et al., 2016), thereby offering a reasonable alternative to the use of isotopes. To a great extent, PCR has replaced most of these approaches, though the S1 NPA and RNase protection assay (RPA) remain popular and reliable techniques. The key advantage that these techniques offer is that they do not require a reverse transcriptase step (see Chapter 9: cDNA: A Permanent Biochemical Record of the Cell and Chapter 10: RT-PCR: A Science and an Art Form, for details).

The S1 NPA (Fig. 18.1) can be performed using double-stranded DNA probes or RNA probes. Hybridization is performed in a buffer containing elevated concentrations of formamide, which maintains a high level of stringency. Keeping the hybridization volume small and the concentration of nucleic acid high drives the hybridization to completion in a few short hours. The formamide-based environment tends to produce a very low background by promoting the formation of RNA:DNA duplexes rather than the

TABLE 18.1 Advantages and disadvantages of nuclease protection assays.

Advantages	What	Why
	More sensitive than Northern blots	No filter; solution-based hybridization favors more complete interaction between probe and target mRNA.
	Tolerant of partially degraded mRNA	Only the region of the mRNA complementary to the probe must remain intact.
	Non-PCR method	An alternative to PCR that can be used to demonstrate the authenticity of data by eliminating PCR bias.
	Flexible experimental design	Probes can be DNA or RNA; hybridization conditions can be modified; isotopic and nonisotopic labeling methods supported.
Disadvantages	Size of native mRNA not revealed	Only the section of the mRNA that is protected by the probe survives the nuclease degradation step.
	Non-PCR method	Is not comparable to the amplication power of PCR, thereby limiting assay sensitivity.
	Isotopic labeling can be messy	Numerous manipulations of the sample, pre- and posthybridization, generate a fair amount of radioactive waste.
	Enzymatic degradation can be difficult to control	S1 nuclease is an aggressive enzyme; RNase protection assay may result in high background if not performed properly.
	Optimization can be time-consuming	New probes often require greater optimization than other molecular biology methods.

renaturation of complementary strands of DNA, and it is for this reason that the use of single-stranded probes is clearly preferred. The hybridization temperature must be slightly below the melting temperature (T_m) of the hybrid molecules that are expected to form. Because of the nuclease-mediated, post-hybridization removal of single-stranded regions, including mismatches and overhangs in DNA:DNA, RNA:DNA, and RNA:RNA structures, failure to identify this "window of hybridization" will very likely result in suboptimal probe:target duplex formation and thus yield misleading data.

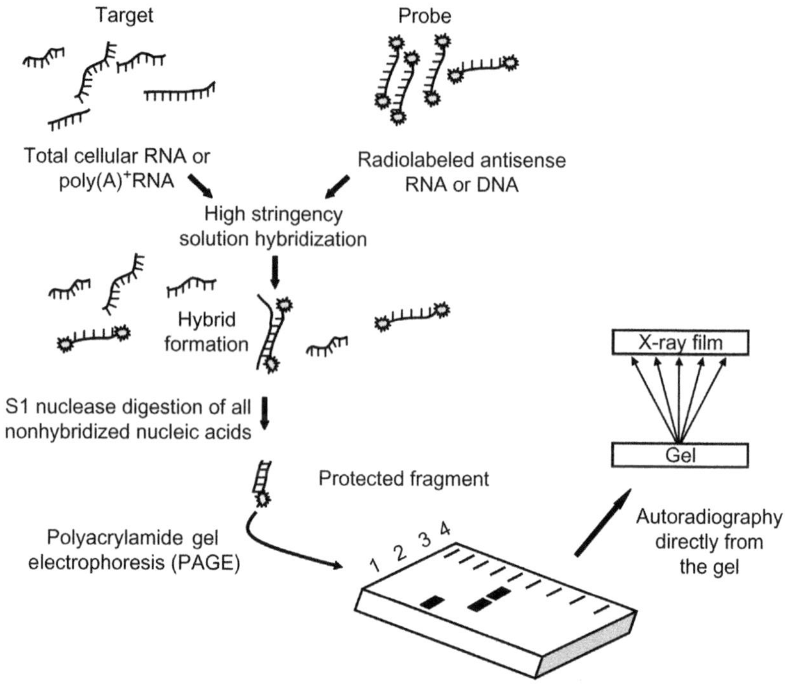

FIGURE 18.1 The S1 nuclease assay for the quantification of specific RNA species. Purified RNA is hybridized in solution with a labeled probe sequence to form thermodynamically stable hybrid molecules. Any RNA or probe molecules that do not participate in the formation of hybrid molecules are removed by digestion with the single-strand-specific nuclease S1. The size and abundance of protected RNAs are deduced by electrophoresis of the intact hybrid molecules, followed by autoradiography performed directly from the gel. Lane 1, undigested probe; lanes 2 and 3, experimental samples; lane 4, molecular weight standards. The mechanics of this assay are nearly identical to RPA.

Nuclease S1 (Ando, 1966; Vogt, 1973) is a naturally occurring glycoprotein isolated from the filamentous fungus *Aspergillus oryzae*, is relatively inexpensive, and was once used interchangeably with mung bean nuclease for RNA initiation and exon-intron mapping assays (Weaver and Weissman, 1979). The S1 nuclease is a multifunctional, zinc-dependent metalloenzyme. It has a high specificity for single-stranded or imperfectly base-paired polynucleotide structures, manifesting a 3′-phosphomonoesterase activity. Any single-stranded DNA or RNA regions are hydrolyzed *exo*- and endonucleolytically, yielding 5′ phosphoryl-mono- or oligonucleotides. When directed against DNA, the hydrolysis activity of S1 nuclease is about five times faster than is observed for RNA. Optimal nuclease activity is observed under conditions of relatively high ionic strength, between pH 4.0 and 4.5, though most reactions are performed at pH 4.5 to minimize DNA depurination. At moderate-to-high concentrations of enzyme, as required in some applications, a concomitant increase in the ionic strength of the assay environment is needed to minimize the nicking of double-stranded molecules that are often observed at higher enzyme concentrations. Nicking of double-stranded DNA is suppressed at 200 mM NaCl and at a slightly higher than optimal pH, though S1 nuclease is essentially inactive at pH 7.2 and it exhibits thermostable properties up to 60°C–65°C. Moreover, the enzyme is resistant to denaturants, including formamide, sodium dodecyl sulfate (SDS), and urea, thereby enhancing its usefulness.

Working with S1 nuclease presents some challenges as well. It is an aggressive enzyme that, in the earliest days of molecular biology, was partially responsible for the poor overall synthesis of cDNA using the procedures that were prevalent in the early 1980s. In the S1 assay, the nuclease digestion steps, more than any other, are the most likely to require optimization, with a typical range of 100U to 200U S1 nuclease added per mL of reaction volume. Excessive amounts of S1 nuclease activity will result in the destabilization and rapid digestion of hybrid molecules; naturally unstable AT- and AU-rich regions are particularly susceptible to S1 attack. Nicking, if it occurs on one strand, will become evident by the presence of an unexpected lower molecular weight band(s) due to cleavage of the other strand; this is also known to occur at abasic sites. This occurrence can be prevented by reducing the amount of enzyme used for the digest. Further, because of the low pH requirement for optimal activity of this enzyme, some level of acid depurination of double-stranded DNA is an expected consequence of prolonged (excessive) incubation in such an environment. Owing to these difficulties controlling S1 nuclease, many investigators prefer the RPA instead.

The basic approach in the RPA (Fig. 18.2) is quite similar to the S1 assay: a probe that is complementary to RNA transcripts under investigation is used in solution hybridization, followed by digestion of excess probe and all transcripts that are not locked up in a double-stranded configuration. From a mechanical point of view, one should note that in the S1 assay, either a DNA probe or an RNA probe can be used, because the S1 nuclease is single-strand specific. Table 18.2 presents a comparison of the closely related S1 nuclease- and RNase protection techniques. In the RPA, however, digestion is performed with a combination of RNase A and RNase T1, mandating the use of an antisense RNA probe. RNase A is an endoribonuclease that cleaves single-stranded RNA, specifically on the 3′ side of pyrimidine (cytosine and uracil) nucleotides (Py/pN). The resulting degradation products are all classified as 3′ pyrimidine phosphates, meaning that each molecule ends in a pyrimidine with a 3′ phosphate group. RNase T1, also a single-strand-specific endoribonuclease, cleaves the phosphodiester bond at the 3′ end of guanine nucleotides (GpN) and, like RNase A, produces oligonucleotides with terminal 3′ phosphates. The actions of these enzymes are summarized in Table 18.3, and the combination of the two ensures thorough posthybridization degradation of nonbase-paired RNA probe and target molecules.

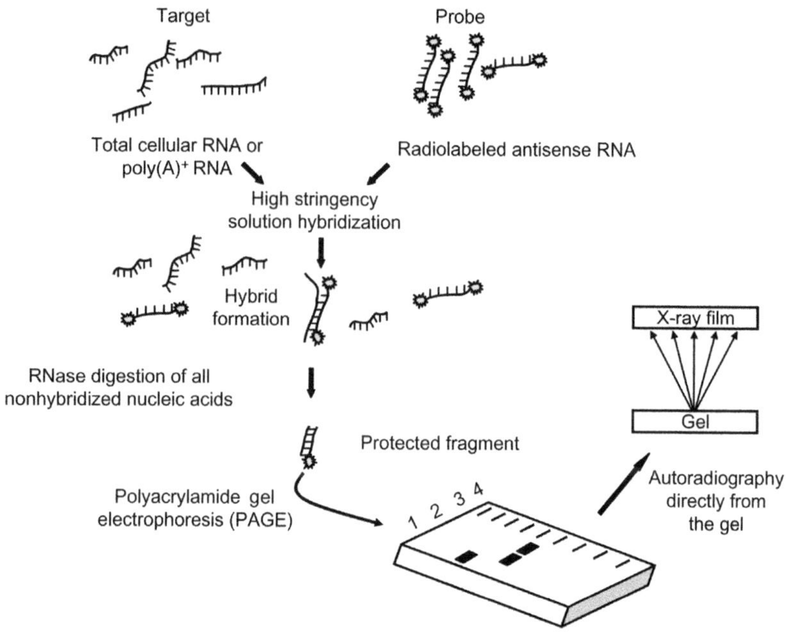

FIGURE 18.2 RNase protection assay for the quantification of specific RNA species. Purified RNA is hybridized in solution with a labeled antisense probe sequence to form thermodynamically stable double-stranded RNA molecules. Any target RNA or probe molecules that remain single-stranded are digested by an RNase cocktail. Following electrophoresis, the size and abundance of protected RNAs are then deduced by autoradiography, performed directly from the gel. Lane 1, undigested probe; lanes 2 and 3, experimental samples; lane 4, molecular weight standards. The mechanics of this assay are nearly identical to RPA.

From a strategy perspective, it is possible to multiplex NPA reactions, that is, to use more than one probe in a single reaction tube. As long as the probes manifest similar thermodynamic behavior and the sizes of the protected fragments are electrophoretically distinguishable, they can be assayed simultaneously, generating a different size band for each protected fragment, all of which appear in a single lane on a gel (Fig. 18.3).

Probe selection

Many of the potential problems associated with spurious hybridization and probe renaturation may be reduced by the use of single-stranded nucleic acid probes. In vitro transcription is the favored method for producing large quantities of uniform length, well-characterized, and high-specific activity RNA probes for nuclease protection analysis. Alternative, though less frequently used, methods for single-strand probe synthesis for this application include oligonucleotide synthesis. Thus, when single-stranded probes are generated,

TABLE 18.2 Comparison of S1 and RNase protection assays.

	S1 nuclease assay	RNase protection assay
Target molecule	mRNA	mRNA
Probe	DNA or RNA	RNA
Nuclease	S1 nuclease	RNase A + RNase T1
Potential for acid depurination	High	Low
Resolution	Reasonably good ability to distinguish between mRNAs that share strong sequence homology	Strong ability to distinguish between mRNAs that share strong sequence homology
Background potential	Low to moderate	Moderate to high
Ability to control enzyme	Very difficult	Moderately difficult
Multiprobe analysis	Yes; resolution less than RPA	Yes, routinely
Average limit of sensitivity	25–50 fg	5–10 fg

the potential for probe renaturation becomes a nonissue, although intramolecular base-pairing can still occur under conditions of relaxed stringency.

Single-stranded probes generated by in vitro transcription often contain at least a small quantity of contaminating DNA template from whence the probe was derived. When present, both the DNA template and contiguous vector sequences can (1) potentially hybridize to full-length probe sequences, protecting them from nuclease degradation, (2) promote artifactual hybridization, and (3) reduce the concentration of available probe. High stringency hybridization will reduce the likelihood of this type of interference, though ideally the transcription vector should be removed by DNase I digestion, followed by probe precipitation to remove the degraded template. Finally, failure to maintain a (ten-fold) molar excess of the probe over target RNA will certainly put the sensitivity of the assay in jeopardy.

Another major difference between the Northern analysis and the NPAs involves the interpretation of the banding distribution observed after electrophoresis and autoradiography (Fig. 18.4). In the Northern analysis, the observed signal is interpreted as corresponding to the native size of the transcript. For example, in humans, the major mRNA produced by transcription of

TABLE 18.3 Comparison of RNase A and RNase T1.

	Typical stock solution	Substrate[a]	Action	Natural source
RNase A	2 mg/mL	Single-stranded RNA	Very stable endoribonuclease that targets pyrimidines (C and U); cleaves the phosphodiester bond 3′ to Py/pN; generates 3′ phosphate group. Example: pApGpCpGpA→ pApGpCp + GpA.	Bovine pancreas
RNase T1	10U/mL	Single-stranded RNA	Endoribonuclease that targets guanine nucleotides; attacks the phosphodiester bond 3′ to GpN; generates 3′ phosphate group. Example: pApApGpCpA→ pApApGp + CpA.	*Aspergillus oryzae*

[a]*Attack of double-stranded RNA occurs upon exposure to excessive quantities of these enzymes for extended periods.*

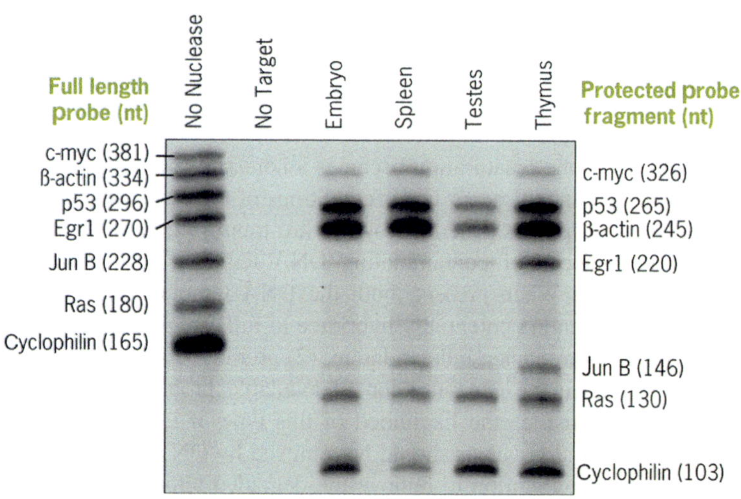

FIGURE 18.3 Ten micrograms of various mouse tissue total RNAs were hybridized overnight with approximately 50,000 cpm each of seven distinct probe transcripts. Nuclease digestion, product separation on a denaturing 6% acrylamide gel, and a 4 h exposure to film at −80°C were used to assess hybridization levels. The radiolabeled probes were synthesized in 5 μL transcription reactions using Ambion's MAXIscript Kit with [α-^{32}P]UTP (800 Ci/mmol, 10 mCi/mL) and gel purified prior to hybridization. The specific activities of the cyclophilin and β-actin probes were reduced 20- and 200-fold, respectively, by adding appropriate amounts of nonradioactive UTP to the transcription reaction. *Courtesy Ambion (now part of Thermo Fisher Scientific).*

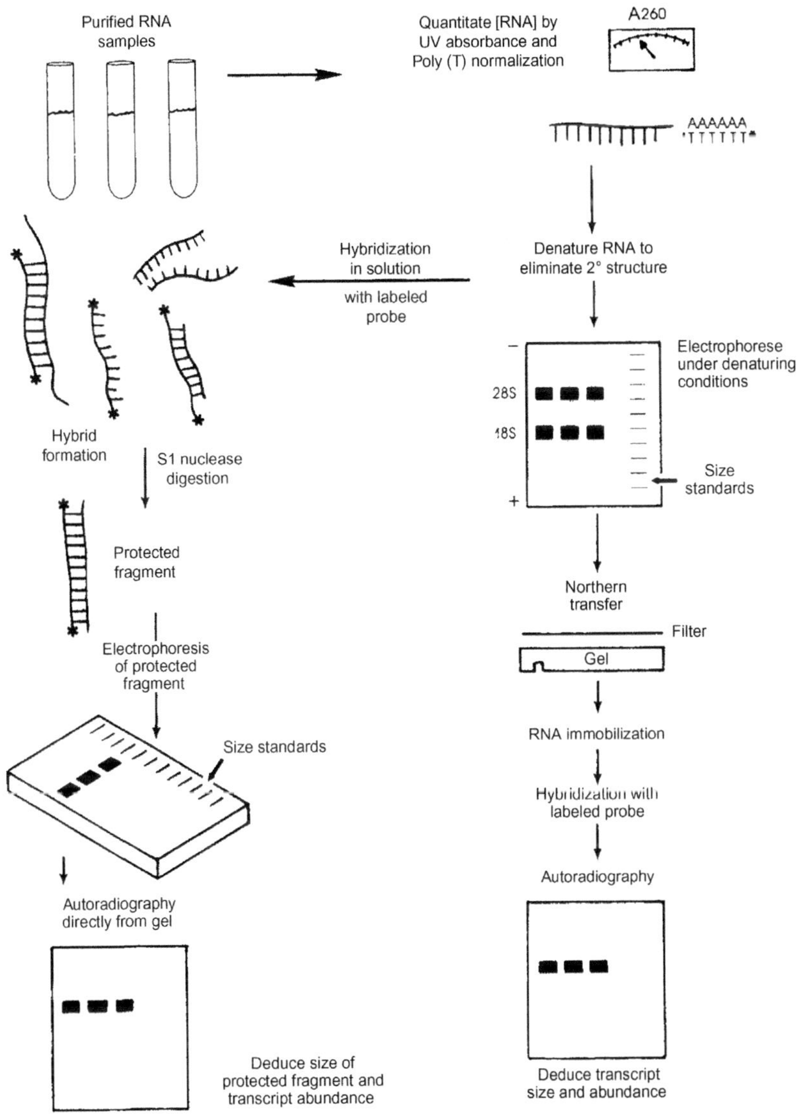

FIGURE 18.4 Quantification of specific RNA transcripts by nuclease protection compared with conventional Northern analysis. The bands generated by Northern blotting represent full-length mRNAs, while bands generated by the S1 or RNase protection assay represent the portion of the mRNA that was "protected" by hybridization to the probe.

the *c-myc* oncogene is 2.4 kb. As such, this transcript is expected to lie between the very abundant ribosomal RNAs (rRNAs), just above the 18S species. In contrast, the resulting hybrid molecule generated by nuclease protection analysis will only be as long as the probe and is usually shorter because nucleotides from both the 5′ and 3′ ends of the probe and target molecules that do not directly participate in duplex formation are digested. The mechanics of the NPAs make these methods tolerant of partially degraded RNA samples as long as the integrity of the section of the mRNA complementary to the probe has not been compromised; recall that a single break in a transcript results in that molecule not being detected by standard Northern analysis. In a well-planned assay, the length of the hybrid (the protected fragment) after nuclease digestion should be observably shorter than nonhybridized or renatured (probe with template) molecules. Under standard assay conditions, electrophoresis of nuclease-treated samples on polyacrylamide gels can easily resolve molecules differing in size by only a few nucleotides.

The manner in which probes are labeled is an important concern, especially when one is attempting to characterize a transcript for the first time. End-labeled probes (5′ or 3′) are *not* preferred and should be avoided, especially in the initial characterization of probe. If the nucleotide containing the (end)-label does not directly participate in the duplex formation, the label will be digested away, rendering the hybrid undetectable. This concern can be quickly alleviated with the use of continuously labeled probes (probes that are labeled along the length of the molecule) and probes that are 100–400 bases tend to work very well.

Single-stranded RNA probes generated by in vitro transcription must be handled using an excellent RNase-free technique. In addition, the high specific activity associated with radiolabeling of these and other probes may result in radiolysis, the phenomenon by which radiolabeled molecules are actually degraded by the decay of the label that they carry. This occurs most often with probes that have been labeled to a very high specific activity ($>1\times10^9$ cpm/μg) and are then stored for more than a few days prior to use. Fragmented probes translate into the increased background and nonspecific hybridization. The investigator is well advised to monitor the integrity of the probe (labeled or not) on a gel prior to each use. It is also prudent to include an aliquot of the undigested probe on the gel along with the experimental samples for direct comparison, a strategy that will reveal probe degradation if present.

Optimization suggestions

Once the type of probe and method of labeling have been established, empirical determination of the hybridization parameters will maximize assay sensitivity and reproducibility. Many NPA reagents are available in kit form; therefore, some of the following optimization procedures may not be necessary. The good news is that an optimization regimen is usually required only once. Tweaking the reaction, however, is almost always required for each new probe.

1. The stable formation of target:probe hybrids must be favored. In the absence of detailed information about the behavior of a particular probe

in a NPA, try hybridization first at 50°C. Then, repeat the assay at various temperatures (± 5°C, ± 10°C, and so forth) to determine the optimum stringency level for each probe. The temperature should be high enough to allow hybrid formation but not so high as to cause the dissociation of duplex molecules—try adjusting the hybridization temperature to 5°C below the T_m.

2. With regard to the thermodynamic behavior of the probe, one should beware of AU-rich domains. The intrinsic instability of these regions can result in incomplete hybridization, thereby contributing to a drastic reduction in signal. To an extent, this difficulty can be alleviated by reducing the formamide concentration in the hybridization buffer, though a loss of probe specificity may result. Following hybridization, incubation at 15°C for 30 minutes in diluted hybridization cocktail *prior* to nuclease degradation permits more complete annealing of AU-rich regions, without favoring nonspecific hybridization that might otherwise accompany an increase in reaction volume under these conditions (Ausserlechner et al., 1998). Following this incubation, nuclease digestion is performed as usual.
3. Hybridization kinetics must be driven to completion and keeping the nucleic acid components of the reaction as concentrated as possible at all times will help. This can be accomplished by the inclusion of carrier RNA (e.g., junk yeast RNA or tRNA). To determine this empirically, remove aliquots at specific intervals and digest each with S1 nuclease as described below. When the hybridization signal intensity no longer increases, the hybridization reaction has gone to completion.
4. When antisense transcripts are used as probes it is important to ensure that they are of uniform length. Nonfull-length transcripts, when present, will produce smaller than expected protected fragments, manifested as more than one band upon autoradiography. If necessary, gel purification of the probe can be performed to isolate the full-length transcripts for the cleanest signal possible. As noted above, it is wise to electrophorese an aliquot of the undigested probe alone, in a lane adjacent to the experimental samples, to convince data critics that the assay is functioning properly.

Potential difficulties

Although the NPAs are straightforward, non-PCR-based methods, they are not without commonplace difficulties. Most of the issues the investigators' experience, especially when working with these assays for the first time, generally fall into one or more of the following categories:

1. *Excessive background*. This is more of a problem when working with a new probe or set of reagents. The excessive background appears as long smears in the lanes of the gel, usually due to incomplete nuclease digestion. It is normal, however, to observe fairly broad smears at the leading

edge of the gel far beyond the location of the protected fragments. Elevated background occurs when there is a delay in moving the samples to the hybridization temperature immediately following the high-temperature RNA denaturation step. This interval affords the opportunity for RNA molecules to regain some degree of secondary configuration, or to base-pair to one another or to nonspecific target sequences inappropriately. This is easily avoided by performing all steps in a thermal cycler which will quickly and automatically move the samples between the denaturation and hybridization temperatures as opposed to physically moving the samples to two (or more) water baths. Background smearing may also occur when the experimental RNA is partially degraded or as a result of probe breakage due to radiolysis or the instability of older probes. Infrequently, smearing in the lanes of the gel may be caused by adding a huge excess, as opposed to a "healthy" excess, of the probe to the hybridization reaction mixture. A large smear at the leading edge of the gel is expected due to the directed degradation of the nonhybridized probe by S1 nuclease or an RNase cocktail. Last, the investigator may wish to check the activity of the S1 nuclease or RNase cocktails to ensure that its use is consistent with the recommendations of the manufacturer. Check the expiration date, too.

2. *No signal.* The common reasons for lack of signal at the end of a NPA include:

a. The probe was not labeled with sufficiently high specific activity. If end-labled, the nucleotide(s) containing the label may have been removed from the hybrid upon nuclease digestion. Although continuously labeled probes (Chapter 17: Nucleic Acid Probe Technology) are less likely to experience this phenomenon, it remains a concern unless it is known with absolute certainty that the actual labeled nucleotide(s) are retained in the protected fragment.

b. The experimental RNA may have been degraded. While it is true that NPAs are more tolerant of *partially* degraded RNA samples, the degradation cannot extend into the region of the RNA which is complementary to the probe. Always assess the integrity of experimental RNA before doing anything with it.

c. Hybridization stringency may have been too high for the type of probe used. The preferred way to remedy this is to lower the hybridization temperature 15°C–20°C in order to determine if the components of the assay will permit any hybridization at all. Then, gradually increase the temperature until the optimal conditions have been determined.

d. The precipitated pellet (sample RNA, probe, carrier RNA) may not have been thoroughly resuspended and, therefore, some or most of the sample was unable to participate in the hybridization reaction. Keeping the pellet *slightly* damp with ethanol will assist in dispersing and dissolving it in the hybridization buffer at the onset of the assay. Also ensure that the nucleic acid pellet was not lost during the post-precipitation ethanol washes.

c. The specific mRNA of interest may be below the level of detection. While the selection of poly(A)$^+$ mRNA prior to S1 analysis may be marginally helpful, it is also potentially counterproductive: the loss of poly(A)$^+$ material because of the intrinsic inefficiency of the selection process may further under-represent low abundance mRNAs of interest and will exclude poly(A)$^-$ species. Increasing the starting mass of total RNA in the assay tube may help. Alternatively, the investigator may be aware of a method for inducing or superinducing the specific gene(s) of interest. The hybridization competence of an RNA sample can be tested using a different probe, ideally a probe for a fairly abundant housekeeping gene, to demonstrate that the sample was at least capable of hybridization to something (anything).

f. Inadvertent use of a "sense," rather than an "antisense," probe. This occurs more often than one might guess.

3. *Background is excellent but there are too many bands.* If this is the only difficulty with this assay, only minimal adjustments should be needed to optimize it. The appearance of multiple, well-defined bands can result from the presence of DNA that is homologous to the probe. cRNA molecules produced by in vitro transcription may be contaminated with plasmid template sequences. A vector or template sequence generally bands closer to the origin (the wells of the gel) because of its much greater size and the fact that it was linearized to support in vitro transcription. Clearly, it is important to destroy the template by DNase I digestion. Alternatively, relaxed hybridization stringency may have tolerated probe hybridization to related transcripts, from the members of the same or a closely related gene family. Depending on the degree of homology and distribution of mismatches, fairly interesting (or frustrating!) banding patterns may be observed. To explore this possibility, conduct an identical hybridization 5°C above the hybridization temperature that resulted in the appearance of multiple bands. This simple procedural change is likely to destabilize mismatched hybrids and reduce or eliminate the phantom bands. Lastly, the appearance of multiple bands may occur when a denatured, double-stranded DNA begins to reanneal. For this reason, many investigators simply prefer to use RNA probes directed against RNA target sequences.

Protocol: transcript quantification by S1 analysis

The most important point to keep in mind is that there is no hard and fast regimen for performing the S1 nuclease assay. The procedure given here is but one of several variations and is a modification of the procedure of

Quarless and Heinrich (1986). With this procedure, it is possible to detect as little as a few hundred femtograms of a specific mRNA.

1. Wear gloves and appropriate protective eyewear. Use RNase-free materials, reagents, and techniques throughout this procedure. Purified RNA samples are particularly susceptible to RNase digestion.
   Note: Be certain to check the concentration, purity, and integrity of the RNA sample(s) before continuing with this protocol.
2. In a microfuge tube, mix 20–30 μg cellular RNA and sufficient carrier RNA to give a final mass of 100 μg. Add approximately 1×10^6cpm of the probe. Adjust the volume of the mixture to 100 μL with sterile H_2O. Be sure to prepare a negative control tube containing carrier RNA and probe only, which should yield no protected fragment.
   Note 1: Beginning with 10 μg total cellular or total cytoplasmic RNA is usually sufficient to quantify many mRNAs, though some investigators begin with as much as 50 μg of RNA. The quantity of S1 nuclease used in this protocol is more than adequate to completely digest all of the non-hybridized RNA and probe remaining in the reaction.
   Note 2: It is very important that the probe is present in at least a 5- to 10-fold molar excess. In most cases, adding 10^6 cpm of radiolabeled probe will fulfill this requirement.
   Note 3: A stock solution of 1 mg/mL yeast RNA for use as carrier can be prepared in 0.1 M sodium acetate, pH 5.2. Store convenient aliquots at −20°C.
3. Coprecipitate sample, carrier, and probe with the addition of 11 μL 3 M sodium acetate, pH 5.2, and 250 μL ice-cold 95% ethanol. Store at −20°C overnight.
   Note: The experimental RNA and carrier can be coprecipitated without the probe and stored in hybridization buffer for up to 1 week at −80°C.
4. Collect precipitate by centrifugation. Decant supernatant and wash pellet with 70% ethanol. Discard ethanol (radioactive waste) and invert tube to dry the pellet for approximately 10 minutes. An optional, second wash of the pellet with 95% ethanol will accelerate the drying process. Maintain the RNA pellet *slightly* damp with ethanol.
   Note: Do not Speed Vac or otherwise allow the pellet to dry to completion; doing so will make the RNA virtually impossible to resuspend.
5. Resuspend the pellet in 20 μL S1 hybridization buffer (80% freshly deionized formamide); 40 mM PIPES (piperazine-1,4-bis(2-ethanesulphonic acid), pH 6.4; 400 mM NaCl; 1 mM EDTA, pH 8.0). Pipette the sample up and down at least 20 times to ensure complete resuspension.
   Note 1: If necessary, the volume of hybridization buffer in which the sample is resuspended may be increased to as much as 30 μL.
   Note 2: Hybridization buffer can be prepared in advance and stored in 1 mL aliquots at −80°C.

Note 3: If experimental RNA and carrier were coprecipitated ahead of time without the probe (not recommended), add the probe to the pellet first and then add hybridization buffer to a final volume of 30 μL.

Note 4: To detect low abundance mRNAs, keeping the RNA concentration near the limit of solubility (5 mg/mL = 100 μg/20 μL) will favor complete hybridization.

Note 5: Hybridization buffer consisting of 80% freshly deionized formamide is a standard formulation; see Appendix G for recommended deionization procedure. If using a DNA probe, in which case the thermodynamic stability of the hybrid will not be as high as in the case of a cRNA probe, a hybridization buffer consisting of as little as 40%–50% formamide may be substituted. Recall that stringency is modulated by temperature (salt and pH, too) and the presence of formamide.

6. Cover the reaction with one small drop of light mineral oil. By preventing evaporation, the reaction will proceed with maximum efficiency.
7. Heat the sample in a water bath at 85°C for 10 minutes to fully denature the RNA sample.

   Note: It is imperative that the sample be fully denatured, as secondary structures will have an adverse effect on the proper formation of hybrids.
8. Pulse centrifuge the tube(s) for 5 seconds to collect the hybridization mixture at the bottom of the tube. *Immediately* place the tube in a prewarmed water bath and incubate the reaction mixture at 50°C or other predetermined temperature, for 3 hours.

   Note 1: Hybridization temperature is a direct function of the type of probe (RNA or DNA) used, probe length, and G + C content, and must be empirically determined. The typical range for the parameters defined in this protocol is 45°C–60°C. In the absence of experience with or information about a particular probe, begin with 50°C.

   Note 2: If carrier RNA was omitted from the reaction tube (step 2), the hybridization interval should be extended to an overnight incubation.
9. Dilute the reaction mixture with the addition of 160 μL sterile H_2O. Add 20 μL of 10 × S1 nuclease buffer that accompanied the purchase of the S1 nuclease. If this pre-made buffer is not available, prepare a generic 10 × S1 buffer (500 mM sodium acetate, pH 4.5; 2.8 M NaCl; 40 mM $ZnSO_4$, 50% glycerol).

   Note: In place of $ZnSO_4$, ZnOAc (zinc acetate) is also used routinely in S1 digestion buffer formulations.
10. Add sufficient S1 nuclease to produce a final enzyme concentration of 200–300 U/mL. The actual volume required will be variable from one lot to the next, and among vendors.

    Note 1: Do not add any more than 10 μL of S1 nuclease to a 200 μL digest because the final glycerol concentration will inhibit the reaction. In the case of fewer units of S1 in the reaction it may be necessary to extend the incubation to ensure complete digestion.

Note 2: This part of the protocol, more than any other, is the step that is most likely to require optimization. Recall that AU-rich regions in the hybrid are particularly susceptible. Nicking, if it does occur, will become evident by the presence of one or more unexpected lower molecular weight bands. If this occurs, the remedy is to reduce the amount of enzyme being used to perform the digest.

11. Pulse centrifuge briefly to collect the entire volume in the bottom of the microfuge tube. Incubate at 37°C for 1 hour.

12. If necessary, transfer the entire reaction volume to a polypropylene microfuge tube. Extract the reaction once with an equal volume of phenol:chloroform:isoamyl alcohol (25:24:1) or phenol:chloroform (1:1) to terminate S1 digestion. Microfuge for 2 minutes at top speed to separate the phases.

Recommended: Rather than using PCI, purification of the protected fragments using a silica-based clean-up kit is a suitable alternative. Be sure that the size of the protected fragment is greater than the molecule weight cut-off of the purification column.

13. Carefully transfer the aqueous (upper) phase to a fresh microfuge tube. Add 1.0 μL 20 mg/mL glycogen (produces a final glycogen concentration of approximately 100 μg/mL). Add 2.5 vol of 95% ethanol to precipitate the hybrids that were protected from nuclease digestion.

14. Incubate tubes on dry ice for 15 minutes or at −20°C for 2 hours.

15. Collect the precipitate by centrifugation. Carefully decant the supernatant and allow the pellet to air dry. If desired, it is acceptable to *briefly* dry down the pellet under vacuum, though it is critical not to allow the sample to dry out.

16. Resuspend the pellet in 5 μL TE buffer (10 mM Tris, 1 mM EDTA, pH 8.0).

17. Add 5 μL of 2 × formamide loading buffer (50% formamide, 2 × TBE buffer, 0.2% bromophenol blue). Heat each sample to 95°C for 3 minutes and then cool on ice for 2 minutes. Electrophorese on a 5%–6% polyacrylamide/8 M urea gel in 1 × TBE buffer at 250 volts, constant voltage until the dye is close to or at the bottom of the gel.

Alternatively: Add an equal volume of 2 × loading buffer (125 mM Tris, pH 6.8; 10% glycerol; 0.2% bromophenol blue) to the sample. Electrophorese on a 5%–6% native polyacrylamide gel or use a percentage appropriate for characterizing the size of the expected protected fragment. Use 1 × TBE (Tris-borate-EDTA) as the electrophoresis buffer. Electrophorese at 250 volts, constant voltage, until the dye front is at or near the bottom of the gel.

18. When electrophoresis is complete, dry down the gel and perform autoradiography at −70°C with an intensifying screen (Chapter 17: Nucleic Acid Probe Technology). Alternatively, a Northern blot-type transfer can be set up (do not attempt to stain or photograph the gel) by which the protected fragment(s) will be eluted from the gel and onto a nylon blotting membrane. Then, wrap the membrane in plastic wrap and perform autoradiography as usual.

Protocol: transcript quantification by RNase protection

As noted indicated in the preceding protocol, there are no hard and fast rules for transcript quantification by nuclease protection. The guiding principle is high stringency hybridization, and the RPA offers greater thermodynamic probe:target stability and discriminatory power via the use of RNA probes than is achievable when using DNA probes in conjunction with, for example, the S1 nuclease assay. The procedure given here is a modification of the procedures of Zinn et al. (1983) and Melton et al. (1984) and is but one of several variations; optimized reagents need to perform the RPA are widely available in kit form and are highly recommend. When optimized, this procedure offers at least 0.1 pg levels of sensitivity, and is often much better. Although the substitution of nonisotopic labeling and detection methods has been reported, radiolabeling offers maximum sensitivity and resolution.

1. Wear gloves and appropriate protective eyewear. Use RNase-free materials, reagents, and techniques throughout this procedure. Purified RNA samples and RNA probes prepared by in vitro transcription are always susceptible to RNase digestion.
2. Prepare the antisense probe by in vitro transcription and then purify it using any commercially available probe purification system or as described elsewhere (e.g., Green and Sambrook, 2014).
3. Isolate RNA from cells or tissues under investigation, using any of the protocols described herein.

   Note: Be certain to check the concentration, purity, and integrity of the sample before continuing with this protocol.
4. In a microfuge tube, mix 20–30 μg cellular RNA and sufficient carrier RNA to give a final mass of 100 μg. Add approximately 1×10^6cpm of the probe. Adjust the volume of the mixture to 100 μL with sterile H_2O. Be sure to prepare a negative control tube containing carrier RNA and probe only, which should yield no protected fragment.

   Note 1: Beginning with 10 μg total cellular or total cytoplasmic RNA is usually sufficient to quantify many mRNAs, though some investigators begin with as much as 50 μg of RNA. The quantity of RNase used in this protocol is more than adequate to completely digest all of the nonhybridized RNA and probe remaining in the reaction.

   Note 2: It is very important that the probe is present in at least a 5- to 10-fold molar excess. In most cases, adding 10^6 cpm of radiolabeled probe will fulfill this requirement.

5. Coprecipitate sample, carrier, and probe with the addition of 11 μL 3 M sodium acetate, pH 5.2 and 250 μL ice-cold 95% ethanol. Store at −20°C overnight.

   Note: The experimental RNA and carrier can be coprecipitated without the probe and stored in hybridization buffer for up to 1 week at −80°C.

6. Collect precipitate by centrifugation. Decant supernatant and wash pellet once with 70% ethanol. Discard ethanol (radioactive waste) and invert tube to dry pellet for about 10 minutes. An optional, second wash of the pellet with 95% ethanol will accelerate the drying process. Maintain the RNA pellet *slightly* damp with ethanol.

   Note: Do not Speed Vac or otherwise allow the pellet to dry to completion; doing so will make the RNA virtually impossible to resuspend.

7. Resuspend the pellet in 20 μL RPA hybridization buffer (80% freshly deionized formamide; 40 mM PIPES, pH 6.4; 400 mM NaCl; 1 mM EDTA, pH 8.0). Pipette the sample up and down at least 20 times to ensure complete resuspension.

   Note 1: If necessary, the volume of hybridization buffer in which the sample is resuspended may be increased to as much as 30 μL.

   Note 2: Hybridization buffer can be prepared in advance and stored in 1 mL aliquots at −80°C.

   Note 3: If experimental RNA and carrier were coprecipitated ahead of time without the probe (not recommended), add the probe to the pellet first and then add hybridization buffer to a final volume of 20 μL.

   Note 4: To detect low abundance mRNAs, keeping the RNA concentration near the limit of solubility (5 mg/mL = 100 μg/20 μL) will favor complete hybridization.

   Note 5: Hybridization buffer consisting of 80% freshly deionized formamide is a standard formulation; see Appendix G for recommended deionization procedure. Recall that stringency is modulated by temperature (salt and pH, too) and the presence of formamide.

8. Cover the reaction with one small drop of light mineral oil. By preventing evaporation, the reaction will proceed with maximum efficiency.

9. Heat the sample in a water bath at 85°C for 10 minutes to fully denature the RNA sample.

   Note: It is imperative that the sample be fully denatured, as secondary structures will have an adverse effect on the proper formation of hybrids.

10. Pulse centrifuge the tube(s) for 5 seconds to collect the hybridization mixture at the bottom of the tube. *Immediately* place the tube in a prewarmed water bath and incubate the reaction mixture at 55°C, or other predetermined temperature, for 3 hours.

    Note 1: Hybridization temperature is a direct function of the probe length and G + C content and must be empirically determined. The typical range for the parameters defined in this protocol is 45°C–60°C. In the absence of experience with or information about a particular RNA probe, begin with 55°C.

Note 2: If carrier RNA was omitted from the reaction tube (step 4), extend the hybridization to an overnight incubation.

11. At the conclusion of the hybridization period, dilute the reaction mixture with the addition of 240 μL sterile H_2O. Add 30 μL of 10× RNase buffer (100 mM Tris, pH 7.5; 3 M NaOAc, 50 mM EDTA, pH 8.0), followed by the addition of 5 μL 2 mg/mL RNase A and 5 μL 10U/μl RNase T1.

    Note 1: If more than 20 μL of RPA hybridization buffer was used in step 7, be sure to increase the volume of H_2O and 10× RNase buffer proportionately.

    Note 2: The final concentration of RNase A should be approximately 50 μg/mL and final concentration of RNase T1 should be approximately 200 U/mL.
12. Pulse centrifuge briefly to collect the entire volume in the bottom of the microfuge tube. Incubate at 37°C for 1 hour.
13. If necessary, transfer the entire reaction volume to a polypropylene microfuge tube. Extract the reaction once with an equal volume of phenol:chloroform:isoamyl alcohol (25:24:1) or phenol:chloroform (1:1) to terminate RNase digestion. Microfuge samples for 2 minutes at top speed to separate the phases.

    Suggestion: Rather than using PCI, purification of the protected fragments using a silica-based clean up kit is a suitable alternative. Be sure that the size of the protected fragment(s) is greater than the molecular weight cut-off of the purification column.
14. Carefully transfer the aqueous (upper) phase to a fresh microfuge tube. Add 1.5 μL 20 mg/mL glycogen and 2.5 vol of 95% ethanol to precipitate the hybrids that were protected from nuclease digestion.
15. Incubate tubes on dry ice for 15 minutes or at −20°C for 2 hours.
16. Collect the precipitate by centrifugation. Carefully decant the supernatant and allow the pellet to air dry. If desired, it is acceptable to *briefly* dry down the pellet under vacuum, though it is critical not to allow the sample to dry out.
17. Resuspend the pellet in 5 μL TE buffer (10 mM Tris, 1 mM EDTA, pH 8.0).
18. Add 5 μL of 2× formamide loading buffer (50% formamide, 2× TBE buffer, 0.2% bromophenol blue). Heat each sample to 95°C for 3 minutes and then cool on ice for 2 minutes. Electrophorese on a 5% to 6% polyacrylamide/8 M urea gel in 1× TBE buffer at 250volts, constant voltage until the dye is close to or at the bottom of the gel. *Alternatively:* Add an equal volume of standard 2× loading buffer (125 mM Tris, pH 6.8; 10% glycerol; 0.2% bromophenol blue) to the sample. Electrophorese on a

5%–6% native polyacrylamide gel or use a percentage appropriate for characterizing the size of the expected protected fragment in 1× TBE electrophoresis buffer. Electrophorese at 250volts, constant voltage, until the dye front is close to or at the bottom of the gel.

19. When electrophoresis is complete, dry down the gel and perform autoradiography at −70°C with an intensifying screen (Chapter 17: Nucleic Acid Probe Technology). Alternatively, a Northern blot-type transfer can be set up (do not attempt to stain or photograph the gel) by which the protected fragment(s) will be eluted from the gel and onto a nylon blotting membrane. Then, wrap the membrane in plastic wrap and perform autoradiography as usual.

Troubleshooting

When difficulties are encountered in the course of a NPA, the troubleshooting suggestions offered here may be helpful for resolving the issue(s).

1. Always use a freshly prepared probe.
2. Be sure to coprecipitate the probe and the test RNA sample(s) the night before performing the assay. This will enable resuspending the pellet in the smallest possible volume of hybridization buffer (ideally 20 μL) and favor assay reproducibility and data reliability.
3. Scrupulously observe the RNase-free strategies described herein and elsewhere.
4. Prior to the onset of the assay, run a minigel to ensure the integrity of the RNA sample.
5. Consider lowering the hybridization stringency in the case of no signal or raising the hybridization stringency if everything "lights up." The precise temperature calculation will also be dependent on the nature of the probe: recall that double-stranded RNA duplexes are more stable thermodynamically than DNA:RNA duplexes and are therefore able to withstand more stringent hybridization than would be tolerated by a single- or double-stranded DNA probe (discussed in Chapter 17: Nucleic Acid Probe Technology).
6. If no signal is detected and the investigator suspects that the transcript is below the level of detection, perform a quick PCR assay, if primers are available. This will answer the question of whether the transcript is present.
7. Incubation with DNase and RNase, in independent reactions, can be used to demonstrate the specificity of the NPA for RNA detection.
8. Reduce the amount of sample loaded onto the gel if bands appear to "smile" or if one band obscures visualization of a nearby band because of signal strength. If possible, only load every other well on the gel. Also, running a gel for a longer time will give a better separation of bands that are too close.

9. In the case of conflicting data, ensure the probe is present in a substantial (ten-fold) molar excess in each assay. This is performed by molecular titration, in which a constant mass of probe is assayed against various quantities of a particular RNA sample. If the probe is present in excess, changing the starting mass of RNA target should change the signal proportionally.

10. S1 assay is an aggressive enzyme that, in excess concentrations, will attack double-stranded molecules. If difficulties are encountered with this component of the assay, vary the amount of S1 nuclease so that the probe is completely degraded without loss of hybridization signal. Avoid the use of homemade preps of S1 or RNase – these require extensive, time-consuming titration in order to be considered useful and, frankly, their unpredictability isn't worth the effort.

References

Ando, T., 1966. A nuclease specific for heat-denatured DNA isolated from a product of *Aspergillus oryzae*. Biochim. Biophys. Acta 114, 158–168.

Ausserlechner, M.J., Sgonc, R., Wick, G., 1998. Chemiluminescence-based RNase protection assays for simultaneous quantification of procollagen mRNAs containing AU-rich regions. BioTechniques 24 (3), 366–368.

Filer, J., Channon, R., Henry, C., Geiss, B., 2019. A nuclease protection ELISA assay for colorimetric and electrochemical detection of nucleic acid. Anal. Methods 11, 1027–1034. Available from: https://doi.org/10.1039/C8AY02729C.

Green, M.R., Sambrook, J. (Eds.), 2014. Molecular Cloning: A Laboratory Manual. Cold Spring Harbor Laboratory Press, Cold Spring Harbor, NY.

Kris, R.M., Felder, S., Deyholos, M., Lambert, G.M., Hinton, J., Botros, I., et al., 2007. High-throughput, high-sensitivity analysis of gene expression in Arabidopsis. Plant. Physiol. 144, 1256–1266.

Li, J., Zhao, Q., Tang, Y., 2016. Label-free fluorescence assay of S1 nuclease and hydroxyl radicals based on water-soluble conjugated polymers and WS2 nanosheets. Sensors 16 (6), 865. Available from: https://doi.org/10.3390/s16060865.

Melton, D.A., Krieg, P.A., Rebagliati, M.R., Maniatis, T., Zinn, K., Green, M.R., 1984. Efficient in vitro synthesis of biologically active RNA and RNA hybridization probes from plasmids containing a bacteriophage SP6 promoter. Nucleic Acids Res. 18, 7035–7056.

Quarless, S.A., Heinrich, G., 1986. The use of complementary RNA and S1 nuclease for the detection of low abundance mRNA transcripts. BioTechniques 4 (5), 434–438.

Rosenau, C., Kaboord, B., Qoronfleh, M.W., 2002. Development of a chemiluminescence-based ribonuclease protection assay. BioTechniques 33, 1354–1358.

Vogt, V.M., 1973. Purification and further properties of single-strand-specific nuclease from *Aspergillus oryzae*. Eur. J. Biochem. 33, 192–200.

Weaver, R.F., Weissman, C., 1979. Mapping of RNA by a modification of the Berk-Sharp procedure: the 5′ termini of 15S β-globin mRNA precursor and mature 10S β-globin mRNA have identical map coordinates. Nucleic Acids Res. 7, 1175–1193.

Xu, W., Peng, L., Li, B., Xie, Z., Tong, C., Liu, B., 2016. A real-time S1 assay at neutral pH based on graphene oxide quenched fluorescence probe. Sens. Bio-Sensing Res. 7, 42–47.

Zinn, K., DiMaio, D., Maniatis, T., 1983. Identification of two distinct regulatory regions adjacent to the human β-interferon gene. Cell 34, 865–879.

Further reading

Noviana, E., Jain, S., Hofstetter, J., Geiss, B.J., Dandy, D.S., Henry, C.S., 2020. Paper-based nuclease protection assay with on-chip sample pretreatment for point-of-need nucleic acid detection. Analytical and Bioanalytical Chemistry 142 (13), 3051–3061. Available from: https://doi.org/10.1007/s00216-020-02569-w.

Chapter 19

Analysis of nuclear RNA

Rationale

A fundamental goal in the study of gene expression is the elucidation of the level(s) at which the genes of interest are being controlled. While potential regulatory points are myriad, they are broadly categorized in eukaryotes as transcriptional in nature or as a result of one or more posttranscriptional events. Examples of the former include transcriptional (promoter) efficiency and transcription rate, while examples of the latter include hnRNA splicing efficiency, nucleocytoplasmic transport, miRNA regulation, and accessibility of mRNA to the protein translation machinery, to name but a few (Fig. 19.1). This mandates direct interrogation of nuclear transcripts independently of those found in the cytoplasm.

Traditional methods used to assay transcription include RT-PCR techniques, Northern analysis, and nuclease protection (Chapters 10, 16, and 18, respectively). When performed correctly, these methods measure steady-state levels of the message, that is, the final accumulation of RNA in the cell and the data are widely interpreted as being equivalent to gene expression. No information is gained, however, about the rate of transcription by which these RNAs were produced. Knowledge of this aspect of gene regulation is important because many transcripts have much longer natural half-lives than others, and because the turnover of many mRNA species is subject to variability in response to various stimuli. Thus, the study of nuclear RNA includes steady-state analysis by RNA isolation directly from the nucleus and an independent assessment of the relative rates of transcription in intact nuclei. Traditional methods are presented here, along with a discussion of suitable contemporary alternatives for global transcriptome analysis.

Transcription rate assays

Steady-state nuclear RNA analysis reveals what types of RNA have accumulated in a cell or organelle and in what amount. Transcripts are easily recoverable from the nucleus through minor modifications of standard RNA isolation procedures found elsewhere in this volume. Characterization of these transcripts commonly involves RT-PCR, though the more traditional nuclease protection assay, or even Northern analysis can be beneficial. To

RNA Methodologies. DOI: https://doi.org/10.1016/B978-0-323-90221-2.00002-3

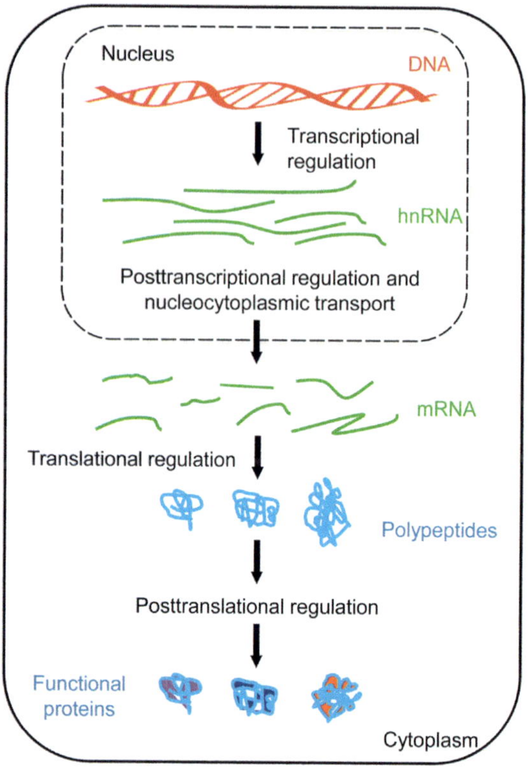

FIGURE 19.1 mRNA biogenesis. Transcription of genes by RNA polymerase II produces hnRNA, the direct precursor of mRNA. Changes in the cellular biochemistry that affect the production of a primary transcript, including transcription factor mediation, are designated as being transcriptional in nature, while downstream events such as splicing, export to the cytoplasm, and stabilizing or destabilizing events are said to posttranscriptional regulatory control points. Steady-state RNA represents the final accumulation of one or more types of RNA in the cell while assessing the relative rate of transcription requires labeling transcripts as they are synthesized in the nucleus.

sequester nuclear transcripts, intact nuclei must be quickly recovered from cells, usually following gentle, hypotonic lysis. Once separated from the cytosolic contents, nuclei are then processed in a manner similar to that used if starting with whole cells.

In another approach, the *rate* of transcription is measured in intact nuclei by incorporating labeled precursor ribonucleotides into RNA transcripts as they are elongated on endogenous chromatin. This widely used technique, known as the nuclear run-on (NRO) assay (Marzluff, 1978) is a sensitive method for measuring transcription rate as a function of cell state (Fig. 19.2) in both animals and plant cells (Kanazawa et al., 2000). It also has the advantage of not supporting the initiation of new transcripts once the nucleus is

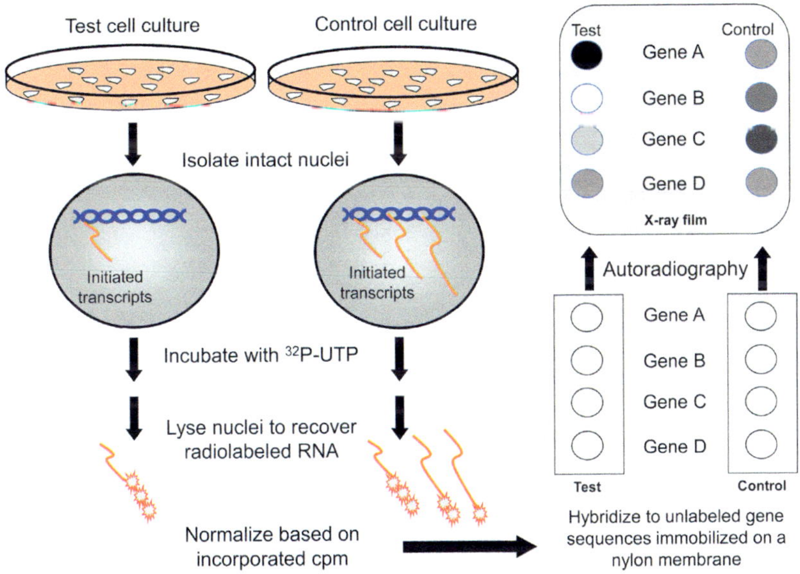

FIGURE 19.2 Nuclear run-on assay. The relative rate of transcription of all genes can be assessed by incubation of intact nuclei with an NTP cocktail containing labeled UTP and then hybridizing those transcripts to unlabeled filter-bound probes. Nonisotopic labeling and detection is an alternative.

isolated from the cell. Labeled RNA (NRO-RNA) is subsequently purified for characterization by any of a number of methods, including hybridization to membrane-bound gene-specific probes, sequencing, array analysis, or other transcriptomics methods. Among the advantages of the NRO assay is that labeling occurs while maintaining the natural geometry of the transcription apparatus and it is a direct measure of the transcription frequency of all genes while at the same time being independent of RNA turnover. Since all RNA being transcribed is labeled, the NRO is a whole transcriptome assay, including the study of lncRNA (Kashi et al., 2016) and miRNA biogenesis (Zhang et al., 2014). A comparison of the traditional Northern analysis, nuclease protection assays, NRO assay, and RT-PCR is presented in Table 19.1.

In contrast to the traditional methods of labeling and detection (described below), these days transcripts are often converted into cDNA and then sent for next-generation sequencing. Nuclear transcription assays have also been shown to be valuable for studies of gene expression knockdown involving RNAi (Khraiwesh, 2011). Other variants include coupling the NRO assay to array analysis (Fan et al., 2011) or recovery by immunoprecipitation. In one approach, for example, the uridine derivative bromouridine (5-bromouridine; BrUrd) functions as a uridine analog, allowing BrUrd nucleotides to be incorporated into nascent RNA molecules in direct proportion to their rate of

TABLE 19.1 Comparison of northern analysis, nuclease protection assay, nuclear run-on assay, and RT-PCR.

	Northern analysis	Nuclease protection assay	Nuclear run-on assay	RT-PCR
Advantages	Provides a qualitative component to RNA analysis. Is compatible with total, cytoplasmic, nuclear, or poly $(A)^+$ RNA. RNA is relatively stable when fixed onto a filter. Supports several rounds of hybridization with different probes. Is able to assess integrity of the sample.	More sensitive than Northern analysis because solution hybridization is more quantitative than filter hybridization. Hybridization volume is very small. Is tolerant of partially degraded RNA. Less handling of RNA than other types of analysis. Can be used for steady-state or transcription rate assays.	Measures transcription rate. Natural geometry of the chromatin is maintained. Able to study of several genes simultaneously. Able to assign gene regulation as transcriptional or posttranscriptional in nature. Supports transcript elongation, and not initiation (this can also be a disadvantage, depending upon the questions being asked).	Provides unparalleled sensitivity. Gold standard. Well-designed primers provide superior resolution. Technology is universally available. Requires very small amounts of starting material. Supersedes many of the classical techniques. Rapid technique that favors productivity.

Disadvantages	Is the least sensitive assay. Some denaturants are toxic. Is a time-consuming process. Characterizes steady-state RNA only. More handling of RNA required.	Protected fragment is smaller than native RNA. Nucleases, especially S1, can be difficult to control. Assay is more sensitive to hybridization parameters than other assays. dsDNA probes can compromise assay sensitivity if reannealing occurs.	Nuclear isolation requires a fair amount of skill. Is a time-consuming, multi-step process. Probe complexity is very large. Unlabeled endogenous RNA can compete with labeled RNA during hybridization.	Much more sensitive to the precise reaction components and conditions than other assays. Exquisitely sensitive to contaminants, especially genomic DNA. Carryover contamination must be controlled. Optimization can be time-consuming and costly.

synthesis (Tani et al., 2012; Roberts et al., 2015). This supports the genome-wide observation of transcriptional activity. BrUrd-labeled RNA can be purified by anti-BrUdr antibody capture on a magnetic bead. The recovered RNA can then be analyzed by qRT-PCR or any other method suitable for the study of RNA.

Another contemporary alternative for the assay of transcription rate is ChIP-seq (Sandoval et al., 2004; reviewed by Tan and Wong, 2019). ChIP, or chromatin immunoprecipitation, is a method for measuring proteins associated with chromatin by crosslinking them, most often with formaldehyde. In the case of a gene transcription rate assay, the precipitation or pulldown is directed at RNA polymerase II occupancy on the gene(s) of interest. Inasmuch as exposure to formaldehyde instantaneously affects crosslinking, this methodology is useful for assay of transcriptional responses in studies involving multiple time points. The pulled-down DNA may then be identified by next-generation sequencing or real-time PCR.

A third alternative to the traditional NRO is a method known as GRO-seq, or global run-on sequencing (Tzerpos et al., 2021) which essentially combines the NRO assay with next-generation deep sequencing to measure nascent RNA genome-wide. Briefly, transcripts in the process of elongation are labeled with BrUrd and then sequestered by magnetic bead antibody capture. Alternatively, the use of the ATP analog N^6-allyadenosine triphosphate (a^6ATP) has been used for labeling (Gao et al., 2022). In either case, the RNA is barcoded for source identification, reverse transcribed, and then sequenced. This method offers sensitive detection of both coding and non-coding transcripts. Traditional transcriptional assay methods require validated probes or primers of known identity; in contrast, all transcripts, known and unknown, are detectable and quantifiable through next-generation sequencing. Information about RNA-seq is presented in Chapter 24, RNA-seq: The Premier Transcriptomics Tool. As with many contemporary methods, outsourcing is readily available for those labs that do not commonly perform these procedures. A comparison of the traditional nuclear run assay, RNA pol II ChIP, GRO-seq is presented in Table 19.2.

The mechanics and reaction conditions of the NRO assay promote the elongation of initiated transcripts and do not support new initiation events; see Zhang and Tijan (2018) for a review pertaining to the measurement of transcript initiation. Thus, the NRO assay is an in vitro extension of what began in vivo. Because the NRO assay yields unspliced precursor RNA, however, the investigator is uncertain as to the natural posttranscriptional fate of these RNA molecules, suggesting that the presence of these transcripts should also be assayed in the cytoplasm. The degree of labeling of any particular RNA species (and hence, the relative transcription rate at a specific genetic locus) is then assessed, in the case of radiolabeling, by autoradiography (Fig. 19.3). For greater precision, the "dots" or "slots" onto which DNA probes are spotted on the hybridization filter can be separated

TABLE 19.2 Transcription rate assay comparison.

	GRO-seq	RNA pol ChIP-seq	Contemporary nuclear run-on assay	Traditional nuclear run-on assay
Starting material	10^3–10^5 cells	10^3–10^5 cells	10^5–10^6 cells	10^5–10^7 cells
Label	5-bromouridine (BrUrd) or a^6ATP	None; crosslinking performed instead	5-bromouridine (BrUrd)	^{32}P-UTP or ^{3}H-uridine
RNA recovery method	Immunoprecipitation	Immunoprecipitation	Immunoprecipitation	RNA purification and precipitation
Antibody specificity	Anti-BrUrd or anti-a^6ATP	Anti-RNA polymerase II	Anti-BrUrd	Not applicable
RNA identification	Next-generation sequencing	Next-generation sequencing	qRT-PCR	Hybridization and quantification by autoradiography or liquid scintillation counting

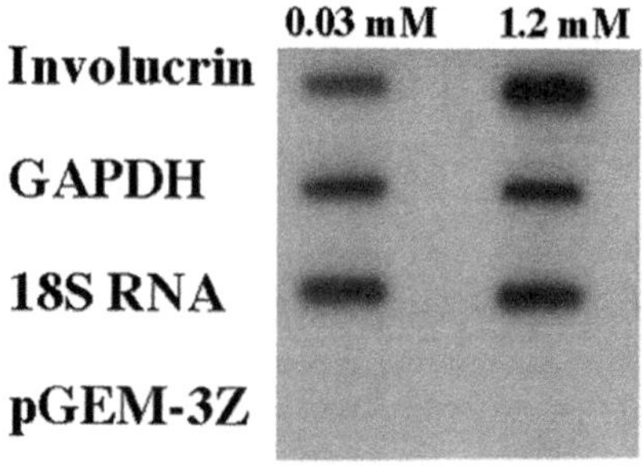

FIGURE 19.3 Nuclear run-on assay using normal human keratinocytes (NHK). Cells were cultured in a medium supplemented with 0.03 or 1.2 mM calcium in the presence of ^{32}P-UTP. The newly synthesized RNA was then hybridized to pGEM-3Z containing involucrin cDNA, GAPDH cDNA (transcription control), and cDNA of 18S RNA (loading control) that were blotted onto a filter. pGEM-3Z was used as the negative control (Ng et al., 1996). *From Ng, D.C., M-J. Su, R. Kim, and D.D. Bikle. (1996). Regulation of involucrin gene expression by calcium in normal human keratinocytes. Front. Biosci. 1, 16–24, and reproduced with permission under the Creative Commons Attribution License 4.0 (CC BY 4.0; https://creativecommons.org/licenses/by/4.0/).*

from each other by cutting the filter and measuring the radioactivity from the hybridized labeled RNA by liquid scintillation counting. Under a defined set of experimental conditions, these data correlate directly with the number of RNA polymerase molecules engaged in transcribing a specific gene and indirectly with the transcriptional efficiency of regulatory sequences associated with the gene(s) of interest. Alternatively, this procedure can be carried out using nonradioactive variants of some of the established and newer procedures (Patrone et al., 2000; Kwei, 2004; Roberts et al., 2015).

Another important advantage of the NRO assay is the ability to simultaneous assay transcriptional activity at multiple genomic loci, all of which are presumably transcribed in the same relative amounts in isolated nuclei as in intact cells. Because a "typical" eukaryotic cell is actively transcribing thousands of different genes at any given point in the cell cycle, the complexity, that is, the total length of different RNA sequences in labeled RNA population is quite large. If used in a hybridization assay, the concentration of the specific transcribed RNA sequences of experimental interest may be only a very small percentage of the total mass of the probe. Despite this, the sensitivity of hybridization assays is quite good because (1) the labeled RNA is single-stranded, so a reduction of effective probe concentration due to double-stranded probe renaturation is a nonissue, (2) the RNA is labeled continuously along its length to a very high specific activity, and (3) the thermodynamic stability of RNA:DNA hybridized is greater than a dsDNA. By tolerating more stringent posthybridization washes, the signal-to-noise ratio is greatly improved. It is important to note that transcripts that had been completed prior to the addition of the label will not be detected.

Relationship to the study of steady-state RNA

Cellular lysis and subsequent RNA isolation performed by any of the methods described in Chapters 4–7 renders steady-state RNA which, as noted above, is best thought of as the net accumulation of RNA in the cell at the moment of lysis. The precise composition of the steady-state RNA population is a function of the prevailing rates and of turnover (degradation) in the cell. One might consider this mnemonic:

relative synthesis + relative degradation = relative abundance

Consequently, steady-state RNA transcripts are not indicative of the rate of transcription at the associated gene locus, the age of the transcripts in the cell, or the rate at which identical transcripts are being degraded. By measuring the rate of gene transcription and comparing it with steady-state data, for example, Northern analysis with cytoplasmic RNA, it is possible to assign observed modulation of gene expression to the transcription level (up- or downregulation) or to a posttranscriptional event. An example of this is presented in Table 19.3. As noted above, the mechanics of transcription rate assays maintain the natural geometry of the chromatin so that the rate data mirrors the in vivo conditions as closely as possible.

A transcription rate assay consists of several distinct phases, each of which has a distinct goal (Table 19.4). The success of the method is almost entirely dependent on the speed with which nuclei can be isolated from intact

TABLE 19.3 Nuclear expression of cellular oncogenes as a function of cell shape.

Cell line	Proliferation phenotype	Capacity for contact inhibition	Relative transcription rate			
			c-ras	*c-myc*	*c-fos*	*c-fes*
CUA-1	Shape dependent	+	30.0	16.0	0.92	0.93
HT-1080	Shape independent	−	0.86	0.67	0.93	0.92
HT-IFN^R	Shape dependent	+	0.38	0.51	0.64	0.66

Notes: Values reflect the abundance of gene-specific nuclear transcripts in flat cells divided by the transcript abundance in rounded cells that had been plated onto the ethanol-soluble polymer poly [HEMA] to control cell shape. Abundance levels were determined from image analysis performed on nuclear run-on autoradiograms (Farrell and Greene, 1992). The obvious shape-dependent difference in *c-ras* and *c-myc* expression patterns in CUA-1 cells stands in stark contrast to the behavior of other these and other the genes in the established cell lines. CUA-1: diploid fibroblasts; HT-1080: fibrosarcoma cells; HT-IFN^R: HT-1080 fibrosarcoma cells treated with interferon, resulting in the transient restoration of shape-dependent proliferation characteristics.

TABLE 19.4 Major steps associated with nuclear transcription assays.

Procedural phase	Goal
Nuclear isolation	Rapid isolation of intact nuclei capable of supporting transcript elongation.
Transcript elongation and labeling	Incorporation of labeled UTP precursors into nascent transcripts while maintaining the natural geometry of the transcriptional apparatus.
Recovery of labeled transcripts	Purification of the newly labeled RNA.
Identification of labeled transcripts	Determine which genes are transcriptionally active under the conditions defined by the experiment. Options include: • Filter hybridization to complementary gene sequences and autoradiographic detection (includes macroarray analysis). • Microarray analysis. • Next generation sequencing. There is no limit to the number of sequences that can be assayed using this approach.
Quantification of transcript abundance	Determination of the relative transcription rates of all genes of interest. Options include: • Comparative autographic signal intensity, which correlates with the degree to which gene-specific transcripts are present. • RT-PCR. • Array analysis, coupled with RT-PCR for greater dynamic range. • In silico evaluation of next-generation sequencing results.

cells and the nascent transcripts radiolabeled. Expedient preparation of the nuclei is by far the most critical step. Failure to produce extensively labeled RNA is usually a direct result of inexperienced handling of the nuclei prior to labeling. Nuclear isolation must also be carried out in such a way as to preserve RNA polymerase activity and nuclear structure during the isolation from cells cultured in vitro or from tissue. Most protocols appearing in the literature vary with respect to the postlabeling RNA purification steps only.

Cells grown in culture may be harvested with a cell scraper or by trypsinization according to standard methods. We have learned the hard way that over-trypsinization greatly reduces the labeling efficiency in this assay. Moreover, when cells are trypsinized, shape distortions of the plasma membrane occur which, theoretically, could alter patterns of RNA synthesis in diploid cells in particular. This could potentially influence the outcome of experiments designed to study elements of the cytoskeleton, signal transduction, or other genes that might be sensitive to the shape of the cell. For this reason, scraping cells from dishes resting on ice is the method least likely to cause distortion of the cellular morphology and is therefore strongly recommended (Farrell and Greene, 1992).

In general, nuclei are isolated from whole cells by incubation in hypotonic lysis buffer featuring the inclusion of the gentle, nonionic detergent NP-40, now known as IGEPAL CA-630 (Price and Penman, 1972; Greenberg, 1988); or in isosmotic sucrose buffer containing Triton X-100 (Marzluff et al., 1973, 1974). Nuclei may then be labeled immediately or stored frozen in cell freezing ampoules (*never* microfuge tubes) in liquid nitrogen; if unfamiliar with methods for the safe handling of liquid nitrogen, liquid nitrogen dewars, or the freezing and thawing of tubes, be sure to seek the guidance of someone who is experienced in this area. This is necessary for the safety of everyone in the lab. Alternatively, nuclei can be stored at −80°C for several months without significant loss of labeling potential. In this lab, nuclei have been successfully labeled after storage for as long as 13 months.

Nuclear run-off versus nuclear run-on assay

A fair amount of confusion exists about the nomenclature associated with the study of in situ transcription. The terms "nuclear run-on assay" and the "nuclear run-off assay" technically refer to two different approaches, though these two terms are sometimes (incorrectly) used interchangeably. The original intent of the term "nuclear run-off" was to describe those in vitro methods by which the individual factors required to support transcription were elucidated, one by one, by adding various components to a DNA template, often a linear restriction fragment with an adjacent promoter. When conditions supporting transcriptional activation are met, the RNA polymerase will "run off" the end of the template, producing transcripts of uniform length. The quantity of RNA produced is therefore an indicator of the efficiency of transcription and what is necessary to support it. The term "nuclear run-on" refers to methods for the identification of genes being expressed at a particular moment under a defined set of conditions. Rapidly isolated nuclei from cells of interest are incubated in the presence of labeled nucleotides under conditions that promote short-term elongation and the concomitant incorporation of the label into transcripts that have been initiated.

Sometime in the 1980s the terms "nuclear run-off assay" and "nuclear run-on assay" started to be used interchangeably. Many peer-reviewed papers were published describing the labeling of RNA in isolated nuclei; some authors called the technology a run-off assay while others referred to the same strategy as a run-on assay. It now seems that these two terms are taken to mean the same thing, namely the elongation and labeling of initiated transcripts in nuclei isolated from cells and tissues and, to an extent, the usage of one term over another seems to be regional in nature. Inasmuch as "NRO assay" is a historically more accurate description of in situ transcription assays, this nomenclature will be used throughout this laboratory guide to refer to the transcription rate assays which follow.

Protocol: nuclear run-on assay

In this laboratory, the best labeling has occurred when cells are maintained as cold as possible until the initiation of the labeling reaction. It is very helpful to fill two or three large pans or trays (the type often used for autoclaving bottles) with ice so that tissue culture flasks or dishes can be set directly on ice for the first three steps in this protocol. Harvesting cells from tissue culture is greatly accelerated when cells are grown in 100 or 150 mm tissue culture dishes rather than in tissue culture flasks. The protocols presented here are improvements on the procedures of Groudine et al. (1981), Nevins and Darnell (1978) and Greenberg (1988).

Harvesting of cells and preparation of nuclei

1. Decant and discard cell growth medium from adherent cells growing in tissue culture.

   Note: While as few as 5×10^6 nuclei may be utilized in a labeling reaction, it is wise, to begin with $3-5 \times 10^7$ nuclei to achieve efficient labeling. Sufficient counts per minute (cpm) must be generated in order to titrate the RNA sample at the hybridization step. As needed, cell number can be determined quickly by hemacytometer counting just prior to lysis.
2. Wash the monolayer once with 10–15 mL ice-cold phosphate-buffered saline (PBS). Decant most of the PBS, leaving approximately 1 mL in the tissue culture vessel.
3. Using a sterile cell scraper, harvest the cells by gentle but rapid scraping.

   Note: For suspension culture cells, centrifuge growth medium at $500 \times g$ for 5 minutes at 4°C in a precooled rotor. Decant and discard the supernatant. Resuspend the cell pellet in 10 mL of ice-cold $1 \times$ PBS and pellet as described above. Remove the supernatant as completely as possible and then proceed to step 5.
4. Transfer the cells to a pre-chilled centrifuge tube (e.g., a 15 or 30 mL nuclease-free Corex glass tube) and collect cells by centrifugation at $500 \times$ g for 5 minutes at 4°C. Remove the supernatant as completely as possible. Make sure that glass centrifuge tubes are supported by the proper rubber adapter sleeves when placed into the centrifuge rotor.

   Suggestion: Transport tubes containing cells and/or nuclei to and from the centrifuge in an ice bucket.

5. Prior to resuspending the cells, loosen the pellet by very gentle vortexing for 5 seconds, which will prevent the clumping of cells. *Slowly* add 4 mL of precooled hypotonic lysis buffer (10 mM Tris, pH 7.4; 10 mM NaCl; 5 mM $MgCl_2$; 0.5% NP-40) while continuing to vortex gently. Incubate on ice for 5 minutes.
6. Collect the nuclei by centrifugation at 500 × *g* for 5 minutes at 4°C.
   Note: If desired, cytoplasmic RNA can be purified from the supernatant for Northern analysis. Store supernatant on ice and continue working up nuclei first.
7. Resuspend the nuclear pellet in 4 mL of the lysis buffer described in step 5. Collect the nuclei by centrifugation at 500 × g at 4°C.
8. Carefully decant and discard the supernatant. Resuspend the nuclei in 200 to 300 μL of prechilled glycerol storage buffer (50 mM Tris, pH 8.0; 30% glycerol; 2 mM $MgCl_2$; 0.1 mM EDTA) by gentle vortexing as described in step 5. The nuclei are now ready for immediate labeling or may be stored in sealed cell freezing ampoules in liquid nitrogen or at −80°C for several months.

Alternative protocol for preparation of nuclei from cell culture

In this procedure, a modification of the procedure of Marzluff and Huang (1984), fragile nuclei can be prepared by cell lysis in isosmotic sucrose buffer and pelleted into a dense cushion of sucrose. Nuclear integrity is maintained by nuclear integrity by preventing nuclei from slamming into the bottom of the centrifuge tube. Inasmuch as the density of nuclei varies from one cell type to the next, tweaking the buffer formulations to accommodate various cell types may be necessary. As in the previous protocol, the recovery of cells from tissue culture is greatly accelerated when cells are grown in 100 or 150 mm tissue culture dishes rather than in tissue culture flasks.

1. Decant and discard cell growth medium from adherent cells growing in tissue culture.
   Note: Begin with $3-5 \times 10^7$ nuclei to generate sufficient cpm to titrate the RNA sample at the hybridization step. As needed, cell number can be determined quickly by hemacytometer counting just prior to lysis.
2. Wash the monolayer once with 10–15 mL cold PBS. Decant most of the PBS, leaving approximately 1 mL in the tissue culture vessel.
3. Using a sterile cell scraper, harvest the cells by gentle but rapid scraping.
   Note 1: Rather than using a cell scraper, cells may be harvested by trypsinization.
   Note 2: For suspension culture cells, centrifuge growth medium at 500 × g for 5 minutes at ambient temperature. Decant and discard the supernatant. Optional: wash the cell pellet one time with an aliquot of

cold PBS and pellet as described earlier. Remove the supernatant as completely as possible and then proceed to step 5.

4. Transfer the cells to a suitable centrifuge tube (e.g., 15 or 30 mL nuclease-free Corex glass tube) and collect cells by centrifugation at $500 \times g$ for 5 minutes at room temperature. Make sure that glass centrifuge tubes are supported by the proper rubber adapter sleeves when placed into the centrifuge rotor. At the conclusion of the centrifugation, remove the supernatant as completely as possible.

   Suggestion: Transport tubes containing cells and/or nuclei to and from the centrifuge in an ice bucket.
5. Resuspend the cells in prechilled sucrose buffer (0.32 M sucrose; 5 mM $CaCl_2$; 3 mM magnesium acetate; 0.1 mM EDTA; 0.1% Triton X-100; 1 mM DTT; 10 mM Tris, pH 8.0) at a density of 1×10^7 cells/mL.

   Note: This buffer can be prepared in advance; however, do not add DTT or Triton X-100 until just prior to use.
6. Homogenize the cells in a Dounce homogenizer (Fig. 6.3) with a maximum of 10 strokes using a tightly-fitting pestle.

   Note: One common cause for poor label incorporation is the use of a loosely fitting pestle and the concomitant failure to liberate a large number of nuclei.
7. Fill the bottom one-third of an RNase-free ultracentrifuge tube with an aliquot of cushion buffer (2 M sucrose; 3 mM magnesium acetate; 1 mM DTT; 10 mM Tris, pH 8.0). Dilute the homogenate with an appropriate aliquot of cushion buffer to produce a sufficient volume to occupy the remaining two-thirds of the ultracentrifuge tubes. Carefully layer the resulting mixture over the sucrose cushion already in place.
8. Collect the nuclei by centrifugation at $30{,}000 \times g$ for 45 minutes at 4°C.

   Note: Be sure to carefully balance the tubes by weight prior to centrifugation.
9. Resuspend the nuclear pellet in storage buffer (40% glycerol; 5 mM magnesium acetate; 0.1 mM EDTA; 5 mM DTT; 50 mM Tris, pH 8.0) at a density of 10^8 nuclei/mL.
10. Immediately seal nuclei in cell freezing ampoules and store in liquid nitrogen or at −80°C for several months. Otherwise, initiate labeling protocol without delay.

Alternative protocol for preparation of nuclei from whole tissue

This is a modification of the procedures of Derman et al. (1981) and Marzluff and Huang (1984).

1. Harvest tissue according to standard procedures. Rapidly mince tissue and rinse with an ice-cold solution consisting of 0.14 mM NaCl and 10 mM Tris, pH 8.0.

2. Homogenize minced tissue in cold lysis buffer (0.32 M sucrose; 5 mM $CaCl_2$; 3 mM magnesium acetate; 0.1 mM EDTA; 0.1% Triton X-100; 1 mM DTT; 10 mM Tris, pH 8.0). Use 5 mL lysis buffer per gram of tissue.
3. *Rapidly* filter homogenate through several layers of sterile cheesecloth. This will facilitate the removal of connective tissue and other debris. Alternatively, a low-speed centrifugation step will do the job.
4. Dounce-homogenize filtrate with 10–20 strokes using a tightly fitting pestle. Remove a 5 μL aliquot of the homogenate and place it on a microscope slide, overlay it with a coverslip and assess the extent of cellular lysis by light microscopy.
5. Add 1–2 vol of prechilled solution of 2.0 M sucrose; 3 mM magnesium acetate; 0.1 mM EDTA; 1 mM DTT; 10 mM Tris, pH 8.0 to the homogenate. Layer the resulting mixture over a cushion of this same buffer that occupies one-third of the volume of an RNase-free ultracentrifuge tube.

   Note: The exact sucrose concentration is a function of the cell type under investigation and should be empirically determined before using valuable starting material. The usual range is 1.8–2.2 M.
6. Collect nuclei by centrifugation at 30,000 × *g* for 45 minutes at 4°C.

   Note: Be sure to carefully balance the tubes by weight prior to centrifugation.
7. Resuspend the nuclear pellet in glycerol storage buffer (25% glycerol; 5 mM magnesium acetate; 0.1 mM EDTA; 5 mM DTT; 50 mM Tris, pH 8.0) at a density of 1×10^8 nuclei/mL.
8. Immediately seal aliquots of nuclei in cell freezing ampoules and freeze in liquid nitrogen or at −80°C for several months. Otherwise, initiate labeling protocol without delay.

Labeling and recovery of transcripts

Note: It is important to have the 2 × reaction buffer prepared before thawing nuclei. Permitting thawed nuclei to sit for just a few minutes on ice will greatly diminish labeling efficiency.

1. Prepare a 2 × reaction buffer (10 mM Tris, pH 8.0; 5 mM $MgCl_2$; 300 mM KCl; 1 mM ATP, 1 mM CTP, 1 mM GTP; 5 mM DTT).

   Note 1: Using an NTP stock solution from any of several vendors will give the greatest reproducibility from one experiment to the next. Because NTP stock solutions prepared in the lab are lacking in quality control, so to speak, the labeling efficiency is likely to be impacted.

   Note 2: If desired, radiolabeled UTP (see step 3) can be added to the 2 × reaction buffer in this step, thereby creating a master mix of sorts.

2. Add 200 μL of the 2 × reaction buffer to each 200 μL aliquot of nuclei. If nuclei were previously stored frozen, add 200 μL of the 2 × reaction buffer to 200 μL of thawed nuclei.

   Note 1: In this laboratory, optimum labeling efficiency is observed when 200 μL of 2 × reaction buffer is added to the ampoule containing the frozen nuclei immediately after placing it on ice to thaw. Do not wash nuclei after thawing, and thaw an aliquot only once.

   Note 2: Do not be alarmed if the mixture of thawed nuclei appears viscous. Although frozen and thawed nuclei may lyse to some extent, transcriptional complexes remain intact and are competent to support the elongation of initiated transcripts. Of course, care should be taken in each step prior to the labeling reaction to minimize the amount of nuclear damage.
3. *Immediately* add 10 μL [α-^{32}P] UTP 800 Ci/mmol, 10 mCi/mL, and incubate for 30 minutes at 26°C with *gentle* shaking.

   Note 1: If nuclei were previously stored at −80°C or in liquid nitrogen, labeling can be carried out directly in the freezing ampoule. Alternatively, freshly prepared nuclei are best labeled in a larger polypropylene tube to better manage the spread of radioactivity that will invariably result from the mechanics of this assay.

   Note 2: In this laboratory, the most reproducible degree of label incorporation was observed when the labeling reaction is shaken at 30−50 rpm on a temperature-controlled orbital platform (e.g., Lab-Line Instruments).

   Note 3: At the conclusion of the labeling period, the use of a commercial RNA product (e.g., TRIzol) is an alternative for RNA recovery.
4. In a separate, RNase-free tube, mix 1 mL of high salt buffer (500 mM NaCl; 50 mM $MgCl_2$; 2.5 mM $CaCl_2$; 10 mM Tris, pH 7.4) with 50 μL of 1 mg/mL RNase-free DNase I, just prior to use. Add 750 μL of the resulting mixture to each tube containing labeled nuclei.

   Note: The high salt buffer will terminate the labeling reaction and disrupt all the remaining intact nuclei. The nuclear disruption greatly increases the viscosity of the solution, because of the liberation of chromatin. Use a silanized (Appendix H), RNase-free Pasteur pipette, a micro pipettor, or a 25-gage needle attached to a 1 mL tuberculin syringe to shear the DNA several times.
5. Incubate this mixture for 5 minutes at 37°C.

   Note: The purpose of this brief treatment with DNase I is to partially fragment, rather than completely degrade, genomic DNA so that the solution is physically more manageable and less likely to trap labeled RNA molecules in the subsequent purification steps. Partially degraded DNA is removed in subsequent steps.
6. Add 200 μL SDS/Tris buffer (5% SDS; 500 mM Tris, pH 7.4; 125 mM EDTA) and 15 μL of 20 mg/mL proteinase K. Incubate for 30 minutes at 42°C to ensure extensive protein hydrolysis.

7. Extract each sample with 1.25 mL of phenol:chloroform:isoamyl alcohol (25:24:1). Centrifuge for 8–10 minutes at 800 × *g* to separate the phases.
8. Transfer the upper aqueous phase to a fresh tube and add, in order:
   2 mL sterile (nuclease-free) H_2O,
   3 mL 10% TCA (trichloroacetic acid) in 60 mM sodium pyrophosphate,
   15 μL 10 mg/mL carrier RNA (e.g., junk yeast RNA).
   Incubate on ice for 45 minutes.
   Note: These steps will result in precipitation of the labeled RNA.
9. Filter the precipitate onto Whatman GF/A glass fiber filters using a vacuum filtration manifold (e.g., Millipore® model 1225). Wash each filter three times with 10 mL aliquots of 5% TCA, 30 mM sodium pyrophosphate.
   Note: Omission of the TCA precipitation step and filtration will usually result in unacceptably high levels of background even after the most stringent posthybridization washes. This is a key step contributing to the quantitativeness of this assay.
10. Transfer each filter to a glass scintillation vial and add 1.5 mL DNase I buffer (20 mM HEPES (free acid), pH 7.5; 5 mM $MgCl_2$; 2.5 mM $CaCl_2$). Add 50 μL 1 mg/mL RNase-free DNase I and incubate for 30 minutes at 37°C.
11. Terminate the digestion with the addition of 40 μL of 500 mM EDTA, pH 8.0, and 70 μL of 20% SDS.
12. Elute the purified RNA from each filter paper by heating the samples to 65°C for 10 minutes. *Carefully* transfer all of the supernatants to a fresh tube.
13. Repeat the elution step by adding 1.5 mL of elution buffer (1% SDS; 10 mM Tris, pH 7.5; 5 mM EDTA) to the scintillation vials containing the filters. Heat to 65°C for 10 minutes and then *carefully* recover the supernatant and combine it with the original supernatant from step 9.
14. Extract the resulting RNA solution one time with an equal volume of phenol:chloroform:isoamyl alcohol (25:24:1), followed by a single extraction with an equal volume of chloroform.
    Note: In both instances, the RNA will partition to the aqueous (upper) phase.
15. Transfer the aqueous material (~3 mL) to a silanized, RNase-free Corex glass tube (or the equivalent).
16. Optional step: Add 650 μL of 1 N NaOH and incubate on ice for no more than 2 minutes and then quickly neutralize the reaction with the addition of 1.5 mL 1 M HEPES (free acid), pH 7.4.
    Note: The limited hydrolysis of RNA tends to improve hybridization efficiency (Jelinek et al., 1974). Incubation for more than 2 minutes runs the great risk of excessive RNA damage by alkaline hydrolysis.

Further, limited hydrolysis reduces the likelihood of high-molecular-weight RNA self-hybridization, in favor of forward, intermolecular hybridization kinetics. If desired, this step and the neutralization step that follows can be eliminated altogether.

17. Precipitate the RNA with the addition of 0.1 vol of 3 M sodium acetate, pH 5.2 (~500 μL) and 2.5 vol of ice-cold 95% ethanol (~15 mL). Store overnight at −20°C.

Preparation of target DNA

DNA sequences corresponding to the genes of interest are best prepared by PCR. In the event that the genes of interest have been cloned into plasmids and if the DNA sequence of the insert is known, the best strategy is to use primers that anneal to the regions of the vector that flank the insert, which will facilitate the recovery of the insert in substantial quantities. This precludes the need to excise the gene (cDNA or gDNA) by restriction endonuclease digestion and gel purification. Any type of dsDNA will need to be denatured prior to being applied to the membrane; in the case of oligonucleotides, however, these molecules can be directly applied without any pretreatment. As the NRO assay is commonly used to simultaneously assess the extent of transcription of a variety of genes, and it is important to include in this assay at least one gene with a purported "housekeeping" function, that is, a gene whose level of expression is not expected to vary under the defined experimental conditions.

One worthwhile blotting strategy is to apply the target DNAs corresponding to the different genes of interest in the same vertical column of a blotting manifold (Appendix J) so that the filter can be cut into narrow strips and hybridized to the extremely heterogeneous RNA probe in a minimum volume of hybridization buffer. The target DNAs are best applied to a nylon membrane so that retention on the membrane is a nonissue; DNA molecules less than 1 kb are poorly retained on nitrocellulose, which also has a limited binding capacity and a proclivity for background. Plan to load 1–2 μg of denatured DNA to each well in a volume of 100–200 μL.

1. Denature linear target DNA with the addition of 0.1 vol of 1 N NaOH. Incubate at room temperature for 30 minutes or at 37°C for 10 minutes.
2. Neutralize the alkaline pH of the mixture by the addition of 10 vol of ice-cold 6× SSC.
3. Store denatured DNA on ice until ready to blot, but for no longer than 15–20 minutes. It is best to complete blotting within 15 minutes after the addition of 6× SSC to preclude the reannealing of the denatured sample.
4. Assemble a Minifold I (dot blot) or Minifold II (slot blot) apparatus (Cytiva) and begin to apply a low vacuum. Pipette 2–5 μg (or other

predetermined mass) of denatured DNA to each well in a volume of 100–200 μL. After the buffer has been pulled through the filter, wash each well with 300 μL 6 × SSC.

Note 1: Be sure to load control wells that contain (a) a corresponding mass of carrier DNA or linear, and (b) buffer only, to assess nonspecific binding to nonhomologous DNA and to the filter itself, respectively.

Note 2: Slot-blotting will concentrate the sample into a smaller area (6 mm^2) than dot-blotting (12.5 mm^2); thus slot-blotting is expected to yield data that are easier to quantify than dot-blotting because the slot configuration permits easier quantitation with image analysis software.

5. Immobilize the target DNA onto the filter paper for subsequent hybridization according to the instructions of the manufacturer. DNA may be cross-linked to nylon membranes using a calibrated UV light source while nitrocellulose (not recommended) is baked for 2 hours at 80°C in vacuo. Alternatively, nylon filters may be baked instead, for as little as 1 hour at temperatures as low as 65°C (vacuum not required). The filter may then be used immediately or stored in a cool, dry location for later use. Details pertaining to nucleic acid immobilization on solid supports can be found in Chapter 16, Northern Analysis.

Preparation of RNA for hybridization

1. Recover the precipitated RNA by centrifugation at 10,000 × g for 30 minutes at 4°C. Carefully decant as much of the ethanol as possible into radioactive waste. Allow the tubes to drain briefly, and then wash once with 1 mL 70% ethanol. Centrifuge, if necessary, and then decant as much of this ethanol wash as possible. Allow the tubes to drain briefly. If excess ethanol remains, carefully wash the RNA once with 1 mL 95% ethanol, centrifuge if necessary, and allow the tube(s) to drain briefly.
2. Resuspend the RNA pellet in 1 mL of TES (*N*-tris(hydroxymethyl) methyl-2-aminoethanesulfonic acid) solution (10 mM TES, pH 7.4; 10 mM EDTA; 0.2% SDS). Shake the sample at 100 rpm on an orbital platform for 30 minutes at room temperature to redissolve the RNA.
3. Remove a 5 μL aliquot from each sample and count for 1 minutes to determine cpm/mL. Normalize the samples with respect to cpm/mL by the addition of TES solution, to provide nearly equal counts per hybridization. In general, the cpm/mL averages about 10^7 cpm/mL per 5×10^7 nuclei when the labeling reaction is efficient.

   Note: The normalization of samples by cpm assumes that the overall level of RNA synthesis is not changing as a function of cell state, but rather that specific genes are being up- or downregulated. The investigator must empirically determine the reasonableness of this assumption for each new set of cell state conditions and also ensure equal numbers of nuclei at the onset of the experiment.

4. Cut the filter containing the immobilized target DNA sequences into strips such that each strip contains one complete set of target DNAs to be hybridized to the labeled RNA probes.
5. Hybridization is most efficiently and reproducibly executed directly in a clean 5 mL scintillation vial. This may be accomplished by coiling the filter strips containing the DNA and carefully pushing each strip to the bottom of individual scintillation vials such that the side of the filter onto which the DNA is fixed is facing the center of the tube.

   Note 1: Hybridization in this fashion will minimize the amount of probe required and will maintain the maximum probe concentration. Moreover, this approach will minimize the radioactive mess that is invariably associated with hybridization in plastic bags.

   Note 2: To ensure that the scintillation vial is RNase-free it is strongly suggested that the vial be soaked in H_2O_2 for a minimum of 20 minutes, and then rinsed thoroughly with sterile H_2O.

   Note 3: Pre-wet the filter strips in sterile H_2O and briefly blot away excess water (Whatman 3MM paper) before placing it into the scintillation vial for hybridization.
6. Mix 1 mL of RNA solution with 1 mL of TES/NaCl (10 mM TES, pH 7.4; 10 mM EDTA; 0.2% SDS; 600 mM NaCl). Add the resulting 2 mL of RNA probe solution to the scintillation vials containing the target DNA filter strips, close the tubes tightly, and hybridize for 24–36 hours at 65°C. Be certain that filters in the vials are completely immersed in hybridization solution.

Note: In this lab, the most efficient method for hybridization is to insert individual scintillation vials into the disposable polystyrene or cardboard racks in which conical 15 mL centrifuge tubes are often packaged. The rack is then placed on the platform of an orbital shaker incubator preheated to 65°C and agitated slowly (50–75 rpm) for the duration of the hybridization.

Posthybridization washes and detection

1. At the conclusion of the hybridization period, wash the filters in batch for 1 hour in an excess volume of 2× SSC, preheated to 65°C.
2. Place filters in individual glass scintillation vials containing 8 mL of 2× SSC and 12 μL of 10 mg/mL RNase A. Alternatively, nonspecific RNA molecules can be digested with a combination of RNase A (5.0 μg/mL) and RNase T1 (5 U/mL) in 2× SSC. Incubate at 37°C for 30 minutes. No shaking is necessary.
3. Wash the filters in batch for 1 hour in an excess volume of 2× SSC at 37°C. *Briefly* blot the filters, ensuring that they do not dry out completely. Plastic-wrap the damp filters and then tape them to a solid support such as Whatman 3MM paper or even a piece of cardboard.

Then, proceed to set up autoradiography (see Chapter 17, Nucleic Acid Probe Technology).

Note: Signal intensity on autoradiograms is most easily quantified through the use of image analysis software, after first ensuring that the observed signal is within the linear range of the film. This can be easily accomplished by preparing multiple X-ray films for varying exposure periods.

4. Following autoradiography, the hybridized RNA probe can be stripped from the membrane-bound target DNA by washing the filters in 0.4 N NaOH at 42°C for 30 minutes, followed by a 60 minutes wash in 1 × SSC, 0.5% SDS at 65°C. This technique removes all hybridized RNA, leaving the efficiency of subsequent hybridization unimpaired.
5. If the filter membrane is not to be used for subsequent hybridization, the magnitude of the observed autoradiographic signal can be precisely quantified by simply cutting the appropriate samples (e.g., the "dots") from the filter membrane strip and counting directly in scintillation fluor.

Protocol: alternative procedure for nuclear run-on assay

Several simplified versions of the traditionally cumbersome and labor-intensive NRO assay has been described (Celano et al., 1989b; Fei and Drake, 1993), a modification of one of which (Celano et al., 1989b) is included here. In this procedure, labeled RNA transcripts are purified rapidly by chaotropic acid–phenol–guanidinium thiocyanate extraction but without TCA precipitation. As always, observe a good RNase-free technique.

1. Isolate nuclei as described above, or elsewhere (Celano et al., 1989a; deBustros et al., 1986).
2. Collect isolated nuclei (up to 5×10^7) in sterile 2.2 mL microfuge tubes in 160 μL of nuclear storage buffer (40% glycerol; 50 mM Tris, pH 8.3; 5 mM $MgCl_2$; 0.1 mM EDTA). Quick-freeze the nuclei in an ethanol-dry ice bath and store at −80°C for up to 1 year. *Do not* store this type of tube in liquid nitrogen. Alternative, aliquots of nuclei can be stored in cell freezing ampoules, tightly sealed, and stored in liquid nitrogen for up to a year.
3. Thaw aliquots of frozen nuclei on ice. For each 200 μL of nuclei, add 25 μL of 10 × run-on buffer (40 mM ATP, GTP, CTP; 5 mM DTT; 50 mM $MgCl_2$; 800 mM KCl) and 20 μL [α-^{32}P]-UTP (200uCi; 3000 Ci/mmol). Incubate for 15 minutes at 26°C.

   Note: Using an NTP stock solution from any of several vendors will give the greatest reproducibility from one experiment to the next. Because NTP stock solutions prepared in the lab are lacking in quality control, so to speak, the labeling efficiency is likely to be impacted.

4. Lyse the nuclei and initiate DNA digestion by adding 12 μL 20 mM $CaCl_2$ and 12 μL 10 mg/mL RNase-free DNase I. Incubate for 5 minutes at 26°C.
5. Add 30 μL of 10× SET (5% SDS; 50 mM EDTA; 10 mM Tris, pH 7.4) and 7 μL 10 mg/mL carrier RNA (e.g., junk yeast RNA). Initiate peptide hydrolysis with the addition of 3 μL 10 mg/mL proteinase K. Incubate the samples at 37°C for 30 minutes.
6. Shear genomic DNA by drawing the lysate carefully and repeatedly through a 25-gage needle positioned on a 1 mL tuberculin syringe or a silanized Pasteur pipette (Appendix H).
7. Extract the RNA from the sample by adding the following to each tube containing labeled nuclear RNA (mix after the addition of each):

   500 μL GTC (4 M guanidinium thiocyanate; 25 mM sodium citrate, pH 7.0; 0.5% sarcosyl; 0.1 M 2-mercaptoethanol),
   80 μL sodium acetate (2 M, pH 4.8),
   800 μL water-saturated phenol,
   160 μL chloroform:isoamyl alcohol (49:1).

   Incubate samples on ice for at least 15 minutes.

   Note: If desired, commercially available formulations for RNA isolation, such as Trizol or TRI Reagent®, can be substituted in this step. Follow the instructions that accompany this product for efficient isolation of RNA from this step forward.
8. Microcentrifuge samples at 12,000 × *g* for 10 minutes, at 4°C if possible.
9. Transfer the upper aqueous phase to a fresh microfuge tube, taking care to avoid the interface. Add an equal volume of isopropanol to this aqueous material to precipitate the RNA. Store sample(s) for at least 1 hour at −80°C or overnight at −20°C.
10. Collect the precipitate RNA by microcentrifugation at 12,000 × g for 10 minutes, at 4°C if possible.

    Note: It is recommended that the RNA pellet be redissolved in 300 μL of guanidinium thiocyanate solution (see step 7) and transferred to a 1.7 mL microfuge tube. The RNA is then reprecipitated by the addition of 300 μL of isopropanol and collected by centrifugation as described above.
11. Wash the pellet two or three times with 70% ethanol and briefly air-dry. If desired, a final wash with 95% ethanol will accelerate the drying process. In either case, do not allow the RNA to dry completely. Dissolve the RNA pellet in 250–500 μL in a solution of 10 mM Tris, pH 7.2; 1 mM EDTA; 0.1% SDS.
12. Determine the activity of the RNA probe by counting a 2 μL aliquot. The total expected activity is approximately 10^7cpm for the assay when starting with 5×10^7 cells.
13. Hybridize labeled RNA to DNA target sequences as described above, or elsewhere.

Protocol: nuclease protection–pulse label transcription assay

The mechanics of the traditional transcription rate assay (label incorporation into nascent transcripts in isolated nuclei) differ markedly from pulse labeling techniques. Subjecting cell cultures to a 5-minute pulse with [^{3}H]-uridine (Nevins and Darnell, 1978) would require a sufficiently high rate of transcription in order to derive meaningful incorporation of label. This is typically not the case with single-copy genes or other sequences that are transcribed at basal levels. This issue, and other concerns pertaining to the inherent difficulties of preparing nuclei capable of adequate label incorporation, have been partially circumvented by coupling the nuclease protection assay with the pulse-labeling of nuclear RNA (Greene and Pearson, 1994).

The pulse-labeling technique involves incubation of tissue culture cells in the presence of radiolabel for ten minutes. At the conclusion of the labeling period, RNA is isolated from the cells for hybridization analysis against cold (unlabeled) gene-specific sequences, and finally incubated with S1 nuclease to remove nonhybridized sequences. In a way, this approach represents the best features of a traditional nuclear transcription rate analysis coupled with a nuclease protection assay. Because of the nature of this technique, cDNA probes for specific genes produce higher fidelity data than oligonucleotide probes. The nuclease-resistant double-stranded hybrids are precipitated and the relative rate of transcription is inferred from the incorporated label, as determined by liquid scintillation counting. As with many transcription assays, this protocol has the advantage of labeling transcripts while maintaining natural nuclear geometry and, in this protocol specifically, maintaining cellular geometry as well. In short, this is a rapid, highly quantitative assay. Moreover, without the requirement for X-ray film, the inherent limitations associated with autoradiography (extended exposure time; nonlinearity of the film) are superseded by scintillation counting. Thus, this assay has the ability to detect subtle changes in transcriptional activity that might otherwise by missed by the more traditional NRO approach.

1. Cells growing in tissue culture are labeled with the addition of 25 μCi/mL [^{3}H]-uridine (specific activity 27.1 Ci/mmol) added directly to the growth medium tissue culture vessel. Incubate for 10 minutes.

   Note: RNA from 10^7 cells is generally required for one assay in duplicate, including all controls (see step 11). At the conclusion of the labeling period, be sure to dispose of radioactive tissue culture media and RNA isolation by-products in a safe manner, according to departmental guidelines.
2. Harvest cells and wash cell pellets twice with ice-cold PBS. All subsequent centrifugations are carried out at 4°C.
3. Resuspend cells (up to 10^7) in 1.8 mL lysis buffer (10 mM Tris, pH 7.4; 10 mM NaCl; 3 mM $MgCl_2$; 0.5% NP-40 [or IGEPAL CA-630]). Transfer the cell suspension to an autoclaved 2.2 mL microfuge tube. Incubate on ice for 3 minutes.

4. Centrifuge at 4000 × g for 5 minutes at 4°C to collect nuclei.
5. Carefully remove and discard supernatant. Resuspend nuclei in 800 μL SDS buffer (0.5% SDS; 1 mM $CaCl_2$; 5 mM $MgCl_2$; 20 mM HEPES, pH 7.5).
6. Add 50U RNase-free DNase I and incubate for 15 minutes at 37°C.
7. Split the nuclear lysate into two microfuge tubes, and then carefully extract each with an equal volume of phenol:chloroform:isoamyl alcohol (25:24:1).
8. Centrifuge to separate the phases. Carefully transfer the upper aqueous phase to a fresh tube.
9. Precipitate RNA with the addition of 0.1 vol 3 M sodium acetate, pH 5.2, and 2.5 vol of 95% ethanol. Store at −20°C for several hours to overnight.
10. Carefully wash sample once with 70% ethanol and centrifuge again if necessary. Briefly air dry the sample but do not allow it to dry out.
11. Resuspend the RNA pellet in 125 μL S1 hybridization buffer (40% deionized formamide; 40 mM PIPES, pH 6.4; 400 mM NaCl; 1 mM EDTA, pH 8.0). Be sure to redissolve the RNA by repeated pipetting. This is a critical step. Do not vortex.

    Note 1: Dissolve the RNA pellet in enough hybridization buffer to allow 20 μL for each hybridization reaction to be performed and run each sample in duplicate.

    Note 2: Appropriate controls to be performed include (1) a tube with no probe and no S1 nuclease; (2) a tube with no probe and 250 U S1 nuclease; and (3) a tube with a probe, but no S1 nuclease.

    Note 3: Deionization of formamide is described in Appendix G.
12. Heat each resuspended RNA sample to 50°C for 2 minutes.
13. Add denatured cold probe (unlabeled) to the samples to a final concentration of 25 ng/μL, or 0.5 μg probe per 20 μL hybridization reaction.

    Note: Be sure to use linearized, denatured DNA. Double-stranded probes can be easily denatured by boiling them for 10 minutes and then plunging them into the ice until further use. Alternatively, the RNA/probe mix can be heated to 65°C for 10 minutes to ensure all-inclusive denaturation at the onset of the hybridization.
14. Hybridize overnight at an experimentally determined temperature. For optimization suggestions, see Chapter 17, Nucleic Acid Probe Technology. If uncertain where to begin, hybridize at 37°C.
15. At the conclusion of the hybridization period, add to each sample:

    160 μL 2 × S1 buffer (0.5 M NaCl; 0.1 M sodium acetate, pH 5.5; 9 mM $ZnSO_4$),
    142 μL H_2O,
    6 μL boiled salmon DNA (10 mg/mL; [Final] ≈ 0.02 μg/mL),
    12 μL S1 Nuclease (25 U/μL; [Final] ≈ 1 U/μL).

    Incubate at 32°C for 1 hour.

    Note: Don't forget about the control samples (step 11, above).

16. Add 80 μL stop buffer (3 M sodium acetate, pH 5.2; 20 mM EDTA, pH 8.0; 40 μg/mL carrier RNA) to each sample.
17. Add 1.0 mL ice-cold 95% ethanol to each tube. Invert to mix thoroughly and then incubate on ice for 20 minutes.
18. Collect precipitate on GF/F glass fiber filters using a vacuum filtration manifold (e.g., Millipore model 1225).
19. Wash each filter twice with 500 μL cold 70% ethanol.
20. Transfer filters to scintillation vials and adds an appropriate volume of fluor. Count samples in triplicate for 5 minutes. The activity remaining on each filter is directly proportional to the extent of hybridization between the probe and target RNA, and hence proportional to the relative rate of synthesis.

Distinguishing among the activities of RNA polymerases

The relative rate of transcription of several genes can be assayed simultaneously using the methods described above. In eukaryotic cells, the collective products of transcription are the result of the transcriptional prowess of multiple RNA polymerases. The activities of these enzymes may be distinguished from each other by using the fungal cyclic octapeptide α-amanitin, from the very poisonous mushroom *Amanita phalloides*. This toxin binds to RNA polymerase molecules and is one to which the eukaryotic RNA polymerases exhibit differential sensitivity (Roeder, 1976). For example, at a concentration of 0.01 μg/mL, RNA polymerase II activity is reduced by 50%. RNA polymerase III is also sensitive to the inhibitory activity of α-amanitin, although a concentration of 25 μg/mL is required to achieve an equivalent 50% inhibition of activity. In contrast, RNA polymerase I exhibits no sensitivity to this peptide. The relatively recently discovered eukaryotic RNA polymerases IV and V, like RNA polymerase I, are also rather insensitive to α-amanitin, as are the RNA-dependent RNA polymerases (RdRPs) that are associated with miRNA expression (refer back to Table 1.6). While α-amanitin is extremely toxic and must be handled with great care, it is a useful tool for determining the amount of RNA synthesis attributable to each polymerase. The inclusion of α-amanitin in one or more labeling reactions is an important and excellent negative control by which hybridization signals are validated as authentic and not artifactual.

In nuclei isolated from cultured cells, the contribution to the total observed RNA synthesis due to RNA polymerase I, RNA polymerase II, and RNA polymerase III is about 50%–70%, 20%–40%, and about 10%, respectively. Clearer definition may be given to any model system by performing three separate labeling reactions in which α-amanitin is added as a reaction buffer component *before* the addition of the nuclei:

1. Reaction 1: no α-amanitin added to labeling reaction.
2. Reaction 2: 1 μg/mL α-amanitin added to the labeling reaction.
3. Reaction 3: 100 μg/mL α-amanitin added to the labeling reaction.

Following the addition of nuclei, the reaction mixtures are briefly incubated on ice and then transferred to the temperature at which the labeling will take place. The difference in the mass of transcription products in Reactions 1 and 2 represents RNA polymerase II activity. The difference between Reactions 2 and 3 represents RNA polymerase III activity. The activity in Reaction 3 is due to RNA polymerase I activity. These differences associated with α-amanitin poisoning may be assessed by dot-blot analysis (Appendix J), by the separation of transcription products on a denaturing agarose gel, or by electrophoresis on a denaturing 10% polyacrylamide gel (Appendix E). Electrophoresis will permit visualization of the size distribution of all transcription products by autoradiographic exposure directly from the gel. Alternatively, the size distribution of transcribed species can be determined by sucrose density gradient centrifugation (Marzluff and Huang, 1984). This is an older method that is not recommended because it is much more time consuming than simple gel electrophoresis and because it requires specialized equipment.

Extraction of nuclear RNA for steady-state analysis

Transcription rate assays facilitate the assessment of gene activity by the elongation of nascent polyribonucleotides in the presence of labeled NTP precursor. Examination of steady-state nuclear RNA is also essential for understanding the processing of the primary transcript. Northern analysis of nuclear RNA, for example, yields quantitative data based on signal intensity of discrete bands, but also provides a qualitative component that cannot be discerned by PCR or nuclease protection, that is, the length of the native transcript. The appearance of a reproducible banding pattern for a single RNA species strongly supports the existence of a primary RNA transcript and, in so doing, attests to the reliability of the NRO assay. A series of bands for the same transcript is associated with the attendant processing of the transcript in eukaryotic cells, that is, the systematic splicing together of coding (exon) sequences and concomitant removal of noncoding (intron) sequences.

To assess the overall size distribution of labeled nuclear transcripts, an aliquot of the reaction mixture can be studied by minigel electrophoresis coupled with in-gel autoradiography. Alternatively, target DNA may be digested, electrophoresed, and Southern blotted onto a filter membrane, followed by attempted hybridization with the heterogeneous population of labeled nuclear RNA (a reverse Northern blot, of sorts). This type of assay is also effective in dot-blot format. A major drawback of this technique, however, is the lack of information about the size of the nuclear RNA molecules themselves, since it is the DNA molecules that are electrophoresed, rather than the RNA. In the case of partial degradation of full-length hnRNA molecules, nonspecific hybridization of the resulting oligoribonucleotides to target DNA molecules may ensue. High-stringency washes, in conjunction with RNase digestion of nonspecific RNA molecules, will alleviate some, but not all, of the background noise, should this occur.

The isolation of steady-state nuclear RNA begins with the preparation of nuclei from whole cells or tissue samples as described previously. Invariably the highest quality RNA always results from the most expedient extraction procedures, that is, those which minimize the interval between cellular disruption and the endpoint of the assay. Inhibition of RNase activity in problematic cells can be accomplished by incorporating chaotropic lysis and purification techniques. Once the nuclei have been isolated from the cytosolic contents, one may elect to purify the RNA from the nuclei as if starting with intact cells, keeping in mind the imperative to control nucleases persists for nuclear RNA isolation. Many of the protocols describing the isolation of nuclear RNA recommend heating the sample in the presence of phenol, accompanied by vigorous shaking to shear the genomic material. However, excellent quality RNA can also be recovered from biological sources through the use of silica-based products described in Chapter 4, RNA Isolation Strategies, a strategy this Author endorses.

Protocol: direct isolation of nuclear RNA

The following protocol is a modification of the procedure of Soeiro and Darnell (1969) in which the use of hot phenol for RNA purification was originally described.

1. Thaw previously frozen nuclei on ice and pellet at 750 × *g* for 3 minutes at 4°C. If starting with freshly prepared nuclei, proceed to step 2.
2. Wash nuclear pellet with 1 mL ice-cold modified RSB/K buffer (10 mM Tris, pH 7.9; 10 mM NaCl; 10 mM $MgCl_2$; 100 mM KCl). Transfer nuclei to an RNase-free polypropylene tube.

   Note: If this extraction is scaled down, it is possible to isolate the RNA in a 2.2-mL microcentrifuge tube. As always, it is best to keep nucleic acids concentrated, working in the smallest possible volumes.
3. Resuspend the nuclear pellet to a concentration of $1-5 \times 10^7$ nuclei/mL in HSB buffer (10 mM Tris, pH 7.4; 500 mM NaCl; 50 mM $MgCl_2$; 2 mM $CaCl_2$), to which RNase-free DNase I has been added to a final concentration of 50U/mL.

   Note: Add RNase-free DNase I just prior to use.
4. Pipette this mixture up and down for 30 seconds at room temperature. This is necessary for lysate homogeneity.

   Note: The objective of this brief treatment with DNase I is to partially fragment, rather than completely degrade, the genomic DNA. The decrease in viscosity increases the manageability of the prep. The DNA will be selectively removed later in this procedure.
5. Add an equal volume of SDS extraction buffer (10 mM Tris, pH 7.4; 20 mM EDTA; 1% SDS).
6. Add an equal volume of phenol saturated with NETS buffer (10 mM Tris, pH 7.4; 100 mM NaCl; 10 mM EDTA; 0.2% SDS).

7. Add an equal volume of chloroform **or** a mixture of chloroform:isoamyl alcohol (24:1). Mix thoroughly and *carefully*.
8. Heat to 55°C for 10 minutes, with periodic shaking.
9. Cool on ice or in an ice-water bath for 5 minutes. Centrifuge at 2500 × g for 3 minutes to separate the phases.
10. Remove the lower organic phase by carefully sliding a sterile silanized Pasteur pipette or microcentrifuge tip along the side of the tube, through the aqueous phase and interphase, and aspirate from the bottom of the tube. Leave the aqueous phase and interphase in the tube and repeat steps 6 and 7.

    Note 1: This approach is especially important when working with small quantities of nuclear RNA, where there is a potential for loss of RNA by entrapment in chromatin.

    Note 2: See Appendix H for instructions on silanizing glassware.
11. Centrifuge at 2500 × g for 3 minutes to separate phases.
12. Carefully recover the aqueous phase, taking care not to disrupt the lower organic phase or interphase, if present. Add an equal volume of chloroform to the aqueous material, mixing thoroughly and carefully. Pulse centrifuge briefly to separate the phases.
13. Transfer the aqueous phase to a new tube. Precipitate the nuclear RNA at −20°C with the addition of 2.5 vol of ice-cold 95% ethanol.
14. Collect the precipitate by centrifugation at 4°C. Wash at least once with 70% ethanol to remove excess salt. Quantify yield and perform quality control testing as described in Chapter 8, Quality Control for RNA Preparations.

Protocol: preparation of nuclear RNA from cells enriched in ribonuclease

The following is a modification of the procedure of Chomczynski and Sacchi (1987). This method is the basis of RNA isolation products TRI Reagent and TRIzol. For those who prefer not to prepare individual reagents in-house, either of the above-referenced products can be substituted with confidence in place of the following protocol.

1. Prepare nuclei as described above; nuclei can be prepared ahead of time and stored frozen. Allow frozen nuclei to thaw on ice for 3–5 minutes.
2. Lyse intact nuclei with the addition of GTC solution (4 M guanidinium thiocyanate; 25 mM sodium citrate, pH 7.0; 0.5% sarcosyl; 100 mM 2-mercaptoethanol). Use 100 μL per 2×10^6 nuclei.

    Note: The volumes indicated for the remaining steps in this procedure assume a starting volume of 1 mL of GTC solution. For scaled-up extractions, remember to increase the volumes of all other reagents appropriately. If less than 800 μL of GTC solution is used, the entire extraction can be carried out in a 2.2 mL microfuge tube.

3. Transfer the homogenate to a disposable *polypropylene tube* (15 mL size works the best) and add, in order:
   a. 0.1 mL 2 M sodium acetate, pH 5.2,
   b. 1.0 mL water-saturated phenol (molecular biology quality),
   c. 0.2 mL chloroform:isoamyl alcohol (49:1).

   Cap tube and mix *carefully* and thoroughly by inversion following the addition of a reagent. Shake tube vigorously for 10 seconds after all reagents have been added.
4. Cool sample on ice for at least 15 minutes and then centrifuge at 4°C to separate the phases.

   Note: Be sure to check the manufacturer's maximum g-force rating before use.
5. Transfer aqueous phase (contains RNA) to a fresh tube and mix with 0.75 vol of isopropanol (~1 mL). Store at −20°C for at least 1 hour to precipitate RNA.

   Note: The RNA should be precipitated in a tube that can withstand the required 10,000 × g centrifugation in the next step. If the volume is too great to fit into one or a few microcentrifuge tubes, then acid-washed Corex glass tubes, used with the proper rubber adapters, are a suitable alternative.
6. Collect precipitate by centrifugation at 10,000 × *g* for 20 minutes at 4°C. Carefully decant and discard supernatant.
7. Completely dissolve RNA pellet in 300 μL of GTC solution (see step 2) and then transfer to an RNase-free 1.5 mL microcentrifuge tube.
8. Reprecipitate the RNA by adding 0.75 vol of ice-cold isopropanol at −20°C for 1 hour.
9. Collect precipitate at top speed in a microcentrifuge for 10 minutes at 4°C. Carefully decant and discard supernatant.
10. Wash pellet with 70% ethanol and centrifuge again. The pellet may be washed once more with 95% ethanol to accelerate air-drying. Do not allow the sample to dry out completely.
11. Redissolve damp RNA pellet in 50 μL of TE buffer, pH 7.5 or 50 μL of sterile H_2O. Incubation at 65°C for 10 minutes may facilitate solubilization. Determine sample concentration (Chapter 8: Quality Control for RNA Preparations); store RNA in suitable aliquots at −80°C. As always, avoid repeated freezing and thawing.

Troubleshooting nuclear RNA analysis

If difficulties are encountered when performing nuclear RNA analysis, consider the following:

1. Streamline the process used for nuclear isolation. This involves keeping the samples cold, unless otherwise noted, and keeping the processing

time to an absolute minimum. As efficiency in this part of the protocol improves, so does the labeling of the RNA.

2. Poor label incorporation may be due to incomplete cell lysis. Cells must be lysed in order for the labeled nucleotide(s) to gain access to the nucleus. Use a drop of trypan blue to assess lysis—the nuclei will look like blue ovals under a microscope.
3. Poor label incorporation may also be due to the incubation temperature. Transcription in isolated nuclei appears to decline abruptly after a brief period at 37°C (Marzluff et al., 1973); for this reason, most investigators perform the labeling step between room temperature and 30°C.
4. If the assay does not appear to be linear, check to ensure that the cold DNA probes immobilized on the filter(s) are in a molar excess relative to the labeled transcripts. That the target DNA is in excess can be determined by (a) varying the amount of input RNA, in which case the hybridization signal should be proportional to the amount of input RNA, (b) varying the amount of target DNA on the filter paper, in which case there should be no effect on the amount of RNA hybridized if the DNA is in excess, and (c) rehybridizing RNA that does not bind to the target DNA or the filter paper, none of which should participate in hybridization.
5. To further investigate the linearity of the assay, make triplicate filter membrane strips, onto which samples of target DNA have been immobilized. Prepare multiple twofold dilutions of the heterogeneous labeled run-on RNA (based on measured specific activity) and then hybridize to individual filter paper strips containing the blotted DNAs. The signal intensity obtained upon hybridization detection should, of course, reflect the twofold dilutions of the labeled RNA.
6. Ensure that all solutions, lab implements, and consumables have been treated to render them RNase-free.
7. Check the date of the last micropipette recalibration. While this may seem obvious, routine maintenance is often overlooked.

References

Celano, P., Baylin, S.B., Casero, R.A., 1989a. Polyamines differentially modulate the transcription of growth-associated genes in human colon carcinoma cells. J. Biol. Chem. 264, 8922–8927.

Celano, P., Berchtold, C., Casero, R.A., 1989b. A simplification of the nuclear runoff assay. BioTechniques 7 (9), 942–945.

Chomczynski, P., Sacchi, N., 1987. Single-step method of RNA isolation by acid guanidinium thiocyanate-phenol-chloroform extraction. Analyt. Biochem. 162, 156–159.

deBustros, A., Baylin, S.B., Levin, M.A., Nelkin, B.D., 1986. Cyclic AMP and phorbol esters separately induced growth inhibition, calcitonin secretion, and calcitonin gene transcription in cultured human medullary thyroid carcinoma. J. Biol. Chem. 261, 8036–8041.

Derman, E., Krauter, K., Walling, L., Weinberger, C., Ray, M., Darnell, J.E., 1981. Transcriptional control in the production of liver-specific mRNAs. Cell 23, 731–739.

Fan, J., Chen, Y., Watkins, T., Dang, C.V., Gorospe, M., Cheadle, C., 2011. Array-based nuclear run-on analysis. Meth. Molec Biol. 809, 505–517.

Farrell Jr., R.E., Greene, J.J., 1992. Regulation of c-myc and c-Ha-ras oncogene expression by cell shape. J. Cell. Physiol. 153, 429–435.

Fei, H., Drake, T.A., 1993. A rapid nuclear run-off transcription assay. BioTechniques 15, 838.

Gao, M., et al., 2022. New chromatin run-on reaction enables global mapping of active RNA polymerase locations in an enrichment-free manner. ACS Chem. Biol. 17, 768–775.

Greenberg, M.E., 1988. Identification of newly transcribed RNA. In: Ausubel, F.M., Brent, R., Kingston, R.E., et al.,Current Protocols in Molecular Biology. John Wiley & Sons, New York, 4.10.1.

Greene, J.J., Pearson, S.L., 1994. Measurement of gene-specific transcription by nuclease protection of pulse-labeled nuclear RNA. J. Biochem. Biophys. Meth. 29, 179–187.

Groudine, M., Peretz, M., Weintraub, H., 1981. Transcriptional regulation of hemoglobin switching on chicken embryos. Mol. Cell. Biol. 1, 281–288.

Jelinek, W., Molloy, G., Fernandez-Munoz, R., Salditt, M., Darnell, J.E., 1974. Secondary structure in heterogenous nuclear RNA: involvement of regions from repeated DNA sites. Mol. Biol. 82, 361–370.

Kanazawa, A., O'Dell, M., Hellens, R.P., Hitchin, E., Metzlaff, M., 2000. Mini-scale method for nuclear run-on transcription in plants. Plant. Mol. Biol. Rep. 18, 377–383.

Kashi, K., Henderson, L., Bonetti, A., Carninci, P., 2016. Discovery and functional analysis of lncRNAs: methodologies to investigate an uncharacterized genome. Biochem. Biophys. Acta 1859 (1), 3–15.

Khraiwesh, B., 2011. Using nuclear run-on transcription assays in RNAi studies. Meth. Mol. Biol. 744, 199–209.

Kwei, K.A., 2004. Transcriptional repression of catalase in mouse skin tumor progression. Neoplasia 6, 440–448.

Marzluff, W.F., 1978. Transcription of RNA in isolated nuclei. Meth. Cell Biol. 19, 317–332.

Marzluff, W.F., Huang, R.C.C., 1984. Transcription of RNA in isolated nuclei. In: Hames, B.D., Higgins, S.J. (Eds.), Transcription and Translation: A Practical Approach. IRL Press, Oxford, p. 89.

Marzluff, W.F., Murphy, E.C., Huang, R.C.C., 1973. Transcription of ribonucleic acid in isolated mouse myeloma nuclei. Biochemistry 12, 3440–3446.

Marzluff, W.F., Murphy, E.C., Huang, R.C.C., 1974. Transcription of the genes for 5 S ribosomal RNA and transfer RNA in isolated mouse myeloma cell nuclei. Biochemistry 13, 3689–3696.

Nevins, J.R., Darnell, J.E., 1978. Steps in the processing of Ad2 mRNA: poly(A)$^+$ nuclear sequences are conserved and poly(A)$^+$ addition precedes splicing. Cell 15, 1477–1493.

Ng, D.C., Su, M.-J., Kim, R., Bikle, D.D., 1996. Regulation of involucrin gene expression by calcium in normal human keratinocytes. Front. Biosci. 1, 16–24.

Patrone, G., Puppo, F., Ceccherine, I., Puliti, A., Ravazzolo, R., 2000. Nuclear run-on assay using biotin labeling, magnetic bead capture and analysis by fluorescence-based RT-PCR. BioTechniques 29, 1012–1017.

Price, R., Penman, S., 1972. A distinct RNA polymerase activity, synthesizing 5.5 S, 5 S, and 4 S RNA in nuclei from adenovirus 2-infected HeLa cells. J. Mol. Biol. 70, 435–450.

Roberts, T.C., Hart, J.R., Kaikkonen, M.U., Weinberg, M.S., Vog, P.K., Morris, K.V., 2015. Quantification of nascent transcription by bromouridine immunocapture nuclear run-on RT-qPCR. Nat. Protoc. 10 (8), 1198–1211.

Roeder, R.G., 1976. Nuclear RNA polymerase. In: Losick, R., Chamberlain, M. (Eds.), RNA Polymerase. Cold Spring Harbor Laboratory Press, Cold Spring Harbor, NY, pp. 283–329.

Sandoval, J., Rodríguez, J.L., Tur, G., Serviddio, G., Pereda, J., Boukaba, A., et al., 2004. RNAPol-ChIP: a novel application of chromatin immunoprecipitation to the analysis of real-time gene transcription. Nucleic Acids Res. 32 (11), e88. Available from: https://doi.org/10.1093/nar/gnh091.

Soeiro, R., Darnell, J.E., 1969. Competition hybridization by "pre-saturation" of HeLa cell DNA. J. Mol. Biol. 44, 551.

Tani, H., Mizutani, R., Salam, K.A., Tano, K., Ijiri, K., Wakamatsu, A., et al., 2012. Genome-wide determination of RNA stability reveals hundreds of short-lived noncoding transcripts in mammals. Genome Res. 22 (5), 947–956.

Tan, K., Wong, K.H., 2019. RNA polymerase II ChIP-seq - a powerful and highly affordable method for studying fungal genomics and physiology. Biophys. Rev. 11 (1), 79–82. Available from: https://doi.org/10.1007/s12551-018-00497-9.

Zhang, Y., Fan, M., Zhang, X., Huang, F., Wu, K., Zhang, J., et al., 2014. Cellular microRNAs up-regulate transcription via interaction with promoter TATA-box motifs. RNA 20, 1878–1889.

Zhang, Z., Tijan, R., 2018. Measuring dynamics of eukaryotic transcription initiation: challenges, insights, opportunities. Transcription 9 (3), 159–165. Available from: https://doi.org/10.1080/21541264.2017.1363017.

Further reading

García-Martínez, J., Aranda, A., Pérez-Ortín, J.E., 2004. Genomic run-on evaluates transcription rates for all yeast genes and identifies gene regulatory mechanisms. Mol Cell 15 (2), 303-13.

Greenberg, M.E., Ziff, E.B., 1984. Stimulation of 3T3 cells induces transcription of the c-fos proto-oncogene. Nature 311, 433–438.

Kofoed, R.H., Betzer, C., Lykke-Andersen, S., Molska, E., Jensen, P.H., 2018. Investigation of RNA synthesis using 5-bromouridine labelling and immunoprecipitation. J. Vis. Exp. Available from: https://doi.org/10.3791/57056.

Larson, D.R., Zenklusen, D., Wu, B., Caho, J.A., Singer, R.H., 2011. Real-time observation of transcription initiation and elongation on an endogenous yeast gene. Science 332 (6028), 475–478.

Marzluff, W.F., 1973. Transcription of ribonucleic acid in isolated mouse myeloma nuclei. Biochemistry 12, 3440–3446.

Marzluff, W.F., Huang, R.C.C., 1975. Chromatin directed transcription of 5 S and tRNA genes. Proc. Natl. Acad. Sci. 72, 1082–1086.

Nevins, J.R., 1987. Isolation and analysis of nuclear RNA. Methods Enzymol. 152, 234–241.

Pontier, D., Yahubyan, G., Vega, D., Bulski, A., Saez-Vasquez, J., Hakimi, M.A., et al., 2005. Reinforcement of silencing at transposons and highly repeated sequences requires the concerted action of two distinct RNA polymerases IV in Arabidopsis. Genes. Dev. 19, 2030–2040.

Sparks, J.D., Collins, H.I., Chirieac, D.V., Cianci, J., Jokinen, J., Sowden, M.P., et al., 2006. Hepatic very-low-density lipoprotein and apolipoprotein B production are increased following in vivo induction of betaine-homocysteine S-methyltransferase. Biochem. J. 395 (2), 363–371.

Wittelsberger, S.C., Kleene, K., Penman, S., 1981. Progressive loss of shape-responsive metabolic controls in cells with increasingly transformed phenotype. Cell 24, 859–866.

Chapter 20

RNA in situ hybridization

Rationale

The method of RNA in situ hybridization (RNA ISH) permits the observation of gene expression within the histological architecture of a specimen. RNA ISH is amenable to freshly isolated, flash-frozen tissue samples, as well as tissues that have been chemically crosslinked in either the recent or distant past. In situ hybridization is based on the same probe-target base-pairing principle that characterizes nearly all molecular biology methods; it alone, however, is the only method that precisely maps transcripts to specific cells within a tissue specimen. Information of this type can be of great importance in the realm of developmental biology as well as for the detection and staging of various diseases. The method is sufficiently sensitive to detect low-abundance transcripts, though it lacks the broad dynamic range associated with other methods including RT-PCR and RNA-seq. A variant of the RNA in situ method is a widely used cytogenetic diagnostic tool of localizing DNA targets on intact chromosomes.

Technical considerations

The term "in situ" means localized, in position, or in its original place or native location. Studying RNA is this manner represents a significant departure from traditional methods for the analysis of RNA which begin with cell or tissue disruption, precluding the ability of the investigator to ascertain the spatial distribution of cells expressing the gene(s) of interest. Thus, in situ studies pertaining to transcription identify spatial and temporal patterns of gene expression, and subcellular target areas may include the nucleus, the cytoplasm, mitochondria, or chloroplasts. There are many variants of RNA (and DNA) in situ hybridization, some of which are described here. When in situ gene expression studies are performed using fluorescent-labeled probes to detect, localize, and quantify one or more targets, the method is commonly known as fluorescence in situ hybridization (Fig. 20.1) or simply FISH. Alternatively, there are well-respected immunochemical probe labeling and chromogenic detection options which are in widespread use.

An in situ hybridization technique was first used to identify amplified rRNA gene loci on intact chromosomes in *Xenopus* oocytes (Gall and Pardue, 1969).

RNA Methodologies. DOI: https://doi.org/10.1016/B978-0-323-90221-2.00046-1

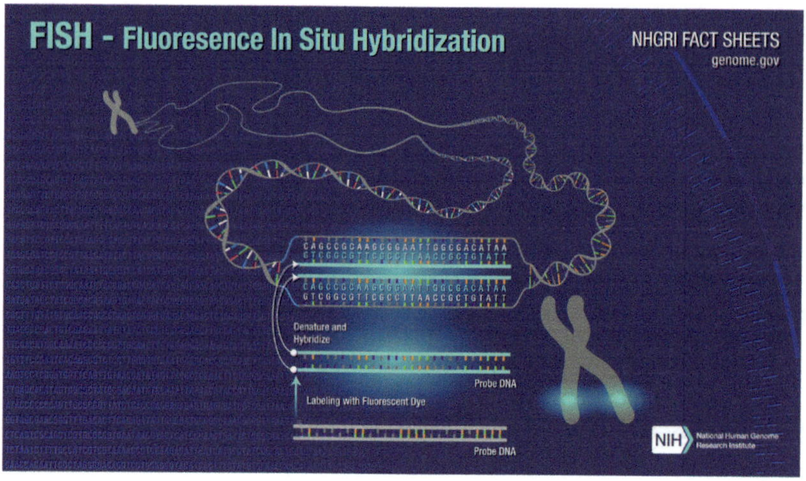

FIGURE 20.1 One method of in situ hybridization. Although the image depicts a gene locus as the target on a metaphase chromosome, the same principle applies to the detection of RNA transcripts in whole cells and tissue sections. *Image courtesy National Human Genome Research Institute (NHGRI; http://www.genome.gov).*

This application utilized a ^{3}H-labeled probe and was an early form of cytogenetics. Almost simultaneously, a similar application was developed to localize specific genes in hamster tissue (Buongiorno-Nardelli and Amaldi, 1970). The technology improved over time, leading to the development of FISH in the late 1970s and early 1980s (Bauman et al., 1980; Langer-Safer et al., 1982). In short order, the in situ method was sufficiently refined so that it could be used to detect specific mRNA species within the architecture of a tissue sample (Singer and Ward, 1982). RNA FISH has become a widely recognized mainstream molecular biology tool which has many applications (Young et al., 2020). A contemporary DNA-target variation of this method, which is currently considered to be the gold standard, utilizes fluorescent-labeled probes to detect specific forms of aneuploidy, changes in chromosome structure, and gene amplification in the diagnosis of cancer (e.g., HER-2 gene; PathVysion Kit, Abbott Labs) and other diseases. The advantages and disadvantages of in situ hybridization are presented in Table 20.1.

As with all methods that involve the processing of samples for RNA analysis, speed, efficiency, and attentiveness to the control of RNase activity are essential for a successful outcome. When performing in situ gene expression studies there is a fair amount of latitude with respect to the nature of the probe (RNA or DNA), the probe labeling method, and the ensuing detection methods. RNA probes are advantageous because double-stranded RNA (dsRNA; RNA probe base-paired to a mRNA target) has much greater thermodynamic stability than a DNA:RNA hybrid (DNA probe base-paired to an mRNA target). Thus, RNA probes, generally referred to as riboprobes, allow

TABLE 20.1 Advantages and disadvantages of in situ hybridization.

Advantages	Disadvantages
Natural histological architecture is preserved.	Fair amount of skill needed to prepare and section tissues.
Specimens can be either flash-frozen tissue or tissue that was previously FFPE.	Toxic crosslinking fixatives such as formaldehyde, formalin, or paraformaldehyde required to maintain RNA integrity and positioning with the sample.
Specific cells within a tissue are identifiable.	Tissue preparation requires either a microtome or cryostat.
Nonradioactive detection methods work as well as radioactive methods antisense, cDNA, or oligonucleotide probes.	Quantification is possible, though the dynamic range is limited.
Transcriptional and translational regulation can be inferred by the presence or absence of mRNA in a cell, compared to the corresponding protein level in that same cell.	Nonabundant RNAs may be difficult to detect.
Whole embryo sectioning is very helpful for understanding gene expression during development.	Location of gene expression is revealed, but not the size of the transcripts being produced.
Archived tissue may support in situ analysis and slides can be stored indefinitely.	Multistep process requiring optimization for each new probe or tissue type.
Useful detecting chromosome-level abnormalities.	Knowledge of histology and microscopy competence are essential.

FFPE, Formalin-fixed and paraffin-embedded.

the assay to be performed with a much higher level of stringency, meaning less background and greater resolution. Owing to their unique properties, antisense cRNA probes are a popular choice for this type of gene expression investigation, though longer cDNA and oligonucleotide probes are also options. Note that oligonucleotides have gained popularity for in situ applications due to the ease with which they are able to penetrate tissue sections and, given their characteristically short nature, they are much more discriminating in terms of target sequence identification than are longer probes.

Sample preparation

In situ studies can be performed in a number of formats, though they all share the major steps of fixation, permeabilization, hybridization, and detection (Table 20.2). While some of the manipulations are similar to other commonplace molecular methods, additional steps are necessary due to the unique processing of intact tissue. The most important and underlying principle behind this technology is that the target transcripts need to be immobilized in such a way that they are readily accessible to the probe all the while maintaining the histological features of the specimen. With regard to the packaging of RNA (and DNA) in the cell, both types of nucleic acids are commonly entangled with, bound to, or otherwise engaged with significant amounts of protein. Therefore, a permeabilization step commonly involving partial protein hydrolysis with proteinase K is an important part of sample preparation. In addition, it is not uncommon to acetylate the tissue sections, the purpose which is to block cationic groups that would otherwise attract negatively charged nucleic acid probes and thereby contribute to the background. Another processing step associated with high-lipid-content tissue, for example, brain, is delipidation (removal of lipids) through immersion in chloroform.

There are two basic ways to approach tissue preparation for gene expressions, namely, using freshly frozen tissue or using tissue that has been chemically fixed and embedded in wax, that is, formalin-fixed and paraffin-embedded (FFPE) tissue. Both methods involve thin-sectioning tissue, requiring either a cryostat for frozen tissue or a microtome for fixed tissue. The sections are immediately immobilized on specially coated microscope slides (e.g., polylysine or gelatin) to ensure that the specimen is not lost as the ensuing processing steps progress. In order to save time and maintain some level of quality control, some investigators prefer to buy pretreated slides (e.g., VWR Superfrost Plus; #48311-703). In the case of chromosome squashes for gene localization or the identification of chromosomal abnormalities, no special pretreatment is needed other than to ensure that the slides are scrupulously clean. This can be accomplished by soaking the slides in or wiping them with alcohol.

In another application, developing embryos from very small animals can also be studied using in situ techniques. In this case, there is a major procedural difference: while tissue and organ sections are immobilized on a microscope slide prior to tissue fixation and permeabilization, insect embryos, as one example, are fixed on a microscope slide as the last step. The embryos are permeabilized and the in situ protocols are completed while the specimens are free-floating in a small vessel such as a microfuge tube, a scintillation vial, or a six- or 24-well cell culture dish. As it is important to transfer as little liquid as possible from one fixative step to the next, it is helpful to work with cassettes from any of several suppliers of ISH equipment and consumables. Alternatively, a homemade "basket" of sorts can be prepared by removing the lower one-fourth of the microfuge tube and then placing a plastic mesh plug at the opening which will allow one fixative solution to drain away before placing the specimen in the next fixative solution. Performing

TABLE 20.2 Experimental options for in situ hybridization.

Parameter	Options	Advantages	Disadvantages
Specimen preparation	Flash frozen tissue prepared on a cryostat	Thin, high-quality tissue sections	Skill and cryostat required
	FFPE tissue prepared on a microtome	Specimen is stable for years	Skill and microtome required; archived tissue may have experienced some level of degradation
	Direct mounting on a microscope slide (whole mount)	Well suited for analysis of metaphase chromosomes, cells, and for small invertebrates	Knowledge of developmental stages and skill in handling very small specimens
Probe type	DNA	Easy to handle	Lower T_m may contribute to background
	RNA	Higher T_m improves specificity and reduces background	Natural instability; susceptible to RNase
	Locked nucleic acids	Very high T_m maximizes specificity	Cost
Label types	Isotopic (^{35}S, ^{33}P, ^{3}H)	High specific-activity signals	Radioactive handling and disposal; radio-decay scatter limits spatial resolution
	Fluorescence	Strong signals with reduced background	Specialized equipment for detection and photodocumentation
	Hapten (digoxigenin; biotin)	Stable after labeling	Lower end of detection sensitivity spectrum
Detection procedures	Autoradiography or microscope slide emulsion	Easy to process; resolution and sensitivity are very good	Radioactivity remains in the specimen (depends on half-life)
	Fluorescence excitation and recording	Low background; sensitivity and resolution are excellent	Specialized equipment needed

(Continued)

TABLE 20.2 (Continued)

Parameter	Options	Advantages	Disadvantages
	Chromogenic/ colorimetric immunochemical approaches	Safe and easy to process	Sensitivity and resolution a bit lower compared to other methods
Experimental controls	Positive controls	Sample known to express target; technical controls to verify functionality of reagents	None; positive controls are imperative
	Negative controls	Use of an antisense or irrelevant control is essential	None; negative controls are imperative

FFPE, Formalin-fixed and paraffin-embedded.

the in situ protocol prior to fixation on a microscope slide is highly advantageous because of the very fragile nature of the specimens. In the case of very small specimens, the investigator may choose to altogether forego permanently fixing them to a slide. Instead, the specimens are examined microscopically in a depression slide with a drop of methyl cellulose, after which the specimens can be returned to storage in PBS at 4°C.

Finally, a convenient and standard method of processing multiple slides at once is to load them into a slide frame which can then be submerged in the recommended series of fixatives specific for the tissue type (Fig. 20.2). Moving the slides in batch ensures a uniform processing time for each step in the series. The movement of the specimens in a slide frame also favors rapid and complete drainage prior to the following step and likewise limits the investigator's contact with toxic chemicals.

Fresh frozen samples

Freshly-isolated tissues and organs produce excellent thin sections for RNA ISH when they are frozen immediately after harvesting. The actual preparation procedure will vary by tissue sample and the primary literature is rife with procedures that are tissue-, organ-, and species-specific. Most tissues section well when they are embedded in Tissue-Tek O.C.T. (optimal cutting temperature) compound, a standard material used to prepare tissue samples for thin sectioning. The sample is then frozen in the cryostat itself rather than being flash-frozen. The Tissue-Tek is also used to attach the frozen

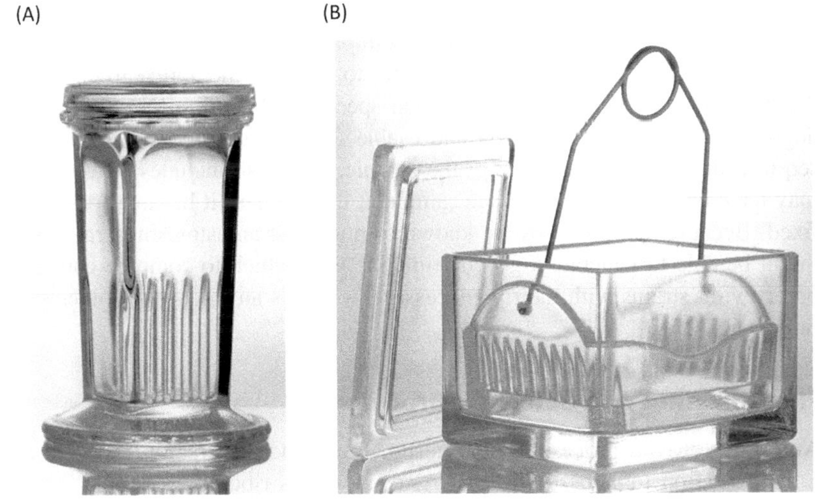

FIGURE 20.2 Coplin jar (A) and staining box (B) are both convenient vessels for processing multiple slides simultaneously for any of a number of procedures, including tissue fixation and staining. These glass containers are easily cleaned, chemically resistant, and easy to make RNase-free for each new application. *Image courtesy Witeg Labortechnik GmbH (http://www.witeg.de).*

tissue to the chuck (the specimen holder), keeping it in place while it is advanced toward the cryostat blade. Fresh tissue samples are generally cooled gradually in O.C.T. medium within the cryostat chamber at −22°C. Ensuring that the spatial distribution of the RNA is maintained during tissue sectioning and in all subsequent steps is an absolute requirement if in situ hybridization experiments are to be meaningful. Flash-freezing, in contrast, should be avoided because it often distorts the spatial arrangement of certain tissue regions, thereby rendering unreliable data with respect to the local distribution of the transcripts within the tissue.

Fixed, paraffin-embedded samples

Tissue samples which may have been collected recently or sometime in the past may have been chemically fixed and then embedded in wax (paraffin) in order to facilitate efficient thin-sectioning on a microtome. Chemical fixation is a preferred method when the slides on which the specimens are mounted are to be stored for an extended period of time. A well-established method for preserving tissue samples is fixation in 10% buffered formalin or 4% paraformaldehyde. The ability to later recover RNA from archival tissue offers an unparalleled opportunity to study unique samples that have been stored for perhaps decades. This prospect is particularly tantalizing because of the substantial archives found in pathology labs and other clinical settings.

Thousands of biopsy and resection specimens are added annually representing various normal and disease-state examples, and providing new insight into disease initiation and progression, not to mention the ability to examine the genomes and transcriptomes of each specimen. While mRNA is usually degraded, miRNA is often intact and usable for at least some level of RNA-seq to profile the biochemistry of the sample. This also includes viruses that may have been present in a tissue sample at the moment it became formalin-fixed. Because some viruses are known to mutate at an astonishing rate, one might be able to generate a genetic timeline with which to compare contemporary viral strains with their predecessors, which is an important contemporary topic.

Hybridization and detection procedures

Antisense (cRNA) probes are in widespread use and are greatly respected as a hybridization tool for RNA ISH. Also known as riboprobes (Table 20.3), they are continuously-labeled RNAs of uniform length which are synthesized by in vitro transcription. In fact, both antisense (probe) and sense (negative control) transcripts can be produced from the same plasmid construction, albeit in different reaction tubes, by cloning the gene sequence of interest between two high-efficiency transcription promoters situated in opposite orientations (refer back to Fig. 17.4 for review). While RNA:RNA probe:target interactions offer enhance thermodynamic stability, RNA probes are still subject to RNase degradation if not handled using proper RNase-free technique. The most common nonradioactive labels include the nucleotide precursors carrying either of the well-known haptens digoxigen (DIG) or biotin, both of which are widely available in commercial systems for probe synthesis and detection. If FISH is to be performed, the detection procedure method will require fluorescence microscopy and a suitable means for image capture. FISH nucleic acid probes carry a fluorophore, for example, Cy3, Cy5, Alexa,

TABLE 20.3 Riboprobes.

Sense RNA	Antisense RNA
+RNA	-RNA
Same sequence as mRNA	Complementary to mRNA; cRNA
Not able to hybridize to mRNA, but is a good negative control	Is able to hybridize to mRNA
Same sequence as the gene coding strand	Same sequence as the gene template strand

Note: The term riboprobe is used generically to refer to an RNA probe, even though Riboprobe is a trademark-protected name (Promega; www.promega.com).

and many others, attached directly to their 5′ end, their 3′ end, or both ends; otherwise, the fluorophore will be introduced during the detection protocol.

As has been a recurrent theme throughout this book, maintaining the integrity of RNA is nonnegotiable in all applications. To this end, some investigators utilize vanadyl ribonucleoside (VRC; also VDR) complexes in the 2–5 mM range either during the sample fixation process (Mistry et al., 2020) or add it directly to the hybridization buffer formulation to protect the RNA in instances when sample integrity is problematic. A description of the uses of VRC along with a synthesis protocol is presented in Chapter 2, Creating a Ribonuclease-Free Environment. VRC as a nuclease inhibitor has fallen out of favor with the RT-PCR crowd because it inhibits reverse transcriptase. Since RNA ISH is directed at transcript localization, VRC is well suited for this application. It should be noted that many hybridization formulations are sufficiently chaotropic such that an additional RNase inhibitor is *usually* not needed.

Detection procedures include immunochemistry which produces a color precipitate, fluorescence, and even autoradiography in the case of a radioactive probe. As with chromogenic or chemiluminescence detection in filter-based assays such as the venerable Northern blot analysis, the same principles apply to in situ studies. Biotinylated probes are localized in a tissue section with streptavidin while DIG is easily localized with anti-DIG antibodies. The streptavidin and the anti-DIG antibodies may exist as an enzyme conjugate that will generate a signal. Alternatively, the streptavidin and the anti-DIG antibodies may in turn be localized with a biotin-enzyme conjugate or secondary antibody-enzyme conjugate, respectively, for signal enhancement by building a type of molecular sandwich and positioning the enzyme a bit further away from the tissue section surface. The more commonly used enzymes supporting chromogenic detection include horseradish peroxidase (HRP) and alkaline phosphatase (AP). In the case of either enzyme, endogenous enzyme activity can contribute to background noise and may need to be suppressed during the preparation of the sample. HRP within a sample can be eliminated by incubating the sample for 30 minutes in 1% H_2O_2 in methanol. Endogenous AP can be neutralized by incubating the specimen in 10.0 mM levamisole 4 × 30 minutes. Common substrates for AP-mediated detection in 5-bromo-4-chloro-3-indolyl-phosphate (BCIP) with nitro-blue tetrazolium (NBT). Common substrates for HRP include diaminobenzidine (DAB) with imidazole. In addition, there are many other substrates which will produce a fluorescence signal.

When performing RNA FISH, autofluorescence is a common problem with lipid-rich specimens. The histochemical stain Sudan Black B can reduce autofluorescence by staining lipids. For this application, it is commonly prepared as 0.1% solution in 70% ethanol or 70% isopropanol and is used as a posthybridization stain prior to the onset of the fluorescence detection procedures. This step may not be needed if a lipid-rich tissue thin section was processed in chloroform early in the tissue fixation process.

There are two mainstream methods utilized to enhance the detection of low abundance transcripts. As noted above, in situ hybridization is a sensitive method, though methods which boost the signal without contributing to the background are always worth considering. The two methods are fluorophore-labeled tyramide detection (Raap et al., 1995; Marras et al., 2019) and bDNA signal amplification (Player et al., 2001).

Tyramine is derived from the amino acid tyrosine, and tyramide is a tyramine derivative. The signal enhancement method is HRP-dependent, with HRP or sometimes poly-HRP conjugated to the secondary antibody used in the detection procedure. Structurally, tyramide is linked to a fluorescent dye, that is, a tyramide fluorophore. In the presence of H_2O_2, the catalytic activity of HRP produces tyramide radicals which begin to quickly bind tyrosine residues in the immediate vicinity of the reaction and which may even include the primary and secondary antibodies. In so doing, large numbers of fluorophores are positioned proximal to the target epitope, thereby producing an enhanced fluorescent signal. Depending on the system used, this method offers signal enhancement as much as 100 × greater than standard detection chemistries (Fig. 20.3).

In the case of bDNA, which is short for branched DNA, a tree-like structure is created in this RNA ISH signal amplification technique. Probes are designed such that one end hybridizes to the target RNA while the other end consists of unique sequences that will base-pair with a preamplifier (Fig. 20.4). In turn, the preamplifier base pairs with numerous amplifier molecules, thereby forming an expansive branched structure which is able to anneal large numbers of fluorescent probes.

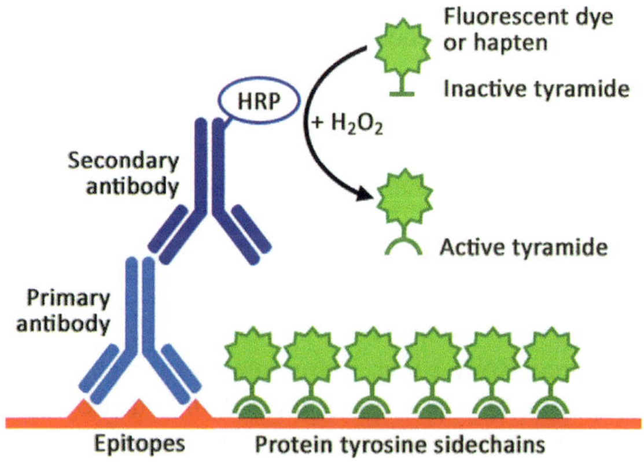

FIGURE 20.3 Tyramine-fluorochrome ISH signal amplification. In the presence of H_2O_2, HRP produces tyramide radicals which bind to tyrosine residues near the site of the reaction, thereby increasing the number of fluorochrome molecules close location where the antisense RNA is base-paired with a target mRNA. *Image courtesy of Biotium (www.biotium.com).*

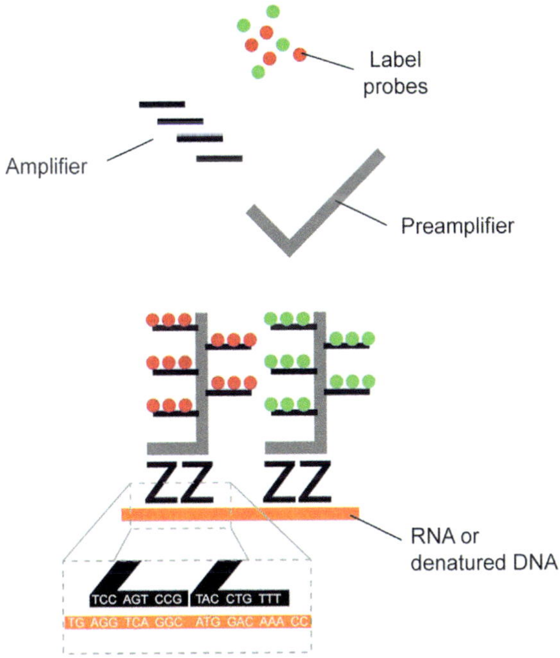

FIGURE 20.4 Branched DNA (bDNA) as a method for signal amplification. One end of a nucleic acid probe (black) is able to base-pair to an RNA target in a specimen, while the other end is complementary to a preamplifier sequence. The subsequent association of the amplifier sequences creates a branched, tree-like structure that is able to bind numerous labeled probe molecules, thereby enhancing signal intensity significantly (**Sannier et al., 2020**). *Figure courtesy Sannier, G., M. Dubé, and D.E. Kaufmann. (2020). Single-cell technologies applied to HIV-1 research: reaching maturity. Front. Microbiol. 11, 297. https://doi.org/10.3389/fmicb.2020.00297; and reproduced with permission under the Creative Commons Attribution 4.0 International License (CC BY 4.0; creativecommons.org/licenses/by/4.0/).*

Positive and negative control considerations

As with all forms of experimentation, the results are credible only to the extent that they are backed with well-considered and well-executed positive and negative controls. In situ methods are no exception. As always, positive controls should be used judiciously so that there is no possible confusion with a positive experimental result. This is especially important when performing in situ methods because although the positive signal localizes gene expression (RNA ISH), unlike PCR methods there is no indication of the size of the target. Contrast this with PCR, in which positive controls are designed to be electrophoretically- or melting curve-distinguishable from authentic experimental amplified products.

Positive controls show that all components of the experiment are functioning properly and that a failure to detect a positive experimental result, if this occurs, was not caused by a technical error. Positive controls may include the same reference or "housekeeping" genes described in Chapter 10, RT-PCR: A Science and An Art Form. To the extent possible, the selection of a gene whose expression is unique or commonly associated with the tissue under investigation is also a wise strategy. In the unlikely event that a suitable positive control cannot be identified for the sample(s) at hand, it is advisable to slot- or dot-blot (Appendix J: Dot blot analysis) a high-dilution series of the labeled probe onto the filter membrane and then perform the detection procedure. This will ensure not only that the probe has been labeled properly, but it will also demonstrate that the detection chemistries are likewise functioning properly. In any event, this is a good standalone test to demonstrate efficient probe synthesis.

Negative controls are likewise an important consideration as a means of assessing background or noise, which is particularly important when assessing hybridization within a tissue section. This includes assessing not only direct probe interaction with nontarget areas of the tissue section but also the extent to which the hapten-specific (digoxigen or biotin) primary and/or secondary antibodies interact with the specimen in a nonspecific manner. One worthwhile strategy is to hybridize the probe, for example, antisense RNA, to one tissue section, and then attempt to hybridize the sense RNA probe to the very next tissue section in order to assess the extent to which the probe preparation interacts nonspecifically with the elements on the slide (Fig. 20.5). Another negative control-type strategy is to pretreat a specimen with RNase which will destroy all target transcripts in the negative control sample, meaning that any signal is a result of probe affinity for something else on the slide. An elevated signal outcome in this particular case may

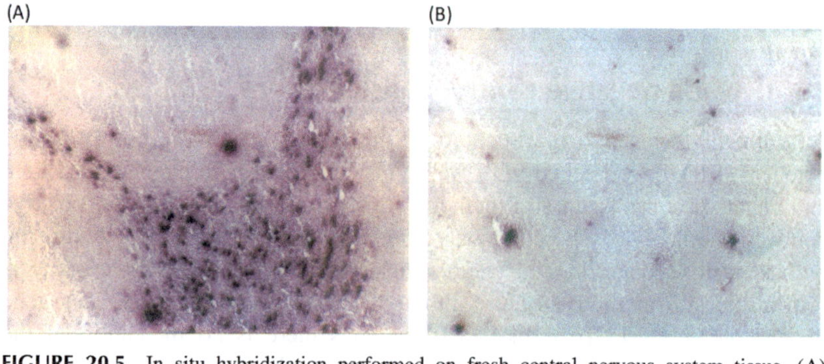

FIGURE 20.5 In situ hybridization performed on fresh central nervous system tissue. (A) Proteolipoprotein expression in rat cerebellum using an antisense RNA probe and chromogenic detection. The probe was labeled with digoxigenin and detected using standard immunochemical methods. (B) Negative control performed using a sense RNA probe (same sequence as the mRNA) directed against the next tissue slice taken from the frozen specimen; this procedure is critical for the assessment of background hybridization.

inform of the need to change the probe, or the protocol, altogether. A parallel assessment with DNase may help determine if DNA is the culprit behind the background signal. One useful tip for eliminating problematic background signal is to perform what might best be described as a sham prehybridization: incubate the hybridization solution (including the probe) briefly with a nonexperimental tissue specimen. In so doing, whatever is in hybridization + probe mixture that has a nonspecific affinity for the specimen will stick to the sham specimen instead. The hybridization cocktail is then transferred to the experimental section for the in situ investigation.

Protocol: preparation of fresh frozen brain tissue for in situ hybridization

Depending on the organ or tissue type being studied, some investigators prefer to perfuse the organism, if applicable, prior to dissection. The necessity or added value of that approach will need to be determined empirically in each laboratory.

1. Prepare freshly-dissected tissue for sectioning by placing the tissue in a sterile appropriately-shaped cryomold containing a few drops of O.C.T compound. Carefully position the tissue specimen as desired and then fill the mold completely with O.C.T., making sure the tissue is completely submerged.

   Note: Freshly-dissected tissue can be snap-frozen prior to embedding it in O.C.T., though some histological distortion may occur. Never plunge the tissue into liquid N_2, which will cause the specimen to crack and make sectioning difficult.
2. Immediately place the sample(s) in an ethanol/dry ice bath or at −80°C. Alternatively, some investigators simply place the specimen inside a precooled cryostat (−22°C) for several hours. The tissue will become embedded in the O.C.T. compound, which will become opaque when it freezes.

   Note 1: Keep in mind that small tissue specimens will freeze more rapidly than large specimens.

   Note 2: Use all recommended safety precautions while working with an ultracold ethanol-dry ice bath.
3. Carefully applying fresh O.C.T. to the chuck and immediately press the frozen embedded sample against it.
4. Section tissue on cryostat and mount sections on precoated (gelatin or polylysine) slides which have been precooled to −20°C. The unused tissue can be stored at −80°C for up to three months or stored indefinitely in liquid N_2.

   Note: Gelatin-coated slides are widely available for purchase. Alternatively, they can be prepared in-house by dipping cleaned slides in a freshly-filtered solution of 0.5% gelatin and 0.05% CrK $(SO_4)_2 \cdot 12H_2O$ five times for about 5 seconds each and then air-drying for 48 hours. Slides that will accept tissue sections prepared on a cryostat should be stored −20°C. Otherwise, the slides may be stored and protected from dust at room temperature.

5. Warm mounted sections to room temperature for 10 minutes.
6. Place the individual slides into a slide frame for a glass Coplin jar (rectangular box shape). This will be used to transfer the slides through the various solutions in steps 7–12.
7. Fix tissue for 5 minutes in 4% paraformaldehyde which has been freshly prepared in 1× PBS.

   Note 1: 1× PBS (per liter) = 8 g NaCl; 0.2 g KCl; 1.44 g Na_2HPO_4; 0.24 g KH_2PO_4; autoclave.

   Note 2: Important: Do not autoclave any solutions containing paraformaldehyde.
8. Wash 2× in PBS.
9. Acetylate for 10 minutes at room temp in freshly prepared 0.25% acetic anhydride prepared in 0.1 M triethanolamine (prepared in saline and pH-adjusted to 8.0).
10. Dehydrate in a graded series of ethanol:
    - **a.** 70% ethanol for 2 minutes;
    - **b.** 80% ethanol for 2 minutes;
    - **c.** 95% ethanol for 2 minutes;
    - **d.** 100% ethanol for 2 minutes.
11. Delipidate CNS tissue for 5 minutes in chloroform.
12. Rinse for 1 minute in 100% ethanol.
13. Allow slides to air-dry completely.
14. Add 200 μL of hybridization buffer (including the probe) on each slide. Carefully apply a cover slip that covers the entire specimen.
15. Hybridize 12–16 hours (overnight) in a well-humidified chamber at 70°C.

    Note: Plastic Nalgene storage or staining boxes with tight-fitting covers work well for this application. Line the inside of the box with Whatman 3 MM paper and wet it with nuclease-free water. Placing broken plastic pipettes, or similar implements, on top of the Whatman paper will create a platform for elevating the slides. Close the box and place in the incubator along with a beaker of water to maintain the humidity level throughout this step.
16. At the end of the hybridization period, carefully transfer the slides, including the coverslips to an excess volume of 50% formamide in 1× SSC and 01% Tween 20. Incubate with very gentle rocking at 65°C for 15 minutes, during which time the coverslips should float away from the slides.

    Note: Do not attempt to pull coverslips away from the slide as this will damage the tissue.

17. Wash samples two more times in the same wash solution as in the previous step for 15 minutes per wash at 65°C.
18. Wash slides twice in 1× MABT for 30 minutes per wash at room temperature.

    Note 1: 5× MAB (per liter) – 58 g maleic acid, 41 g NaCl, adjust to pH 7.5 with approximately 25–30 mL 10 N NaOH).

    Note 2: 1× MABT = 1× MAB with 0.01% Tween 20.
19. Detection procedures are specific to the type of label carried by the probe.

Protocol: RNA in situ hybridization for zebrafish embryos

The following protocol is a modification of the procedures of Thisse and Thisse (2008) and Petko et al. (2009). A special word of thanks to Dr. J. Petko at Penn State York for her input and suggestions pertaining to this protocol. Although the procedure describes the use of antisense DIG-labeled RNA probes, this is just one option. Biotinylation is a suitable alternative, as are DNA probes, including oligonucleotides.

1. Cultivate fish embryos until the desired developmental stage is reached.
2. Fix embryos overnight at 4°C in a freshly prepared solution of 4% paraformaldehyde in 1× PBS. Fixation is easily performed in standard 100 mm or 150 mm Petri dishes.

   Note: The paraformaldehyde solution should be prepared no more than 24 hours in advance. Refer to Appendix K, Useful stock solutions for the molecular biologist, for the formulation of PBS.
3. Carefully remove paraformaldehyde solution and discard it according to departmental guidelines for hazardous waste.
4. If the specimens exhibit residual pigmentation, it should be bleached away for 30 minutes in a 3% H_2O_2 solution in 0.5% KOH at room temperature. Otherwise, proceed to the next step.
5. Rinse fixed embryos once with 1× PBS.
6. Remove PBS as completely as possible and then rinse embryos in 100% methanol (MeOH).
7. Remove MeOH rinse and replace it with a fresh aliquot of 100% MeOH. Incubate specimens for a minimum of 2 hours at −20°C. This is a dehydration step.

   Note: As needed, the embryos may be stored in MeOH at −20°C for several months.
8. Permeabilize and process embryos for prehybridization by moving them through the following buffer series. All incubations are performed at room temperature in convenient sterile 24-well cell culture plates. It is important to transfer as little of a previous solution into the next solution during these steps. Consider using a cassette or a "basket" device

for draining small specimens as described in the "Sample Preparation" section, above.

a. 75% MeOH/25% PBS for 5 minutes.
b. 50% MEOH/50% PBS for 5 minutes.
c. 25% MeOH/75% PBS for 5 minutes.
d. 1 × PBS containing 0.1% Tween 20 for 5 minutes.
e. Repeat step 8d, above, with a fresh aliquot of PBS + 0.1% Tween 20;
f. Proteinase K (10 μg/mL in PBS + 0.1% Tween 20).

   Note: The incubation time in proteinase K typically ranges from 3 to 15 minutes and will be a function of the development stage of the embryo. A good guideline is to incubate for 5 minutes for every 24 hours post fertilization (hpf).

g. 4% paraformaldehyde prepared in 1 × PBS for 20 minutes.

   Note: The paraformaldehyde solution should be prepared no more than one day in advance and stored at 4°C.

h. Perform four washes in 1 × PBS + 0.1% Tween. Perform each wash for 5 minutes.

9. Carefully transfer the embryos to a sterile 1.5 mL microfuge tube. Multiple embryos may be placed in a single tube.
10. Remove any residual PBS solution as completely as possible.
11. At 700 μL hybridization solution at incubate at 70°C for 2–6 hours.

    Note 1: Hybridization solution = 50% deionized formamide; 5 × SSC, 0.1% Tween 20; 50 μg/mL heparin; 500 μg/mL carrier tRNA; adjust to pH 6.0 with 1 M citric acid solution (up to 460 μL per 50 mL of hybridization solution).

    Note 2: Upon completion of this prehybridization step, the embryos may be stored tightly sealed for weeks in the hybridization solution at −20°C.

12. Remove the hybridization solution as completely as possible and replace it with a fresh 200 μL aliquot of hybridization solution to which 50 ng antisense DIG-labeled RNA probe has been added. Allow the hybridization to proceed 12–16 hours (overnight) at 70°C.
13. At the conclusion of the hybridization period, transfer the embryos and hybridization solution to sterile 24-well cell culture plates.
14. Perform all of the following stringency washes at 70°C. Important: the first four washes (a–d) use a modified hybridization solution (*50% deionized formamide; 5 × SSC, 0.1% Tween 20; adjust to pH 6.0 with 1 M citric acid solution with up to 460 μL per 50 mL of hybridization solution)* which contains neither heparin nor carrier tRNA. These washes will remove any nonhybridized or imperfectly hybridized probe.

    a. 100% hybridization solution for 10 minutes.
    b. 75% hybridization solution in 2 × SSC for 10 minutes.
    c. 50% hybridization solution in 2 × SSC for 10 minutes.

d. 25% hybridization solution in 2 × SSC for 10 minutes.
e. 2 × SSC for 10 minutes.
f. 0.2 × SSC for 30 minutes.
g. 0.2 × SSC for 30 minutes.

15. Transfer 24-well plates to an orbital shaker or platform. Perform all of the following incubations at room temperature at 40–50 rpm.
a. 75% 0.2 × SSC in PBS containing 0.1% Tween 20 for 10 minutes.
b. 50% 0.2 × SSC in PBS containing 0.1% Tween 20 for 10 minutes.
c. 25% 0.2 × SSC in PBS containing 0.1% Tween 20 for 10 minutes.
d. PBS containing 0.1% Tween 20 for 10 minutes. While this last wash is in progress, begin to prepare the blocking buffer for the detection protocol.

16. Add freshly prepared blocking buffer (1 × PBS containing 0.1% Tween 20; 2% sheep serum; 2 mg/mL bovine serum albumin (BSA, Fraction V)) to the wells for a fresh 24-well cell culture plate. Immediately add embryos. Incubate for 1 hour at room temperature on an orbital platform at 40–50 rpm.

17. Remove the blocking buffer and replace it with a fresh aliquot of the same buffer to which the anti-DIG antibody (1/10,000 dilution) has been added just prior to this step. Incubate for 12–16 hours (overnight) at 4°C on an orbital platform at 40–50 rpm.

18. At the conclusion of the incubation period, remove the antibody solution and perform the following washes at room temperature on an orbital shaker platform.
a. PBS containing 0.1% Tween 20 for 10 minutes. Then, repeat this step four more times.
b. AP buffer for 5 minutes. Then, repeat this step two more times.
Note: AP buffer = 100 mM Tris · Cl, pH 9.5; 50 mM $MgCl_2$; 100 mM NaCl; 0.1% Tween-20.

19. Transfer the embryos to a sterile 24-well cell culture dish. Remove any residual AP buffer.

20. Add 700 μL BM purple or a mixture of BCIP/NBT.
Note: BM purple (Roche 11442074001) is one example of an alternative chromogenic substrate to the traditional pairing of BCIP and NBT. It is preferred by some investigators because of lower background levels and because it does not require any premixing.

21. Maintain plates completely in the dark while the color developing process proceeds. Monitor the progress of the color development every 10 minutes.

22. Stop the development process by transferring the embryos to a sterile 1.5 mL microcentrifuge tube and washing them twice for 5 minutes each with 1.0 mL PBS.

23. Fix embryos again by incubating in 4% paraformaldehyde solution (prepared in 1 × PBS) for 20 minutes.

24. Wash embryos three times for 5 minutes each in 1.0 mL PBS.
25. Place embryos on a depression slide in a drop of methyl cellulose for microscopic examination. For more permanent storage, the specimen can be fixed to the slide in ClearMount.

Protocol: RNA in situ hybridization whole mount for arachnid embryos

The following protocol is a modification of the procedures of Tautz and Pfeifle (1989), Damen and Tautz (1998), Klinger and Gergen (1993), and as described in Gruzin et al. (2020). A special word of thanks to Dr. J. Petko at Penn State York for her input and suggestions pertaining to this protocol. A typical outcome of this procedure is shown in Fig. 20.6.

1. Incubate spider or other insect eggs to the desired developmental stage according to standard protocols.
2. Transfer eggs to a clean glass scintillation vial or similar vessel.
3. Fix specimens in 1 mL 2 × PEMS (200 mM PIPES; 2 mM EDTA; 2 mM $MgSO_4$; pH 6.9), 1 mL 37% formaldehyde, and 3 mL heptane.
4. Incubate samples for 4 hours at room temperature on an orbital platform (40–50 rpm) or other devices at will keep samples in a gentle motion.
5. Carefully remove the fixative solution and discard according to departmental guidelines.
6. Wash eggs five times with PEMS-t (2 × PEMS; 0.1% Tween-20) and then transfer to a 2.0 mL microfuge tube.

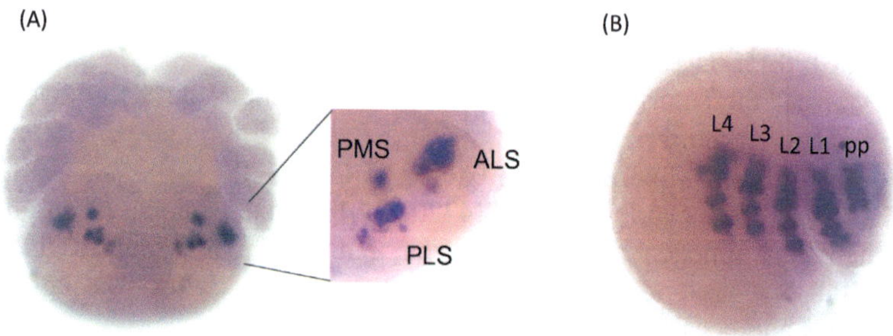

FIGURE 20.6 RNA ISH expression analysis of Doublesex (Dsx) transcription factor genes in the common house spider. (A) Ventral view showing mRNA expression pattern of Dsx1 in embryonic growth stage 13 in the posterior medial spinneret (PMS), anterior lateral spinneret (ALS), and in the posterior lateral spinneret (PLS). (B) Lateral view showing mRNA expression pattern of Dsx2 in embryonic growth stage 13 in legs 1–4 (L1-L4) and the pedipalps (pp) (**Gruzin et al., 2020**). *Image courtesy Gruzin, M., M. Mekheal, K. Ruhlman, M. Winkowski, and J. Petko. (2020). Developmental expression of doublesex-related transcripts in the common house spider, Parasteatoda tepidariorum. Gene Expr. Patterns 35, 119101. https://doi.org/10.1016/j.gep.2020.119101; and reproduced with permission from Elsevier.*

7. Dehydrate the eggs by incubating them at room temperature through the following graded methanol (MeOH) series. It is important to transfer as little of a previous solution into the next solution during these steps. Consider using a cassette or a "basket" device for draining small specimens as described in the "Sample Preparation" section, above.
   a. 25% MeOH/PEMS-t for 5 minutes.
   b. 50% MeOH/PEMS-t for 5 minutes.
   c. 75% MeOH/PEMS-t for 5 minutes.
   d. 100% MeOH for 5 minutes.
   e. 100% MeOH (fresh aliquot) for 1 hour. At the conclusion of this incubation, store this tube at −20°C for a minimum of 12–16 hours (overnight).
8. Remove samples from −20°C storage and allow to warm to room temperature. In preparation for proteinase permeabilization, incubate the eggs at room temperature and then rehydrate as follows.
   a. 50% MeOH/PBS-T (1 × PBS containing 0.02% Tween 20) for 5 minutes.
   b. 25% MeOH/PBS-T for 5 minutes.
   c. 10% MeOH/PBS-T for 5 minutes.
   d. PBS-T for 5 minutes.
   e. PBS-T (fresh aliquot) for 5 minutes

   Note 1: Depending on the model species, it may be necessary remove any remaining membranes from the eggs (e.g., devittelinization) immediately prior to this step.

   Note 2: Refer to Appendix K, Useful stock solutions for the molecular biologist, for PBS formulation.
9. Fix embryos in 5% formaldehyde in PBS-T for 20 minutes at room temperature.
10. Rinse embryos with PBS-T for 5 minutes. Then, repeat this step twice.
11. Incubate embryos in 15 mg/mL proteinase K at room temperature for 4 minutes.

    Note: This permeabilization step may require optimization due to possible differences in the developmental stage of the specimens.
12. Rinse embryos with PBS-T.
13. Wash embryos in PBS-T for 5 minutes. Then, repeat this step.
14. Fix embryos in 5% formaldehyde in PBS-T for 20 minutes at room temperature.
15. Wash embryos in PBS-T for 5 minutes. Then, repeat this step twice.
16. Incubate the embryos in a 1:1 mixture of PBS-T:HYB-B for 5 minutes at room temperature.

    Note: HYB-B = 50% freshly deionized formamide; 5 × SSC; 0.02% Tween 20; adjust to pH 6.5 with 1 N HCl.
17. Incubate the embryos in 500 μL HYB-B solution at 65°C for 5 minutes.
18. Incubate the embryos in 500 μL of HYB-A at 65°C for 1 hour.

Note 1: HYB-A = 50% freshly deionized formamide; 5 × SSC; 0.2 mg/mL denatured carrier DNA; 0.1 mg/mL tRNA or carrier RNA; 50 μg/mL heparin; 0.02% Tween 20; adjust to pH 6.5 with 1 N HCl.

Note 2: Prehybridization and hybridization temperatures may need to be optimized depending primarily on the length and G + C content of the probe. See Chapter 17, Nucleic Acid Probe Technology, for details.

19. Remove 455 μL of the HYB-A solution from the tube.

20. Add 5 μL of the RNA probe and then incubate at 65°C for 12–16 hours (overnight). The final volume of probe plus residual HYB-A solution from the previous step should be 50 μL.

21. At the conclusion of the hybridization period, add 1.0 mL of HYB-B solution that has been prewarmed to 70°C. Quickly mix the two solutions in the tube and maintain tubes at 65°C throughout.

Note: It is important the temperature be maintained at 65°C to ensure that the probe does not interact with nontarget sequences due to the reduction in stringency that accompanies falling temperatures.

22. Carefully remove as much of the solution from the tube as possible. Then, add a fresh, prewarmed 65°C aliquot of HYB-B to the embryos and incubate at 65°C for 10 minutes.

23. Carefully remove as much of the solution from the tube as possible. Then, add a fresh, prewarmed 65°C aliquot of HYB-B to the embryos and incubate at 65°C for 30 minutes.

24. Carefully remove as much of the solution from the tube as possible. Then, add 500 μL of a 1:1 mixture of HYB-B:PBS-T and incubate at room temperature for 10 minutes.

25. Wash the embryos twice with PBS-T for 1 minutes per wash.

26. Wash the embryos twice with PBS-T for 15 minutes per wash.

27. Begin the antibody detection by incubating the embryos in PBS-T containing 1% casein or a commercially available blocking reagent (e.g., *Roche 1096176*) for 30 minutes at room temperature.

28. Remove the blocking solution from the previous step and immediately replace it with anti-DIG antibody that has been diluted 1:2000 in PBS-T. Incubate for 4 hours at room temperature.

29. At the conclusion of the incubation period, remove the antibody solution and was the embryos twice in PBS-T for 10 minutes per wash.

30. Wash the embryos twice in PBS-T for 30 minutes per wash.

Note: If elevated background has been a problem in the past, consider perform one or two additional 30-min washes before moving forward.

31. Wash to embryos in PBS-T at 4°C for 12–16 hours (overnight).

32. Initiate the staining protocol by washing the embryos three times in staining buffer for 5 minutes per wash at room temperature.

Note: Staining buffer = 100 mM Tris · Cl, pH 9.5; 50 mM $MgCl_2$; 100 mM NaCl; 0.02% Tween 20). In some cases, $MgCl_2$ has been known to contribute to the spotty background, particularly when a membrane is involved (e.g., Northern analysis). If the background persists in this application, try reducing the $MgCl_2$ concentration.

33. Remove the staining buffer from the previous step and immediately replace it with as fresh aliquot of staining buffer to which 20 μL BCIP/NBT has been added per mL.

    Note 1: BCIP and NBT are widely available (e.g., Roche 11681451001) for this type of application.

    Note 2: Handle BCIP and NBT according to the manufacturer's safety recommendations.

34. Allow the staining to proceed in the dark at room temperature for as many as several hours, but monitor staining progress every 15 minutes. If performing the staining at the end of the day, place the samples in the refrigerator overnight.
35. Arrest the staining process by washing the embryos several times in PBS-T.
36. Store embryos in sterile lab-grade water at 4°C until ready for microscopic examination.

Spatial transcriptomics

The explosive growth of innovations pertaining to the detection of transcripts coupled with RNA-seq improvements led to the development of spatial transcriptomics, a method by which single, multiple, or global gene expression is quantified in a tissue section (Ståhl et al., 2016). The importance of this rapidly evolving tool is evident inasmuch as it was designated as "Method of the Year" for 2020 by the journal *Nature Methods* (Marx, 2021). Briefly, tissue sections are prepared and mounted on glass slides to which highly specialized capture probes have been attached, an example of which is shown in Fig. 20.7. The 5′ end of each probe is oligo(dT)-tailed, followed by a barcode that denotes the precise position of each spot on the slide, followed by a specialized sequence that will support amplification, and finally a cleavage sequence at the end of the probe close to the point where it is immobilized on the slide. Once the tissue section is fixed onto the slide, it is important to image the preparation so that gene expression data at the end of the experiment can be histologically aligned with the original sample. The cells are permeabilized in such a way that the mRNA from each cell is accessible to the probes in the spot directly beneath it. cDNA is then synthesized directly on the slides, which creates the sequencing library. The library is cleaved from the slide and evaluated by massively parallel sequencing and data analysis.

In situ hybridization tips for success

As with all molecular techniques, attentiveness to the subtleties of the hands-on components can make the difference between spectacular success and an

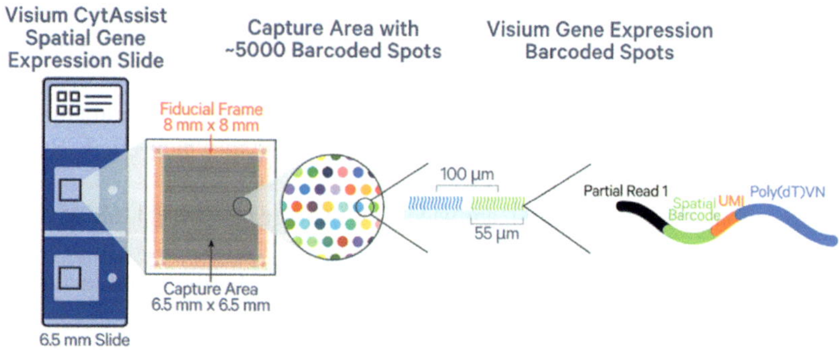

FIGURE 20.7 Method of mRNA capture from thin-sectioned tissue for precision mapping of gene expression. Polyadenylated transcripts are affinity-captured onto glass slides directly from fixed and permeabilized tissue samples. During cDNA synthesis a spatial barcode is added for localizing each transcript that is subsequently identified by RNA-seq. *Image © 10× Genomics, Inc. (http://www.spatialtranscriptomics.com) and reproduced here with permission.*

abysmal failure. When suboptimal results occur, it may be helpful to consider the following:

1. Observe scrupulous RNase-free methods not only during the preparation of the sample and all related implements and chemicals but during all procedural steps as well.
2. Skillful tissue preparation and processing are essential to success. Be attentive to recommendations for specimen processing because the protocol for processing one type of tissue may be wholly inadequate for a different tissue type. For example, fresh-frozen tissue at the wrong cutting temperature will severely impact the quality of the thin section and therefore the quality of in situ study.
3. Pay attention to the thickness of the tissue sections which, ideally will be in the 8–10 μm range.
4. Consider the nature of the probe. Because dsRNA (RNA probe, mRNA target) is more stable than a DNA:RNA hybrid (DNA probe, mRNA target) more stringent hybridization and washing conditions can be used. This has the effect of minimizing random cross-hybridization. By decreasing the background, the signal-to-noise ratio can be greatly improved. For an added level of probe stability consider increasing the T_m of the probe by the inclusion of one or more locked nucleic acid (LNA) nucleotides. Although this will add to the cost of the probe, it may very well solve the problem.
5. Consider altering the stringency of the post-hybridization washes. This includes (a) increasing the stringency when the background is too high and/or when using tightly hybridizing RNA probes, or (b) decreasing the stringency if there is reason to believe that the probe may have been destabilized in the washing steps; this is a greater concern when using

DNA probes. The most common methods of controlling stringency include raising or lowering the temperature, or lowering the T_m of the probe:target duplex by adding formamide (expect about 0.75°C T_m reduction for each 1% formamide).

6. Be certain to use freshly deionized formamide for all hybridization applications. This process is described in Appendix G, Deionization of formamide, formaldehyde, and glyoxal.
7. Check the pH of all solutions that are prepared in-house.
8. Labeling and detection options offer both opportunities and challenges. Both fluorescent labeling and immunochemical staining are excellent choices and, in the case of the latter, no specialized equipment is needed (e.g., laser or fluorescence detector/imager). Generally speaking, the use of biotinylated probes is usually not the best choice because, being a naturally occurring vitamin, its presence in a tissue sample may compromise assay sensitivity by contributing to background. Digoxigenin (DIG), in contrast, is a well-regarded hapten for in situ hybridization applications. When included in an in vitro transcription reaction, DIG-labeled nucleotides (e.g., DIG-11-UTP) will produce DIG-labeled probes of uniform length. Probes labeled in this manner are later detected through the use of anti-DIG antibodies. If radiolabeling is to be used, ^{3}H or ^{35}S are the preferred isotopes; the high-energy emission of ^{32}P will cause a significant loss of resolution at the time specimen is examined.
9. In cases where the background remains elevated after optimizing the other procedural steps, consider performing controlled nuclease digestion to remove any residual probe carryover. Performed with an RNase A + RNase T1 cocktail for RNA probes or using S1 nuclease for either DNA or RNA probes, this is the same principle that is used in the respective nuclease protection assays described in Chapter 17, Nucleic Acid Probe Technology, in which single-stranded nucleic acids, and not double-stranded probe:target molecules, are susceptible to *endo*- and exonucleolytic cleavage.

References

Bauman, J.G.J., Wiegant, J., Borst, P., Van Duijn, P., 1980. A new method for fluorescence microscopical localization of specific DNA sequences by in situ hybridization of fluorochrome-labelled RNA. Exp. Cell Res. 128 (2), 485–490.

M.Buongiorno-Nardelli, M., Amaldi, F., 1970. Autoradiographic detection of molecular hybrids between rRNA and DNA in tissue sections. Nature 225, 946–948. Available from: https://doi.org/10.1038/225946a0.

Damen, W.G., Tautz, D., 1998. A Hox class 3 orthologue from the spider *Cupiennius salei* is expressed in a Hox-gene-like fashion. Dev. Genes. Evol. 208 (10), 586–590. Available from: https://doi.org/10.1007/s004270050218.

Gall, J.G., Pardue, M.L., 1969. Formation and detection of RNA-DNA hybrid molecules in cytological preparations. Proc. Natl. Acad. Sci. 63 (2), 378–383.

Gruzin, M., Mekheal, M., Ruhlman, K., Winkowski, M., Petko, J., 2020. Developmental expression of doublesex-related transcripts in the common house spider, *Parasteatoda tepidariorum*. Gene Expr. Patterns 35, 119101. Available from: https://doi.org/10.1016/j.gep.2020.119101.

Klinger, M., Gergen, J.P., 1993. Regulation of runt transcription by *Drosophila* segmentation genes. Mechanisms Dev. 43 (1), 3–19. Available from: https://doi.org/10.1016/0925-4773(93)90019-T.

Langer-Safer, P.R., Levine, M., Ward, D.C., 1982. Immunological method for mapping genes on *Drosophila* polytene chromosomesD. C Proc. Natl. Acad. Sci. 79 (14), 4381–4385.

Marras, S.A.E., Bushkin, Y., Tyagi, S., 2019. High-fidelity amplified FISH for the detection and allelic discrimination of single mRNA molecules. Proc. Natl. Acad. Sci. USA 116 (28), 13921–13926. Available from: https://doi.org/10.1073/pnas.181446311.

Marx, V., 2021. Method of the year: spatially resolved transcriptomics. Nat. Meth. 18, 9–14. Available from: https://doi.org/10.1038/s41592-020-01033-y.

Mistry, R.M., Singh, P.K., Mancini, M.G., Stossi, F., Mancini, M.A., 2020. Single cell analysis of transcriptionally active alleles by single molecule FISH. J. Vis. Exp. 163. Available from: https://doi.org/10.3791/61680.

Petko, J.A., Kabbani, N., Frey, C., Woll, M., Hickey, K., Craig, M., et al., 2009. Proteomic and functional analysis of NCS-1 binding proteins reveals novel signaling pathways required for inner ear development in zebrafish. BMC Neurosci. 10, 27. Available from: https://doi.org/10.1186/1471-2202-10-27.

Player, A.N., Shen, L.P., Kenny, D., Antao, V.P., Kolberg, J.A., 2001. Single-copy gene detection using branched DNA (bDNA) in situ hybridization. J. Histochem. Cytochem. 49 (5), 603–612. Available from: https://doi.org/10.1177/002215540104900507.

Raap, A.K., Van de Corput, M.P.C., Vervenne, R.A.M., Van Gijlswijk, R.P.M., Tanke, H.J., Wiegant, J., 1995. Ultra-sensitive FISH using peroxidase-mediated deposition of biotin-or fluorochrome tyramides. Hum. Mol. Genet. 4 (4), 529–534. Available from: https://doi.org/10.1093/hmg/4.4.529.

Sannier, G., Dubé, M., Kaufmann, D.E., 2020. Single-cell technologies applied to HIV-1 research: reaching maturity. Front. Microbiol. 11, 297. Available from: https://doi.org/10.3389/fmicb.2020.00297.

Singer, R.H., Ward, D.C., 1982. Actin gene expression visualized in chicken muscle tissue culture by using in situ hybridization with a biotinated nucleotide analog. Proc. Natl. Acad. Sci. U. S. A. 79 (23), 7331–7335. Available from: https://doi.org/10.1073/pnas.79.23.7331.

Ståhl, P.L., Salmén, F., Vickovic, S., Lundmark, A., Navarro, J.F., Magnusson, J., et al., 2016. Visualization and analysis of gene expression in tissue sections by spatial transcriptomics. Science 353, 78–82. Available from: https://doi.org/10.1126/science.aaf2403.

Tautz, D., Pfeifle, C., 1989. A non-radioactive in situ hybridization method for the localization of specific RNAs in *Drosophila* embryos reveals translational control of the segmentation gene hunchback. Chromosoma 98, 81–85.

Thisse, C., Thisse, B., 2008. High-resolution in situ hybridization to whole-mount zebrafish embryos. Nat. Protoc. 3 (1), 59–69. Available from: https://doi.org/10.1038/nprot.2007.514.

Young, A.P., Jackson, D.J., Wyeth, R.C., 2020. A technical review and guide to RNA fluorescence in situ hybridization. PeerJ 19 (8), e8806. Available from: https://doi.org/10.7717/peerj.8806.

Further reading

Bruno, L., Ronchini, M., Gagliardi, O., Corinti, T., Chiappetta, A., Gerola, P., et al., 2015. Analysis of AtGUS1 and AtGUS2 in *Arabidopsis* root apex by a highly sensitive TSA-

MISH method. Int. J. Development. Biol. 59 (4–5–6), 221–228. https://doi.org/10.1387/ijdb.140195LB.

Chen, K.H., Boettiger, A.N., Moffitt, J.R., Wang, S., Zhuang, X., 2015. Spatially resolved, highly multiplexed RNA profiling in single cells. Science 348 (6233), aaa6090. Available from: https://doi.org/10.1126/science.aaa6090Eng2019.

Eng, C.-H.L., M. Lawson, Q. Zhu, R. Dries, N. Koulena, Y. Takei, et al. Transcriptome-scale super-resolved imaging in tissues by RNA seqFISH. Nature 568 (7751), 235–239. https://doi.org/10.1038/s41586-019-1049-y.

Hauptmann, G. (Ed.), 2015. In situ Hybridization Methods. Human Press, New York, NY.

Liehr, T. (Ed.), 2017. Fluorescence In Situ Hybridization (FISH): Application Guide. 2nd edition Springer Publishing Company, New York, NY.

Nielsen, B.S. (Ed.), 2014. In situ Hybridization Protocols. 4th edition Humana Press, New York, NY.

Xia, C., Babcock, H.P., Moffitt, J.R., Zhuang, X., 2019. Multiplexed detection of RNA using MERFISH and branched DNA amplification. Sci. Rep. 9, 7721. Available from: https://doi.org/10.1038/s41598-019-43943-8.

Chapter 21

Array analysis of gene expression

Rationale

The venerable gene expression tools at hand in the early days of the biotechnology revolution are low-throughput in that they are limited to the study of one or a few genes at a time. Microarrays first appeared in the early 1990s (Fodor et al., 1991, 1993) and came about as a result of improvements both in robotic liquid handling for precision spotting and subsequent advances in fluorescence labeling and detection technology. They were first used predominantly for the assay of mRNA expression from various biological sources. The astonishing density of unique sequences that can be placed on a microarray qualifies it as a device for high-throughput and highly inclusive analysis of gene expression. This aligned well with the nearly ubiquitous interest in global or whole-transcriptome analysis. Later microarray applications included the measurement of gene copy number, profiling the single nucleotide polymorphisms (SNPs) that characterize a sample, comparing the genome structure and composition of two different organisms or cell types, detection of alternative splicing at exon-intron junctions, and examining DNA methylation patterns. Microarray technology then expanded to include protein microarrays (reviewed by Huan et al., 2019), commonly known as protein biochips, proteome microarrays, and antibody (Ab) microarrays. Much like their nucleic acid microarray counterpart, the miniaturization of protein microarrays affords unparalleled high-throughput sensitivity with low sample consumption. Information about protein arrays is presented at the end of this chapter.

Microarray expression data are archived in any of a number of online repositories, a notable example of which is the Gene Expression Omnibus (GEO; http://www.ncbi.nlm.nih.gov/geo). In the mind of some, the heyday of the microarray has come and has gone because of the decreasing cost of outsourcing whole transcriptome sequencing. This may be true of some of the original types of microarrays that were manufactured early, though technological improvements and platform innovations are responsible for the resurgence of interest that some newer microarray platforms now enjoy.

RNA Methodologies. DOI: https://doi.org/10.1016/B978-0-323-90221-2.00036-9

What is a microarray?

A microarray is a device in which single-stranded DNA sequences are immobilized on a solid support and which are able to participate in hybridization with complementary sequences. The principle is a readily recognized theme at the very heart of molecular biology. The surface of a microarray consists of a collection of many thousands of different gene segments which are nicely organized into rows and columns and which have been tethered to a glass slide, a silicon wafer, a plastic substrate (Fig. 21.1), or to silica microbeads. It is the reproducible pattern of sequences that defines any particular array and supports parallel hybridization. With microarrays, note that the traditional use of the terms "probe" and "target" is reversed, that is, the probes are the sequences fixed in place on the surface of the array.

Although microarrays are considered high-throughput tools because they allow the measurement of many genes simultaneously, they are not especially quantitative. Candidate genes are generally identified by high-throughput microarray analysis, after which the behavior of these genes may be more fully characterized, at least at the transcriptional level, by real-time

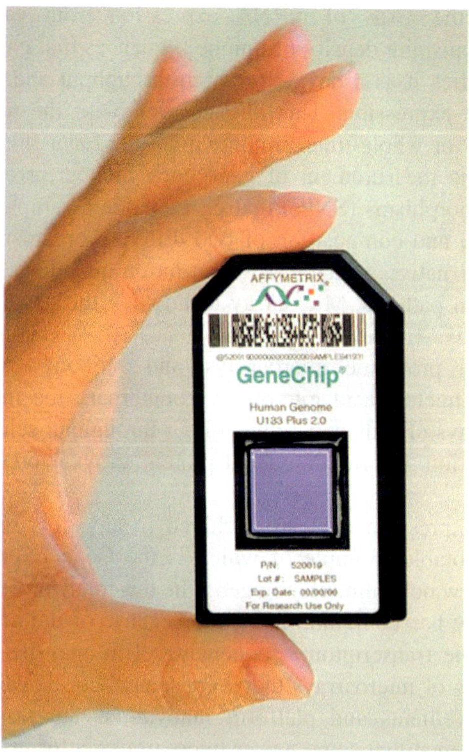

FIGURE 21.1 The GeneChip Human Genome U133 Plus 2.0 Array. *Courtesy of Affymetrix.*

PCR or digital PCR. By comparing microarray and quantitative PCR data, one is able to paint a more complete picture of the behavior of a gene.

The basic principle for assaying gene expression with a microarray involves working with cDNA from two sources. Most often, the cDNA molecules are labeled with the popular dyes Cy3 and Cy5, which fluoresce green and red, respectively, upon laser scanning during the detection process. For example, control cDNA would be labeled with Cy3 and cDNA from a treated or test sample would be labeled with Cy5. These labeled target sequences are mixed together in a small volume of hybridization buffer (200 μL) and co-hybridized overnight to the probes printed on the microarray (Fig. 21.2).

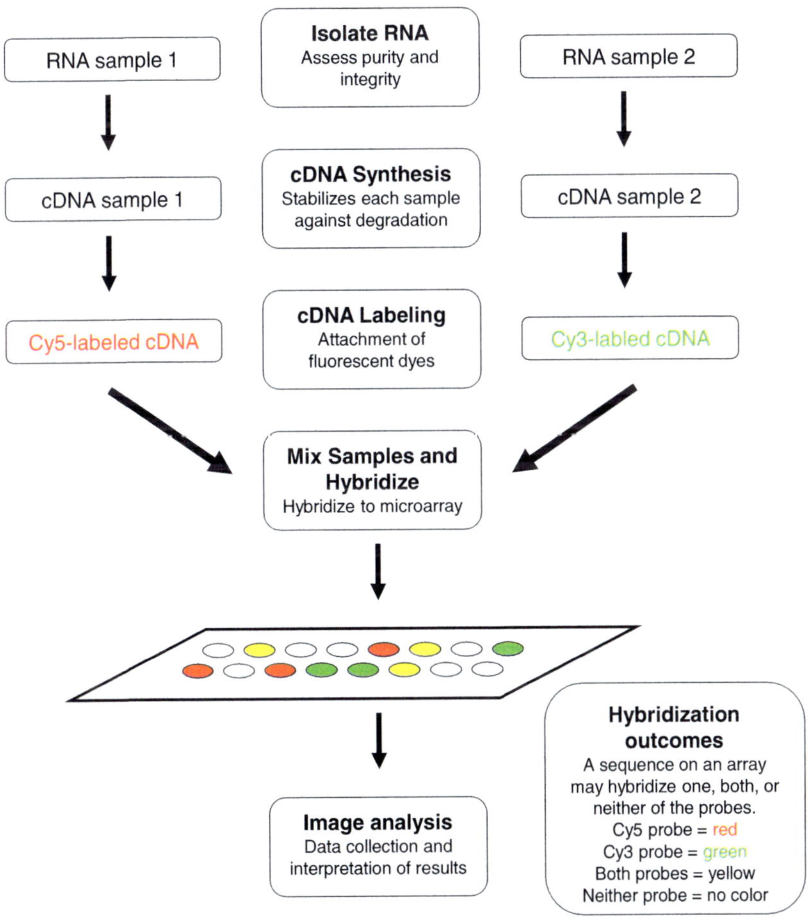

FIGURE 21.2 Major steps for microarray analysis. cDNAs are synthesized, labeled and hybridized to an array. Fluorescence detection or autoradiography coupled with image analysis provides a great deal of information about patterns of gene expression in the samples under investigation.

If Cy3-labeled cDNA and Cy5-labeled cDNA hybridize to the same spot on a microarray, the laser-induced fluorescence of both Cy3 and Cy5 makes the spot appear yellow. The quantity of the fluorophore-labeled cDNA populations retained by hybridization and a fluorescence ratio-based analysis of each spot on the microarray provides information about the source and (within limits) the abundance of particular transcripts and, by extension, whether a gene is up- or downregulated in the context of the experiments being performed.

When two different cDNA populations are individually labeled with Cy5 and Cy3 and then co-hybridized to the target in the manner described above, the method is known as a dual-channel array. This is currently the more widely used protocol and has the advantage of requiring only one array to perform an experiment in which two samples are being compared simultaneously. In contrast, a one-channel microarray screening involves labeling the sample with a single fluorophore and then hybridizing that one sample into a microarray. The investigator typically compares relative fluorescence intensities among the various probe locations on the array, some of which are designed as controls, reference genes, or genes known to exhibit baseline expression. The single-channel approach is a common method in a diagnostic setting in which, for example, an SNP profile for an individual is generated. These data can also be used to determine the best drug regimen when a patient presents with a particular illness or disease.

Historically there are three microarray formats that have been used for nucleic acid-type investigations – the short oligonucleotides (25-mer) microarrays (by far the most common), cDNA or gDNA microarrays, and chemically-synthesized longer oligonucleotide (60- to 70-mer) arrays. Note that this does not include microfluidics lab-on-a-chip products, biosensors, or similar clinical devices. Each spot or well on the array typically contains picomole quantities of a unique DNA sequence. The longer oligonucleotide platform is viewed by some as the preferred format because it offers greater specificity than 25-mer oligonucleotides because of greater tolerance for mismatches and more sequence T_m-consistency from spot to spot. Beyond 70 bases, there does not appear to be any appreciable gain in sensitivity that would justify the added cost of synthesizing longer oligonucleotides, and the likelihood of incorrect bases increases as the length of the oligonucleotide increases (Villaseñor-Park and Ortega-Loayza, 2013; Barrett and Kawasaki, 2003). In contrast, the use of shorter oligonucleotide probes is more economical and the sequences can be applied at a much higher density, in this format, each gene is represented in its entirety, albeit in sections and in multiple places on the array. Rarely used cDNA arrays often consist of highly characterized segments of PCR products.

A significant advance in microarray screening was the development of bead-based microarrays (e.g., BeadChips; http://www.illumina.com). Rather than tethering the probes to planar glass or plastic substrates, these devices consist of microwells etched onto the array surface. Each microwell houses a single silica microbead (3μm) that is coated with many copies of a gene locus-specific oligonucleotide that acts as a probe. When present in the assay, labeled complementary RNA (cRNA) sequences representing the experimental sample base-pair to the sequences on the microbead. Following standard washing and staining steps, the data provide a gene expression profile related to the sample.

For SNP analysis, a unique feature of the BeadChip platform is the design of the oligonucleotide probes attached to the beads, which hybridize to the sample genomic DNA one base short of the locus of interest (Fig. 21.3). Allelic identity is determined by a single base extension step using chain-terminating dideoxynucleotides (ddNTPs). The ddNTP bases carry either a biotin label (C and G) or a dinitrophenyl (DNP) label (A and T). After base extension, the identity of the incorporated base is determined using green-fluorescent streptavidin (biotinylated base detection) or red-fluorescent anti-DNP antibodies. The use of secondary fluorescent-labeled antibodies is useful for boosting the signal by creating a molecular

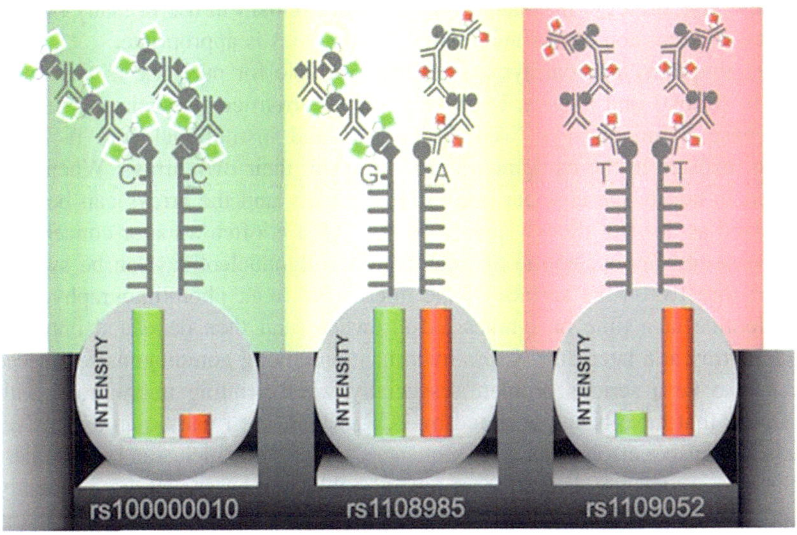

FIGURE 21.3 Contemporary Infinium BeadChip array. Each bead is coated with many copies of a unique oligonucleotide that hybridizes its complementary genomic sequence, stopping one nucleotide short of the SNP position being interrogated (not shown). The oligonucleotide is extended one base (only) and is complementary to the next base on the hybridized genomic sequence. As shown, the oligonucleotides attached to the beads in the three wells were extended with cytosine only (homozygote), guanine or adenine (heterozygote), or thymine only (homozygote). The newly added bases have different detection strategies (see text for details) which produce a characteristic fluorescence signal for each allele (from left to right: green, yellow, or red). *Figure courtesy of Illumina (www.illumina.com) and reproduced with permission.*

sandwich-type structure through successive rounds of antibody layering. Laser excitation of the fluorescence label permits easy detection of signal intensity from each well and, in the case of heterogeneity, through fluorescence ratio measurements. Inasmuch as these arrays consist of multiethnic genome-wide content, they are useful for studies examining population genetics, pharmacogenomics, profiling epigenetic changes such as methylation, and basic research and development applications.

More efficient and comprehensive transcriptome profiling methods, including RNA-seq, have emerged as practical alternative platforms. To that end, the cost of whole transcriptome sequencing continues to decline and is easily outsourced, which is particularly attractive to laboratories with limited funding or that have already invested in microarray technology. RNA-seq offers many advantages, particularly in the realm of uncharacterized transcripts or when working with an organism for the first time because microarrays are limited to the detection of predefined genes or transcripts in a hybridization-dependent manner. However, the microarray approach to the analysis of gene expression remains useful because the relative ease and immediacy with the resulting data can be managed. In some laboratories, RNA-seq data are used to validate microarray data, and vice versa, which is a very wise strategy. Inasmuch as microarrays remain in use in many laboratories, a discussion of this and related technologies is appropriate.

High-quality microarrays are readily available for purchase in any number of formats and themes. Because of the recurrent cost associated with the purchase of microarrays, well-funded labs and institutional core facilities have invested in the equipment needed to print their own arrays. When prepared in-house, the cost per array drops quickly and the arrays can be customized as needed, though quality control (QC) is often a major concern. In the case of oligonucleotide microarrays, the oligonucleotides can be synthesized directly on the surface of the microarray (e.g., photolithography) one nucleotide at a time or synthesized elsewhere and then deposited onto the microarray at a later time. Cross-hybridization among homologous sequences can also be a serious problem, especially when printing microarrays with cDNAs that have been generated by PCR, which itself can be an expensive, labor-intensive effort. The gene sequences selected for inclusion on commercial microarrays are derived from several public databases that are frequently updated such as NCBI Genome Workbench, RefSeq, Ensembl, and others. The arrays most often include well-annotated and characterized genes; this precludes the need to do any cloning or sequencing precisely because the identity of the gene is already known. Microarrays may also contain uncharacterized DNA sequences of obscure function and, if prepared in-house, custom sequences.

Commercial microarrays are designed to be used only once, and attempts to strip off previously hybridized sequences have produced lackluster robustness. In contrast, nylon macroarrays (described below) can be stripped and reused. Succinctly, microarrays represent the ability to gather qualitative and semiquantitative information about cellular biochemistry on a mass scale owing to genome project sequencing data and the newer bead-based array devices have rapidly become mainstream research tools.

Microarrays are commonly prepared with sequences representing either a unique tissue source or as multiple-tissue devices. This latter approach is analogous to the multiple-tissue Northern blots that were popular in the early years of molecular biology: RNA from different tissues or organisms was blotted onto a single membrane for direct side-by-side comparison. In a way, multiple-tissue microarrays are also a high-tech, high-throughput extension of in situ hybridization in which gene expression is assigned to specific cell types; keep in mind, of course, that arrays do not provide any histological information. Microarrays that represent specific themes are referred to as panels; popular panels include cancer, apoptosis, cell signaling, growth and development, cytokines, cytoskeleton, and others. Specialized microarrays, as well as microarrays printed with broad-ranging sequences, are designed to provide investigators with as much latitude as possible in designing their experiments.

What is a heat map?

As a high-throughput method, a lot of data results from microarray work. Heat maps are useful for reporting and summarizing data that often represent many samples, many genes, or both. A heat map, also referred to as a heat-map matrix, is a graphical, two-dimensional representation of data (in this case from microarray experiments) in which the magnitude of each data point is color-coded. For example, the columns within a heat map often represent different samples, which may include biological or experimental replicates, while the rows represent different genes and the associated color reflects departures from baseline or control samples, for example, changes in gene expression or relative abundance. Regarding the color scheme, red (warmer color) usually indicates high (or increased) expression while green (cooler color) usually indicates low (or decreased) expression. In addition to summarizing data, heat maps are useful for detecting trends.

In the construction of a heat map, it is possible to associate common occurrences. For example, a heat map may reveal patterns of gene expression shared by patients with a common illness; alternatively, the behavior of a particular gene across several tissues, organs, or organisms may become evident. The simplest way to associate gene expression similarities that become evident during the assembly of a heat map is by hierarchical clustering in which a tree-like structure known as a dendrogram groups genes with similar levels of expression into clusters. Small clusters are then merged into progressively larger clusters until all of the genes under investigation are eventually merged into a single cluster. Genes are clustered to gain a sense of expression similarity, and the clustering of individual samples, on the other axis, may likewise provide valuable information. The resulting dendrogram structures are typically shown along the axes of the heat map for unambiguous interpretation of the data (Fig. 21.4). Although the volume of microarray data may seem daunting, visually representing it in this way is helpful for understanding the temporal behavior of genes as well as naturally occurring or induced variances.

What microarrays can do

The most obvious advantage of microarray-based profiling of gene expression is the ability to assay very large numbers of genes (thousands) in a single hybridization experiment. As a result, microarray data have been used to implicate specific genes and gene relays in the onset and progression of various diseases or cellular dysfunctions. A well-planned microarray experiment asks questions about the behavior of families or groups of genes as a way of elucidating pathways associated with a specific phenotype. Microarrays provide a means for large-scale comparison of various categories of genes.

Other methods in molecular biology likewise offer opportunities to assay all up- and down-regulated sequences simultaneously (Table 21.1), a strategy known as global analysis of gene expression. The clear advantages of microarrays are the speed and efficiency with which huge amounts of data are acquired, processed and archived, and the immediate identification of genes in which expression is observed to fluctuate. Unlike differential display or suppression subtractive hybridization (SSH), the modulated sequences don't have to be cloned or sequenced because the genes on the microarray are already known and have been mapped to a precise spot on the array.

How do microarrays compare to cDNA libraries? Microarray data provide a genetic snapshot of a cell or tissue undergoing any type of physiological or biochemical change, much like a collection of cDNA from a biological source and, as indicated above, the identity of microarray sequences and probable physiological significance has in most cases been defined. It is important to note that traditional cDNA libraries may contain sequences that are not present on the microarray, perhaps representing unknown genes, or mapped genes of unknown function, including small and long noncoding RNAs (depending upon when and how the library was synthesized). Microarray technology is also complementary to and compatible with other contemporary tools available to the molecular biologist, including real-time PCR, RNAi, and transgenic studies. The technology is not intended to replace other time-honored molecular methods but rather to add a new dimension to structural and functional gene expression studies.

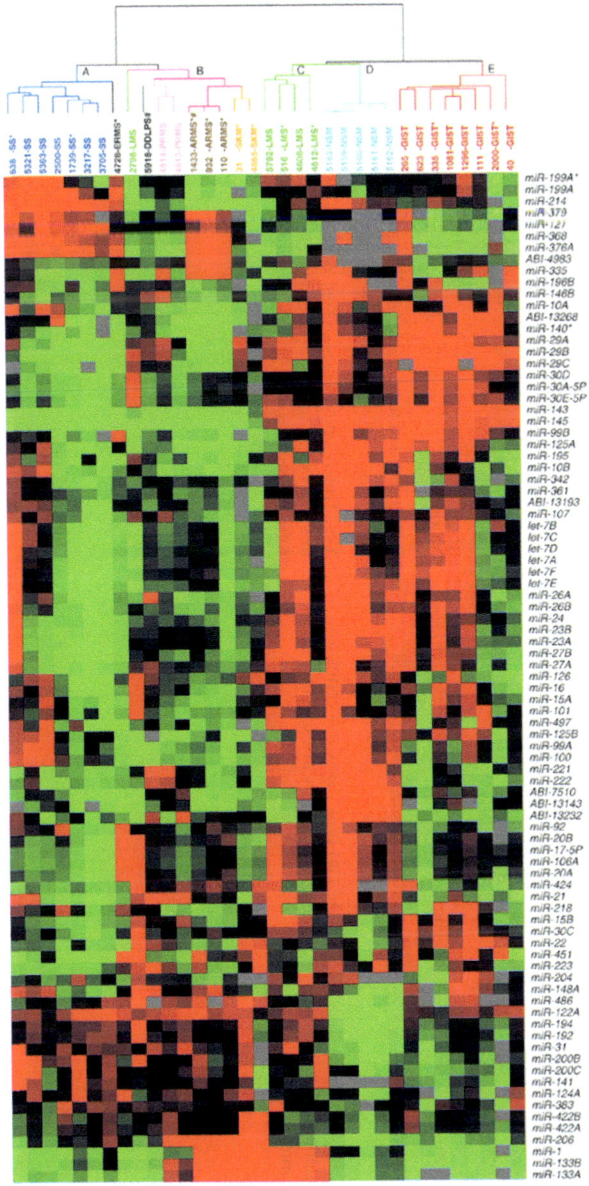

FIGURE 21.4 A heat map with hierarchical clustering of genes (y-axis) and samples (x-axis). Red usually indicates a higher level of expression, green represents a lower level of expression, and the shade of red or green is an indicator of how high or how low the level of expression is. The objective behind clustering is to produce groups that behave in a similar manner. This particular image shows a hierarchical clustering analysis of 27 sarcomas, five normal smooth muscles, and two normal skeletal muscle tissues; each row represents the relative levels of expression for a single miRNA and each column shows the expression levels for a single sample (Subramanian et al., 2008). *Reprinted by permission from Subramanian, S., W.O. Lui, C.H. Lee, I. Espinosa, T.O. Nielsen, M.C. Heinrich, et al., (2008). MicroRNA expression signature of human sarcomas. Oncogene 27, 2015–2026. Macmillan Publishers Ltd.*

TABLE 21.1 Comparison of gene expression analysis techniques.

	Microarrays	Real-time PCR	SSH[a]	RNA-seq
High throughput analysis	Yes	No	No	Yes
Assay many genes simultaneously	Yes	Limited capacity	Yes	Yes
Quantitativeness	Moderate	The best	Moderate	High
Cost	Moderate	Low to moderate	Low to moderate	Moderate to high; declining
New gene discovery	No – can only assay sequences that are on the microarray	No – can only assay the genes for which the primers are specific	Yes! This is one of the goals of subtraction-based methods	Yes! Any RNA regardless of function will be sequenced
Global analysis	Yes	No	Yes	Yes
Side-by-side comparison of two (or more) cell populations	Yes	Not easily	Yes	Yes
Reference	Universal Reference RNA	Housekeeping genes, cRNA, or a control sample	Driver cDNA	Reference genomes are available; if not, de novo sequencing

Common fluorescence labeling options	Cy3, Cy5	6-FAM, TAMRA (or SYBR Green) and others	Not applicable	Platform dependent
Detection options	Fluorescence detection	Fluorescence detection	Gel electrophoresis	Platform dependent
Qualitativeness	Moderate	Moderate	Limited	High
Special equipment and consumables	Requires expensive chips, microarray reader, and analysis software	Requires a real-time instrument, fluorescence-labeled probes (or SYBR Green), and analysis software	No special equipment	Next-generation sequencing and compatible analysis software

[a] *Suppression subtractive hybridization.*

What microarrays cannot do

DNA array technology, considered to be an important tool for the global assay of changes in gene expression, is not without intrinsic limitations. First, microarrays can assay only the gene sequences that have actually been printed onto it. As such, microarrays are not tools for new gene discovery. The difference between microarray analysis compared to methods such as SSH and RNA-seq is that these alternative methods render all genes that are differentially expressed and are not limited to the assay of a predefined set of genes.

Second, microarrays are semiquantitative at best. The rather narrow dynamic ability to detect increases or decreases (two- to fourfold) is most commonly a result of background issues and/or signal saturation. Comparatively speaking, microarrays also have a sensitivity floor that is related to the fluorescence signal itself; in cases of low abundance transcripts, the weak fluorescence emission from hybridized, labeled cDNA at a particular spot on the microarray may be obscured altogether by background fluorescence. If microarray data do indicate a difference in gene expression, then subsequent analysis of those gene(s) by real-time PCR will add an entirely new quantitative dimension to the behavior of that gene under the conditions defined by the investigation. Microarrays are also limited in terms of the qualitative information that they can furnish. A positive signal from a microarray gene sequence, therefore, indicates transcription from a particular locus, though the microarray may or may not allow recognition of alternative transcript splicing or the use of multiple transcriptions start sites. In fact, direct observation of full-length transcript size is possible only by Northern analysis. Alternative start sites can be detected by 5′ RACE, though detection is not guaranteed by microarray analysis. Further, because the cost of microarray technical replicates may be prohibitively high for some labs, in many instances a particular sample may be assayed only once.

Third, limiting amounts of RNA starting material may actually preclude profiling a sample by microarray analysis. This is true especially when only a few cells are available, such as in laser capture, cell sorting, when dealing with microscopically dissected tissues, or when RNA is extracted from archived clinical (FFPE; formalin-fixed, paraffin-embedded) samples. Depending on the sample, one strategy is to amplify the RNA in a linear manner (Eberwine method; Chapter 10: RT-PCR, a science and an art form) before preparing it for hybridization. For example, cDNA can be synthesized from RNA using downstream primers that incorporate a T7 or an SP6 promoter. This will allow the cDNA to be transcribed in vitro in a subsequent step, thereby producing much more RNA (cRNA) that can be labeled with a hapten or a fluorescence tag; otherwise, the newly transcribed RNA can be used to make more cDNA that can be labeled. This strategy will not distort the natural abundance relationships among the various transcripts in the sample, even if two or more rounds are needed to produce a suitable RNA mass for labeling. PCR, in contrast, is a process resulting in the exponential amplification of template sequences, though it is well known to be biased in favor of some sequences over others. Consequently, the true abundance relationships among transcript species in high-complexity samples become increasingly distorted with each passing cycle.

Fourth, and somewhat ironically, some of the occasional ambiguity pertaining to data analysis is a result of the fact that microarrays produce so much data all at once. As such, direct side-by-side comparisons are not always cut and dry, and most users agree that microarray data are not perfect. Nor are microarray experiments immune from interference due to contaminants and poor technique in the laboratory. The major sources of variability among experiments are (1) the biological source itself; (2) the quality of the RNA at the onset; and (3) user-dependent phenomena. Nonreproducibility problems can be further compounded by *Mycoplasma* contamination, which has been the bane of cell biologists for years and has been shown to alter microarray data when RNA is isolated from a contaminated cell culture (Miller et al., 2003).

Major steps in microarray analysis

The basic steps performed in a typical microarray analysis are delineated below. Keep in mind that the method of labeling, method of detection, control RNA, and other parameters are subject to change due to the preferences of each laboratory as well as the microarray platform itself.

1. **Acquisition of the microarray system**. As described above, microarrays are readily available from a number of sources. Of course, an investigator may be limited by the fact that the organism being studied is not yet represented on commercially available microarrays, and therefore must be prepared in-house.
2. **Isolation of high-quality RNA**. Methods of the isolation of high-quality RNA and the assessment thereof are described in detail in Chapter 4, RNA isolation strategies, Chapter 5, Isolation of polyadenylated RNA, Chapter 6, The truth about tissues, and Chapter 7, Going Green: RNA and the molecular biology of plants. In general, RNA intended for microarray analysis requires an A_{260}/A_{280} greater than 1.8, a RIN >7, and a RIN range among all RNA samples of <1.5. The amount of RNA required is protocol-dependent, ranging anywhere from as little as 500 ng to as much as 10 μg.
3. **cDNA synthesis**. Due to the inherent chemical instability of RNA in general, mRNA from the samples to be profiled is converted into cDNA much as it is for other applications. However, reverse transcription of RNA in the presence of nucleotides carrying a fluorescent label is known

to be inefficient and skewed in favor of Cy3 incorporation. The method of indirect fluorescent labeling has circumvented this problem. The amino allyl method of labeling involves the incorporation of the amine derivative of dUTP, 5-(3-aminoallyl)-2′-deoxyuridine 5′-triphosphate, in a nucleotide cocktail consisting of dATP, dCTP, dGTP, amino allyl-dUTP. Comparatively speaking, the incorporation of this highly reactive nucleotide by reverse transcriptase is very efficient. After the sample is reverse-transcribed, a fluorescent dye is attached to purified cDNA in a subsequent reaction.

4. **Linkage of fluorescent dyes**. Following the amino allyl labeling reaction, the $-NH_2$ groups incorporated into the newly synthesized cDNA are exposed to the reactive forms of Cy3 and Cy5, N-hydroxysuccinimide esters, which bind to the modified amino allyl nucleotides. This is known as the coupling reaction and is directly responsible for linking each dye to the cDNA. The labeling efficiency and degree of label incorporation are commonly assessed by scatter plot analysis (Fig. 21.5); the tighter the scatter plot, the more likely it is that the now-labeled target sequences will perform in a microarray assay.

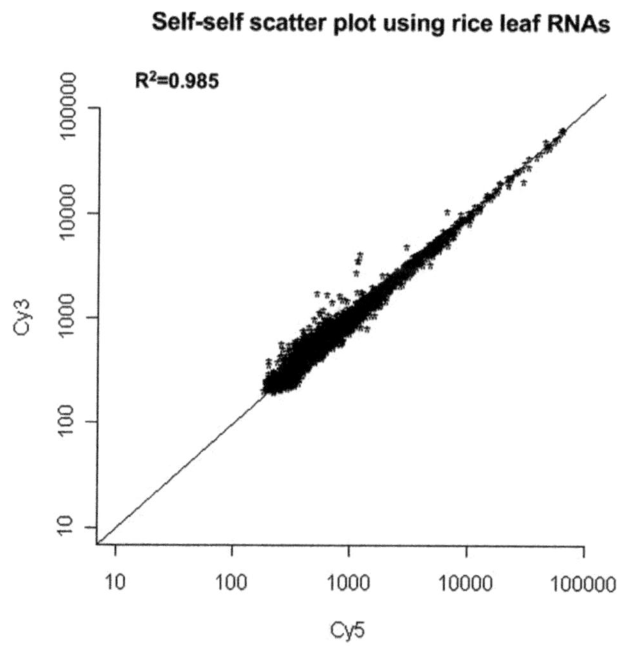

FIGURE 21.5 Scatter plot analysis is used to compare two data sets, in this case to demonstrate equivalent probe labeling efficiencies. The fluorescence intensity from Cy3-labeled target sequences is plotted against the same from Cy5-labeled target sequences. The closer the correlation coefficient (r) is to 1.0, the better the targets will perform during hybridization to a microarray. *Courtesy of Dr. Pam Ronald, University of California - Davis.*

5. **Purification of Cy-labeled target molecules**. The purpose of this very important step is to remove any unincorporated dye from the now-labeled cDNA. If the dye removal procedure is inefficient, the background on the microarray will be extremely high. One reason microarray data sometimes does not appear to be reproducible pertains directly to the quality of the cDNA and the overall labeling and clean-up efficiencies. High-quality cDNA comes from high-quality RNA, and careful attention must be given to the removal of unincorporated labels and other reaction components.
6. **Microarray hybridization**. The labeled cDNA molecules are exposed directly to the array. Maintaining a minimum volume of hybridization buffer has the effect of increasing the effective target molecule concentration and favoring forward hybridization kinetics. Hybridization is characteristically performed under stringent conditions, followed by stringent washing to remove nonhybridized and imperfectly hybridized molecules.
7. **Acquisition of data from the microarray slide or the filter**. The processed microarray is placed in a scanner, or "chip reader," which scans it with a laser to promote the fluorescence signal, accompanied by image capture. Image analysis software unique to the microarray system is then used to manage the images generated on different color channels and interpret the resulting data.
8. **Determination of the biological significance and relevance of data**. This is user dependent.

Reference RNA

Universal reference RNA (URR) is widely available and an indispensable part of many contemporary techniques, including microarray analysis, RNA-seq, and other approaches when a consistent baseline sample is needed for data set comparisons. Considered by some to be "control RNA," a better way to think of URR is in terms of it being an additional control or as a surrogate control in situations where a control RNA for a particular experimental model is not immediately obvious, or readily accessible, as in the study of disease progression. URR is very high-quality RNA that is produced by pooling RNA from a number of different sources, in some cases as many as ten cell lines, representing (1) various tissues, some of which are difficult to obtain and (2) various abundance categories of RNA. Stringent QC tests for URR typically certify that the characteristics of each lot include $A_{260}/A_{280} \geq 1.8$ and RIN often > 9, certified as being free of genomic DNA, certified as being RNase-free ($RNase^{-}$), manifesting at least a 2:1 fluorescence ratio of the 28 S:18 S rRNAs upon electrophoresis, and QC-tested in a microarray experiment. It is routinely shipped on dry ice as an ethanol precipitate or, less commonly, dissolved in sterile water. A high level of URR quality assurance is absolutely necessary for assay optimization and, perhaps more importantly, assay standardization. A standard functionality descriptor of URR is known as gene coverage, which is the number of microarray sequences that hybridize to it. Gene coverage is important for comparison purposes among and between the genes under investigation. The typical coverage range is 85%–99%.

Microarray experiments are performed using one of two basic formats:

1. **Pairwise Design**. Samples 1 and 2 are labeled and hybridized onto the same microarray. For example, label sample 1 with Cy3 (green) and sample 2 with Cy5 (red) and then hybridize both to the same microarray. If there are four samples, then the investigator must perform all hybridizations involving all permutations:

Sample 1 + Sample 2	Sample 2 + Sample 3
Sample 1 + Sample 3	Sample 2 + Sample 4
Sample 1 + Sample 4	Sample 3 + Sample 4

for a total of six hybridizations. The ratio analysis performed is Sample X ÷ Sample Y and, of course, must be performed on all six microarrays.

2. **URR Design**. This approach reduces the required number of hybridizations because each labeled sample (Cy3), in this case Samples 1–4, are individually hybridized to the microarray along with labeled (Cy5) URR. This format requires only four hybridizations, and the ratio analysis is Sample 1/URR, Sample 2/URR, Sample 3/URR, and Sample 4/URR. One of the major advantages of the universal reference design approach is that individual samples can be compared directly because each sample is compared to a single common control. Thus, the investigator is able to cross-reference the data to all other experiments being performed in the laboratory; when using this approach to data analyses, it is strongly recommended that microarrays from the same lot be utilized when comparing closely related data sets. Prior to the availability of URR researchers were not always able to compare microarray data directly due to laboratory-specific protocols. Because URR is manufactured on an industrial scale, large numbers of aliquots representing the same lot are available for long-term studies and are simultaneously available to collaborating laboratories that are geographically dispersed.

What is a macroarray?

DNA sequences can be printed on nylon, too. In contrast to their microarray counterparts, the spots containing the DNA sequences are much larger on macroarrays. The diffusion associated with the porosity of a nylon filter limits the number of cDNAs that can be spotted on the filter to the hundreds, rather than the hundreds of thousands. Macroarrays, therefore, are really not high-throughput tools; rather they are better classified as a "convenience."

Like some types of microarrays, macroarrays are often themed, for example, cancer or apoptosis, and are very useful for labs that have an interest in only a few genes or that do not have the equipment or resources to perform microarray analysis. They are also ideal as a teaching tool. Target sequences used to hybridize macroarrays are commonly labeled with ^{32}P or ^{33}P and evaluated by autoradiography. The DNA sequences on macroarrays are printed in duplicate, side-by-side so that a true positive hybridization event will produce nearly identical signals from both duplicate spots. Later, if handled properly, the hybridized target molecules can be removed by a high stringency wash for reuse. In this lab, we have also performed macroarray analysis using a nonradioactive hapten label and chemiluminescence detection with very good results.

The major mechanistic difference between macroarray screenings vs microarray analysis is that the detection of changes in gene expression requires a side-by-side comparison of two arrays, in which the investigator looks for increases and decreases in hybridization signal intensity (Fig. 21.6). This is quite different from the fluorescence-based analysis used in conjunction with microarrays.

Macroarray advantages include (1) the geometric arrangement of sequences on the array; (2) the ability to assess the up- or downregulation of multiple gene sequences simultaneously; (3) the fact that no special equipment is necessary to determine the location and extent of hybridization on the array; (4) the ability to the visually inspect the X-ray film or other image detection modality (Chapter 15: Photodocumentation and image analysis) for

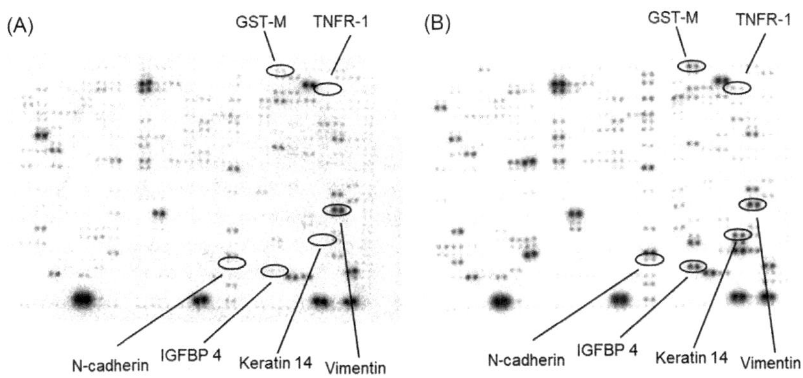

FIGURE 21.6 Differential expression in murine mesothelioma cell lines. The two filters represent expression in normal cells (A) and tumor cells (B). Briefly, cDNA target molecules were synthesized from normal and tumor mRNA in the presence of ^{32}P. Following hybridization and stringent washing, the data were generated by phosphorimager exposure and digital image analysis. Target genes on the filter are printed in pairs, and cDNA abundance is directly related to signal strength. Differential gene expression was later confirmed by Northern analysis. *Courtesy of Dr. A. Kane, Brown University, Department of Pathology and Laboratory Medicine.*

changes in gene expression; and (5) the relatively low cost of macroarrays, which is particularly helpful for labs with limited funding resources.

Macroarray disadvantages include first and foremost a lack of sensitivity associated with the detection of changes in gene expression, certainly in comparison with any PCR-based methods. Thus, macroarrays, like microarrays, are not well suited for assaying highly abundant of very low abundance transcripts. Another disadvantage relates to the limited number of sequences that can be printed on a macroarray. The classification of macroarrays as a nonglobal method should not be a concern for a lab that is interested in only a few genes.

Applications

Commercialization of microarray technology developed from a marriage between biochemistry and the semiconductor manufacturing industry. Because of the intrinsic value of this industry to the scientific community, the number of vendors grew from one (Affymetrix; now part of Thermo Fisher Scientific) to several, including Agilent, GE Healthcare (now Cytiva), Illumina, and others. Succinctly, anyone who is studying gene expression can take advantage of microarray technology because it provides a global snapshot of gene expression, at least with respect to the sequences printed on the array. Replicates can be expensive and, consequently, a resulting lack of data points may present challenges for meaningful statistical analysis. There are also upper and lower limits of detection associated with the use of microarrays. Real-time PCR remains the gold standard method for accurately quantifying changes in gene expression that are suggested by microarray data. Real-time PCR is also a method that validates microarray data. Microarrays reveal which genes are being modulated, and real-time PCR accurately reveals the magnitude of the modulation (by how much the expression of the gene has changed). Thus, these two methods complement each other nicely (Fig. 21.7). Microarray technology is useful for the study of disease initiation and progression, developmental phenomena, SNP profiling, all areas of functional genomics, the response of cells to drugs, and to laboratories performing basic research. Microarrays in one form or another are now also used for diagnostic testing in a clinical setting.

It is important to keep in mind that RNA, believe it or not, does not always tell the whole story. Gene expression is also about translation and the production of functional proteins by the acquisition of the proper subset of posttranslational modifications. As such, protein chips can be used to discover which types of molecules (RNA, DNA, other proteins) bind to the immobilized cadre of proteins, providing clues about how these molecules interact in the cell, for example, protein–protein, protein–nucleic acid, or protein–metabolite. Antibody (Ab) microarrays, also known as antibody chips, constitute an analogous method for the simultaneous expression assay

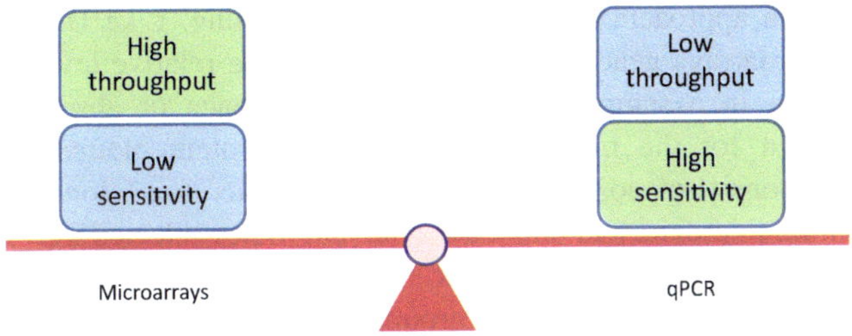

FIGURE 21.7 Arrays are relatively low sensitivity and high-throughput while real-time PCR, which complements array data, is a very high-sensitivity and low-throughput method. The combined use of these methods provides overall balance and context to studies examining various dimensions of gene expression.

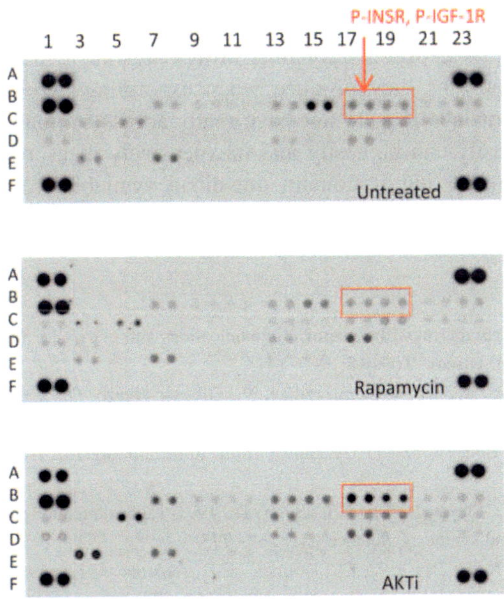

Fold induction P-RTK in BT-474

P-RTK	AKTi	Rapa
HER3	12.1	4.6
HER4	3.2	1.0
INSR	2.4	0.5
IGF-IR	2.3	0.6
FLT-3	6.1	1.2
MSPR	2.9	1.5
Mer	1.7	2.0
Eph A1	2.4	0.7
Eph A7	3.1	0.5
MuSK	1.3	2.0
MET	0.2	1.0

FIGURE 21.8 Human phospho-receptor tyrosine kinase antibody arrays. Experimental samples and a control membrane show the abundance of a variety of proteins in each of the three samples. The changes in protein expression can be measured by densitometric image analysis or estimated by visual inspection. The accompanying table reflects the fold-change in the expression of specific receptor proteins of interest, specifically in response to exposure of BT-474 cells to rapamycin and a protein kinase B inhibitor (AKTi). The asymmetric placement of positive controls helps to maintain the orientation of the filter as well as demonstrates that varying signal intensities elsewhere on the array represent authentic changes in protein expression levels (Chandarlapaty et al., 2011). *From Chandarlapaty, S., A. Sawai, M. Scaltriti, V. Rodik-Outmezguine, O. Grbovic-Huezo, V. Serra, et al., (2011). AKT inhibition relieves feedback suppression of receptor tyrosine kinase expression and activity. Cancer Cell 19, 58–71. doi 10.1016/j.ccr.2010.10.031; Figure 5B. Reproduced with permission from Elsevier.*

of proteins, an approach known as proteome profiling. Like DNA microarrays, Ab microarrays generate data pertaining to the relative levels of a particular protein in a sample and not just its presence or absence and are therefore ideal for the multiplex screening of protein biomarkers. These microarrays consist of dozens to hundreds of various monoclonal antibodies printed on the microarray in duplicate. The antibodies selected may represent a particular theme, such as a cancer panel, and the antibodies commonly used represent various functional groups as well as subcellular locations throughout the cell. Detection is ideally based on Cy3 and Cy5 labeling for higher-density arrays; nonfluorescent labels generally cause an unsatisfactory loss of sensitivity. Antibody arrays can also be assayed using HRP or alkaline phosphatase-based chemiluminescence methods when the array is printed on a membrane, rather than a chip (Fig. 21.8). In general, these arrays are semiquantitative with typical sensitivity comparable to an ELISA. Moreover, the observed sensitivity of this type of assay is dependent on a number of factors, perhaps one of the most important of which is the amount of sample to which the array is exposed. Proteome arrays are much more valuable for the qualitative profiles they produce when examining two or more samples side-by-side. For labs that do not have ready access to image analysis software, signal intensity can be easily and inexpensively measured using Imagej, a free densitometer public-domain download available from NIH (imagej.nih.gov).

References

Barrett, J.C., Kawasaki, E.S., 2003. Microarrays: the use of oligonucleotides and cDNA for the analysis of gene expression. Drug. Discov. Today 8, 134–141.

Chandarlapaty, S., Sawai, A., Scaltriti, M., Rodik-Outmezguine, V., Grbovic-Huezo, O., Serra, V., et al., 2011. AKT inhibition relieves feedback suppression of receptor tyrosine kinase expression and activity. Cancer Cell 19, 58–71. Available from: https://doi.org/10.1016/j.ccr.2010.10.031.

Fodor, S.P., Read, J.L., Pirrung, M.C., Stryer, L., Lu, A.T., Solas, D., 1991. Light-directed, spatially addressable parallel chemical synthesis. Science 251, 767–773.

Fodor, S.P., Rava, R.P., Huang, X.C., Pease, A.C., Holmes, C.P., Adams, C.L., 1993. Multiplexed biochemical assays with biological chips. Nature 364, 555–556.

Huan, Q., Wang, F., Tao, S., 2019. Proteome microarray technology and application: higher, wider, and deeper. Expert. Rev. Proteom. 16 (10), 815–882. Available from: https://doi.org/10.1080/14789450.2019.1662303.

Miller, C.J., Kassem, H.S., Pepper, S.D., Hey, Y., Ward, T.H., Margison, G.P., 2003. Mycoplasma infection significantly alters microarray gene expression profiles. BioTechniques 35, 812–814.

Subramanian, S., Lui, W.O., Lee, C.H., Espinosa, I., Nielsen, T.O., Heinrich, M.C., et al., 2008. MicroRNA expression signature of human sarcomas. Oncogene 27, 2015–2026.

Villaseñor-Park, J., Ortega-Loayza, A.G., 2013. Microarray technique, analysis, and applications in dermatology. J. Investig. Dermatol. 133 (4), 1–4.

Further reading

Hacia, J.G., Fan, J.B., Ryder, O., Jin, L., Edgemon, K., Ghandour, G., et al., 1999. Determination of ancestral alleles for human single-nucleotide polymorphisms using high-density oligonucleotide arrays. Nat. Genet. 22, 164–167.

Rao, M.S., Van Vleet, T.R., Ciurlionis, R., Buck, W.R., Mittelstadt, S.W., Blomme, E.A.G., et al., 2019. Comparison of RNA-seq and microarray gene expression platforms for the toxicogenomic evaluation of liver from short-term rat toxicity studies. Front. Genet. 9, 1–16. Available from: https://doi.org/10.3389/fgene.2018.00636.

Chapter 22

Subtractive and nonsubtractive methods for the analysis of gene expression

Rationale

The exquisite orchestration of gene expression at the transcriptional level and beyond is responsible for the differentiation of cells and tissues and, ultimately, for conferring the phenotype of the organism. It has long been the goal of many molecular biologists to be able to identify which genes are expressed in common among two or more different cell types and, just as importantly, which genes are expressed differentially in a side-by-side comparison. Such insight would have profound ramifications on understanding the molecular basis of everything.

A logical focus, initially, was on the mRNA component of the transcriptome and the manner in which the mRNA complement changed in response to experimental stimuli or biochemical cues. It is important to note that noncoding RNAs, both short and long, are equally important for understanding the modulation of gene expression. Prior to the development of PCR, several variations of a method known generically as subtraction hybridization (Sargent and Dawid, 1983; Hedrick et al., 1984; Fornace and Mitchell, 1986; Sargent, 1987; Orr et al., 1992; Lisitsyn et al., 1993; Zhang et al., 1995) were utilized as a means of removing sequences shared by two transcript populations. Methods that support the physical separation of differentially expressed sequences include chromatography through hydroxyapatite (Bernardi, 1971; Sargent, 1987), biotinylation coupled with streptavidin capture and modified agarose chromatography (Welcher et al., 1986), PCR-based magnetic bead capture (Sharma et al., 1993), biotinylation coupled with magnetic bead separation (e.g., purification of poly(A)$^+$ mRNA, other macromolecules, and sometimes whole cells) (Rodriguez and Chader, 1992; Lambert and Williamson, 1993; López-Fernández and del Mazo, 1993; Coche et al., 1994), biotinylation coupled with phenol extraction (Travis and Sutcliffe, 1988), cross-hybridization of entire libraries (Kulesh et al., 1987; Cochran et al., 1987), and asymmetric PCR (Houge, 1993). As a historical note, hydroxyapatite $[(Ca_5(PO_4)_3OH)_2]$, an insoluble crystalline form of

RNA Methodologies. DOI: https://doi.org/10.1016/B978-0-323-90221-2.00033-3

calcium phosphate, was once used extensively for the purification of nucleic acids other macromolecules. The unique nature of this matrix permits the physical partitioning of single-stranded nucleic acids from double-stranded nucleic acids (cDNA:mRNA and DNA:DNA) in a salt-dependent manner. Some of these methods are still in use though most have been abandoned in favor of more contemporary whole-transcriptome approaches, including direct NGS (Chapter 24: RNA-seq: The Premier Transcriptomics Tool) or the alternative techniques described below. Owing to the labile nature of RNA, these methods characteristically involve working with cDNA.

Essential issues

Regardless of the method utilized, there are terms and concepts common to all methodologies that must be clearly understood to ensure that an investigation aimed at differential expression is well-designed. The *tester* (also known as the target) is the cDNA population containing induced sequences that are to be identified. The *driver* is the control or reference population of cDNA sequences against which the tester is being compared. During the subtraction hybridization, the driver is always present in large molar excess (often 10-fold or more) over the tester. This will ensure first-order hybridization kinetics, resulting in complete hybridization and concomitant removal of all sequences common to both populations. To that end, it is also helpful to add a fresh driver to the hybridization cocktail after several hours. The overall efficiency of subtractive methods depends on maintaining high stringency conditions while favoring efficient forward hybridization kinetics, both of which are dependent upon salt concentration, hybridization temperature, hybridization time, and the presence of excess driver throughout.

Keeping in mind that phenotype changes occur when genes are upregulated or downregulated, it is necessary to perform a second subtraction in which the tester and the driver are reversed. For example, in the forward subtraction, cDNA from sample A is the tester while cDNA from sample B is the driver. Then, in the reverse subtraction, cDNA from sample B is used as the tester and cDNA from sample A becomes the driver. Performing the forward and reverse subtractions will identify sequences that have been induced and those that have been repressed, thereby giving a more complete look at the biochemistry associated with the expression of genes from sample A and from sample B.

Postsubtraction, the remaining sequences must be screened using one or more alternative methods to confirm their differential expression status. These methods may include but are not limited to Northern analysis or reverse Northern blot (Chapter 16: Northern Analysis), nuclease protection (Chapter 18: Quantification of Specific mRNAs by Nuclease Protection), and other quantitative PCR-based methods (Chapter 10: RT-PCR: A Science and an Art Form and Chapter 11: Quantitative PCR Techniques); RNA-seq can also be a valuable complementary method. In any case, the mere presence of a sequence in a subtracted population does not guarantee the differential status of the corresponding gene; although infrequent, false positives are occasionally observed.

Subtractive methods

The terms gene subtraction and subtraction hybridization refer to a global search for up- and/or downregulated sequences, rather than examining a few specific genes using techniques described in previous chapters in this text and elsewhere. Subtraction hybridization methods also differ from both antibody-mediated immunopurification methods in which a transcript is sequestered because of its association with a protein. Nucleic acid-based subtraction is attractive because the endpoint of the assay is an enriched population of sequences associated with a change in cellular phenotype. As with most contemporary reverse transcriptase (RT)-PCR procedures, neither RNA size selection nor poly(A)$^+$ selection are requirements for efficient subtraction because of the cleverly designed primers and cDNA synthesis strategies. By linking sequence subtraction to the amplification power of PCR, even cDNAs representing low-abundance mRNAs can be detected when they are up- or downregulated. Poly(A)$^+$ selection may, in fact, compromise the goals of the subtraction methodology by further underrepresenting rare transcripts.

The ability to discriminate between sequences expressed uniformly vs those which are modulated under the conditions defined by an experiment suggests some intriguing possibilities. For example, the following are some of the comparisons that colleagues and collaborators have dreamed up over the years:

1. Quiescent cells versus senescent cells.
2. Precursor cells versus terminally differentiated cells.
3. Infected tissue versus uninfected tissue.
4. Physiological changes in the kidney of certain types of fish allow these animals to osmoregulate as they migrate from salt water into fresh water.
5. Benign tumors versus metastatic tumors.
6. Diploid cells versus established cell lines.
7. Apples versus oranges, literally.

The take-home lesson, then, is that a comparison can be made between any two populations for any reason at all, resulting in the capture of sequences that differ.

Suppression subtractive hybridization

The method of nucleic acid subtraction coupled with suppression PCR (Diatchenko et al., 1996; Siebert et al., 1995) has become a favored technique for the isolation of differentially expressed gene sequences. This approach has the advantage of being a sensitive assay, and the unique design of the nucleic acid adapters needed to support this assay is the key to the system (Clontech PCR-Select cDNA Subtraction System; Takara). Moreover, the mechanics of the assay (Fig. 22.1) favor normalization of differentially expressed sequences, meaning that the end of the assay produces an enriched pool of differentially expressed sequences that are present at similar concentrations, regardless of their abundance level in the original biological material. This situation is highly desirable during the gene identification process because in a nonnormalized library abundant cDNAs are more likely to be retrieved than cDNAs of lesser abundance. When sequencing clones from a normalized pool, there is an equal probability of retrieving cDNAs representing high- or low-abundance transcripts, thereby providing a much more comprehensive snapshot of the variety of up- or downregulated sequences. The magnitude of the up- or downregulation may be discerned later through qPCR methods. When performed correctly, sequencing even hundreds of clones usually renders very few redundant sequences.

In this laboratory the suppression subtractive hybridization (SSH) method has been used in several applications, with outstanding results (Bassett et al., 2014; Liu et al., 2011; Norelli et al., 2009; Bassett et al., 2006). While technically a bit more challenging than some of the older methods, SSH greatly enriches in favor of low-abundance transcripts and, in the mind of this Author, generates a more complete profile of gene expression than related methods. One can expect that any sequence which differs in abundance by four- to fivefold will be sequestered easily by the subtraction process, though it is not uncommon to identify genes with as little as a 1.5-fold difference compared to the driver population. This method is commonly used when global sequencing or microarray analysis is not accessible to the investigator.

Succinctly, the methodology (Fig. 22.2) involves creating two tester subpopulations; half of the cDNA is ligated to one type of adapter while the other half of the cDNA is ligated to a different adapter. Adapters are not ligated to the driver. To prepare the cDNA for adapter ligation, it is first cut with the restriction enzyme *Rsa* I, a four-base cutter that prepares blunt-ended restriction fragments. This precludes the need for end-repair of the cDNA, which would be yet another opportunity to lose a precious sample. The uniqueness of the adapter ligation approach lies in the fact that the adapters do not exhibit 5′ phosphate groups, meaning that (1) the cDNA itself will be the phosphate group donor, (2) only one strand of each adapter will be joined to each cDNA, and (3) stretches of identical sequences within the adapters will facilitate the annealing of primers and supporting PCR amplification after a simple filling-in reaction has been performed.

Two independent hybridization reactions are performed, each of which contains one of the adapter-ligated cDNA subpopulations mixed with an aliquot of the driver. The first (primary) hybridization requires 6–8 hours.

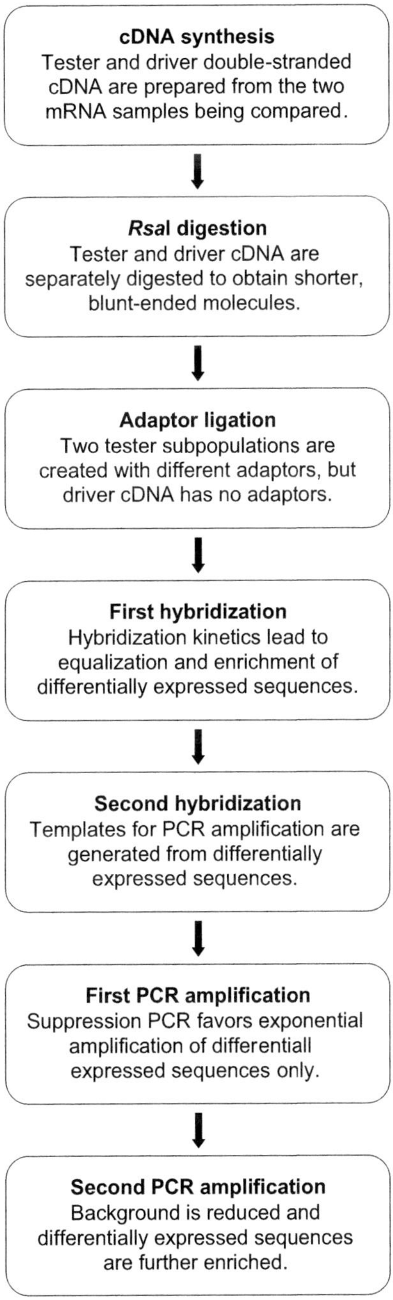

FIGURE 22.1 Overview of the PCR-Select cDNA Subtraction system. *Courtesy Clontech Laboratories (now Takara Bio USA).*

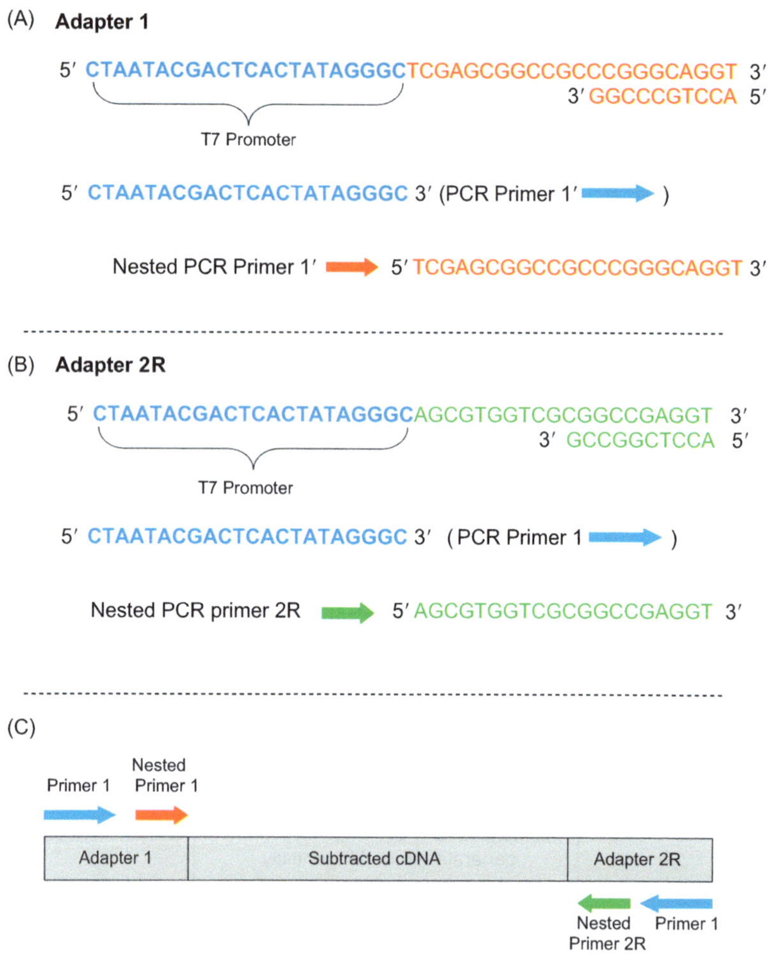

FIGURE 22.2 Preparation of adaptor-ligated tester cDNAs for hybridization and PCR. To perform the forward and reverse subtractions, each tester cDNA must be ligated to the appropriate adaptors as shown. (A) Structure of adaptor 1; (B) Structure of adaptor 2R; (C) PCR amplification of differentially expressed sequences is favored only when molecules contain adaptor 1 sequences on one end and adaptor 2R sequences on the other end. See text for details. *Courtesy Clontech Laboratories (now Takara Bio USA).*

Beyond 12 hours in the first hybridization the single-stranded tester molecules have a tendency to base pair intramolecularly and with each other, making the formation of double-stranded cDNAs very difficult in the second hybridization. At the end of the first hybridization period, the two separate hybridization reactions are mixed together for a longer, secondary hybridization which may proceed overnight. At the end of the secondary hybridization, the remaining single-stranded tester subpopulations (Fig. 22.3) now act as templates for PCR

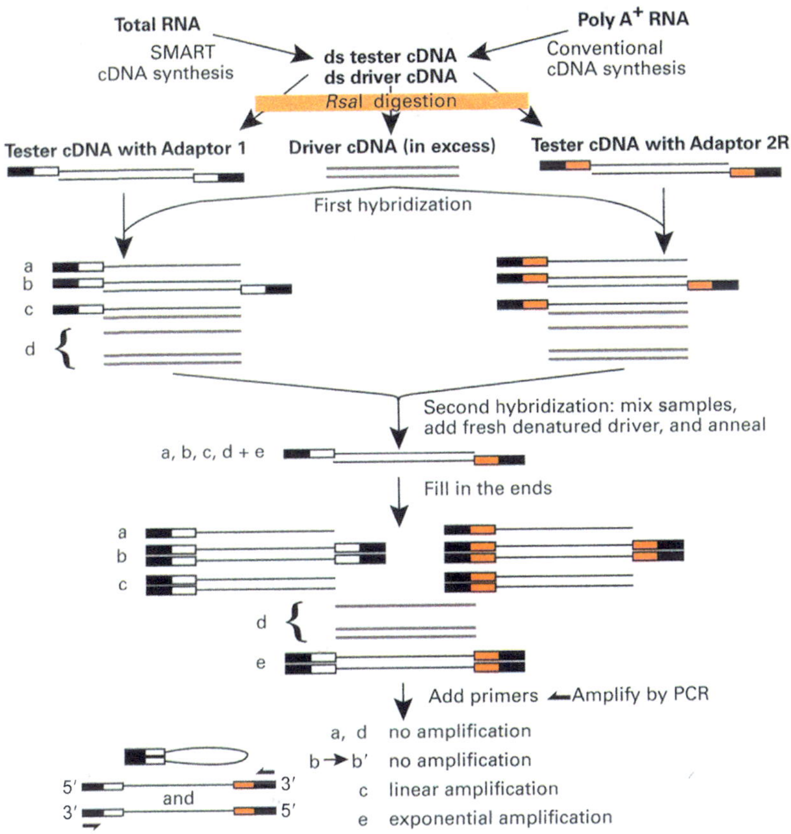

FIGURE 22.3 Subtraction-suppression procedure showing how the various cDNAs in the reaction, both tester and driver, are able to interact with each other. *Courtesy Clontech Laboratories (now Takara Bio USA; PCR-Select cDNA Subtraction system).*

amplification of the differentially expressed sequences that they represent. The recommended hybridization periods are more than adequate for the removal of sequences shared by both nucleic acid populations.

PCR is performed in two stages using nested primers; performing the secondary (nested) amplification is absolutely essential for diluting out low-molecular-weight junk and also generating an adequate mass of larger, useful cDNAs for identification purposes. When performed correctly, the primary and secondary PCR reactions render an enriched, equalized population of subtracted sequences; the suppression phenomenon that occurs (Fig. 22.4) limits the amplification of sequences which have identical primer binding sites on both ends. In this lab, when an attempt was made to clone the subtracted PCR products after a single PCR amplification using only the outer primers, the results were wholly unsatisfactory.

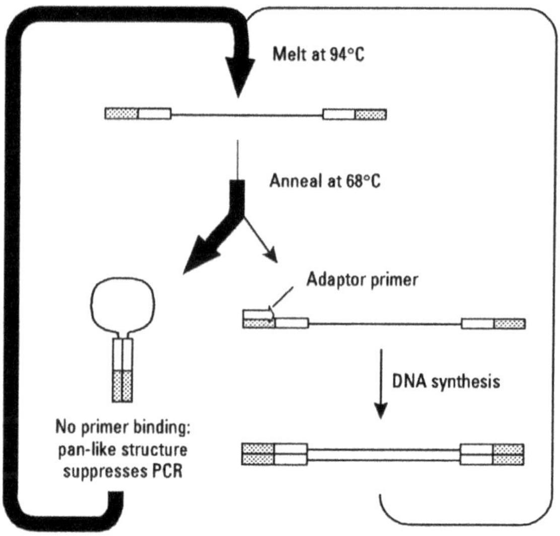

FIGURE 22.4 Suppression PCR. During PCR, nonspecific amplification is efficiently suppressed by the formation of a pan-like structure, while specific amplification of double-stranded cDNAs with different adaptors at both ends proceeds normally. *Courtesy Clontech Laboratories (now Takara Bio USA).*

Following the primary (outer primer) and secondary (inner primer) amplifications, the enriched pool of differentially expressed sequences (Fig. 22.5) is now ready for identification and, most importantly, a demonstration that the sequences which survived the subtraction are really differentially expressed. In this lab, we have adopted the following semiquantitative routine, at least for an initial assessment. The subtracted cDNAs are ligated to plasmids, followed by transformation into *Escherichia coli* and plating. One hundred (or more) clones are picked and plasmid preps are performed. Any insert found to be less than 200 base pairs is ignored; this can be ascertained by using vector-specific primers which will amplify the insert (only) so that it can be examined on an agarose gel. Clones that meet or exceed the size criteria are sent for sequencing. Then, gene sequence-specific primers are synthesized and used with an aliquot of the tester cDNA and, separately, with an aliquot of the driver cDNA. If the identified gene is truly modulated, there should be a stronger band in the tube with the tester cDNA, compared to the product (if any) when using the driver cDNA as a template (Fig. 22.6). In some cases, there will be a tester-only PCR product while in other cases, there will be a weaker, though obvious, band produced by the driver sample; this confirms differential expression status of the genes being investigated. If the bands are of near equal intensity, it will be necessary to use real-time PCR to assess the level of modulation. If the band on the gel from the driver cDNA is stronger than the band produced from the tester cDNA, this is probably a false positive.

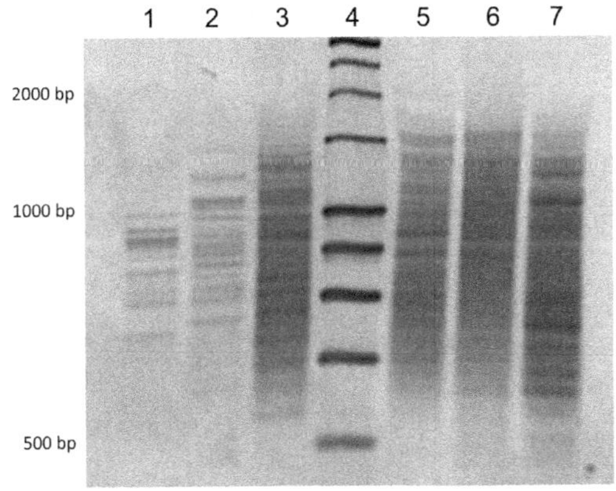

FIGURE 22.5 Typical results of subtraction-suppression PCR. Six RNA populations representing precisely controlled environmental conditions were examined for differential expression. The samples under investigation were paired as the "A" reactions, the "B" reactions, and the "C" reactions, and a forward (AF, etc.) and reverse (AR, etc.) subtraction was performed for each. The subtracted products were then PCR amplified; note the fairly uniform density of each band, which demonstrates the normalization process among the differentially expressed sequences, regardless of their natural relative abundance in the cell. The bands present in each lane represent differentially expressed cDNA sequences that were subsequently cloned for sequencing and further experimentation in order to confirm differential expression. Lane 1, A-forward; Lane 2, A-reverse; Lane 3, B-forward; Lane 4, molecular weight standard; Lane 5, B-reverse; Lane 6, C-forward; Lane 7, C-reverse.

As an alternative, a form of older-style bacterial colony hybridization can be performed instead of the approach described above. As plasmid preps and sequencing can be time-consuming and costly, a rapid, preliminary screening to confirm differential status can be helpful in this regard. The colony hybridization method first uses an aliquot of labeled tester as a probe, followed by a screening replica plated colonies with an aliquot of labeled driver. Colonies that light up with the tester probe, but not with the driver probes are assumed to be differentially expressed. At that point individual clones can be plasmid-prepped and sequenced. One way for efficiently managing relatively small numbers of clones (<500) involves picking the colonies with sterile toothpicks and transferring them to a 96-well plate containing 200 μL sterile LB medium supplemented with 10% glycerol and the appropriate antibiotic in each well. After incubating the plates at 37°C overnight, a replica-plating tool (Fig. 22.7) is used to inoculate a fresh agar dish such that the clones grow in a geometric pattern on the agar plate, and each clone can be easily traced back to a specific well. The clones that grow after replica plating in this manner are quite a bit larger than the normal size of colonies that results from standard plating methods after a routine bacterial transformation. These larger colonies produce quite unambiguous signals upon hybridization to a probe, though fewer colonies can be grown per plate.

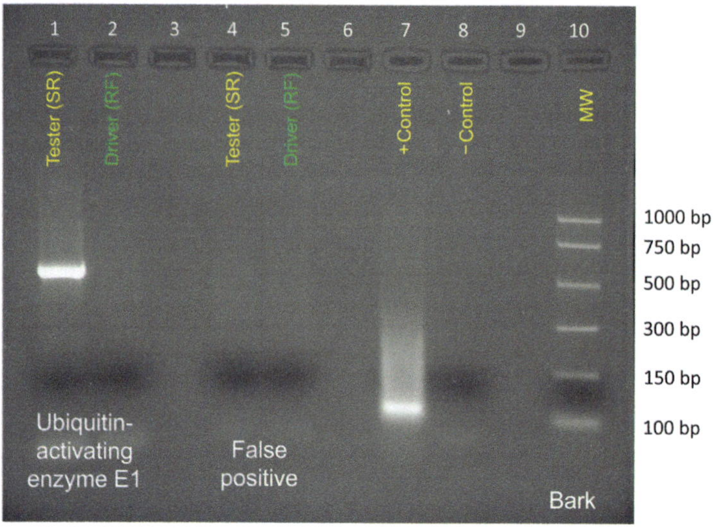

FIGURE 22.6 Confirmation of differential expression by semiquantitative PCR. Gene-specific primers are mixed with an aliquot of tester cDNA and with an aliquot of driver cDNA. A stronger band in the tester lane is good evidence that the gene is upregulated. Lane 1, tester [sample SR]; Lane 2, driver [sample RF]; lane 3, empty; lane 4, tester [sample SR]; lane 5, driver [sample RF]; lane 6, empty; lane 7, positive control, lane 8, negative control, lane 9, empty; lane 10, MW size standard. Differential expression was confirmed for ubiquitin-activating enzyme E1 (lanes 1 and 2); false positives also occur periodically (lanes 4 and 5). Original RNA samples were recovered from bark under drought and drought-recovery conditions.

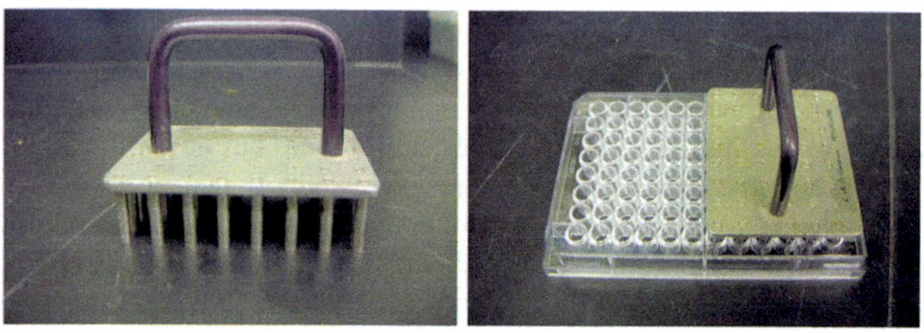

FIGURE 22.7 Bacterial colony replica-plating tool. The stainless-steel prongs fit precisely into 48 of the 96 wells of a standard microtiter plate into which individual bacterial clones have been grown (usually 200 μL per well). After dipping the prongs into the bacteria-containing wells, the instrument is gently placed on the surface of a standard 1.5% LB agar/antibiotic plate for a few seconds and quickly removed. The plate is then incubated as usual. The bacterial clones will grow on the plate with a very precise geometry; by comparison with a typical bacterial colony, this plating method produces a much larger target area for hybridization. Because of the geometry and strong signals from positive clones, it will be very easy to recover positive clones from the master 96-well plate. The 96-well format is also an excellent way to archive and store clones for extended periods.

Granted, the method is somewhat cumbersome and is practical only when working with a small number of clones. If the colonies were organized into a glycerol-supplemented medium in 96-well plates, these plates can then be sealed with Parafilm and then stored at −80°C with great confidence in the long-term viability of these clones.

The SSH method overall requires a fair amount of skill and can be time-consuming relative to other methods, though the potential benefits are extraordinary. One of the nice things about the system is that the subtractions can (and should) be performed in both directions, leading to the capture of both up- and downregulated sequences. The first- and second hybridizations very effectively normalize the enriched population of differentially expressed sequences such that one has nearly an equal chance of sequencing a cDNA from an abundant transcript or a low-abundance transcript. Best of all, SSH is truly a method for new gene discovery, since *any* differentially expressed sequence can be profiled, regardless of its identity. This is an advantage compared to microarray screening in which the only sequences that can be assayed are those that are printed on the array. SSH is an excellent alternative, or complementary technique, to gene expression profiling when an array is not available representing the organism of interest or when transcriptome sequencing is not available to the investigator.

As mentioned above, the complete SSH procedure usually takes four to five days to complete. There are numerous manipulations that must be performed, many of which involve generating the cDNA prior to the actual subtraction itself. The normalization of sequences which occurs in the two hybridizations is efficient, but higher abundance sequences sometimes show an occasional level of redundancy greater than is observed among other sequences of lower natural abundance. Also, it is very important to confirm that the subtracted sequences are, in fact, differentially expressed, preferably using a non-PCR-based method. Overall, the approach is excellent and is a valuable research tool.

Troubleshooting

1. Nonnegotiables
   a. High-quality RNA:
      i. Use RNase-free technique throughout the procedures.
      ii. Use guanidinium thiocyanate- or guanidinium HCl-based lysis buffers and procedures. NP-40 lysis buffers can also be used depending on the biological source, though most protocols recommend the use of a chaotropic guanidinium lysis reagent.
      iii. Double precipitation with guanidinium lysis buffer can be helpful.
      iv. Treat samples with RNase-free DNase.

v. Wash RNA pellets thoroughly and extensively with 70% ethanol prepared in nuclease-free H_2O to remove excess salt; cationic salts bind dNTPs quantitatively and can cause havoc in the cDNA synthesis and PCR reactions.
vi. Properly store purified RNA.

b. Efficient reverse transcription into cDNA, and PCR amplification:
i. Use an equal mass of RNA per cDNA reaction (e.g., 200–300 ng).
ii. RTs lacking endogenous RNase H activity maximize efficiency.
iii. Always work from a cDNA master mix.
iv. Always work from a PCR master mix.
v. Use a hot-start PCR method, ideally in conjunction with an enzyme blend. This will ensure high processivity, specificity, fidelity, and reproducibility.

2. Common problems
a. Heavy smearing in one or more lanes:
i. Optimize Mg^{2+} concentration (usually not required).
ii. Reduce the number of PCR cycles.
iii. Decrease the amount of enzyme in the reaction.
iv. Slightly increase PCR annealing temperature.

b. No product:
i. Check the integrity of the starting RNA!
ii. Add an RNase inhibitor (e.g., RNasin) to the cDNA synthesis reaction.
iii. Test the components of the RT and PCR reactions, especially the enzymes.
iv. Run *all* recommended control reactions.
v. Check off each component as the reactions are pipetted together.
vi. Check PCR cycle parameters.

c. Low yield:
i. Incomplete homogenization or lysis of samples could be the cause.
ii. Ensure that the final RNA pellet is incompletely redissolved.
iii. If $A_{260/280} < 1.65$, the sample is probably contaminated with protein.
iv. Check for RNA degradation:
1. Tissue was not immediately processed/frozen after removing from the animal. Consider the use of RNAlater.
2. Purified RNA was stored at −20°C instead of −80°C.
3. Aqueous solutions or tubes were not RNase-free.

d. DNA contamination:
i. DNase-treatment should be standard procedure.

e. Nonreproducibility of data:

- **i.** Use only thin-walled tubes for PCR (200 μL work well). Avoid thick-walled tubes for all PCR applications.
- **ii.** Consider hot-start PCR.
- **iii.** Prepare a single, large-volume master mix per cell type and aliquot for each reaction.
- **iv.** Match T_m of upstream and downstream primer pairs as closely as possible.
- **v.** Consider enzyme blends designed for long-range PCR.
- **vi.** Nonreproducibility can occur when RNA is not isolated by the same method or on the same day. Biological and technical replicates are very important.
- **vii.** Optimize and run reactions in the same thermal cycler (not just the same model number).
- **viii.** Use a thermal cycler with a heated lid and avoid using mineral oil *except* to overlay the hybridization reactions, which are only a few microliters.

3. General suggestions

- **a.** Keep denaturation steps as short as possible.
- **b.** Keep denaturation temperature as low as possible (90°C–92°C).
- **c.** Elongation temperature can be reduced to 68°C.
- **d.** Extend elongation time for each successive cycle (5–20 seconds/cycle).
- **e.** Accurate pipetting is critical for reproducibility. Micropipettors should be recalibrated at regular intervals. Use micropipette barrier tips to avoid PCR carryover contamination.
- **f.** In some cases, very low-abundance mRNA can be superinduced by adding a subcytotoxic concentration of cycloheximide (10 ng/mL) to the cell culture growth medium and incubating for 4–6 hours before cell lysis and RNA isolation. When the protein translation apparatus is inhibited in this manner, mature mRNAs tend to accumulate in the cytoplasm. However, some cells, especially cultures of normal diploid cells, will not tolerate this treatment.
- **g.** Transcript subtraction approaches are quite sensitive to the quality of the starting material. The highest quality RNA is generally prepared in guanidinium-containing buffers.
- **h.** Silanized microcentrifuge tubes are helpful in minimizing losses during the various manipulations because both cDNA and mRNA adhere to the inside of microfuge tubes. Silanizing is especially helpful when only very small quantities of irreplaceable samples are available (see Appendix H: Silanizing Centrifuge Tubes and Glassware for protocol).

Nonsubtractive methods

An alternative strategy by which comparatively profile gene expression in two or more samples involves coupling standard methods for cDNA synthesis to PCR-based amplification using large combinations of short primers. In so doing, transcripts common to the RNA populations under investigation are manifested as PCR products of identical size when compared side by side using polyacrylamide gel electrophoresis (PAGE) when identical primer sets are used. Differential gene expression is indicated when a band in one lane is *absent* or of *different abundance* in the corresponding location of the adjacent lane(s). This strategy for gene expression profiling is based on selective amplification rather than transcript subtraction.

mRNA differential display

The mRNA differential display process is governed by exclusive patent rights granted to GenHunter Corporation (Nashville, TN). As such, GenHunter is the only authorized vendor of kits for performing differential display PCR. Visit http://www.genhunter.com for more information. The most scientifically correct terminology to describe this alternative method for the identification of up- and downregulated genes is mRNA differential display (Liang and Pardee, 1992; Liang et al., 1992; Liang et al., 1993; Mou et al., 1994; Liang et al., 1995; Zhao et al., 1995), though it has also been referred to as differential display PCR, DDPCR or the more cumbersome DDRTPCR. When using this shorthand, be certain to use uppercase letters. Writing "ddPCR" strongly suggests that the writer is referring to dideoxy sequencing. Other methods that have evolved from this general approach, but with subtle variations in primer design, are the amplified restriction fragment length polymorphism technique (Vos et al., 1995), and RNA arbitrarily primed PCR (RAP-PCR) (Welsh et al., 1992; McClelland and Welsh, 1994). Although subtly different from one another, each of these techniques essentially accomplishes the same end albeit with varying levels of coverage with respect to transcript characterization.

The strategy behind DDPCR is quite simple: incorporation of traditional synthesis of cDNA from purified RNA, followed by amplification of the products of the first-strand cDNA reaction. Unlike traditional cDNA synthesis, *several* cDNA reactions are performed for each sample, each of which is supported by a unique combination of downstream primers and upstream primers, which, in any given reaction tube, will support the amplification of only a subset of all possible PCR products. Upon PAGE analysis, DDPCR products manifest themselves as being highly abundant, not at all abundant, and everywhere in between. The data make the greatest sense and are easiest to interpret when all of the PCR products generated using the same downstream primer are electrophoresed on a single gel. "State A" PCR reactions are loaded adjacent to "State B" PCR reactions, in order of upstream primer used, an example of which is shown in Table 22.1. This permits the direct side-by-side comparison of the products from each pair of reactions. Products of identical molecular weight will be observed when cells in state

TABLE 22.1 Gel loading strategy for mRNA differential display.

Lane	cDNA source	Downstream primer	Upstream primer
1	State A	3	1
2	State B	3	1
3	State A	3	2
4	State B	3	2
5	State A	3	3
6	State B	3	3
7	State A	3	4
8	State B	3	4
9	State A	3	5
10	State B	3	5
11–48 *Same strategy*			

PCR products generated using identical primer combinations are loaded into adjacent lanes, permitting the direct side-by-side comparison of the variety of cDNAs amplified. Bands of equal molecular weight in adjacent lanes correspond to sequences transcribed in both cell types. Differences in intensity, including the complete absence of a band in one of the lanes, suggest differential modulation of specific genes. Standard-sized, older-style DNA sequencing gels favor excellent resolution and a 48-tooth comb permits loading several upstream- and downstream primer combinations for both cell types.

A and cells in state B are producing the same transcript (Fig. 22.8); the appearance of band without a corresponding band in the adjacent lane suggests differential expression (Fig. 22.9). Conveniently, 48-tooth combs used for DNA sequencing permit the loading of multiple primer combinations, using a single downstream primer, on the same gel (e.g., 2 cell samples × 1 downstream primer × 24 upstream primers). The use of a molecular weight marker is not required on this gel, the purpose of which is simply to highlight the differences in each reach pair. Molecular weight standards are more valuable following band retrieval and subsequent characterization. Taken together, these steps constitute a structured method for simultaneous examination of all transcribed sequences.

Virtually any model in which experimental manipulation causes changes in gene expression is a suitable candidate for evaluation using mRNA differential display. The method affords a side-by-side comparison of mRNA from different sources and is excellent for discerning both the induction and repression of genes. Examples include, but certainly are not limited to, the study of gene modulation (1) in the presence of pharmacological concentrations of growth

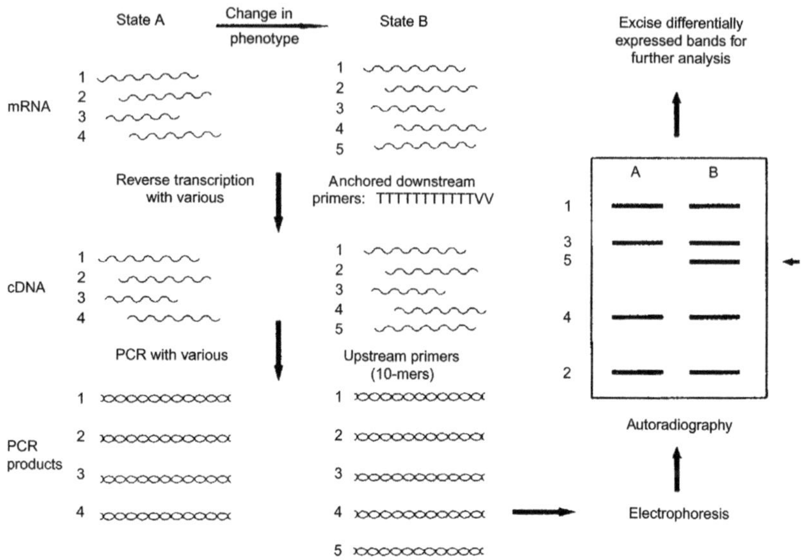

FIGURE 22.8 Schematic showing the general method used for mRNA differential display. Transcripts produced in both cell types are displayed simultaneously as PCR products when identical primer combinations are used. The presence of a PCR product from only one sample strongly suggests the differential transcriptional activity of the corresponding gene.

factors or drugs, (2) when comparing quiescent to senescent cells, (3) as a function of seasonal changes in plants, (4) in response to the expression of transfected foreign genes, (5) as a function of differentiation, (6) for the isolation of new members of multigene families, and (7) for the discernment of transcriptionally regulated sequences vs those that are posttranscriptionally regulated (relates to a method of RNA preparation – refer to Fig. 1.13).

The major steps in DDPCR are as follows:

1. **RNA isolation**. Identification of modulated sequences by DDPCR begins with the isolation of very high-quality RNA. In particular, the isolation of RNA by guanidinium-acid-phenol is one of the better, more widely accepted, purification methods for analysis by DDPCR. Investigators may wish to explore the protocols found in Chapters 4–7 or use one of the numerous commercial guanidinium-containing reagents to accomplish this isolation. Of great importance is the removal of all traces of contaminating genomic DNA (gDNA) by DNase-treating the sample. Remember that primers will base pair to any complementary molecule, DNA or RNA. The efficient removal of DNA is validated by the no-RT control reaction, described later.
2. **cDNA downstream primer design**. The standard approach to cDNA synthesis intended to support analysis by DDPCR is to perform several cDNA synthesis reactions for each unique RNA sample, reverse

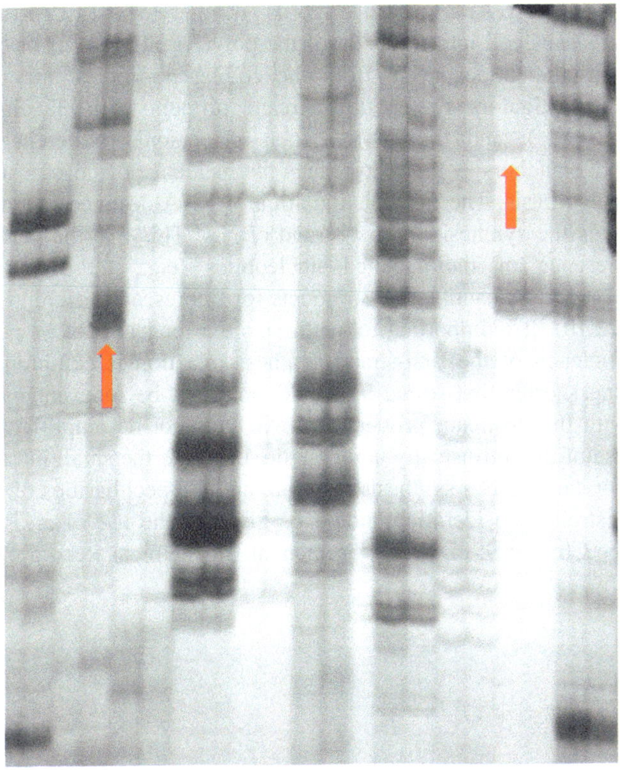

FIGURE 22.9 mRNA differential display. Typical appearance of PCR products following electrophoresis and autoradiography. Identical bands appearing in adjacent lanes represent genes transcribed in common and with similar abundance levels. Notice the differing intensities of the various product pairs, directly reflecting the relative abundance of the original RNA. Red arrows indicate the presence of a band in one lane, while the band is missing or of reduced intensity in the adjacent lane, suggesting that the corresponding gene is expressed differentially when comparing the two samples.

transcribing only a fraction of all possible messages into cDNA in any one reaction tube. Thus, each RT reaction is *intended* to generate subpopulations of cDNA so as to resolve discrete bands by electrophoresis, rather than producing a glorious smear. The key to accomplishing this lies with the use of anchored downstream primers which often have a generic structure (Liang et al., 1994):

$$5' \text{TTTTTTTTTTTTV or } 5' \text{TTTTTTTTTTTVV}$$

where V = A, C, or G.

The 3′ nucleotide degeneracy ensures the eventual reverse transcription of all possible mRNAs, albeit in different reaction tubes. Further, the 3′ nucleotide anchor ensures that the primer will base pair at the 5′-most end of the poly(A)$^+$ tail: an oligo(dT) primer (5′ TTTTTTTTTTTTT) can base pair well into the body of the poly(A) tail, dozens or even hundreds of nucleotides away from the coding portion of the transcript. Some investigators prefer downstream primers with 3′ dinucleotide anchors (VV), and this is a matter of personal preference; single-nucleotide anchors (V) tend to reduce the incidence of false positives.

3. **cDNA synthesis**. The first manipulation of RNA destined for differential display is the synthesis of first-strand cDNA. This may be performed in the presence of radiolabelled nucleotides (e.g., ^{33}P-dATP), though GenHunter also manufactures products for fluorescent differential display (Fig. 22.10). It is important to synthesize the longest cDNAs possible: if the efficiency of the first-strand synthesis reaction is compromised, the resulting shorter first-strand molecules may not be long enough to base pair with the upstream primers and will thus not be amplified by PCR. Because of the intrinsically labile nature of RNA, the small mass of RNA used for each cDNA synthesis reaction, and the peculiarities of this technique, strong consideration should be given to the use of an RNase H$^-$ RT (see Chapter 9: cDNA: A Permanent Biochemical Record of the Cell

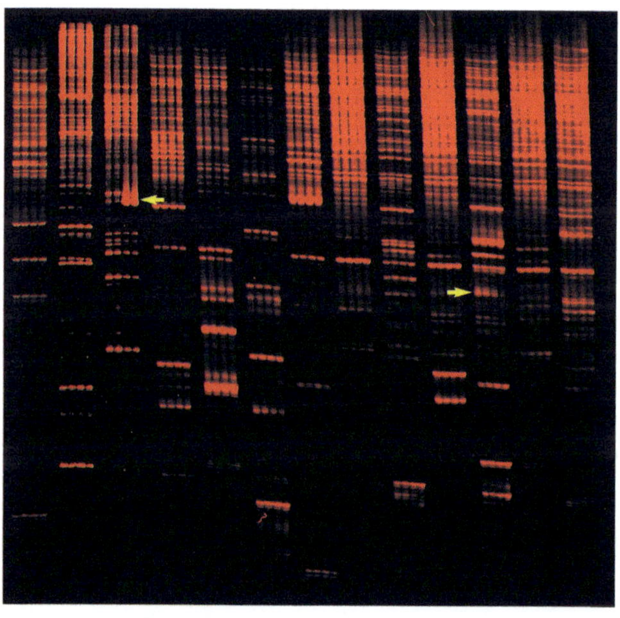

FIGURE 22.10 Identification of differentially expressed genes via 5′ rhodamine end-labeled primers. The yellow arrows indicate a clear increase (left) and decrease (right) by the change in the abundance of the PCR products in the immediately adjacent lanes. Fluorescent mRNA differential display (e.g., RNAspectra Red System) is an excellent alternative to traditional radiolabeling. *Courtesy GenHunter Corporation; www.genhunter.com.*

and Chapter 10: RT-PCR: A Science and an Art Form for details). Upon completion, the first-strand cDNA should be stored in appropriate aliquots; ideally, the second-strand synthesis reactions should be performed as soon as possible.

4. **PCR amplification**. The design of the upstream primers is generally less structured than that of the downstream primers. The upstream primers are typically 10-mers or 13-mers and should be thought of as what this Author likes to refer to as "semi-nonspecific." Given the relatively relaxed primer annealing conditions, and the fact that the primers are rather short, only seven or eight of the bases in any given upstream primer are expected to anneal. While this relatively relaxed stringency may at first be unsettling, it is a necessary component of the DDPCR approach: the idea is to be able to amplify *any* transcripts that nature can generate without having prior knowledge of their exact base nucleotide composition. Thus, it is not necessary to worry about all possible 10-mer (4^{10}) or 13-mer (4^{13}) combinations. Precisely because of the nature of the primers, hot-start PCR may be helpful.

   The upstream primers support the synthesis of the second-strand cDNA and, in conjunction with the downstream primer, will support the amplification of the cDNA by PCR. Inasmuch as there are a potentially large number of PCR reactions that could be assembled, it may not be practical to run all possible primer combinations simultaneously. Instead, one may store some of the first-strand cDNA synthesis reactions at $-80^\circ C$. Should interesting PCR products be observed by DDPCR with the first set of samples, the bands can be fully characterized, and the remaining PCR reactions, representing different primer pairs, may be performed at a later date.

   The importance of suitable controls cannot be overstated because no molecular biology technique is perfect. First, it is incumbent upon the investigator to always run a no-RT (RT^-; RT-minus) control whenever RNA is reverse transcribed. Failure to generate product is evidence that the RNA prep was free of contaminating, competing genomic DNA template material. If PCR product is generated in an RT^- control reaction, the investigator must DNase-treat the sample (with RNase-free DNase I) before moving forward. One should assume that all RNA samples are tainted with genomic DNA, and it is a tremendous time saver to DNase-treat all RNA samples at the time of isolation rather than having to backtrack.

   Second, be certain to run control reactions with only one primer, to show that the primers are not self-priming. One method that is useful in this regard is to kinase (5′ end label) the downstream primers. Thus, the only PCR products visible by autoradiography are those generated with an upstream *and* downstream primer, as opposed to upstream primers alone.

Third, a simple way to ensure the authenticity of a differentially amplified product is to assay two concentrations of the same RNA sample. A true positive, one which will generate a signal with a least two starting concentrations of the same RNA, can be further investigated. This precludes wasting time investigating artifactual variations, that is, differences in signal intensity due to the way the samples were pipetted together or the way the gel was loaded.

5. **Pilot study**. In view of the potential for technical difficulties, it is strongly recommended that a pilot study, or at least a pilot set of reactions, be developed before actually investing labor and materials on samples that may well be irreplaceable, especially if reproducibility is an issue. For example, it can be helpful to electrophorese a small aliquot of the completed reactions through a 3% agarose gel and then stain it with SYBR Green or ethidium bromide (Fig. 22.11) before making the effort to run and analyze the DDPCR products on a polyacrylamide gel. While it is true that these fluorescent dyes and a 3% gel offer very limited sensitivity and resolving power, the objective here is to simply demonstrate that the reactions actually worked before moving forward.

   Typical DDPCR reactions generate usable products that range from 75 to 800 bp, while products larger than 800 bp begin to show signs of compression. Clearer definition may be given to the larger molecular weight products by double loading the gel, as is often done for older-style, gel-based DNA sequencing purposes. Radiolabelled PCR products are visualized by autoradiography. It is usually a good idea to identify internal

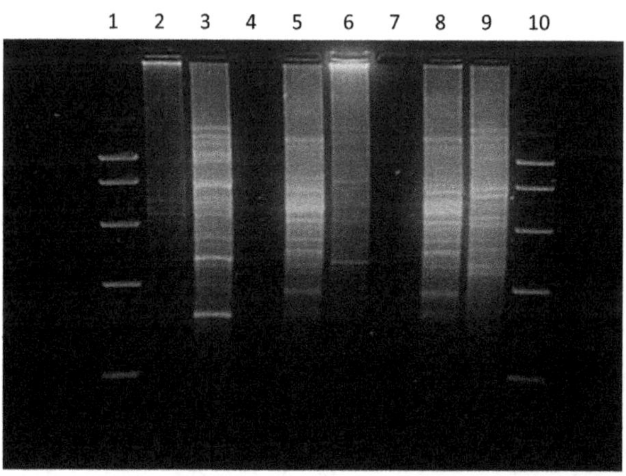

FIGURE 22.11 Appearance of DDPCR products after first round amplification when stained with ethidium bromide. Aliquots of six samples were electrophoresed in a 3% agarose gel supplemented with 1 μg/mL ethidium bromide. This diagnostic can be used as a preliminary indicator of the efficiency of the reactions. Notice that, even with this very insensitive method of visualization, some differences are obvious among the samples.

markers within a given set of reactions electrophoresed on the same gel; that is, look for the superabundant sequences in the same relative position on the gel each time the experiment is performed. Very short exposure to X-ray film will show these strong signals only, and not the lesser-strength signals: this is a good way to make a type of fingerprint of the reaction and is a way to check efficiency each time a set of samples is assayed because these abundant "landmark" sequences should show up each time an experiment is repeated. Differential display PCR is at least semiquantitative, reflecting the complexity of the RNA starting material: signal intensities of electrophoretically resolved products are in proportion to the abundance of the original mRNA template material. Finally, bands of interest are cut out of the acrylamide gel using the autoradiogram as a guide. Gel slices are heated in sterile H_2O to liberate the PCR product(s) after which it is reamplified using the same primer combination that supported its synthesis in the DDPCR reactions. These products are then cloned into a suitable vector for sequencing, confirmation of differential expression, and any of a variety of other downstream applications. There is a limit to the sensitivity of DDPCR, which might be described as semiquantitative at best. Subsequent analyses, such as real-time PCR (Chapter 10: RT-PCR: A Science and an Art Form) or nuclease protection (Chapter 18: Quantification of Specific mRNAs by Nuclease Protection) are used to measure changes in gene expression in a much more quantitative manner.

6. **Confirmation of differential expression**. After DDPCR is run once using all primer combinations, a second run is conducted using only those primer pairs that show differences in the first run. To an extent, this approach favors the elimination of errors due to micropipetting or gel issues. Of course, the exact experimental milieu that defined the first run should be duplicated as closely as possible to afford an opportunity to reproduce the data.

With respect to biological replicates, one should not expect, however, that any two experiments will be exactly reproducible because (a) cell culture conditions are impossible to duplicate precisely and (b) tissue samples from different patients are exactly that—biochemically different organisms. While many similarities are to be expected, many differences should not be unexpected. To maximize the reproducibility of different runs from the same sample, the investigator should aliquot the RNA during the initial isolation procedure such that the RNA is precipitated in several aliquots at $-80°C$, and only one of which is centrifuged and used as cDNA template material. Upon completion of the entire DDPCR protocol, the investigator will then have access to a fresh aliquot of RNA that was isolated on the same day, by the same person, and that represents precisely the same biochemical status of the cells.

Reliable methods for assessing whether a sequence purported to be differentially expressed using the mRNA differential display method truly represents a biochemical change in the cell include Northern analysis (Chapter 16: Northern Analysis), ribonuclease protection assay (Chapter 18: Quantification of Specific mRNAs by Nuclease Protection), nuclear runoff assay (Chapter 19: Analysis of nuclear RNA), real-time PCR (Chapter 10: RT-PCR: A Science and an Art Form), or even RNA-seq whole-transcriptome analysis (Chapter 24: RNA-seq: The Premier Transcriptomics Tool). When possible, the confirmation is all the more convincing when demonstrated using a non-PCR method.

The isolation of differentially modulated sequences opens many possibilities for gaining a clearer understanding of this aspect of gene regulation in the cell. Isolated DNA sequences can be reamplified repeatedly, constituting a limitless source of a template and/or nucleic acid probe material. Well-characterized sequences may also be cloned into expression vectors in order to observe the effects of overproduction of a particular protein. The gene sequence can also be mutated (site-directed mutagenesis) to determine the detriment to the cell when a misfolded or truncated protein is produced or transiently silenced via RNA interference (Chapter 13: RNA Interference and Gene Editing). The possibilities are endless and mind-boggling.

There are clear advantages and disadvantages associated with the DDPCR method which are summarized in Table 22.2. Succinctly, the requirement for small input RNA mass is nice, though it is strongly recommended that a larger mass of RNA be purified and stored as an ethanol precipitate for reference purposes at a later date, particularly when the time comes to repeat the experiment and, hopefully, reproduce the data. Moreover, the selection of poly(A)$^+$ RNA is neither required nor recommended, which may actually serve to further under-represent low-abundance transcripts.

Inasmuch as phenotypic change unquestionably involves several genes, another key advantage of DDPCR is that the mechanics of the assay favor the simultaneous display of all transcripts present in the cell at the moment of cell lysis. The speed and sensitivity of PCR facilitate the detection of very low-abundance transcripts in a relatively short time frame. A most important concept here is that differences in gene expression are manifested in the form of PCR products, though the investigator has no idea what these PCR products are until they have been sequenced. In this way, DDPCR is clearly a tool for new gene discovery and, as such, overcomes an important limitation associated with the use of microarrays.

RNA used for differential display is often prepared by guanidinium-acid-phenol extraction, which is an excellent method for RNA recovery from a variety of biological sources. Recall, however, that such chaotropic reagents yield mRNA from the cytoplasm mixed with unspliced hnRNA upon cellular disruption. Thus, there exists the potential that a sequenced PCR product may show both exon and intron sequences as well as splice variants from the same gene locus. In addition, it is often the case that two different size cDNAs, corresponding to the same transcript, are identified on the same gel.

TABLE 22.2 Advantages and disadvantages of mRNA differential display.

Advantages
• The method is simple, fast, sensitive method for detecting specific transcripts.
• The method is versatile and semiquantitative.
• It is a systematic approach to the identification of differential gene expression.
• Poly(A)$^+$ selection is neither required nor recommended.
• There is a simultaneous display of all up- and downregulated genes.
• Microgram quantities of RNA are required (often less than 3 μg).
• Data are reproducible, though this is user-dependent.
• Tool for new gene discovery; not limited by sequences preprinted on a microarray.
Disadvantages
• PCR products need to be sequenced for identification; compare to microarrays in which the sequences printed on the array are already known.
• Very pure samples of RNA are required.
• The reaction is very sensitive to contaminants, especially genomic DNA.
• The location of known sequences on the resulting gel cannot be predicted.
• Large numbers of PCR reactions are necessary to fully compare a pair of samples.
• Incidence of false positives can be high because of fairly low-stringency annealing and other variables.
• False positives may also be observed after reamplification.
• Purportedly "positive" cDNAs are sometimes contaminated with unrelated cDNA sequences (identifiable through bioinformatics analysis).
• The traditional format favors the amplification of sequences near the 3′ end of transcripts.

This occurs because in both cases all of the molecules of a specific transcript are reverse transcribed by the same downstream primer, though two or more upstream primers (different reaction tubes) find complementary sites on the cDNA, thereby supporting amplification by PCR. Although this may seem disadvantageous, it should be viewed as a means of maximizing the chances that any given transcript will be detectable.

Troubleshooting

Among the more common complaints engendered by the use of DDPCR is the seemingly nonreproducible nature of the reactions, and there appear to be limitless theories promulgated to explain this. As with all PCR-based methods, the assay is exquisitely sensitive to the precise concentration of all reaction components; thus, accurate pipetting and the use of master mixes for both cDNA synthesis and PCR are essential. The DDPCR technique works very nicely when conditions are optimal. The following suggestions outline what may be done to minimize aggravation, maximize reproducibility and productivity, and reduce user-induced error.

1. Nonnegotiables
   A. High-quality RNA:
      i. Use RNase-free technique throughout the procedures.
      ii. Use guanidinium thiocyanate- or guanidinium HCl-based lysis buffers and procedures. NP-40 lysis buffers can also be used depending on the biological source, though most protocols recommend the use of a chaotropic guanidinium lysis reagent.
      iii. Double precipitation with guanidinium lysis buffer can be helpful.
      iv. Treat samples with RNase-free DNase I.
      v. Wash RNA pellets thoroughly and extensively with 70% ethanol prepared in nuclease-free H_2O to remove excess salt; cationic salts bind dNTPs quantitatively and can cause havoc in the cDNA synthesis and PCR reactions.
      vi. Properly store purified RNA.
   B. Efficient reverse transcription into cDNA, and PCR amplification:
      i. Use equal mass of RNA per cDNA reaction (e.g., 200–300 ng).
      ii. RTs lacking endogenous RNase H activity (RNase H^-) maximize cDNA synthesis efficiency and increase yield.
      iii. Always work from a cDNA master mix.
      iv. Always work from a PCR master mix.
2. Common problems
   A. Heavy smearing in one or more lanes:
      i. Optimize Mg^{2+} concentration; decrease or increase incrementally by 0.1 mM.
      ii. Reduce the number of PCR cycles to as few as 30.
      iii. Decrease the amount of enzyme in the reaction.
      iv. Slightly increase PCR annealing temperature.

      Note: In some cases, smearing may be reduced if cycles 1–5 are annealed at 40°C and cycles 6–40 are annealed at 45°C. Be aware, however, that this strategy may reduce the variety of bands by as much as 25%–30%.
   B. No product:
      i. Check the RNA!
      ii. Add an RNase inhibitor (e.g., RNasin).
      iii. Test the components of the RT and PCR reactions, especially the enzymes.

- iv. Check off each component as the reactions are pipetted together.
- v. Check PCR cycle parameters.

C. Low yield:
- i. Consider that incomplete homogenization or lysis of samples may have occurred.
- ii. Check to see if the final RNA pellet is incompletely redissolved.
- iii. Determine if $A_{260/280} > 1.65$. A lower ratio suggests protein or other contaminants.
- iv. Check for RNA degradation:
  - a. Tissue was not immediately processed/frozen after removing from the animal.
  - b. Purified RNA was stored at −20°C instead of −80°C.
  - c. Aqueous solutions or tubes were not RNase-free.

D. DNA contamination:
- i. DNase-treatment should be standard procedure.

E. Nonreproducibility of data:
- i. Use thin-walled tubes (200 μL work well); for thick-walled tubes, extend cycle parameters.
- ii. Consider hot-start PCR (described in Chapter 10: RT-PCR: A Science and an Art Form).
- iii. Prepare a single, large-volume master mix per cell type and aliquot for each reaction and be attentive to the master mix composition.
- iv. Verify primer combinations used with each sample.
- v. Match T_m of upstream and downstream primer pairs as closely as possible.
- vi. Consider enzyme blends designed for long PCR.
- vii. Nonreproducibility can occur when RNA is not isolated by the same method or on the same day. Biological and technical replicates are very important.
- viii. Optimize reactions in the same thermal cycler (not the same model number).
- ix. Use a thermal cycler with a heated lid and avoid using mineral oil to overlay the reactions.

3. General suggestions
   - A. Keep denaturation steps as short as possible.
   - B. Keep the denaturation temperature as low as possible (90°C−92°C).
   - C. The elongation temperature can be reduced to 68°C.
   - D. The elongation time can be extended for each successive cycle (5−20 seconds/cycle).
   - E. Accurate pipetting is critical for reproducibility. Micropipettors should be calibrated at regular intervals thereafter. Use micropipette barrier tips to prevent PCR carryover contamination due to aerosol formation.

F. The mRNA differential display technique is especially sensitive to the exact input mass of RNA. As with most RT-PCR reactions "less is more," meaning that it is counter-productive to overload the reaction with starting material (200–300 ng of RNA is optimal).

G. Unlike many PCR-associated techniques, the differential display seems to be quite sensitive to the quality of the starting material. The highest quality RNA is generally prepared in guanidinium-containing buffers.

H. Silanized microcentrifuge tubes are helpful in minimizing losses during the various manipulations because cDNA and PCR products are present in picogram quantities (or less) prior to reamplification (see Appendix H: Silanizing Centrifuge Tubes and Glassware).

References

Bassett, C.L., Baldo, A.M., Moore, J.T., Jenkins, R.M., Soffe, D.S., Wisniewski, M.E., et al., 2014. Genes responding to water deficit in apple (*Malus x domestica* Borkh.) roots. BMC Plant. Biol. 14, 182. Available from: https://doi.org/10.1186/1471-2229-14-182.

Bassett, C.L., Wisneiwski, M.E., Arlip, T.S., Norelli, J.L., Renaut, J., Farrell Jr., R.E., 2006. Global analysis of genes regulated by low temperature and photoperiod in peach bark. J. Am. Soc. Hort. Sci. 131, 551–563.

Bernardi, G., 1971. Chromatography of nucleic acids on hydroxyapatite columns. Meth Enzymol. 21, 95–139.

Coche, T., Dewez, M., Beckers, M.C., 1994. Generation of an unlimited supply of subtracted probe using magnetic beads and PCR. Nucleic Acids Res. 22, 1322–1323.

Cochran, B.H., Zumstein, P., Zullo, J., Rollins, J., Mercola, M., Stiles, C.D., 1987. Differential colony hybridization: molecular cloning from a zero data base. Meth. Enzymol. 147, 64–85.

Diatchenko, L., Lau, Y.F., Campbell, A.P., Chenchik, A., Moqadam, F., Huang, B., et al., 1996. Suppression subtractive hybridization: a method for generating differentially regulated or tissue-specific cDNA probes and libraries. Proc. Natl. Acad. Sci. 93 (12), 6025–6030.

Fornace, A.J., Mitchell, J.B., 1986. Induction of B2 RNA polymerase III transcription by heat shock: enrichment for heat shock induced sequences in rodent cells by hybridization subtraction. Nucleic Acids Res. 14 (14), 5793–5811.

Hedrick, S.M., Cohen, D.I., Nielsen, E.A., Davis, M.M., 1984. Isolation of cDNA clones encoding T-cell specific membrane-associated proteins, Nature, 308. pp. 149–153.

Houge, G., 1993. Simplified construction of a subtracted cDNA library using asymmetric PCR. PCR Meth. Appl. 2 (3), 204–209.

Kulesh, D.A., Clive, D.R., Zarlenga, D.S., Greene, J.J., 1987. Identification of interferon-modulated proliferation-related cDNA sequences. Proc. Natl. Acad. Sci. 84, 8453–8457.

Lambert, K.N., Williamson, V.M., 1993. DNA library construction from small amounts of RNA using paramagnetic beads and PCR. Nucleic Acids Res. 21, 775–776.

Liang, P., Pardee, A.B., 1992. Differential display of eukaryotic messenger RNA by means of the polymerase chain reaction. Science 257, 967–971.

Liang, P., Bauer, D., Averboukh, L., Warthoe, P., Rohrwild, M., Muller, H., et al., 1995. Analysis of altered gene expression by differential display. Meth. Enzymol. 254, 304–321.

Liang, P., Averboukh, L., Pardee, A.B., 1993. Distribution and cloning of eukaryotic mRNAs by means of differential display: refinements and optimization. Nucleic Acids Res. 21, 3269–3275.

Liang, P., Averboukh, L., Keyomarsi, K., Sager, R., Pardee, A.B., 1992. Differential display and cloning of messenger RNAs from human breast cancer vs mammary epithelial cells. Cancer Res. 52, 6966–6968.

Liang, P., Zhu, W., Zhang, X., Guo, Z., O'Connell, R.P., Averboukh, L., et al., 1994. Differential display using one-base anchored oligo-dT primers. Nucleic Acids Res. 22, 5763–5764.

Lisitsyn, N., Lisitsyn, N., Wigler, M., 1993. Cloning the differences between two complex genomes. Science 259, 946–951.

Liu, J., Wisniewski, M., Droby, S., Norelli, J., Hershkovitz, V., Tian, S., et al., 2011. Increase in antioxidant gene transcripts, stress tolerance and biocontrol efficacy of *Candida oleophila* following sublethal oxidative stress exposure. FEMS Microbiol. Ecol. 80, 578–590.

López-Fernández, L.A., del Mazo, J., 1993. Construction of subtractive cDNA libraries from limited amounts of mRNA and multiple cycles of subtraction. BioTechniques 15 (4), 654–658.

McClelland, M., Welsh, J., 1994. RNA fingerprinting by arbitrarily primed PCR. PCR Methods Appl. 4, S66–S81.

Mou, L., Miller, H., Li, J., Wang, E., Chalifour, L., 1994. Improvements to the differential display method for gene analysis. Biophys. Res. Commun. 199, 564–569.

Norelli, J.L., Farrell Jr., R.E., Bassett, C.L., Baldo, A.M., Lalli, D.A., Aldwinckle, H.S., et al., 2009. Rapid transcriptional response of apple to fire blight disease revealed by cDNA suppression subtractive hybridization analysis. Tree Genet. Genomes 5, 27–40.

Orr, S.L., Gese, E., Hood, L., 1992. A new approach to understanding T cell development: the isolation and characterization of immature $CD4^-$, $CD8^-$, $CD3^-$ T cell cDNAs by subtraction cloning. Mol. Biol. Cell 3, 761–773.

Rodriguez, I.R., Chader, G.J., 1992. A novel method for the isolation of tissue-specific genes. Nucleic Acids Res. 20, 3528.

Sargent, T.D., 1987. Isolation of differentially expressed genes. Meth. Enzymol. 152, 423–432.

Sargent, T.D., Dawid, I.B., 1983. Differential gene expression in the gastrula of *Xenopus laevis*. Science 222, 135–139.

Sharma, P., Lönneborg, A., Stougaard, P., 1993. PCR-based construction of subtractive cDNA library using magnetic beads. BioTechniques 15 (4), 610–611.

Siebert, P.D., Chenchik, A., Kellogg, D.E., Lukyanov, K.A., Lukyanov, S.A., 1995. An improved PCR method for walking in uncloned genomic DNA. Nucleic Acids Res. 23, 1087–1088. Available from: https://doi.org/10.1093/nar/23.6.1087.

Travis, G.H., Sutcliffe, J.G., 1988. Phenol emulsion-enhanced DNA-driven subtractive cDNA cloning: isolation of low-abundance monkey cortex-specific mRNAs. Proc. Natl. Acad. Sci. 85, 1696–1700.

Vos, P., Hogers, R., Bleeker, M., Reijans, M., Van De Lee, T., Hornes, M., 1995. AFLP: a new technique for DNA fingerprinting. Nucleic Acids Res. 23, 4407–4414.

Welcher, A.A., Torres, A.R., Ward, D.C., 1986. Selective enrichment of specific DNA, cDNA, and RNA sequences using biotinylated probes, avidin, and copper-chelate agarose. Nucleic Acids Res. 14, 10027–10044.

Welsh, J., Chada, K., Dalal, S.S., Cheng, R., Ralph, D., McClelland, M., 1992. Arbitrarily primed PCR fingerprinting of RNA. Nucleic Acids Res. 20, 4965–4970.

Zhang, M., Zou, Z., Maass, N., Sager, R., 1995. Differential expression of elafin in human normal mammary epithelial cells and carcinomas is regulated at the transcriptional level. Cancer Res 55, 2537–2541.

Zhao, S., Ooi, S.L., Pardee, A.B., 1995. New primer strategy improves precision of differential display PCR. BioTechniques 18 (5), 842–850.

Further reading

Freeman, F.M., 2019. Differential Display of Gene Expression. Neuroscience Methods. CRC Press, pp. 209–214.

Gupta, V., et al., 2021. Recent Advances in OMICS Technologies. In: Kumar, A., Kumar, R., Shukla, P., Pandey, M.K. (Eds.), Omics Technologies for Sustainable Agriculture and Global Food Security, 1. Springer, Singapore, doi.org/10.1007/978-981-16-0831-5_1.

Lohman, J., Schickle, H., Bosch, T., 1995. REN display, a rapid and efficient method for nonradioactive differential display and mRNA isolation. BioTechniques 18, 200–202.

Chapter 23

Transcriptomes and bioinformatics

Rationale

The completion of the human genome project generated massive amounts of data that provided information about the structure and organization of the genetic blueprint of the cell. Among the benefits of this monumental undertaking was a first-time-ever ability to associate genome anatomy with genome physiology from an ultrastructure perspective as well as an ability to perform sequence-wise and structure-wise comparisons with other organisms. In the past, positional cloning (i.e., identification of a gene by finding its position on a genetic map) was one of the few imprecise tools available for zeroing in on genes that, when mutated, were responsible for the onset and progression of various diseases. People at risk for various genetic disorders were often identified by the presence of a restriction fragment length polymorphism (RFLP), meaning that DNA from normal, carrier, and afflicted individuals would be cut differently by one or more restriction enzymes; the major shortcoming of that method is that not all genetic diseases have an associated RFLP. With the development first of PCR-based methods and now next-generation sequencing, entire genomes have been sequenced from bacteria, viruses, yeast, plants, and animals, with new organisms being added continually to the ranks. Thus, in the postgenomics era, elucidation of genome function is the next sought-after prize, and the ready availability of genome information from globally available databases has greatly facilitated progress toward that goal.

It is not surprising, then, that technologies concerned with the function of the genome (i.e., regulatory and coding functions of RNA and the associated cognate proteins) are of immense, if not frenzied, interest. New and emergent disciplines and subdisciplines require new names. It became the norm to add the suffix "-omics" to a noun that has something to do with the flow or expression of genetic information, or analysis thereof, to produce a name for the discipline and this trend continues. Although one could reasonably guess the subject matter associated with the new disciplines, a short working definition for several of these terms is provided in Table 23.1 for the convenience of the reader.

RNA Methodologies. DOI: https://doi.org/10.1016/B978-0-323-90221-2.00034-5

TABLE 23.1 The "-omics."

Study of the	Is known as	And examines the	And which (is)
Genome	Genomics	DNA content of a cell	Static[a]
Transcriptome	Transcriptomics	RNA content of a cell	Changes constantly
miRNome	miRNomics	miRNA content of a cell	Changes constantly
Proteome	Proteomics	Protein content of a cell	Changes constantly
Epigenome	Epigenetics	Inventory of chemical modifications of DNA that regulate gene expression	Changes constantly
Epitranscriptome	RNA epigenetics	Inventory of posttranscriptional chemical modifications of RNA that influence its structure, function, and stability	Changes constantly
Epitome[b]	Epitopics (also, epitope mapping)	Antigenic residues found in proteins (epitope mapping)	Changes constantly
Glycome	Glycomics	Carbohydrate content of a cell	Changes constantly
Metabolome	Metabolomics	Inventory of all of the metabolites found in a biological sample	Changes constantly
Exome	Exomics	Exon content of genes and the genome, which directly impacts the proteome	Static, but subject to splicing variability

Note: Well-orchestrated gene expression and associated regulatory events at all levels result in cells, tissues, and ultimately organisms that have specific, recognizable characteristics.
[a]*Mutations, telomere length, and epigenetic modifications notwithstanding.*
[b]*"Epitome" is also the name of a database containing comprehensive epitope information.*

Genomics, transcriptomics, and proteomics are intimately linked, obviously, as professed by the central dogma of molecular biology. It is important to keep in mind, however, that DNA sequence information is only part of the story because not all genomic DNA has the capability to be "functional." Recall also that RNA, although a commonly examined parameter of gene expression, does not tell the whole story either because of the enormous turnover of hnRNA in the nucleus, the fact that mRNAs in the cytoplasm are not guaranteed to be translated, and that many RNAs are noncoding regulatory molecules. Because genomes are stable but the RNA and protein composition of the cell are not, it makes perfect sense to study both RNA and protein expression data to gain a more complete picture of the physiology of the cell under the prevailing conditions (Table 23.2).

Bioinformatics is a unique blend of basic biology, genetics, and information sciences. The heart of bioinformatics is the use of digital technology to analyze DNA, RNA (and cDNA), and protein sequence information. Inasmuch as the amount of new nucleic acid and protein sequence information continues to be deposited into various databases worldwide at an astonishing pace, informatics platforms need to efficiently interface molecular and clinical data, especially in light of the growing interest in personalized medicine. It is no exaggeration that the drug discovery process has been accelerated by taking advantage of revolutionary informatics tools, including high-speed and cloud-based "big data" analysis.

Essential vocabulary

The contemporary bioinformatics practitioner is familiar with the unique and descriptive vocabulary associated with this discipline. The use of many of the following terms is commonplace in the laboratory, in presentations made at society meetings and professional conferences, and in published work. A short list of some of the essential terms is presented here is the benefit of the novice who is beginning to work in this area.

Accession Number: A numeric identifier of a nucleic acid sequence or an amino acid sequence. The accession number acts as an electronic card catalog of sorts, facilitating expedient recovery of an entry from a database.

TABLE 23.2 Epigenetics and gene expression.

Molecule	Example	Purpose
DNA	DNA methylation (e.g., formation of 5-methylcytosine and 5-hydroxymethylcytosine)	Transient or permanent gene silencing by blocking transcription
RNA	Modification of mRNA by cytosine base acetylation (N^4-acetylcytidine)	Stability of mRNA increases translation
Protein	Extensive variety of posttranslational modifications of polypeptides (e.g., acetylation of histones) confers full functionality	Histone protein acetylation relaxes nucleosome structures, giving transcription factors and RNA polymerase access to the chromatin

Epigenetic modifications influence gene expression at many levels, often acting in concert.

Assembly: A set of sequences that represent a model of a (usually haploid) genome. For example, the current human reference genome assembly is designated as GRCh38.p14, which stands for Genome Reference Consortium Human Build 38. P14 stands for patch release 14. The reported haploid reference genome has a total sequence length of 3,099,441,038 (http://www.ncbi.nlm.nih.gov).

Bioinformatics: The marriage of computers, information technology, and biotechnology through which users are able to sort through and archive enormous quantities of nucleic acid and protein sequence information. This also extends into the realm of image informatics, where data extracted from complex gel and microarray images are quantified, interpreted, and archived.

Bit score (also, S score): A statistical value related to the alignment of a sequence provided by the user with the match(es) retrieved from the database. The higher the bit score, the better the match.

Basic local alignment search tool (BLAST) software: Perhaps the best known and most widely used bioinformatics resource. A query may be entered as either a FASTA sequence or, if known, as an accession number. Variants include:

- *BLASTn*: Searches a nucleotide database using a nucleotide query to find a sequence match. Referred to as a "standard nucleotide blast," it is the most straightforward type of search.
- *BLASTp*: Searches a protein database using an amino acid query to find a sequence match.
- *BLASTx*: Searches a protein database using a translated nucleotide query to find a match.
- *tBlastn*: Searches a translated nucleotide database using an amino acid query to find a sequence match.
- *tBlastx*: Searches translated nucleotide databases using a translated nucleotide query.

Chemogenomics: Chemical manipulation of a genome with organic compounds to study the consequences for the cell when cellular pathways are disturbed. Widely used for drug discovery.

Contig: A group of clones representing overlapping regions of a genome (contiguous pieces).

Digital object identifier (doi): A unique combination of letters, numbers, and sometimes symbols that points to the permanent location of an article on the internet. Not to be confused with an accession number.

Expect value (E value): The number of matches with the same S score that one would expect to find by chance. Low E-values (e.g., 1e-30) indicate a strong match.

Epigenetics: Refers to the covalent modification of RNA or DNA nucleotides, resulting in structural and/or functional changes to the polynucleotide without altering the base sequence. May also refer to covalent, posttranslational polypeptide modifications.

Expressed sequence tag (*EST*): A unique segment of cDNA with a base sequence identical to at least part of the coding region of a gene and generally used as a landmark for mapping. Succinctly, an EST is a partial cDNA sequence, 200–500 nts, often representing the 3′ end of the cDNA.

FASTA: A standard bioinformatics text format by which nucleotides, amino acids, or sequences of either are represented by a single letter. For example, A = adenine, M = methionine, and R = purine. FASTA is an acronym for fast adaptive shrinkage threshold algorithm.

FASTQ: A standard bioinformatics text format used to store the results of a sequencing run as well as the associated quality score.

Functional genomics: The study of the function of the genome as a whole rather than using the one-gene-at-a-time approaches associated with traditional molecular biology.

Genome: The entirety of chromosomal DNA found in a cell. Depending upon the organism, it may be useful to distinguish nuclear genomic DNA from organellar genomes (e.g., mitochondrial and chloroplast chromosomes).

Genomics: The study of the structure and organization of genes based in DNA sequencing data.

Metabolome: The overall chemical composition, that is, all metabolites, in a cell or in a biological sample.

Metabolomics: Compound profiling; the systematic study of metabolite levels in a biological sample and the associated chemical processes. The goal is to produce a fingerprint that reflects the biochemical status of a cell or an organism.

Next-generation sequencing (*NGS*): Any of a variety of non-Sanger high-throughput sequencing methods used to elucidate base sequence identify of genomic or cDNA populations. NGS is characterized by a massively parallel format which facilitates simultaneous sequencing of millions of unique DNA strands. NGS is often described as sequencing by synthesis owing to the consecutive addition of fluorescent nucleotides to growing DNA molecules during the replication of the templates.

Pangenome: An entire inventory of all genes found in all strains of a species.

Proteome: The full complement of translation products (proteins) produced by a cell at a particular time under a particular set of conditions.

Proteomics: The study of the dynamic protein composition of a cell.

Reference genome: A DNA database containing the precise order of bases representing an ideal, that is, a consensus, genome of an organism. The database is assembled in silico from the results of individual sequencing reactions that almost ways represent a number of different donors from the same species. Reference genomes, also referred to as reference assemblies, are commonly represented as haploid even though the organism may be diploid or exhibit polyploidy.

Redundancy: A situation in which sequence is found in multiple places in one or more related databases.

Transcriptome: The complete set of mRNA molecules produced by a particular cell under a particular set of conditions.

Transcriptomics: The all-inclusive analysis of an organism's transcripts and how the expression of those transcripts changes as conditions change inside the cell change.

Transcriptomes and transcriptomics

The full complement of mRNAs that a cell is producing is known as its transcriptome. In the broadest sense, this entire book is about the characterization of the transcriptome because of the numerous methods presented for the isolation and characterization of RNA. The history of transcriptional assessment looks something like this:

Northern analysis → Nuclease protection → End-point PCR → Real-time PCR → RNA-seq

Between the development of the milestone techniques listed above, additional methodologies, some more quantitative than others, emerged and have been used to answer very specific questions. For example, microarray analysis (Chapter 21) is a contemporary method for assessing which genes are expressed under a defined set of conditions and how expression changes in response to the experimental challenge. In a way, microarray analysis might be thought of as a glorified (and costlier) form of dot-blot analysis in that known DNA probe sequences immobilized on solid support hybridize fluorophore-labeled cDNA molecules that represent the RNA population in the cell at the moment of lysis.

Transcript complexity and abundance within a sample is mirrored by cDNA sequencing data. One source of expression data is ESTs (Adams et al., 1991), which are partial cDNA sequences. EST analysis can be used to profile mRNA expression in cell culture, in tissue samples, in the form of cDNA libraries, or in any biological source. A method known as serial analysis of gene expression (SAGE) (Velculescu et al., 1995), involves the analysis of a large population of very short (9–10 bp) cDNA molecules, produced by cutting much larger cDNA molecules with *Nla* III, ligating adapters, and then cutting the cDNA with an appropriate restriction enzyme (e.g. *BSM* FI) (Fig. 23.1). The resulting modified cDNAs are referred to as tags, each of which represents a defined position within a unique transcript. The tags are ligated together into concatemers, which can be readily sequenced. The frequency with which a specific tag is identified is related to the abundance of the corresponding mRNA; it is considered to be a high-throughput method for analysis of gene expression and has the advantage of not requiring any prior knowledge of the genes that will be analyzed. Collectively, SAGE and other EST analyses have produced a vast body of knowledge pertaining to gene expression in many different organisms. By comparison with SAGE, traditional EST analysis requires sequencing of significantly longer cDNAs. One thing that has become abundantly clear from these various

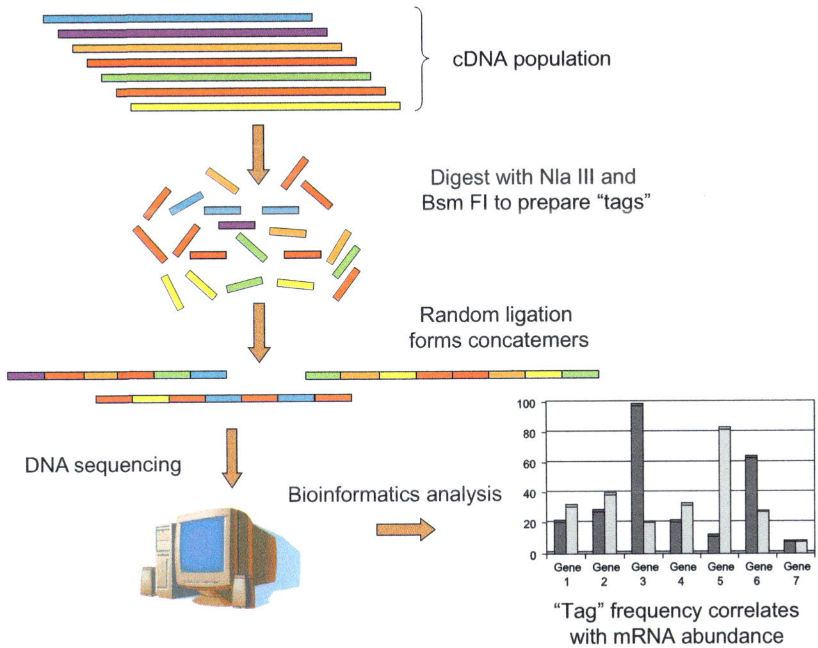

FIGURE 23.1 Serial analysis of gene expression. The sequential actions of cutting cDNA into fragments, ligating adapters, cutting the adapter-ligated cDNA, and then randomly ligating the fragments (tags), produce concatemers that are finally sequenced. The frequency with which each tag appears is in direct proportion to the frequency of an mRNA species in the original sample.

forms of expression profiling is that many genes are capable of producing more than one transcript as a result of alternative splicing and many also show variation in the choice of transcription start site (TSS). Each of the various transcripts has the potential to produce a different protein isoform.

The epitranscriptome (epigenetics of RNA)

The complexity of transcribed RNA is significantly more extensive than anyone could have imagined even as recently as 20 years ago. We now know about alternative splicing, alternative TSSs, both short and long noncoding RNA, and intracellular strategies for both the posttranscriptional enhancement and suppression of gene expression. In addition, much has come to light about the covalent modification of RNA. RNA epigenetic modifications were discovered decades ago and only more recent investigations have shown that epitranscriptomic changes to RNA are the direct counterpart of the well-known epigenetic modifications commonly observed throughout the genome; they may also be considered an indirect counterpart of the posttranslational modifications that are widely observed among polypeptides.

More than 170 epigenetic modifications are known to occur in RNA (Boccaletto et al., 2017). The prevalence of each modified RNA nucleotide varies in abundance and location compared to the others and, in many cases, the function remains unclear. Many of these modifications were discovered through the combined use of liquid chromatography, mass spectrometry, and direct sequencing. Examples of some of the better-characterized mRNA modifications are presented in Table 23.3. The RNA epigenomics database, known as MODOMICS, can be accessed at https://iimcb.genesilico.pl/modomics/.

TABLE 23.3 Select mRNA epigenetic modifications. Grozhik and Jaffrey, 2017; Wiener and Schwartz, 2021

Nomenclature	Symbol	Probable mRNA prevalence[a]	Overall prevalence in RNA
N^6-methyladenosine	m^{6}A	High (most highly abundant modification in mRNA)	High
Pseudouridine	Ψ	Low	Medium to high, primarily in rRNA and tRNA; most commonly modified nucleotide after m^{6}A
5-Methylcytosine	m^{5}C	Low to medium	Medium
N^1-methyladenosine	m^{1}A	Very low overall	High in tRNA
N^4-acetylcytidine	ac^{4}C	Not confirmed	Very low
N^7-methylguanosine	m^{7}G	Abundant in 5′ mRNA cap; also present internally in mRNA	Unknown
8-Oxoguanosine	8-oxo-G	Forms in response to oxidative stress	Unknown
N^6,2′-O-dimethyladenosine	m^{6}A$_m$	Uniquely found at the first transcribed nucleotide of mRNA	High prevalence in eukaryotic mRNA; has an impact on mRNA stability
2′-O-methylated nucleotides (N$_m$)	A$_m$, C$_m$, G$_m$, U$_m$	Found at the second transcribed nucleotide of mRNA; also found internally in yeast RNA	High in tRNA and rRNA

[a]*The relative abundance of many modified RNA nucleotides is not clear-cut and is often a matter of debate owing to mapping approach issues.*
Sources: Mauer et al. (2017); Linder and Jaffrey (2019); Boccaletto et al. (2022); Grozhik and Jaffrey (2017); Wiener and Schwartz (2021); MODOMICS database (http://genesilico.pl/modomics/modifications).

Many epitranscriptomic-modified nucleotides are known to occur with a consensus motif, for example:

−G−m^6A−C−[N^6-methyladenosine]

and

−G−G−G−T−Ψ−C−[Pseudouridine]

In contrast, the positioning of other types of modified RNA nucleotides appears to be random or directed by some nonmotif-driven mechanism. It would appear that cellular and tissue differentiation in terms of both form and function is driven by the cellular response to both internal and external cues. Many of these cues are posttranscriptional in nature and involve epitranscriptomic events; what is very clear is that m^6A and other modified RNA nucleotides are powerful regulators of gene expression at the posttranscriptional level. It is likewise important to acknowledge the roles of translational, and posttranslational regulatory events such as protein splicing and other forms of maturation, which themselves may be tied to patterns and prevalence of epitranscriptomic nucleotides.

Most epigenetic modifications of RNA are found in transcripts of the noncoding variety, to wit, snRNA, tRNA, and rRNA; the modifications are often of considerable importance in the tertiary structure(s) that these molecules form and their concomitant functions. RNA epigenetic modifications are also found in mRNA and were originally observed in the eukaryotic 5′ cap (7-methylguanosine). Modified nucleotides are now known to exist in the structural portion of mRNA, including the abundant N^6-methyladenosine (m^6A) nucleotide, where their presence is functionally associated with normal growth and development. Conversely, cellular anomalies involving the presence or absence of modified bases may lead to the disease state. In addition to cellular RNAs, m^6A is widely observed in various RNA viruses. The prevalence of epitranscriptomic modifications varies greatly; generally speaking, they are found more commonly among rRNA, tRNA transcripts, and other noncoding transcripts (Machnicka et al., 2013) while of much lesser prevalence in mRNA, as noted above. Though some epigenetic RNA modifications are rare, it is very likely they exert important regulatory control over various cellular functions.

With respect to mRNA methylation, there are two discrete domains, namely, those associated in or near the 5′ cap and those found within the structural portion of the mRNA. Among these two domains, there are presently five known methylation patterns associated with mRNA, namely, m^6A (internal mRNA nucleotides), m^7G (5′ cap), N_{m1} (2′-O-methylation of first transcribed mRNA nucleotide; usually m^6A_m), N_{m2} (2′-O-methylation of the second transcribed mRNA nucleotide), and m^6A_m (N^6 methylation associated with A_m in the first transcribed mRNA nucleotide). These mRNA epigenetic modifications exert considerable influence over maturation, splicing, stability, decay, intracellular trafficking, enhanced resistance miRNA, translation, and may even contribute to overall homeostasis, particularly in T-cells (Li et al., 2017). The prevalence of many other epigenetic modifications has not been determined with high resolution due, in part, to the difficulty, and variegated methods for detecting modifications, and the high number of false positives associated with some mapping procedures. Unlike the well-documented reversibility of DNA epigenetic modification, the reversibility of at least some RNA epigenetic changes remains obscure. For example, the formation of m^6A is widely acknowledged to be reversible methylation, while the fate of other modified nucleotides is less certain. If it is eventually determined that RNA epigenetic modifications are more recalcitrant to reversal than their DNA counterparts, one could easily envision the cell ridding itself of unnecessary or unneeded epitranscriptomic-modified nucleotides by virtue of the relatively short half-life of cellular RNA: epitranscriptomic modifications are eliminated by natural or accelerated decay of the RNA, particularly mRNA. RNA modification patterns are also known to the altered by the presence of viral RNA (Lichinchi et al., 2016).

RNA–chromatin, RNA–RNA, and RNA–protein interactions

RNA and proteins are each fascinating biomolecules in their own right, though it is rare that one or the other is not interacting with other macromolecules. As noted in Chapter 1, an RNA molecule has an intrinsic ability to form impressive secondary and tertiary structures, often through intramolecular base-pairing. Similarly, with proteins, the formation of the correct tertiary structure is a nonnegotiable requirement for normal functionality. Both RNA and protein molecules are well known for their ability to interact with each other, and with similar molecules for any variety of biochemical regulatory activities. Thus, there is a great deal of interest in the ability to detect these interactions. There are many methods for detecting RNA–RNA and RNA–protein interactions, the choice of which is governed by the questions being asked and the experimental goal; for review see Ramanathan et al. (2019). While the broad range and details of these methods are beyond the scope of this volume, it is important to briefly describe some of the options here.

1. *Nitrocellulose filter binding*: Nitrocellulose is well-known among experienced life-science investigators as a blotting filter membrane upon which proteins may be immobilized for Western blot analysis. Because of the affinity that this substrate has for proteins, it is easy to envision how proteins, and any RNAs bound to them would be captured on the filter, while unbound nucleic acids would not be retained and simply pass through the filter. In terms of capturing proteins and the RNAs with which they may have formed complexes, nitrocellulose does not work well with crude cell lysates because of the high levels of nonspecific retention that will occur, resulting in unacceptable levels of background noise. Rather, nitrocellulose filter binding is better suited for work in which one or a few purified proteins are mixed with a nucleic sample, either RNA or DNA, to determine what type of molecular interactions take place.

2. *Cross-linking*: The methods used to isolate RNA can be quite harsh. The same is true of many protein isolation methods. This means that any natural interactions, including binding between molecules, will almost certainly be lost during the course of macromolecule extraction from the cell. To preclude this from occurring, one strategy is to crosslink macromolecules so that they cannot easily separate, allowing the investigator an opportunity to study them following recovery from the cell. Traditional cross-linking agents include psoralen, a calibrated UV light source, and formaldehyde. The cross-linking immunoprecipitation (CLIP) method (Fig. 23.2) involves antibody capture of the protein of interest, after cross-linking it to all of the RNA molecules with which it was interacting (Jensen and Darnell, 2008). This method is particularly effective when coupled with next-generation sequencing. A newer cross-linking option, PAR-CLIP (Garzia et al., 2017) utilizes photoactivatable-ribonucleoside-enhanced cross-linking. After incubating cells in the presence of a photoreactive nucleoside such as 4-thiouridine, cross-linking at 365 nm increases the efficiency of the cross-linking process and also reduces the background signal.

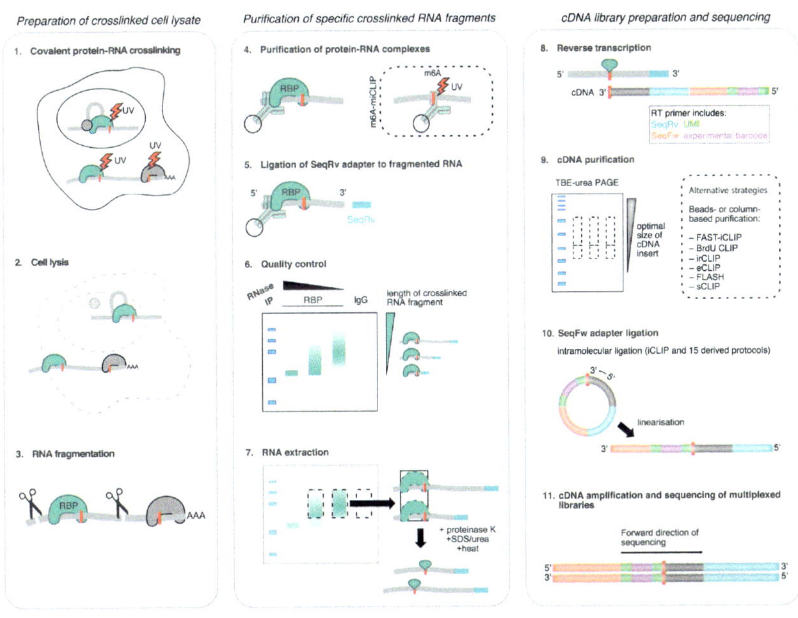

FIGURE 23.2 Overview of CLIP workflow. The major steps include cross-linking proteins with the engaged RNA, RNA recovery and purification, and then RNA identification by sequencing or other methods such as RT-PCR (Lee and Ule, 2018). *Reproduced from Lee, F.C. Y. Ule, J., 2018. Advances in CLIP technologies for studies of protein-RNA interactions. Mol. Cell 69 (3), 354–369. https://doi.org/10.1016/j.molcel.2018.01.005, with permission from Elsevier.*

3. RNA–ChIP, that is, the RNA–chromatin immunoprecipitation assay, is a standardized method of detecting the interaction of RNA and proteins by "pulling down" crosslinked nucleic acid–protein complexes by antibody-mediated immunoprecipitation. Since the inception of this method two decades ago, RNA-ChIP has become a mainstream method for mapping and quantifying the extent to which proteins and RNA interact and mediate each other's function(s). Contrary to the sound of its name, RNA–ChIP does not necessarily require a physical chip. In some variations of the protocol, RNA–protein complexes are sequestered using widely utilized and very efficient magnetic bead technology. As shown in Fig. 23.3, the major steps in this process include cross-linking the sample(s), DNase treatment to remove genomic DNA contaminants, sequestration of the target complexes with antibody-couple magnetic beads, elution, and quantification of the transcripts of interest by PCR. More traditional methods to assay the cognate RNAs, such as Northern analysis or nuclease protection lack the sensitivity to provide reliable data, though interestingly enough, an immuno-Northern blot can be performed by utilizing antibodies against naturally or artificially modified nucleosides (Mishima et al., 2015).

Among the keys to the successful outcome of an RNA-ChIP or other pull-down assay are excellent RNase-free technique (naturally) and a well-characterized antibody for precision pull-down.

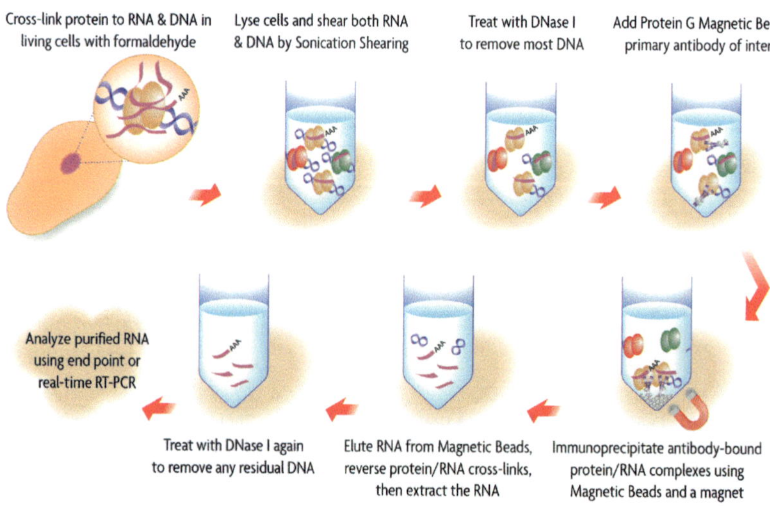

FIGURE 23.3 Flow chart of the RNA-ChIP process. *Courtesy Active Motif (http://www.activemotif.com).*

Aptamer biology

Aptamers are small, single-stranded RNA or DNA molecules which, by virtue of their exact base sequence, are able to fold into a characteristic three-dimensional shape (refer to Fig. 26.3). This, in turn, allows aptamers to bind to one or more ligands, much like epitope recognition by an immunoglobulin molecule. The smaller size of aptamers allows them to penetrate cells and histological structures more easily than antibodies and the high precision of aptamers offers tremendous potential for diagnostic and therapeutic applications.

The method by which aptamers are commonly developed is a process known as SELEX, or systematic evolution of ligands by exponential enrichment. It is a repetitive, stepwise methodology for identifying a nucleic acid sequence for which a protein has a binding preference based on the shape into which an aptamer folds (Tuerk and Gold, 1990; Ellington and Szostak, 1990). In the SELEX approach, an aptamer library consisting of up to 10^{14} unique, random 60–80 nts single-stranded sequences are generated and mixed with a ligand. Aptamers with no affinity for the ligand are washed away. The bound aptamers are recovered from the ligand, PCR amplified (the sequences at the ends of the aptamers are constant), and then exposed to the ligand again. By performing this process repeatedly, eventually only the aptamer(s) with high affinity and specificity for the ligand will be retained. The aptamer is then sequenced so that it can be synthesized again, as needed. Potentially useful targets of aptamers include specific membrane-localized or subcellular structures for both diagnostic and therapeutic purposes—a type of smart bullet. Macugen (Pfizer), for example, is an aptamer drug that can be used to treat macular degeneration. Further, aptamers have the potential to be modified by the addition of various functional groups (Aptabodies; http://www.aptagen.com) to confer drug-like properties.

Bioinformatics

Vast new amounts of novel nucleic acid- and protein-sequencing data are produced daily, most of which eventually become part of the public domain. The National Center for Biotechnology Information (NCBI; Bethesda, Maryland) was established for the expressed purpose of making these data available to anyone wishing to access it, and at the same time provide the proper software tools to support queries into the database. Similarly, the European Molecular Biology Laboratory (EMBL) (Heidelberg, Germany), and the DNA Data Bank of Japan (DDBJ) (Mishima, Japan) coordinate their efforts with NCBI to provide ready access to this massive quantity of data to investigators worldwide. Collectively, NCBI, EMBL-EBI (European Bioinformatics Institute), and DDBJ constitute the International Nucleotide Sequence Database Collaboration (INSDC). All three organizations accept sequencing data and have standardized the format for submissions. The well-known Sequin tool that was previously used for submitting eukaryotic or prokaryotic sequences to GenBank was retired in January 2021. It was replaced by the Genome Workbench Submission Wizard which can be accessed at http://www.ncbi.nlm.nih.gov/tools/gbench. As of August 2022, Genbank contained approximately 1.5 trillion base pairs of data representing over 239 million entries. Check out the statistics anytime at ftp.ncbi.nih.gov/genbank/gbrel.txt. Since the inception of GenBank, the number of bases it contains has doubled approximately every 18 months.

A variety of other specialized search tools and databases are readily accessible from these central portals and Table 23.4 lists some important bioinformatics websites. These allow the user to search not only for nucleic acid and protein sequence information but scientific literature as well.

Search for genes—have a BLAST!

The amount of data that is available through GenBank is impressive, to the point that it can even be frightful. More and more, bench scientists are becoming in silico savvy, a necessary skill if one is to be able to mine what has become known as "big data." The NCBI home page is one resource with which the contemporary molecular biologist should have familiarity. While navigation may seem overwhelming during the first few visits, a number of tutorials, videos, and documentation materials can be readily accessed at http://www.ncbi.nlm.nih.gov/home/learn. Still, other tutorials are readily available on YouTube.

The ever-increasing wealth of information resulting from genome, cDNA, and protein sequencing represents a daunting challenge to investigators who wish to utilize this information: how does one efficiently search and recover specific items from databases consisting of hundreds of billions of bases? Perhaps the best-known resource is the Basic Local Alignment Search Tool software, better known as BLAST, a powerful and versatile tool that searches databases for nucleic acid or amino acid sequences that have high-scoring matches with the input sequence. A BLAST query result will also show all matching sequence alignments and report the statistical significance of those alignments. The BLAST program can be accessed directly at http://www.blast.ncbi.nlm.nih.gov, and many other internet sites have links to BLAST.

The process for performing a BLAST search is simple, and the most straightforward method for nucleic acid analysis is to select "nucleotide BLAST" (BLASTn). Users are asked to either manually enter or cut-and-paste the desired search string, that is, the query (the nucleotide sequence being investigated). Note that one may enter more than one query in a single search, each preceded by a right arrow and the number of the query,

TABLE 23.4 Important bioinformatics websites.

National Center for Biotechnology Information (NCBI)	http://www.ncbi.nlm.nih.gov	GenBank and related bioinformatics tools
European Molecular Biology Laboratory (EMBL)	http://www.embl.org	EBI core databases
RNAcentral	https//rnacentral.org	Noncoding RNA sequence database
MODOMICS	https://iimcb.genesilico.pl/modomics/	RNA epitranscriptome database
Ensembl	http://www.ensembl.org	Browser for vertebrate genomes
The Arabidopsis Information Resource (TAIR)	http://www.arabidopsis.org	*Arabidopsis* database and information resource (TAIR)
DNA Data Bank of Japan (DDBJ)	http://www.ddbj.nig.ac.jp	DNA Data Bank of Japan
UniProt[a]	http://www.uniprot.org	Protein sequence database
PeptideAtlas	http://www.peptideatlas.org	Mass spectrometry proteomics data
The European Bioinformatics Institute (EMBL-EBI)	http://www.ebi.ac.uk	Microarray functional genomics data and links to many other databases
Codon Usage Database (Kazusa)	http://www.kazusa.or.jp/codon/	Codon utilization tables

[a]*UniProt is a central protein sequence repository. It was created by the merger of Swiss-Prot, PIR-PSD, and TrEMBL.*

followed by the query itself. For example, three separate queries would be entered as

>seq1
TGTCGTAAATCCGTACGTCTACC...
>seq2
CTAGTACCATCAATCATCCAGAGGCAT...
>seq3
ATGCTGATCGCTATCTACCA...

The user then has the option of searching an entire query sequence or portions thereof by entering the first and last nucleotides in the "FROM" and

"TO" windows, respectively. The user is also asked to select a database to be searched. Search tip: search the smallest database that is likely to contain the sequence of interest. This strategy will greatly reduce the "noise" among the search results. As needed, the search can be narrowed down to a model organism. In contrast, selecting the default nonredundant database "nr" will perform the most comprehensive search possible. If available, the user may also enter the accession number to examine that single entry. Finally, click on the BLAST icon to begin the search process.

Depending on a number of variables, the search may require anywhere from seconds to minutes, or longer, to complete. The BLAST software will provide an indication of the approximate time needed to complete the search and will also provide a request ID (RID) number, which is a reference number for the transaction. When the search has been completed, another window will then open in which the results of the BLAST search will be presented. When the search results are finally available, there will be two scores assigned to each sequence: (1) the *S score* (bit score), a statistical value related to the alignment (number of matches) between the sequence provided by the user and the match(es) retrieved from the database; and (2) the *Expect value* (E score), which is the number of alignment matches with the same S score that one would expect to find by chance. Thus, for any BLAST search, the higher the S score and the lower the E value (should be much lower than 1.0), the better the match.

Most BLAST searches return multiple matches, particularly when the query is run against the nr database. Clicking on "Taxonomy" will reveal a list of organisms that matched the query and, if desired, a phylogenetic "Distance tree of results" can also be generated. In the event that a DNA sequence is not available for performing a BLAST search, the NCBI website has other extensive, very impressive resources with which one can search for genes and proteins. In the event that assistance is needed interpreting the results of a search, one may contact the BLAST help desk via https://support.nlm.nih.gov/support/create-case, or by sending an email directly to blast-help@ncbi.nlm.nih.gov. Be sure to have the RID that is associated with the search; RIDs are typically retained on the NCBI server for 36 hours.

To become familiar with the proper and expedient use of the BLAST programs, one can go to any of the above-referenced websites at any time and practice using the various information tools. A convenient way to approach a first-time practice session is to enter either a known sequence, perhaps related to a gene being studied in the lab, or the PCR primers being used to amplify portions of the gene. Alternative, one can simply make up a sequence for the purpose of gaining familiarity with the search process. Also noteworthy is the Primer-BLAST tool which can be used to design primers for your own template. Give it a try at http://www.ncbi.nlm.nih.gov/tools/primer-blast. It is enlightening to visit the NCBI website and explore the possibilities!

References

Adams, M.D., Kelley, J.M., Gocayne, J.D., Dubnick, M., Polymeropoulos, M.H., Xiao, H., et al., 1991. Complementary DNA sequencing: expressed sequence tags and human genome project. Science 252, 1651–1656.

Boccaletto, P., Stefaniak, F., Ray, A., Cappannini, A., Mukherjee, S., Purta, E., et al., 2022. MODOMICS: a database of RNA modification pathways. 2021 update. Nucleic Acids Res. 50 (D1), D231–D235. Available from: https://doi.org/10.1093/nar/gkab1083.

Boccaletto, P., Machnicka, M.A., Purta, E., Piatkowski, P., Baginski, B., Wirecki, T.K., et al., 2017. MODOMICS: a database of RNA modification pathways. 2017 update. Nucleic Acids Res 46, D303–D307. Available from: https://doi.org/10.1093/nar/gkx1030.

Ellington, A.D., Szostak, J.W., 1990. In vitro selection of RNA molecules that bind specific ligands. Nature 346, 818–822.

Garzia, A., Meyer, C., Morozov, P., Sajek, M., Tuschl, T., 2017. Optimization of PAR-CLIP for transcriptome-wide identification of binding sites of RNA-binding proteins. Methods 118–119, 24–40. Available from: https://doi.org/10.1016/j.ymeth.2016.10.007.

Grozhik, A.V., Jaffrey, S.R., 2017. Epitranscriptomics: shrinking maps of RNA modifications. Nature 551, 174–176.

Jensen, K.B., Darnell, R.B., 2008. CLIP: crosslinking and immunoprecipitation of in vivo RNA targets of RNA-binding proteins. Meth. Mol Biol 488, 85–98.

Lee, F.C.Y., Ule, J., 2018. Advances in CLIP technologies for studies of protein-RNA interactions. Mol. Cell 69 (3), 354–369. Available from: https://doi.org/10.1016/j.molcel.2018.01.005.

Li, H.B., Tong, J., Zhu, S., Batista, P.J., Duffy, E.E., Zhao, J., et al., 2017. m^6A mRNA methylation controls T cell homeostasis by targeting the IL-7/STAT5/SOCS pathways. Nature 548, 338–342. Available from: https://doi.org/10.1038/nature23450.

Lichinchi, G., Zhao, B.S., Wu, Y., He, C., Rana, T.M., 2016. Dynamics of human and viral RNA methylation during Zika virus infection. Cell Host Microbe 20 (5), 666–673. Available from: https://doi.org/10.1016/j.chom.2016.10.002.

Linder, B., Jaffrey, S.R., 2019. Discovering and mapping the modified nucleotides that comprise the epitranscriptome of mRNA. In: Cech, T.R., Steitz, J.A., Atkins, J.F. (Eds.), RNA Worlds: New Tools for Deep Exploration. Cold Spring Harbor Laboratory Press, Cold Spring Harbor, NY. Available from: http://doi.org/10.1101/cshperspect.a032201.

Machnicka, M.A., Milanowska, K., Oglou, O.O., Purta, E., Kurkowska, M., Olchowik, A., et al., 2013. MODOMICS: a database of RNA modification pathways-2013 update. Nucleic Acids Res 41, D262–D267.

Mauer, J., Luo, X., Blanjoie, A., et al., 2017. Reversible methylation of m^6A_m in the 5′ cap controls mRNA stability. Nature 541, 371–375. Available from: https://doi.org/10.1038/nature21022.

Mishima, E., Jinno, D., Akiyama, Y., Itoh, K., Nankumo, S., Shima, H., et al., 2015. Immuno-Northern blotting: detection of RNA modifications by using antibodies against modified nucleosides. PLOS One . Available from: https://doi.org/10.1371/journal.pone.0143756.

Ramanathan, M., Porter, D.F., Khavari, P.A., 2019. Methods to study RNA–protein interactions. Nat Meth. 16, 225–234. Available from: https://doi.org/10.1038/s41592-019-0330-1.

Tuerk, C., Gold, L., 1990. Systematic evolution of ligands by exponential enrichment: RNA ligands to bacteriophage T4 DNA polymerase. Science 249, 505–510.

Velculescu, V.E., Zhang, L., Vogelstein, B., Kinsler, K.W., 1995. Serial analysis of gene expression. Science 270, 484–487.

Wiener, D., Schwartz, S., 2021. The epitranscriptome beyond m^6A. Nat Rev Genet 22, 119–131. Available from: https://doi.org/10.1038/s41576-020-00295-8.

Further reading

Bartee, D., Nance, K.D., Meier, J.L., 2022. Site-specific synthesis of N^4-acetylcytidine in RNA reveals physiological duplex stabilization. J. Am. Chem. Soc. 144 (8), 3487–3496. Available from: https://doi.org/10.1021/jacs.1c11985.

Besenbacher, S., Liu, S., Izarzugaza, J.M.G., Grove, J., Belling, K., Bork-Jensen, J., et al., 2015. Novel variation and de novo mutation rates in population-wide de novo assembled Danish trios. Nat. Commun. 6, 5969. Available from: https://doi.org/10.1038/ncomms6969.

Darnell, R.B., Ke, S., Darnell Jr., J.E., 2018. Pre-mRNA processing includes N6 methylation of adenosine residues that are retained in mRNA exons and the fallacy of 'RNA epigenetics'. RNA 24, 262–267.

Engel, M., Chen, A., 2017. The emerging role of mRNA methylation in normal and pathological behavior. Genes Brain Behav 17 (3), e12428. Available from: https://doi.org/10.1111/gbb.12428.

Gokhale, N.S., Horner, S.M., 2017. RNA modifications go viral. PLOS Pathogens . Available from: https://doi.org/10.1371/journal.ppat.1006188.g001.

Klungland, A., Dahl, J.A., Greggains, G., Fedorcsak, P., Filipczyk, A., 2017. Reversible RNA modifications in meiosis and pluripotency. Nat. Meth. 14, 18–22.

Mauer, J., Jaffrey, S.R., S. R., 2018. FTO, m6Am, and the hypothesis of reversible epitranscriptomic mRNA modifications. FEBS Lett 592, 2012–2022.

Pace, C.N., Scholtz, J.M., 1998. A helix propensity scale based on experimental studies of peptides and proteins. Biophys. J. 75, 422–427.

Schaefer, M., Kapoor, U., Jantsch, M.F., 2017. Understanding RNA modifications: the promises and technological bottlenecks of the 'epitranscriptome'. Open Biol 7, 17007. Available from: https://doi.org/10.1098/rsob.170077.

Zhao, B.S., Nachtergaele, S., Roundtree, I.A., He, C., 2018. Our views of dynamic N6-methyladenosine RNA methylation. RNA 24, 268–272.

Chapter 24

RNA-seq: the premier transcriptomics tool

Rationale

RNA-seq is an amalgam of standard molecular biology techniques and computational bioinformatics that are designed to identify and quantify transcriptional activity in the cell. The nomenclature, RNA-seq, comes from "RNA sequence analysis" and the technology is a widely regarded asset within the scientific community. The utility and popularity of this transcriptomics tool are based on the high-throughput and massively parallel character of NGS (next-generation sequencing) and the unique characteristics of third-generation sequencing which are exemplified by RNA-seq. RNA-seq has revolutionized and invigorated the mining of the transcriptome for its secrets by providing much greater latitude in experimental design and providing an understanding of the transcriptome at heretofore unknown levels of detail. This represents an unprecedented opportunity to further refine the current models of gene expression. In addition to the coding, noncoding, and intergenic RNAs that are already known, it is likely that there is a treasure trove of unannotated transcripts awaiting characterization.

RNA-seq produces a quantitative and a qualitative profile of each sample. The quantitative profile reflects the expression level of each gene and permits the discernment of abundance relationships among RNAs in the sample across a wide dynamic range. The qualitative data that are generated are essential for observing structural alterations in transcripts as a consequence of transcriptional or posttranscriptional events. While earlier molecular biology hybridization-, PCR-, or microarray-based methods were limited to the study of only one or a few transcripts in a single experiment, RNA-seq will assay whatever transcripts are present in the starting material. Further, no a priori knowledge or designation of target sequences is needed.

The goals of nucleic acid sequencing include demonstrating collinearity with a previously identified protein, assigning a nucleic acid sequence to a multigene family based on some level of homology, or annotation of sequences previously not observed. The resulting base sequence identity of each unique RNA can also be used to predict the manner in which the molecule will assume secondary- and tertiary-folded structures, a relatively new

RNA Methodologies. DOI: https://doi.org/10.1016/B978-0-323-90221-2.00045-X

area known as the structurome. These data can be used to make similar predictions for any encoded peptide(s) as well. Without a doubt, RNA-seq is a versatile and powerful method for transcriptome analysis that has revolutionized the world of RNA biology.

Essential vocabulary

Barcoding: The addition of a six-base sequence to all of the cDNA from a specific sample. Adding a unique barcode to each sample allows multiple samples to be mixed together and sequenced simultaneously (multiplex sequencing).

Base calling: The assignment of the identity of each consecutive base in a sequencing reaction.

Coverage (also known as *sequencing depth*): The number of times each template nucleotide is sequenced.

ddNTP: A dideoxynucleotide, that is, a DNA nucleotide with a 2′-H and 3′-H.

de novo genome assembly: The alignment of each molecule in a sequencing reaction based on sequence overlap and in the absence of a reference sequence (reference transcriptome or reference genome).

dRNA-seq: Direct sequencing of RNA molecules, for example, Nanopore sequencing, without chemical modification or creating cDNA by reverse transcription. May also stand for differential RNA-seq.

ERCC: The External RNA Controls Consortium is an independent ad hoc group that developed a series of RNA transcripts designed to be spiked into a RNA-seq reaction in order to measure performance using a particular protocol on a particular platform. Using the ERCC standardized transcripts is particularly important when performing scRNA-seq.

Genome: The entirety of DNA found in the cell. The term usually refers to nuclear DNA, rather than mtDNA or cpDNA, unless otherwise specified. Mutations and epigenetic modifications notwithstanding, the genome is considered to be static.

Genomics: The comparative study of the structure, organization, and function of the genome.

Paired-end sequencing: The sequencing of both ends of a template by performing sequencing reactions initiating from each end of the molecule; that is, sequencing both strands using a unique adapter on each end.

Read: One of perhaps millions of short molecules that are created by RNA or cDNA fragmentation and then sequenced.

Read depth: The total number sequencing reads per sample, for example, how many molecules were produced through the sequencing-by-synthesis reactions. This may range from 10,000,000 to well in excess of 30,000,000 reads per sample.

Read length: The number of bases sequenced per molecular template, that is, sequence reads, which typically range from 50 to 200 for RNA-seq.

Redundancy: A situation in which homologous sequences appear together in the same data set/sample; essentially, two or more transcripts/genes perform the same function.

Reference genome: A DNA database containing the precise order of bases representing an ideal, that is, a consensus, genome of an organism. The database is assembled in silico from the results of individual sequencing reactions that almost always represent a number of different donors from the same species. Reference genomes, which are often referred to as reference assemblies, are commonly represented as haploid even though the organism may be diploid or exhibit polyploidy.

RNase H: An enzyme that cleaves the RNA strand in an RNA:DNA hybrid.

RNA-seq: RNA sequencing.

Sanger sequencing: A sequencing process by which the order of bases in molecules is determined via chain termination reactions mediated by labeled dideoxynucleotides.

scRNA-seq: Single-cell RNA-seq.

Single read: Sequencing of only strand of DNA. Produces large amounts of data quickly and economically. This maybe is sufficient to draw accurate conclusions pertaining to gene expression when working with well-annotated reference genomes.

Tag-seq: A one-time lower cost alternative to RNA-seq which focuses on 3′-end sequencing, thereby reducing redundancy and cost.

Transcriptome: The entirety of RNA being produced by a cell at any particular moment, which includes all coding and noncoding RNA species. Since the cell responds quickly to changes in its local environment, the transcriptome is in constant flux.

Transcriptomics: The comparative study of the RNA composition of the cell.

Overview of RNA-seq

Nucleic acid sequencing enjoys a colorful past. Technically speaking "RNA sequencing" first occurred in the mid-1960s when the first tRNA was sequenced (Holley et al., 1965) using enzymatic cleavage methods and a lot of analytical detective work. The 1970s saw the development of inferential chemical cleavage methods for DNA sequencing (Maxam and Gilbert, 1977) and the "plus and minus" method of withholding a nucleotide to deduce the identity of the next base in a growing chain (Sanger and Coulson, 1975). By the early 1980s, cDNAs (and the RNAs they represented) were being identified using dideoxy chain termination methods (Sanger and Nicklen, 1977) by running four parallel reactions per sample, with each reaction featuring all four dNTPs plus one lower-concentration ddNTP that was randomly incorporated by a DNA polymerase into a growing chain, thereby terminating it. The reaction products, and accompanying radioactive mess, were run on long polyacrylamide gels in four adjacent lanes, creating a step-like series of sequencing products which differed in size by a single base. Later iterations included the use of four different fluorescent ddNTPs which enabled a single-tube sequencing reaction accompanied by the analysis of sequencing products by capillary electrophoresis. An improvement in productivity accompanied the development of serial analysis of gene expression (SAGE; Velculescu et al., 1995) and related methods. All of the previous sequencing methods described above are collectively referred to as first-generation methods.

What may be thought of as second-generation sequencing, more commonly known as NGS, began with the rollout of pyrosequencing (Nyrén and Lundin, 1985), a technology that, at the time, was commonly associated with 454 Life Sciences. This method and the instrumentation platform that supported it quickly began to be referred to as a massively parallel approach to sequencing because of the large number of sequencing reactions that could be supported side by side each time the instrument was run. The technique featured PCR reactions in an oil-water emulsion (emPCR) which relied on adapter-coated beads to support sequencing, ideally with one DNA molecule base-pairing to one bead and which was then amplified in its own emulsion microreactor. This later evolved into a sequencing variant developed by Solexa (which was acquired by Illumina) known as "bridge amplification," a term used to describe the manner in which bridge or inverted U-shaped structures form as newly synthesized molecules in the sequencing reaction interact with adjacent oligonucleotides that are distributed as a lawn on the surface of a flowcell. The efficacy of the massively parallel format to sequencing was the impetus behind the formation of other companies that featured variations on this theme, including the now-legacy SOLiD platform which was best classified as sequencing by ligation rather than sequencing by synthesis.

Third-generation sequencing commonly involves single-molecule sequencing which, with that ability, eliminates the need for any type of DNA amplification which is the hallmark of most of the second-generation methods. Common third-generation sequencing platforms include real-time fluorescence detection using nanostructures known as zero-mode waveguides (Pacific Biosciences SMRT Cell technology) during a sequencing synthesis reaction (Eid et al., 2009); and nanopore sequencing (Oxford Nanopore Technologies) in which the movement of each base through a channel with an applied voltage causes a characteristic distortion in ionic flow that identifies the base. Third-generation sequencing methods typically produce very long reads that are associated with NGS methods.

A major impediment for many laboratories has been the cost of whole transcriptome sequencing. Technical improvements have paid two important dividends, namely, greater quantities of data with a higher level of accuracy, and a significant decline in the cost per reaction. Fortunately, the cost per reaction continues to decline and it is expected that sooner, rather than later, even modestly-funded laboratories will be able to utilize RNA-seq. When access to the technology does become available, investigators who are accustomed to older-style Sanger sequencing are surprised to learn that NGS methods in general, and RNA-seq in particular, produce relative short reads, most often between 200 and 400 bases. Although RNA-seq has largely supplanted microarray profiling, research involving the use of microarrays continues to be somewhat productive, owing in no small measure to the investment made in microarray equipment and consumables, as well as comfortable familiarity with the data analysis processes. One worthwhile strategy is to use RNA-seq data to validate microarray data, and vice versa. Likewise, whole-transcriptome RNA-seq results can be validated by droplet digital PCR, the outcome of which usually demonstrates a strong correlation between these two complementary methods. Given the cost of instrumentation, the cost and often labile nature of consumables, and balancing the need for this type of data, sequencing is one of the most outsourced services utilized by the molecular biology community.

The reasons for performing RNA-seq are highly variable depending on the questions being asked in an experimental setting. Some of the more common applications include:

1. Whole transcriptome profiling;
2. Biomarker discovery;
3. Novel coding and noncoding transcript discovery;
4. RNA subset profiling (e.g., mRNA, miRNA, scRNA);
5. Allelic expression profiling;
6. QTL (quantitative trait loci) analysis;
7. RNA isoform identification;
8. Mapping exon-intron boundaries;
9. Mapping transcription start sites (TSS);
10. Identification of SNPs, indels, or altered polyadenylation patterns;
11. Elucidation of gene expression pathways;
12. Validation of sequence identity;
13. Confirmation of genetically-engineered sequence modifications;
14. Confirmation of data generated by other means.

When identified, previously uncharacterized sequences are assigned an accession number and uploaded to any of a number of worldwide sequencing databases. If a previously unknown sequence is deemed as proprietary or of some potential commercial value, it may not be immediately shared with the scientific community. Naturally-occurring sequences cannot, however, be patented (Association for Molecular Pathology vs Myriad Genetics, Inc.). In contrast, cDNA and artificial constructs thereof are patentable insofar as they meet certain criteria of originality.

The amount of information from RNA-seq greatly exceeds that which is derived from microarrays or other transcriptomics assays and is often in the order of gigabytes (GiB) per sample. The depth and coverage data that are produced by RNA-seq is one of its most important strengths (Table 24.1). Unfortunately, that strength had also been a serious weakness due to the lack of computational bioinformatics tools to handle the enormous data sets until the arrival of "big data" methods and cloud-based computing. While the data processing situation has improved considerably compared to just a few short years ago, challenges remain. Tag-seq, a lower-cost alternative to RNA-seq, arose as a modification of SAGE sequencing and which supports multiplex analysis (Rozenberg et al., 2016). Tag-seq realizes cost saves by focusing on the 3′ ends of sequencing template molecules and is much less data-intensive than RNA-seq.

NGS refers to any of a variety of non-Sanger high-throughput sequencing methods used to elucidate base sequence identification of genomic or cDNA populations. The common thread among NGS systems is sequencing by

TABLE 24.1 RNA-seq advantages and disadvantages.

Advantages	Disadvantages
Best method of whole transcriptome profiling	Cost of equipment
Detection of both coding and noncoding transcripts supported	Per reaction cost is declining, but is still prohibitive for some investigators
Reliable method for new transcript discovery, alternative splicing, and SNP identification	Sequencing library preparation and workflow are complicated
Wide dynamic range	Massive amount of data generated
No prior knowledge of sequences required	Access to cloud-based or big-data computing required
Effective for biomarker discovery	Data analysis is time-consuming
Easy to outsource RNA-seq if the technology is not available in-house	High level bioinformatics fluency required
Can be performed on individual cells	Analysis of organisms with highly-repetitive sequences can be challenging
Can be performed on RNA recovered from formalin-fixed, paraffin-embedded (FFPE) samples	Lack of cross-platform standardization creates data interpretation and experimental reproducibility issues

synthesis (SBS), an enzymatic process by which a new molecule of DNA is synthesized from a user-provided template one base at a time. Each base is modified in some way (e.g., carries a different color fluorescent tag) so that generates a characteristic signal upon excitation. It is a rapid, accurate, and systematic means of securing detailed information, in the case of RNA-seq, about patterns of gene expression. NGS is characterized by a massively parallel format which facilitates simultaneous sequencing of millions unique DNA strands. Very specialized and sensitive equipment and data analysis software are non-negotiables. Several sequencing platforms are in use and each of which has unique technological features. Examples of other applications not directly related to RNA-seq include the study of protein-DNA interactions (ChIP-seq), DNA methylation analysis (Methyl-seq), whole genome analysis (WGS), de novo sequencing (sequencing of a new genome for which there is no reference genome), and sequencing of one or a few specific genes or cloned gene fragments (directed sequencing).

RNA-seq workflow

Investigators planning to perform RNA-seq for the first time would be wise to first visit the RNA-seqlopedia. Housed at http://rnaseq.uoregon.edu, this excellent resource is maintained by the Cresko lab at the University of Oregon. Also, an excellent series of tutorials, RNA-seq Bioinformatics, compiled collaboratively by the Griffith Lab is accessible at https://rnabio.org. A very good way to keep abreast of new lab innovations and developments within the industry is the visit the RNA-seq blog located at http://www.rnaseqblog.com. The website http://www.biostars.org is also a very good resource, though one needs to create an account in order to access the posts and contribute to the discussions.

There are several discrete components associated with an RNA-seq investigation, and there are numerous permutations of each (Fig. 24.1). An important first step, as with any investigation, is to define the question(s) being asked. The goals must be clearly defined, as well as the objectives, strategies, and tactics that are to be used to achieve each goal. For example, in-advance considerations include the use of technical and biological replicates, the desired depth of sequencing, and the degree of transcriptome coverage. Advance planning will keep the investigation on track, on target, and on budget. It will also help each member of the lab team understand their unique role and shared responsibilities. This is especially important for RNA-seq because of the cost involved. It is important to understand that there may be significant protocol variability depending on the nature of the RNA subpopulation and, in particular, the NGS platform. As with all good laboratory experimental practices, there are important quality control checks that much be performed at each step to ensure a high level of confidence in the results. This includes having high confidence in the quality of the sequencing reads.

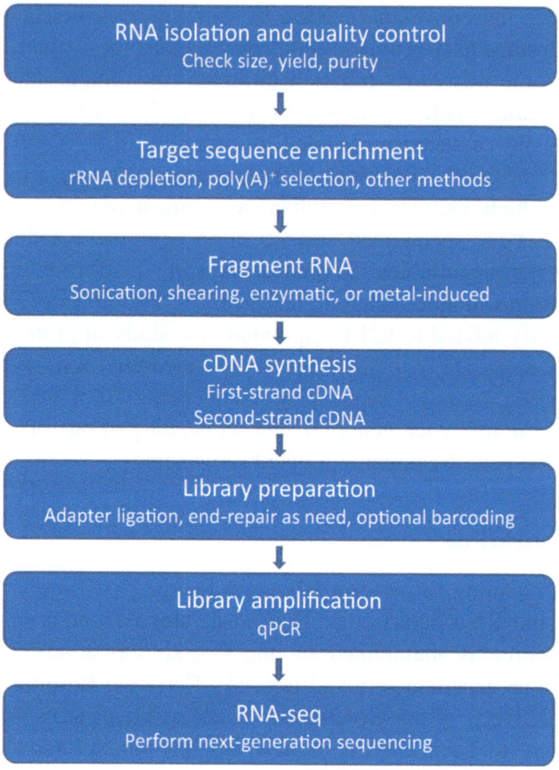

FIGURE 24.1 In vitro workflow for RNA-seq. There is no single one-size-fits-all methodology. The precise protocol followed is largely dependent upon the exact questions being asked and the available instrumentation.

RNA isolation and quality control

As described in earlier chapters, working with subpopulations of RNA is often useful for answering questions that cannot be addressed using total cellular RNA. Due to the rather sensitive nature of RNA-seq, the best RNA isolation strategy involves the use of a commercial product or kit which has been optimized for nucleic acid recovery. This is strongly preferred over the use of homemade reagents, particularly when working with rare or irreplaceable samples. With respect to RNA recovery from biofluids, there is no single preparation or procedure that will support RNA-seq due to the significant differences among various types of body fluid matrices. Saliva, as one example, is a difficult biofluid to examine due to the high abundance of bacteria in the mouth and the low abundance of extracellular RNA (exRNA), make the process fraught with challenges. As with most applications in molecular biology, the condition and age of the biological source, and the quality of the RNA starting material all have a direct impact on the quality of an ensuing sequencing library.

With respect to other quality control measures, the concentration of the sample must be known and, while standard spectrophotometric analysis can be performed (OD_{260}; A_{260}:A_{280} ratio), measuring nucleic acid concentration by fluorometric assay is considered to be a superior approach particularly when dealing with low sample concentrations and minute quantities. In addition, the size distribution of the sample is another important attribute. One may either run a gel (Chapter 14: Electrophoresis of RNA) or profile the RNA using a Bioanalyzer chip (Chapter 8: Quality Control for RNA Preparations) to generate an RNA integrity number (RIN) which, ideally, should be $\geq$ 7.0. Between RIN 6.0 and 7.0, the sample quality is highly questionable and below 6.0 it should not be used. Two exceptions: (1) In the case of RNA recovered from FFPE, the RNA sample will have experienced heavy degradation and the resulting RIN would be almost meaningless. A novel metric for determining the integrity of RNA isolated from FFPE tissues has been described, using an area under-the-curve analysis, rather than RIN, for determining the probable utility of the sample (Chung et al., 2016). (2) The calculation of RIN for a sample is based on mammalian RNA, in which the 28 S:18 S fluorescence ratio is ideally at least 2.0. In some organisms, for example, yeast it is normal for the major rRNA subunit ratio to be less than 2.0, and sometimes considerably so. If the interpretation of the sample RIN is uncertain, one suggestion is to heat-denature the sample with urea and run a minigel to ensure minimal smearing near the leading edge of the gel. When present, low molecular weight smearing suggests that the integrity of the sample has been compromised by RNase or perhaps by improper handling or storage.

RNA enrichment

With regard to the synthesis of an RNA-seq library, there are two common approaches. First, library construction may proceed with total RNA as the starting material, sometimes referred to as "total RNA-seq," which is also known as whole transcriptome sequencing. Libraries constructed by total cellular RNA have all coding and noncoding transcripts represented among their membership, including the highly abundant rRNA species. While total RNA-seq libraries will provide comprehensive coverage, the fact that cytoplasmic and mitochondrial rRNAs constitute nearly 90% of the sample will add to the cost of the sequencing and may result in missing rare transcripts. The second approach involves enrichment of one or more RNA subpopulations by removal of transcript species such as the abundant rRNAs that may not be of immediate experimental interest (Table 24.2). The most common enriched populations of RNA used for RNA-seq applications include:

TABLE 24.2 RNA-seq library preparation strategies.

Strategy	RNA population to be sequenced	How performed
Total RNA	All transcripts isolated from a biological source	No selection method required
Poly $(A)^+$ selection	Polyadenylated mRNA; emphasis on coding RNA	Oligo(dT) selection using magnetic beads
rRNA depletion (ribo-depletion)	All transcripts except for rRNA; emphasis on noncoding	RNase H-mediated rRNA/ oligonucleotide cleavage
Targeted transcript reduction	All transcripts except for abundant sequences that were reduced or altogether eliminated prior to NGS	Not-so-random (NSR) primers or pseudo-random (PS)-primers; sequence-specific hybridization to oligos on magnetic beads; anti-DNA:RNA hybrid antibodies
RNA sequestered by immunoprecipitation (RIP)	RNAs specifically associated with particular protein(s)	Antibody mediated pull-down of specific protein-RNA complexes
Size Fractionation	Pre-determined size range of transcripts; miRNA	Gel electrophoresis/elution or silica spin columns

1. poly$(A)^+$ RNA, isolated using oligo(dT)-tailed magnetic beads. This method will enrich in favor of polyadenylated mRNA but will eliminate poly$(A)^-$ mRNA in addition to tRNA and rRNA. Procedures for poly$(A)^+$ selection are described in detail in Chapter 5, Isolation of Polyadenylated RNA.
2. rRNA-depleted RNA (total RNA minus the rRNA). Ribo-depletion is viewed as a desirable quality for RNA samples destined for RNA-seq because, unless removed, the highly abundant rRNAs will consume a high percentage of the sequencing reads, thereby putting the detection of low or very low abundance transcripts in peril. Preparation of an $rRNA^-$ sample can be accomplished either by poly$(A)^+$ selection (above), or enzymatically by annealing complementary oligodeoxynucleotides to both cytoplasmic and mitochondrial rRNA species and then degrading the resulting RNA:DNA hybrid molecules through incubation with RNase H. In other cases, rRNA can be reduced by hybridization capture using complementary rRNA-specific oligonucleotides linked to paramagnetic beads. The poly$(A)^+$ selection methodology is easy, though the resulting subpopulation excludes nonadenylated mRNAs and many noncoding transcripts. Elimination of rRNA, in contrast, is more involved

than poly(A)$^+$ selection, but the RNA that will be used to construct the cDNA library retains important categories of noncoding RNA as well as the poly(A)$^-$ mRNA species.

Fractionated poly(A)$^+$ or rRNA$^-$ subpopulations should be tested using rRNA primers to demonstrate a limited or inability of the sample to support PCR amplification of an rRNA product. This should be done concurrently using primers directed toward a reference (housekeeping) gene (e.g., actin, GAPDH, transferring receptor; see Chapter 10, RT-PCR: A Science and An Art Form, for other suggestions) to show that the sample is able to support the amplification of something (anything) and, in so doing, provide evidence that the sample integrity was maintained during the enrichment procedure.

Regardless of the method used, it is important to purge contaminating gDNA from the sample by incubation with DNase I (see Appendix C: Removal of DNA from an RNA sample, for suggestions). This is particularly important if the rRNA will be targeted for enzymatic depletion with RNase H because gDNA is sometimes able to participate in hybridization to complementary, non-rRNA species, thereby targeting these unintended hybrids for immediate destruction by RNase H. This would have the undesirable effect of altering the profile of the starting material. Regardless of methodology, removal of gDNA should be a standard component of RNA purification anyway, if the data are to be considered as credible

3. RNA immunoprecipitation (RIP) is the enrichment of mRNAs encoding a specific protein by capturing the entire ribonucleoprotein complex (mRNA, ribosome, nascent protein) during the process of translation. Known as a form of RNA pull-down, the components of the complex are subsequently dissociated, and the mRNA continues in the RNA-seq workflow.
4. Target-enriched RNA, represents a gene-specific subset of one or a few RNA species that are of immediate interest to the investigator. The efficiency of transcriptome coverage during sequencing is dependent on the ability to detect rare transcripts, or common transcripts that are present at exquisitely low abundance. As noted above, a common difficulty in achieving this goal is the prevalence of ribosomal RNAs and highly abundant RNAs that are specific to a particular cell type (e.g., hemoglobin in reticulocytes). Given the preponderance of sequence information about genes and their respective transcripts, one strategy is to computationally design primers that will not recognize highly abundant transcripts, thereby greatly reducing their prevalence among the ensuing sequencing data. Known as not-so-random (NSR) primers (Armour et al., 2009; Arnaud et al., 2016; reviewed by Fitzpatrick et al., 2021), the process involves priming for cDNA synthesis with a "modified" pool of random primers that do not contain any exact-match sequences to the abundance transcripts that the investigator wishes to exclude from

sequencing. While an enticing and effective strategy, NSR primers can add significant cost when either the depletion target is changed or when the priming strategy is altered by adding a linker or other 5′ tail, as one example. A modification of the NSR approach takes advantage of the fact that reverse transcriptase will dependably tolerate as many as two mismatches during priming (Mizuno et al., 1999). This method also requires the generation of an organism- and target-specific target set of pseudo-random (PS) primers (Arnoud et al., 2018) though far fewer primers are needed for the target depletion strategy to be successful.

RNA fragmentation

RNA template material for cDNA synthesis is typically fragmented prior to reverse transcription. This action tends to create a fairly uniform size distribution among the resulting cDNAs. Fragmentation is desirable because the ability to sequence molecules, that is, the read length, using some of the current NGS platforms is limited. Note that in some cases, fragmentation of molecules is performed after the synthesis of cDNA, rather than fragmenting the RNA. While there are several protocols, the operative term is "controlled" fragmentation. Methods include sonication, enzymatic methods, mechanical shearing, and metal-induced fragmentation. The latter method is popular and is mediated by heating RNA to 95°C for less than 5 minutes in the presence of Mg^{++} at a quasi-alkaline pH; this option is the basis of some kit-based procedures for library preparation for RNA-seq. After heating and cooling, the fragmented RNA is cooled on ice as quickly as possible, ethanol precipitated, and resuspended in sterile H_2O. A kinasing reaction follows to prepare for adapter ligation. The enzyme T4 polynucleotide kinase will add 5′ phosphate groups *and* it will remove 3′ phosphoryl groups that may have been generated by the fragmentation process. If any spike-in, exogenous, or control RNA is to be included as part of the sequencing experimentation, it needs to be added to the sample prior to the fragmentation step. To that end, the ERCC RNA standards (often 90 + transcripts ranging from 250 to 2000 bases and covering six logs of dynamic range) are well-suited RNA-seq performance indicators, as are other established standards such as Universal Reference RNA. Finally, depending on the nature of the primers that will support reverse transcription and the intended sequencing platform, the entire procedure may be RNA fragmentation-free.

cDNA synthesis

The manner in which cDNA is synthesized is consistent with well-established strategies used for other nonsequencing applications, that is, by reverse transcription. Owing to the fact that fragmentation produces randomized molecules with unknown 3′ and 5′ end sequences, cDNA synthesis is readily accomplished through random priming. Following the synthesis of both the first- and second-strand cDNAs, the ends of the ds cDNA are either blunt-ended with a DNA repair enzyme (e.g., T4 DNA polymerase) and/or A-tailed. These steps are designed to support adapter ligation by creating predictable end structures on each molecule.

A contemporary modification to cDNA synthesis in the RNA-seq library preparation is the inclusion of dUTP in the second-strand synthesis reaction. dUTP will function as a dTTP analog and, following adapter ligation (below), will facilitate the removal of the second-strand cDNA by enzymatically (uracil-N-glycosylase) targeting uracil-contain nucleotides. This approach to cDNA synthesis is useful when it is important to be able to distinguish the first-strand cDNA from the second, primarily for asymmetric adapter/barcode labeling. It is important to maintain the "strandedness" of sequencing templates to avoid ambiguity at the time of data analysis.

Library amplification

Immediately preceding NGS, the cDNA from each sample is generally modified by the ligation of adapters to the ends of the blunt/A-tail cDNA from the previous step. In so doing the 5′ and 3′ ends consist of known sequences. This facilitates the very simple process of adding barcode and PCR primer sequences to the ends of each molecule. Barcoding permits RNA samples from multiple sources to be sequenced simultaneously; during the postsequencing data analysis, the reads can then be grouped by barcode so that the biological source of each is easily distinguishable. It is also important to be confident that the adapter ligation step was successful because the adapters are key to the quality and inclusiveness of the sequencing data. Further, the placement of known sequences at the end of each molecule will facilitate reamplification in the event the cDNA yield from the preceding steps is low. In view of the known amplification bias of PCR, one should increase the mass of RNA by linear amplification methods rather than using PCR to generate more cDNA for library construction.

Next-generation sequencing

There are numerous widely-available platforms for performing high-throughput NGS and ensuing analysis of the resulting data. Examples of platforms that support RNA-seq are shown in Table 24.3. Without a doubt, NGS has revolutionized the ability to profile gene expression in that it is possible to sequence an entire transcriptome or genome, rendering hundreds of millions of data points, in a few short hours, rather than the days or weeks required at the advent of automated sequencing. RNA-seq is also a valuable tool for the profiling nonmodel organisms, meaning those for which there is no reference transcriptome or reference genome. One must also be aware

TABLE 24.3 Examples of RNA-seq platforms.

Company	Platform	Process	Website
Illumina	NGS	Synthesis of a complementary DNA strand on flowcell, one nucleotide at a time, accompanied by a unique fluorescence signal which corresponds to the added base. Also known as sequencing by synthesis.	http://www.illumina.com
Ion Torrent	NGS	Synthesis of a complementary DNA strand, one nucleotide at a time and recorded as a voltage change caused the release of protons, and therefore pH, when a new nucleotide is added to the DNA template. Also known as sequencing by synthesis.	http://www.thermofisher.com
Pacific Biosciences	Third-generation sequencing (single molecule technique)	Sequencing by synthesis at the bottom of a nanowell in which DNA polymerase is immobilized; amplification and sequencing results reported in real-time via zero-mode waveguide nanoarrays and the single-molecule real-time (SMRT) process.	http://www.pacb.com
Oxford Nanopore Technologies	Third-generation sequencing (single molecule technique)	The passage of single-stranded RNA through a protein nanopore; characteristic disruptions in the current flow are directly related to the identity of each base as it moves through the pore.	http://www.nanoporetech.com
Singular Genomics	G4	Sequencing by synthesis; proprietary enzymes, nucleotides, and image detection technology; short run time and high data output.	https://singulargenomics.com

(Continued)

TABLE 24.3 (Continued)

Company	Platform	Process	Website
SOLiD	Legacy technology	No longer supported	ThermoFisher
454	Legacy technology	No longer supported	Roche
Helicos	Legacy technology	No longer supported	Helicos BioScience

that, albeit rare, sequencing errors do occur; rather clever strategies have been developed to identify and correct them, an example of which is known as error-correct high-throughput sequencing or more commonly, maximum-depth sequencing (MDS; Jee et al., 2016). Perhaps the biggest difficulty encountered when performing RNA-seq is the immense amount of data and the required storage space. The time-consuming data analysis that is associated with each sequencing run is an art form in itself. Interestingly, technology has improved to the point that RNA-seq can be performed on a single cell (Tang et al., 2009; Islam et al., 2011; Ramsköld et al., 2012; Hashimshony et al., 2012).

RNA-seq data analysis

Don't try to learn this on your own. The best advice is to begin with online or YouTube tutorials and to seek the guidance/mentorship of someone who is highly experienced in this particular aspect of bioinformatics. Online courses that describe RNA-seq data analysis approaches and options, online tutorials, and in-person workshops and professional development opportunities likewise abound to learn some of the contemporary computational bioinformatics approaches that are tied to RNA-seq analysis. The variety of data analysis tools continues to expand, with products associated with specific platforms as well as online RNA-seq analysis solutions. With respect to the former, many software providers offer a free trial period in which to evaluate their specific products for compatibility with the needs of the lab team. Be sure to visit https://bioformatics.ca for a schedule of bioinformatics workshops and other resources that are particularly useful for investigators working in this disciple.

While each propriety data analysis product differs in unique ways, including the manner in which the data are analyzed and ultimately presented, there are some standard tasks that are associated with the processing of

RNA-seq raw data regardless of the source and intended use (Fig. 24.2). Examples of these tasks include trimming, which refers to the removal of adapter sequences, chimera detection, alternative transcript detection (including alternative TSS, exon skipping, and intron retention) sequence alignment with a reference genome (if available), de novo assembly (when appropriate), transcript identification, transcript quantification, data comparison across samples, normalization, and quality control analysis.

With respect to data output, it is important to note that for each molecule present in the NGS reaction, it is likely that the entire molecular may not be sequenced. When the read length is expected to be short, one strategy is to perform sequencing from both ends of the molecular (e.g., paired-end sequencing) in order to produce as much coverage as possible. In so doing, it will be much easier to unambiguously identify the transcript. A good indicator of the accuracy of the data is the percentage of reads that map to the reference transcriptome or genome. As noted above, the amount of output data are not insignificant, ranging from 150 million to as many as 250 million

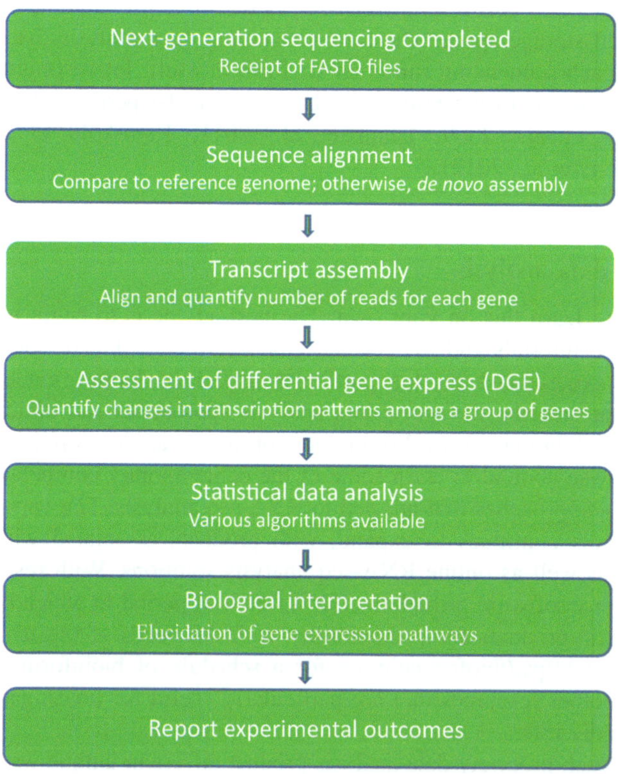

FIGURE 24.2 In silico workflow for RNA-seq. Several computational bioinformatics data suites are available to manage and to mine the large amount of data generated by RNA-seq.

reads per lane on some NGS platforms. With respect to the identification of genes being expressed differentially, one must quantify the number of reads associated (aligned) with a specific gene, and then compare the observed level of expression across the various samples under investigation. In order to perform these data analyses, investigators rely on a suite of software applications that performs the requisite data manipulations. These are all examples of how computational bioinformatics algorithms are used to answer sequencing questions and extend to assist in biological systems modeling.

One of the better-known collections of RNA-seq data analysis tools is known as "Tuxedo," consisting of applications need to interpret the data. The major steps associated with the assay of gene expression include:

1. Quality control, which is a function of the sequencing reaction and the instrument in which it was run. Raw data from a sequencing run is commonly delivered in.fastq (FASTQ) format.
2. Alignment (through the use of TopHat2 or, more recently, HISAT2), in which the transcripts represented by each read are assigned to specific genes. A preferred location at which to access reference genomes, when available, is Ensembl (http://www.ensembl.org). From the Ensembl website, the available FASTA files contain the entirety of the genome information; one should also download the gene annotation (GTF) files which represent previously identified and mapped sequences. As noted above, alignment may be preceded by trimming, an in silico procedure by which the adapter sequences are eliminated and thereby improving the FASTQ data read quality.
3. Transcript assembly (through the use of Cufflinks or, more recently, StringTie or HTSeq), in which the number of reads aligned to each gene is determined.
4. Assessment of differential gene expression (through the use of Cuffdiff or, more recently, Ballgown or DESeq 2, and other algorithms), in which a determination is made of one or more genes are expressed differentially amount a group of samples.

RNA-seq variations

RNA-seq offers a different approach to the assay of gene expression compared to more traditional methods such as microarray analysis and qPCR (Table 24.4). Much like PCR, in which the numerous variations of the method involving the clever use of primers all revolve around the themes of sample denaturation followed by primer annealing and extension, many iterations of RNA-seq have been described, which share the major steps of RNA isolation and enrichment and some form of NGS or third-generation sequencing. The following are a few examples of RNA-seq variations that are designed to sharpen the focus on one or more RNA species or populations.

TABLE 24.4 Comparison of gene expression assays.

	qPCR	Microarrays	Sanger Sequencing	RNA-seq
Principle	Hybridization and amplification	Hybridization	Dideoxy sequencing	NGS and third-generation sequencing
New gene discovery	Limited; application dependent	No	No	Yes
Direct sequence identity	No; requires follow-up sequencing	Tentative, based on hybridization	Yes	Yes
Background signal / noise	Low-to-medium	High	Low	Low
Standardized protocols	Yes	Yes	Yes	No
Throughput	Low	High	Low	High
Detection of alternative splicing, TSS, or allelic expression patterns	Sequence and primer-dependent	Limited to sequences that are already known and present on the array	Yes	Yes
Dynamic range	Low-to-medium	Low; limited by fluorescence scanner	Low	Very high
Resolution	Low-to-medium	Low-to-medium	High; single base	High; single base
Required informatics	Standard	Elevated	Standard	Very high

CaptureSeq

The emphasis of CaptureSeq (Mercer et al., 2012; De Paoli-Iseppi et al., 2021) is a subpopulation of cDNAs (and the transcripts they represent) that are complementary to and therefore able to hybridize to capture probes prior to the sequencing library amplification by PCR. The capture probes can be bound to an array or immobilized on paramagnetic beads. cDNAs that are unrelated to the capture probes are washed aware. The eluted cDNAs which then move on to NGS are highly enriched, making CaptureSeq ideal for the characterization of transcript isoforms, previously uncharacterized TSSs, and other novel phenomena associated with one or a few gene loci. The method requires are fairly large amount of starting material, generally five or more micrograms of total RNA to produce an adequate amount of cDNA for probe capture.

DropSeq

It doesn't take much imagination to envision the plethora of information that would result from RNA-seq performed on a single cell, sometimes referred to as scRNA. DropSeq (Macosko et al., 2015) is precisely such a method in which single cells are sequestered into individual drops through the use of a microfluidics device in a manner similar to droplet digital PCR. In addition to the cell, the droplet also contains lysis buffer and oligo (dT)-barcoded primers commonly attached to a microbead. Inasmuch as each bead contains a unique barcode, this allows the pooling of beads from all droplets for cDNA synthesis and subsequent massively parallel sequencing. The resulting data are unique to each cell due to the original primer barcode on the bead. The method requires access to a microfluidics device to produce the droplets and, because of the manner in which the cDNA is synthesized on oligo(dT) beads, the assay is limited to mRNA, and only those which are polyadenylated.

First described more than a decade ago, scRNA-seq has become an important research tool due to procedural improvements and technological advances in microfluidics. This has been invaluable in advancing our understanding of polyclonal tumor heterogeneity as one practical example; different regions of the same tumor often exhibit marked differences in genetic composition. scRNA-seq methods, of which DropSeq is an example, permit a detailed examination of the contributions of individual cells that constitute any type of histological structure.

CEL-Seq

Another method that involves the assay for minute quantities of RNA starting material is CEL-Seq (Hashimshony et al., 2012). Like DropSeq, CEL-Seq involves barcoding and the addition of T7 transcription promoter to the cDNA as it is synthesized from the RNA found in a single cell. When completed the barcoded cDNA from each sample (cell) is pooled together and the sample can be amplified in a linear manner by T7-based in vitro transcription. Linear amplification eliminates the amplification bias associated with PCR amplification by which the true abundance relationships among transcripts are quickly distorted. This is followed by sequencing. During the data analysis, the barcode identifies the source of the each read, thereby assigning them to the same sample group. Unlike DropSeq, the partitioning of cells is performed in microfuge tubes, that is, cloning by limiting dilution, rather than encapsulating each cell in a microfluidics-generated droplet.

TIF-seq

Transcript isoform sequencing (TIF-seq) is an RNA-seq adaptation that focuses on the identification of transcript isoforms (Pelechano et al., 2013; Wang et al., 2020). This is accomplished by sequencing full-length mRNAs (only) via their respective cDNA counterparts in a manner that fully captures the 5′- and 3′-most sequences of each template. The method of library preparation involves exclusion of nonfull-length mRNAs via phosphatase treatment to remove exposed 5′ phosphate groups, removal of the 5′ cap from full-length mRNA via tobacco acid pyrophosphatase (leaves a residual 5′ phosphate to which a 5′ adapter is ligated), and 3′-end amplification of the cDNA using a barcoded primer with an oligo(dT) tract that will enable base-pairing to only those template molecules endowed with a poly$(A)^+$ tract. The process also involves template circularization and subsequent fragmentation to minimize the chances of intermolecular interactions which would produce misleading data. The manner in which the 5′ end of full-length mRNA is prepared is nearly identical to the preparation of transcripts for RLM-RACE as described in Chapter 10, RT-PCR: A Science and An Art Form.

Global run-on sequencing

Global run-on sequencing (GRO-seq) is an innovative approach to RNA-seq that is coupled with measuring the rate of gene transcription (Chapter 19: Analysis of Nuclear RNA) rather than measuring steady-state transcript levels, which the other RNA-seq methods do. The method of nuclear run-on transcript labeling is a well-established method dating back to the early years of molecular biology; GRO-seq was developed (Core and Lis, 2008; Core et al., 2008) to provide a comprehensive snapshot of the coding and noncoding transcriptome from a different perspective.

In the GRO-seq method, transcripts in the cell are labeled with bromouridine (Br-U; 5-bromo-5′-uridine) in a manner not unlike the radiolabeling of initiated transcripts in the classical forms of the transcription rate assay. Following RNA isolation and controlled hydrolysis, the Br-U-labeled is captured by anti-Br-U antibodies fixed on the surface of a magnetic or other types of bead. The eluted RNA is used to synthesize cDNA and the ensuring NGS identifies the relative transcriptional activity at the respective loci. All active loci are identifiable without any prior information, including bidirectional (sense and antisense) transcription orientations. This method is limited to cell culture because of the need to incubate the isolated nuclei in the presence of labeled nucleotides. Depending on the lysis methods, the assay may include new transcription initiation events rather than being limited to the labeling of only those transcripts that were initiated at the moment of cell lysis.

cP-RNA-seq

The most common chemical structure associated with the 3′ end of RNA and DNA molecules is -OH associated with the ribose or deoxyribose, respectively, of the last nucleotide. When polynucleotides, including both coding and noncoding transcripts, are cleaved by the action of an RNase, it is not uncommon to observe a 3′ phosphate (3′-P) or a cyclic phosphate (cP; $2' - 3'$-cyclic phosphate) (Fig. 24.3). In view of the manner in which RNA-seq is performed, it is easy to understand why molecules with 3′-P or cP ends would not be detectable due to the absence of a 3′-OH group. One might dismiss the presence of these molecules as a reasonable consequence of natural RNA turnover until one considers newer evidence that cP-RNAs are involved in breast, prostate, and other hormone-associated malignancies (Honda et al., 2015) and probably other regulatory pathways. Given the likely importance of these RNAs, cP-RNA-seq was developed (Honda et al., 2016) to prepare these specialized RNAs for RNA-seq by first removing the phosphate group at the 3′ end, an example of which is shown in Fig. 24.4. Subsequent studies have associated elevated levels of what was once considered to be junk RNA with infection, neurodegeneration, and other cellular functions associated with the immune system (reviewed

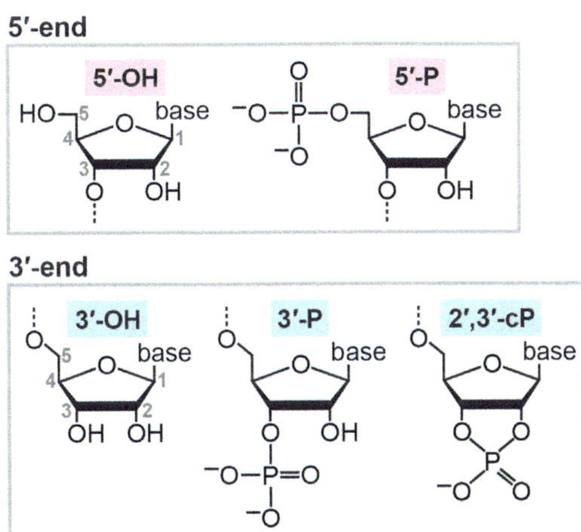

FIGURE 24.3 Chemical structures of RNA termini. *Image courtesy Shigematsu, M., Kawamura, T., and Kirino, Y. (2018). Generation of 2′,3′-cyclic phosphate-containing RNAs as a hidden layer of the transcriptome. Front. Genet. 9, 562. https://doi.org/10.3389/fgene.2018.00562 and reproduced with permission under the Creative Commons Attribution 4.0 International License (CC BY 4.0; creativecommons.org/licenses/by/4.0/).*

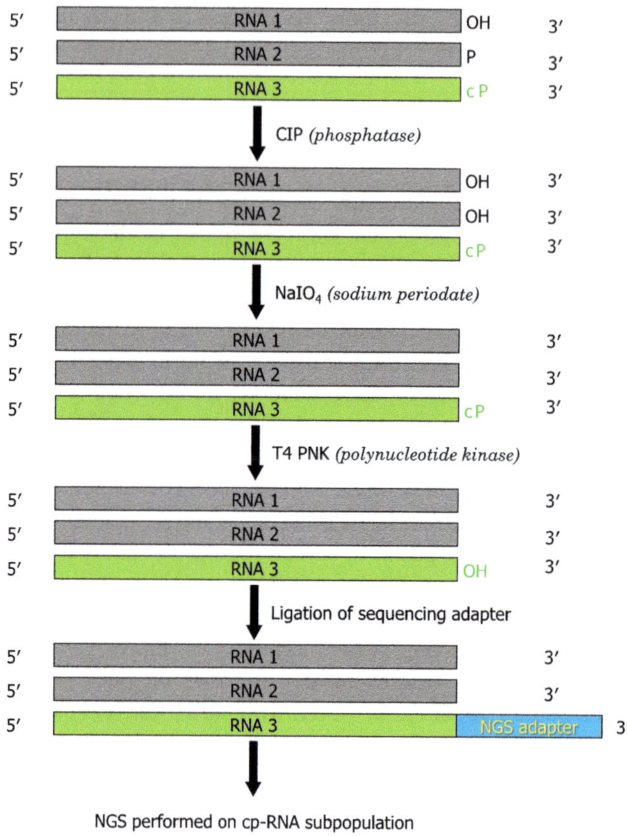

FIGURE 24.4 Preparation of cp-RNA for RNA-seq. The stepwise manipulation of 3′-end structures removes 3′-OH and 3′ phosphates, where present, from RNA molecules and ultimately produces a population of transcripts in which only RNAs that originally had a cyclic 3′ phosphate (RNA 3) are able to ligate a suitable adapter and then move forward to NGS. The resulting sequencing data reflect the cp-RNA pool in the original biological source.

by Shigematsu and Kirino, 2022). With this in mind cp-RNA molecules have considerable potential as molecular biomarkers.

References

Armour, C.D., Castle, J.C., Chen, R., Babak, T., Loerch, P., Jackson, S., et al., 2009. Digital transcriptome profiling using selective hexamer priming for cDNA synthesis. Nat. Method 6 (9), 647–649.

Arnoud, O., Kato, S., Poulain, S., Plessy, C., 2018. Targeted reduction of highly abundant transcripts using pseudo-random primers. BioTechniques 60, 169–174.

Chung, J.-Y., Cho, H., Hewitt, S.M., 2016. The paraffin-embedded RNA metric (PERM) for RNA isolated from formalin-fixed, paraffin-embedded tissue. BioTechniques 60 (5), 239–244.

Core, L.J., Lis, J.T., 2008. Transcription regulation through promoter-proximal pausing of RNA polymerase II. Science 319 (5871), 1791–1792. Available from: doi.org/10.1126/science.1150843.

Core, L.J., Waterfall, J.J., Lis, J.T., 2008. Nascent RNA sequencing reveals widespread pausing and divergent initiation at human promoters. Science 322 (5909), 1845–1848. Available from: doi.org/10.1126/science.1162228.

De Paoli-Iseppi, R., Gleeson, J., Clark, M.B., 2021. Isoform age- splice isoform profiling using long-read technologies. Front. Mol. Biosci. 8, 767743. Available from: https://doi.org/10.3389/fmolb.2021.767743.

Eid, J., Fehr, A., Gray, J., Luong, K., Lyle, J., Otto, G., et al., 2009. Real-time DNA sequencing from single polymerase molecules. Science 323 (5910), 133–138. Available from: https://doi.org/10.1126/science.1162986.

Fitzpatrick, A.H., Rupnik, A., O'Shea, H., Crispie, F., Keaveney, S., Cotter, P., 2021. High throughput sequencing for the detection and characterization of RNA viruses. Front. Microbiol. 12, 1–25. Available from: https://doi.org/10.3389/fmicb.2021.621719.

Hashimshony, T., Wagner, F., Sher, N., Yanai, I., 2012. CEL-seq: single-cell RNA-seq by multiplexed linear amplification. Cell Rep. 2, 666–673.

Holley, R.W.J., Apgar, G.A., Everett, J.T., Madison, M., Marquisee, S.H., Merrill, J.R., et al., 1965. Structure of a ribonucleic acid. Science 147, 1462–1465. Available from: https://doi.org/10.1126/science.147.3664.14.

Honda, S., Loher, P., Shigematsu, M., Palazzo, J.P., Suzuki, R., Imoto, I., et al., 2015. Sex hormone-dependent tRNA halves enhance cell proliferation in breast and prostate cancers. Proc. Natl. Acad. Sci. USA 112 (29), E3816–E3825. Available from: https://doi.org/10.1073/pnas.1510077112.

Honda, S., Morichika, K., Kirino, Y., 2016. Selective amplification and sequencing of cyclic phosphate–containing RNAs by the cP-RNA-seq method. Nat. Protoc. 11, 476–489. Available from: doi.org/10.1038/nprot.2016.025.

Islam, S., Kjällquist, U., Moliner, A., Zajac, P., Fan, J.-B., Lönnerberg, P., et al., 2011. Characterization of the single-cell transcriptional landscape by highly multiplex RNA-seq. Genome Res. 21, 1160–1167.

Jee, J., Rasouly, A., Shamovsky, I., Akivis, Y., Steinman, S.R., Mishra, B., et al., 2016. Rates and mechanisms of bacterial mutagenesis from maximum-depth sequencing. Nature 534, 693–696.

Macosko, E.Z., Basu, A., Satija, R., Nemesh, J., Shekhar, K., Goldman, M., et al., 2015. Highly parallel genome-wide expression profiling of individual cells using nanoliter droplets. Cell 161 (5), 1202–1214. Available from: https://doi.org/10.1016/j.cell.2015.05.002.

Maxam, A.M., Gilbert, W., 1977. A new method for sequencing DNA. Proc. Natl. Acad. Sci. 74 (2), 560–564.

Mercer, T.R., Gerhardt, D.J., Dinger, M.E., Crawford, J., Trapnell, C., Jeddeloh, J.A., et al., 2012. Targeted RNA sequencing reveals the deep complexity of the human transcriptome. Nat. Biotechnol. 30 (1), 99–104. Available from: https://doi.org/10.1038/nbt.2024.

Mizuno, Y., Carninci, P., Okazaki, Y., Tateno, M., Kawai, J., Amanuma, H., et al., 1999. Increased specificity of reverse transcription priming by trehalose and oligo-blockers allows high-efficiency window separation of mRNA display. Nucl. Acids Res. 27 (5), 1345–1349.

Nyrén, P., Lundin, A., 1985. Enzymatic method for continuous monitoring of inorganic pyrophosphate synthesis. Anal. Biochem. 151 (2), 504–509.

Pelechano, V., Wei, W., Steinmetz, L.M., 2013. Extensive transcriptional heterogeneity revealed by isoform profiling. Nature 497 (7447), 127–131. Available from: https://doi.org/10.1038/nature12121.

Ramsköld, D., Luo, S., Wang, Y.C., Li, R., Deng, Q., Faridani, O.R., et al., 2012. Full-length mRNA-seq from single-cell levels of RNA and individual circulating tumor cells. Nat. Biotechnol. 30, 777–782.

Rozenberg, A., Leese, F., Weiss, L.C., Tollrian, R., 2016. Digital gene expression analysis with sample multiplexing and PCR duplicate detection: a straightforward protocol. BioTechniques 61 (1), 26–32.

Sanger, F., Coulson, A.R., 1975. A rapid method for determining sequences in DNA by primed synthesis with DNA polymerase. J. Mol. Biol. 94 (3), 441–448.

Shigematsu, M., Kawamura, T., Kirino, Y., 2018. Generation of 2′,3′-cyclic phosphate-containing RNAs as a hidden layer of the transcriptome. Front. Genet. 9, 562. Available from: https://doi.org/10.3389/fgene.2018.00562.

Shigematsu, M., Kirino, Y., 2022. Making invisible RNA visible: discriminative sequencing methods for RNA molecules with specific terminal formations. Biomolecules 12 (5), 611. Available from: doi.org/10.3390/biom12050611.

Tang, F., Barbacioru, C., Wang, Y., Nordman, E., Lee, C., Xu, N., et al., 2009. mRNA-seq whole-transcriptome analysis of a single cell. Nat. Methods 6, 377–382.

Velculescu, V.E., Zhang, L., Vogelstein, B., Kinsler, K.W., 1995. Serial analysis of gene expression. Science 270, 484–487.

Wang, J., Li, B., Marques, S., Steinmetz, L.M., Wei, W., Pelechano, V., 2020. TIF-seq2 disentangles overlapping isoforms in complex human transcriptomes. Nucleic Acids Res. 48 (18), e104. Available from: doi.org/10.1093/nar/gkaa691.

Further reading

Barbacioru, C., Wang, Y., Nordman, E., Lee, C., Xu, N., Wang, X., et al., 2009. mRNA-seq whole-transcriptome analysis of a single cell. Nat. Methods 6, 377–382.

Besenbacher, S., Liu, S., Izarzugaza, J.M.G., Grove, J., Belling, K., Bork-Jensen, J., et al., 2015. Novel variation and de novo mutation rates in population-wide de novo assembled *Danish trios*. Nat. Commun. 6, 5969. Available from: https://doi.org/10.1038/ncomms6969.

Brooks, A.N., Hughes, A.L., Clauder-Münster, S., Mithcell, L.A., Boeke, J.D., Steinmetz, L.M., 2022. Transcriptional neighborhoods regulate transcript isoform lengths and expression levels. Science 375 (6584), 1000–1005. Available from: https://doi.org/10.1126/science.abg016.

Bussotti, G., Leonardi, T., Clark, M.B., Mercer, T.R., Crawford, J., Malquori, L., et al., 2016. Improved definition of the mouse transcriptome via targeted RNA sequencing. Genome Res. 26 (5), 705–716. Available from: https://doi.org/10.1101/gr.199760.115.

Lan, F., Haliburton, J., Yuan, A., Abate, A.R., 2016. Droplet barcoding for massively parallel single-molecule deep sequencing. Nat. Commun. 7, 11784. Available from: https://doi.org/10.1038/ncomms11784.

Nagalakshmi, U., Wang, Z., Waern, K., Shou, C., Raha, D., Gerstein, M., et al., 2008. The transcriptional landscape of the yeast genome defined by RNA sequencing. Science 320 (5881), 1344–1349. Available from: https://doi.org/10.1126/science.1158441.

Saint-Léger, A., Bello, C., Dans, P.D., Torres, A.G., Novoa, E.M., Camacho, N., et al., 2016. Saturation of recognition elements blocks evolution of new tRNA identities. Sci. Adv. 2 (4), e1501860. Available from: https://doi.org/10.1126/sciadv.1501860.

Stark, R., Grzelak, M., Hadfield, J., 2019. RNA sequencing: the teenage years. Nat. Rev. Genet. 20, 631–656. Available from: https://doi.org/10.1038/s41576-019-0150-2.

Tzerpos, P., B. Daniel, and L. Nagy. 2021. Global run-on sequencing (GRO-Seq). In: Borggrefe, T., Giaimo, B.D. (eds) Enhancers and Promoters. Methods in Molecular Biology 2351. Humana, New York, NY. Available from: https://doi.org/10.1007/978-1-0716-1597-3_2.

Wang, Z., Gerstein, M., Snyder, M., 2009. RNA-seq: a revolutionary tool for transcriptomics. Nat. Rev. Genet. 10, 57–63.

Zappia, L., Theis, F.J., 2021. Over 1000 tools reveal trends in the single-cell RNA-seq analysis landscape. Genome Biol. 22, 301. Available from: https://doi.org/10.1186/s13059-021-02519-4.

Chapter 25

RNA biomarker discovery and validation

Rationale

A major research and development goal of many in the medical community, and many organizations in the biopharma sector, is the expedient identification of biological molecules that are reliably indicative of the threat, presence, and progression or regression of a disease state. Such molecules, known as molecular biomarkers, can be DNA, RNA, proteins, metabolites, or even circulating tumor cells or cell fragments. With the ability to assay quantitatively for the presence of a biomarker, it becomes easier for the medical community to diagnose a disease state earlier and with greater accuracy. Given the currently high level of sensitivity associated with many molecular assays, biomarker screening has the potential to signal a potential health problem well before the onset of any clinical manifestation. With early detection, medical intervention may come earlier and, hopefully, a significantly better prognosis for the patient. Note that biomarkers are not limited to molecules; biomarkers can also be physiological manifestations, as described below.

It makes perfect sense to interrogate the transcriptome for novel biomarkers because it is the RNA expression profile of a cell that drives its phenotype. For this reason, molecular biomarkers are occasionally referred to as "signature molecules." While many very well-respected assays measure proteins using a variety of platforms, it is important to remember that protein abundance does not always correlate directly with RNA abundance and that nucleic acid screening methods, for example, RT-PCR, are often more sensitive than protein detection assays. Moreover, protein assays are simply not possible in cases where an RNA biomarker is a noncoding transcript. Though a great deal of the information currently known about the role of RNAs as biomarkers was distilled from a very large amount of high-throughput sequencing data, much more information is yet to come.

Biomarkers defined

To understand the potential usefulness of biomarkers, it is important to understand what is meant by a "sign" and how it differs from a "symptom."

RNA Methodologies. DOI: https://doi.org/10.1016/B978-0-323-90221-2.00038-2

A sign is an indicator of a medical condition that is observed from outside the body; it is objective, reproducible, and quantifiable, such as one's blood pressure. In contrast, a symptom is an indicator of health that is reported by the patient based on their own perception, for example, "I have a headache." An important difference is that one cannot fake a sign. A biomarker, then, is an objective, clinically relevant sign that reflects the biological condition of a cell or an organism. As objective indicators, biomarkers are clear signs of health status, or of disease progression or disease regression. Examples of traditional biomarkers are shown in Table 25.1.

There are many ways in which biomarkers are valuable. As summarized in Table 25.2, biomarkers can be prognostic (suggesting whether a disease is likely to develop), diagnostic (what disease or form of disease is afflicting a patient), or predictive (what drug or therapy is best for a patient, based on that patient's unique biochemistry, and whether the disease is likely to return following some form of therapeutic regimen). Thus, biomarkers can also be important tools for identifying at-risk individuals and designing preventative therapies in an asymptomatic population.

With respect to biomarker discovery when the marker is the target of a drug (pharmacotoxicology), the organ-on-a-chip (OOAC) platform (wyss.harvard.edu) is becoming increasingly important for biomarker discovery. With these devices, it is possible to test drugs as an alternative to the use of animals. Briefly, advances in microfabrication and the ability to create an in

TABLE 25.1 Examples of traditional biomarkers.

Biomarker	Indicates
Cardiac enzymes in the blood	Heart damage, especially associated with a heart attack
Body mass index	Obesity
Intraocular (eyeball) pressure	Glaucoma
Blood glucose	Pancreatic function; diabetes
Blood cholesterol	Risk of heart disease
Blood pressure	Risk of heart attack or stroke
Bone density	Risk of osteoporosis
QT interval (ECG)	Ventricular function; may indicate adverse drug reaction
Prostate specific antigen	Prostate cancer
Viral RNA	Infection (virus-specific)

TABLE 25.2 Uses of biomarkers.

Early diagnosis	Early detection is best, allowing the maximum amount of time for medical intervention.
Patient stratification	Determine which patients are likely to respond to a therapeutic regimen vs those whose genetic makeup renders them nonresponders, in which case another form of medical intervention should be sought.
Disease staging	Early stage, late stage, and spreading of the disease; condition of affected organs.
Disease prognosis	Indicative of the outlook for the patient in terms of time to recovery, level of recovery, or likelihood of recovery.
Pharmacogenomics	Tailor-made therapy based on one's genetic makeup. Sometimes known as theranostics (therapy + diagnostics).
Pharmacodynamics	Determination of the optimal dose for a patient as a genetically unique individual.
Pharmacotoxicology	Determination of consequences of toxic exposure to a drug, which is important for biomarker discovery especially when the biomarker is the target of a drug.

vivo-type environment through innovative tissue engineering as led to the design of devices that replicate tissue- and organ-level functionality (Fig. 25.1). The unique qualities of the OOAC replicate structural and functional relationships among the various types of cells that reside in these microfluidic devices such that responses to external stimuli and biochemical cues that are observed depict and predict in vivo behaviors. While simpler than the true organ that the OOAC represents, this innovative approach represents a greatly enhanced model by comparison with two-dimensional cell culture methods in which to observe and test cell behaviors. One of the areas that are highly likely to benefit from OOAC technology is biomarker discovery. In that capacity, OOACs are becoming valuable platforms for disease modeling and personalized medicine.

Biomarkers may also be classified as follows:

1. Biomarkers of risk, including germ-line mutations. For example, inheriting the mutated form of BRCA-1 places a person at higher risk for developing breast cancer.
2. Biomarkers of exposure, an example of which are observable chromosomal changes in peripheral lymphocytes that occur due to exposure to intense or sustained levels of ionizing radiation.
3. Biomarkers of early detection, in which abnormal levels of a blood protein such as prostate-specific antigen or circulating nucleic acids (CNAs) often signal trouble.

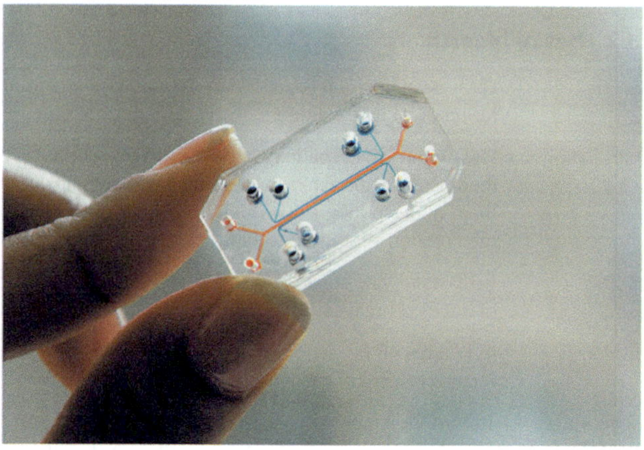

FIGURE 25.1 Organ-on-a-chip device. These microchips recapitulate the microarchitecture and functions of living human organs. As such, they offer a potential alternative to animal testing as well as nearly unlimited potential for biomarker discovery. The hollow microfluidic channels are lined by living human cells interfaced with a human endothelial cell-lined artificial vasculature, and mechanical forces can be applied to mimic the physical microenvironment of living organs. *Courtesy the Wyss Institute at Harvard University; wyss.harvard.edu.*

4. Biomarkers of prognosis and treatment; for example, BRCA-1 carriers have a better prognosis than BRCA-2 carriers. Similarly, patients with breast cancer caused by amplification of the *Her-2* gene respond to the drug Herceptin, while patients with other forms of breast cancer do not, thereby requiring alternative forms of medical intervention. There are many other examples.

Characteristics of useful biomarkers

With each passing day, there is a growing overlap between biomarker identification and drug development. As a direct consequence, many of the historically formidable barriers to biomarker discovery are conquered. While no biomarker is perfect, favorable characteristics include:

1. Patient specimen is easy to obtain and easy to assay. Examples include saliva, urine, blood, or minimally-invasive surgical biopsy.
2. The biomarker is reliable across ethnic groups and in both males and females.
3. The biomarker is cost-effective to assay.
4. The biomarker is quantifiable; the ability to measure variability is helpful in documenting how a patient responds to treatment, and the biomarker should be indicative of the probability of recurrence.
5. Biomarkers must be validated and provide clinically useful information. An ideal biomarker should be an indicator of prognosis.

6. Biomarkers should reflect the efficacy, toxicity, or safety profile of a drug.
7. Biomarkers should be indicators of the type of therapy that should be given to a patient based on their genetic makeup. Thus, biomarkers have an increasingly important role in personalized medicine.

For the purpose of comparison, Table 25.3 provides examples of biomarkers and the diversity of assay methods that are used to quantify their manifestation.

miRNA biomarkers

Any aberration in transcription or posttranscriptional gene regulation is fully capable of disrupting the function of genes associated with maintaining the health of the cell or organism-level homeostasis overall. This includes changes in miRNA biogenesis and functionality which, as described in Chapter 12,

TABLE 25.3 Diversity of biomarkers.

Biomarker category	Example	Assay method
Nonmolecular physiological indicators	Blood pressure, body mass index, blood cholesterol, and medical imaging (e.g., X-ray and MRI).	Physical examination; laboratory testing
Gene expression	Dysregulation	PCR
Single nucleotide polymorphism	Asymptomatic carrier of a single base change	Microarray or next generation sequencing
Chromosomal abnormalities	Gene duplications, deletions, or translocations	Fluorescence in situ hybridization (FISH)
Antibodies	Viral infection	Western blot or ELISA
Histopathology	Invasive tumor growth	Microscopic examination
Elevated serum protein - disease	PSA, serum creatinine, blood glucose	ELISA, Western blot, or other device
Elevated serum protein - injury	Cardiac enzymes following a heart attack	ELISA, Western blot, or other device
Nucleic acids	miRNA; circulating nucleic acids (CNA)	RNA-seq; PCR

miRNA and Other Noncoding RNAs, may include splicing defects, Drosha- or Dicer-specific dysfunctions, nuclear egress, or the catalytic activities associated with RISC, particularly related to the Argonaute proteins. For example, it is known that altered expression of Dicer is a prognosis predictor for certain types of malignancies (reviewed by Vergani-Junior et al., 2021).

It appears that miRNAs have important regulatory roles in each of the areas associated with the malignant phenotype, including angiogenesis, abrogation of proliferation controls, capacity for immortalized growth, invasive growth into normal surrounding tissue, and metastasis, among other properties. It is quite clear, given a preponderance of the evidence, that miRNA expression patterns are reliable indicators of cellular functionality and, perhaps more importantly, perturbation of molecular pathways. Of great interest, therefore, is the scrutiny of miRNAs expressed under certain conditions, circumstances, or pathologies for possible biomarker discovery (Table 25.4).

miRNA is easily purified from animal and plant tissue, cell culture, blood, cerebrospinal fluid, urine, fungi, and formalin-fixed paraffin-

TABLE 25.4 Examples of miRNAs and disease detection.

miRNA profile	Purpose	Potential advantages
miRNAs circulating in blood	• Early detection • Disease progression • Disease regression	May replace need for biopsy
miRNA in saliva	• Evidence of a sustained concussion[a] • Biomarker for nasopharyngeal carcinoma	Rapid test result that may suggest further treatment
miRNA SNPs	• Determining disease predisposition or carrier status • SNPs commonly found in miRNA genes	An alternative to SNP microarrays
miRNA expression signatures	• Distinguish normal tissue from abnormal or diseased tissue • Determination of disease subtype • In the case of cancer, establish the site of the primary tumor	A possible alternative to histopathology studies, expensive bloodwork and other lab tests, and imaging studies

[a]*Johnson, J.J., A.C. Loeffert, J. Stokes, R.P. Olympia, H. Bramley, and S.D. Hicks. 2018. Association of salivary microRNA changes with prolonged concussion symptoms.* JAMA Pediatr. ***172(1)****, 65–73. https://doi.org/10.1001/jamapediatrics.2017.3884*

embedded (FFPE) archived tissues. The ability to easily collect miRNAs from biofluids in a noninvasive manner is particularly helpful because the profile of CNA, including tumor RNA, is a useful diagnostic indicator. Blood plasma is known for elevated levels nucleases, and freely circulating miRNA, or any other type of RNA, would not be expected to remain intact for very long in that environment. miRNAs are surprisingly stable in biofluids in general, however, owing to their packaging as they move about the body. Their protective "outwear" includes association with Argonaute complexes (involved in gene silencing), encapsulation within lipid-bound structures, or association with other protective proteins. It is therefore very possible that the detection of key indicator miRNAs may soon replace moderately- to heavily-invasive surgical biopsies as a preferred diagnostic tool. In general, as tumor staging progresses, so does the concentration of cell-free nucleic acids in the blood. To this end, some early successes included detecting incidences of lung cancer and ovarian cancer (Taylor and Gercel-Taylor, 2008, 2010; Rabinowits et al., 2009) in the search for appropriate biomarkers. The expression of miR-1, as another example, which has anatomical and physiological implications for developing cardiac and other muscle tissue, has taken on added value because of its importance as a heart attack biomarker following a cardiac infarction event (D'Alessandra et al., 2010). In the screening for potential circulating biomarkers, there is a need to consider lifestyle choices, too, such as smoking status, BMI, and geographic location. Other characteristics that must be considered are disease history, age, sex, and ethnicity.

One might envision therapeutic strategies to be directed at either the knockdown of overactive miRNA expression or the induction or replacement of miRNAs that have been downregulated. In the case of the former, it is possible that miRNAs in excess abundance could be controlled through the introduction of antisense miRNAs, sometimes call antagomiRs (Krützfeldt et al., 2005; Czech, 2006,) to silence the endogenous miRNAs by base-pairing with them. In the case of the latter, restoring normal levels of a downregulated miRNA type is possible by transfection or through the use of a gene construction that has been put into a cell to take the place of the endogenous gene locus that is failing to produce adequate levels of its encoded miRNA.

The therapeutic potential of miRNA faces one major impediment: inefficient delivery vehicles. To address this, miRNAs have been used to create triple helices as a means of shrinking tumors (Conde et al., 2016). The notion is that a triple helical structure should exhibit enhanced stability, thereby increasing transfection efficiency. Triple helices are possible when Hoogsteen hydrogen bonds (Hoogsteen, 1963) and standard Watson–Crick base pairing allow three single-stranded molecules to interact. The nature of the partner strands could, for example, deliver a two-pronged miRNA therapeutic approach, for example, a tumor suppressor-associated miRNA and a miRNA species targeting another component of a proliferation pathway in the case of a patient with cancer.

A final thought: inasmuch as RNA dysregulation can be the cause of a disease, small RNA transcripts also have the potential to cure the disease.

Circulating RNA

Nucleic acids were first shown to be blood-borne more than 40 years ago (Leon, et al., 1977). Later, it was shown that some of the CNAs may portend the onset of cancer or other diseases. Thus, liquid biopsies, for example, blood samples are important as diagnostics because they are easily acquired and minimally invasive. In addition to rendering biomarkers of potential importance, a blood sample can provide a comprehensive overview of tissue and organ functionality whereas a more traditional invasive biopsy only provides information about the specific site from which it was acquired.

There are several ways for RNA to enter the circulatory system where they can be detected as early or late-stage biomarkers. First, RNAs are often present in apoptotic bodies that emerge as a cell dismantles itself. Second, some RNAs travel throughout the circulatory system in the form of ribonucleoprotein assemblies which make them resistant to the plethora of RNases that would otherwise degrade them. Third, nucleic acids in general often travel in lipid-bound extracellular vesicles, including exosomes (Fig. 25.2) which are small membrane-bound vesicles <150 nm that bud off the plasma membrane of various cells types, including tumor cells, (Larrea et al., 2016; Zhang et al., 2015). The vesicles commonly transport myriad transcript types, including mRNA and miRNA, circRNA, and various proteins. Comparatively speaking, cell-free RNA (cfRNA) exhibits significantly greater stability when protect by an exosome membrane. Exosomes are easily recovered from blood, urine, and saliva, thereby making them attractive biomarker candidates.

One thing that is apparent regarding tumor cells: they release significant amounts of RNA into the blood, often just larger amounts of normal transcripts, and in a tissue-specific manner; even at low abundance, it is possible to make quantitative measurements. To be clear, however, one is able to detect circulating RNAs from the normal cells of cancer patients as well as healthy individuals who do not have cancer. How and why CNAs are liberated from normal cells is somewhat nebulous, and is an area of great interest. There is speculation that tumor cells may use RNAs stabilized in or on a carrier to signal normal cells in the neighborhood in an attempt to promote angiogenesis, prevent normal cells from attempting to suppress them in some manner, or even induce normal cells to acquire the malignant phenotype. The importance of CNAs is not to be underestimated and as noted above, exosome transport is an important means of intercellular communication. In addition to exosome transport vehicles, whole tumor cells are also known to travel through the circulatory system, which is the calling card of metastasis.

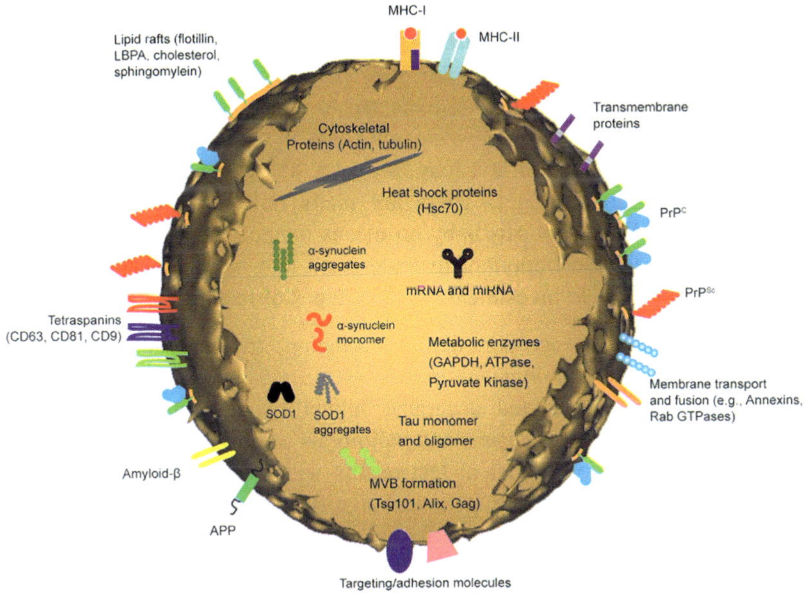

FIGURE 25.2 Exosomes are small membrane-bound vesicles, typically less than 150 nm in diameter, which shares many structural and topological similarities with the plasma membrane. They contain mRNA and miRNA as well as a vast array of different proteins depending on their host cell, and they are probably present in all body fluids (Bellingham et al., 2012). *Image courtesy Bellingham, S.A., B. B. Guo, B.M. Coleman, and A.F. Hill. (2012). Exosomes: vehicles for the transfer of toxic proteins with neurodegenerative diseases? Front. Physiol. 3 https://doi.org/10.3389/fphys.2012.00124; and reproduced with permission in accordance with the Creative Commons License (CC BY 4.0; https://creativecommons.org/licenses/by/4.0/).*

Identification of biomarkers for research and diagnostic applications

Generally speaking, molecular biomarkers link together epidemiology, ecology, and molecular biology. Ideally, one could envision a standard panel of biomarkers being identified which could be used to determine the direction a pharmaceutical company takes with respect to the development of a new drug, including whether an investment should be made in developing the drug at all. Then, a rapid diagnostic device, a theranostic (therapy + diagnostic), consisting of various biomarkers on a microarray or in a microfluidics chip format could become standard fare with respect to determining what drugs, and in what dose, a patient should be given; the theranostic approach is a prime example of a useful device for personalized medicine. Think of this also in terms of patient stratification for earlier identification of drug responders from nonresponders. Given the complexity of the gene expression machinery, a comprehensive, personalized biomarker profile for a patient is likely to include both DNA and RNA components, as dysregulation may involve changes in gene structure and gene expression at the onset and during the progression of the disease. Because of the demand for personalized medicine, the worldwide market for molecular biomarkers is exploding, with the global biomarker market expected to approach $150 billion by the year 2028 (http://www.grandviewresearch.com).

Biomarker development is a long, arduous, and expensive road to travel and involves multiple stages. There are many milestones associated with these stages, and this synopsis is by no means intended to be comprehensive or complete. It is also important to understand that biomarker development requires a considerable investment of many types of resources.

1. *Discovery*: Next-generation deep sequencing, transcriptome tools, and other -omics methodologies.
2. *Validation*: Demonstrated accuracy, reliability, reproducibility, performance of technical replicates, and validity as a surrogate end-point.
3. *Clinical trials*: Correlation of test results with phenotype, results within and among patient cohorts, statistical significance of results.
4. *FDA approval*: Demonstrated effectiveness, patient safety, designation as a medical device in vitro diagnostic device) or laboratory test; interpretation of increased or decreased biomarker levels (refer to the FDA's Biomarker Qualification Program; http://www.FDA.gov).
5. *Acceptance*: Commercialization, health insurance buy-in, patient care guidelines, high-level confidence by the medical community as well as ease of use in hospitals.

The techniques that are commonly used for molecular biomarker discovery may focus on the three facets of gene expression, to wit, the gene itself, the mRNA that the gene produces by the process of transcription, and the mRNA-encoded protein that is synthesized by the process of translation and, of increasing interest, the metabolome.

DNA approaches

The sequencing of the complete complement of human chromosomal DNA signaled the beginning of the era of genomics. While the data set is complete, the function of much of what is in that data set remains unknown. One thing that is clear, however, is the considerable nonallelic variability in humans that is defined by a change in a single base that is known as a single-nucleotide polymorphism (SNP). SNPs are more common than one might imagine in the human genome, occurring on the order of approximately one every 1300 nucleotides, and often in intergenic regions. The most commonly observed SNP is a cytosine to thymine ($C \rightarrow T$) base change.

Many of the mapped SNPs do not alter the phenotype and are innocuous, though sometimes they are not; sickle-cell anemia and cystic fibrosis come to mind immediately in the case of the latter. SNPs are identified through

DNA sequencing or by microarray analysis (SNP Chip) screening of genomic DNA. When discovered in a patient sample, the number and location of SNPs may indicate that a patient is at risk for a particular disease. If the patient already has a disease at the time of screening, SNP analysis may help identify the disease subtype. It is important to note, however, that while SNPs are useful, many diseases are polygenic in nature and many involve more than one SNP. Two handy resources are the Allele Frequency Database (ALFRED), located at alfred.med.yale.edu, and NCBI's dbSNP database located at http://www.ncbi.nlm.nih.gov/SNP.

RNA approaches

Transcriptomics is the study of the expressed RNA content of the cell. Unlike the static nature of the genome, the transcriptome is dynamic, changing constantly in response to both extra- and intracellular cues. These changes include up- and downregulation (abundance), alternative splicing, alternative transcription start sites, and epigenetic variations of RNA (epitranscriptome; reviewed by Wiener and Schwartz, 2021). Various traditional and contemporary methods may be used for assaying differential gene expression, including microarray screening, whole transcription profiling (RNA-seq), and other high-throughput techniques. The richness of RNA-seq data lends itself to biomarker identification which correlate with specific diseases. The various forms of next-generation deep sequencing have revealed an astonishing variety RNAs which have shown great potential as biomarkers and in many cases have already been designated as such.

Noncoding RNA (ncRNA) is synthesized by transcription, as are other RNAs, though they are commonly observable in extracellular locations as well, including biofluids. Many of the array-based screening methodologies require nanogram quantities of RNA, or less, or a very small number of cells, and it is common to find ncRNAs printed on commercial microarrays along with traditional coding sequences. Previously archived FFPE samples, while not ideal, are also capable of rendering useful ncRNA information.

Among the small noncoding RNAs (sncRNAs) are the well-known microRNAs (miRNAs; miRs), the over- or underexpression of many of which correlates with various human diseases or pathologies. miRs continue to have great potential as biomarkers, owing to their well-characterized post-transcriptional role in the stability and functionality of their cognate mRNAs. The number of known miRs increases monthly (visit http://www.miRBase.org), as more and more of these powerful regulatory molecules are uncovered across the eukaryotes. Given the capability to create siRNAs to mimic miRNAs, or act as decoys, this avenue of research may turn out to be very productive in terms of developing personalized, nucleic acid-based therapies. It is very likely that, as the number of small and large RNAs continues to grow, the reasons why some cancer patients are naturally resistant to certain drugs or therapies will almost certainly be uncovered. Owing to miRNA redundancy and cooperation, meaning that a particular transcript can be regulated by several different miRNAs, this avenue of biomarker discover is all the more intriguing.

A long noncoding RNA (lncRNA) is longer than 200 nts, while a small (short) noncoding RNA (sncRNA) may be as long as 200 nts. Diversified in structure and variety, lncRNAs also vary greatly in terms of the abundance range they display and are established RNA biomarkers. There are many thousands of known lncRNAs, most of which are of unknown function. The long ncRNA group in general is widely believed to hold enormous potential as RNA biomarkers. For example, PCA3 is a lncRNA that is an effective reliable biomarker for the diagnosis of prostate cancer; found only in prostate tissue and is easily detected in urine, this transcript is present at elevated levels in the disease state. Many such lncRNAs and mRNA splice variants over which they often have influence have been identified by transcriptomic-associated assays, though the significance of many of these transcripts remains something of a puzzle.

Another class of RNA, long intergenic noncoding RNA (lincRNA), has been identified in several organisms, including animals (from several thousand loci in mammals alone), plants, and some protozoans (reviewed by Ulitsky and Bartel, 2013). Although the typical abundance level for a lincRNA transcript is usually less by a factor of ten than a typical mRNA (Sigova et al., 2013), there is much greater variability in lincRNA expression levels among tissues. Several functions have been proposed for lincRNAs, including cotranscriptional regulation, base-pairing with other transcripts, acting as molecular decoys, and forming requisite RNA protein complexes in support of other cellular functions. For a review of lncRNA and lincRNA, see Chen and Shan (2020).

Noncoding RNAs are not the only type of established biomarker transcripts; mRNAs are among the ranks as well. Examples include (1) the telomerase mRNA, an important biomarker because reactivation of this gene is a key factor, in maintaining telomere length, in the abrogation of cellular senescence, (2) a splice variant of the androgen receptor, AR-V7, is used to predict the likely efficacy of antiandrogen receptor drugs (Li et al., 2013), and (3) sustained, elevated expression of the HER-2 oncogene in certain forms of breast cancer (Slamon et al., 1987; 1989). A few examples of known RNA biomarkers in the sncRNA, lncRNA, and mRNA categories are presented in Table 25.5.

At this juncture, qPCR, dPCR, and RNA-seq are in widespread use for the detection of all types of RNA, regardless of the source. These methods offer the investigator an ability to detect even minute quantities of RNA, and many up- and downregulated transcripts associated with a host of malignant diseases have been cataloged. Fortunately, what is known about miRNA, lncRNA, and coding mRNA abundance profiles is accessible in public databases so that expression trends may become discernable over time.

TABLE 25.5 Examples of RNA biomarkers.

	Transcript	Source	References
Small noncoding RNA ≤200 nt	miR-1	Elevated in acute myocardial infarctions	Liu et al. (2015); Ai et al. (2010)
	miR-10b	Overexpressed in some forms of breast cancer	Ma, et al. (2007), Heneghan et al. (2010)
	miR-22	Overexpressed in some forms of ovarian cancer	Li et al. (2010)
	miR-155	Elevated in whole blood in sepsis and is indicative of severity and prognosis	Reviewed by Chen et al. (2021)
	miR-497	Significantly downregulated in osteosarcoma primary tumor tissue	Pang et al. (2016)
	SNORD 33/66/76	Three snoRNA biomarkers for lung cancer	Liao et al. (2010)
	Exosomal RNA (exoRNA)	Various transcripts, including miRNA	Reviewed by Chu et al. (2020)
Long noncoding RNA (>200 nt)	H19	miRNA precursor; epigenetic imprinting; elevated in some cancers	Tanos et al. (2004), Takeuchi et al. (2007)
	HOTAIR	Gene silencing; breast cancer and pancreatic cancer biomarker	Rinn et al. (2007), Gupta et al. (2010)
	MALAT-1	Prostate, lung cancer biomarker (upregulated)	Ren et al. (2013), Ji et al. (2003)
	PCA3	Prostate cancer (upregulated as much as 100-fold)	Hessels and Schalken (2009), Salagierski et al. (2010), Gunelli et al. (2021)
mRNA	$BIRC_5$	Baculoviral IAP (inhibitor of apoptosis) Repeat-Containing 5 mRNA is elevated in colorectal cancers patients with a poor prognosis	He et al. (2021)
	HER-2	Elevated oncogene expression in certain types of breast cancer; used to determine treatment	Llombart-Cussac et al. (2017), Slamon et al. (2001)
	hTERT	Human telomere reverse transcriptase is detected in >90% of all cancers	Hannen and Bartsch (2018)

Protein approaches

Proteomics profiling methods include various traditional and contemporary methods for assaying differential protein expression, including Western analysis, ELISA, protein and antibody chip technologies, and other protein assay techniques. This is the era of in silico proteomics, meaning that powerful computers are able to predict protein 2°, 3°, and 4° structures, as well as function, based on amino acid sequence (1° structure). Also included is cataloging the comprehensive variety of posttranslational modifications to which eukaryotic proteins are subjected. These data sets are valuable assets for rational drug design in that valid drug targets may be identified. In some cases, biomarkers may themselves be the target of a drug, particular if the biomarker has a known signaling function.

Metabolomics approaches

The term metabolome refers to the overall metabolite composition of body fluid, a cell, or other biological samples of interest. Metabolites are byproducts of metabolism, and commonly include amino acids, glucose, acetate, ethanol, methanol, urea, pyruvate, creatine, and many other types of molecules. Hundreds of thousands of different metabolites have been cataloged and are readily accessible through the Human Metabolome Database (https://hmdb.ca) or the Metabolomics Workbench (http://www.metabolomicsworkbench.org). Metabolomics has become quite relevant in biomarker identification. As noted above, ideal matrices for biomarker development include blood, saliva, and urine because the collection of a specimen is simple and noninvasive, economical, and generally free of risk or complications; these biofluids also contain large sets of metabolites. Since everything that enters the body is destined for metabolic rearrangement in one form or another, a metabolite profile accuracy reflects both cellular- and organism-level physiology. Methods used for metabolite profiling commonly include mass spectrometry (MS), gas-liquid chromatography (GLC), high-performance liquid chromatography (HPLC), and nuclear magnetic resonance (NMR).

Biomarker issues and shortcomings

If there is one area that is most in need of fine-tuning with respect to biomarker discernment, it is the lack of standardization with respect biomarker assays in a clinical setting. There is, in fact, no general agreement about what, exactly, constitutes a valid biomarker. This is particularly concerning since many diseases involve the deregulation of several genes, and not just one. Exacerbating this situation is the fact that SOPs, including sample collection methods, often vary widely from lab to lab. Refinements in this area would be beneficial at numerous levels, particularly for intractable forms of cancer in which detection and diagnosis come only when the disease is in an advanced stage; at that point, intervention is often only marginally helpful.

The source and processing of a specimen is impactful on the overall sensitivity of an assay. In the case of blood, for example, the utility of the relatively small amount of recoverable RNA is often compromised by the excessive quantity of protein found in blood plasma, including immunoglobulins, albumin, transferrin, and many others. Differences in RNA purification efficiency are widely believed to be the cause of the observed variability among microarray assay results; procedures which favor the removal of impurities or contaminants in the fewest steps narrow that variability. False negatives are also common.

Regarding circulating RNA, there is no consensus about what population or subpopulation of transcripts to use to normalize a sample. Ideas include endogenous miR transcripts, 5 S rRNA, and even spiking a sample with an external control transcript. Ideally, uniformly expressed miRs could be used, but there are reports that even the "housekeeping miRs" are subject to abundance variation, just like many of the mRNAs that were once considered to be expressed constitutively with minimal modulation. Adding to this, the magnitude of miR variability in human populations has not yet been determined.

In one novel application, a panel of nineteen blood biomarkers was developed that is being used by some to prognosticate just how gracefully a person will age (Sebastiani et al., 2017). Markers used for this study included standard blood and lipid biomarkers, as well as markers of inflammation and frailty; these data were correlated with perceived risk for diabetes, cardiovascular disorders, and cancer. There appeared to be a standard biomarker signature shared by most of the nearly 5000 participants in this study, though several other signatures were also observed. The biomarker signature for an individual suggests whether that person is expected to have more or fewer health issues as they age. Among the minority biomarker signatures are disease-free aging, risk of heart disease, and risk of dementia. This study also demonstrated that the dysregulation of one of the panel biomarkers was able to alter the expression pattern of some of the other known biomarkers. For a recent review, see Hartmann et al. (2021).

The idea of a clinically relevant biomarker panel is not new. OncotypeDX (http://www.oncotypedx.com) has developed panels of genes that are assayed in order to generate a recurrence score, that is, the likelihood that cancer will return, usually after five years. Screening panels are available for breast, colon, and prostate cancer; because they involve measuring the expression of multiple gene loci, they are often referred to as in vitro diagnostic multivariate index assays, or IVDMIA (FDA docket number

FDA-2006-D-0233). The specimen and pathology reports are generated in a clinical setting, though molecular biology is performed in a centralized laboratory in order to maintain stringent quality control over the technical aspects of the screening. While the prospects for early intervention are intriguing, the potential for discriminatory practices based on a patient's "signature" does give one pause.

A new discipline known as multiomics (also, multiomics) is emerging. Multiomics is the synthesis of information gathered from the multiple "omes" (e.g., the transcriptome, the proteome, and the metabolome) to describe a biological phenomenon. These studies may focus on the behavior of single cells or may occur at the organism-level Large-scale studies of this nature, also described as multimodal platforms, often involve big data computational methods, and have great potential in the area of biomarker identification. Many software applications, most notably the bioconductor project (http://www.bioconductor.org) have been developed to manage a growing and a diverse number of multiomic databases (Table 25.6).

Finally, intellectual property issues also remain an impediment to progress in this area. Recently, the entire issue of the patentability of naturally-occurring molecules, including DNA, RNA, and proteins has been called into question by court rulings (reviewed by Harrison, 2014a,b). While this issue is less stringently regulated abroad, the final resolution is likely to have far-reaching ramifications.

TABLE 25.6 Useful multiomics resources.

Name	Website	Multiomic focus
LinkedOmics	http://www.linkedomics.org	Cancer
OASIS	https://atlasgeneticsoncology.org	Cancer cell lines and tumor tissue
ZikaVR	https://bioinfo.imtech.res.in/manojk/zikavr/	Zika virus
Ecomics	http://prokaryomics.com/	*Escherichia coli*
ProteomicsDB	http://www.proteomicsdb.org	Human proteome
MaizeGDB	https://maizegdb.org	*Maize*
KEGG	http://www.genome.jp/kegg	Plants, animals, and microorganisms
Lipid Maps	http://www.lipidmaps.org	Lipidomics
PRIDE	http://www.ebi.ac.uk/pride	Proteomics
FlyBase	https://flybase.org	*Drosophila*

References

Ai, J., Zhang, R., Li, Y., Pu, J., Lu, Y., Jiao, J., et al., 2010. Circulating microRNA-1 as a potential novel biomarker for acute myocardial infarction. Biochem. Biophys. Res. Commun. 391, 73–77. Available from: https://doi.org/10.1016/j.bbrc.2009.11.005.

Bellingham, S.A., Guo, B.B., Coleman, B.M., Hill, A.F., 2012. Exosomes: vehicles for the transfer of toxic proteins with neurodegenerative diseases? Front. Physiol. 3. Available from: https://doi.org/10.3389/fphys.2012.00124.

Chen, H., Shan, G., 2020. The physiological function of long-noncoding RNAs. Non-coding RNA Res. 5 (4), 178–184. doi.org/10.1016/j.ncrna.2020.09.003.

Chen, M., Wang, F., Xiqa, H., Yao, S., 2021. MicroRNA-155: regulation of immune cells in sepsis. Mediators Inflamm. 2021. Available from: https://doi.org/10.1155/2021/8874854.

Chu, Y.L., Li, H., Ng, P.L.A., Kong, S.T., Zhang, H., Lin, Y., et al., 2020. The potential of circulating exosomal RNA biomarkers in cancer. Expert. Rev. Mol. Diagn. 20 (7), 665–678. Available from: https://doi.org/10.1080/14737159.2020.1745064.

Conde, J., Oliva, N., Atilano, M., Song, H.S., Artzi, N., 2016. Self-assembled RNA-triple-helix hydrogel scaffold for microRNA modulation in the tumour microenvironment. Nat. Mater. 15, 353–363.

Czech, M.P., 2006. MicroRNAs as therapeutic targets. N. Engl. J. Med. 354 (11), 1194–1195.

D'Alessandra, Y., Devanna, P., Limana, F., Straino, S., Di Carlo, A., Brambilla, P.G., et al., 2010. Circulating microRNAs are new and sensitive biomarkers of myocardial infarction. Eur. Heart J. 31 (22), 2765–2773.

Gunelli, R., Fragalà, E., Fiori, M., 2021. PCA3 in prostate cancer. In: Salvi, S., Casadio, V. (Eds.), Urinary Biomarkers. Methods in Molecular Biology, Vol 2292. Humana, New York, NY, https://doi.org/10.1007/978-1-0716–1354-2_9.

Gupta, R.A., Shah, N., Wang, K.C., Kim, J., Horlings, H.M., Wong, D.J., et al., 2010. Long non-coding RNA HOTAIR reprograms chromatin state to promote cancer metastasis. Nature 464, 1071–1076.

Hannen, R., Bartsch, J.W., 2018. Essential roles of telomerase reverse transcriptase hTERT in cancer stemness and metastasis. FEBS Lett. 592, 2023–2031. Available from: https://doi.org/10.1002/1873-3468.13084.

Harrison, C., 2014a. Patenting natural products just got harder. Nat. Biotechnol. 32, 403–404.

Harrison, C., 2014b. Patent watch: Australian court upholds Myriad's gene patent. Nat. Rev. Drug Discov. 13, 805.

Hartmann, A., Hartmann, C., Secci, R., Hermann, A., Fuellen, G., Walter, M., 2021. Ranking biomarkers of aging by citation profiling and effort scoring. Front. Genet. 12, 0.3389/fgene.2021.686320.

Heneghan, H.M., Miller, N., Lowery, A.J., Sweeney, K.J., Newell, J., Kerin, M.J., 2010. Circulating microRNAs as novel minimally invasive biomarkers for breast cancer. Ann. Surg. 251, 499–505. Available from: https://doi.org/10.1097/SLA.0b013e3181cc939f.

Hessels, D., Schalken, J.A., 2009. The use of PCA3 in the diagnosis of prostate cancer. Nat. Rev. Urol. 6, 255–261.

He, J., Wu, F., Han, Z., Hu, M., Lin, W., Li, Y., et al., 2021. Biomarkers (mRNAs and non-coding RNAs) for the diagnosis and prognosis of colorectal cancer – from the body fluid to tissue level. Front. Oncol. 29. Available from: https://doi.org/10.3389/fonc.2021.632834.

Hoogsteen, K., 1963. The crystal and molecular structure of a hydrogen-bonded complex between 1-methylthymine and 9-methyladenine. Acta Crystallogr. 16, 907–916.

Ji, P., Diederichs, S., Wang, W., Böing, S., Metzger, R., Schneider, P.M., et al., 2003. MALAT-1, a novel noncoding RNA, and thymosin beta4 predict metastasis and survival in early-stage non-small cell lung cancer. Oncogene 22, 8031–8041. Available from: https://doi.org/10.1038/sj.onc.1206928.

Krützfeldt, J., Rajewsky, N., Braich, R., Rajeev, K.G., Tuschl, T., Manoharan, M., et al., 2005. Silencing of microRNAs in vivo with 'antagomirs'. Nature 438, 685–689.

Larrea, E., Sole, C., Manterola, L., Goicoechea, I., Armesto, M., Arestin, M., et al., 2016. New concepts in cancer biomarkers: circulating miRNAs in liquid biopsies. Int. J. Mol. Sci. 17, 627.

Leon, S.A., Shapiro, B., Sklaroff, D.M., Yaros, M.J., 1977. Free DNA in the serum of cancer patients and the effect of therapy. Cancer Res. 37, 646–650.

Liao, J., Yu, L., Mei, Y., Guarnera, M., Shen, J., Li, R., et al., 2010. Small nucleolar RNA signatures as biomarkers for non-small-cell lung cancer. Mol. Cancer 9, 198. Available from: https://doi.org/10.1186/1476-4598-9-198].

Liu, X., Fan, Z., Zhao, T., Cao, W., Zhang, L., Li, H., et al., 2015. Plasma miR-1, miR-208, miR-499 as potential predictive biomarkers for acute myocardial infarction: an independent study of Han population. Exp. Gerontol. 72, 230–238. Available from: https://doi.org/10.1016/j.exger.2015.10.011.

Li, Y., Chan, S.C., Brand, L.J., Hwang, T.H., Silverstein, K.A., Dehm, S.M., 2013. Androgen receptor splice variants mediate enzalutamide resistance in castration-resistant prostate cancer cell lines. Cancer Res. 73, 483–489.

Li, J., Liang, S.H., Lu, X., 2010. Potential role of ezrin and its related microRNA in ovarian cancer invasion and metastasis. Zhonghua fu Chan ke za zhi 45, 787–792.

Llombart-Cussac, A., Cortés, J., Paré, L., Galván, P., Bermejo, B., Martínez, N., et al., 2017. HER2-enriched subtype as a predictor of pathological complete response following trastuzumab and lapatinib without chemotherapy in early-stage HER2-positive breast cancer (PAMELA): an open-label, single-group, multicentre, phase 2 trial. Lancet Oncol. 18 (4), 545–554. Available from: https://doi.org/10.1016/S1470-2045(17)30021-9.

Ma, L., Teruya-Feldstein, J., Weinberg, R.A., 2007. Tumour invasion and metastasis initiated by microRNA-10b in breast cancer. Nature 449, 682–688. Available from: https://doi.org/10.1038/nature06174.

Pang, P.C., Shi, X.Y., Sun, K., 2016. miR-497 as a potential serum biomarker for the diagnosis and prognosis of osteosarcoma. Eur. Rev. Med. Pharmacol. Sci. 20 (18), 3765–3769.

Rabinowits, G., Gerçel-Taylor, C., Day, J.M., Taylor, D.D., Kloecker, G.H., 2009. Exosomal microRNA: a diagnostic marker for lung cancer. Clin. Lung Cancer 10 (1), 42–46.

Ren, S., Wang, F., Shen, J., Sun, Y., Xu, W., Lu, J., et al., 2013. Long non-coding RNA metastasis associated in lung adenocarcinoma transcript 1 derived miniRNA as a novel plasma-based biomarker for diagnosing prostate cancer. Eur. J. Cancer 49, 2949–2959.

Rinn, J.L., Kertesz, M., Wang, J.K., Squazzo, S.L., Xu, X., Brugmann, S.A., et al., 2007. Functional demarcation of active and silent chromatin domains in human HOX loci by noncoding RNAs. Cell 129 (7), 1311–1323.

Salagierski, M., Verhaegh, G.W., Jannink, S.A., Smit, F.P., Hessel, D., Shalken, J.A., 2010. Differential expression of PCA3 and its overlapping PRUNE2 transcript in prostate. Prostate 70, 70–78.

Sebastiani, P., Thyagarajan, B., Sun, F., Schupf, N., Newman, A.B., Montano, M., et al., 2017. Biomarker signatures of aging. Aging Cell Jan. 6. Available from: https://doi.org/10.1111/acel.12557.

Sigova, A.A., Mullen, A.C., Molinie, B., Gupta, S., Orlando, D.A., Guenther, G.G., et al., 2013. Divergent transcription of long noncoding RNA/mRNA gene pairs in embryonic stem cells. Proc. Natl. Acad. Sci. 110 (8), 2876–2881.

Slamon, D.J., Clark, G.M., Wong, S.G., Levin, W.J., Ullrich, A., McGuire, W.L., 1987. Human breast cancer: correlation of relapse and survival with amplification of the HER-2/neu oncogene. Science 235, 177–182.

Slamon, D.J., Godolphin, W., Jones, L.A., Holt, J.A., Wong, S.G., Keith, D.E., et al., 1989. Studies of the HER-2/neu proto-oncogene in human breast and ovarian cancer. Science 244, 707–712.

Slamon, D.J., Leyland-Jones, B., Shak, S., Fuchs, H., Paton, V., Bajamonde, A., et al., 2001. Use of chemotherapy plus a monoclonal antibody against HER2 for metastatic breast cancer that overexpresses HER2. N. Engl. J. Med. 344 (11), 783–792. Available from: https://doi.org/10.1056/NEJM200103153441101.

Takeuchi, S.W.K., Hofmann, Tsukasaki, K., et al., 2007. Loss of H19 imprinting in adult T-cell leukaemia/lymphoma. Br. J. Haematol. 137 (4), 380–381.

Tanos, V.I., Ariel, D., Prus, N., De-Groot, Hochberg, A., 2004. H19 and IGF2 gene expression in human normal, hyperplastic, and malignant endometrium. Int. J. Gynecol. Cancer 14 (3), 521–525.

Taylor, D.D., Gercel-Taylor, C., 2008. MicroRNA signatures of tumor-derived exosomes as diagnostic biomarkers of ovarian cancer. Gynecol. Oncol. 110, 13–21.

Taylor, D.D., Gercel-Taylor, C., 2010. MicroRNA signatures of tumor-derived exosomes as diagnostic biomarkers of ovarian cancer. Gynecol. Oncol. 116, 153.

Ulitsky, I., Bartel, D.P., 2013. lincRNAs: genomics, evolution, and mechanism. Cell 154 (1), 26–46.

Vergani-Junior, C.A., Tonon-da-Silva, G., Inan, M.D., Mori, M.A., 2021. DICER: structure, function, and regulation. Biophys. Rev. 13, 1081–1090. Available from: https://doi.org/10.1007/s12551-021-00902-w.

Wiener, D., Schwartz, S., 2021. The epitranscriptome beyond m6A. Nat. Rev. Genet. 22, 119–131.

Zhang, X., Yuan, X., Shi, H., Wu, L., Qian, H., Xu, W., 2015. Exosomes in cancer: small particle, big player. J. Hematol. Oncol. 8, 83. Available from: https://doi.org/10.1186/s13045-015-0181-x.

Further reading

Bae, J., Yang, S.-H., Kim, A., Kim, H.G., 2022. RNA-based biomarkers for the diagnosis, prognosis, and therapeutic response monitoring of prostate cancer. Urologic Oncol. 40 (3), 105. Available from: https://doi.org/10.1016/j.urolonc.2021.11.012. e1–105.e10.

Buder, A., Tomuta, C., Filipits, M., 2016. The potential of liquid biopsies. Curr. Opin. Oncol. 28 (2), 130–134.

Condrat, C.E., Thompson, D.C., Barbu, M.G., Bugnar, O.L., Boboc, A., Cretoiu, D., et al., 2020. miRNAs as Biomarkers in disease: latest findings regarding their role in diagnosis and prognosis. Cells 9 (2), 276. Available from: https://doi.org/10.3390/cells9020276. PMID: 31979244.

Goossens, N., Nakagawa, S., Sun, X., Hoshida, Y., 2015. Cancer biomarker discovery and validation. Transl. Cancer Res. 4 (3), 256–269.

Hoeng, J., Bovard, D., Peitsch, M. (Eds.), 2019. Organ-on-a Chip. Academic Press, San Diego, CA, ISBN 9780128172025.

Leung, C.M., de Haan, P., Ronaldson-Bouchard, K., Kim, G.-A., Ko, J., Rho, H.S., et al., 2022. A guide to the organ-on-a-chip. Nat. Rev. Meth. Primers 2, 33. Available from: https://doi.org/10.1038/s43586-022-00118-6.

Neves, A.F., Araújo, T.G., Biase, W.K.F.S., Meola, J., Alcântara, T.M., Freitas, D.G., et al., 2008. Combined analysis of multiple mRNA markers by RT-PCR assay for prostate cancer diagnosis. Clin. Biochem. 41, 1191–1198.

Peng, Y., Croce, C.M., 2016. The role of microRNAs in human cancer. Signal. Transduct. Target. Ther. 1, 15004. Available from: https://doi.org/10.1038/sigtrans.2015.4.

Quinn, J.J., Chang, H.Y., 2016. Unique features of long non-coding RNA biogenesis and function. Nat. Rev. Genet. 17, 47–62.

Neves, A.F., Araújo, T.G., Biase, W.K.F.S., Meola, J., Alcântara, T.M., Freitas, D.G., et al., 2008. Combined analysis of multiple mRNA markers by RT-PCR assay for prostate cancer diagnosis. Clin. Biochem. 41, 1191–1198.

Chapter 26

Functional genomics strategies

Rationale

The completion of the original human genome sequencing project was a milestone achievement, opening the door for more physical and functional studies. The development of various forms of next-generation sequencing (NGS) has facilitated rapid sequencing of the complete genomes of many other organisms, rendering prodigious amounts of data. Very little information is gleaned from sequencing data alone, however, as to how individual genes and the genome as a whole actually function. The postgenomics era in which the scientific community finds itself has been dubbed the era of functional genomics: investigators now have the script but need to interpret its meaning. Functional genomics is all about various forms of mostly high-throughput, global analyses of gene expression; as a discipline it distills down to really one goal, to wit, making sense out of a LOT of data. While deciphering and interpreting that data was once believed to require years, big data and cloud computing have accelerated the process. One important discovery that has emerged is that most of the genome is transcribed as noncoding RNAs, among which the miRNA subpopulation appears to have an astounding ability to modulate mRNA and therefore protein levels in the cell.

There are two types of investigative work within the sphere of functional genomics. First, experimental methods are used to monitor changes in gene expression through the direct assay of RNA and protein abundance. There are many useful methods, though they tend to be time-consuming and can be expensive. Second, in silico analysis, that is, bioinformatics, is supremely useful for performing homology searches, predicting gene function from DNA sequence information, and managing and archiving large amounts of sequencing data associated with the species being interrogated. These data have already made significant contributions to all areas of modern biology and to human health in particular.

Functional genomics defined

An appropriate way to think about functional genomics is to compare it to structural genomics. A good analogy is a relationship between anatomy and

RNA Methodologies. DOI: https://doi.org/10.1016/B978-0-323-90221-2.00030-8

physiology. Anatomy is the study of a biological structure while physiology is the study of biological function. The manner by which a cell, a tissue, or an organ function is directly related to its structure. A major function of the genome is the expression of the information that is encoded within it, beginning with the engagement of the regulatory (promoter) elements with RNA polymerase and the entire company of ancillary factors which support transcription. Whereas the genome is indicative of what a cell *can do*, functional genomics is concerned with what a cell is *actually doing* at a particular moment and is, therefore, a logical extension of structural sequencing. Functional genomics is a holistic study of the entire genome through which one endeavors to understand phenotype by first understanding genotype. The notion of functional genomics for modeling the consequences of gene behaviors is not new. Many of the methods described below have been in use for decades.

The concept of a functional genome is more about a multitiered, global analysis of gene expression, and less about the gene-by-gene assessment methods associated with traditional molecular biology. Whereas transcriptomics inventories RNA variety and abundance, and how that inventory changes in an experimental context, "functional genomics" is a more inclusive term. One would be quite remiss if proteins were not also a part of the evaluation of genome functionality. Given the dynamic nature of the transcriptome, a logical next step after examining transcript abundance is to examine the regulatory circuits associated with the synthesis of the cognate polypeptide(s). Remember, for instance, that the presence of a particular mRNA in the cytoplasm does not guarantee its translation which, from a functional genomics perspective, is important. In addition to monitoring the process of translation, proteomics profiling also takes into account posttranslational modifications, protein–protein interactions, and protein turnover. Proteomics, in turn, is a major platform for drug discovery because of the link between genes, the protein products of genes, and diseases that arise from nonfunctional proteins or from proteins that have acquired "unauthorized functions." These types of data led to the development of metabolomics and later, systems biology. Taken together, observations at each of these levels of gene expression proffer a better understanding of the physiology of the cell.

Importance of functional genomics approaches

As noted above, structure imparts function. It is important to understand that, while the genome is static in the sense that the semiconservative manner in which DNA is replicated ensures genetic continuity, the genome is also subject to mutations which can easily change the function of one or more genes. For example, single-nucleotide polymorphisms (SNPs; base positions in a gene that are subject to change) are often associated with unique characteristics and, therefore, are interesting in a functional genomics context. Thus, qualitative data are just as important as quantitative data. Inasmuch as SNPs often identify candidate genes associated with the onset and progression of the disease, cataloging them is of great importance in the development and advancement of pharmacogenomics, known more casually as personalized medicine. The SNP database is accessible at http://www.ncbi.nlm.nih.gov/projects/SNP.

The cell uses combinatorial strategies at the DNA, RNA, and protein levels to create genetic diversity among the cells of a single organism. At the DNA level, for example, antibody genes and T-cell receptor genes are well known for their ability to undergo somatic recombination in lymphocytes. These nongerm-line rearrangements allow cells to produce highly-specific antibodies against just about any antigenic structure to which the organism is exposed. In the RNA world, posttranscriptional splicing allows the inclusion or exclusion of specific exons from a particular mRNA, thereby producing transcript variants from the same gene locus. Further, the selection of alternative transcription start sites can support the inclusion of a transit sequence at the amino terminus of the encoded protein that will allow the protein to move into various organelles. In the protein world, immature polypeptides require a certain set of posttranslational modifications to acquire their intended function(s). In some cases, protein functionality also requires the formation of a quaternary structure, meaning that two or more polypeptide subunits are required to interact in a precise manner in order to activate the protein; the commonly used example of hemoglobin requires the association of two α chains and two β chains to function correctly in the erythrocyte. Understanding these and other cellular and subcellular phenomena are highly relevant in the realm of functional genomics.

Commonly used functional genomics approaches

This book presents a wide variety of classical and contemporary molecular biology techniques, some of which are reiterated in this chapter so as to convey to the reader a sense of the options that are available. The following list is by no means meant to be all-inclusive but rather is intended to highlight functional genomics methodologies and strategies for understanding the function, directly or indirectly, of the genome. Note that these methods are not unique to functional genomics; rather the methods are often used in various combinations in order to gain a holistic vision of genome-wide functionality.

1. *RNA-seq* is a powerful NGS tool that has gained widespread acceptance. As detailed in Chapter 24, RNA-seq: The Premier Transcriptomics Tool, RNA-seq is a global RNA profiling method that is capable of probing the transcriptome at a heretofore unknown depth. The

massively parallel nature of RNA-seq has revolutionized and invigorated mining of the transcriptome by virtue of the quantitative (transcript amount) and qualitative (transcript variety) data that it generates. Given the power of RNA-seq, it is likely that many as-of-yet-to-be-detected transcript species may soon be added to the ranks of annotated members of the transcriptome. While the expense associated with performing RNA-seq has prevented some laboratories from taking full advantage of this technology in the past, there has been a continuous decline in the per-reaction cost. This will increase demand for RNA-seq as the functional genomics tool of choice because of the much greater dynamic range that it offers in comparison with microarrays and related methods. RNA-seq is also used to validate microarray data and vice versa.

2. *CRISPR technology* represents a precision tool for gene editing or gene inactivation. One may learn a great deal about the function of a gene by determining the consequences when a wild-type gene is intentionally mutated, when a mutated gene is edited to return it to the wild-type state, or when a gene is permanently disabled. As the CRISPR technology continues to be refined, scaling up and multiplexing-type capabilities are quickly becoming mainstream tools. See Przybyla and Gilbert (2022) for a recent review of CRISPR methods as they pertain to the realm of functional genomics.
3. *Genome-wide association studies (GWAS)* is an investigative process with the goal of inventorying genetic variation, especially SNPs, across large populations of individuals in order to identify genetic markers associated with the disease. For example, this approach to gene profiling has implicated the complement factor H gene in blindness. Similar findings have linked specific genes to elevated risk for Parkinson's disease, obesity, heart disease, type 2 diabetes, Crohn's disease, and schizophrenia to name a few examples. The usual methodology involves a direct, side-by-side comparison of DNA from patients with a disease with DNA from volunteer donors without the disease (Fig. 26.1). Genomic variations are most often detected by microarray analysis or NGS. These data are then uploaded and archived in the database of Genotypes and Phenotypes (dbGaP), an NCBI website located at http://www.ncbi.nlm.nih.gov/gap which can be accessed by anyone. See Uffelmann et al. (2021) for a review of this important initiative.
4. *ChIP*, or chromatin immunoprecipitation, is a valuable technique for studying the manner in which proteins and nucleic acids interact (Fig. 26.2). Briefly, proteins and the chromatin sequences to which they are bound are crosslinked (e.g., UV light or formaldehyde) such that they are stably joined. After random fragmentation of the DNA, an antibody is used to sequester the protein for which it has specificity, which also pulls down the genomic DNA region to which the protein is bound.

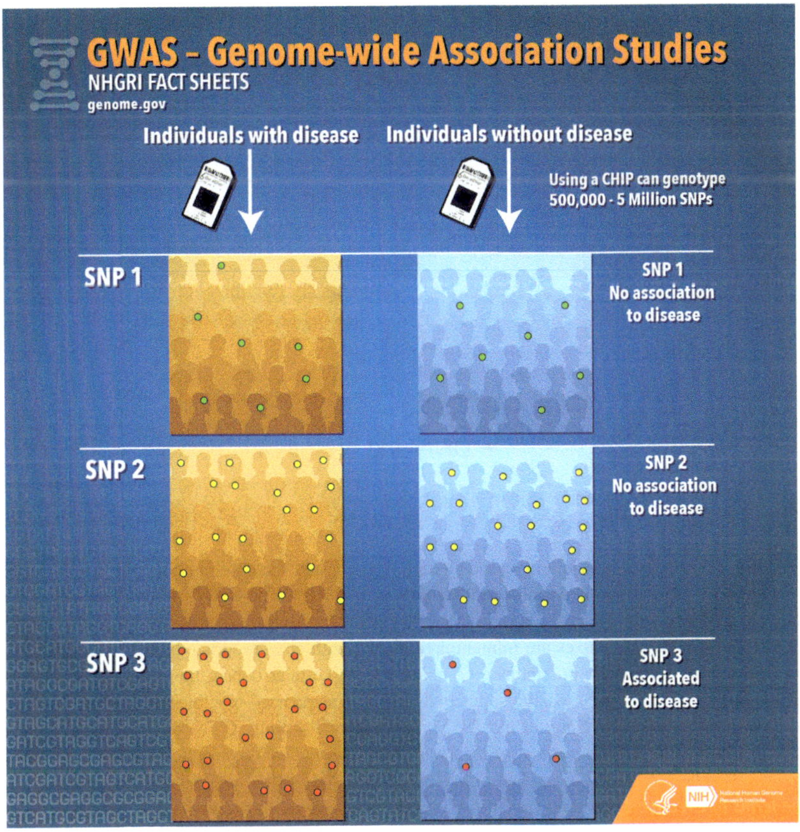

FIGURE 26.1 Genome-wide association study. DNA from healthy donors and DNA from those with an illness or disease are compared side-by-side by microarray (chip) analysis. A significant difference in the observed frequency of an SNP in patients manifesting a disease, when compared with the asymptomatic population, is often a strong sign that the gene variation is in some way associated with the initiation or progression of the disease. *Image courtesy National Human Genome Research Institute (NHGRI; http://www.genome.gov).*

The DNA is then sequenced to identify the chromosomal regions and the specific nucleotide sequence with which the protein was interacting. This type of work has been particularly valuable in the study of transcription factors and the nature of the gene promoter regions for which they have an affinity.

5. *Microarray analysis* is a high-throughput method for the simultaneous assay of very large numbers of genes, as described in Chapter 21, Array Analysis of Gene Expression. A global method for the analysis of gene expression, microarrays (gene chips) consist of solid support upon which hundreds of thousands of individual gene sequences, referred to as probes, have been applied. The high level of sensitivity is due to the

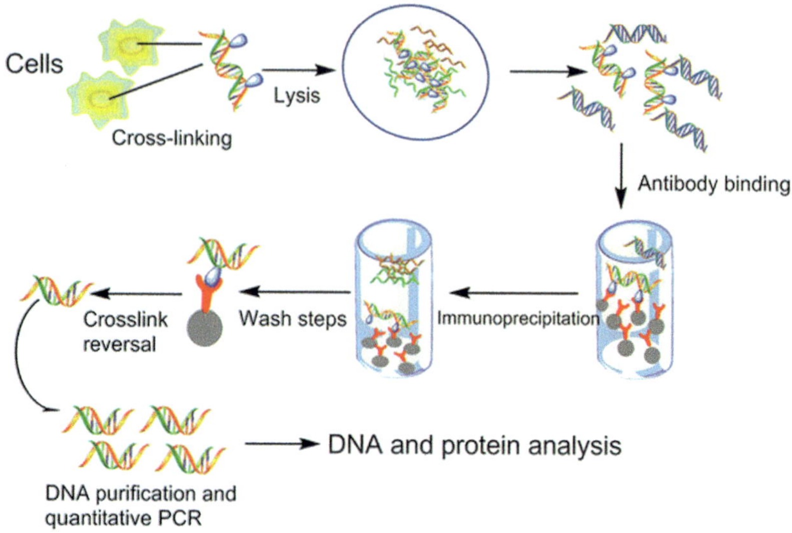

FIGURE 26.2 Chromatin immunoprecipitation. Protein and the chromatin to which it is bound in living cells are crosslinked with formaldehyde or UV light. The DNA–protein complexes are then sheared into approximately 500 bp DNA fragments by sonication or by enzymatic digestion. The DNA–protein complexes are immunoprecipitated with an appropriate protein–specific antibody. After separating the precipitated complexes from cellular debris, the sequestered DNA is released by reversal of the crosslinks, purified, and then sequenced or subjected to other analyses to determine its identity (Song et al., 2015). *Image courtesy Song, C., S. Zhang, and H. Huang. (2015). Choosing a suitable method for the identification of replication origins in microbial genomes. Front. Microbiol. 6, 1–19. https://doi.org/10.3389/fmicb.2015.01049 and reproduced according to the terms of Creative Commons (CC by 4.0; https://creativecommons.org/licenses/by/4.0/).*

fact that it is a fluorescence-based platform. Microarray applications include SNP profiling, detection of alternative RNA splicing patterns, direct comparison of genomes from different organisms, changes in gene expression patterns as a result of some type of experimental manipulation, and measurement of gene copy number variation (e.g., Johansson et al., 2015). Microarrays, by comparison, are prepared by spotting cDNA onto a nylon filter and represent a less sophisticated and less expensive method for performing gene expression studies. Though infrequently used, macroarrays are useful when the investigator wishes to assay only a particular subset of genes and the arrays are usually themed (e.g., cancer, cell signaling, apoptosis, growth and development). Macroarrays are typically limited to no more than a few hundred sequences due in part to the limiting, porous nature of the membrane. As the detection methodology is chemiluminescence or autoradiography, the macroarray method is not nearly as sensitive as its microarray counterpart.

6. *Aptamers* are short RNA or DNA molecules which have ligand binding capabilities. Depending on their exact nucleotide sequence, aptamers are able to fold into various three-dimensional shapes (Fig. 26.3) enabling them to bind to specific membrane-localized or subcellular target ligands for both diagnostic and therapeutic purposes—a type of smart bullet. Many investigators consider aptamers to be a really low-cost alternative or complementary platform to antibodies. There is significant growing interest in modifying aptamers by the addition of various functional groups (e.g., Aptabodies; http://www.aptagen.com) to confer drug-like properties. Aptamers that are synthesized from locked nucleic acids (LNAs; Chapter 10: RT-PCR: A Science and an Art Form) exhibit enhanced specificity due to greater thermodynamic stability, meaning that they are well-suited for a variety of molecular biology applications. For example, LNAs possess highly discriminatory diagnostic capabilities allowing them to distinguish among various bacterial strains in a clinical specimen. The RNA-ligand interaction database (http://RNAligands.ccbr.utoronto.ca) is an excellent resource for mapping RNA secondary and tertiary structures to ligands for which they have known binding affinity. When designed properly, folded RNA molecules and their cognate ligands have the potential to intentionally impact normal and dysregulated biological functions, and to interfere with microbial replication.
7. *Nanobiotechnology* is an area in which remarkable advances are being made for single molecule analysis and improved bioavailability of

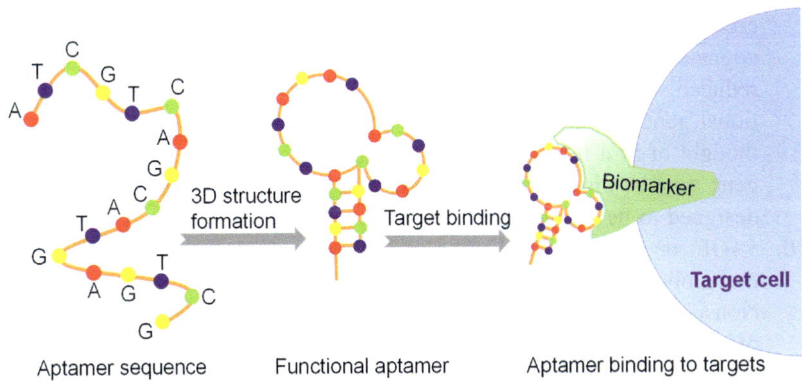

FIGURE 26.3 Aptamers comprising judiciously chosen oligonucleotide sequences form functional three-dimensional structures, and they bind to their targets with high specificity and affinity (Yang et al., 2018). *Image courtesy Yang, S., H. Li, L. Xu, Z. Deng, W.Han, Y. Liu, et al. (2018). Oligonucleotide aptamer-mediated precision therapy of hematological malignancies. Mol. Ther. Nucleic Acids 13, 164–175. https://doi.org/10.1016/j.omtn.2018.08.023 and reproduced according to the terms of Creative Commons (CC by 4.0; https://creativecommons.org/licenses/by/4.0/).*

medicine. The technology represents a unique blend of chemistry, biology, and physics. Nanofabrication is the science of making useful nanometer-size objects that literally involves manufacturing at the atomic level. In conjunction with lab-on-a-chip microfluidic devices, nanosensors are well-suited for the detection of extremely small quantities of DNA and RNA for disease and pathogen detection. Thus, large numbers of genes can be assayed from incredibly minute, highly differentiated cell populations or tissue samples in order to characterize genome functionality.

8. *Gene transfer* is a process by which gene sequences are transferred into cells for over-expression or ectopic expression. Alternatively, a gene construction may be transferred into cells for the purpose of silencing (see RNAi, below). Then, any of a variety of functional assays may be performed to determine the consequences for the cell. Gene transfer may be transient or stable, and the most useful constructions feature genes that are under the control of inducible promoters, meaning that they can be up- or downregulated as required by the experimental protocol. Examples of inducible promoters include the well-known tetracycline (tet) regulatory (tet-on and tet-off) systems (for reviews, see Bertram et al., 2021; Navabpour et al., 2020) for controlling gene expression in mammalian cells.
9. *RNAi*, or RNA interference, is a molecular biology tool for targeted gene silencing. As such, it is an important interrogation method for understanding gene function. As discussed in Chapter 13, RNA Interference and Gene Editing, this naturally occurring phenomenon is now sufficiently well-understood such that it can be used to target specific mRNAs for immediate destruction. In so doing, synthesis of the cognate protein is downregulated because the template for translation is reduced, if not altogether eliminated, making RNAi an important functional genomics tool. The science behind the technology might best be thought of as a form of reverse genetics, meaning that the function of a gene may be discovered by silencing it. The silencing of a gene can be designed to be transient or permanent.
10. *SAGE*, or serial analysis of gene expression, is a nonarray method for the global analysis of gene expression (Chapter 22: Subtractive and Nonsubtractive Methods for the Analysis of Gene Expression) and it offers greater sensitivity than microarray platforms. Briefly, cellular RNA is converted into cDNA which is then digested into small fragments (<20 nts) known as tags. The tags are randomly ligated together to form larger concatemers, and the concatemers are then sequenced. The frequency with which each tag is detected by sequencing is directly related to the abundance of the corresponding mRNA in the sample. Bioinformatics methodologies are used to identify the tags and associate them with the biochemistry of the original biological source. Inasmuch

as this is a sequencing methodology, it is capable of detecting novel transcripts. SAGE is a good way to get a lot of sequencing data quickly and affordably.

11. *Real-time PCR* is the premier method for quantitative measurement of DNA and RNA (RT-PCR). This method, described in detail in Chapter 10, RT-PCR: A Science and an Art Form and Chapter 11, Quantitative PCR Techniques, revolutionized research involving all facets of the life sciences. RT-PCR is a versatile mainstream functional genomics tool that is widely regarded as the gold standard for the assay of transcript abundance. Real-time PCR should be viewed as a complementary partner technique to RNA-seq, rather than a competitor methodology. Digital PCR has emerged as a sensitive alternative platform for quantifying gene expression, though this technology is not (yet) in widespread use.
12. *Bioinformatics* represents a marriage between biotechnology and information technology. As discussed in Chapter 23, Transcriptomes and Bioinformatics, bioinformatics is a result of the "information revolution" which enables the practitioner to efficiently organize, manage, and mine the immense amounts of nucleic acid and protein sequencing data that are already available in various online repositories. Well-written data mining software, which this Author considers to be an electronic art form, is needed to recover specific types of data and to identify patterns that are biologically meaningful with respect to gene functionality. There is an exigent and growing need for bioinformatics specialists because of the immense amount of new sequencing information that is added to these databases every day. Good places to begin bioinformatics-type queries are the National Center for Biotechnology Information (NCBI; http://www.ncbi.nlm.nih.gov), the DNA Data Bank of Japan (http://www.ddbj.nig.ac.jp), and the European Bioinformatics Institute (http://www.ebi.ac.uk).
13. *Epigenetics* refers to changes in gene expression that are *not* the result of a change in the base sequence of a gene (i.e., changes in phenotype that are not the result of changes in genotype). Examples of epigenetic changes (Fig. 26.4) include reversible chemical modifications such as DNA methylation, which correlates with the suppression of gene expression, and histone acetylation, which correlates with the induction of gene expression. Epigenetic changes may or may not be passed to descendant daughter cells. The areas of stem cell formation, cancer, and aging are among the research foci with the greatest interest in how epigenetics guides genes' functionality.
14. *Two-dimensional gel electrophoresis* is the classical tool for examining the proteome. Proteins are first separated by isoelectric point (pI), which is the pH at which a protein has no net charge. Then, proteins are separated based on mass by performing electrophoresis in a direction

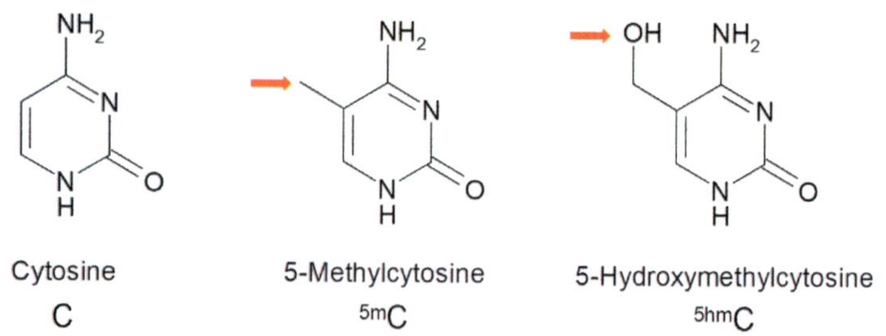

FIGURE 26.4 One of many examples of an epigenetic modification. The methylation of cytosine involves the addition of chemical groups without changing the identity of the base. The red arrows indicate the changes compared to natural cytosine. Epigenetic modifications occur in DNA and in RNA.

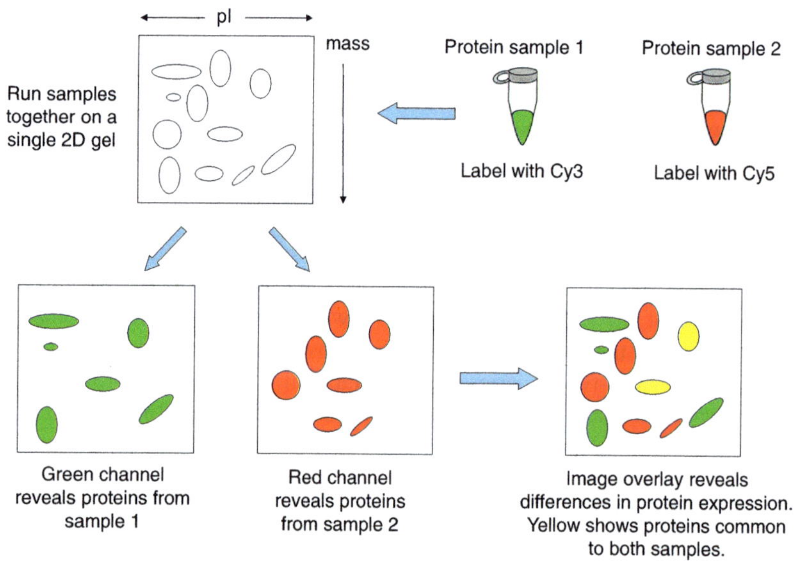

FIGURE 26.5 Fluorescent two-dimensional gel electrophoresis. Each sample is labeled with a different fluorescent dye fold by protein chromatography, first by isoelectric point (pI) and then, in a perpendicular direction by mass. *Green*, *red*, and *yellow* fluorescence signals identify the source of the protein, that is, sample 1, or sample 2, or both samples, respectively.

perpendicular to the direction of the first-dimension separation (Fig. 26.5). This method permits the simultaneous separation of thousands of proteins, each of which appears as a unique spot on a polyacrylamide gel. The size of each protein spot is proportional to the abundance of the protein. The protein localized in each spot can be recovered from the gel and sequenced to determine its identity.

15. *Protein arrays* are analogous to DNA microarrays. Colloquially known as "protein chips," protein arrays consist of various proteins that have been immobilized on a glass slide or other solid support. The purpose of the array is to determine which ligands in a biological sample bind to the proteins on the slide (refer back to Fig. 21.7). The process is rapid and generates a lot of data from a single experiment. There are four variant protein array formats which differ in terms of what is actually immobilized on the slide, namely (1) proteins, for binding other macromolecules in a sample, including other proteins for the study of protein–protein interactions, (2) antibodies (known as antibody arrays), for the detection of specific antigens in a sample, (3) whole cell lysates representing various experimental conditions, for binding specific antibodies that recognize their cognate antigen in the lysate, and (4) affinity arrays, for the capture of any molecules to which a ligand has some type of natural biological affinity. Affinity arrays may consist of a variety of immobilized targets including, specific nucleic acid sequences, antibodies, and macromolecule complexes such as nucleic acid + protein, protein + protein, and so forth.
16. *Mass spectrometry* or simply "mass spec" is an analytical tool which, among other things, can be used to elucidate the chemical structure of various compounds, including proteins. There are several variants of this method, though the basic principle involves ionizing the sample and measuring its migration through a magnetic field, which will be a function of its charge-to-mass ratio and which is related to its relative abundance. For example, proteins can be partially digested with trypsin, and the resulting peptide fragments analyzed by mass spec in order to discern the amino acid sequence of the original protein.
17. *Metabolomics* refers to the measurement of large sets of metabolites in cells. This is a tool of increasing importance in the world of functional genomics because the induction or repression of even a single gene has the potential to result in one or more metabolic disturbances. Metabolomics queries the overall metabolite composition of biological fluids; blood, saliva, and urine are ideal matrices because these biofluids are easy to collect. Metabolomics has also become quite relevant in biomarker identification because any observed changes may reflect some type of change or abnormality in cellular function. Common metabolomics measurement methods include nuclear magnetic resonance, high-performance liquid chromatography, gas–liquid chromatography, and mass spectrometry. The human metabolome database is accessible at https://hmdb.ca.

Relationship of functional genomics approaches to classical molecular biology

Is there really a relationship between functional genomics and traditional molecular biology? The answer is a resounding "yes!" Although there is an implied association of functional genomics with all things "high throughput" and "high tech," even a modest molecular biology laboratory is able to perform meaningful research in this sphere with little more than standard lab equipment (thermal cycler, electrophoresis equipment, dedicated micropipettors) and access to the internet. As needed, outsourcing DNA sequencing projects is commonplace and much more economical than only infrequently performing sequencing in-house with labile reagents.

Keep in mind that there are methods for the global analysis of gene expression that do not involve RNA-seq or microarrays. Suppression subtractive hybridization (SSH; Chapter 22: Subtractive and Nonsubtractive Methods for the Analysis of Gene Expression) is a tremendously useful tool for subtracting sequences that are common to two samples, for example, test versus control. The remaining sequences represent an enriched population of uniquely expressed RNAs which are then cloned and sequenced to determine the identity of each modulated gene. While SSH is not a high-throughput method, it is certainly a powerful method. One major advantage of SSH over microarray analysis is that SSH is truly a tool for new gene discovery in that *any* sequence that is up- or downregulated can be identified by this method. In the case of microarray analysis, one is limited to the assay of genes that have already been discovered *and* which have been printed on the array.

Similarly, mRNA differential display (Chapter 22: Subtractive and Nonsubtractive Methods for the Analysis of Gene Expression) has an aim much like that of SSH, though the mechanics of these processes are quite different. Succinctly, cDNAs representing two or more biological sources are PCR-amplified in parallel using specific primer pairs. The judicious selection of primers ensures that only a small percentage of all cDNAs are amplified in a particular reaction tube which, upon electrophoresis, produces well-resolved sharp bands in each lane that profile active genes in each sample.

In addition to real-time PCR-based techniques, RNA-seq, and microarray analysis, there is additional information that can be garnered through some of the time-honored molecular biology methods presented in earlier chapters in this book. For example, the size of the native (full length) transcript can be determined by Northern analysis, assuming that the RNA species is sufficiently abundant. Nuclease protection will provide an added level of assay sensitivity. If nothing else, these non-PCR methods are very convincing means for confirming results that were generated by PCR.

Classical genetics can also play a role in the realm of functional genomics. One example of such an approach involves selective breeding of organisms that exhibit a phenotype of interest. For example, mutagenesis screenings have played an important role in discerning the role of specific genes in organism-level growth and development. Briefly, random germline mutations can be fairly easily induced in a population of animals (or plants) by exposure of some of the organisms to a compound with nucleic acid mutagenizing properties. For example, zebrafish exposed to a xenobiotic would then be mated with wild-type zebrafish. Upon examination of the F_1, one might infer that any new phenotypes that emerge are due to xenobiotic-induced dominant mutations and which are also heterozygous at the particular genetic locus. Further breeding, including a backcross with the "exposed" parent, is necessary to identify any xenobiotic-induced recessive mutations, which are expected to appear in a Mendelian manner. Candidate genes responsible for phenotype variations or a disease pathology can be identified by positional cloning (Fig. 26.6), which involves classical linkage and mapping analysis as well as DNA sequencing, the ultimate identification tool, in

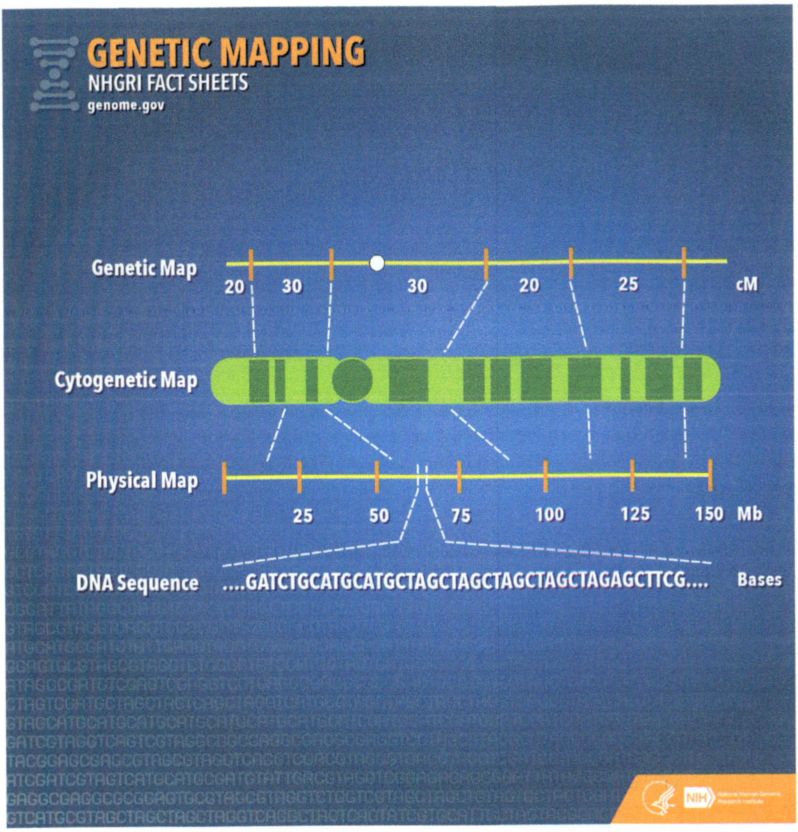

FIGURE 26.6 Genetic mapping. Shown are relationships between gene locus positioning via genetic, cytogenetic, and physical mapping, compared with gene identity determination by direct DNA sequencing. *Image courtesy National Human Genome Research Institute (NHGRI; http://www.genome.gov).*

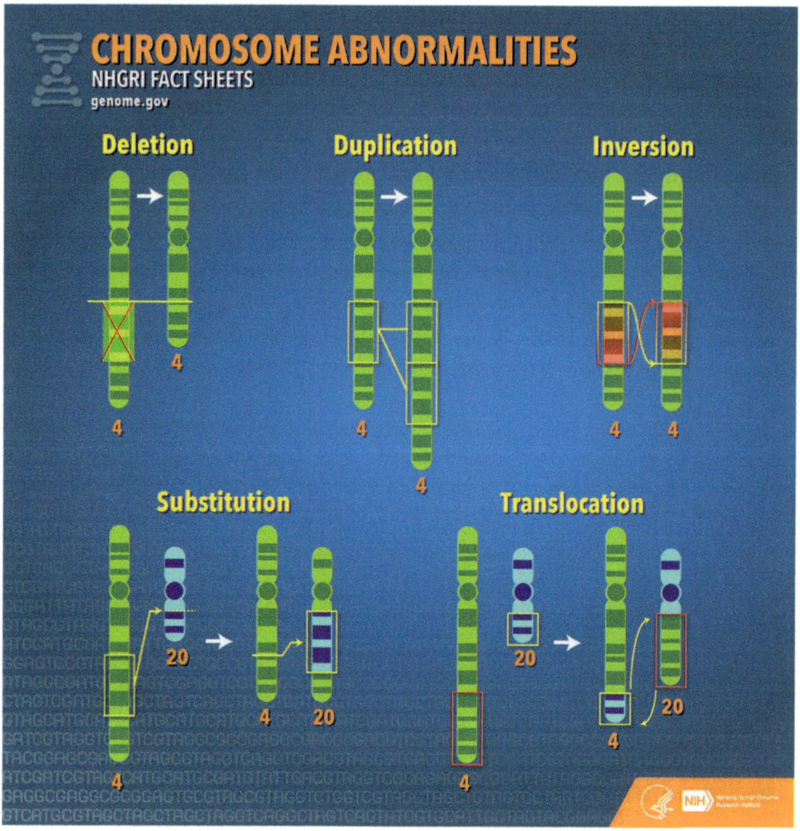

FIGURE 26.7 Examples of gross changes in chromosome structure which may impact genome function. The abnormality may be inherited or may be acquired at some point during the life of the organism. *Image courtesy National Human Genome Research Institute (NHGRI; http://www.genome.gov).*

order to discern the precise nature of the underlying mutation. Finally, examples of gross morphological changes to chromosome structure (Fig. 26.7) are often observed upon karyotype analysis.

References

Bertram, R., Neuman, B., Schuster, C.F., 2021. Status quo of tet regulation in bacteria. Microb. Biotechnol. 15, 1101–1119. Available from: https://doi.org/10.1111/1751-7915.13926.

Johansson, M.M., Van Geystelen, A., Larmuseau, M.H.D., Djurovic, S., Andreassen, O.A., Agartz, I., et al., 2015. Microarray analysis of copy number variants on the human Y chromosome reveals novel and frequent duplications overrepresented in specific haplogroups. PLOS ONE 10 (8), e0137223. Available from: https://doi.org/10.1371/journal.pone.0137223.

Navabpour, S., Kwapis, J.L., Jarome, T.J., 2020. A neuroscientist's guide to transgenic mice and other genetic tools. Neurosci. Biobehav. Rev. 108, 732–748.

Przybyla, L., Gilbert, L.A., 2022. A new era in functional genomics screens. Nat. Rev. Genet. 23, 89–103. Available from: https://doi.org/10.1038/s41576-021-00409-w.

Song, C., Zhang, S., Huang, H., 2015. Choosing a suitable method for the identification of replication origins in microbial genomes. Front. Microbiol. 6, 1–19. Available from: https://doi.org/10.3389/fmicb.2015.01049.

Uffelmann, E., Huang, Q.Q., Munung, N.S., de Vries, J., Okada, Y., Martin, A.R., et al., 2021. Genome-wide association studies. Nat. Rev. Meth. Prim. 1, 59. Available from: https://doi.org/10.1038/s43586-021-00056-9.

Yang, S., Li, H., Xu, L., Deng, Z., Han, W., Liu, Y., et al., 2018. Oligonucleotide aptamer-mediated precision therapy of hematological malignancies. Mol. Ther. Nucleic Acids 13, 164–175. Available from: https://doi.org/10.1016/j.omtn.2018.08.023.

Further reading

Baldo, A.M., Norelli, J.L., Farrell Jr., R.E., Bassett, C.L., Aldwinckle, H.S., Malnoy, M., 2010. Identification of genes *differentially expressed during interaction of resistant and susceptible apple cultivars (*Malus x *domestica*) with *Erwinia amylovora*. BMC Plant Biol. 10, 1. Available from: https://doi.org/10.1186/1471-2229-10-1.

Bocsanczy, A.M., Phillips, J.G., Korban, S.S., Dardick, C.D., Bassett, C.L., Wisniewski, M.E., et al., 2011. Apple (*Malus* x *domestica*) transcriptome in response to the compatible pathogen *Erwinia amylovora* and the incompatible pathogen *Pseudomonas syringae*. Acta Horticult. 896, 237–241.

Gardiner, S.E., Norelli, J.L., de Silva, N., Fazio, G., Peil, A., Malnoy, M., et al., 2012. Putative resistance gene markers associated with quantitative trait loci for fire blight resistance in *Malus robusta* 5 accessions. BMC Genet. 13, 25. Available from: https://doi.org/10.1186/1471-2156-13-25.

Malnoy, M., Borejsza-Wysocka, E.E., Baldo, A.M., Aldwinckle, H.S., Carlisle, C.M., Bowatte, D., et al., 2008. Identification and mapping of genes differentially expressed during interaction of resistant and susceptible apple cultivars (*Malus* x *domestica*) with *Erwinia amylovora*. Acta Horticult. 793, 195–199.

Norelli, J.L., Bassett, C.L., Lalli, D.A., Wisniewski, M.E., Farrell Jr., R.E., Baldo, A.M., et al., 2008. Transcriptional response of *Malus* to *Erwinia amylovora* infection. Acta Horticult. 793, 189–194.

Norelli, J.L., Lalli, D.A., Bassett, C.L., Wisniewski, M.E., Gardiner, S.E., Celton, J.M., et al., 2009. Using functional genomics to identify molecular markers for fire blight resistance (*Erwinia amylovora*) in apple (*Malus*). Acta Horticult. 839, 415–420.

Norelli, J.L., Farrell, R.E., Leder, E.H., Bassett, C.L., Baldo, A.M., Malnoy, M., et al., 2006. Using functional and applied genomics to identify genes that confer either resistance or susceptibility to fire blight. Phytopathology 96, S86 (Abstract).

Norelli, J.L., Farrell Jr., R.E., Bassett, C.L., Baldo, A.M., Lalli, D.A., Aldwinckle, H.S., et al., 2009. Rapid transcriptional response of apple to fire blight disease revealed by cDNA suppression subtractive hybridization analysis. Tree Genet. Genomes 5, 27–40.

Palazzo, A.F., Lee, E.S., 2015. Non-coding RNA: what is functional and what is junk? Front. Genet. 6, 2. Available from: https://doi.org/10.3389/fgene.2015.00002.

Chapter 27

A few RNA success stories

The protocols and accompanying discussion in the preceding chapters of this text illustrate various methods for the expedient isolation of RNA from a variety of biological sources and its subsequent analysis. The judicious application of various combinations of these techniques constitutes a systematic method for understanding the role of the products of transcription in the regulation of gene expression. A typical RNA-based experimental design includes assessing steady-state levels of RNA, assessing the levels of related regulatory molecules such as miRNAs, and assessing the modulation of these elements in response to naturally occurring biological changes of experimental challenge. Because experiments involving RNA are almost certainly asking gene expression questions, these techniques are commonly performed to assign observed changes to the transcriptional level or due to some posttranscriptional event(s). Therefore, assays providing the greatest sensitivity and resolution are always desirable. Real-time PCR, miRNA profiling, and the emergence of next-generation sequencing, particular RNA-seq, have facilitated transcriptome profiling the cellular biochemistry at levels inconceivable even 10 years ago and these methods continue to offer great potential.

The era of RNA therapeutics has arrived. Many years of investigating the multifarious facets of RNA by the scientific community have brought to fruition a number of very important products and processes. Unlike traditional drugs which are directed toward the active sites of proteins, including many membrane-embedded receptors, RNA "medicines" directly impact the flow of genetic information in the cell in order to decrease or increase, the production of a particular protein. Depending on the nature of the molecular intrusion, an impact is usually a posttranscriptional event. A revolution involving the refinement of RNA technologies for the treatment of illness and disease is in full view and well underway.

Nucleic acids as pharmaceuticals

DNA therapeutics have enjoyed some notable successes. In order to function, these double-stranded molecules must either make their way into the nucleus of a cell or be placed there directly. One of the most common means of

RNA Methodologies. DOI: https://doi.org/10.1016/B978-0-323-90221-2.00041-2

doing so is through the use of reprogrammed viral vectors (usually adenovirus) in which the genome of the virus has been genetically engineered to carry an exogenous gene or gene segment of some pharmaceutical value. Alternatively, the introduction of a foreign gene sequence may be introduced in the form of plasmids, which are covalently closed circular molecules. In either case, the encoded DNA must first be transcribed by the nuclear machinery of the host cell to produce the encoded RNA, followed by the production of the RNA-encoded peptide(s), this time using the host cell's translation apparatus. The general idea is to provide a cellular platform for the production of the normal counterpart of abnormal or missing proteins. These approaches are not without shortcomings, however. Plasmids, for example, may be expelled from the nucleoplasm, resulting in a concomitant loss in its therapeutic value. While the adenovirus sequences currently use favor stable integration into the chromosome of the host, there is also the chance, however small, that integration into a critical off-target region of the chromosome may compromise or otherwise interfere with another normal cellular function. In a related application, viral vectors can be used to introduce sequences that intentionally alter the phenotype and physiology of the cell. Finally, the CRISPR-Cas9 system is at the heart of contemporary gene editing by directly changing a target DNA sequence, though it may also be considered as a form of RNA therapeutics owing to the central role played by the two RNA molecules (tracrRNA and crRNA) which guide this high-precision process, as described in Chapter 13, RNA Interference and Gene Editing.

RNA vaccines and therapeutics

The notion of RNA as a therapeutic is not a novel idea. It makes perfect sense that using RNA which encodes a viral (or bacterial) antigenic structure would be an efficient way to provoke an immune response. Of necessity, the entire world was recently treated to an amazing display of the utility of RNA vaccines to address and hopefully suppress the devasting health consequences for people who are exposed to the severe acute respiratory syndrome coronavirus 2 (SARS-CoV-2), thereby contracting COVID-19 disease (Table 27.1). A coronavirus is so named because the term "corona" means "crown," which the spike proteins on the surface of coronavirus particles are thought to resemble. It is against these antigenic spike proteins that COVID-19 vaccines are designed to induce a robust immune response.

Several pharmaceutical companies independently produced vaccines to remediate the current public health crisis with entrepreneurial, record-breaking speed (Table 27.2), three of which are approved for use in the United States at the time of this writing. The mRNA strategy used in the Pfizer-BioNtech and Moderna vaccines is similar: vaccination leads to

TABLE 27.1 COVID-19 essential vocabulary.

Term	What it stands for	What it is
COVID-19	Coronavirus disease 2019	The proper name of a disease
SARS-CoV-2	Severe acute respiratory syndrome coronavirus 2	A virus that first emerged late in the year 2019 that causes COVID-19
Coronavirus	The name of a family of viruses, more properly known as Coronaviridae	Any one of several types of single-strand, (+) sense RNA viruses
SARS-CoV	Severe acute respiratory syndrome coronavirus	The virus that causes SARS
SARS	Severe acute respiratory syndrome	The proper name of a disease
MERS-CoV	Middle East respiratory syndrome coronavirus	The virus that causes MERS
MERS	Middle East respiratory syndrome	The proper name of a disease

the synthesis of the virus spike protein via the natural translation machinery in the cell when the cognate mRNA is injected intramuscularly. In the case of the Johnson & Johnson vaccine, however, a different strategy is used to build immunity to SARS-CoV-2. Rather than being an RNA vaccine, the Johnson & Johnson vaccine is a modified adenovirus (common cold virus). This virus acts as a vector that carries the genetic information for the spike protein and, as a DNA vaccine, requires that chromosomal integration, transcription, and translation all occur. In any case, the immune system then recognizes the newly synthesized spike protein as a nonself antigen and immediately mounts an immune response to protect the vaccinated individual against the advent of exposure to SARS-CoV-2.

mRNA in a vaccine need only cross the plasma membrane and enter the cytoplasm to be effective; inasmuch as they are excluded from the nucleus, there is no interaction with chromosomal DNA (Fig. 27.1). The half-life of mRNA is short in animal cells and the natural mRNA turnover and recycling processes that affect levels of endogenous mRNA in the cytoplasm are likewise influencing vaccine mRNA stability. Although the Pfizer-BioNTech mRNA (4284 bases) differs from the Moderna mRNA (4004 bases), they both share the requisite landmarks (e.g., 5′ cap for ribosome recognition; chemically modified nucleotides for stability and reduced immunogenicity; poly$(A)^+$ tail for stability) to ensure their efficient translation (Fig. 27.2). Inasmuch as every person is a biochemically unique

TABLE 27.2 Examples of vaccines against COVID-19.

Manufacturer	Alternative or Brand Name	Classification	How it Works
Pfizer, Inc. and BioNTech [BNT162b2]	COMIRNATY	mRNA vaccine	Lipid nanoparticle-encapsulated mRNA encoding the SARS-CoV-2 spike protein which, when made in the host cell, triggers an immune response
ModernaTX, Inc.[mRNA-1273]	Spikevax	mRNA vaccine	Lipid nanoparticle-encapsulated mRNA encoding the SARS-CoV-2 spike protein which, when made in the host cell, triggers an immune response
Johnson & Johnson [JNJ-78436735]	Janssen	DNA vaccine	Modified type 26 adenovirus carrying genetic instructions (DNA) for making the SARS-CoV-2 spike protein
Novax [NVX-COV2373]	Nuvaxovid and Covovax; not currently authorized for use in the United States	Antigenic protein vaccine	Vaccine contains a nanoparticle-bound SARS-CoV-2 spike protein and an adjuvant to enhance the immune response
Oxford-AstraZenica [AZD1222]	Vaxzevria; not currently authorized for use in the United States	DNA vaccine	Modified adenovirus carrying the SARS-CoV-2 spike protein code

organism, the magnitude of any short-term, vaccine-induced side-effects is highly variable from one person to the next. RNA vaccines in general are able to induce cell-mediated immunological pathways as well as humoral immunity which, of course, includes the production of antibodies (Martinon et al., 1993; Petsch et al., 2012).

With new advances in the medical field, new worries are sure to follow. Regarding the safety of RNA vaccines, it is important to keep three things in mind. First and as noted above, the mRNA component of the vaccine is quickly broken down by the cell by normal mRNA surveillance processes.

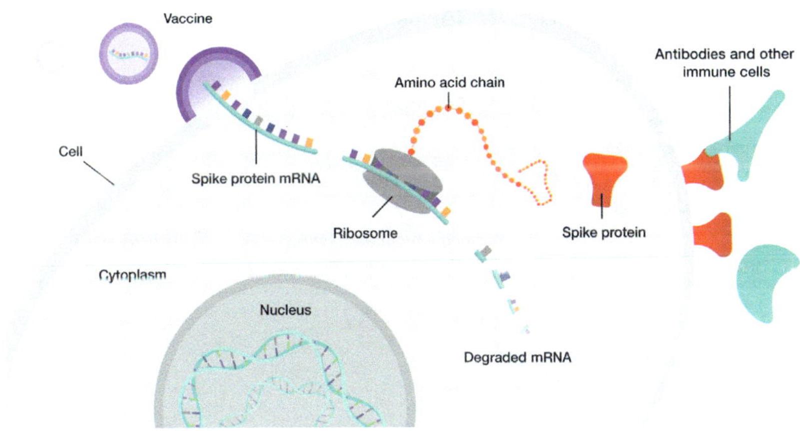

FIGURE 27.1 RNA vaccines deliver functional mRNA molecules encapsulated in specialized lipid nanoparticles (purple). This mRNA is translated in the cytoplasm to produce, as shown, the SARS-CoV-2 spike protein. The RNA is then quickly degraded by the natural turnover machinery that resides in all cells. The mRNA never enters the nucleus and has no interaction with the chromosomes of the vaccine recipient. The synthesized spike proteins act as powerful antigens which provoke an immune response. *Image courtesy National Human Genome Research Institute (http://www.genome.gov).*

Second, the mRNA contains only a small segment of sequences corresponding to the SARS-CoV-2 genetic material, so there is no way that new viruses can be made in the cells of a vaccine recipient. Nor do RNA vaccines contain any virions (live or dead). Third, the mRNA present in the vaccine is incapable of entering the cell nucleus, thereby preventing direct or indirect interaction with the DNA of the vaccine recipient. Concerns about this type of cellular change have reached a new level of fervor with the rollout of the COVID-19 mRNA vaccines because many people outside the medical community did not/do realize that the mRNA in vaccines does not interact with the chromosomes of the vaccine recipient.

Perhaps the most important consideration when RNA is to be used as a therapeutic is the method by which the transcripts will be stabilized and delivered to the cells of the vaccinated individual. The current state of the art is the use of modified lipid nanoparticles (Fig. 27.3); the mostly cationic (and other) lipids are commonly arranged as a lipid monolayer (and sometimes as a bilayer) encapsulating its cargo. This is in contrast to liposomes which are always characterized by a lipid bilayer, tend to be greater in size than nanoparticles, and largely mimic the architecture of the plasma membrane of the cell. In many cases, the lipid nanoparticles carry polyethylene glycol on the surfacc which functions as a transient barrier to highly prevalent blood proteins, including nucleases from absorbing across the surface. This strategy increases the lifetime of the lipid nanoparticles as they move

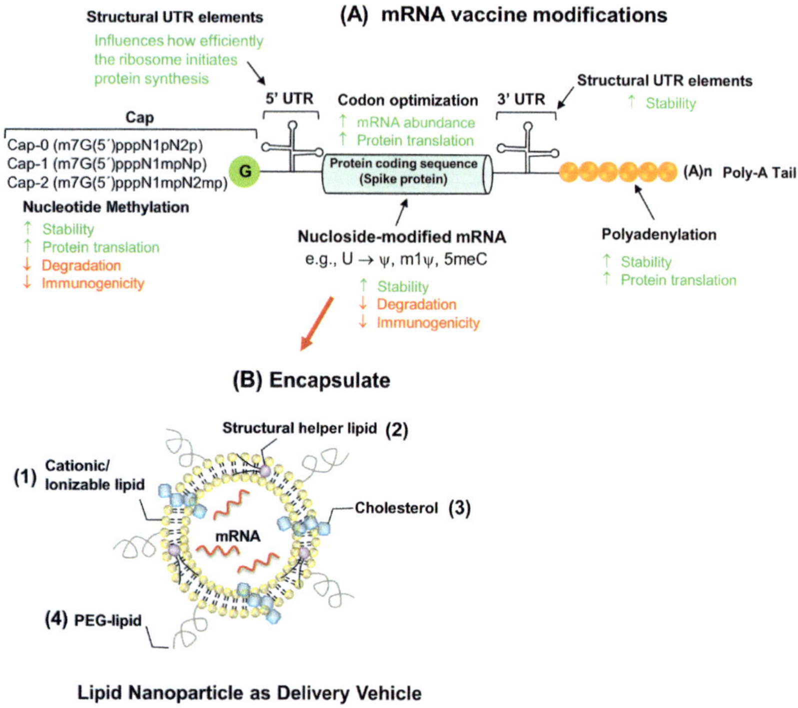

FIGURE 27.2 Structural motif of vaccine mRNA. (A) Vaccine mRNA modifications were designed to enhance stability, lower immunogenicity, and maximize translation efficiency. This includes the use of pseudouridine (ψ), which is a uridine analog. (B) mRNA is encapsulated in customized lipid nanoparticles for efficient entry into the cells of the vaccinated individual and for protection of the mRNA against nuclease and other inhibitory proteins (Granados-Riveron and Aquino-Jarquin, 2001). *Image courtesy Granados-Riveron, J.T. and G. Aquino-Jarquin. (2001). Engineering of the current nucleoside-modified mRNA-LNP vaccines against SARS-CoV-2. Biomed. Pharmacother. 142, 111953. https://doi.org/10.1016/j.biopha.2021.111953 and reproduced with permission from Elsevier.*

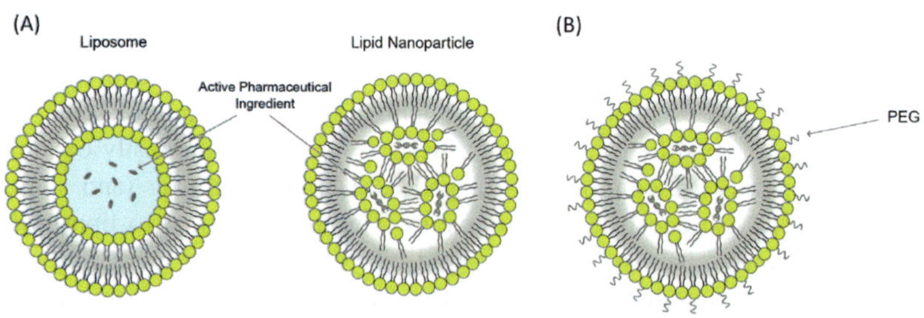

FIGURE 27.3 RNA delivery vehicles. (A) Comparative structure of liposomes and lipid nanoparticles (LNPs). (B) PEGylated LNP structure stabilizes nucleic acid cargo, extends its life in the bloodstream, and facilitates release from the endosomes, i.e., endosome escape, following endocytic entry into the cell. *Images courtesy of Exelead (www.exeleadbiopharma.com) and reproduced with permission from Merck KGaA, Darmstadt, Germany and its affiliates.*

through the circulatory system. In addition, the formulation of lipid nanoparticles with polylactic acid or poly(lactic-*co*-glycolic acid) polymers has been shown to facilitate the release of the contents of lipid nanoparticles in a controlled manner (Gao et al., 2013).

Media coverage shortly after Food and Drug Administration (FDA) approval of the vaccines frequently expounded on the need to store and transport mRNA vaccines at ultracold temperatures owing to the natural labile nature of the RNA molecules; subsequent data presented to the FDA indicated that less extreme storage conditions may be sufficient to maintain the stability of the vaccine, which would simplify logistics and lower the cost of storage and transport. Given the demonstrated efficacy of mRNA vaccines for combatting COVID-19 and the solving of many logistical transportation and distribution issues, it is likely that the scientific community will soon witness explosive R&D investment and productivity in the area of RNA therapeutics, including the potential to use mRNA vaccines for the conferral of passive immunity in a situation where the mRNA directly encodes a well-characterized antibody.

Beyond mRNA vaccines, there are two effective strategies involving the use of RNA in general as a therapeutic, either (1) introducing into a host cell either a full-length or shortened mRNA which encodes a needed protein or (2) preventing an mRNA that is produced naturally within the cell from becoming translated either by blocking access to it or by destroying it directly via an RNAi-type pathway (Table 27.3). Sometimes referred to as mRNA replacement therapy, providing a patient with the normal counterpart of a missing or defective mRNA is a logical approach to the treatment of any of a number of metabolic disorders and there are several mRNA candidates currently in clinical trials. One of the more innovative approaches to mRNA replacement therapy involves the use of self-replicating mRNA, which is a genetically engineered variant of single-stranded RNA alphavirus. Known for its naturally elevated capacity for replication, the (+) sense genome is modified such that it contains the encoded antigen sequences downstream from other regulatory elements necessary for replication, including viral replicase genes (reviewed by Lundstrom, 2018). The net result is a situation in which more of the encoded antigenic protein is made from a much lower mRNA inoculum. As in all cases involving mRNA in the cell, these transcripts will likewise be turned over in a relatively short period of time owing to their natural labile nature. Alternatively, one could envision using antisense RNA to form a dsRNA region with the retroviral genome, thereby targeting it for immediate destruction. This is the same principle miRNA- and siRNA-mediated control of mRNA stability and translational repression; dsRNA in mammalian cells is quickly targeted for cleavage.

A longstanding challenge has been conquering the sophisticated strategies enabling viruses to evade the immune system. Fortunately, the development of rapid test kits and PCR-based assays facilitate the identification of persons

TABLE 27.3 RNA therapeutic strategies.

RNA molecule type	Action	Result	Domain
mRNA	Introduces new encoded protein template	Translated by ribosomes to produce the encoded protein which then has a specific structural, functional, or antigenic role	mRNA vaccines and other therapeutics
Aptamer	Folds into a characteristic 3D shape that is a result of its unique sequence	Mimics antibodies by binding to or blocking ligands such as cytoplasmic proteins or transmembrane receptors that exhibit a complementary shape	Ligand binding
miRNA	Base-pairs to complementary sequences on mRNA	Blocks translation (repression) or mediates destruction of the mRNA through the formation of a region of dsRNA	RNAi
siRNA	Base-pairs to complementary sequences on mRNA	Blocks translation (repression) or mediates destruction of the mRNA through the formation of a region of dsRNA	RNAi
Antisense oligonucleotides (ASO; DNA or LNA)	Base-pairs to complementary sequences on mRNA	Blocks translation (repression) or destabilizes mRNA most often through steric interference or RNase H-mediated cleavage	ASO technology

who have an infection, allowing them to isolate themselves as soon as possible to contain the spread of illness. As the molecular biology of pathogens and the genetic basis of disease are unraveled and become better understood, advances in RNA-based therapies will continue to move forward.

RNA biobanking

The Precision Medicine Initiative (PMI) was a federal initiative to develop exacting, personalized medicine by collecting, studying, banking, and occasionally distributing samples from vast numbers of tissue specimens. The concluding report produced by the Working Group Advisory Committee became the basis of the current iteration of this initiative which is now known as the "All of Us" research program (https://allofus.nih.gov).

The Mayo Clinic was designated in 2018 as the location of the biobank repository for all samples collected as part of the All of Us PMI. As the name suggests, a biobank is a well-maintained and properly annotated (bio)repository where various types of biological specimens (biospecimens) are maintained for long-term investigation. Participation in this program is purely voluntary and anyone who wishes to participate can register online at the website noted above. Anyone who chooses to participate will have full access to their own data as well as summary information pertaining to all participants in aggregate, and likewise, have access to related findings that may result from the analysis of all of the contributed specimens. As a federally funded program, all data is readily accessible to the scientific research community. The recruitment of participants began in 2017. The cohort size is expected to be at least one million participants who donate biological material (e.g., blood, urine, saliva) and likewise share diet, lifestyle, and genetic information. Anyone in the United States may volunteer to participate and contribute a biospecimen; the goal is to have the population, diverse as it is, represented as completely as possible. Moreover, having large numbers of participants will greatly increase the likelihood that rare disorders will be included in the accessioned biocollection. The notion is that, through big data analysis, it should be possible to determine the proper medicine and dose to prescribe for any illness or disease for each biochemically unique individual. The development of personalized medicine in this manner is known as pharmacogenomics. Biobanks can also be virtual, often set up to share critical time-sensitive information in the face of a world health crisis. Visit the World Health Organization website (http://www.who.int) for more information and for some interesting statistics.

Rather than extracting RNA from its biological source, it is commonly stored within the tissue at a biobanking facility, though some biobanks offer cell separation and nucleic acid isolation as a service. Per the conventional wisdom, tissue samples are most often maintained at −80°C. Newer studies, however, suggest that cryopreservation below −130°C greatly extends the "shelf life" of the archived samples. Further, the concentration of previously isolated RNA appears to influence sample stability, even under ultracold storage conditions; more concentrated samples show far less degradation after more than four years (Olivieri et al., 2014) than samples maintained at lower concentrations. In addition to the many nationally and internationally acclaimed biobanks, there are also commercial enterprises that are driving the biobanking industry forward. One immediate need is for the development of standardized procedures for handling and storage of biospecimens in order to ensure that consistent and meaningful data is shared within the informatics infrastructure. Although biobanks have been around since the 1990s, the biobanking of RNA, without a doubt, has tremendous untapped potential.

RNA reprogramming

The technology that has come to be known as RNA reprogramming is used to induce the reversion of differentiated adult cells back to pluripotent stem cells. The process is best thought of as dedifferentiation, and the resulting cells are referred to as induced pluripotent stem cells (iPSCs; Fig. 27.4). Skin fibroblasts in particular quite readily lend themselves to reprogramming. All adults have stem cells in every tissue, the primary purpose of which is to maintain and, as needed, repair the tissues in which they reside. These cells remain quiescent until they are called into action.

This process and the potential benefits are intriguing for a number of reasons. First, it provides an opportunity to study the entire process of growth and development from the beginning. Second, depending on the exact culturing conditions, it is possible to guide the differentiation of iPSCs into any one of several types of tissue. Thus, the potential exists to generate healthy tissue and then return it to the donor in the event of illness or disease without the risk of rejection. Third, stem cells produced by RNA reprogramming are an excellent platform for discerning the effects of various drugs on the stem cell donor, an important step forward for personalized medicine. Fourth, and most importantly, generating stem cells through RNA reprogramming completely eliminates the ethical concerns associated with the use of embryonic stem cells.

The earliest attempts to induce dedifferentiation involved the viral delivery of iPSC-inducing genes and the direct introduction of DNA. A significant improvement in this technology accompanied the demonstration that artificially synthesized RNA can be used for reprogramming (Warren et al., 2010; reviewed by Bailly et al., 2022). In so doing, one precludes the possibility of inducing changes to the genome through the unintentional introduction of foreign DNA. A subsequent significant innovation was the development of polycistronic, self-replicating RNA molecules containing the key reprogramming genes (Yoshioka et al., 2013). The products of the reprogramming genes are all transcription factors. Individually, each transcription factor influences cellular physiology in many ways. When used

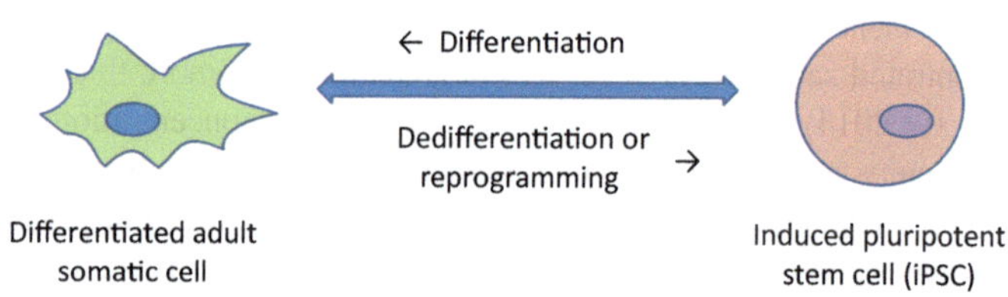

FIGURE 27.4 RNA reprogramming. Adult somatic cells, for example, skin fibroblasts, can be induced to revert back to a dedifferentiated state with the introduction of the appropriate subset of RNAs and miRNAs. In so doing, the cells re-acquire the attributes of pluripotent stem cells, which can subsequently be guided to differentiate into any of a number of tissue types.

collaboratively they are referred to as Yamanaka factors (Takahashi and Yamanaka, 2006) because they were shown to induce pluripotency from both mouse and human adult fibroblasts. Working together, these gene products are directly responsible for the reversion in phenotype and maintenance in a much less differentiated state.

1. OCT4, Octomer-binding transcription factor 4, involved in the self-renewal of stem cells;
2. KLF4, Kruppel-like factor 4, which regulates telomerase in stem cells;
3. SOX2, Sex determining region Y-Box 2, a transcription factor that maintains stem cell pluripotency;
4. c-MYC, a well-known, key transcription factor regulator of proliferation and metabolism.

The overall efficiency of the established RNA reprogramming process was low until it was demonstrated that primary human fibroblasts could be efficiently induced to re-acquire the dedifferentiated phenotype by supplementing the reprogramming mRNAs with the miRNA-302/367 cluster (Kogut et al., 2018). The cluster consists of five similar miRNAs that were earlier demonstrated to have the ability as a group to induce pluripotency in various mouse and human cell types (Anokye-Danso et al., 2011) and which are known to be highly expressed in stem cells.

miRNA	Sequence	Accession no.
miRNA-302a	5′ UAAGUGCUUCCAUGUUUUGGUGA	MIMAT0000684
miRNA-302b	5′ UAAGUGCUUCCAUGUUUUAGUAG	MIMAT0000715
miRNA-302c	5′ UAAGUGCUUCCAUGUUUCAGUGG	MIMAT0000717
miRNA-302d	5′ UAAGUGCUUCCAUGUUUGAGUGU	MIMAT0000718
miRNA-367	5′ AAUUGCACUUUAGCAAUGGUGA	MIMAT0000719

Interestingly, the miR302/367 cluster is a target of OCT4 and SOX2, which makes perfect sense in the capacity of these gene products as transcription factors. In addition, miRNAs are easily transfected into cells and demonstrate a much greater dedifferentiation efficiency, by as much as two orders of magnitude, compared to the classical transcription factor approach. The combining of the miRNA cluster and the RNA reprogramming genes in the dedifferentiation protocol further improved the dedifferentiation efficiency. The appropriate subset of RNAs can be introduced into cells by lipid-based transfection and, as one might imagine, there are many optimized products for RNA reprogramming from various vendors.

Trans-splicing: mRNA repair

Eukaryotic genes in higher plants and animals consist of coding regions known as exons that are separated by intervening, noncoding intron regions.

Early in the course of gene expression, transcription results in the synthesis of large, immature pre-mRNA molecules consisting of both exon and intron sequences from a specific gene locus. Subsequent removal of introns and the ligation of exons from the same RNA molecule ensues via the well-orchestrated formation of a spliceosome, that is, an RNA splicing complex, so that all of the coding information in a mature mRNA is contiguous prior to the onset of translation. This type of splicing is known as *cis*-splicing because the exons from a single pre-mRNA molecule are ligated together; it is the predominant form of transcript processing to remove intronic sequences. In contrast, *trans*-splicing involves the precision joining of exons from two different pre-RNA molecules, resulting in the formation of a hybrid (chimeric) RNA molecule. Like *cis*-splicing, *trans*-splicing is a naturally occurring process in eukaryotes, including humans, which occurs in a spliceosome-mediated manner, albeit ordinarily at a much lower frequency (Lei et al., 2016; Lasda and Blumenthal, 2011; Hastings, 2005). *Trans*-splicing may proceed in an intergenic manner, involving transcripts from two distinct gene loci; alternatively, intragenic *trans*-splicing involves variant transcripts produced from the same gene locus. The controlled and directed application of this natural phenomenon portends opportunities to produce an astonishing array of designer proteins both in vitro and in vivo, not to mention the potential to repair defective mRNAs.

With respect to the use of *trans*-splicing as a therapeutic platform, this spliceosome-mediated process can be harnessed to preempt cellular damage caused by the formation of aberrant proteins by fixing or "reprogramming" defective pre-mRNA molecules so that normal proteins are produced (Puttaraju et al., 1999; Wally et al., 2012) and the technology has enjoyed some successes (reviewed by Hong et al., 2020). In a more drastic application, *trans*-splicing could be utilized to induce processes such as apoptosis in, or to deliver an intracellular toxin to, cells that have experienced abrogation of normal proliferation controls. The attractiveness of *trans*-splicing is that it has the potential to be effective even if less than 100% of all defective mRNAs are "repaired." Exons can be repaired regardless of their position in the transcript: near the 5′ end, near the 3′ end, or somewhere in the middle of the transcript. An example of this method is shown in Fig. 27.5 in which a 3′ exon carrying a mutation is replaced with the wild-type nucleotide sequence. This is not an example of gene therapy in the traditional sense, however, in that *trans*-splicing has the potential to be effective even when a mutation that is harbored in the DNA persists. There are also instances in which *trans*-splicing can be mediated by Group 1 ribozymes, though this method is limited to modifications involving only the 3′-most exon in a transcript (reviewed by Lee et al., 2018).

In another application of this technology, gene expression can be effectively knocked down by *trans*-splicing a stop codon into an endogenous RNA that will result in the cessation of protein synthesis when this

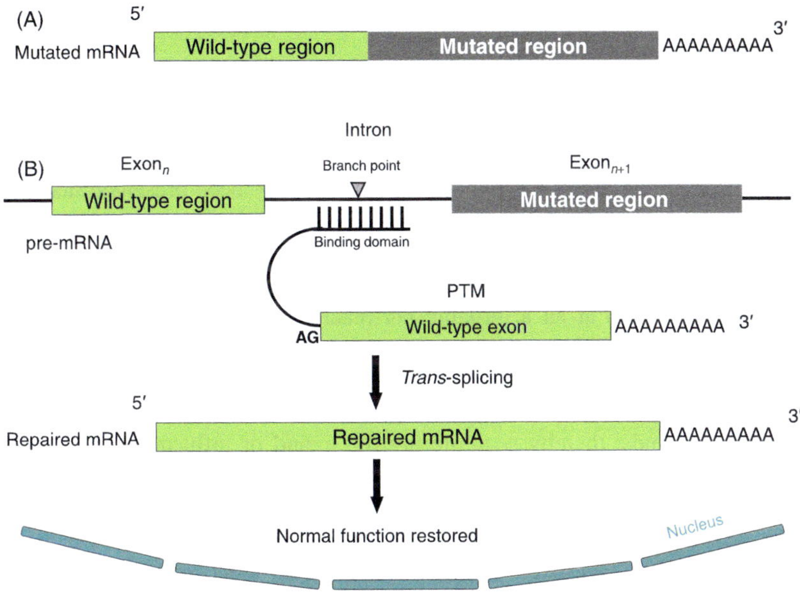

FIGURE 27.5 *Trans*-splicing joins together RNA from two different transcription units. (A) Mutant mRNA, capable of encoding a mutant protein. (B) *Trans*-splicing mRNA repair begins by targeting the pre-mRNA, which is produced in the cell from an endogenous gene locus. The RNA that will be spliced into the mutant pre-mRNA is referred to as the PTM (pre-*trans*-splicing molecule) and is likewise produced inside the cell following transfection of a PTM-encoding transcription vector. A splicing event involving the PTM binding domain and an intron that lies upstream of the mutation produces a repaired mRNA that is able to direct the synthesis of a wild-type protein. Notice that the PTM provides the requisite "AG" splicing acceptor site. This is a superb example of RNA therapy, rather than gene therapy.

molecular stop sign is encountered. Interestingly, the net result is the same as RNA interference (RNAi; see Chapter 13, RNA Interference and Gene Editing); RNAi causes the destruction of the mRNA template so that it cannot be translated, and *trans*-splicing a stop codon into an mRNA causes an abrupt premature end to protein synthesis. Most truncated proteins in the cell have little or no function and will end up being recycled. *Trans*-splicing may well represent an alternative to RNAi, particular when the exact RNAi sequences for gene silencing are elusive or the target cells are recalcitrant. As with RNAi, the process is completely and easily reversible since the DNA remains unchanged in both instances.

Although useful and innovative, the *trans*-splicing technology does have the potential to induce off-target effects; this is a concern any time nucleic acids are subjected to gene editing procedures. The greatest *trans*-splicing successes have been observed in vitro and it is unclear how well this technology will translate into success in vivo applications. Further, inasmuch as RNA repair is not a permanent solution to genetic defects, the effects of sustained delivery as well as practical economic considerations are at this time unclear.

Other RNA innovations

Some of the other RNA success stories have already been presented, namely RNA-seq (Chapter 21, Array Analysis of Gene Expression) and RNA aptamer technology (Chapter 24, RNA-seq: The Premier Transcriptomics Tool). RNA-seq in all of its permutations has revolutionized mapping the transcriptome and doing so at an affordable per-reaction cost. Aptamers are single-stranded RNA (or DNA) molecules consisting of usually less than 100 bases. Each unique aptamer, by virtue of its base sequence, is able to fold into a characteristic three-dimensional shape that will allow it to bind to any of a variety of ligands, thereby sequestering the ligand or influencing its natural function. Aptamers can also be modified so that they can carry drugs to specific organs and tissues, owing to the shapes that they assume. To date, the only FDA-approved aptamer drug sold under the name Macugen (Pegaptanib, OSI Pharmaceuticals) is used to treat macular degeneration by binding to VEGF (vascular endothelial growth factor). Many other aptamer-type drugs are in clinical trials as of this writing. Aptamers are much more economical to develop and manufacture than antibodies and exhibit low intrinsic immunogenicity. It is the belief of many that as the technology behind aptamer development improves, they may one day replace antibodies altogether. Among the downside of aptamers are their sensitivity to nucleases (unless encapsulated or 2′-fluoro-substituted), as well as their rapid removal from the body via the renal clearance system.

A typical experiment?

There is no such thing as a typical experiment. Before investing countless hours and thousands of dollars investigating a particular problem with various expensive kits and high-tech approaches, a simple initial characterization of a new system is always in order (Fig. 27.6). The nature of a "quick and dirty" assessment is a direct function of the model being proposed. For example, an abundant transcript might easily be assayed from large numbers of samples or across multiple time points by something as blatantly unsophisticated as dot blot analysis. If transcripts of interest are not detected, it may be necessary to modify the experimental design. Experimental modifications could include (1) selecting an alternative method of RNA isolation, (2) selecting an alternative biological source, (3) selecting a more sensitive assay (most common), or (4) induction of the genes of interest by chemical or physiological stimulation. At the same time, there is nothing wrong with attempting to generate preliminary data using supersensitive real-time PCR

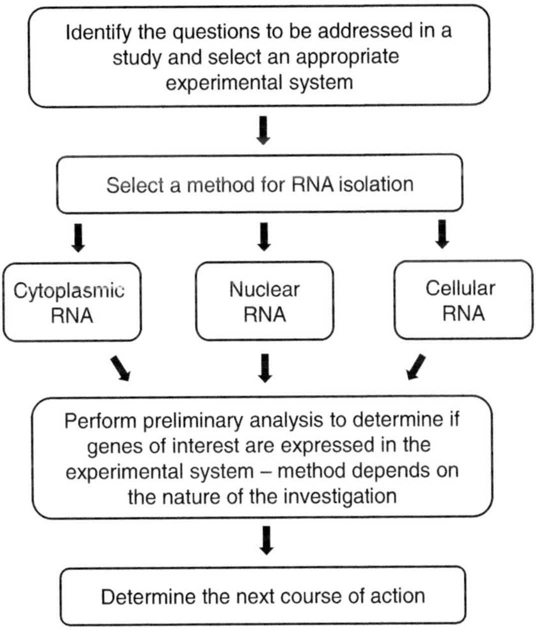

FIGURE 27.6 Preliminary gene expression analysis is intended to determine whether a selected experimental system is appropriate for reaching experimental goals. In-house expertise, equipment, and the nature of the questions being asked will suggest the most reasonable "quick and dirty" approaches. For the novice, the most important piece of information to be gleaned from preliminary characterization is whether the investigator is capable of isolating RNA of sufficiently high quality to support the intended experiments.

methods; the choice really depends on the nature of the starting material, the questions being asked, and the resources at hand.

In the course of an initial characterization, analysis of total cellular *and* total cytoplasmic RNA may be useful because many transcribed RNAs are degraded in the nucleus (hnRNA) and therefore neither mature nor appear in the cytoplasm in the form of mRNA. Further, if a sequence is observed in cellular RNA samples but not in cytoplasmic RNA samples, this suggests some type of posttranscriptional regulatory mechanism in action, possibly related to nucleocytoplasmic transport (transcript is retained in the nucleus) or miRNA-mediated regulation (e.g., a transcript is targeted for destruction). If the initial assessment yields encouraging data, a more detailed examination of the system is in order. As needed, the size of the transcript and an estimate of its abundance can be demonstrated by performing the time-honored Northern analysis. Recall, however, that a great deal of contemporary research involves the assay of genes that are undetectable by standard blot analysis and RT-PCR may be needed. The more sensitive quantitative assays of transcription are those based on solution hybridization, such as in

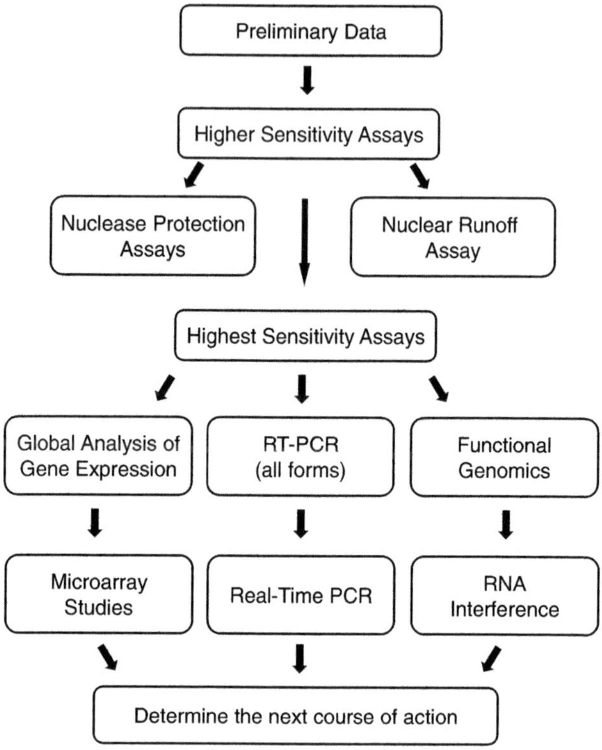

FIGURE 27.7 The nuclease protection assays and nuclear transcription assays offer greater sensitivity than filter-based assays (e.g., Northern analysis). For maximum experimental latitude, sensitivity, resolution, and complementation of techniques, microarray screening, real-time PCR, gene suppression by RNA interference, RNA-seq, and other functional genomics tools are all state-of-the-art methods.

nuclease protection and the various microarray- and PCR-based methods (Fig. 27.7); with the elimination of the filter, many problems related to assay sensitivity disappear.

Molecular scrutiny of a model system does not necessarily stop at the transcriptional characterization of one or a few genes. Having assayed the prevalence of transcriptional products, an investigator may wish to begin to characterize the molecular basis of observed variations from control or baseline expression. Gene expression is, after all, only partly about the biogenesis of RNA. It is also possible that observed changes in gene expression might be a result of a structural change of rearrangement in the gene locus itself, or even point mutations. It is entirely possible that an aberration within the coding portion of a locus, or the flanking sequences that influence its expression, constitutes at least one underlying basis for the up- or downregulation observed upon transcript quantification. These possibilities can be explored by microarray analysis or by direct gDNA sequencing.

Superior sensitivity and resolution are accolades attributed to two newer methods: digital PCR and RNA-seq. The revolutionary digital PCR approach to the amplification of cDNA and genomic DNA sequences is among the surest and most sensitive methods to assay transcription in virtually any experimental context, and is perhaps soon to displace standard real-time PCR as the "gold standard." Although RT-PCR can be a stand-alone technique in the assay of gene expression, it is wise to seek corroborating data using one or more of the previously described non-PCR-based methods. In the case of RNA-seq, this method has been greatly enhanced by datamining refinements in the field of bioinformatics.

Sensitivity issues

This laboratory guide presents a number of methodologies for the isolation and subsequent characterization of RNA. It should be abundantly clear that each of these methods offers distinct advantages and disadvantages, most often concerning issues of sensitivity and what, exactly, the data represent. In the literature, the term *relative abundance* is often used to describe the observed prevalence of specific transcripts when comparing samples. One of the samples is designated as the control or baseline sample and assigned a value of 1.0 and to which the other samples are compared. For example, it might be reported that the abundance of *c-myc* oncogene mRNA in exponentially proliferating, subconfluent cultures of diploid fibroblasts is fifty-fold higher compared to the abundance of *c-myc* mRNAs in contact-inhibited cultures of the same. Thus, statements can be made about the increase or decrease in the prevalence of transcripts without having knowledge of the absolute mass of these transcripts in the cell. Be aware, however, that a perceived relative abundance is directly related to the quality of the starting material and to the relative sensitivity of the assay at hand; evaluating the same set of samples using a battery of different methods will often produce startlingly different results.

If desired, however, the starting mass of a sample can be accurately gauged by performing real-time PCR. The relative sensitivity of several common transcription assay methods was described in Chapter 11 as "Farrell's RNA Sensitivity Index," in which RNA-based assays are ranked from the highest sensitivity; at present, the closest estimate of the absolute number of copies of a particular transcript in a sample comes from digital PCR, though this requires specialized equipment. Keep in mind that techniques that rank lower on the sensitivity index nonetheless provide useful information. Consider, for example, Northern analysis which provides information about the full size of an mRNA, a datum not associated with methods of higher sensitivity ranking unless RNA-seq is performed.

What to do next

The observation that one or more RNAs of interest have been induced or repressed, or otherwise up- or downregulated, may be the basis for a decision to invest additional laboratory resources to further study the cellular response. Gene expression is, by its very nature, multifaceted and the experimental design and associated timeline will be determined by the particular questions being asked and the available tools. Moreover, unexpected results may suggest a new research direction which may include one or more of the following areas:

1. Transcription start site (TSS) mapping of specific genes. As the TSS is subject to change, often in a tissue-specific manner, a single genetic locus may produce multiple transcript variants.
2. Using RNAi to discover the precise function of a gene.
3. Performing RNA-seq in order to better understand the behavior of the genes of interest in the context of global patterns of gene expression. When discovered to be associated with a particular phenotype, response, or condition, a sequence can be annotated with that information and added to GenBank or to the database of Genotypes and Phenotypes (dbGaP; http://www.ncbi.nlm.nih.gov/gap) for the benefit of all.
4. Assessing the splicing pattern of hnRNA, often from RNA-seq data, to identify alternative mRNA transcripts arising from the same genetic locus.
5. Examining flanking DNA sequences, in vitro or *in silico*, to better characterize the entire gene, particularly a gene's promoter region.
6. Analyzing the gene itself and its flanking sequences for possible protein binding sites via gel shift assays, DNA footprinting, or other proteomics assays.
7. Determining the exon–intron structure of the loci under investigation.
8. Performing site-directed mutagenesis in the structural portion of the gene, the regulatory sequences, or both, to discover the consequences to the cell (or an organism) of over- and underproduction of a wild-type protein or an improperly folded variant. This may also be achievable through RNAi.
9. Performing in situ studies to determine the histological distribution of cells expressing genes of interest.
10. Developing PCR primers for diagnostic applications.
11. Studying putative regulatory elements by ligation to reporter genes.
12. Performing homologous recombination of the gene under investigation with undifferentiated cells in order to examine the role of the locus in differentiation and development.
13. Since RNA doesn't tell the whole story, including Western analysis, immunochemical, or other proteomics techniques will provide a more complete understanding of gene expression under a specific set of conditions.

14. Performing high-level protein expression in prokaryotic systems and eukaryotic systems for clinical or biophysical studies.
15. *Responsibly* producing transgenic animals to facilitate in vivo production of life-saving pharmaceuticals.

Where to turn for help

In this age of technology, bioinformatics, social media, email, and all types of online tools, it is not difficult to perform a literature search or do a great deal of background reading without leaving one's desk. PubMed is a very quick way to access a great deal of scientific literature. This resource can be accessed indirectly from a number of websites worldwide, by going to the NCBI website (*pubmed.ncbi.nlm.nih.gov*) or by simply typing *pubmed.gov* into a web browser. Further, most universities and government agencies subscribe electronically to large numbers of journals for easy access. Other extensive offerings are easily accessible from the CDC and the NIH, and there are YouTube videos that demonstrate how to efficiently navigate the NCBI website.

There are a number of different Internet-based bulletin boards where scientists with common interests can postspecific questions about challenges they are experiencing in the hope of an answer from the readership. It is very important, however, to not believe everything you read, especially on the Internet. It is this Author's opinion, though, that nothing is better for collaboration and assistance than beginning with a smile and firm handshake at the conference. Professional development should be a priority for every researcher regardless of their role, stature, or rank in an organization or research institution.

Every biotech vendor has a website containing extensive and comprehensive product data including product descriptions, QA and QC procedures, lot numbers, levels of purity, protocols, user guides, applications notes, safety data sheets (SDS), opportunities for real-time "chat," and a wealth of other information. When the notion comes to mind that a product may be just what is needed to increase the productivity of the laboratory, it is very easy to print out the user guide. Should the need arise to speak with someone, most biotech companies have easily accessible technical services departments, all of which can be reached via email or by calling them directly for in-person assistance (Appendix O). One should take full advantage of tech services and not hesitate to request replacement materials when a product does not perform in the laboratory as advertised. Moreover, one should not be shy about approaching a company to request a significant discount on the first-time purchase of a new kit or reagent. This will provide the laboratory with the opportunity to try something new at a minimal cost. Clearly, if the kit works well, additional kits and related products will be purchased, and all the vendors know this. One should be polite, clear, and direct in making the request.

If setting up a new lab, looking to expand the current lab operation, or simply trying to identify which products are available from which vendors, there are two excellent strategies. First, many vendors offer new lab discount programs, new lab start-up guides, and many other online resources. By first approaching organizations with whom your institution has a purchasing history or already-in-place institutional (competitive) pricing, there is a good chance of finding what you need to fit the available budget. Second, Linscott's Directory of Immunological and Biological Reagents (http://www.linscottsdirectory.com) is another valuable online resource for searching for kits, antibodies, biologicals, and related materials. Of course, one can always go to a search engine and enter the name of the product of interest. Don't reinvent the wheel or try to do this on your own. Reaching out to someone who is already affiliated with an operational laboratory and utilizing online resources will save a lot of time and aggravation.

Remember the take-home lesson: you are limited only by your own ingenuity.

References

Anokye-Danso, F., Trivedi, C.M., Juhr, D., Gupta, M., Cui, Z., Tian, Y., et al., 2011. Highly efficient miRNA-mediated reprogramming of mouse and human somatic cells to pluripotency. Cell Stem Cell 8 (4), 376–388. Available from: https://doi.org/10.1016/j.stem.2011.03.001.

Bailly, A., Milhavet, O., Lemaitre, J.-M., 2022. RNA-based strategies for cell reprogramming toward pluripotency. Pharmaceutics 14 (2), 317. Available from: https://doi.org/10.3390/pharmaceutics14020317.

Gao, W., Hu, C.M.-J., Fang, R.H., Zhang, L., 2013. Liposome-like nanostructures for drug. Deliv. J. Mater. Chem. B 1, 6569–6585.

Granados-Riveron, J.T., Aquino-Jarquin, G., 2001. Engineering of the current nucleoside-modified mRNA-LNP vaccines against SARS-CoV-2. Biomed. Pharmacother. 142, 111953. Available from: https://doi.org/10.1016/j.biopha.2021.111953.

Hastings, K., 2005. SL trans-splicing: easy come or easy go? Trends Genet. 21, 240–247.

Hong, E.M., Ingemarsdotter, C.K., Lever, A.M., 2020. Therapeutic applications of trans-splicing. Br. Med. Bull. 136 (1), 4–20. Available from: https://doi.org/10.1093/bmb/ldaa028.

Kogut, I., McCarthy, S.M., Pavlova, M., Astling, D.A., Chen, X., Jakimenko, A., et al., 2018. High-efficiency RNA-based reprogramming of human primary fibroblasts. Nat. Commun. 9, 745. Available from: https://doi.org/10.1038/s41467-018-03190-3.

Lasda, E.L., Blumenthal, T., 2011. Trans-splicing. Wiley Interdiscip. Rev. RNA 2, 417–434.

Lee, C.H., Han, S.R., Lee, S.W., 2018. Therapeutic applications of group I intron-based trans-splicing ribozymes. Wiley Interdiscip. Rev.: RNA 9 (3), e1466. Available from: https://doi.org/10.1002/wrna.1466.

Lei, Q., Li, C., Zuo, Z., Huang, C., Cheng, H., Zhou, R., 2016. Evolutionary insights into RNA trans-splicing in vertebrates. Genome Biol. Evol. 8, 562–577. Available from: https://doi.org/10.1093/gbe/evw025.

Lundstrom, K., 2018. Self-replicating RNA viruses for RNA therapeutics. Molecules 23 (12), 3310. Available from: https://doi.org/10.3390/molecules23123310.

Martinon, F., Krishnan, S., Lenzen, G., Magné, R., Gomard, E., Guillet, J.G., et al., 1993. Induction of virus-specific cytotoxic T lymphocytes in vivo by liposome-entrapped mRNA. Eur. J. Immunol. 23 (7), 1719–1722. Available from: https://doi.org/10.1002/eji.1830230749.

Olivieri, E.H., de Andrade, F.L., Pereira, R.G., Carvalho-Mota, L.D., Campos, A.H., Carraro, D. M., 2014. Biobanking practice: RNA storage at low concentration affects integrity. Biopreserv Biobank 1, 46–52. Available from: https://doi.org/10.1089/bio.2013.0056.

Petsch, B., Schnee, M., Vogel, A., Lange, E., Hoffman, B., Voss, D., et al., 2012. Protective efficacy of in vitro synthesized, specific mRNA vaccines against influenza A virus infection. Nat. Biotechnol. 30, 1210–1216. Available from: https://doi.org/10.1038/nbt.2436.

Puttaraju, M., Jamison, S.F., Mansfield, S.G., 1999. Spliceosome mediated RNA trans-splicing as a tool for gene therapy. Nat. Biotechnol. 17, 246–252.

Takahashi, K., Yamanaka, S., 2006. Induction of pluripotent stem cells from mouse embryonic and adult fibroblast cultures by defined factors. Cell 126 (4), 663–676. Available from: https://doi.org/10.1016/j.cell.2006.07.024.

Wally, V., Murauer, E.M., Bauer, J.W., 2012. Spliceosome-mediated trans-splicing: the therapeutic cut and paste. J. Invest. Dermatol. 132, 1959–1966.

Warren, L., Manos, P.D., Ahfeldt, T., Loh, Y.H., Li, H., Lau, F., et al., 2010. Highly efficient reprogramming to pluripotency and directed differentiation of human cells with synthetic modified mRNA. Cell Stem Cell 7, 618–630.

Yoshioka, N., Gros, E., Li, H.R., Kumar, S., Deacon, D.C., Maron, C., et al., 2013. Efficient generation of human iPSCs by a synthetic self-replicative RNA. Cell Stem Cell 13 (2), 246–254. Available from: https://doi.org/10.1016/j.stem.2013.06.001.

Further reading

Damase, T.R., Sukhovershin, R., Boada, C., Taraballi, F., Pettigrew, R.I., Cooke, J.P., 2021. The limitless future of RNA therapeutics. Front. Bioeng. Biotechnol. 9. Available from: https://doi.org/10.3389/fbioe.2021.628137.

Hou, X., Zacs, T., Langer, R., Dong, Y., 2021. Lipid nanoparticles for mRNA delivery. Nat. Rev. Mater. 6, 1078–1094. Available from: https://doi.org/10.1038/s41578-021-00358-0.

Sandbrink, J.B., Shattock, R.J., 2020. RNA vaccines: a suitable platform for tackling emerging pandemics? Front. Immunol. 11, 608460. Available from: https://doi.org/10.3389/fimmu.2020.608460.

Epilogue

A few pearls of wisdom

To suggest that research and interest in all things RNA has grown spectacularly over the last few years would hardly be a stretch of one's imagination! It has been amazing to watch as new roles for RNA unfold owing to the development of increasingly sophisticated molecular tools. One thing that is clear is that RNA is an ancient molecule, predating the arrival of DNA on the primeval scene, and capable of fulfilling the roles that are currently associated with proteins and DNA. Perhaps the biggest eye-opening revelation has been the multifaceted functional diversity of RNA, particularly the noncoding products of transcription. Working successfully with RNA requires skill and attentiveness to detail.

With all its diverse applications, the essential character of molecular biology is the judicious use of fewer than two dozen standard reactions. Upon gaining proficiency using these techniques, many of which have been presented here, the research potential is limitless. Molecular biology *is not* about kits; rather, it is about planning, determining how best to use the tools at one's disposal to get the job done and, in the final analysis, it's all about nucleic acid quality control. A thoughtful, well-considered strategy and the implementation of the correct techniques, in the correct order, will ultimately govern the outcome of an investigation. To this end, keep in mind this Author's definition of the five Rs of molecular biology:

1. Rapid
2. Representative
3. Reproducible
4. RNase-free
5. Reliable

Most importantly, take every opportunity to attend a scientific conference and, if possible, make a presentation there. Direct interaction with colleagues working in the same discipline can be a valuable source of information, inspiration, consolation, and encouragement. While it is certainly true that one learns (a lot) from one's mistakes, there is no point in reinventing the wheel. If someone else has already solved the same technical problem that you are experiencing or has developed a modified experimental approach, learning about the solution to the problem or about methodological alternatives sooner, rather than later, can be invaluable.

The following is a list of suggestions, some obvious and others not so obvious, that may help streamline the day-to-day activities in the lab and avoid some of the difficulties that may otherwise be encountered.

1. Get rid of powdered gloves. The powder can inhibit PCR and other enzymatic reactions, increase the background on blotting membranes used for nucleic acid hybridization, and it causes an all-around mess.
2. Prepare recipe cards that delineate the exact formulation of commonly used buffers and solutions. Having a card at hand will save a great deal of time when reagents are in short supply or when they must be prepared quickly in the middle of an experiment. Alternatively, write the formulations directly in this book in Appendix K under "notes."
3. Read the label. Just as one would not want to administer the wrong medication, the use of the wrong reagent will put an immediate end to one's experiment. In fact, read the label twice.
4. Be sure to label tubes unambiguously. Even seasoned investigators often need to be reminded of the importance of being able to associate a particular tube with a particular entry in a lab notebook. Consider, for example, using different color markers to label sample and reaction tubes, and record the color of the ink in the notebook. In the case of complicated or multistep protocols, writing the step number on the top of the tube, and other salient information on the side of the tube is very helpful in terms of organization and tube retrieval from the freezer. It also helps to store unrelated tubes in different storage boxes in the freezer.
5. All stock and working solutions in the lab should be labeled with a predetermined expiration date, in addition to the date of preparation. Be sure to check the expiration date of everything before using it! A sad consequence of the use of out-of-date and even worse, ancient solutions, in the lab is the unintentional introduction of RNase from microorganisms that may have taken up residence in the bottles.
6. As needed, RNA samples can be "cleaned up" by redissolving them in the lysis buffer that accompanies spin column RNA isolation kits. The sample is then processed as if working with a whole cell lysate. The results of this type of cleanup are remarkable. This method is particularly useful for changing the buffer in which the RNA is dissolved and for concentrating the RNA *without* having to precipitate the sample.
7. Running a gel is the single best diagnostic that can be used to assess the integrity of a nucleic acid preparation. A small aliquot of sample electrophoresed on a minigel (e.g., 7×8 cm) will reveal the probable usefulness of the sample in downstream applications. This can be

performed at several points in a series of experiments to check for degradation of RNA, improper denaturation, and to confirm the efficient synthesis of cDNA.

8. If denaturing RNA with formaldehyde for any application, be aware that the sample will quickly become degraded if the formaldehyde pH is below 4.0. Always deionize a fresh aliquot of formaldehyde (and formamide) just prior to use (see Appendix G).
9. If the loading buffer containing bromophenol blue changes from blue to green when added to a sample, there has been a radical pH change in the tube, and the sample integrity itself is at risk.
10. When RNA is to be recovered from a lysate that also contains DNA, organic extraction at pH below 6.0 will drive DNA to the aqueous-organic interface. Keep in mind, however, that samples isolated in this manner are almost certainly tainted with genomic DNA. DNase I treatment is necessary, especially for RT-PCR applications.
11. If denaturing RNA with dimethyl sulfoxide (DMSO) and glyoxal, especially for dot blot analysis, be aware that DMSO dissolves nitrocellulose. Use a nylon blotting membrane instead.
12. When preparing to wash precipitated RNA pellets, be sure to dilute 95%–100% ethanol down to 70%–75% using nuclease-free water.
13. Wash nucleic acid pellets with 70% ethanol to remove excess salt. This is necessary because nucleic acid molecules and salt form aggregates in a solution that demonstrates dramatically reduced solubility in alcohol. Thus, the precipitate at the bottom of the tube is both salt and nucleic acid. Washing the pellet with 70% ethanol will remove much of the salt; however, it is not sufficiently aqueous to redissolve nucleic acids. After two or three washes, do a final wash with 95% ethanol to facilitate rapid drying of the pellet.
14. RNA pellets do not adhere to the side of microfuge tubes as firmly as DNA pellets. When decanting an alcohol supernatant, keep the pellet in sight at all times. If afraid of losing a pellet when decanting, pour off the supernatant into another microfuge tube so that the pellet will not be lost if it slides out of the tube.
15. When pipetting microliter volumes, it can be helpful to pipette individual reagents onto the inside wall of the microfuge tube, close to, but not at the bottom. The individual reagents can then be mixed by pulse centrifugation. This avoids the problem of inaccurate volume delivery to the tube due to capillary action in the tip of the micropipette. It also shows convincingly that small aliquots have actually been delivered into the tube.
16. Always test enzyme and reaction components when opening a new package or kit. Non-functional enzyme translates into one thing: no results!
17. Get in the habit of making up master mixes for cDNA- and PCR-related applications. This is *especially* important for quantitative analyses and

should be a standard operating procedure. Even subtle variations in microliter reaction volumes or nanomolar concentrations can have a strong negative impact on the accuracy of the resulting data.

18. When preparing a master mix for multiple reactions, be sure to multiply the "per reaction" volumes by at least one more than the total number of required reactions. This standard procedure will ensure an adequate volume of master mix in the event of subtle micropipetting errors. Although critical for the preparation of cDNA master mixes and PCR master mixes, it is a good all-around practice whenever multiple samples are involved.

19. Pay attention to the volume of enzyme add to a reaction or master mix. Recall that molecular biology enzymes are often stored in 60% glycerol. The volume of a reaction occupied by the enzyme should not exceed 5% of the total volume. Beyond this amount, the enzymes will not function optimally.

20. Consider the use of REDTaq (Millipore Sigma) or REDTaq master mix to ensure not only that the enzyme has been added to the reaction tube, but to also ensure that the enzyme/master mix has been properly mixed within the reaction tube. The red dye in RedTaq also functions well as a gel loading buffer.

21. Nucleic acid sequences are always written $5' \rightarrow 3'$, from left to right. This applies to RNA and DNA. This is especially important when ordering oligonucleotides: the sequence of the $5'$ primer and the $3'$ primer must both be written $5' \rightarrow 3'$.

22. If reporting the sequence of dsDNA, the convention is to write the sequence of the coding strand $5' \rightarrow 3'$; it is not necessary to report the sequence of the template strand because it should be intuitive that the two strands are complementary to each other and are base-paired to each other in an antiparallel manner.

23. Do not store primers below a concentration of 20 μM. Primers diluted below this concentration are often unreliable in later experiments.

24. Do not store primers or templates at excessively high concentrations, which is an excellent way to promote amplicon anarchy (massive carry-over contamination) in the lab.

25. Always keep tubes containing RNA samples on ice unless the protocol specifically dictates otherwise. When reading the protocol, let the samples rest on ice until the next course of action has been discerned. Also, be sure to prechill centrifuge tubes when carrying out a sequential extraction regimen so that aliquots are always delivered to a cold tube. Remember that cool temperatures partially inhibit nuclease activity.

26. When a protocol specifies placing a heat-denatured sample on ice immediately after heating, it is a good idea to have an ice bucket close to the heating block or thermal cycler. Plunging the tube into ice, and then traveling back to one's work area will maintain the sample in a denatured state for as long as possible.

27. Take logistical measures to keep stock solutions of RNA concentrated. This will facilitate UV spectrophotometric measurements, and subsequent dilution, as needed.

28. It is mechanistically easier to recover small RNA yields in minimal volumes from 1.7 mL or 0.5 mL microfuge tubes than from larger tubes (e.g., 2.0 mL tubes) due to the shape of the bottom of the tubes.

29. Do not allow nucleic acid pellets to dry out completely unless explicitly specified in the protocol. RNA and DNA pellets that are *slightly* damp with ethanol will dissolve more readily than those that are completely dry.

30. Spend the extra money and purchase molecular biology-grade reagents. The use of these highly purified, nuclease-free materials will favorably increase the yield, stability, and utility of nucleic acid preps for all downstream applications. Plus, each item has a lot number for record-keeping purposes that will keep the quality assurance and the quality control people happy.

31. In addition to RNase, keep in mind that heavy metals can cause RNA degradation when present for extended periods. If this is a suspected problem, filtering buffers through Chelex (Bio-Rad) can help. In addition, the inclusion of 8-hydroxyquinoline, which chelates heavy metals and is itself a partial RNase inhibitor, may optimize RNA purification efficiency.

32. In the course of extraction with organic solvents to deproteinate nucleic acid samples, complete recovery of the aqueous phase without disturbing the protein interface is often difficult. To maximize the recovery of aqueous material after most of it has been removed, it helps to tilt the tube to a 45-degree angle, causing a "bubble" of the remaining aqueous material to form up against the side of the tube (Fig. 1). At this point, one may slide the micropipette tip along the upper inside wall of the tube, penetrate the aqueous bubble, and remove most of the remaining aqueous volume without disturbing the protein interphase or organic material. *Caution*: Use great care when handling tubes containing organic solvents, wear appropriate eye protection and skin protection, and use common sense.

33. When purifying nucleic acids with phenol, always perform a final extraction with chloroform alone or a mixture of chloroform:isoamyl alcohol (24:1). Due to the high solubility of phenol in chloroform, even trace amounts of carryover phenol will be removed in the final chloroform extraction. This is a necessary step because quinones, the oxidation products of phenol, can compromise the integrity of the nucleic acid sample.

34. Do not use polystyrene plasticware or thin-wall PCR tubes with organic solvents. The plastic will begin to dissolve immediately in the presence of phenol or chloroform, resulting in catastrophic failure.

35. Be sure to check for chemical compatibility with centrifuges tubes, especially when organic solvents are involved. Also be sure to check for β-mercaptoethanol incompatibility with ultracentrifuge tubes (e.g., polyallomer).

36. When collecting nucleic acid precipitates by centrifugation, be certain not to exceed the recommended *g*-force for a particular type of tube. Attention to this detail is especially important when scaling up or scaling down RNA isolation procedures, as significant *g*-forces may be required. For smaller-scale procedures, polypropylene microfuge tubes are used almost exclusively. Properly cushioned, nuclease-free Corex glass tubes (or the equivalent) are handy for larger-scale preps because of their ability to withstand the *g*-force needed to pellet precipitated RNA.

37. Sodium dodecyl sulfate (SDS) is often added to buffers to control nuclease activity. In many applications, it may be desirable to eliminate it because it precipitates out of solution at cold temperatures (the sodium salt of SDS is especially insoluble) and in high-salt environments. Moreover, even traces may be incompatible with downstream applications, especially PCR. Other types of RNase inhibitors, such as RNasin, may be used instead.

38. When preparing buffers and reagents that include EDTA, be sure to use the disodium salt (Na_2-EDTA) rather than the tetrasodium salt (Na_4-EDTA). This is important because of the great sensitivity of many molecular biology enzymes to salt concentration. Buffers prepared with Na_4-EDTA will initially be more alkaline than those made with Na_2-EDTA. The subsequent addition of HCl to establish pH 7–8 will greatly increase the NaCl concentration in the final buffer, and likely have a negative influence on the outcome of an experiment.

39. The ubiquitous instruction to store labile materials in "suitable aliquots" is quite vague at best, particularly because different applications require different quantities of RNA or DNA. If the precise definition of a suitable aliquot is unknown, purified RNA is best stored in the precipitated form at −80°C until such a determination can be made. It is unwise to repeatedly thaw and refreeze RNA samples, as this will dramatically reduce the storage life of the material. If a redissolved sample must be stored, aliquot it and then store it at −80°C so that an aliquot is thawed only once.

40. Do not discount the importance of key UV absorbance ratios as indicators of the purity of nucleic acid samples. Be sure to examine A_{260}/A_{280}, A_{260}/A_{230}, and A_{260}/A_{240}, as described in Chapter 8. Deviant ratios may be indicative of protein contamination, phenol contamination, excess salt, incomplete solubilization of the RNA pellet, or other issues.

41. In the case of membrane-based hybridization assays, and contrary to popular belief, it is not detrimental to hybridize more than one membrane in the same hybridization chamber. As long as there is sufficient buffer to keep the filters from sticking together and the filters are free to move in the chamber with the motion of the incubator (orbital shaker incubators work well), simultaneous hybridization of two or more blotting membranes should proceed without any impediment.

42. If a particular hybridization buffer formulation continually produces elevated or splotchy background on the X-ray film, think about performing a sham hybridization with the hybridization buffer first (with or without the probe). In other words, incubate the hybridization buffer with a sheet of blotting membrane with nothing on it; if the probe will be present with the sham filter, be sure to prehybridize/block it first. The idea is that anything in the hybridization buffer that will stick nonspecifically to the experimental membrane will stick to the blank membrane instead. Then, transfer the hybridization buffer to the bag or container with the experimental membrane. Sometimes, the improvement is astonishing.
43. Define the linear range of X-ray films when preparing to record experimental data by autoradiography or chemiluminescence. For example, preparing dilutions of at least one sample will facilitate the accurate interpretation of data. Remember that changing the exposure time for X-ray film, just like altering image analysis exposure time and aperture settings digitally, will minimize problems associated with the limited linear response of film in general and highlight the differences among the samples.
44. The detection limit for nucleic acids is strongly tied to the specific activity of the probe. Attentiveness to this parameter is important.
45. When performing Northern analysis, permanently record the location of size standards on X-ray film by labeling a very small amount of probe that will hybridize to the size markers. This will eliminate guesswork when comparing them to experimental samples.
46. Three-hole punch dried X-ray films and place them in a three-ring notebook for organizational and storage purposes. Place a sheet of paper between films to prevent them from becoming scratched.
47. All chemicals should be considered potentially hazardous. Individuals should be trained in good laboratory techniques and safety practices by their respective departments. In a molecular biology setting, there are many inherent dangers associated with what is considered "the daily routine." Always follow the safe handling, containment, and personal safety recommendations from the manufacturers of chemicals and equipment.
48. Be sure to keep and maintain easy access to the Safety Data Sheet (SDS; *formerly* Material Safety Data Sheet, MSDS) that accompanies hazardous chemicals and reagents. It is important for all members of a lab team to become familiar with the potential health hazards associated with each of these compounds, even if they are not working directly with them. Each SDS also contains appropriate first aid measures in the event of accidental exposure.
49. Do not store microfuge tubes in liquid nitrogen, ever.

50. Using a mixture of oligo(dT) and random primers goes a long way toward improving first-strand cDNA synthesis, both in terms of size and yield.
51. Failure to inactivate reverse transcriptase after first-strand synthesis (one-tube PCR notwithstanding) is a big no-no.
52. Learn to use WebLogo. This is a free, web-based tool for creating graphical representations of nucleic acid or protein sequence alignments, consensus sequences, and other uses. Visit weblogo.berkeley.edu.

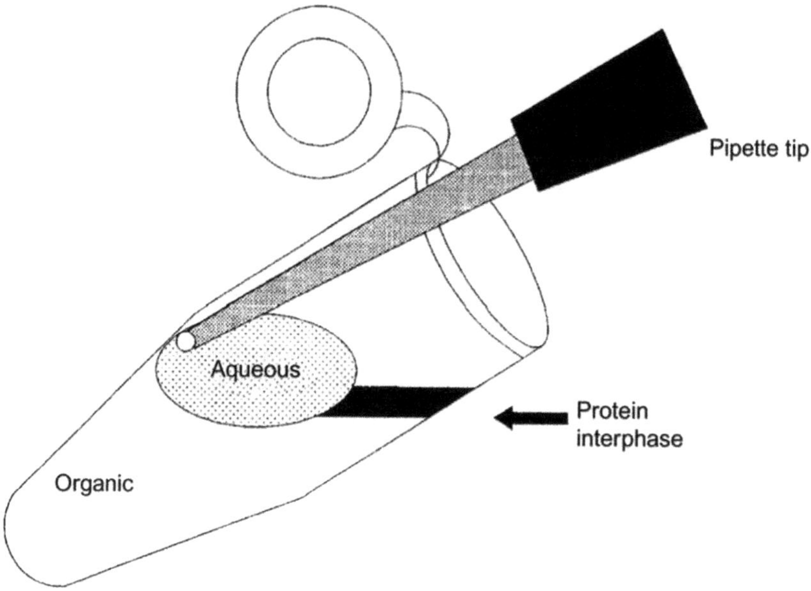

FIGURE 1 Technique for the quantitative recovery of aqueous material during an organic extraction. Be sure to wear appropriate eye and skin protection.

53. Never mouth pipette anything and always wear safety glasses.
54. Wear appropriate attire in the lab at all times: no jeans with holes, no shorts, no short skirts, no long hair unless it is pulled back, no face piercings or dangling jewelry, no sandals, no open-toe shoes. Wool, cotton, and silk are the best fabrics for clothing because they are less flammable than synthetic fibers that may melt if ignited. And wear a lab coat.

Appendix A

Maintaining complete and accurate records

The absolute necessity of maintaining accurate records of experiments performed and the resulting data are indisputable. Perhaps no other facet of scientific inquiry has the potential to have the greatest impact on research than the *art* of excellent recordkeeping. Many graduate students have been amazed by the intralaboratory fallout that results from the failure to produce well-written, easy-to-interpret documentation of one's research-related activities. Not only are accurate, meaningful records an important component of continuous productivity, but also such records are often the only verification that the work has been completed, as well as the order in which each tantalizing bit of data was generated. Sustained funding and successful laboratory audits often depend precisely on the quality of research records. These issues also have legal implications that can be far-reaching.

Students who begin their research experience as undergraduates have a distinct advantage when they enter graduate school or otherwise become involved in an industrial research setting. Laboratory investigation at the most rudimentary level, which may involve little more than "grunt" work, is often the first exposure that students have to the imperative for keeping a hand-written or electronic notebook. Although requests from more senior members of the lab may seem inane at times, invariably the time will come when that inane information must be recalled. It is the opinion of this Author that it is better to write down too much than to leave anything to memory.

The head of the laboratory is the party with the primary responsibility for breaking bad habits that may be imported into the lab and as well as preventing new members of the lab from acquiring any bad habits with respect to recordkeeping or poor lab technique. What follows is a list of guidelines that one would reasonably expect to be adhered to in a research setting, all of which are part of good laboratory practice. The list is not intended to be all encompassing, and it should be modified to suit the local clientele and the nature of the research enterprise.

1. Record each facet of each experiment as it is being performed. Do not trust details to memory and do not wait until the end of the day to write everything down.
2. Do not write procedural information or data on scraps of paper. All too often this type of stationery is misplaced or accidentally discarded. It is also possible that data might be transferred to record books, written or electronic, in an incorrect order.
3. If possible, laboratory records should be maintained on a shared network drive and the data backed up daily. Electronic laboratory notebooks are becoming increasing commonplace and well-funded laboratories are leaning toward automated data collection technology.
4. Clarity is of the utmost importance. It is useless putting procedures into flamboyant prose garnished with elaborate punctuation if the interpretation is unclear. The level of detail recorded in a notebook should be sufficient to allow the reader to reproduce the experiment from the notebook entry alone. A well-written lab notebook is itself a work of art.
5. Don't worry about experiments that did not work out as expected or did not work out at all; it is imperative that everything is documented. Often, useful information can be gleaned from scrutiny of a string of failures. Just like everything else in life, we all learn from our mistakes in the laboratory.
6. A professional lab notebook is a bound (tamper-evident) journal, has a sequential number on each page, and includes a table of contents.
7. Never tear or otherwise remove pages from a lab notebook.
8. If data are entered manually into laboratory notebooks, entries should be made in blue or black ink only. This is an industry standard. Of late, pens with fluorescent- and pastel-colored ink have become stylish; these are difficult to read, unprofessional for everything, and should be avoided.
9. If one's handwriting is poor, then one should print instead.
10. Put the date on each page of the notebook as procedures and data are entered.
11. At the beginning of each new experiment, include a sentence or two that provides a rationale for the experiment and be sure to reference or otherwise call attention to previous experiments or samples that have relevance to the business at hand.
12. Never "white-out" or scratch out entries in laboratory notebooks. If changes are to be made, a single line should be made through the incorrect entry, initialed by the author, and a brief expository note describing the reason for the change should be included.
13. Be sure to record lot numbers for each new reagent, kit, or other chemistry that is used to support the ongoing research. Many suppliers now provide all relevant data on adhesive strips on the spec sheets that

accompany their products; these strips can be easily applied to the current page of one's notebook for future reference. Expiration dates should also be recorded.

14. Figures, tables, photographs, and equations should be assigned a number in the notebook. These data should subsequently be referred to by that number elsewhere in the notebook.
15. Laboratory records should be reviewed periodically by someone associated with the project, but not directly associated with the experiments and data recorded, to verify the completeness of these records and the ease of comprehension. Further, suggestions as to how recordkeeping might be improved can be proffered, as needed.
16. There is no substitute for a well-motivated investigator.

Appendix B

Converting mass to moles

Each investigator has a preferred method for preparing and labeling stock solutions. Some people prefer to describe solutions in μg/μL (mass per volume) while others prefer to describe everything in terms of nmols (a mole quantity) or μM (molarity; e.g., μmoles/L). It is important that persons performing molecular biology methods be able to quickly convert mass to moles, moles to mass, μg/μL to molarity, and molarity to μg/μL. The conversions can often be performed in one's head without a calculator by adherence to a few simple mnemonics. Remember, practice makes perfect.

With respect to PCR, the most important thing to recall is that most protocols require equimolar amounts (same number of molecules) of the upstream and downstream primers. In other words, in the reaction tube there is a 5′ primer molecule for every 3′ primer molecule. Thus, adding a 1 μL of 100 μg/μL stock solution of each primer to a master mix may not do the trick, particularly when the length or base composition of the primers differ. Therefore, primer stock solutions are best labeled in terms of their molarity. In this laboratory, primers are prepared as 200 μM master stock solutions, an aliquot of which is diluted down to produce a working stock of 10 μM (1:20 dilution) for routine preparation of PCR master mixes. The final concentration of each primer in a PCR tube is 0.1 μM, meaning that a simple 1:100 dilution of the 10 μM stock is all that is required (e.g., a 200 μL master mix requires 2 μL of the 10 μM primers to achieve a 0.1 μM final concentration). Good advice: keep primers and other stock solutions at a concentration that is easily divisible by some multiple of 10.

Regarding the concentration conversions, the first and perhaps the easiest thing to remember is that μM = pmol/μL. Second, to convert from molar concentration to mass per volume (e.g., μM → μg/μL) the molecular weight of the material must be known. Third, when performing a conversion of any type, keeping track of the units will usually prevent making a mistake. In the case of working with primers, which is a common application, the molecular weight can be determined in advance using any of the free online primer design tools. When primers are delivered, the specification sheet that accompanies the product also provides molecular weight information. The following scenarios exemplify the types of conversions one might routinely perform.

Scenario 1

Consider the oligonucleotide sequence GCTGTCGATCGATTCGATCGCTAGA, a 25-mer with a molecular weight of 7673 g/mol. A new tube of this primer arrives in the lab from the manufacturer. The spec sheet accompanying the lyophilized primer indicates that the tube contains 247 μg of the primer. You add 247 μL H_2O to produce a 1 μg/μL concentration. You then realize that you should express the concentration in terms of molarity. To make the conversion, perform the following calculation:

1 μg	10^6 μL	g	mol	10^6 μmol
μL	L	10^6 μg	7673 g	mol

By multiplying all numerators and all denominators and, of course, keeping track of the units, one finds that a 1 μg/μL stock solution of this primer has a concentration of 130 μM.

Scenario 2

You have 247 μL of a 130 μM primer stock solution and need to dilute all of it down to 100 μM to conform to the standard operating procedure in the lab. To make the dilution, perform the following calculations:

130 μM (initial concentration) ÷ 100 μM (final concentration) = 1.3 (dilution factor)
247 μL (initial volume) × 1.3 (dilution factor) = 321 μL (final volume to prepare)
321 μL (final volume to prepare) − 247 μL (initial volume) = 74 μL (amount of water to add)

Thus, adding 74 μL H_2O to the 247 μL of 130 μM stock will produce 321 μL of a 100 μM solution.

Scenario 3

The new tube of a primer arrives in the lab from the manufacturer. The accompanying spec sheet indicates the properties of the primer to be 4.17 nmol per OD_{260} (optical density at 260 nm) and that the tube contains 3.7 ODs of material. You want to make a 50 μM stock solution and need to figure out how much water to add to the tube. To complete your task, perform the following calculation:

3.7 OD	4.17 nmol	μmol	L	10^6 μL
	OD	10^3 nmol	50 μmol	L

By multiplying all numerators and all denominators and, of course, keeping track of the units, one finds that the result is 309 μL. Thus, the addition of 309 μL to the tube described above will produce a 50 μM stock solution. Note that the contents of the fourth column (L ÷ 50 μmol) in the calculation table pertain to the final desired 50 μM concentration.

Scenario 4

The principal investigator of the lab hands you a tube containing an oligonucleotide with MW = 5933 g/mol at a concentration of 150 μM and tells you to remove 20 μg from the tube for an experiment. To complete your task, perform the following calculation:

L	10^6 μmol	mol	g	10^6 μL	20 μg
150 μmol	mol	5933 g	10^6 μg	L	

By multiplying all numerators and all denominators and, of course, keeping track of the units, one finds that the result is 22.5 μL. Therefore, a 22.5 μL aliquot from the 150 μM stock solution described above contains a mass of 20 μg of the oligonucleotide.

Important: Regardless of one's mathematical prowess, *always* check the final concentration spectrophotometrically whenever a dilution is made or when lyophilized material is resuspended. Measuring A_{260} is highly desirable because the actual content of primer tubes often differs from what is indicated on the label. See Chapter 10, RT-PCR: A Science and an Art Form, for details.

Nucleotide anhydrous molecular weights

Single-stranded DNA		Single-stranded RNA	
dAMP	313.2 g/mol	AMP	329.2 g/mol
dCMP	289.2 g/mol	CMP	305.2 g/mol
dGMP	329.2 g/mol	GMP	345.2 g/mol
dTMP	304.2 g/mol	TMP	—
dUMP	290.2 g/mol	UMP	306.2 g/mol
dIMP	314.2 g/mol	IMP	330.2 g/mol

Average MW of a nucleotide ≈ 330 g/mol.
Average MW of a nucleotide base pair ≈ 660 g/mol.
The molecular weights shown in the table above are approximate and refer to the corresponding nucleoside monophosphate when it is part of a polynucleotide. If phosphorylation is observed at the beginning of the molecule, add 79 g/mol for a 5′ monophosphate and 159 g/mol for a 5′ triphosphate. The data presented here are for estimation purposes only.
Note: I = inosine; corresponding base is hypoxanthine, Hx.

Appendix C

Removal of DNA from an RNA sample

Laboratory procedures for the isolation of RNA often require the removal of contaminating DNA. The most common reason for "DNasing" a sample is to destroy any potentially competitive genomic DNA templates prior to RT-PCR. For example, the popular acid-phenol method (Chapter 4, Ribonucleic Acid Isolation Strategies) for separating RNA from DNA, while efficient, is not efficient enough to ensure that only RNA remains in the aqueous environment; as one approaches the organic phase with the micropipette tip, the likelihood that DNA will be carried over increases. This will almost certainly compromise the quantitative nature of RNA-based assays because probes and primers often have difficulty distinguishing between RNA targets and DNA targets when both are present. While silica-based isolation strategies yield purer RNA, these samples, too, may be tainted with contaminating DNA. Regardless of the isolation method, it is unwise to use RNA for any RT-PCR or other quantitative application without first treating it with DNase because even one molecule of genomic DNA can produce misleading data.

The most commonly used enzyme for the degradation of contaminating DNA molecules is the endonuclease DNase I, for which both ssDNA and dsDNA are substrates. As there is no easy, cost-effective way to prepare RNase-free DNase I in the lab because of its labile nature, molecular biology-certified enzyme formulations are widely available and well worth the purchase price. Many commercial preparations of RNase-free DNase I are certified by the manufacturer to harbor no intrinsic RNase activity and their use is strongly recommended. In general, the final enzyme concentration is typically 1–2 U/μg of RNA or, if the sample contains less than 1 μg, use 1U DNase I/10 μL reaction. DNase I is supplied in a glycerol storage buffer, and one should therefore not exceed more than 1 μL enzyme/10 μL reaction volume. The sample(s) are then incubated at 37°C for up to 30 minutes.

At the conclusion of the incubation period, DNase is usually heat-inactivated; no sample clean-up is necessary and the residual denatured DNase I is ignored altogether. If required, DNase I can be removed by phenol:chloroform extraction (not recommended) or, more conveniently, with the use of a silica spin column which will help keep the sample concentrated. Applications that benefit from additional clean-up post-DNase treatment include, but are not limited to, degradation of DNA template molecules following in vitro transcription reactions, the study of DNA–protein interactions by DNase I footprinting, and other situations where the RNA must be free of DNA polynucleotides.

Finally, it is important to note that DNase I has strict cofactor requirements for Ca^{2+} and Mg^{2+} (or Mn^{2+}). Due to the sensitivity of PCR to divalent cation concentration, the Mg^{2+} in the DNase I reaction buffer must be taken into account if the intended use of the RNA is an RT-PCR assay due to the influence that Mg^{2+} exerts on PCR efficiency and specificity.

Protocol: digestion of DNA

1. Use RNase-free plasticware throughout this procedure. Reaction assembly template:

RNA	≤1 μg
10× reaction buffer	1 μL
Nuclease-free H_2O	to 9 μL
RNase-free DNase I	1 μL; final (0.1U/μL)

Note 1: DNase I typically will show optimal activity in the presence of 5–10 mM $MgCl_2$ and 1–2 mM $CaCl_2$. These cofactors and other reaction components are provided by the 10× reaction buffer that accompanies the purchase of the enzyme.
Note 2: Remember to keep the sample as concentrated as possible.
2. Incubate the mixture for 15–30 minutes at 37°C. As needed, incubations up to 1 hour are also acceptable.
Note: Keep in mind that DNase I is strongly inhibited in buffers containing EDTA, SDS, and other denaturants or if heated above 37°C.
3. Terminate the reaction by heating the sample to 65°C for 10 minutes. Recommended: arrest the activity of DNase I by the addition of EGTA, pH 8.0 to a final concentration of 2 mM prior to heating the sample.
Note: The addition of EDTA is not recommended for samples intended for RT-PCR. Some “stop buffer” formulations contain EGTA instead. Be sure to check prior to use.
4. Optional: If sufficient RNA is available, assess its integrity by electrophoresing an aliquot of the sample under denaturing conditions (Chapter 8: Quality Control for RNA Preparations).
5. The RNA is now ready for immediate use in RT-PCR or other applications. Alternatively, store RNA in convenient aliquots at −80°C.

Appendix D

Removal of RNA from a DNA sample

Laboratory procedures for the isolation of high-molecular-weight genomic DNA or plasmid DNA, of necessity, require the removal of all contaminating RNA without compromising the integrity of the DNA. Most kits used to isolate DNA already incorporate an "RNasing" step as part of their standard procedure. Alternatively, the use of commercial DNase-free RNase products, certified by the manufacturer to harbor no intrinsic DNase activity, is strongly recommended. The industry standard for removing RNA from a DNA prep involves incubating the sample with a cocktail of DNase-free RNases (usually RNase A and RNase T1), which does the job nicely. The final enzyme concentrations, which vary among manufacturers, are typically 1–100 μg/mL RNase A and 0.1–1.0 U/μL RNase T1. The use of only one RNase is not recommended because incomplete RNA digestion will almost certainly occur, rendering bands that will be plainly visible on a gel. Upon addition of the RNase A/RNase T1 mixture to a DNA prep, the sample should be incubated for up to 30 minutes at 37°C. In this laboratory, DNA preps are routinely incubated with RNase at 45°C to minimize any intrinsic DNase activity that may be present in the sample itself. At the conclusion of the incubation period, RNase can be completely removed through the use of a silica spin column or, if absolutely necessary, by phenol:chloroform extraction. Depending on the intended application, it may be possible to ignore the residual RNase altogether.

Crude, homemade preparations of RNases, if not specially treated, are widely known to harbor significant levels of DNase activity as well. In all cases, purging of all intrinsic DNase activity from these RNase stock solutions must be carried out before use and, truthfully, it is hardly worth the effort. However, because RNase preparations can be made DNase-free in the most basic molecular biology laboratories, the following procedure pertains to the preparation of the homemade version only. As noted above, commercial RNase preparations are strongly preferred, widely available, and ready for immediate use.

Protocol: removal of DNase activity from homemade RNase stock solutions

1. Prepare a 10 mg/mL stock solution of RNase A, RNase T1 (1000 U/μL), or both in TE buffer (10 mM Tris-Cl, 1 mM EDTA, pH 7.5–8.0) and 15 mM NaCl. Alternatively, stock solutions can be prepared in sterile H_2O.
2. Heat RNase stock solutions to 90°C for 10–15 minutes. Cool on ice and store at −20°C in suitable aliquots.
   Note: Boiling is not recommended, even though the RNase will maintain considerable activity; at 90°C DNase activity will be eliminated quickly without compromising the RNase activity of the preparation.

Protocol: digestion of RNA

1. Add DNase-free RNase to the sample to a final concentration of 10 μg/mL (RNase A) and/or 1 U/μL RNase T1.
2. Incubate for 10–30 minutes at 37°C. Alternatively, the sample may be incubated at 45°C to inactivate any intrinsic DNase activity.
3. Purify the sample with a silica spin column. Alternatively, terminate RNase digestion by extraction with an equal volume of phenol:chloroform (1:1) or phenol:chloroform:isoamyl alcohol (25:24:1). Concentrate DNA by salt and alcohol precipitation, or use directly for restriction endonuclease digestion, PCR, or other applications.
4. Store purified genomic DNA in convenient aliquots at 4°C. Plasmid DNA can be stored for extended periods at −20°C; small-aliquot storage will preclude multiple freeze-thaw cycles.

Appendix E

Electrophoresis: principles, parameters, and safety

Electrophoresis is the technique by which mixtures of charged macromolecules, such as nucleic acids and proteins, are rapidly resolved in an electric field. Nucleic acids exhibit a net negative charge at any pH used for electrophoresis; in contrast, the net charge of protein species is governed by the pH of the milieu or by the inclusion of an anionic surfactant such as sodium dodecyl sulfate. Electrophoresis produces a qualitative and semiquantitative profile of a sample, which is useful for assessing its integrity, its purity and, to a lesser extent, its relative mass. Thus, electrophoresis is a powerful technique for concentrating like-species of RNA and DNA because the parameters governing electrophoresis favor co-migration of molecules of equivalent size. Electrophoresis is efficient and quick, and in most instances, the investigator can recover specific bands in a variety of ways for subsequent characterization.

The earliest applications of this technique involved macromolecular chromatography in sucrose. Refinements in the methodology extended electrophoresis into starch gels and then later into polyacrylamide, in which the structure of the gel itself influences the migration characteristics of the molecules under investigation. It was quickly realized that gels with large pores were required to efficiently electrophorese high-molecular-weight nucleic acids. At first, a low percentage (2.5%) polyacrylamide gels were utilized, though the instability of low percentage polyacrylamide matrices precluded their widespread use. Agarose, a linear polysaccharide extracted from seaweed, was later added to acrylamide gels to give them enhanced physical strength. Currently, the two standard matrices for the electrophoretic profiling of nucleic acids are agarose and polyacrylamide slabs, and the electrophoresis of RNA usually being conducted under denaturing conditions. The choice of one matrix over the other depends on the size range of the molecules of interest as well as the heterogeneity of the sample. Less frequently, RNA may be characterized by two-dimensional electrophoresis, a contemporary example of which is the posttranscriptional mapping of tRNA (Antoine and Wolff, 2020).

Theoretical considerations

(Adapted, in part and with permission, from Hoefer Pharmacia Biotech Application Guide)

When a molecule is placed in an electric field, the force exerted on it (F) is dependent upon the net charge of the molecule (q) and the strength of the field (E/d) into which it is placed. Therefore,

$$F = E/d \cdot (q) \tag{E.1}$$

in which E is the potential difference between the electrodes and d is the distance between them. A frictional force (F) opposes the migration toward an electrode and is dependent upon the size and shape (radius, r) of the molecule and the viscosity of the medium (η) through which it passes; v is the velocity at which the molecule is moving:

$$F = 6\pi r\eta\nu \tag{E.2}$$

By combining Eqs. (E.1) and (E.2), we see that

$$E/d\cdot(q) = 6\pi r\eta\nu \tag{E.3}$$

and by algebraic rearrangement, one can express the velocity with which a molecule moves in an electric field as

$$\nu = (Eq)/d6\pi r\eta \tag{E.4}$$

Take-home lesson 1: What may be distilled from Eqs (E.1)–(E.4) is that the velocity at which a molecule moves is proportional to the field strength and net charge, while the velocity is inversely proportional to the size of the molecule and the solution viscosity (stiffness of the gel). Succinctly, larger molecules move more slowly than smaller molecules.

Next, according to Ohm's Law, the electrical current (I, amperes or amps) is directly proportional to the voltage (E, volts) and inversely proportional to resistance (R, ohms). Thus,

$$I = E/R \tag{E.5}$$

In addition, power (P, watts), a measure of the amount of heat produced, is the product of voltage (E, volts) and current (I, amps), and can be expressed as

$$P = EI \tag{E.6}$$

Substitution of E with the product $I\,R$ from Eq. (E.5) permits the expression of Eq. (E.6) as

$$P = I^2R \tag{E.7}$$

When performing electrophoresis, either, voltage, current, or power is always held constant. The consequences of an increase in resistance during the run (due to electrolyte depletion or temperature fluctuation) differ in the following ways:

1. When the constant current mode is selected, in which case velocity is directly proportional to the current, heat is generated, and the velocity of the molecule is maintained.
2. When the constant voltage mode is selected, there is a reduction in the velocity of the charged molecules, though no additional heat is generated during the run. This is the most common format for routine lab work.
3. When the constant power mode is selected, there is an unpredictable and nonuniform reduction in the velocity of the molecules, but there is no associated change in heating.

Take-home lesson 2: Temperature regulation is a *primary* concern throughout the electrophoretic process, particularly when running an agarose gel. Overheating will very quickly cause the gel matrix to begin to melt, so maintaining a constant temperature is very important.

The reactions that occur at the electrodes mounted inside the electrophoresis unit, which permit the passage of current from the cathode to the anode, are as follows:

Cathode reactions

$$2e^- + 2H_2O \rightarrow 2OH^- + H_2$$

$$HA + OH^- \leftrightarrows A^- + H_2O$$

Anode reactions

$$H_2O \rightarrow 2H^+ + \tfrac{1}{2}O_2 + 2e^-$$

$$H^+ + A^- \leftrightarrows HA$$

These reactions describe the electrolysis of water in the electrophoresis buffer, resulting in the production of hydrogen at the cathode and molecular oxygen at the anode. For each mole of hydrogen produced, a one-half mole of oxygen is produced. By direct observation of the cathode, one can see about twice as many gas bubbles as are produced at the anode. While this is certainly not the best way to ascertain whether the electrodes have been connected properly, it is an indicator that the current is flowing.

The electrophoresis of RNA can be performed in any of several ways although there are several aspects that these variations all share. Typically, RNA is denatured in a relatively small volume (20 μL or less), applied into the wells of a horizontal agarose gel slab, and electrophoresed for an experimentally determined period. Once the optimal parameters have been determined in a particular laboratory setting, establishing a standard protocol will favor the reproducibility of data. The central considerations are (1) heterogeneity of the sample, (2) the required level of resolution, and (3) whether the procedure is to be analytical or preparative.

Agarose gel electrophoresis

Agarose is a highly purified form of agar, a polysaccharide derived from seaweed. As with all reagents for molecular biology, there are several grades, many of which contain at least trace amounts of anions such as pyruvate and sulfate which can contribute to electroendosmosis (EEO). One should only use very high-quality molecular biology-grade agarose that is certified to be low in contaminants which could contribute to EEO and completely free of nuclease activity. These contaminants may interfere with downstream applications of nucleic acids (especially RNA) upon recovery from the gel. The nontoxic nature of agarose, absent the fluorescent dyes needed to visualize nucleic acids, facilitates ease in handling, gel preparation, and disposal. Low-melting- or low-gelling-temperature agaroses are also available for specialized applications, such as recovering discrete molecular-weight species directly from the gel.

When the experimental design calls for the blotting or transfer of RNA out of the gel after electrophoresis, as in the traditional Northern analysis (or blotting of DNA for Southern analysis), the matrix of choice is agarose. Agarose gels are traditionally poured and electrophoresed as a horizontal slab. The slab configuration is desirable because the gel is fully supported by the casting tray beneath it. Horizontally positioned agarose gels are completely submerged in electrolyte buffer, albeit with the minimum volume necessary to completely cover the gel. This facilitates reasonable dissipation of heat when the gel is run at room temperature and at sufficiently low voltages. Some vertical apparatuses accommodate the weaker nature of agarose gels by the inclusion of a frosted glass plate to which the gels may cling.

The concentration of agarose (the "stiffness" of the gel) profoundly influences the rate of migration, and the mobility of a molecule of any given size differs as the concentration of agarose changes. This is because the pore size of an agarose gel is determined by the concentration of the agarose in the gel, which creates a molecular meshwork not unlike a tangled web. The migration rate is linear over most of the length of the gel, although very large or very small molecules may end up outside its linear range. When this occurs, erroneous size determinations based on comparisons to size standards that fall *within* the linear portion of the gel may occur. In addition, there is a certain amount of smearing of low-molecular-weight species frequently observed at the dye front, that is, the so-called "leading edge" of the gel. This is especially noticeable in gels consisting of less than 1% agarose. The usual working concentration of agarose is between 1.0% and 1.5%, although higher or lower concentrations may be adapted to specific circumstances, for example, the use of 2.5%–3% agarose gels for the analysis of end-point PCR products.

Polyacrylamide gel electrophoresis

Important: Acrylamide in the unpolymerized (monomeric) form is a dangerous, potent neurotoxin that is absorbed readily through the skin. It should be handled with great care when preparing polyacrylamide gels, both prior to and after polymerization, and in full accordance with all safety precautions recommended by the manufacturer. Be sure to seek the assistance of someone who has experience in the proper preparation of gels and handling of acrylamide. A mask rated for working with acrylamide and gloves should be worn when handling the acrylamide in both solid and liquid form. It is the policy of many laboratories to purchase precast polyacrylamide gels and dispense with the toxic precursors altogether. Never assume that the polymerized form no longer poses a potential health hazard. To that end, precast gels should also be handled with gloves.

Polyacrylamide gels offer much greater resolving power than agarose gels. In forming the gel, acrylamide monomers polymerize into linear chains that are covalently joined by a crosslinker. It is the crosslinker that actually holds the structure together, creating a gel matrix that acts as a molecular sieve. The most common crosslinker is *N,N'*-methylene-bis-acrylamide, or *bis* for short. Other crosslinkers whose special properties aid in solubilizing polyacrylamide are also used occasionally. In polyacrylamide gel electrophoresis (PAGE), the molecular sieving of nucleic acid molecules as they pass through the gel is a direct function of the concentration of the acrylamide as well as the ratio of acrylamide to bis in the gel. As oxygen inhibits polymerization, the monomeric mixture must be de-aerated just prior to use, most commonly by purging the mixture with inert gas or by evacuating the mixture with a vacuum. Better yet, buy precast polyacrylamide gels if at all possible. They are well worth the investment, have passed quality control by the manufacturer, and have a lot number that can be used for recordkeeping.

The pore size in a polyacrylamide gel may be predetermined by (1) adjusting the total percentage of acrylamide; or (2) varying the amount of crosslinker added to the acrylamide to induce polymerization. When there is a wide molecular weight in a sample, the researcher may prepare a pore gradient gel in which the pore size is larger at the top of the gel than at the bottom. Thus, the gel becomes more restrictive as the electrophoretic run progresses.

Polyacrylamide gels are typically poured between glass plates, are much thinner than agarose gels, and offer greater resolving power; agarose gels, however, allow the investigator to separate nucleic acid molecules over a much greater size range (Table E.1) owing to the large size of the pore of an agarose gel. Polyacrylamide, in contrast, makes a small-pore gel and can be used to separate smaller- to average-size proteins and oligonucleotides. In general, the stiffer the gel, that is, the more agarose or polyacrylamide makes up the matrix, the greater its resolving power. Agarose gels are used to

TABLE E.1 Useful range of agarose and polyacrylamide gels in the electrophoresis of nucleic acids.[a].

Matrix type	Percentage (w/v)	Useful range (bases)
Agarose[b]	0.6	1000–20,000[c]
	0.8	600–10,000
	1.0	500–9000
	1.2	300–6000
	1.5	200–3000
Acrylamide[d]	3.5	100–1500
	5.0	80–500
	8.0	60–400
	12.0	40–200
	15.0	25–150
	20.0	6–100

[a]*Table derived, in part, from data presented by Sambrook and Russell (2001); Ogden and Adams (1987); and Farrell (1993).*
[b]*Agarose gels consisting of as much as 2.5%–3.0% agarose may be used to observe short products (100–500 bp) of the polymerase chain reaction.*
[c]*Larger DNA molecules can be resolved efficiently by pulse-field electrophoresis.*
[d]*Gel composition = acrylamide: N,N′-methylene-bis-acrylamide, 19:1.*

electrophorese nucleic acid molecules as small as 150 bases to more than 50,000 bases (50 kb), depending on the concentration of the agarose and the precise nature of the applied electric field (constant or pulse-field). Polyacrylamide gels, in contrast, are routinely used to separate nucleic acid molecules as small as 5–10 bases to as many as 500–600 bases. The resolving power of such gels is evident when one considers that polyacrylamide gels used for older-style DNA sequencing resolve nucleic acid molecules that differ in size by one base. Conveniently, premade gels may now be purchased from a number of suppliers and are available for many gel box designs and formats. Polyacrylamide is almost always electrophoresed in a vertical configuration, cast either as a slab or as a cylinder and usually mounted between two buffer chambers containing separate electrodes so that the only electrical connection between the two chambers is through the gel.

As the preparation of polyacrylamide gels is beyond the scope of this volume, the reader is referred to Westermeir (2016), Green and Sambrook (2012), and "A Guide to Polyacrylamide Gel Electrophoresis and Detection" at http://www.bio-rad.com (bulletin #6040) for technical and procedural information.

Molecular size range of sample

Nucleic acid molecules migrate at a rate inversely proportional to the $\log_{10}$ of the number of bases (base-pairs for double-stranded DNA and DNA:RNA hybrids) that make up the molecule. This is true because the larger the molecule, the greater the frictional coefficient, or drag, that these molecules experience as they are pulled through the pores that constitute the geometry of the matrix. Size determinations and all associated calculations are performed automatically through the use of image analysis software. In the absence of this technology, a size calibration curve for electrophoretically separated RNA can be easily constructed by plotting the $\log_{10}$ nucleotides of at least two molecular weight standards against the distance (in centimeters) that these standards migrated from the origin, that is, the well into which they were loaded.

Nucleic acid conformation

The electrophoretic behavior of RNA is also influenced by the presence of various types of formidable secondary structures which nondenatured RNA molecules naturally form due to intramolecular base-pairing. This is analogous to the influence that the shape of DNA molecule, for example, a linear or covalently closed circle, has on its electrophoretic mobility. To encourage identical species of RNA to co-migrate, the electrophoresis of RNA is routinely carried out under denaturing conditions (Chapter 14: Electrophoresis of RNA). These conditions favor disruption of secondary structure such that the separation of molecules is based exclusively on the size (number of bases) of the molecule.

Applied voltage

The efficient dissipation of heat is a key concern in all electrophoresis applications. When working with agarose gels, one must be certain that the applied voltage does not result in the overheating of the gel. Gels run at relatively low voltage and can be cooled as a direct function of the gel box configuration. Gels run at excessive voltages will manifest a noticeable loss of resolution and the gel will actually melt if the system becomes too hot. In some cases, it may be necessary to continually pump cool water around the electrophoresis chamber to maintain adequately low temperatures and generate flat, high-resolution bands, and equipment so designed is outfitted with the required plumbing. Agarose gels should never be electrophoresed above 5 V/cm of the distance between the electrodes; of course, gels may be run at much lower voltages. In general, the slower the gel runs, the better will be the resolution (2–10 kb range). If agarose gel electrophoresis is maintained within these parameters, it should not be necessary to cool the system. In contrast, vertical gels tend to dissipate heat more rapidly than horizontal gels, and gel boxes with an aluminum backing conduct heat more rapidly and efficiently than a glass of comparable thickness.

Ethidium bromide

Caution: Ethidium bromide is a powerful mutagen. Handle and dispose of it using great care and following the directions of the manufacturer. See Appendix F for disposal details.

Unlike DNA, which naturally exists in double-stranded form and the migration of which can be profoundly affected by the presence of ethidium bromide during electrophoresis, RNA structure is not intrinsically influenced by the addition of ethidium bromide to an agarose gel and running buffer. When used, ethidium bromide (10 mg/mL stock) is most often diluted to a final concentration of 0.5 μg/mL. The dye does, however, retard the rate of sample migration by about 15%. It is usually just as easy and far less messy to stain (and destain) the gel at the conclusion of electrophoresis. Of even greater concern is the fact that ethidium bromide can interfere with the transfer efficiency of RNA and DNA during Northern blotting and Southern blotting, respectively. If ethidium bromide is added directly to the gel during preparation, be sure to add it to the electrophoresis buffer, too, because as the RNA runs from − to +, the ethidium bromide runs from + to −, thereby depleting the dye from the gel (Fig. 14.9). Adding ethidium bromide to the electrophoresis buffer is not necessary, however, if the electrophoresis is short, for example, in a quick check for size, mass, or integrity. Problems occur only when the nucleic acid and the ethidium bromide pass each other on the way to opposite electrodes. For a more thorough discussion of the merits of ethidium bromide as well as other staining alternatives, see Chapter 14, Electrophoresis of RNA.

SYBR dye family

Caution: SYBR Green and SYBR Gold should be handled with the same caution as any nucleic acid binding dye, following the directions of the manufacturer. See Appendix F for disposal details.

SYBR Green I, SYBR Green II, SYBR Gold, and SYBR Safe dyes are popular for use in nucleic acid electrophoresis and other applications. The SYBR dyes offer advantages over ethidium bromide, such as greater sensitivity, and lower background; these differences are discussed in greater detail in Chapter 14, Electrophoresis of RNA. It is not advisable to add SYBR Green or SYBR Gold to molten agarose when preparing the gel because of the negative effect that the dye has on the migration pattern of both DNA and RNA. Gels run in the presence of SYBR Green show bands that are fuzzy, highly irregular in shape, and often indistinguishable from other bands in the vicinity. This phenomenon seems to occur at all concentrations of SYBR Green and at all voltages. In this laboratory, samples are electrophoresed first, followed by staining with either SYBR Green or SYBR Gold. When staining is performed after electrophoresis, the results are outstanding.

Base composition and temperature

Unlike polyacrylamide gels, the migration characteristics of nucleic acids in agarose are not influenced by their respective base compositions. This applies both to single-stranded and double-stranded nucleic acid molecules. Further, the electrophoretic behavior of nucleic acids in agarose gels does not change appreciably between 4°C and room temperature, although the overheating of gels is a matter of great concern, particularly at high voltages. Nucleic acid samples are routinely electrophoresed in agarose gels at room temperature.

Field direction

Conventional electrophoresis is accomplished by separating nucleic acid molecules in an electric field of constant direction. The classical size limitations associated with electrophoretic methods in the past, at least in the case of genomic length DNA molecules, have been overcome through the use of pulse-field gel electrophoresis (see Neoh et al., 2019 for a recent review). This technique is performed by forcing DNA molecules to periodically reorient from one electric field direction to another. This approach permits efficient, reproducible separation of DNA molecules ranging in size from 0.1 to 10 megabases, and even entire chromosomes; it is, therefore, a widely respected method for bacterial genotyping. The technology of alternating field direction is not commonly applicable to the electrophoresis of RNA, due to the naturally smaller size of these molecules.

Types of gel boxes

Agarose gels are routinely cast and electrophoresed as horizontal slabs in order to fully support them from beneath. Gel box chambers of diverse designs and quality are universally available, and it is definitely worth investing in a solid, durable gel box that will withstand normal laboratory wear and tear for several years. Because agarose does not exhibit the toxic qualities associated with monomeric acrylamide, it is very easy to cast an agarose gel on the bench top (or in the fume hood if toxic denaturants are added) and perform the electrophoresis.

In practice, the open ends of the gel casting tray must be sealed to retain molten agarose until the matrix has solidified. Some designs incorporate built-in rubber gaskets that seal the casting tray when inserted into the

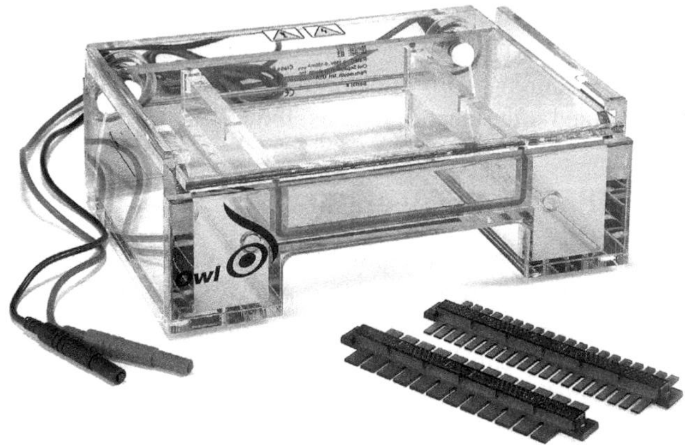

FIGURE E.1 Medium-size gel electrophoresis chamber featuring gasketed tray. Note the positioning of the tray and slots for the combs. After the agarose has solidified, the casting tray is rotated 90 degrees to align the wells formed by the comb to the direction of the electric field. *Courtesy Owl Separation Systems (now part of Thermo Fisher Scientific).*

chamber 90 degrees to the intended direction of electrophoresis (Fig. E.1). This design precludes the taping requirement and frequent leaks associated with older-style trays. When the gel has solidified, the casting tray is rotated into the correct orientation to align it with the direction of the current flow that will occur when the power supply is connected.

Electrophoresis gel boxes (tanks) are often supplied with a variety of combs that can be used to generate different numbers of wells and of different dimensions. The teeth of typical combs generally are 1-, 1.5-, 2-, or 3-mm wide and will produce ten to twenty wells per gel. It may also be desirable to purchase gel boxes with buffer outlets or ports which, if necessary, can accommodate buffer circulation. For example, it is critical to prevent the formation of a pH gradient when performing electrophoresis using phosphate or another buffer that does not at all resist changes in pH, such as when electrophoresing glyoxylated RNA (Chapter 14, Electrophoresis of RNA).

Safety considerations in electrophoresis

Safety is a primary consideration when performing electrophoresis. Never attempt to manipulate the equipment (gel, gel box, leads, power supply) while conducting electrophoresis or while the power supply is plugged into an electrical outlet. Nor should any attempt ever be made to modify the design of either a gel box or power supply; any alterations present a serious risk to both the operator and unsuspecting colleagues in the laboratory. Well-designed systems make it impossible to open or remove the lid while maintaining the flow of electrical current. For example, some designs feature leads mounted in the gel box cover; when the cover slides off, electrical contact with the power supply is broken before the interior of the tank becomes accessible. This permits the safe inspection of the progress of the electrophoresis; the lid must be replaced to restore electrical connections and resume electrophoresis. Moreover, it is wise to disconnect the leads directly from the power supply as well as the gel box when accessing the tank for an inspection of the system. If the power supply is equipped to accommodate two sets of leads from two different gel boxes, be sure to turn the power supply off before connecting the second set of leads (another gel box) to the same power. Be sure to leave the power supply off until the leads have been connected to the power supply and the gel box cover is in place. After all, connections have been made properly, the power supply may be turned on again. Be sure to turn the power supply off before attempting to disconnect any of the leads. Last, *never* take hold of both leads simultaneously; there is a possibility that he who does so may complete the circuit, resulting in electrocution.

Maintenance of electrophoresis equipment

Electrophoresis is a daily ritual in many molecular biology laboratories. While good quality electrophoresis gel boxes and power supplies can represent an investment of thousands of dollars, it is an unfortunate, if not potentially dangerous, practice to ignore the upkeep of the equipment. Basic preventative maintenance of electrophoresis apparatuses is cost-effective and ethically compelling.

The proper maintenance and use of electrophoresis gel boxes and power supplies are important, though often overlooked part of lab safety. It is clearly the responsibility of the operator to inspect wires, connections, and electrodes on a regular schedule to prevent potentially life-threatening injuries. Perhaps the most common problem is the exposure of bare copper wire at the junction of the power cords, electrode leads, and plug ends. This situation poses a severe shock hazard. Replacement cords are usually available at a nominal cost from the manufacturer or distributor, and it is well worth ordering a few extras when the equipment is initially purchased. In addition, all power cords, insulation, electrodes, connection nuts, and gaskets should be scheduled for intra-laboratory inspection on a weekly or biweekly basis. For a suggested regimen for electrophoresis apparatus maintenance, see Landers (1990).

References

Antoine, L., Wolff, P., 2020. Mapping of posttranscriptional tRNA modifications by two-dimensional gel electrophoresis mass spectrometry. In: Arluison, V., Wien, F. (Eds.), RNA Spectroscopy. Methods in Molecular Biology, 2113. Humana, New York, NY. Available from: https://doi.org/10.1007/978-1-0716-0278-2_8.

Farrell Jr, R.E., 1993. RNA Methodologies. A Laboratory Guide for Isolation and Characterization. Academic Press, San Diego, CA.

Green, M.R., Sambrook, J., 2012. Molecular Cloning: A Laboratory Manual, fourth edition. Cold Spring Harbor Laboratory Press, Cold Spring Harbor, NY.

Landers, T., 1990. Electrophoresis apparatus maintenance. Focus 12 (2), 54.

Neoh, H., Tan, X., Sapri, H.F., Tan, T.L., 2019. Pulsed-field gel electrophoresis (PFGE): a review of the "gold standard" for bacteria typing and current alternatives. Infection, Genet. Evol. 74, 103935. Available from: https://doi.org/10.1016/j.meegid.2019.103935.

Ogden, R.C., Adams, D.A., 1987. Electrophoresis in agarose and acrylamide gels. Methods Enzymol. 152, 61–87.

Sambrook, J., Russell, D.W., 2001. Molecular Cloning: A Laboratory Manual, third edition. Cold Spring Harbor Laboratory Press, Cold Spring Harbor, NY.

Westermeir, R., 2016. Electrophoresis in Practice: A Guide to Methods and Applications of DNA and Protein Separation, fifth edition Wiley-VCH Publishers, Weinheim, Germany.

Appendix F

Disposal of ethidium bromide and SYBR Green solutions

Ethidium bromide (EtBr), SYBR Green, and similar dyes used to stain nucleic acids in gels exhibit varying degrees of mutagenicity; as such, they should be handled with care while wearing gloves. Inasmuch as gel-staining dyes are in widespread use, attentiveness to the containment of these dyes is of paramount importance. As intercalating agents and nucleic acid-binding dyes, they should be treated as toxic waste; the proper processing and disposal of these materials also include preventing environmental contamination. Only after proper treatment can ethidium bromide and SYBR Green-tainted wastes be disposed of with minimal apprehension.

The salient issues surrounding the mutagenicity of ethidium bromide (MacGregor and Johnson, 1977) and the treatment of ethidium bromide waste have been described at length (Lunn and Sansone, 1987, 1990; Bensaude, 1988). The early practice of adding bleach to ethidium waste is not recommended because such treatment while reducing the mutagenicity of ethidium bromide, converts the dye into another mutagenic compound (Quillardet and Hofnung, 1988). Instead, one of the following protocols should be adopted as standard lab policy and applied to the handling of SYBR dyes as well. These protocols suggest the handling of working concentrations of ethidium bromide (0.5–1.0 μg/mL) and SYBR Green (1–4 ×); at higher concentrations, these dyes can be diluted down and processed as described in Protocol 1, Protocol 2, Protocol 3, or Protocol 4.

Protocol 1

The Whatman Extractor (10448031) is a one-step filtration method for the removal of ethidium bromide or SYBR Green from gel-staining solutions (Fig. F.1). The device is capable of filtering 10 L of electrophoresis buffer containing EtBr, with greater than 99% removal (data not shown). The filter is an activated carbon matrix that should be disposed of according to

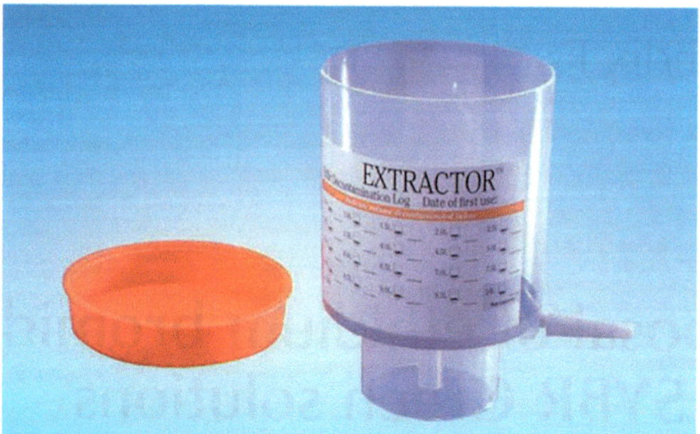

FIGURE F.1 Extractor filtration device for the removal of ethidium bromide and SYBR Green from electrophoresis- and staining buffers. *Photograph originally courtesy Schleicher & Schuell Bioscience.*

departmental guidelines, whereas the filtrate can be discarded safely down the drain. This is a self-contained, cost-effective method for dealing with a common waste product in the molecular biology laboratory. Although the manufacturing of this item has been discontinued, it remains available through various distributors.

Ancillary protocol

Quantitative assay to detect residual EtBr in filtrate (Menozzi et al., 1990).

1. Prepare a solution of 100 μg/mL salmon sperm DNA in running buffer or staining buffer. Prepare dilutions of EtBr in the range of 0–500 ng/mL to prepare a standard curve.
2. Add salmon sperm DNA to 1 mL aliquots from each volume of the filtrate to a final concentration of 100 μg/mL.
3. Read standards and unknowns in a fluorimeter (excitation 526 nm, emission 586 nm). Plot the standard curve and read the EtBr concentration of the unknown from the standard curve. This assay is usually linear ($r = 0.999$) for EtBr concentrations as low as 4 ng/mL.

Protocol 2

An EtBr GreenBag (Ethidium Bromide Disposal Kit; MP Biomedicals, 112350-200) is a device reminiscent of a tea bag that is used to bind and concentrate up to 10 mg of EtBr from contaminated solutions (Fig. F.2). Place one GreenBag into a container with a solution consisting of up to 10 mg EtBr. The container should be gently agitated on an orbital shaker platform up to 50 rpm for 12–16 hours, ensuring that the liquid does not splash out of the container. Alternatively, a magnetic stir bar/plate can be

FIGURE F.2 GreenBag for the removal of EtBr from electrophoresis buffers and staining solutions. Each GreenBag has an EtBr binding capacity of up to 10 µg EtBr. *Photograph courtesy MP Biomedicals.*

used, though care should be taken to avoid more than one revolution per second so as not to damage the GreenBag. At the conclusion of the treatment period, the liquid may be safely discarded in a drain. A used GreenBag should be disposed of according to departmental procedures for the handling of solid hazardous waste.

Protocol 3

Solutions containing working concentrations of ethidium bromide, SYBR Green, and related dyes can be decontaminated by adding about 3 g of Amberlite XAD-16 (Sigma Cat. No. XAD-16) for each 100 mL of solution (Joshua, 1986; Lunn and Sansone, 1987). Amberlite resin is a nonionic, polymeric absorbent and is manufactured by the Dow Chemical Company or one of its affiliates. The solution may be shaken intermittently for 12–16 hours at room temperature, after which it is filtered with Whatman No.1 filter paper or the equivalent. The filter and Amberlite resin should be treated as toxic waste and disposed of according to in-house laboratory policy. The filtrate can then be safely discarded.

Protocol 4

Add 100 mg of activated charcoal (e.g., Sigma Cat. No. C-2889) to each 100 mL of ethidium bromide or SYBR Green at the standard working concentration (Bensaude, 1988). The resultant mixture may be shaken intermittently for 1 hour at room temperature, after which it is filtered with Whatman No.1 filter paper or the equivalent. The filter and charcoal should be treated as toxic waste and disposed of according to in-house lab policy. The filtrate can then be safely discarded.

References

Bensaude, O., 1988. Ethidium bromide and safety-readers suggest alternative solutions. Letter to editor. Trends Genet 4, 89–90.

Joshua, H., 1986. Quantitative absorption of ethidium bromide solutions from aqueous solutions by macroreticular resins. BioTechniques 4 (3), 207–208.

Lunn, G., Sansone, E.B., 1987. Ethidium bromide: destruction and decontamination of solutions. Analyt. Biochem. 162, 453.

Lunn, G., Sansone, E.B., 1990. Degradation of ethidium bromide in alcohols. BioTechniques 8 (4), 372–373.

MacGregor, J.T., Johnson, I.J., 1977. In vitro metabolic activation of ethidium bromide and other phenanthridinium compounds; mutagenic activity in *Salmonella typhimurium*. Mutation Res 48, 103–108.

Menozzi, F.D., Michel, A., Pora, H., Miller, A.O.A., 1990. Absorption method for rapid decontamination of solutions of ethidium bromide and propidium iodide. Chromatographia 29, 167–169.

Quillardet, P., Hofnung, M., 1988. Ethidium bromide and safety-readers suggest alternative solutions. Trends Genet 4, 89.

Appendix G

Deionization of formamide, formaldehyde, and glyoxal

Formamide, glyoxal, and formaldehyde are utilized for a variety of molecular biology applications. The rapid oxidation of these reagents mandates deionization prior to use for optimal nucleic acid denaturation and hybridization efficiency. The most common format by far for deionization is the batch approach, that is, directly mixing deionization resin with the material to be deionized, though a resin-filled chromatography column can be prepared instead. In either case, be sure to use only molecular biology-grade resins that are certified as being nuclease-free.

Deionization of commonly used molecular biology reagents can be achieved by the addition of the reagent to a small mass of mixed bed resin (e.g., Sigma-Aldrich; M-8032). This mixed anionic and cationic exchange resin is strongly acidic and strongly basic, facilitating rapid deionization. Briefly, for each 10 mL of formamide or formaldehyde to be deionized, add 1 g of mixed bed resin. The deionization of glyoxal is similar, though the resin:volume ratio is different, which each mL glyoxal requiring 1.0 g mixed bed resin. Swirl the mixture every 2–3 minutes or maintain it in a sealed tube on an orbital platform at approximately 50–75 rpm. As the capacity of the resin is reached, its color will change from blue to gold. The deionization process is recommended to proceed for 1 hour and may require the use of more than one aliquot of the resin. In this lab the deionization process is often complete within 30 minutes; much depends on the age and condition of the starting material, and the optimal denaturation time will need to be determined empirically. In the case of formamide deionization, the color change occurs rapidly, even before the deionization capacity of the resin is reached. At the conclusion of the deionization period, remove the resin by filtration or by centrifugation. Store deionized glyoxal in tightly sealed, convenient aliquots at −20°C; deionized formamide and formaldehyde should be used shortly after deionization. Once opened, the unused portion of a previously deionized aliquot should be discarded. As needed, the pH of the deionized material can be checked by spotting a small aliquot on pH paper.

Alternatively, formamide, glyoxal, and formaldehyde (Fig. G.1) can be deionized with AG 501-X8 mixed bed resin (Bio-Rad). AG 501-X8 resin is a 1:1 equivalent mixture of AG 1-X8 resin (OH^-) and AG 50W-X8 resin (H^+). It is recommended that 1 g of AG 501-X8 resin be added for each 20 mL of material to be deionized, except for glyoxal which requires 1 g resin/1−2 mL glyoxal. Deionization requires about 1 hour, although the mixture can be left overnight, and a blue-to-gold color change occurs as the capacity of the (anion exchange) resin is reached. When deionizing formamide, the color change occurs rapidly, even before the deionization capacity of the resin is reached. Then the resin can be removed by filtration; formaldehyde and formamide should be used immediately, while deionized glyoxal can be stored in frozen aliquots. As noted above, the pH of the deionized material can be checked by spotting a small aliquot on pH paper.

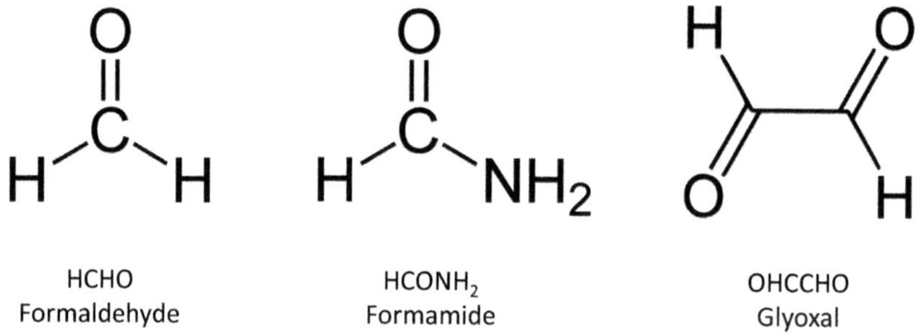

FIGURE G.1 Molecular structure of the common RNA denaturants formaldehyde, formamide, and glyoxal.

Appendix H

Silanizing centrifuge tubes and glassware

Nucleic acids have a natural affinity for glassware and plasticware commonly used for RNA or DNA isolation and ensuing procedures for analysis. Consequently, some of the sample is lost because it "sticks" to the inside wall of an untreated tube. This can be especially problematic when working with low concentrations of nucleic acids because the loss of sample may underrepresent low abundance transcripts, thereby calling into question the results of quantitative assays. Sample loss can be minimized by silanizing (siliconizing) the interior of glass tubes and certain types of plastic tubes prior to use. Although some loss is inevitable, silanizing increases the hydrophobicity of the treated surface and nucleic acid recovery is greatly improved. In addition to following protocol, low adhesion microcentrifuge tubes may be purchased from any of a number of vendors.

Protocol

Sigmacote® (Sigma-Aldrich; Cat. No. SL2) is a very useful reagent for the rapid silanizing of glassware. Sigmacote is a solution of a chlorinated organopolysiloxane in heptane.[1] It forms a microscopically thin, water-repellent covalent film on surfaces to which it is exposed. Sigmacote is reusable if it is kept free of moisture. Store at 4°C.

1. Be sure to follow the manufacturer's instructions for the safe handling of Sigmacote.
2. Fill articles to be coated with Sigmacote. Avoid dipping, as tubes will become very slippery if the outside becomes silanized, and it will be difficult to write on them.
   Note: Tubes should be cleaned and thoroughly dry before coming into contact with Sigmacote.

---

1. Sigma-Aldrich product information sheet.

3. Drain and allow articles to air dry in a chemical fume hood. Drying is usually complete in 5–10 minutes.
4. Prior to use, rinse Sigmacote-treated tubes with nuclease-free water to remove any silanization by-products (i.e., traces of HCl) that may have formed. To strengthen the hydrophobic film, Sigmacote-treated glassware can be dried in an oven up to 100°C for 30 minutes.

Appendix I

Centrifugation as a mainstream tool for the molecular biologist

Centrifugation is an important separation technique based on the fact that objects moving in a circular path are subjected to an outward-directed force. The magnitude of this force, commonly expressed in terms of the earth's gravitational force [relative centrifugal force (RCF) or the "number times *g*"] is a direct function of the radius of rotation and the angular velocity, i.e., $F = \omega^2 r$, where ω = angular velocity (radians) and r = radius (cm) of rotation. Centrifugation is an indispensable procedure for the separation of whole cells, organelles, virus particles, and macromolecules (including RNA and DNA) from a solution, based on their shape, size, and density. As such, centrifuges are in constant use in most laboratories. Improper handling and maintenance, however, can cause permanent injury or worse. It is the responsibility of the operator of the centrifuge, regardless of its size or capacity, to ensure that the centrifuge instrument and the centrifuge rotor are operated within their respective design limits. This includes derating the rotor, which is described in the manual that accompanied its purchase. For excellent overviews of centrifugation and applications, see Mikkelsen and Cortón (2016) and Rickwood and Graham (2015).

Types of centrifuges

Centrifuges vary widely with respect to application, number of samples that can be accommodated, rotor configuration, temperature, maximum speed, and footprint in the laboratory. Table I.1 summarizes the basic types of centrifuges commonly found in a biological research setting. Often, larger and more sophisticated instrumentation is shared among multiple laboratories in a department or may otherwise be available in a core facility. Owing to the numerous applications for which a centrifuge is used, routine cleaning (including decontamination, as needed) and regular maintenance protocols

TABLE I.1 Types of centrifuges.

Instrument type	Attributes and applications
Benchtop clinical centrifuges	• Usually operate below 3000 rpm and at ambient temperature
High-speed centrifuges	• Operate between 20,000 and 25,000 rpm and are usually refrigerated
Benchtop microcentrifuges (microfuges)	• Operate up to 14,000 rpm (12,000 × *g*) • Are sometimes equipped with refrigeration • Are indispensable for processing nucleic acid samples in small volumes
Ultracentrifuges	• Operate up to 500,000 × *g* (75,000 rpm with $r = 8$ cm) • Permit fractionation of subcellular organelles • Permit fractionation of molecules based strictly on density • Capable of operating across a wide temperature range • Rotor operates in a vacuum (air friction severe above 40,000 rpm) • Floor model and benchtop configurations are available

Notes: For general information purposes only. Consult user's guide that accompanied the centrifuge and the rotor for proper use and important safety information.

are necessary to ensure the proper functioning and the safety of all users and those in the vicinity of the instrument.

Rotors

Centrifuge rotors are designed to accommodate the specific requirements for a given type of separation protocol. Parameters for consideration include required RCF, the volume of the samples, sample viscosity, shape of gradient (e.g., linear or step), and choice of differential or density gradient separation. A fixed angle rotor (Fig. I.1) is one in which the sample is maintained at a defined angle during the centrifugation period. In contrast, swinging bucket rotors (Fig. I.2) allow the holder or bucket into which the sample is placed to swing outward to a position 90 degrees relative to the axis of rotation. Vertical rotors (Fig. I.3) maintain sample tubes upright throughout the run, parallel to the axis of rotation. The choice of fixed angle, swinging bucket, or vertical rotor is completely dependent on the intended application. Fixed angle rotors are most compatible with differential centrifugation techniques,

FIGURE I.1 Fixed angle rotor (JA-18). Samples are held in a fixed position during centrifugation. *Courtesy of Beckman Coulter Life Sciences.*

FIGURE I.2 Swinging bucket rotor (SW40Ti). Buckets holding the samples swing outward 90 degrees with respect to the axis of rotation. *Courtesy of Beckman Coulter Life Sciences.*

while efficient gradient-based separations are supported by swinging buckets and vertical rotors.

A useful method by which to correlate performance among rotors is a comparison of their respective k or k' factors, values used to compare efficiency among rotors in a particular application. The k factor estimates the time required to pellet a particle when the sedimentation coefficient of the particle is known; the k' factor is indicative of the time required to move a zone of particles to the bottom of a centrifuge tube. Succinctly, the lower the k factor, the more efficient the rotor. The k factor, k' factor, and the other unique characteristics of each rotor are often the basis for comparative shopping.

FIGURE I.3 Vertical-tube rotor (VTi 65.2). Samples are held in a fixed, vertical position during centrifugation. *Courtesy of Beckman Coulter Life Sciences.*

Applications

(Adapted, in part, from Griffith, 1986).

In the molecular biology laboratory, centrifugation fractionation techniques are utilized at several levels during the isolation of nucleic acids, including the pelleting of whole cells prior to cell lysis, phase separation during nucleic acid extraction, gradient formation during isopycnic separation, and collection of salt and alcohol-precipitated material. Be sure to use only those high-strength tubes that can withstand the required RCF. In the case of consumable plasticware, this information accompanies the product when it is shipped. Corex glass tubes or other high-strength glass centrifuge tubes (e.g., Kimble Chase numbers 45500-15 and 45500-30) are used at RCFs greater than can be tolerated by polypropylene products. High-strength glass tubes must be placed in a suitable rubber adapter to support them during centrifugation. Ultracentrifugation, however, requires tubes that are designed specifically for that application. Table I.2 is intended to be a general guide only; the precise RCF maxima, as well as solvent compatibility for each product, are described by the manufacturer, and those specifications should be adhered to without exception. Note that exposing centrifuge tubes to chemicals with which the tubes are incompatible is a recipe for tube failure, not to mention the possible danger to lab personnel and damage to the centrifuge and rotor. Always check for chemical compatibility before loading or pretreating centrifuge tubes in any way.

Differential centrifugation

Differential centrifugation is the simplest, most straightforward centrifugation technique for sample fractionation. In this method, centrifuge tubes

TABLE I.2 Selected centrifuge tubes for common molecular biology applications.

Type of plastic	Tube size	Typical[a] maximum RCF	Common centrifuge (s)
Polypropylene	1.5 mL microfuge	15,000 × g	Microcentrifuge
	15 mL conical	3,000 × g	Clinical centrifuge
	50 mL conical	5,000 × g	Clinical centrifuge
Polystyrene	15 mL conical	1,500 × g	Clinical centrifuge
	50 mL conical	1,800 × g	Clinical centrifuge
Corex glass[b]	15 mL	12,000 × g	Clinical or high-speed centrifuge
	30 mL	12,000 × g	Clinical or high-speed centrifuge
Oak Ridge tubes[c]	10–50 mL sizes	50,000 × g (tube must be 80% full)	High-speed centrifuge
Polyallomer	Thin wall	100,000–500,000 × g	Ultracentrifuge
	Thick wall	70,000–175,000 × g	Ultracentrifuge

Important: this table is intended to be a general guide only.
[a]*Maximum tolerable RCF varies widely by tube design, tube size, manufacturer, and actual volume within the tube. Be sure to check the specifications and safety recommendations for the specific tubes in use in the lab.*
[b]*High-strength glass centrifuge tube.*
[c]*Oak Ridge tubes are made of traditional polycarbonate or the newer polypropylene copolymer (PPCO).*

containing a homogeneous sample mixture are subjected to brief centrifugation (usually less than 30 minutes). At the conclusion of the centrifugation period, the pellet at the bottom of the tube includes all material sedimented during the run. It should be noted that the pellet is also contaminated with anything that was localized at or close to the bottom of the tube from the onset. The liquid resting on top of the sedimented material is known as the supernatant, which contains nonsedimented material. The supernatant is usually recovered/removed by decanting though in the case of a very firm pellet the supernatant can be removed carefully by aspiration. Should further fractionation be required, centrifugation of the supernatant in a fresh tube for a longer period than the initial centrifugation, or at greater RCFs, will produce a new pellet and supernatant.

Density gradient centrifugation–sedimentation velocity

A more sophisticated preparative technique is density gradient centrifugation, in which sample particles move through a density gradient to achieve separation. The density of the gradient is lowest at the top of the tube and increases toward the bottom of the tube. In one type of density gradient centrifugation, known as sedimentation velocity, or rate zonal centrifugation, a sample is carefully layered on top of a preformed gradient. When subjected to a centrifugal force, particles in the sample begin to move downward through the gradient in discrete zones; the rate of zone movement is governed by the sedimentation rate of individual particles within a zone. This type of separation is centrifugation is usually performed using a relatively shallow gradient at low speeds for a short time (compared to buoyant density centrifugation). Successful sedimentation velocity centrifugation requires the following conditions:

1. The density of particles in the sample must be greater than the density of the gradient at every point throughout the gradient.
2. Centrifugation must be terminated before any of the separated material (specifically the zone of greatest density) reaches the bottom of the tube. This method was once widely used to separate RNA molecules by size in a typical 5%–20% sucrose gradient. The RNA will be distributed throughout the gradient based on size, and the method can be considered a form of RNA enrichment.

Density gradient centrifugation–isopycnic technique

Another type of separation through a density gradient is known as isopycnic, buoyant density, or density equilibrium centrifugation. In this application, particles in a sample move through a gradient only to the point at which the density of the gradient is equal to the density of the particle and at which point particles of identical density accumulate and float, often producing a band-like appearance. Unlike sedimentation velocity centrifugation, extending the period of centrifugation will not result in the continued downward migration of the sample through the gradient. The gradient itself must be very steep, and at its greatest point, it must exceed the density of the particles of interest or else they will sediment at the bottom of the tube. Often, very dense components of a sample do pellet while other components migrate only to their isopycnic positions. Commonly used materials for this type of separation include cesium chloride (CsCl), cesium sulfate (Cs_2SO_4), and cesium trifluoroacetate (CsTFA). Cesium chloride, the salt used for the classical studies of semiconservative replication (Meselson et al., 1957; Meselson and Stahl, 1958), is able to establish gradients with densities ranging up to about 1.8 g/cm^3. Cs_2SO_4 and CsTFA are able to form a gradient twice as steep as that which is achievable with the chloride salt and have been used in place of CsCl for various molecular biology applications.

A gradient for buoyant density centrifugation need not be preformed. As the increasing centrifugal force is applied during ultracentrifugation, the gradient material redistributes, due in part to the intrinsic density of cesium salts. These self-forming gradients traditionally require several hours at ultracentrifugation speeds to produce isopycnic macromolecular banding, though the required centrifugation time can be significantly shortened by using a micro-ultracentrifuge. During the run, components of the sample will either sediment or float to their isopycnic locations in the linear gradient that forms, based only on their respective densities. This methodology has long since been replaced by much more rapid and less cumbersome kit-based methods for nucleic acid purification.

References

Griffith, O.M., 1986. Techniques of Preparative, Zonal, and Continuous Flow Ultracentrifugation. Beckman Instruments, Inc., Palo Alto, CA.

Meselson, M., Stahl, F.W., 1958. The replication of DNA. in E. coli. Proc. Natl. Acad. Sci. 44, 671–682.

Meselson, M., Stahl, F.W., Vinograd, J., 1957. Equilibrium sedimentation of macromolecules in density gradients. Proc. Natl. Acad. Sci. 43, 581–588.

Mikkelsen, S.R., Cortón, E., 2016. Centrifugation, Bioanalytical Chemistry, second ed. John Wiley & Sons, Hoboken, NJ, pp. 325–348.

Rickwood, D., Graham, J., 2015. Centrifugation techniques. eLS 1–7. Available from: https://doi.org/10.1002/9780470015902.a0002704.pub2.

Appendix J

Dot blot analysis

The isolation of high-quality RNA from cells and tissue samples is the first and most critical step in the evaluation of any model for the study of gene expression. The subsequent analysis of RNA can be a time-consuming and expensive road to travel. When evaluating a new model system, cell, tissue type, or experimental regimen for the first time, dot blot analysis is a simple technique that allows the investigator to make "quick and dirty" statements about the biochemistry of the sample, at least from a gene expression perspective. In dot-blotting, RNA samples are applied directly onto the surface of a membrane under vacuum through a multiwell dot blot filtration manifold, after which the samples are immobilized, followed by hybridization against sequence-specific probes. The method of dot blotting is very easy and does not require any particular skills, other than the requirement of maintaining a good RNase-free technique throughout the protocol.

Dot blots permit rapid detection of the relative amount of a particular RNA species in a sample without an electrophoresis component. It is important to understand from the onset that dot blots provide an *estimate* of the amount of target in a sample. The degree of hybridization between the sample and the hybridization probe is assessed by measurement of the signal from each "dot" on the filter, most often recorded on X-ray film. All types of nucleic acid molecules can be dot blotted, including poly$(A)^+$ RNA, total cellular or total cytoplasmic RNA, genomic DNA, cDNA, PCR products, and even oligonucleotides. Dot blotting can also be used for protein analysis when suitable antibodies are available. When dilutions are needed, especially for titration purposes, the design of the dot blotter (Fig. J.1) is ideal. The dot blot apparatus is easily cleaned and, if handled with care, will last for years.

Advantages and disadvantages

Among the most obvious advantages of the dot-blotting technique is the speed with which samples are prepped for nucleic acid hybridization: a denatured sample is applied directly to a membrane. The method is rapid and facilitates

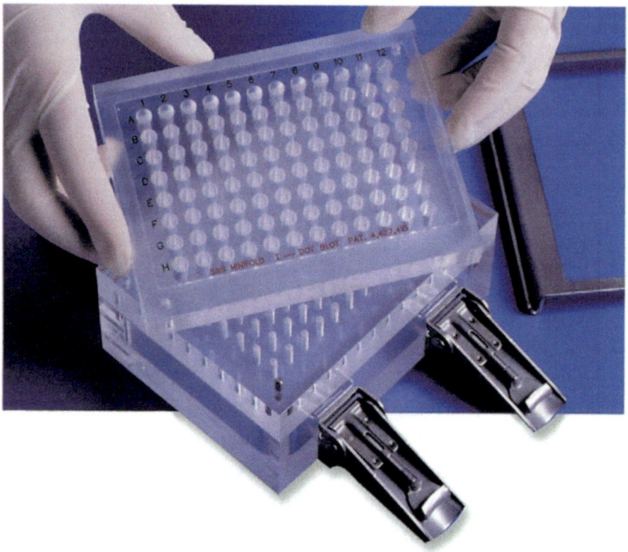

FIGURE J.1 Minifold I dot blot apparatus. Sample dilutions are applied under vacuum directly to the surface of the filter membrane. Depending on the format, the sample is concentrated into either 12.5 mm^2 "dots" or 2 mm^2 "spots." *Courtesy of Schleicher & Schuell Bioscience (now part of Cytiva).*

the screening of many samples simultaneously. For example, the dot blot format is ideal for preparing in-lab panels of RNA from various tissues, different organisms, or entirely different animal or plant species. Moreover, the dot blot format is ideal for analyzing the dilutions of a sample. The ensuing dot blot data may well suggest additional experimental methods or assessing true transcript abundance relationships among the samples through the use of techniques associated with much higher levels of sensitivity.

Dot blotting is occasionally used for confirmation-type screening experiments involving, for example, further characterization of cloned, purportedly differentially expressed genes. The method is particularly well suited when the endpoint is either a yes-or-no answer or when there are moderate differences in gene expression expected among the samples. In the case of a reverse dot blot, a number of unlabeled (cold) probes are dot-blotted onto a filter in advance, and then hybridized with labeled RNA. This allows the investigator to screen for the expression of multiple genes simultaneously. An example of this screening approach is described in detail in Chapter 19 in association with the assay of nuclear transcripts. The formatting and uniform spacing of each sample lend themselves to very easy image analysis (Chapter 15: Photodocumentation and image analysis).

The disadvantages of performing dot blots are closely related to the mechanics of how the assay is performed. Dot-blotting only renders quantitative data because the sample is applied directly to a membrane; as there is no gel, it is not possible to assess the size of the hybridizing transcripts. The strength of the hybridization signal correlates only with the abundance of a transcript. With respect to lower-abundance transcripts, it is useful to consider the area into which samples are concentrated by the manifold. In addition to a dot blot approach, a closely related variant known as a slot blot may increase the sensitivity of the assay, at least marginally. The slot configuration often generates a higher signal-to-noise ratio because of the smaller surface area into which the sample is concentrated (6 mm^2 for slots vs 12.5 mm^2 for dots). Thus, the slot blot configuration may be helpful when working with low mass quantities or in the assay of very low abundance transcripts. Investigators also find the slot blot format useful for following changes in gene expression across multiple time points (Fig. J.2).

In practice, however, it really does not make a great deal of difference in how the sample is configured because of the rather low overall sensitivity of this assay. As with all nucleic acid assays in which a membrane is involved, the quantitativeness of the dot blot assay is compromised because the act of physical immobilization onto the surface of a filter membrane renders some of the molecules incapable of nucleic acid hybridization. Further, dot blot assays do not allow the investigator to discern how many different-sized transcripts in an individual dot have managed to hybridize into the probe. This could be problematic if variable-length transcripts are being produced at the same genetic locus (e.g., alternative transcription start sites). The potential for high background due to even minor cross-hybridization to the abundant rRNA species mandates the inclusion of excellent positive and negative controls.

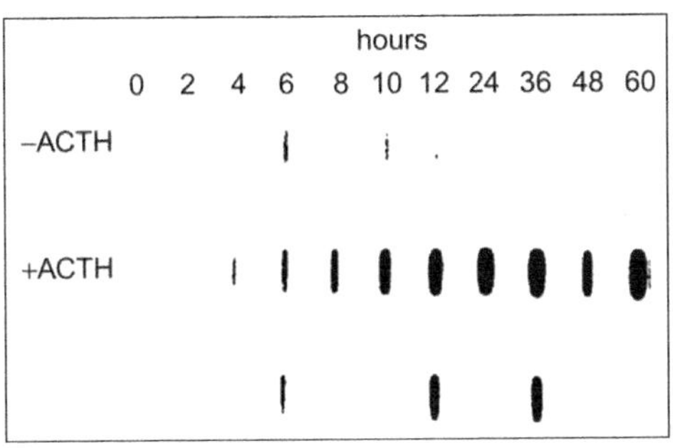

FIGURE J.2 Slot-blot autoradiogram of the temporal pattern of ACTH-mediated increases in P-450$_{SCC}$ mRNA in bovine adrenocortical cells. Cells were harvested and the RNA was extracted at the intervals indicated. Twenty micrograms of RNA were applied per sample to nitrocellulose using the Manifold II slot-blotter and hybridized to a nick-translated pBSCC-2 insert. The three lower samples are duplicates of those above them. *Photograph courtesy of Dr. Maliyakal John. John, M.E., et al. (1984). Identification and characterization of cDNA clones specific for cholesterol side-chain cleavage cytochrome P-450. Proc. Natl. Acad. Sci. 81, 5628–5632.*

Appropriate positive and negative controls

Dot blot analysis yields data which are semi-, semiquantitative at best. To be considered reliable, dot blot analysis must include reliable positive and negative controls, needed to demonstrate hybridization specificity and to gauge nonspecific probe binding to the filter membrane. For example, applying dilutions of rRNA or tRNA to unused wells would show the degree of cross-hybridization of the probe to the nonpoly(A)$^+$ component of the sample, especially the ribosomal RNAs. Further, dot-blotting aliquots of the popular λ-*Hind III* digested- or ΦX174-*HaeIII*-digested bacteriophage genomes (used as DNA molecular weight standards on agarose gels) should yield no observable hybridization to probe molecules if the hybridization and posthybridization washes are conducted with adequate stringency. If molecular weight DNA or RNA standards are not available, *any* DNA that is unrelated to the probe will do. Applying nothing but buffer to at least one well will demonstrate a lack of buffer-associated signal and the effectiveness of blocking the filter during the prehybridization step. Positive controls might include dilutions of cDNA complementary to the transcripts of interest (the probe itself), which, depending on the dilution prepared, may well yield the strongest signal on the filter. Because of the very strong signals usually generated by positive controls, it is best to prepare a substantial dilution of the positive control material before applying it to the filter. Since strong signals from dot blots tend to bleed over and obscure signals from proximal wells, it is also best to avoid positioning experimental samples in those wells immediately adjacent to the positive control. Moreover, good internal dot blot controls are always in order: one should observe equally intense signals from wells into which duplicate samples are applied. When attempting dot blot analysis for the first time or with a new system, it is strongly suggested that dilutions of the positive control target material be made in order to determine the linear range of the assay. For example, it would be useless, quantitatively speaking, if the hybridization signals were too intense to be accurately measured on X-ray film—recall that all films exhibit a rather narrow linear range in which to observe differences among the samples.

Limitations of the data

The greatest limitation of data generated by dot-blotting is the lack of a qualitative aspect because there is no gel and, because of the mechanics of the assay, the compromised sensitivity, compared to more mainstream techniques for the assay of gene expression. The relatively narrow linear range of the dot blot assay is due to (1) the ease with which one can exceed the binding capacity of the filter (spot saturation), (2) fixation of the sample directly onto the membrane, and (3) the intrinsic limitations of X-ray film (which can be partially overcome by phosphorimaging; see Chapter 15:

Photodocumentation and image analysis). Samples that generate strong signals may literally burn out the film so that the true abundance differences among the dots are completely obscured. In order to impart a more quantitative aspect to the assay after autoradiography, the dots themselves can be cut from the filter and dropped directly into a scintillation cocktail, allowing the investigator to associate counts per minute (cpm) with the visual data on the X-ray film. While cut-up filters obviously cannot be used again, the dot blotter instrument is ideally suited for the rapid preparation of replica filters.

As suggested above, there really is no method by which to assess cross-hybridization within a particular spot on the filter, though well-designed negative and positive controls can provide fairly convincing, though indirect, evidence of the resolution of the assay. Image analysis software can only analyze that which the user provides, and it is incumbent upon the investigator to at least try to minimize ambiguity wherever possible. For these reasons, dilutions of each sample are strongly recommended when performing this assay for the first time with uncharacterized material, in order to assess the dynamic range of the sample. Dilutions of the samples are arranged either vertically or horizontally, a geometry that is quite simple to scan and digitize. One should also be attentive to the recommended salt concentration when making dilutions because samples suspended in high-salt buffers are notorious for clogging membranes when applied to the well of the dot-blot manifold.

Lastly, because of the relative ease with which investigators can make cDNA, perform RT-PCR, run gels, and perform nucleic acid blot hybridization, dot blotting may not be the best choice at the onset of the study. Dot blotting is a "graphical" assay: a panel of dots representing hybridized samples is nice to look at but does little more than organize semiquantitative data. If an assay with an array-type format is desired, then blotting and probing the samples is in order. For more quantitative approaches, the investigator should select alternative techniques. Interestingly, some investigators have used dot blot analysis to validate PCR-derived data using a non-PCR assay, at least for fairly abundant transcripts.

Protocol: RNA dot blots

Protocol adapted, in part, from Schleicher & Schuell (1995).

1. *Wear gloves!* Be sure that all reagents are purged of nuclease activity prior to contact with RNA samples.
2. Purify RNA according to any of the protocols presented in this book or elsewhere. RNA should be dissolved in sterile H_2O or TE buffer, pH 7.5 (10 mM Tris-Cl, 1 mM EDTA, pH 7.5). Plan to apply 1–10 µg of RNA per dot/slot in a volume of 100–200 µL per well. Dilute only RNA that will be used for this assay. As always, stock RNA should be stored as concentrated as possible at −80°C for future experimentation.

3. For each 100 μL of diluted RNA, add 60 μL 20× SSC buffer and 40 μL 37% formaldehyde stock solution to the RNA sample. Incubate at 60°C for 15 minutes to denature the sample.

   Note: Formaldehyde (HCHO) must be freshly deionized prior to use to remove formic acid and, in so doing, drive the pH above 4.0. Formamide ($HCONH_2$) must also be deionized prior to use. Formaldehyde and formamide are both toxic and should be handled with care according to the manufacturer's safety data sheets. See Appendix H, Silanizing centrifuge tubes and glassware, for deionization protocol.
4. While the RNA is denaturing, prewet the nylon or other filter in autoclaved water for 5 minutes. Be sure to handle the filter as little as possible and do so only with gloves. Equilibrate the filter in 6× SSC just prior to use and also saturate two sheets of absorbent blotting paper (Whatman #GB003, or the equivalent) in 6× SSC.

   Note: The ionic strength required is dependent on the chemical composition and surface charge of the filter. As always, it is best to follow the manufacturer's instructions for equilibrating any filter prior to any application.
5. Place the saturated sheets of GB003 blotting paper on the filter support plate of the filtration manifold. Place the nylon filter on top of the blotting paper and clamp the sample faceplate into position. When correctly assembled, the O-rings are facing down, in direct contact with the filter, and the wells are facing up.
6. Apply a low vacuum to the dot blot device and ensure that the residual buffer from the prewetting step is being drawn through the membrane. Wash individual wells with 500 μL of 6× SSC.
7. Apply samples into the wells in a volume of 100–400 μL per well. Dilutions, if required, are made in 6× SSC.
8. After all of the sample has been pulled through the nylon filter, wash each well with an additional 300 μL of 6× SSC. When this wash aliquot has been pulled through the membrane, disconnect the vacuum source, and remove the membrane from the manifold.
9. Immobilize the RNA onto the filter membrane according to the instructions provided by the manufacturer, usually by UV crosslinking (see Chapter 14: Electrophoresis of RNA, for immobilization strategies). If the filter will not be probed right away, store it in a cool, dry place, out of direct light.
10. Perform prehybridization blocking, probe hybridization, and posthybridization stringency washes (parameters described in Chapter 17: Nucleic acid probe technology).
11. Perform detection protocols appropriate for the method by which the probe was labeled (discussed in Chapter 17: Nucleic acid probe technology).

Protocol: DNA dot blots

Protocol adapted, in part, from Schleicher & Schuell (1995).

The ability to perform rapid nucleic acid analysis by dot-blotting also extends into the realm of DNA characterization. DNA dot blotting is identical to the methods prescribed for RNA dot blots, the only exception being the method for double-stranded DNA denaturation. The same considerations described above for RNA also apply to the design of DNA dot blots.

1. *Wear gloves!* Be sure that all reagents are purged of nuclease activity prior to contact with DNA samples.
2. Purify DNA according to any of the widely available standard protocols for genomic DNA isolation or cDNA synthesis. A convenient genomic DNA stock concentration for most routine assays is 1.0 mg/mL. DNA should be dissolved in TE buffer, pH 8.0 (10 mM Tris-Cl, 1 mM EDTA, pH 8.0). Plan to apply 5 μg of DNA per dot/slot in a volume of 100–200 μL per well and dilute the DNA sample as needed in TE buffer or sterile H_2O.

   Note: It may be necessary to add even greater amounts of DNA per well to be able to detect single copy sequences.
3. For each 100 μL DNA in TE buffer, add 10 μL 2 N NaOH and incubate at 37°C for 10 minutes. Add 40 μL 20 × SSC buffer and then place on ice if not prepared to apply to the membrane immediately. As an alternative to the addition of 20 × SSC, one may add an equal volume of 2 M NH_4OAc, pH 7.0, and likewise place the sample on ice.

   Note: It is always best to apply denatured samples to the filter as soon as possible so as to preclude possible reannealing of the denatured, single-stranded molecules.
4. While the DNA is denaturing, prewet the nylon or other filter in autoclaved water for 5 minutes. Be sure to handle the filter as little as possible and do so only with gloves. Equilibrate the filter in 6 × SSC just prior to use and then saturate two sheets of absorbent blotting paper (Whatman #GB003, or the equivalent) in 6 × SSC.

   Note: The ionic strength required is dependent on the chemical composition and surface charge of the filter. As always, it is best to follow the manufacturer's instructions for equilibrating any filter prior to any application.
5. Place the saturated sheets of GB003 blotting paper on the filter support plate of the filtration manifold. Place the nylon filter on top of the blotting paper and clamp the sample faceplate into position. When correctly assembled, the O-rings are facing down, in direct contact with the filter, and the wells are facing up.
6. Apply a low vacuum to the dot blot device and ensure that the residual buffer from the prewetting step is being drawn through the membrane. Wash individual wells with 500 μL of 6 × SSC.

7. Apply samples to the wells in a volume of 100–400 μL per well. Dilutions, if required, are made in 6 × SSC.
8. When the entire sample has been pulled through the nylon filter, wash each well with an additional 300 μL of 6 × SSC. After the wash aliquot has been pulled through the membrane, disconnect the vacuum source, and remove the membrane from the manifold.
9. Immobilize the DNA onto the filter membrane according to the instructions provided by the manufacturer, usually by UV crosslinking (see Chapter 14: Electrophoresis of RNA, for immobilization strategies). If the filter will not be probed right away, store it in a cool, dry place, out of direct light.
10. Perform prehybridization blocking, probe hybridization, and posthybridization stringency washes (parameters described in Chapter 17: Nucleic acid probe technology).
11. Perform detection protocols appropriate for the method by which the probe was labeled (discussed in Chapter 17: Nucleic acid probe technology).

References

Schleicher & Schuell. (1995). Nucleic acid dot/slot-blots onto S&S transfer media. Nucleic acid dot/slot-blots onto S&S transfer media, Blotting, Hybridization, and Detection, 6th ed. Schleicher & Schuell, Keene, NH, pp. 20–22.

Further reading

Biasini, A., Marques, A.C., 2020. A protocol for transcriptome-wide inference of RNA metabolic rates in mouse embryonic stem cells. Front. Cell Dev. Biol. 8 (97). Available from: https://doi.org/10.3389/fcell.2020.00097.

Nagarajan, A., Janostiak, R., Wajapeyee, N., 2019. Dot blot analysis for measuring global N^6-methyladenosine modification of RNA. In: Wajapeyee, N., Gupta, R. (Eds.), Epitranscriptomics. Methods in Molecular Biology, vol 1870. Humana Press, New York, NY. Available from: https://doi.org/10.1007/978-1-4939-8808-2_20.

Sabalza, M., Yasmin, R., Barber, C.A., Castro, T., Malamud, D., Kim, B.J., et al., 2018. Detection of Zika virus using reverse-transcription LAMP coupled with reverse dot blot analysis in saliva. PLOS ONE 13 (2), e0192398. Available from: https://doi.org/10.1371/journal.pone.0192398.

Appendix K

Useful stock solutions for the molecular biologist

Whenever possible, investigators should take advantage of widely available molecular biology-grade reagents and stock solutions to support their research. These products are certified as nuclease-free, have a lot number for recordkeeping purposes, and are well worth the investment. When it does become necessary to prepare stock solutions in-house, the following are among the more commonly used formulations.

NOTES

AE Buffer

50 mM sodium acetate,
10 mM EDTA,
Adjust pH to 5.3; autoclave.

Ampicillin (1000×)

100 mg/mL in H_2O,
Filter-sterilize (0.22 μm),
Store aliquots at −20°C.

Denhardt's Solution (50×)

1% BSA (fraction V bovine serum albumin),
1% Ficoll 400,
1% PVP (polyvinylpyrrolidone),
Filter-sterilize (0.22 μm); store aliquots at −20°C.

DNA (heterologous)

10 mg/mL in water,
Boil and shear with 18-gauge needle. Store at −20°C,
Boil aliquots again just prior to use.

NOTES

$EDTA\text{-}Na_2$ (500 mM)

EDTA will not dissolve below pH 7.5,
Adjust to final pH of 8.0 with 10N NaOH or NaOH pellets,
Autoclave; store at room temperature.
$EDTA\text{-}Na_2$ = Ethylenediaminetetraacetic acid; disodium salt.

Ethidium Bromide (EtBr)

10 mg/mL in sterile water,
Store foil wrapped at 4°C.
CAUTION: *EtBr is a powerful carcinogen.*

GTC (also known as "solution D")

4 M guanidine thiocyanate,
5 mM sodium citrate, pH 7.0,
0.5% sarkosyl,
Store in a foil-wrapped bottle at 4°C,
Add 100 mM β-ME to a suitable aliquot just before use.
CAUTION: *Do not autoclave.*
CAUTION: *Extremely chaotropic agent.*

Hybridization Buffer

Many alternative formulations are available, depending upon application.
Consult Chapter 17, Nucleic acid probe technology, for details.

Luria–Bertani (LB) Medium

Per liter: 10 g tryptone,
5 g yeast extract,
10 g NaCl,
Adjust pH to 7.4 with 1N NaOH,
Autoclave; aliquot using aseptic technique.

Loading Buffer (10×) for Agarose Electrophoresis

50% glycerol,
50 mM EDTA, pH 8.0,
0.20% bromophenol blue,
Store 1 mL aliquots at −20°C.
Optional: Include 10 mM Tris, pH 7.5 in the formulation of the 10× loading buffer.
Many alternative formulations are possible.

NOTES

MOPS Buffer (10×)

400 mM MOPS, pH 7.0 (MOPS = 3-[*N*-morpholino]propanesulfonic acid),
100 mM sodium acetate,
10 mM Na_2-EDTA, pH 8.0,
Filter-sterilize only; store at room temperature out of direct sunlight.
[Variation: Prepare 10× MOPS buffer as 200 mM MOPS, 20 mM NaOAc, 10 mM EDTA.]

NaCl (saturated)

Weigh sufficient NaCl to make a 6 M solution in water,
Autoclave as needed.

NaOH (10N)

Add 40 g NaOH pellets to 60 mL H_2O,
Increase volume to 100 mL.C.
CAUTION: *Highly exothermic;* **do not** *autoclave even after cooling. Extremely caustic alkali.*

NP-40 RNA Lysis Buffer

140 mM NaCl,
1.5 mM $MgCl_2$,
10 mM Tris, pH 8.5,
0.5% NP-40 (now sold as IGEPAL CA-630).
Optional: Add RNase inhibitor just prior to use.

PBS

Per liter: 8 g NaCl,
0.2 g KCl,
1.44 g Na_2HPO_4,
0.24 g KH_2PO_4,
Aliquot and autoclave.
PBS = Phosphate buffered saline, Ca^{++}/Mg^{++}*-free.*

Phosphate Buffer (10mM)

Electrophoresis buffer; described as $Na_xH_xPO_4$,
Per liter: 39 mL 1 M NaH_2PO_4,
61 mL 1 M Na_2HPO_4,
Adjust to pH 7.0,
Adjust volume to 1 L,
Aliquot and autoclave.

NOTES

SDS (10%)

Add 10 g SDS to 80 mL sterile water. Mix by gentle swirling.
CAUTION: *Do not breathe SDS powder.*
Increase volume to 100 mL,
Do not autoclave,
Store at room temperature.
SDS = Sodium dodecyl sulfate.

SSC (20×)

3 M NaCl,
0.3 M Na_3-citrate,
Adjust pH to 7.0 with 1N HCl,
Autoclave; store at room temperature.
SSC = Saline sodium citrate.

SSPE (20×)

3 M NaCl,
0.2 M $NaH_2PO_4 \cdot H_2O$,
0.02 M Na_2-EDTA,
Adjust pH to 7.4,
Autoclave; store at room temperature.
SSPE = Saline sodium phosphate-EDTA.

TAE (50×)

Per liter: 242 g Tris base,
100 mL 0.5 M Na_2-EDTA, pH 8.0,
57.1 mL glacial acetic acid [***CAUTION***: *Handle acid with care*],
Autoclave; working concentration is 1× TAE.
TAE = Tris-acetate-EDTA solution.

TBE (20×)

Per liter: 121 g Tris base,
61.7 g sodium borate,
7.44 g Na_2-EDTA,
Autoclave; working concentration is 1× TBE.
TBE = Tris-borate-EDTA solution.

TE Buffer (traditional)

10 mM Tris, pH 7.5,
1 mM EDTA,

NOTES

Autoclave; store at room temperature.
Note: Adjust pH to 8.0 for DNA storage.

TE Buffer (modified)

10 mM Tris, pH 7.5,
0.1 mM EDTA,
Autoclave; store at room temperature.
Note: Adjust pH to 8.0 for DNA storage.

TE-9 Buffer

500 mM Tris, pH 9.0,
20 mM EDTA,
10 mM NaCl,
Autoclave; store at room temperature.

Tetracycline (1000 ×)

12.5 mg/mL in 70% ethanol,
Filter-sterilize (0.22 μm),
Store aliquots at −20°C.

Tris (1M)

Weigh out sufficient Tris to prepare a 1-M stock solution.
Dissolve in approximately two-thirds of the required amount of water.
Adjust to desired pH. Add water to achieve final volume.
Note: Tris base and Tris · Cl will be pH-adjusted differently.
Autoclave as needed.

Tris/SDS buffer

100 mM NaCl,
1 mM EDTA,
10 mM Tris, pH 8.5,
Autoclave; then add 0.5% SDS,
Store at room temperature.

Water (DEPC-treated; RNase-free)

Add DEPC to laboratory grade (deionized/distilled) water to a final concentration of 0.05%–0.1% (v/v).
Mix thoroughly on an orbital shaker for several hours or stir rapidly on a magnetic stirrer for at least 30 minutes in a chemical fume hood.
Autoclave to destroy DEPC.
CAUTION: *DEPC is a carcinogen.*

NOTES

Follow all of the safety instructions recommended by the manufacturer.
Do not add DEPC to any solution which contains an amine group.
See Chapter 2, Creating a ribonuclease-free environment, for details.

YPD Medium

Used as a yeast growth medium,
Per liter: 10 g yeast extract,
20 g peptone,
20 g dextrose,
Aliquot and autoclave.

Appendix L

Genomes and proteomes

Genomes and genomics

Sequencing a single gene produces a lot of information. Sequencing an entire genome produces so much information that it could not possibly be organized, analyzed, archived, or made readily accessible to other investigators without very powerful, high-speed computers, outstanding software for the recovery of desired information from a database, and users who have a good understanding of the science behind the database entries. One such function of bioinformatics, which is a unique blend of information sciences and biology, is aptly known as "data mining." It is worth noting that many of the technologies described in this book as RNA methods also have direct applicability in the realm of genomics. Students who are proficient in these areas can expect exciting careers in diverse areas over the next several years; majoring in biology or life science and minoring in information technology, or vice versa, is a very good start.

Among the goals of genomics is the elucidation of the precise stretches of DNA that constitute discrete genes. One of the primary methods used to accomplish this goal is to examine sequencing data for an open reading frame (ORF) which is a long run of nucleotides that is *not* punctuated by a nonsense (stop) codon. Bona fide ORFs also manifest the requisite start codon (ATG; less frequently GTG), and eventually one of the three stop codons TAA, TAG, or TGA. In eukaryotic cells, an ORF may be punctuated internally with coding and noncoding domains (exons and introns, respectively), which are easily recognized by the presence of the highly conserved 5′ donor (GT) and 3′ acceptor (AG) splice junctions as well as sequence conservation immediately surrounding the splice sites. Recall that these "interrupted genes" are found in eukaryotic cells, and not in prokaryotic cells, the latter of which contains an absolute economy of genetic information. Some viruses, for example, the phage ΦX174, even exhibit two genes occupying the same stretch of DNA, accomplished by utilizing two different overlapping reading frames. This is not an entirely unexpected finding because (1) the small genome of this virus (5375 nts) must encode all of the proteins necessary to direct the synthesis of progeny virions and (2) viruses, in general, have gone through astronomical numbers of replications resulting in genomes that are sleek and maximally efficient.

The days of Sanger (dideoxy) sequencing are quickly coming to end, being rapidly overshadowed by the prevalence and ease of access to next-generation sequencing (NGS) and even third-generation sequencing platforms. The process of NGS is often described as sequencing by synthesis and as being massively parallel in nature, meaning that millions of sequencing reactions are ongoing simultaneously, thereby generating hundreds of gigabases of sequence data in a matter of a few hours. This is fundamentally different from the mechanics of Sanger sequencing. NGS also differs from Sanger sequencing in that no cloning or subcloning is required. Gone are the days of working with phagemids, cosmids, yeast artificial chromosomes and BACs (bacterial artificial chromosomes). The NGS "reads" or newly synthesize strings of nucleotides are then digitally reassembled, as overlapping fragments, by comparison with a known or reference genome (a process known as resequencing) or simply by the assembly of overlapping fragments when no reference genome is available (de novo sequencing). In one variant of NGS, multiplexing is facilitated by unique barcoding each sample, meaning that a specific string of bases (the barcode) is ligated to the end of each member of a library. In this way, recognition of the barcode sequence during data analysis identifies the sample source. These methods are within the realm of structural genomics because of the emphasis on genome structure and organization with a goal of understanding the encoded proteins (Table L.1). In contrast, functional genomics (including RNA-seq) is all about profiling the expression patterns of the genes resident in a cell.

A new area of enormous interest, especially in the microbiology community, is known as pangenomics. Owing to the rise of high-throughput NGS, it has become apparent there is considerable variability in the distribution of genes within a single species, typically identified at the subspecies level. The pangenome, therefore, is the inventory of all genes found in all strains of species. This is not a matter of the members of a species showing the equivalent of allelic variation among a common set of genes; rather, pangenomics represents the fact there are complete subsets of genes that are present in one organism but missing altogether in another isolate of that same species. In the realm of pangenomics, genes are classified as follows:

- Core genes which, as the name implies, are linked to housekeeping functions and tied closely to viability. These genes are found in each individual in a species.
- Variable genes are those found in some members of species but not others. It is easy to speculate that nonshared genes are related in some way to the ecology of the organism.
- Unique genes are found in one member of a species or are uniquely shared among members of a particular strain, only.

TABLE L.1 Examples of genomics classifications.

Nomenclature	Description	Questions addressed
Structural genomics	Gene structure and organization and how it relates to the three-dimensional architecture of the proteins that the genes encode	With the DNA sequence of a gene elucidated, which is the probable folded structure of the cognate protein?
Functional genomics	Profiling gene expression patterns in response to various intracellular or extracellular stimuli, with emphasis on transcription and translation	In what way does the overall genome function when the cell is challenged, signaled, or otherwise prompted?
Comparative genomics	Examining evolutionary relatedness of two of more species based on genomic similarities, differences, and mutations	How do the DNA sequences, including genes, noncoding, and intergenic regions, from one organism align with those from another organism or from a reference genome of the same species?
Epigenomics	Mapping the entirety of epigenetic modifications distributed across the genome	How do these reversible modifications impact gene expression?
Metagenomics	Sequencing all of the DNA recovered from an environmental sample such as soil or water, with emphasis given to the identification of unculturable organisms that are present in the sample	What is the structure and membership of a microbial community?
Pangenomics	Determining gene composition among individual members of the same species, which may be variable among members of the same species	What are the essential core genes, the variable genes, and the unique genes associated with some, but not all, members of a species?

More recent sequencing data have revealed unexpected levels of variability in genome structure and composition, far beyond intragene differences, in *Saccharomyces*, *Arabidopsis*, and *Maize*. There is also growing evidence of pangenomic variability in mammals, including humans (Sherman and Salzberg, 2020; Besenbacher et al., 2015; Li et al., 2010; Alkan et al., 2009). Needless to say, these revelations have given pause to many, to the extent that the entire concept of a reference organism needs to be reconsidered.

Proteomes and proteomics

Proteins are immensely complex when compared to the DNA and RNA sequences that encode them. It has been well established that the 21,000 + protein-coding genes in the human genome encode 1,000,000 + proteins due to variant mRNA transcription start sites, alternative splicing and, especially, post-translational processing of nascent polypeptides. As is the case with transcription, translation is hardly at the static process, and it changes to support the physiologic status of the cell. Mirroring posttranscriptional RNA biogenesis, posttranslational protein processing refers to the extensive covalent modifications and related events commonly associated with natural protein maturation. Posttranslational events are commonly reversible and include glycosylation, acetylation, isoprenylation, phosphorylation, methylation, sulfation, proteolytic processing, complex protein–protein interactions, sumoylation (transient attachment of small ubiquitin-like modifier proteins to other proteins that changes their function), and ubiquitination, to name but a few. Proteomics is important in its own right because amino acid sequence data can be used to predict folding (Table L.2) of the protein, and folding can be used to predict function. Keep in mind that the final shape of a protein molecule is determined by the type and extent of post-translational modifications to which it is subjected. Thus, the dynamic nature of the proteome adds a level of complexity to proteome profiling in eukaryotic cells.

TABLE L.2 Select amino acid folding propensities.

	α-Helix formation	β-Sheet formation
Favorable tendency	Alanine	Isoleucine
	Arginine	Phenylalanine
	Glutamic acid	Tryptophan
	Leucine	Tyrosine
	Methionine	Valine
Unfavorable tendency	Asparagine	Aspartic acid
	Glycine	Asparagine
	Proline	Glycine
		Glutamic acid
		Proline

Notes: Examples of certain amino acids that are associated with specific types of intramolecular protein folding. These classifications are generalities only; the role of amino acids in 2° protein structure varies greatly depending on the local conditions, including side-chain interactions.
Source: Compiled from data published by Pace and Scholtz (1998), Farzad et al. (2008), and Zvelebil and Baum (2008).

Proteomics as a discipline has an important role in discerning gene function owing to the fact that dysregulation is the basis of various pathologies. As such, proteomics has evolved into the global study of patterns of protein expression in cells and tissues, rather than the analysis of one protein at a time. Without a doubt, protein informatics has become a specialized, formidable subdivision of bioinformatics. It is interesting to note that, although only twenty amino acids are used during protein synthesis, more than 150 amino acids exist in nature. It would seem that the constraining variables are the tRNA molecules to which amino acids are added when they are "charged" by an aminoacyl tRNA synthetase. Current speculation suggests that within the anticodon loops, there are physical space and shape limitations through which the translation apparatus is able to distinguish among tRNAs carrying different amino acids. This threshold limitation appears to have blocked the evolution of greater variety among the amino acids during protein synthesis (Saint-Léger et al., 2016).

Challenges also persist in the realm of quantitative proteomics. The dynamic nature of proteins presents many challenges with respect to developing an accurate proteome catalog. Not only are proteins subject to transient posttranslational modifications, the three-dimensional shape, and therefore its functionality, are also subject to change depending on the environment in which the protein finds itself and to which it is exquisitely sensitive. For example, an antibody may be unable to recognize an epitope when its shape changes as a consequence of posttranslational modifications affecting adjacent amino acids (e.g., the H3 K9S10 paradigm in which lysine (K) 9 is subject to acetylation and serine (S) 10 is subject to phosphorylation in the histone H3 nucleosome protein).

The advent of supercomputing came with the expectation that in silico proteomics would become increasingly prevalent as a tool for predicting the folding and function of novel therapeutic peptides; great progress has been made in this area (Bidwell, 2021; Simeon and Chen, 2018). Proteomics constitutes a systematic means for correlating protein function and abundance with genome function in normal cells as well as the disease state. As it facilitates characterization of many proteins simultaneously, it has emerged as a major platform for drug discovery. One goal in the proteomics community is to develop a robust protein equivalent to RNA-seq because transcription, at the end of the day, transcription is only a proxy for translation. This would enable the study of large numbers of protein samples simultaneously and this prospect is of great interest to the biopharma industry because, by 2027, the proteomics market is forecast to be approximately $50 billion.

The cell type (plant, animal, yeast, microbial) from which proteins are to be extracted will determine the form of cellular disruption at the onset. Methods include mechanical lysis, enzymatic lysis, homogenization, detergent permeabilization, and grinding. All of these techniques have been described extensively elsewhere, and instructions regarding the precise method to use accompany the purchase of each protein isolation reagent or kit from the various biotech vendors that specialize in protein isolation and characterization.

Methods for protein identification include older methods, newer methods, and older methods that are now enjoying a resurgence. First and foremost, proteins may be characterized by Western analysis. Succinctly, proteins are electrophoresed through polyacrylamide gels, electroblotted onto nitrocellulose or PVDF membranes, and then incubated with antibodies to determine the size and abundance of a particular protein in a sample (Fig. L.1). This method is most often used to assay one protein at a time.

Two-dimensional gel electrophoresis is one of the premier tools for taking a first look at the proteome. In this method, proteins are first separated by isoelectric point (pI; the pH at which the protein has no net charge), followed by electrophoresis perpendicular to the direction of separation of the first dimension; in the second dimension, proteins are further chromatographed based on size. The two-dimensional gel method permits the simultaneous separation of thousands of proteins, each of which appears as a unique spot upon visualization (Fig. L.2). A two-dimensional gel is a fingerprint of the sample; the size of each spot is proportional to the abundance of the protein. Like RNA, proteins are present in the cell at varying abundance levels; therefore, it is important to ensure that superabundant proteins do not mask the presence of low abundance proteins that may resolve in proximity under two-dimensional electrophoresis conditions. A number of strategies have been devised to fractionate the proteins in a whole-cell lysate in order to maximize the resolution of the proteome (Corthals et al., 2000; Zuo and Speicher, 2000). These methods, which are in widespread use, are based upon an orderly fractionation scheme so as to simplify the analysis of the sample. Specialized software for two-dimensional gel image analysis is

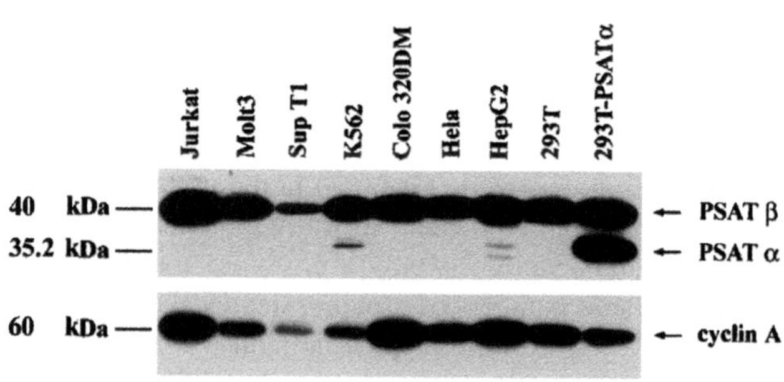

FIGURE L.1 Western analysis involves immunological detection of a protein following electrophoresis and blotting. Expression of phosphoserine aminotransferase (PSAT)β, PSATα, and cyclin A (control) in various human tumor cell lines. The intensity of the band in each lane is indicative of the abundance of the protein in the sample. *Reproduced with permission from Baek, J.Y. et al., 2003. Biochem. J. 373, 191–200. © The Biochemical Society (http://www.biochemj.org).*

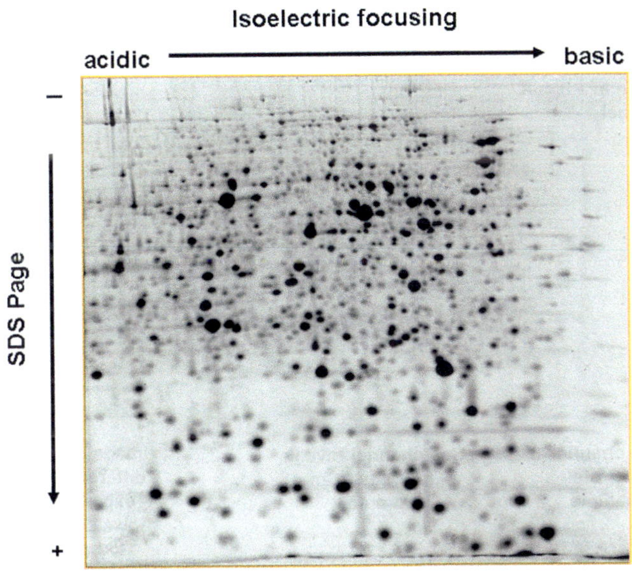

FIGURE L.2 Traditional two-dimensional protein electrophoresis. Proteins are first separated by an isoelectric point and then, perpendicular to the direction of the first separation, separated by mass. The resulting distribution of "spots" is a proteomic fingerprint of the sample. Much like Western analysis, the intensity of each spot correlates with the abundance of the protein at that location. Proteins in individual spots of interest are then cut out of the gel for identification by mass spectrometry. *Gel image courtesy of Sjouke Hoving and Hans Voshol, Novartis Institutes for BioMedical Research. Generated as part of the Fixing Proteomics Campaign.*

readily available for automating the identification and quantification of unique spots on the gel, that is, image informatics, and provides an electronic means for archiving the image.

One popular format for the detection of protein expression involves labeling proteins from a biological source with Cy3 and Cy5 dyes, and then hybridizing these labeled proteins to the arrays which contain the antibodies. This dual labeling scheme can also be applied to the two-dimensional gel format described above (Fig. L.3). For example, proteins in one sample can be labeled with Cy3 and proteins in another sample (control vs experimental) can be labeled with Cy5. The two protein populations are mixed together and run on a single two-dimensional gel. Laser excitation of Cy3 will reveal "green" labeled proteins while laser excitation of Cy5 will reveal the location of "red" labeled proteins. One of the nice things about digital image analysis is the ability to not only overlay images but also to subtract what two or more images have in common. Thus, image subtraction (common locations of red and green fluorescence) will immediately reveal the location(s) of proteins uniquely expressed in each population. This method precludes the requirement for comparing X-ray films from two different gels side-by-side to find differences, as was done extensively in the past.

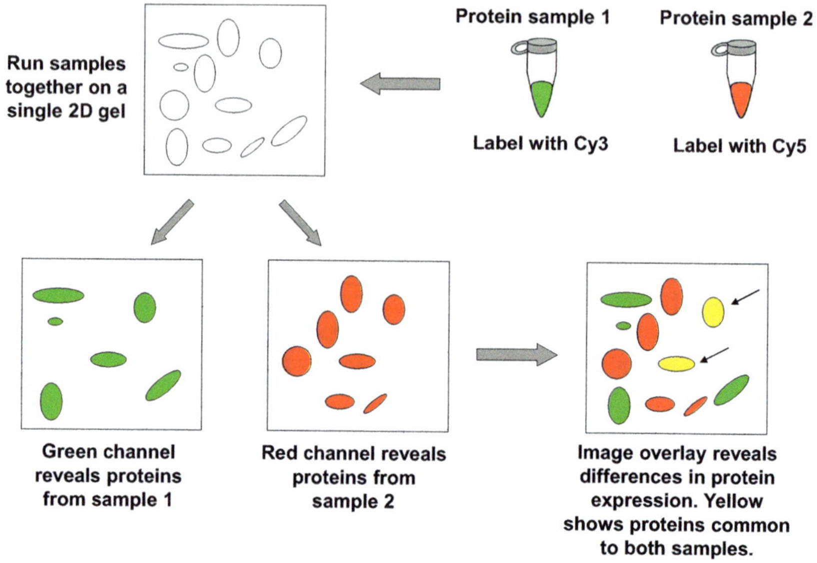

FIGURE L.3 Fluorescence two-dimensional protein analysis. Two protein samples are individually labeled with Cy3 or Cy5 and then electrophoresed together on one gel. Laser excitation reveals the locations of Cy3- and Cy5-labeled proteins. By digitally overlaying the red- and green-channel images, differences in protein expression among the samples are revealed.

Next, unique proteins are recovered from individual spots on the two-dimensional gel. These proteins are digested with proteolytic enzymes, rendering much smaller peptides that are identified by mass spectrometry, a popular form of which is known as matrix-assisted laser desorption ionization-time of flight (MALDI-TOF) analysis. Developed in the 1980s, "MALDI analysis" has become an important tool in the area of proteomics. Because it is the DNA that encodes any and all possible proteins, MALDI-derived amino acid sequence info can be used to match the protein under investigation with the nucleotide sequence associated with a specific genetic locus. However, physically mapping a protein to a gene does not simultaneously identify its function. Elucidation of protein function is often a difficult task because a particular protein may function in a collaborative manner with other proteins or may be a component of much larger complex.

At present there are a number of strategies being employed to discover the function of specific proteins, including the yeast two-hybrid system, various permutations on immunoaffinity capture, nuclear magnetic resonance, crystallization, capillary electrophoresis, high performance liquid chromatography, and protein microarrays that are used to detect protein–protein, protein–lipid, and protein–nucleic acid interactions. Much of the data already generated in this area can be accessed through the protein–protein interaction databases;

for example, the Molecular Interaction Database (http://www.ebi.ac.uk/intact) and the Biomolecular Interaction Network Database (http://www.bioinformatics.ca) are two noteworthy examples. Antibody microarrays are now manufactured, consisting of hundreds of monoclonal antibodies representing a spectrum of biological functions, including "themed" arrays for various infectious diseases, cancer, and apoptosis.

The number of proteomics databases is increasing rapidly as is the amount of information that they contain. The professional organizations related to proteomics are good starting points for novices in the discipline, including UniProt (http://www.uniprot.org), the Human Proteome Organization (http://www.hupo.org), and the British Society for Proteome Research (http://www.bspr.org).

It is worth mentioning in this brief discussion of proteomics that peptides, like DNA and RNA oligonucleotides, can be made-to-order using amino acids with various posttranslational modifications. Applications involving the use of customized peptides include epitope mapping for immunology applications, identification of active peptides for new drugs, identification of enzyme substrates and inhibitors, characterization of protein–protein interactions, characterization of protein–nucleic acid interactions, and development of novel antibodies for vaccine development. This avenue of research is likely to continue to gain momentum as assay resolution and sensitivity continue to improve.

References

Alkan, C., Kidd, J.M., Marques-Bonet, T., Aksay, G., Antonacci, F., Hormozdiari, F., et al., 2009. Personalized copy number and segmental duplication maps using next-generation sequencing. Nat. Genet. 41, 1061–1067.

Besenbacher, S., Liu, S., Izarzugaza, J.M.G., Grove, J., Belling, K., Bork-Jensen, J., et al., 2015. Novel variation and de novo mutation rates in population-wide de novo assembled Danish trios. Nat. Commun. 6, 5969. Available from: https://doi.org/10.1038/ncomms6969.

Bidwell, G., 2021. Novel protein therapeutics created using the elastin-like polypeptide platform. Physiology 36, 367–381. Available from: https://doi.org/10.1152/physiol.00026.2021.

Corthals, G.L., Wasinger, V.C., Hochstrasser, D.F., Sanchez, J.C., 2000. The dynamic range of protein expression: a challenge for proteomic research. Electrophoresis 21, 1104–1115.

Farzad, F., Gharaei, N., Pezeshk, H., Marashi, S.A., 2008. β-Sheet capping: signals that initiate and terminate β-sheet formation. J. Struct. Biol. 161, 101–110.

Li, R., Li, Y., Zheng, H., Luo, R., Zhu, H., Li, Q., et al., 2010. Building the sequence map of the human pan-genome. Nat. Biotechnol. 28, 57–63.

Pace, C.N., Scholtz, J.M., 1998. A helix propensity scale based on experimental studies of peptides and proteins. Biophys. J. 75, 422–427.

Saint-Léger, A., Bello, C., Dans, P.D., Torres, A.G., Novoa, E.M., Camacho, N., et al., 2016. Saturation of recognition elements blocks evolution of new tRNA identities. Sci. Adv. 2 (4), e1501860. Available from: https://doi.org/10.1126/sciadv.1501860.

Sherman, R.M., Salzberg, S.L., 2020. Pan-genomics in the human genome era. Nat. Rev. Genet. 21, 243–254.

Simeon, R., Chen, Z., 2018. In vitro-engineered non-antibody protein therapeutics. Protein Cell 9, 3–14. Available from: https://doi.org/10.1007/s13238-017-0386-6.

Zuo, X., Speicher, D.W., 2000. A method of global analysis of complex proteomes using sample prefractionation by solution isoelectrofocusing prior to two-dimensional electrophoresis. Analyt. Biochem. 284, 266–278.

Zvelebil, M., Baum, J.O., 2008. Obtaining secondary structure from sequence. Understanding Bioinformatics. Garland Science, New York, pp. 413–462.

Further reading

Adams, M.D., Kelley, J.M., Gocayne, J.D., Dubnick, M., Polymeropoulos, M.H., Xiao, H., et al., 1991. Complementary DNA sequencing: expressed sequence tags and human genome project. Science 252, 1651–1656.

Mair, F., Erickson, J.R., Voillet, V., Simoni, Y., Bi, T., Tyznik, A.J., et al., 2020. A targeted multi-omic analysis approach measures protein expression and low-abundance transcripts on the single-cell level. Cell Rep. 31, 107499. Available from: https://doi.org/10.1016/j.celrep.2020.03.063.

Tettelin, H., et al., 2005. Genome analysis of multiple pathogenic isolates of *Streptococcus agalactiae*: implications for the microbial 'pan-genome'. Proc. Natl. Acad. Sci. USA 102, 13950–13955.

Zou, Y., et al., 2019. 1,520 reference genomes from cultivated human gut bacteria enable functional microbiome analyses. Nat. Biotechnol. 37, 179–185.

Appendix M

Common SI prefixes

Factor	Prefix	Symbol
10^{24}	yotta	Y
10^{21}	zetta	Z
10^{18}	exa	E
10^{15}	peta	P
10^{12}	tera	T
10^{9}	giga	G
10^{6}	mega	M
10^{3}	kilo	k
10^{2}	hecto	h
10^{1}	deka (also deca)	da
10^{-1}	deci	d
10^{-2}	centi	c
10^{-3}	milli	m
10^{-6}	micro	μ
10^{-9}	nano	n
10^{-12}	pico	p
10^{-15}	femto	f
10^{-18}	atto	a
10^{-21}	zepto	z
10^{-24}	yocto	y

	Common SI base units		
	Name	**Symbol**	**Imperial/US system conversions**
Amount of a substance	mole	mol	not applicable
Length	meter	m	feet = m × 3.28084
Mass	kilogram	kg	pounds = kg × 2.20462
Temperature	Celsius	°C	°F = (°C × 1.8) + 32
Temperature	Kelvin	°K	°F = (°K−273.15) × 1.8 + 32
Time	sec	s	not applicable
Volume	liter	L	gallons = L × 0.264172

SI Units = *Système International d'Unités* (International System of Units). Of note are the updated definitions of SI units that were internationally adopted on May 20, 2019. See https://doi.org/10.1515/ci-2019-0108 for details.

Appendix N

Common abbreviations

A	adenine
Å	angstrom (10^{-10} meter)
Ab	antibody
Ac	acetyl group
amol	attomole
ARE	adenine/uracil-rich element
BAC	bacterial artificial chromosome
BMP	bitmap
BSA	bovine serum albumin
Bq	becquerel
bp	base pair
C	cytosine
cDNA	complementary DNA
ChIP	chromatin immunoprecipitation
circRNA	circular RNA
COVID-19	Coronavirus disease 2019
cRNA	complementary (antisense) RNA
cpRNA	chloroplast RNA
CRISPR	clustered regular interspaced short palindromic repeats
C_T	threshold cycle
ctRNA	countertranscript RNA
Da	Dalton
DGE	differential gene expression
DNA	deoxyribonucleic acid
DNase	deoxyribonuclease
dNTP	a cocktail of nucleotides containing equimolar amounts of dATP, dCTP, dGTP, dTTP
doi	digital object identifier
dpm	disintegrations per minute
dsDNA	double-stranded DNA
dsRNA	double-stranded RNA
DTT	dithiothreitol
EDTA	ethylenediaminetetraacetic acid
ELISA	enzyme-linked immunosorbent assay

EtBR	ethidium bromide
fg	femtogram (10^{-15} gram)
G	guanine
GIF	graphics interchange format
GLP	good laboratory practices
GMP	good manufacturing practices
gRNA	guide RNA
h	hour
hnRNA	heterogeneous nuclear RNA
HUGO	Human Genome Organization
HUPO	Human Proteome Organization
ISO	International Organization for Standardization
JPEG	joint photographic expert's group
Kb	kilobase
LCR	ligase chain reaction
lincRNA	large intergenic noncoding RNA
LNA	locked nucleic acid
LOD	limit of detection
LOH	loss of heterozygosity
lncRNA	long noncoding RNA
μg	microgram (10^{-6} gram)
μl	microliter (10^{-6} liter)
m	meter
mAb (*also*, moAb)	monoclonal antibody
MCS	multiple cloning site
MDx	molecular diagnostics
min	minutes
miRNA (*also* miR)	microRNA
mol	mole
mRNA	messenger RNA
MSDS	*see SDS*
mtDNA	mitochondrial DNA
mtRNA	mitochondrial RNA
NaOAc	sodium acetate
ncRNA	noncoding RNA
ng	nanogram (10^{-9} gram)
nm	nanometer (10^{-9} meter)
NGS	next generation sequencing
NTP	a cocktail of nucleotides containing equimolar amounts of ATP, CTP, GTP, UTP
nt	nucleotide
OD	optical density
PBS	phosphate-buffered saline
PCI	phenol:chloroform:isoamyl alcohol (often in 25:24:1 ratio)
PCR	polymerase chain reaction
pg	pictogram
pH	$-\log[H^+]$
poly(A)$^+$ RNA	polyadenylated RNA

PNG file	portable network graphic file
PTSG	posttranscriptional gene silencing
Pu	purine (adenine, guanine)
PVSA	polyvinylsulfonic acid
Py	pyrimidine (cytosine, thymine, uracil)
qPCR	quantitative polymerase chain reaction
q.s.	*quantum sufficiat, quantum sufficit, quantum satis*
RCF	relative centrifugal force
RIN	RNA integrity number
rDNA	recombinant DNA
RNA	ribonucleic acid
RNAi	RNA interference
RNase	ribonuclease
RNA-seq	RNA sequencing
RPA	ribonuclease protection assay
rRNA	ribosomal RNA
RT-PCR	reverse transcriptase polymerase chain reaction
S	Svedberg sedimentation unit
SAGE	serial analysis of gene expression
SAM	S-adenosyl-methionine
SBS	Sequencing by synthesis
SDS	safety data sheet (*formerly*, MSDS)
SDS	sodium dodecyl sulfate
shRNA	short hairpin RNA
siRNA	short interfering RNA
SNP	single nucleotide polymorphism
SNR	signal-to-noise ratio
snRNA	small nuclear RNA
SOP	standard operating procedure
SSC buffer	saline sodium citrate buffer
ssDNA	single-stranded DNA
SSPE buffer	saline sodium phosphate EDTA buffer
s, sec	second
T	thymine
Taq	*Thermus aquaticus*
TIFF	tagged image file format
TMB	3,3′,5,5′-tetramethylbenzidine
Tris	tris(hydroxymethyl)aminomethane
tRNA	transfer RNA
TSS	transcription start site
U	uracil
UV	ultraviolet
WGA	whole genome amplification
WT, +	wild type
YAC	yeast artificial chromosome
1°	primary
2°	secondary
3°	tertiary
4°	quaternary

Appendix O

Select suppliers of equipment, reagents, and services

AirClean Systems
2179 East Lyon Station Road
Creedmoor, NC 27522
800.849.0472
http://www.aircleansystems.com

Agilent Technologies
5301 Stevens Creek Blvd
Santa Clara, CA 95051
877.424.4536
http://www.agilent.com

Analytik Jena US
2066W. 11th Street
Upland, CA 91786
http://www.uvp.com
800.452.6788

Aptagen
250 North Main Street
Jacobus, PA 17407
717.278.2436
http://www.aptagen.com

BD Biosciences
2350 Qume Drive
San Jose, CA 95131
877.232.8995
http://www.bdbiosciences.com

Beckman Coulter Life Sciences
5350 Lakeview Parkway S. Drive
Indianapolis, IN 46268
800.742.2345
http://www.beckman.com

Bellco Glass, Inc.
340 Edrudo Road
Vineland, NJ 08360
800.257.7043
http://www.bellcoglass.com

Bertin Corporation
2096 Gaither Road, Suite 230
Rockville, MD 20850
844.723.7846
http://www.bertin-instruments.com

Bio-Rad Laboratories*
1000 Alfred Nobel Drive
Hercules, CA 94547
800.424.6723
http://www.biorad.com

BioSpec Products, Inc.
P.O. Box 788
Bartlesville, OK 74005
800.617.3363
http://www.biospec.com

Biotium, Inc.
46117 Landing Parkway
Fremont, CA 94538
800.304.5357
http://biotium.com

Carestream Health, Inc.
150 Verona Street
Rochester, NY 14608
877.747.4357
http://www.carestreamhealth.com

Center for Professional Innovation & Education, Inc.
7 Great Valley Pkwy
Suite 295
Malvern, PA 19355
610.648.7550
http://www.cfpie.com

* **Bio-Rad**® is a registered trademark of Bio-Rad Laboratories, Inc.

Cole-Parmer
625 East Bunker Court
Vernon Hills, IL 60021
800.323.4340
http://www.coleparmer.com

Corning Life Sciences
836 North Street
Building 300, Suite 3401
Tewksbury, MA 01876
800.492.1110
http://www.corning.com

Cytiva
100 Results Way
Marlborough, MA 01752
800.526.3593
http://www.cytivalifesciences.com

Diversified Biotech, Inc.
65 Commerce Way
Dedham, MA 02026
800.796.9199
http://www.divbio.com

Drummond Scientific Company
500 Park Way
Box 700
Broomall, PA 19008
800.523.7480
drummondsci.com

DWK Life Sciences, LLC
1501 North 10th Street
Millville, NJ 08332
800.225.1437
http://www.dwk.com

Eastman Kodak Co.
343 State Street
Rochester, NY 14650
800.242.2424
http://www.kodak.com

Eppendorf North America
175 Freshwater Blvd.
Enfield, CT 06082
800.645.3050
http://www.eppendorf.com

Exelead
6925 Guion Road
Indianapolis, IN 46268
317.347.2800
http://www.exeleadbiopharma.com

GenHunter Corporation
624 Grassmere Park Drive
Suite 17
Nashville, TN 37211
800.311.8260
http://www.genhunter.com

Hoefer, Inc.
84 October Hill Road
Suite 10
Holliston, MA 01746
800.272.2775
http://www.hoeferinc.com

Illumina, Inc.
5200 Illumina Way
San Diego, CA 92122
858.202.4500
http://www.illumina.com

Interchim
211 Bis avenue J.F. Kennedy
BP 1140
03100 Montluçon Cedex
FRANCE
+ 33 4 70 03 88 55
http://www.interchim.com

JustRight Technology
821 Frederick Road, Suite 201
Catonsville, MD 21228
410.747.2001
http://www.justright.com

Kinematica USA
155 Keyland Court
Bohemia, NY 11716
631.750.6653
http://www.kinematicausa.com

Laronde
A Flagship Pioneering Company
55 Cambridge Parkway, Suite 800E
Cambridge, MA 02142
617-868-1888
http://www.laronde.bio

Lofstrand Labs, Ltd.
7961 Cessna Avenue
Gaithersburg, MD 20879
800.541.0362
http://www.lofstrand.com

Lonza Rockland, Inc.
191 Thomaston Street
Rockland, ME 04841
800.341.1574
bioscience.lonza.com

Lumigen, Inc.
22900W. 8 Mile Road
Southfield, MI 48033
877.586.4436
http://www.lumigen.com

Media Cybernetics, Inc.
1700 Rockville Pike
Suite 240
Rockville, MD 20850
800.263.2088
http://www.mediacy.com

Millipore Sigma
(Sigma Aldrich Chemical Co.)
3050 Spruce Street
St. Louis, MO 63103
800.325.3010
http://www.sigmaaldrich.com

Mirus Bio LLC
5602 Research Park Blvd.
Suite 210
Madison, WI 53719
888.530.0801
http://www.mirusbio.com

Molecular Cloning Laboratories (MCLAB)
320 Harbor Way
South San Francisco, CA 94080
888.625.2288
http://www.mclab.com

Molecular Research Center, Inc.
5645 Montgomery Road
Cincinnati, OH 45212
888.841.0900
http://www.mrcgene.com

MP Biomedicals
9 Goddard
Irvine, CA 92618
800.854.0530
http://www.mpbio.com

New England Biolabs
240 County Road
Ipswich, MA 01938
800.632.5227
http://www.neb.com

Nikon Instruments, Inc.
1300 Walt Whitman Road
Melville, NY 11747
800.526.4566
https://www.microscope.healthcare.nikon.com/

Omni International
935 Cobb Place Blvd.
Kennesaw, GA 30144
800.776.4431
http://www.omni-inc.com

PerkinElmer, Inc.
940 Winter Street
Waltham, MA 02451
800.762.4000
http://www.perkinelmer.com

Phoseon Technology
7425 NE Evergreen Parkway
Hillsboro, OR 97124
503.439.6446
http://www.phoseon.com

Promega Corporation
2800 Woods Hollow Road
Madison, WI 53711
800.356.9526
http://www.promega.com

Qiagen
19300 Germantown Road
Germantown, MD 20874
800.426.8157
http://www.qiagen.com

Roche Life Science
9115 Hague Road
Indianapolis, IN 46250
800.262.1640
http://www.lifescience.roche.com

Takara Bio USA, Inc.
2560 Orchard Parkway
San Jose, CA 95131
800.662.2566
http://www.takarabio.com

10X Genomics
6230 Stoneridge Mall Road
Pleasanton, CA 94588
800.709.1208
http://www.10xgenomics.com

Teledyne Isco
4700 Superior Street
Lincoln, NE 68504
800.228.4373
http://www.teledyneisco.com

Thermo Fisher Scientific
168 Third Avenue
Waltham, MA 02451
800.678.5599
http://www.thermofisher.com

Tricontinent Scientific, Inc.
12740 Earhart Avenue
Auburn, CA 95602
800.937.4738
http://www.tricontinent.com

VWR International
Radnor Corporate Center
Building One, Suite 200
100 Matsonford Road
Radnor, PA 19087
800.932.5000
http://www.vwr.com

Worthington Biochemical Corporation
730 Vassar Avenue
Lakewood, NJ 08701
800.445.9603
http://www.worthington-biochem.com

Zeiss Microscopy, LLC
One North Broadway
White Plains, NY 10601
800.233.2343
http://www.zeiss.com

Zymo Research Corporation
17062 Murphy Avenue
Irvine, CA 92614
888.882.9682
http://www.zymoresearch.com

Address, contact information, and company ownership are correct to the best of my knowledge.

—Author

Appendix P

Trademark citations

AccuTaq™ is a trademark of the Sigma-Aldrich Co. LLC.
Affymetrix® is a registered trademark of Thermo Fisher Scientific Inc.
Amberlite® is a registered trademark of The Dow Chemical Company.
Amersham™ is a trademark of Cytiva.
Amicon® is a registered trademark of Merck KGaA.
Amplifluor® is a registered trademark of Merck KGaA.
Ampligen® is a registered trademark of AIM ImmunoTech, Inc.
AmpliTaq® is a registered trademark of Roche Molecular Systems, Inc.
Aptabody™ is a trademark of Aptagen, LLC.
AptaTaq™ is a trademark of Roche Diagnostics GmbH.
ArrayScript™ is a trademark of Thermo Fisher Scientific, Inc.
BeadBeater™ is a trademark of BioSpec Products, Inc.
Benchkote™ is a trademark of Cytiva.
Bioanalyzer® is a registered trademark of Agilent Technologies.
BioMax® is a registered trademark of Carestream Health, Inc.
Bio-Rad® is a registered trademark of Bio-Rad Laboratories, Inc.
BLAST® is a registered trademark of the National Library of Medicine.
Centricon® is a registered trademark of Merck KGaA.
CDP-Star® is a registered trademark of Thermo Fisher Scientific
Chelex® is a registered trademark of Bio-Rad Laboratories, Inc.
Clontech® is a registered trademark of Takara Bio USA, Inc.
Corning® is a registered trademark of Corning, Inc.
Coomassie® is a registered trademark of Interchim, S.A.
CSPD® is a registered trademark of Applera Corporation.
Cy3® is a registered trademark of Cytiva.
Cy5® is a registered trademark of Cytiva.
Deep Vent® is a registered trademark of New England BioLabs.
Dynabeads® is a registered trademark of Thermo Fisher Scientific, Inc.
eAMV™ is a trademark of Sigma-Aldrich Co. LLC.
EastTides® is a registered trademark of PerkinElmer.
ECL™ is a trademark of Cytiva.
E-Gel® is a registered trademark of Thermo Fisher Scientific, Inc.
ELIMINase® is a registered trademark of Decon Labs, Inc.
EpiScript™ is a trademark of Lucigen.

Eppendorf® is a registered trademark of Eppendorf-Netheler-Hinz GmbH
Expand™ is a trademark of Roche Diagnostics GmbH.
Extractor® is a registered trademark of Whatman, Inc.
FAM™ is a trademark of Thermo Fisher Scientific, Inc.
FastPrep™ is a registered trademark of MP Biomedicals, LLC.
Ficoll® is a registered trademark of Cytiva.
FORMAzol® is a registered trademark of Molecular Research Center, Inc.
Gel-Pro® is a registered trademark of Media Cybernetics, Inc.
GelStar™ is a trademark of DuPont Nutrition USA, Inc.
GenBank® is a registered trademark of the US Department of Health and Human Services.
GeneAmp® is a registered trademark of Applied Biosystems, LLC.
GenElute™ is a trademark of Sigma Aldrich Co. LLC
Hybond® is a registered trademark of Cytiva.
Hyperfilm™ is a trademark of Cytiva.
IGEPAL® is a registered trademark of Rhodia Operations.
JumpStart™ is trademark of the Sigma Aldrich Co. LLC
Kimwipes® is a registered trademark of the Kimberly-Clark Worldwide, Inc.
Lab-Line™ is a trademark of Thermo Fisher Scientific, Inc.
LightCycler® is a registered trademark of Roche Diagnostics GmbH.
LNA® is a registered trademark of Qiagen GmbH.
Lumi-Phos® is a registered trademark of Lumigen, Inc.
Millipore® is a registered trademark of Merck KGaA.
Nalgene® is a registered trademark of Thermo Fisher Scientific
NanoDrop™ is a trademark of Thermo Fisher Scientific.
Nonidet™ is a trademark of Shell Chemical Company.
NuSieve® is a registered trademark of Lonza Sales AG.
Nytran™ is a trademark of Cytvia.
Parafilm® is a registered trademark of Bemis Company, Inc.
pBluescript® is a registered trademark of Agilent Technologies.
pGEM® is a registered trademark of the Promega Corporation.
Plexiglas® is a registered trademark of Röhm GmbH.
Polaroid® is a registered trademark of PLR IP Holdings, LLC.
PolyA Spin™ is a trademark of New England BioLabs, Inc.
Polytron® is a registered trademark of Kinematica AG Company.
Precellys® is a registered trademark of Bertin Technologies.
PrimeScript™ is a trademark of Takara.
Pronase® is a registered trademark of Merck KGaA.
ProtectRNA™ is a trademark of Sigma-Aldrich Biotechnology LP.
ProtoScript® is a registered trademark of New England BioLabs, Inc.
Protran® is a registered trademark of Cytiva.
PYREX® is a registered trademark of Corning, Inc.
Qiagen™ is a trademark of Qiagen.
REDTaq® is a registered trademark of the Sigma-Aldrich Corporation.

Riboprobe® is a registered trademark of Promega Corporation.
RNA*later*® is a registered trademark of Ambion, Inc.
RNAprotect® is a registered trademark of Qiagen GmbH.
RNase AWAY® is a registered trademark of Thermo Fisher Scientific
RNAse-OFF™ is a trademark of Takara Bio USA.
RNaseZap™ is a trademark of Thermo Fisher Scientific
RNasin® is a registered trademark of Promega Corporation.
Rubbermaid® is a registered trademark of Rubbermaid Commercial Products.
Saran™ Wrap is a trademark of S. C. Johnson & Son.
Scorpions® is a registered trademark of Qiagen Manchester, UK.
SeaKem® is a registered trademark of DuPont Nutrition USA, Inc.
Sephadex® is a registered trademark of Cytiva.
Sepharose® is a registered trademark of Cytiva.
SpeedVac® is a registered trademark of Thermo Fisher Scientific, Inc.
Sigmacote® is a registered trademark of Sigma-Aldrich Company, LLC.
SUPERase•In™ is a trademark of Thermo Fisher Scientific
SuperScript® is a registered trademark of Thermo Fisher Scientific Inc.
SYBR® is registered trademark of Thermo Fisher Scientific
SYPRO® is registered trademark of Thermo Fisher Scientific
TAMRA™ is a trademark of Thermo Fisher Scientific
TaqMan® is a registered trademark of Roche Molecular Systems
TaqStart™ is a trademark of Takara Bio USA.
Teflon® is a registered trademark of the Chemours Company FC, LLC
TET™ is a trademark of Thermo Fisher Scientific
TEX™ is a trademark of Thermo Fisher Scientific
Tissue-Tek® is a registered trademark of Sakura-Finetek.
TriPure™ is a trademark of Roche.
TRI Reagent® is a registered trademark of Molecular Research Center, Inc.
TriSure™ is a trademark of Bioline Reagents Limited.
Triton™ is a trademark of Dow Chemical Company or an affiliated company of Dow.
Trizma® is a registered trademark of Sigma-Aldrich Co. LLC.
TRIzol® is a registered trademark of Molecular Research Center, Inc.
Tween® is a registered trademark of Croda International, PLC.
Typhoon™ is a trademark of Cytiva.
Ultrafree® is a registered trademark of Merck KGaA.
Vent® is a registered trademark of New England BioLabs.
VIC® is a registered trademark of Thermo Fisher Scientific, Inc.
Whatman® is a registered trademark of Cytiva.
YouTube™ is a trademark of Google, Inc.
Ziploc® is a registered trademark of S.C. Johnson & Son, Inc.

All product names, logos, and brands are the property of their respective owners. These trademarks and registered trademarks are accurate to the best of my knowledge.

-Author

Glossary

Abundance The prevalence of a particular RNA, or class of RNA molecules, in the cell. The relative amount of a particular messenger RNA (mRNA) species, compared to other samples, is frequently expressed in terms of relative abundance.

Accession Number A numeric identifier of a nucleic acid sequence (genomic DNA, cDNA, or RNA) or an amino acid sequence (protein). The accession number acts as an electronic card catalogue of sorts, facilitating expedient recovery from a database.

Acetylation Introduction of an acetyl group ($-COCH_3$) to a molecule.

Adduct A molecule to which a chemical group or other molecule has been covalently added, that is, a chemical addition product.

Adenine A purine that ordinarily base-pairs with thymine in DNA or uracil in RNA.

Adenosine A nucleoside containing the base adenine.

Affinity chromatography A physical separation of molecules based on a biological activity or a biological structure. Examples include base-pairing between the poly(A) tail of some eukaryotic mRNAs with oligo(dT), and the capture of oligohistidine fusion proteins on nickel resin columns.

Algorithm A series of mathematical operations that are applied in a defined sequenced; often used for data mining purposes.

Alkylation Modification of a molecule by addition of an alkyl group (carbon and hydrogen atoms only, arranged in a linear chain, with the generic formula C_nH_{2n+1}). Alkylation of nucleic acids is often the causative event in the mutation of a gene.

Allele One of two or more alternative forms of a gene.

***Alu* repeat sequence** One of several hundred thousand sequences, about 300 base pairs in length, found dispersed throughout the primate genome. These repetitive sequences are so named because of the characteristic *Alu* restriction enzyme sites (AGCT) that they contain.

α-Amanitin A bicyclic octapeptide toxin derived from the very poisonous mushroom *Amanita phalloides*. This fungal product differentially inhibits eukaryotic RNA polymerases, and RNA polymerase II is especially sensitive.

Amber codon The translation termination codon UAG.

Amino acids The building blocks of proteins.

Ampicillin A beta-lactam antibiotic that is used to select resistant bacteria in molecular cloning. Ampicillin interferes with bacterial cell wall synthesis by inhibition of the enzyme transpeptidase.

AmpliTaq Recombinant form of the naturally occurring thermostable *Taq* DNA polymerase from the organism *Thermus aquaticus*.

Amphoteric A compound which can function as either a base or an acid. Proteins, for example, are amphoteric compounds because their net charge is pH dependent. This influences their electrophoretic behavior.

Angstrom (Å) 10^{-10} meters

Anion An atom or molecule that has a negative charge.

Annealing The base-pairing of complementary polynucleotides to form a double-stranded molecule.

Antigen Any molecule or portion thereof that can induce the synthesis of an antibody or otherwise provoke an immune response.

Antiparallel The manner in which two complementary polynucleotides base-pair; the 5′ and 3′ ends of each molecule are reversed in relation to each other, so that the 5′ end of one strand is aligned with the 3′ end of the other strand. Antiparallel base pairing accompanies the formation of double-stranded DNA (DNA:DNA), double-stranded RNA (RNA:RNA), and DNA-RNA hybrids (DNA:RNA).

Antisense RNA (also, − RNA or minus RNA) Any RNA molecule that is complementary to the naturally occurring sequence of mRNA. Antisense RNA is capable of hybridization with mRNA on a filter, in situ, or in solution, thereby forming extremely stable double-stranded RNA molecules.

AOI (area of interest) A contiguous group of pixels within an image, arranged in any polygonal shape, which is used to isolate this pixel subset from the rest of the image.

α Particle One of the three forms of natural radiation consisting of two protons and two neutrons.

Aptamer An RNA or DNA oligonucleotide that, depending on base composition, is able to bind to a specific ligand (target molecule) by folding into a characteristic three-dimensional shape.

Argonaute A highly conserved family of proteins that are essential components of gene silencing via RNA interference (RNAi). Among the multiple functions of the argonaute proteins is a role in the formation of the RNA-induced silencing complex (RISC) which can induce mRNA degradation or translation repression.

Array *See* microarray and macroarray.

Attenuation Method of regulation of transcription termination in certain prokaryotic operons. The literal meaning of attenuation is a loss in strength, signal, or intensity.

Autoradiograph (also autoradiogram) A photographic record of the spatial distribution of radiation in an object or specimen. It is made by placing the object very close to a photographic film or emulsion.

Autoradiography The process by which the decay of radiolabeled molecules is recorded on X-ray film or other photographic emulsion.

Autosomes Non-sex chromosomes. In humans, chromosome pairs 1–22 are known as the autosomes.

Bacteriophage A virus that infects and is propagated in a bacterial host. In the biotechnology world, bacteriophages have been used extensively for cloning purposes, most commonly for the construction of DNA libraries. Among the best characterized and widely exploited are genetically modified derivatives of the wild type λ bacteriophage.

Band The visible accumulation of molecules that have an identical molecular weight, as observed following electrophoresis. The greater the intensity of the band, the greater is the number of molecules that have co-located to that particular position.

Base pairing The formation of hydrogen bonds between the nitrogenous bases of two nucleic acid molecules.

Becquerel (Bq) A standard unit of radiodecay. One Bq = one disintegration per second (dps).

β Particle An elementary particle emitted from a nucleus during radioactive decay, and which has a single electrical charge; it is one of the three forms of natural radiation. A negatively charged β particle is identical to an electron, and β particle emission is a form of ionizing radiation. A positively charged β particle is known as positron.

Betaine PCR chemical adjunct that reduces secondary structure in GC-rich regions, thereby improving amplification of DNA. The standard stock solution is 5 M and is used at a working concentration of 0.8–1.6 M. Betaine is also known as *N,N,N*-trimethylglycine.

Biobank A repository where biospecimens (often millions) are stored for research purposes. A major goal is to eventually understand the underlying basis of disease (biochemical, risk factors, lifestyle) and generate a personalized medicine profile for each person based on their genetic makeup.

Bioinformatics The marriage of computers, information technology, and biotechnology. Bioinformaticists are multidisciplinary practitioners who are able to retrieve, sort, and archive enormous amounts of nucleic acid and protein sequence information.

Biological replicate A series of biological samples, from different sources, that are assayed in parallel in order to detect any statistically significant variability (i.e., noise versus true differences) among the ensuing results. Biological replicates are necessary ensure the veracity of experimental outcomes.

Biotechnology (*see also* genetic engineering) The deliberate use of microorganisms, plant-, and animal cells to produce useful materials, such as food, medicine, and other compounds. Most practioners consider biotechnology to be just as much an art as much as it is a science.

Biotin A small water-soluble vitamin, also known as vitamin B7 and vitamin H, which can be used to label molecules, including nucleic acids for a variety of purposes, including non-isotopic hybridization detection. Alternatively, biotinylated molecules may be isolated by affinity capture using streptavidin.

Bit The smallest unit of information recognized by a computer. A bit, short for binary digit, has a value of 0 or 1. In an image, the number of bits in a pixel is directly related to its available color palette.

Bit depth The number of bits used to represent one pixel value, such as intensity of color. Also referred to as *pixel depth* and *bits-per-pixel (BPP)*.

Bitmap A two-dimensional array of image pixels used to represent or store an image. Each cell in the array contains a value that describes a portion of an image in terms of its color.

BLAST (Basic Local Alignment Search Tool) A well-known and widely utilized bioinformatics resource that allows users to search for specific nucleotide and protein sequences in GenBank via the internet. BLAST searches also reveal the statistical significance of matches, if any, between the input data and any sequences in the database.

BMP (Bitmap) A digital image file format.

BPP (Bits per pixel) Describes the depth of an image, that is, the number of bits that describe a pixel in terms of its color. Bi-level images are only 1 BPP while true color images are 24 BPP.

Brightness The amount of white in an image. The brighter the image, the more white it contains. As brightness is increased, each color in the image is shifted more toward white.

Buoyant density A measure of the ability of a substance to float in a standard fluid. For example, differences in the buoyant densities of RNA and DNA allow them to be separated by ultracentrifugation in a gradient of cesium chloride (CsCl) or cesium trifluoroacetate (CsTFA).

CAAT box A conserved regulatory sequence, CAAT or CCAAT, located about 75 base pairs upstream (−75) from the transcription start site of eukaryotic genes. The CAAT box is a commonly observed part of the eukaryotic promoter.

Canonical base pairs The commonly observed hydrogen bonding between adenine and thymine, between cytosine and guanine, and between adenine and uracil. *See* Chargaff's Rule.

Cap The structure found at the 5′ end of eukaryotic mRNA molecules, including most eukaryotic viral mRNAs, consisting of an inverted (5′−5′) linkage between the first two nucleotides. The 5′-cap is added enzymatically to the 5′ end of hnRNA immediately after transcription while the transcript is still in the nucleus and exhibits a characteristic level of methylation. The 5′ cap not only stabilizes the 5′ end of the mRNA by resisting exonucleolytic degradation, but it also has a role in enhancing translation as well. *See* Chapter 1 for details.

Casting tray The portion of the gel box into which the molten agarose is poured. Gel casting trays are either stationary or removable, depending on the model of the electrophoresis unit. The ends of the tray must be sealed, either by taping or through the use of rubber gaskets, until the gel has solidified completely.

Cation An atom or molecule that has a positive charge.

CCD (charge coupled device) Equipment designed for digital image capture.

cDNA (complementary DNA) DNA synthesized in vitro by reverse transcriptase from an RNA template. cDNA may be synthesized as single-stranded or double-stranded molecules, as needed. cDNA represents a permanent biochemical record of the cellular biochemistry and also provides a means by which that record can be propagated indefinitely.

Central dogma The long-held belief that the flow of genetic information in all cells begins with the transcription of DNA (the blueprint) into an mRNA intermediary (the message), followed by synthesis of the cognate protein (the manifestation of gene expression). The universality of the central dogma was disproven upon discovery of retroviral reverse transcriptase which seemingly violates the central dogma.

Cesium chloride (CsCl) A dense salt used for isopycnic separation of nucleic acids. CsCl is used to partition macromolecules based on differences in buoyant density via ultracentrifugation. The most common applications include purifying plasmids from bacterial cell lysates and pelleting eukaryotic RNA with concomitant banding of cellular DNA at a lower density location above the pellet.

Cesium trifluoroacetate (CsTFA) A dense salt used for isopycnic separation of nucleic acids via ultracentrifugation. CsTFA is capable of forming steeper gradients than CsCl. Consequently, RNA can either be banded or pelleted depending on the preference of the investigator.

Chaotropic Biologically disruptive, that is, causing cellular chaos. Chaotropic lysis buffers disrupt cellular and subcellular membranes and destroy enzymatic activity on contact.

Chargaff's Rule The predication that under ordinary circumstances, the formation of canonical base pairs (A::T, A::U, C:::G) is strongly favored over the formation of non-canonical base pairs.

Chemiluminescence A non-isotopic hybridization detection technique. Chemiluminescence is the production of visible light by an enzyme-mediated chemical reaction.

Chromatid In metaphase chromosomes, one of two plainly visible distinct subunits. Sister chromatids are held together by a common centromere until they separate during anaphase.

Chromatin The complex of genomic DNA and protein found in the nucleus of a cell during interphase. At the onset of cell division, the chromatin condenses into recognizable chromosomes during prophase.

Chromatin immunoprecipitation (ChIP) Method by which a protein and the RNA or DNA bound to it is isolated using an antibody. The associated nucleic acid is then sequenced to determine in what manner the protein interacts with it and the precise nucleotides involved.

circRNA (circular RNA) RNA molecules that exist in circular form as a consequence of the back-splicing of linear transcript, often formed during the process of intron excision from a precursor molecule. Broadly abundant in mammalian cells, circRNA expression is tissue specific and involves in the regulation of numerous biological pathways.

Cistron A discrete region or specific sequence of DNA. The term cistron is an older name for what is now commonly called a gene.

Clade A group of genes or organisms that appear to share a common ancestor as suggested by phylogenetic analysis.

Clone (*noun*) A collection of genetically equivalent cells or molecules.

Clone (*verb*) A series of manipulations designed to isolate and propagate a specific nucleic acid sequence, or cell, for characterization, storage, or further amplification.

Clone bank Older terminology for what is currently known as a DNA library (genomic or cDNA).

Coding strand In double-stranded DNA, the strand that, when transcribed, has the same sequence as the resulting RNA (except for the substitution of uracil for thymine). The coding strand alone is reported when publishing a gene sequence and is always written $5' \rightarrow 3'$, from left to right.

Codon A group of three nucleotides found in the coding portion of mRNA. The precise order of nucleotides specifies the identity and placement of an amino acid into a nascent peptide during translation. Punctuation (start and stop) is also codon-specified.

Colinearity The manner in which the linear order of nucleotides, read $5' \rightarrow 3'$, dictates the order of amino acids in the resulting protein, read from the amino terminus to the carboxy terminus. Succinctly, the order of codons specifies the order of amino acids.

Competitive PCR A pre-real-time PCR method for quantification of transcript abundance. The addition of an exogenous competitor DNA template to the same reaction tube as an experimental sample (often with a nearly identical sequence) creates competition for primers, dNTPS, and the "attention" of the polymerase. By varying the amount of the competitor DNA, a dilution will be identified in which the concentration of target and competitor are equivalent, resulting in equimolar amplification and from which the amount of starting template in the experimental sample can be inferred.

Condensate (RNA condensate) A membraneless region in cell or within an organelle in which, compared to the surrounding area, there is accumulation of one or more RNA species and their attendant binding proteins.

Consensus sequence A composite nucleotide or amino acid sequence that has been deduced from the examination and alignment of multiple molecules that share this sequence similarity. Succinctly, a consensus sequence is essentially the "average" of several similar sequences. In cases where individual nucleotides or amino acids differ from the majority of the sequences being considered it is customary to indicate the ambiguity by either assigning a percentage (A_{90}, meaning that adenosine occurs 90% of the time in a particular position) or by listing the observed permutations ($ATCG^{C}/_{A}TGTAT$).

Conserved sequence A DNA sequence which exhibits little or no variation from one organism to the next.

Constitutive gene expression Continuous interaction of RNA polymerase with the promoters of specific genes not subjected to additional levels of regulation. Such genes are often expressed continually at low or basal levels and are sometimes referred to as genes with housekeeping functions. In contrast, facultative genes are transcribed only when the encoded protein is needed.

Contig A contiguous stretch of DNA that has been assembled in silico from a series of smaller overlapping genomic or cDNA segments.

Contrast The sharpness of an image. The higher the contrast in an image, the larger is the difference between the darkest and lightest areas in an image.

Coomassie Brilliant Blue Commonly used compound to stain proteins in polyacrylamide gels.

Cosmid A molecular biology cloning vector in which the λ bacteriophage *cos* sequences are joined to a plasmid. Thus, after ligation of an insert (foreign) DNA, the cosmid is packaged into λ bacteriophage for infection into a bacterial host, after which the cosmid begins acting like a plasmid. Cosmids are able to accept much larger inserts than traditional plasmids.

Countertranscript RNA (ctRNA) Another name for antisense RNA that is encoded in the same genomic region, but in the opposite orientation, as its functional sense RNA (mRNA) counterpart. Consequently, sense mRNA and ctRNA are able to form areas of dsRNA and, in so doing, suppress transcription initiation elsewhere. ctRNA is often discussed in the context of prokaryotic gene expression.

C_q (quantification cycle) The cycle in real-time PCR when accumulating product crosses a predetermined threshold level and becomes visible to the detection system. C_q is a more contemporary term replacing C_t as the basic unit of measurement term in real-time PCR.

CRISPR A novel RNA editing system, properly known as Clustered Regular Interspaced Short Palindromic Repeats. This system functions in unison with an endonuclease known as CRISPR ASsociated Protein 9 and is simply known as the CRISPR Cas9 system. This is an efficient, versatile platform for gene knock-in, gene knockout, and site-directed base changes and is in widespread use.

cRNA (complementary RNA) Usually another name for antisense RNA, that is, a transcript which is complementary to mRNA. Occasionally, however, cRNA can also mean a transcript with a similar sequence to mRNA and which can act as a competitor in a cDNA synthesis and amplification reaction.

C_t (Threshold cycle) The cycle in real-time PCR when accumulating product crosses a predetermined threshold level and becomes visible to the detection system. C_t is the basic unit of measurement in real-time PCR. In more recent literature, the designation C_q (quantification cycle) is used in place of C_t.

ctDNA Circulating tumor DNA. Blood-borne nucleic acid is useful as a biomarker for an increasing number of medical conditions.

Curie (Ci) A standard unit for measuring radioactivity in a sample. One curie is equivalent to 3.7×10^{10} Becquerel (i.e., disintegrations per second).

Cy3 (indocarbocyanine) A fluorescent water-soluble dye (absorption 550 nm; emission 570 nm) often used in combination with Cy5 to label nucleic acids or proteins for various applications, including microarray assays. Both Cy3 and Cy5 are more stable, brighter fluorophores than other traditional fluorescent compounds and produce less background.

Cy5 (indodicarbocyanine) A fluorescent water-soluble dye (absorption 650 nm; emission 670 nm) often used in combination with Cy3 to label nucleic acids or proteins for various applications, including microarray. Both Cy5 and Cy3 are more stable, brighter fluorophores than other traditional fluorescent compounds and produce less background.

Cycloheximide A drug that inhibits protein synthesis in eukaryotic cells. It has no effect on prokaryotic cells. Cycloheximide functions by interfering with protein elongation during translation.

Cytidine A nucleoside containing the base cytosine.

Cytogenetics That branch of genetics focusing on the structure and behavior of chromosomes, especially during mitosis and meiosis.

Cytoplasm The cellular contents found between the plasma membrane and the nuclear membrane.

Cytosine A pyrimidine that usually base-pairs with guanine.

Dark genome Genes and their products about which there is little if any understanding of their biological significance or function.

de novo Made from scratch (from metabolites) within a cell, as opposed to being partially degraded and recycled. This term often refers to certain biochemical pathways, such as de novo nitrogenous base synthesis. The term de novo has different implications in different disciplines.

Degenerate codons Two or more codons that specify the same amino acid.

Deep sequencing A next-generation sequencing strategy in which a specific genomic region is sequenced repeatedly, hundreds or thousands of times, to ensure complete and accurate coverage of the regions of interest, or the whole genome. In this context, the term "coverage" refers to the number of times that a specific nucleotide appears in the sequencing data.

Denaturation (of nucleic acids) Conversion of DNA or RNA from a double-stranded form to a single-stranded form. This can mean dissociation of a double-stranded molecule into its two constituent single strands, or the elimination of intramolecular base-pairing that is commonly associated with RNA.

DNA (deoxyribonucleic acid) A polymer of deoxyribonucleoside monophosphates, assembled by a DNA polymerase. In vivo, DNA is produced by the process known as replication. DNA can also be synthesized using a variety of in vitro enzymatic methods, such as PCR, or by a series of chemical reactions in an oligonucleotide synthesizer.

Dicer A dsRNA-specific type III endonuclease that produces short interfering RNA (siRNA) and microRNA (miRNA) cleavage products from longer templates. These shorter molecules are intimately associated with the RNA interference (RNAi) pathway in the cell.

Dideoxynucleotide A modified DNA nucleotide that exhibits a 3′−H, rather than the usual 3′−OH. Incorporation of a dideoxynucleotide prevents the further elongation of nascent strands of DNA during sequencing reactions due to the absence of the requisite 3′−OH needed for phosphodiester bond formation with what would be the next nucleotide in the chain.

Diethyl pyrocarbonate (DEPC) A chemical used to purge reagents of nuclease activity. Nuclease-free H_2O and other reagents are readily available from most vendors, thereby precluding the need to ever use DEPC in the laboratory. DEPC is carcinogenic and, if used, should be handled with extreme care.

Differential Display PCR (DDPCR) *Also* mRNA differential display. A method for identification of uniquely transcribed sequences among two or more RNA populations. DDPCR is a PCR-based method that utilizes large combinations of short primers to ensure amplification of all cDNA species within a sample. Electrophoretic comparison of the PCR products of each reaction shows products of identical molecular weight when a transcript is common to the biological samples under investigation; a band in only one lane is observed if gene expression has been induced or repressed. Detection of up- and downregulated genes is possible using this method.

Differentiation The process of biochemical and structural changes by which cells become highly specialized both in form and function.

Digital PCR (dPCR) A contemporary PCR platform which offers greatly enhanced sensitivity in comparison with real-time PCR. The digital PCR process is not dependent on an observed C_q value. Rather, the process involves partioning the sample, thereby favoring large numbers of parallel amplification reactions, some of which contain a template molecule(s) while others do not. The ratio of positive to negative amplification reactions is used to calculate, with high precision, the quantity of template material at the onset of the assay without the need for comparison to a reference standard.

Diploid Having two complete sets of chromosomes (two of each chromosome). Compare with *haploid* (one of each chromosome) and *triploid* (three of each chromosome).

Directional cloning Unidirectional insertion of a DNA molecule into a vector accomplished by placement of different sequences or restriction enzyme sites at the ends of double-stranded cDNA or genomic DNA molecules. The vector would also need to be similarly asymmetric.

DMSO (dimethyl sulfoxide) A reagent used in conjunction with glyoxal to denature RNA prior to electrophoresis. DMSO can also be used to lower the T_m (melting temperature) of DNA duplexes to support long-range PCR.

DNA polymerase An enzyme capable of synthesizing DNA from a DNA template. There are five known DNA polymerases in *E. coli* and ten different DNA polymerases (excluding terminal transferase) are known in eukaryotic cells. Each polymerase has a specific role in DNA replication or repair, and exhibits a characteristic combination of distinct activities, namely: $5' \rightarrow 3'$ polymerase, $5' \rightarrow 3'$ exonuclease, and $3' \rightarrow 5'$ exonuclease.

doi Digital Object Identifier. This is a unique combination of letters, numbers, and sometimes symbols that point to the permanent location of an article on the internet.

Dot blot A array-type, membrane-based technique for the quantification of specific RNA or DNA sequences in a sample. The sample is applied directly onto a filter by vacuum filtration through a manifold in either a dot or slot configuration (*see also* slot blot). Dot blot analyses lack the qualitative component associated electrophoresis and are considered low sensitivity assays.

Dounce homogenizer A device for mechanical tissue disruption that consists of a glass tube with an accompanying tightly fitting pestle. This is a hand-operated device that is generally less disruptive than electronic homogenizers.

Downstream Sequences in the 3′ direction (further along in the direction of expression) from some reference point. For example, the initiation codon is located *downstream* from the 5′ cap in eukaryotic mRNA.

Drop-seq A form of RNA-seq by which the individual cells are captured in separate microdroplets, barcoded for source identification, and then subjected to next-generation sequencing.

Drosha A nuclear, dsRNA-specific ribonuclease involved in miRNA biogenesis. Drosha, which forms a complex with the nuclear enzyme Pasha, is specifically involved in the maturation of pri-miRNAs into pre-miRNAs.

Duplex A double-stranded molecule or portion thereof that forms by the base-pairing of two complementary polynucleotides. This term refers to both inter- and intramolecular structures.

Eberwine linear amplification A widely used process for the linear amplification of antisense RNA (cRNA) from cDNA templates, most often for microarray-based applications. This technique minimizes the distortion of abundance relationships among transcripts in the original sample because it lacks the amplification bias inherent in PCR.

Electroendosmosis (EEO) The diffusion of water and solutes dissolved in it toward the cathode in a porous matrix such as an agarose gel. This is exacerbated during electrophoresis by the presence of contaminates, resulting in a net result is loss of resolution. For this reason, only highly purified, low EEO agarose should be used for nucleic acid electrophoresis.

Electropherogram A graphical tracing or plot of the results of a DNA sequencing reaction. Each peak of the electropherogram represents a fluorescence signal associate with a particular base in the sequencing reaction.

Electrophoresis A type of chromatography in which macromolecules such as proteins or nucleic acids are resolved through a matrix based on their size and net charge.

Electrophoretogram A photograph of a gel made after electrophoresis, which records the spatial distribution of macromolecules within the gel. An electrophoretogram is widely referred to as a "picture of the gel."

ELISA (enzyme-linked immunosorbent assay) An assay that detects an antigen–antibody complex via an enzyme reaction, thereby producing a visible result such as a color change (chromogenic detection) or measurable fluorescence.

End-point PCR A form of DNA amplification in which PCR products are analyzed, usually by electrophoresis, after the completion of a set number of cycles. Contrast with real-time PCR in which product accumulation is measured after every cycle as amplification proceeds.

Enhancer element A short regulatory DNA sequence that can increase the activity of some eukaryotic promoters, resulting in elevated levels of transcription. Enhancers are *cis*-acting elements usually exert their influence independently of orientation and proximity to target promoters.

Enrichment Any manipulation of RNA that results in an increase in the statistical representation of one or more RNA subpopulations as a percentage of all RNAs synthesized by the cell. For example, selection of poly(A)$^+$ RNA results in an enrichment of mRNA. An enrichment strategy may also involve the physical manipulation of cells before lysis to superinduce one or more species of RNA. Serum stimulation of quiescent cells, for example, results in an induction (enrichment) of proliferation-specific message as cells reenter the cell cycle.

Epidemiology The discipline that is concerned with the occurrence and distribution of a disease in a population. Epidemiological studies are performed with the intention of developing treatment plans as well as ways to avoid a disease or other health-related issue altogether.

Epigenetics The study of the regulation of gene expression due to the addition or removal of chemical groups to RNA or DNA, thereby transiently altering its properties, including the capacity to support transcription. While some epigenetic modifications are inheritable, they do not alter the base sequence of the molecule.

eRNA (efferent RNA) Any transcript that is exported from the nucleus into the cytoplasm. Contrast with transcripts that are retained in the nucleus.

eRNA (endless RNA) A type of circular RNA with an internal ribosome entry site (IRES) that supports sustained translation, thereby producing one or more desirable or necessary proteins, or molecular therapeutics.

eRNA (enhancer RNA) A transcript that arises from a gene enhancer element in eukaryotic cells by RNA polymerase II. eRNA transcripts are nonadenylated and typically less than 2000 nt. The production of eRNA tends to correlate with elevated levels of transcription of other genes in the vicinity. There is speculation that RNA polymerase II may first engage the enhancer RNA region and then travel along the DNA to a nearby gene promoter.

EST (expressed sequence tag) Partial sequences from selected cDNA clones which match at least a portion of the coding region of a gene. Public databases contain immense amounts of sequencing data useful for the identification of ESTs. Assay of ESTs is one strategy for new gene discovery as well as quantifying gene expression.

Ethidium bromide (EtBr) A planar, intercalating agent used to visualize nucleic acids, both DNA and RNA. This dye emits a bright orange fluorescence when UV irradiated, allowing gels to be photographed for future reference. The standard ethidium bromide stock solution is 10 mg/mL in water and the standard staining concentration is 0.5–1.0 μg/mL.

Exosomal RNA (exRNA) *See* Extracellular RNA

Exosome Membrane-bound structure that is originates from the cell membrane. Exosomes typically entrap one or more types of molecules for travel throughout the body, often being secreted or excreted.

Extracellular RNA (exRNA) RNA molecules that are found outside the cell in which they were transcribed. These molecules are not only found in extracellular spaces but are also secreted into and may be assayed in various body fluids, including breast milk, urine, blood, and saliva. exRNAs may have regulatory roles including intercellular communication. exRNAs also have potential as noninvasive biomarkers. Also known as exosomal RNA.

Exon A portion of a eukaryotic gene represented in the mature mRNA molecule. Exons may or may not be translated. Exons are separated from each other by intervening introns in hnRNA molecules in the nucleus. When splicing occurs, exons are joined together upon removal of the introns, which are recycled.

FASTA A bioinformatics text format for representing nucleotide or amino acid sequences using a single-letter symbol according to standard conventions.

File format The specific organization method by which an image is stored on a disk, based on image class, compression type, and halftone pattern. TIFF, BMP, JPEG, and GIF are examples of file formats.

Formaldehyde (HCHO) A commonly used denaturant of RNA and inhibitor of RNase. It should be handled with care in a chemical fume hood. Formaldehyde is a liver carcinogen and, as a known teratogen, should be avoided by expectant mothers.

Formamide ($HCONH_2$) A commonly used organic solvent/denaturant used to lower the melting temperature (T_m) of double-stranded duplexes; its inclusion in hybridization reactions maintains stringency at a temperature lower than might otherwise be required. It is also used with formaldehyde to denature RNA before electrophoresis. Formamide is a carcinogen and, as a known teratogen, should be avoided by expectant mothers.

Fosmid An f-factor cosmid. Cosmids and fosmids are similar in structure to plasmids, but have the ability to carry significantly larger inserts, often on the order of up to 45–50 kb of DNA. The upper limit for many plasmid-borne inserts is about 10 kb DNA. Because of their great size a bacterial cell is able to acquire and replicate only one fosmid, making fosmids less desirable cloning vehicles compared to plasmids. Historically, fosmids were used to support genomic DNA sequencing. The use of fosmids has been largely supplanted through the development and implementation of bacterial artificial chromosomes (BACs) and yeast artificial chromosomes (YACs) for cloning purposes.

Frame grabber Hardware unit, essentially a card inserted into one of the slots in the back of a computer that permits the use of a video camera to import an image directly into image analysis software.

Functional genomics Study of the function of genes and gene interactions by examination of the levels of RNA (and sometimes protein) in the cell and how these levels are modulated. Succinctly, it is the study of how the genome responds to change. The principal tools are genome sequencing and RNA-seq. *See also* transcriptomics.

Gamma A non-linear logarithmic contrast correction factor used to adjust the contrast in dark areas of an image.

γ Radiation The most energetic and penetrating of the three forms of natural radiation. Unlike α and β subatomic particles, γ rays consists solely of electromagnetic radiation.

gDNA *See* Genomic DNA.

Gel box The apparatus in which the electrodes are mounted and in which an electrophoretic separation is carried out. Also known as an "electrophoresis chamber."

Gene The unit of heredity. A gene is a discrete sequence of chromosomal DNA that ultimately encodes the instructions for the synthesis of a polypeptide.

Genetic engineering *Also* recombinant DNA technology, biotechnology, molecular biology. The technology by which genes, or pieces thereof, are moved about, sometimes from one organism to another, for characterization or expression. Genetic engineering has grown into an industry with far-reaching implications in virtually all areas of science and society.

Genome The entirety of chromosomal DNA found in a cell. In some applications, it may be useful to distinguish nuclear genomic DNA from the mitochondrial genome.

Genome editing Site-directed modification of DNA, most commonly via CRISPR technology, to modify a DNA sequence by adding bases, removing bases, or intentionally disabling a gene.

Genomic DNA (gDNA) Chromosomal DNA. The mitochondrial genome is designated as mtDNA, the chloroplast genome is designated ctDNA, and nuclear DNA is generically referred to as genomic DNA.

Genomics Study of the structure, organization, and functions of genes. Genomics is intimately linked to data derived from DNA sequencing.

Genotype The actual collection of genes (and alleles) present in a cell or organism.

GIF (graphics interchange format) A bit-mapped color graphics format that supports 256 colors. Digital images of company logos, line art, and related applications are frequently stored as GIF files.

GLP (good laboratory practice) A established set of principles and best-practice procedures which are used to ensure the quality and reliability of data pertaining to nonclinical investigations and to research and development.

GMP (good manufacturing practice) A established system of quality control by which products are made through the consistent application of stringent industry standards. GMP pertains mostly to pharmaceuticals, cosmetics, medical devices, and food and beverages and, in the United States, is under FDA authority.

Glycobiology Study of the structure and biological functions of sugars, polysaccharides (glycans) and carbohydrates.

Glyoxal (OHCCHO) A reagent that has been used to denature RNA prior to electrophoresis. It is used in conjunction with DMSO and is less popular than the formaldehyde denaturing system.

Gray level The brightness value assigned to a pixel in gray scale images. In an 8-bit image, this value ranges from 0 to 255, specifically from black, through shades of gray, to white.

gRNA (Guide RNA) An RNA molecule that (1) directs naturally occurring single-base substitutions in the cytoplasm or mitochondria, or (2) that homes in on a complementary target sequence in the CRISPR Cas9 editing process.

Guanine A purine that usually base-pairs with cytosine.

Guanosine A nucleoside containing the base guanine.

Haploid Having one complete set of chromosomes (one of each chromosome, as found in gametes). See also *diploid* and *triploid.*

Hapten A small molecule, not antigenic by itself, that can act as an antigen when conjugated to a larger molecule, usually a protein.

Heat Map A graphical two-dimensional data display in which data values are often represented in color. This is a common method for summarizing the up- and downregulation of a number of genes from a number of samples.

hnRNA (heterogeneous nuclear RNA) Precursor mRNA or pre-mRNA; the primary product of eukaryotic transcription by RNA polymerase II.

Histone proteins DNA binding proteins that facilitate supercoiling of chromatin into recognizable chromosomes. Histone proteins are among the most highly conserved among all eukaryotic genes and are at the core of nucleosome formation. Histone mRNAs are sometimes assayed as housekeeping genes.

Hogness box (also Goldberg-Hogness box; TATA box) Older term for the conserved TATAAA-motif located approximately 25 bp upstream (−25) from the transcriptional start site of eukaryotic genes. The Hogness box is a *cis*-acting regulatory component of many eukaryotic promoters for RNA polymerase II.

Holoenzyme A complete, biochemically active enzyme (as opposed to an incomplete enzyme complex or a cleavage product).

Hot-start PCR A chemical or procedural modification that prevents synthesis of any PCR product until after an initial period of high temperature denaturation. Hot-start PCR is frequently used to maximize the sensitivity of PCR by eliminating the synthesis of nonspecific product in the earliest amplification cycles.

Housekeeping gene A gene that is expressed, at least theoretically, at constant levels in all cells and the products of which are required to maintain normal cell function or viability. Because of their purported invariance, transcription assays often include housekeeping genes to demonstrate that an *overall* change in gene expression has *not* occurred as a result of an experimental manipulation.

HUGO (Human Genome Organization) A nonprofit international organization that coordinates scientific issues related to human genetics.

HUPO (Human Proteome Organization) An international nonprofit organization that coordinates scientific issues related to the proteomics and the mapping of proteome.

Hybridization The formation of hydrogen bonds between two nucleic acid molecules that demonstrate some degree of complementarity. The specificity of hybridization is a direct function of the stringency of the system in which the hybridization is being conducted; imperfect or mismatched hybridization is known to occur under low stringency conditions.

Hydrogen bonding The highly directional attraction of an electropositive hydrogen atom to an electronegative atom such as oxygen or nitrogen. This is the manner of interaction of complementary bases during nucleic acid hybridization. Hydrogen bonds are relatively weak and are easily broken by increasing stringency. *See also* base pairing.

Hydroxylation Chemical reaction involving the addition of one or more –OH groups to a target molecule, most often by enzymes known as hydroxylases. Protein hydroxylation is usually directed at proline residues and is one several posttranslational modifications that can influence protein folding and function.

Hyperpolymer A collection of labeled probe molecules that have hybridized to a target sequence and with each other, in an overlapping fashion, such that a tail of labeled molecules is formed and attached to the target sequence at one end. This phenomenon occurs when random breaks persist in the backbone of a nucleic acid probe molecules, often as a consequence of the labeling reaction itself. The net effect is an amplification of the signal, which translates into shorter detection time, although a slight loss in resolution is observed.

IGEPAL CA-630 A nonionic detergent commonly used to gently solubilize the plasma membrane while leaving most of the membrane bound organelles intact. Replaces Nonidet P-40 (NP-40), which is no longer sold.

Image A two-dimensional array of data elements, each of which is related to a color or an intensity. This information can be stored in any of several file formats and can be evaluated using image analysis software.

Imagej Free (public domain) densitometric software for image analysis available from the National Institutes of Health (imagej.nih.gov).

Image analysis An electronic method for the digital capture and storage of an image, accompanied by measurement of parameters such as molecular weight, mass, relative abundance, and optical density of various objects in the image (e.g., bands on a gel).

Imprintome All of the genes that organism possesses that are marked or "imprinted" in such a way that their germline origin can be determined. Methylation of DNA sequences, a common imprint mark, regulates gene expression and may selectively favor the expression of one allele over another.

Indel An insertion or deletion of a single base in one genome, relative to another. An indel is not a base substitution but rather that actual loss, or acquisition of a base. Indels become apparent during sequence alignments.

Inosine (I) An unusual purine nucleoside containing the base hypoxanthine. Inosine is sometimes used in the synthesis of degenerate or "guess-mer" oligonucleotide primers because it is capable of base-pairing with adenosine, cytidine, thymidine, and uridine nucleosides.

In silico Through the use of a computer. For example, using bioinformatics software to analyze large amounts of DNA sequence information.

In situ hybridization A method for localizing gene expression within the architecture of tissue. RNA or DNA probes are used for hybridization to mRNA present within a very thin tissue section. By maintaining natural histological geometry, changes in gene expression can often be mapped to a few cells within a tissue. In situ hybridization can also refer to identifying gene loci on an intact metaphase chromosomes, as in the FISH (fluorescence in situ hybridization).

Internal ribosome entry site (IRES) A specific RNA nucleotide sequence that serves as a ribosome recruiting site within the body of an mRNA, most often polycistronic prokaryotic mRNAs. An IRES is a ribosome recruiting alternative to the 5′ cap that is commonly observed in eukaryotic mRNAs.

Intron An intervening DNA sequence that interrupts the coding sequences (exons) of a gene. Introns are transcribed and represented in hnRNA, though they are spliced out in a systematic, high-precision manner during the RNA maturation.

Ionic detergent A compound often used to disrupt the natural three-dimensional folding of a molecule, particularly proteins, for either inactivation in an assay or as a means of more fully characterizing that molecule. Ionic detergents are also used to disrupt biological membranes to facilitate isolation of cellular components. Ionic detergents, examples of which include SDS and sarkosyl, are reliable inhibitors of RNase.

Ionizing radiation Any radiation that displaces electrons from atoms or molecules, resulting in the formation of ions. α, β, and γ radiation are all forms of ionizing radiation.

Isoprenylation Linkage of lipid groups to cysteine residues. This is one of several possible posttranslational modifications associated with protein maturation.

ISO (International Organization for Standardization) A worldwide management standard that serves to unify business and industry practices with a goal of process and quality improvement at all levels of operation.

Isotope One of two or more atoms with the same atomic number but different atomic weights, that is, different atomic forms of the same element.

JPEG (joint photographic experts group) A widely used electronic compression method used to digitize color images and photographs.

Kanamycin An aminoglycoside antibiotic used to select resistant bacterial cells in molecular cloning. Kanamycin exerts its effect by binding to the small ribosomal subunit and interfering with translation.

Klenow enzyme The large fragment of *E. coli* DNA polymerase I, generated by cleavage of the holoenzyme with the protease subtilisin or obtained by cloning. The Klenow fragment retains the $5' \rightarrow 3'$ polymerase and $3' \rightarrow 5'$ exonuclease activities but lacks the often troublesome $5' \rightarrow 3'$ exonuclease activity associated with the intact enzyme. *See also* DNA polymerase.

Lane The linear path that a molecule follows out of the well during electrophoresis when an electric field is applied.

Leader sequence The non-translated portion of mature mRNA located between the 5′ cap and the initiation codon. This region is also known as the 5′-UTR (untranslated region).

Library A collection of clones that partially or completely represent the complexity of gene sequences from a specific biological source. A library consists of cDNA or genomic DNA molecules that have been ligated to a suitable vector (e.g., plasmid or bacteriophage) and which can be propagated indefinitely. Archived DNA library members may be selected or retrieved from the library by nucleic acid hybridization or, in the case of expression vectors, by antibody recognition. Isolated clones are then definitively identified by DNA sequencing.

Ligase Any enzyme that joins together nucleic acid molecules. Ligases may be thought of as "molecular paste." T4 DNA ligase, for example, repairs the breaks produced by restriction endonucleases (molecular scissors) in dsDNA and, in so doing, is able to join together two unrelated DNA molecules. T4 RNA ligase is used to join single-stranded RNA or single-stranded DNA molecules.

Ligase Chain Reaction (LCR) A DNA amplification technique that was developed primarily to identify point mutations. The most common form of LCR involves the ligation of two primer molecules which flank a putative mutation site, resulting in a single, longer molecule which is then able to support subsequent amplification. If a point mutation has occurred, then the two shorter molecules will not be ligated because of the mismatch, thereby preventing amplification.

lincRNA Long intergenic noncoding RNA. Transcripts without open reading frames, typically longer than 200 nts, which appear to have a regulatory impact on various cellular processes, including growth and development, and disease progression.

Linear RNA Amplification *See* Eberwine linear amplification.

lncRNA Long noncoding RNA. lncRNAs are typically longer than 200 nts and are believed to impact near all cellular processes, directly or indirectly. lncRNA and lincRNA (above) gene expression is observed in a highly tissue-dependent manner.

Loading buffer A dense glycerol- or sucrose-based solution that is added to an RNA, DNA, or protein sample prior to electrophoresis in order to increase the density of the sample. This allows the investigator to load the wells of the gel after the gel has been completely submerged in running buffer. Loading buffers routinely contain a colored dye, such as bromophenol blue, the position of which can be used to track the progress of the electrophoresis. Colloquialisms for loading buffer include "sample buffer" and "blue juice."

Locked nucleic acid (LNA) A modified RNA nucleotide with an extra bridge joining the 2′ oxygen and 4′ carbon atoms. The inclusion of one or more LNAs in oligonucleotide primers or probes significantly increases the T_m, thereby reducing the incidence of nonspecific product formation while increasing the discriminatory power of the primers.

Locus (plural is loci) The precise location of a particular gene, and any possible alleles, on a chromosome.

Long-Range PCR Modified PCR amplification procedure by which very large (genomic length) PCR products are generated. Reaction modifications usually include a more alkaline reaction buffer, higher nucleotide concentrations, a pair of thermostable enzymes, primers with a higher T_m, and a modified cycling profile compared to standard reaction parameters.

Macroarray A nylon filter onto which multiple gene sequences have been immobilized. Macroarrays are able to hold only a small fraction of the number sequences that can be printed on a microarray due to the porous nature of the macroarray filter. Macroarrays are often themed (e.g., cancer, apoptosis, signal transduction, mouse, human) and lack the sensitivity associated with microarray technology.

MALDI-TOF (matrix-assisted laser desorption ionization-time of flight) A form of mass spectrometry used to sequence peptide fragments for proteomics applications.

Marker A generic term that can refer to any gene or gene product of interest in an experiment. In addition, "marker" can refer to molecular weight standards loaded onto a gel.

MDx Molecular diagnostics

Meiosis Cell division responsible for the production of haploid gametes (sex cells).

Melting temperature *See* T_m.

Mendel (Father Gregor Mendel) A Catholic priest who, in 1865, first postulated the notion of the gene. Mendel is universally recognized as the father of classical genetics, also known as Mendelian genetics in his honor.

Methylene blue Commonly used tracking (loading) dye used to monitor the progress of nucleic acid electrophoresis. Methylene blue is a well-known DNA binding dye and is also used for other applications.

mRNA (messenger RNA) The mature product of RNA polymerase II transcription. In eukaryotic cells, mRNA is derived from hnRNA. In conjunction with the protein translation apparatus, mRNA is capable of directing the synthesis of the encoded polypeptide(s).

Metabolome All of the metabolites (small, non-polymerized compounds and monomers) that are present in a biological sample and are able to participate in general metabolic reactions. The composition of the metabolome at any particular moment is a chemical fingerprint of the organism.

Meta-analysis A form of statistical analysis in which data from multiple experimental repeats or from multiple independent studies are combined to address a specific question.

Metabolomics The science of determining the variety, number, and abundance of low molecular weight compounds (metabolites) in a sample or in a cell or tissue sample. There is a growing body of evidence that the metabolome is indicative of the role of genes with as-of-yet unknown function(s). Some of the methods used in metabolomics (and metabonomics, below) include nuclear magnetic resonance (NMR), gas chromatography (GC), liquid chromatography (LC), two-dimensional polyacrylamide gel electrophoresis (2D-PAGE), and mass spectrometry.

Metabonomics The science of metabolomics in humans and other higher animals only. More specifically, metabonomics attempts to profile and compare the overall metabolic response of cells to drugs, the onset of specific diseases, or changes in the local environment by examining data from several individuals.

Metagenome The entirety of genomic DNA isolated from an environmental sample, usually representing a diverse array of organisms.

Methylation The addition of a $-CH_3$ (methyl) group to a variety of substrates.

Methylmercuric hydroxide (also methyl mercury) An extremely toxic reagent that was used in the past to denature RNA prior to electrophoresis or for other molecular biology applications. Methylmercuric hydroxide should always be avoided in favor of other less dangerous denaturation options.

Microarray A silicon, glass, or polymer-coated chip printed with tens of thousands of different DNA sequences for high-throughput global analysis of gene expression.

miR microRNA or miRNA

miRNA (microRNA) A very short (22–25 nt) abundant transcript that plays a role in the regulating gene expression by either transiently suppressing translation of an mRNA molecule or by directing its cleavage.

miRNome The entire population of miRNA transcripts being expressed at any given moment in the cell.

Mismatch A position in a double-stranded molecule where the nucleotides do not base-pair. In order for non-complementary base pairs to be tolerated, the temperature of annealing must be sufficiently below the melting temperature (T_m); at the T_m only perfectly matched duplexes are stable. The location and context of mismatching have profound ramifications with respect to primer annealing and the identity of the resulting product(s) when performing PCR.

Mitosis Cell division in somatic cells. Mitosis is responsible for producing genetically identical diploid copies ($2n \rightarrow 2n$) of a parental cell.

Molecular beacons Fluorescent hybridization probes that are used in certain real-time PCR applications.

Molecular biology The discipline dealing with the study of gene organization and gene expression at the RNA, DNA, and protein levels. Examples of molecular biology techniques include making temporary or permanent changes to the structure and function of nucleic acid molecules, isolating specific DNA or RNA sequences for further analysis, and studying the types of interactions that occur among these molecules. Synonymous with biotechnology, genetic engineering, and recombinant DNA methodology.

Monocistronic Term used to describe mRNA molecules that encode only one polypeptide. Monocistronic mRNAs are commonly associated with eukaryotic cells.

MOPS 3-[*N*-morpholino]propanesulfonic acid; a component of a buffering system used in formaldehyde-agarose gel electrophoresis of RNA. 10 × MOPS Buffer = 400 mM MOPS, pH 7.0; 100 mM sodium acetate; 10 mM Na_2–EDTA, pH 8.0.

MSDS (Material Safety Data Sheet; ***also*** Safety Data Sheet) A document that describes many of the physical and chemical properties of a substance, along with instructions for proper handling, storage, containment, and disposal. An MSDS also contains toxicity data and first aid information, in the event of exposure to or ingestion of the substance. MSDS information should be maintained on file in the laboratory, with ready access at all times.

Multiomics The study of multiple 'omes' (e.g., the transcriptome, the proteome, and the metabolome) to describe a biological phenomenon. Large-scale studies involving multiple omics approaches, also described as multimodal platforms, often involve big data computational method, and have great potential in the area of biomarker identification. Many software applications have been developed to manage a growing number of multiomic databases.

Multiplex PCR Use of multiple primer pairs for the amplification of one or more targets within the same reaction.

NASBA (nucleic acid sequence-based amplification) An isothermal process of "self-sustained sequence replication" used primarily for molecular diagnostics.

Nascent protein A protein molecule which is elongating but has not yet been completely synthesized by the process of translation.

Nascent RNA An RNA molecule which is elongating but has not yet been completely synthesized by the process of transcription.

Next Generation Sequencing (NGS) Also known as massively parallel sequencing or second-generation sequencing. NGS is a contemporary, high-throughput, non-Sanger-type DNA sequencing approach in which millions of independent DNA molecules are sequenced in parallel. The method precludes the need for cloning into a sequencing vector and produces billions of base reads in just a few hours. The methodology is also widely used for whole transcriptome sequencing

Nonidet P-40 Nonionic detergent P-40, now sold as IGEPAL CA-630. Detergent commonly used to gently solubilize the plasma membrane while leaving most of the membrane-bound organelles intact.

Normalization Any number of measurements or manipulations made to allow meaningful comparison among samples in a series. Examples of normalization in a molecular biology typically involves ensuring equal amounts of template material in an assay or standardizing data against baseline data.

Northern analysis (also Northern blotting) A technique for transferring electrophoresed RNA from an agarose gel onto a filter paper for subsequent immobilization and hybridization. The information gained from Northern analysis is used to assess the expression of specific genes qualitatively and quantitatively. The Northern analysis is characteristically lacking in sensitivity by comparison with other transcription assays.

Nuclear run-on assay A method for labeling nascent RNA molecules in intact nuclei as they are being transcribed. In some circles, this technique is also known as the nuclear runoff assay. The rate at which specific RNAs are being transcribed can then be inferred from amount of label incorporation. *Compare with* steady-state RNA.

Nuclease protection assay A method primarily used for quantifying transcript abundance. In general, hybridization between probe and target RNA takes place in solution, followed by nuclease digestion of all non-hybridized molecules or portions thereof. Nucleic acid molecules locked up in a double-stranded configuration are relatively safe, that is, protected, from nuclease degradation. The undigested RNA:RNA or RNA:DNA hybrids are then electrophoresed or precipitated for quantitation. The nuclease protection assay is more sensitive than Northern analysis.

Nucleoside A molecule consisting of a five-carbon sugar (ribose or deoxyribose; pentose) linked to a nitrogenous base (adenine, cytosine, guanine, thymine, or uracil).

Nucleotide A molecule consisting of a five-carbon sugar (ribose or deoxyribose; pentose), a nitrogenous base (adenine, cytosine, guanine, thymine, or uracil), and at least one phosphate group. Nucleotides are the building blocks used to assemble both DNA and RNA.

Ochre codon The translation termination codon UAA.

Oligonucleotide A short, artificially synthesized, single-stranded DNA molecule that can function as a nucleic acid probe or a PCR primer. *Oligonucleotide* can also refer to a short fragment of RNA.

Opal codon The translation termination codon UGA.

Ortholog A gene in one organism that has a similar function to a gene found in another organism. Identifying orthologs, known more formally as orthologous genes, implies some degree of evolutionary relatedness, suggesting that the orthologs derive from a common ancestral gene.

P-Body A stress-associated, cytoplasmic granular structure into which translationally repressed mRNA and mRNA decay-associated proteins are directed. Also known as a "processing body." The role of these structures is not entirely clear, as there is evidence that some mRNAs in P-bodies may be recycled, becoming translationally active. Suggests a role for storing mRNA as well as degrading it.

Palindromic sequence A short, double-stranded DNA sequence in which the order of bases is identical when each strand is read 5′ to 3′.

Pangenome The inventory of all genes found in all members of a biological clade; each member organism will not have all of the genes in the pangenome but rather a subset which confers the unique characteristics of that ancestral descendent.

Pasha A nuclear enzyme (known as DGCR8 in humans) which forms a complex with Drosha, another nuclear enzyme, and which is responsible for processing of pri-miRNA molecules.

PCI Standard laboratory symbol for a mixture of phenol:chloroform:isoamyl alcohol (25:24:1).

PCR (polymerase chain reaction) A primer-mediated enzymatic process for the systematic exponential amplification of minute quantities of specific genomic or cDNA sequences. The technique revolutionized molecular biology since it was first described in 1987. It has the advantage of being a very sensitive technique that can be performed quickly in the lab. PCR is commonly used to isolate and amplify one sequence from a heterogeneous pool, quantify gene expression, produce novel templates for gene expression studies, prepare DNA fingerprints, demonstrate familial relatedness, and study DNA–protein interactions to name but a few applications.

Pentose A five-carbon sugar, either ribose or deoxyribose. A pentose sugar is one of the components of a nucleotide and a nucleoside.

Pharmacogenetics That discipline concerned with the analysis of one or a few genes in response to drugs, and how biochemical differences that exist among individuals might affect the efficacy of a drug.

Pharmacogenomics That discipline concerned with the interactions of every part of the genome and, therefore, the response of a patient to drugs. The use of this term is more inclusive than the term *pharmacogenetics*, which is more concerned with the response of individual genes. The pharmacogenomics discipline focuses on cellular pathway interactions with a goal of creating a personalized drug regimen.

Pharmacokinetics The study of absorption, processing, and clearance of drugs from the body over a particular timeline.

Phosphate-buffered saline (PBS) An isotonic salt solution frequently used to remove residual growth medium from cell culture plasticware or from cell pellets following centrifugation. One liter of calcium-magnesium-free PBS is prepared by mixing 8 g NaCl, 0.2 g KCl, 1.15 g Na_2HPO_4, 0.2 g KH_2PO_4. PBS should then be autoclaved and aseptically handled to maintain tissue culture sterility.

Phosphorimaging A filmless process by which radioactive decay events, fluorescence emission, and chemiluminescence may be detected and electronically analyzed. Phosphorimagers provide a much greater linear response than X-ray film.

Phosphorylation The covalent addition of a phosphate group to a target molecule by the action of a kinase. Phosphorylation is a widely used mechanism for controlling the function of enzymes and proteins, and it is a major posttranslational modification both in eukaryotic and prokaryotic cells.

Photodocumentation A method for preserving the image of a stained gel immediately after electrophoresis. Virtually all gel images are now captured and stored electronically though older media, including thermal paper and X-ray film, persist. *See* image analysis.

Photon A particle that has no mass but is able to transfer electromagnetic energy.

Pixel Picture element, that is, a single point in a graphic image. Pixels are the smallest units of image digitization.

Plasmid A covalently closed double-stranded DNA molecule capable of autonomous replication. Prokaryotic and eukaryotic plasmids are variegated and are in widespread use. Plasmids can accept DNA inserts from any other biological source, though the insert size is usually limited to about 10 kb. Plasmids generally contain a variety of selectable markers and ancillary sequences for characterization, and sometimes expression, of the insert DNA.

Poly(A) tail A tract of adenosine residues enzymatically added to the 3′ terminus of mRNA by nuclear poly(A) polymerases. The addition of the poly(A) tail involves cleavage at the 3′ terminus of the primary transcript followed by polyadenylation. The length of the poly(A) tail correlates with the stability of that particular RNA in the cytoplasm, and the poly(A) tail shortens as an mRNA ages.

Polyadenylation signal A highly conserved six-base motif (AAUAAA) instrumental in the efficiency of polyadenylation. The splicing events that precede the actual polymerization of adenosine residues to the 3′ end of eukaryotic mRNA molecules are controlled, in part, by the polyadenylation signal. AAUAAA is functionally linked to downstream GU-rich and U-rich regions on the native transcript.

Polycistronic Term used to describe mRNA molecules that contain coding regions representing more than one polypeptide. Commonly observed in prokaryotes.

Polymorphism A genetic trait that, within a species, is subject to variation. Examples of human polymorphisms include eye color, blood type, cutaneous fingerprints, and DNA fingerprints.

Polysome An mRNA molecule that is being translated simultaneously by several ribosomes.

Polytron An electronic tissue disruption instrument. Polytron homogenization is much harsher than using a Dounce homogenizer and well-suited for breaking up both soft and fibrous tissue.

Posttranscriptional regulation Any event occurring after transcription termination and that influences one of the subsequent steps involved in the ultimate expression of that gene. Posttranscriptional regulatory events include, but are not limited to, intron removal, polyadenylation, nucleocytoplasmic transport, stability in the cytoplasm, and engagement by the protein translation apparatus.

Posttranslational regulation Any event occurring after synthesis of the primary peptide that influences the subsequent maturation of a protein. Posttranslational regulation usually refers to the events that modify a peptide, such as glycosylation, methylation, and hydroxylation, leading to full functionality.

pre-miRNA The immediate precursor of miRNA. pre-miRNAs arise from the processing of pri-miRNA transcripts.

Precursor RNA (also hnRNA or pre-mRNA) An unspliced RNA molecule; the primary product of transcription produced by eukaryotic RNA polymerase II.

Pribnow box (also Pribnow-Schaller box) The consensus sequence TATAAT that is part of the transcriptional promoter for prokaryotic genes. It is centered about 10 base pairs upstream (−10) from the gene transcription start site. The Pribnow box itself is especially important in the binding of RNA polymerase.

pri-miRNA Primary RNA. This is the immediate product of transcription of a miRNA gene and the immediate precursor pre-miRNA.

Primer A short nucleic acid molecule that, upon base pairing with a complementary sequence, provides a free 3′–OH which supports PCR or any of a variety of other primer extension-dependent reactions.

Probe Labeled DNA or RNA molecules used to hybridize to complementary sequences in a on a filter paper (e.g., Northern or Southern analysis), or in solution (e.g., real-time PCR or nuclease protection assay). Probe specificity can be controlled by raising or lowering stringency. Antibodies can also function as probes, as in Western analysis or when screening an expression vector-based library.

Processivity The ability of an enzyme such as DNA polymerase to remain bound to its substrate (template) and efficiently produce a long product (complementary strand) before disengaging. For example, *Taq* DNA polymerase is said to be a highly processive enzyme because it travels great distances along a template before dropping off, though this particular thermostable enzyme is also well-known for its high error rate.

Promoter A DNA sequence associated with a particular locus to which RNA polymerase and other initiation-associated proteins bind to initiate transcription. Promoters typically consist of several regulatory elements involved in initiation, regulation, and efficiency of transcription.

Proteome The full complement of proteins produced by a cell at a particular time. Proteome maps are generated and assessed by 2D electrophoresis and other techniques designed to identify, quantify, and characterize the products of translation.

Proteomics The study of the protein composition of a cell under certain biochemical conditions and how the assortment of proteins changes in response to natural or experimental stimuli. Proteomics constitutes a systematic means for correlating protein function, abundance, and variety genome function in normal cells as well as the disease state.

Purine The chemical parent compound of the bases adenine and guanine.

Pyrimidine The chemical parent compound of cytosine, guanine, and uracil.

Quinone Oxidation product of phenol. Quinones compromise the quality of RNA preparations by crosslinking nucleic acid molecules and breaking phosphodiester bonds. Avoid quinone-induced damage by always preparing and working with fresh phenol. Quinones characteristically impart a pink tint to phenol solutions, indicating that it is time to dispose of that particular aliquot.

q.s. Abbreviation for the Latin words *quantum satis*, meaning "sufficient quantity." The inclusion of q.s. in a protocol or formulation means that the investigator will add enough water to achieve the specified final volume in a reaction tube.

RACE (rapid amplification of cDNA ends) A PCR method for cloning the 5′ end (5′ RACE) or 3′ end (3′ RACE) of transcripts. RACE can be performed using a number of formats, the most popular of which is RNA-ligase mediated RACE (RLM-RACE) for mapping transcription start sites.

Radiochemical A chemical containing one or more radioactive atoms.

Radiolysis The physical breakage of DNA or RNA probe that occurs when these probe molecules are radiolabeled to extremely high specific activity. Hence, the radioactive decay of the labeled nucleotides is responsible for damaging the probe itself.

Radionuclide An unstable isotope of an element that decays spontaneously and, in so doing, emits radiation.

Real-time PCR A high-precision method for measuring PCR product accumulation as it is produced in each cycle, rather than measuring final product mass when all cycles have been completed (i.e., an end-point measurement). Real-time PCR is widely regarded as the gold-standard PCR assay.

Recombinant DNA technology *see* Molecular Biology.

Reference genome A database that represents the genomic makeup of at least one, but often several, individual(s) of a particular species. The data is collected from multiple sequencing experiments and then assembled into the correct linear order based on alignment of overlapping nucleotide patterns.

Renaturation (also reannealing) The reassociation of complementary strands of DNA or RNA following high stringency denaturation.

Restriction endonuclease A DNA modification enzyme that cuts (restricts) double-stranded DNA at predictable locations. Thought of as the "molecular scissors," restriction endonucleases most often cut substrate DNA at palindromic 4-, 6-, or 8-base sequences.

Retrovirus A virus with an RNA genome (usually single-stranded) that, upon infection of a suitable host cell, propagates via conversion of its genetic material into double-stranded DNA using its own reverse transcriptase, which is carried within the virion.

Reverse Northern Blot A method in which DNA probes immobilized on a filter, often in a dot blot configuration, are hybridized to labeled mRNA or cDNA. "Reverse Northerns" are commonly used to confirm differential gene expression data produced by other assays, for example RT-PCR.

Reverse transcriptase (RT) Also known as RNA-dependent DNA polymerase. A retroviral enzyme that supports the synthesis of a complementary DNA (cDNA) molecule from an RNA template. Many types of natural and genetically modified reverse transcriptase are available, including the traditional AMV (avian myeloblastosis virus) and MMLV (Moloney-murine leukemia virus) varieties, while other forms of reverse transcriptase lack a natural background RNase H activity, thereby making them desirable for cDNA synthesis reactions. Some of the thermostable polymerases that support PCR, such as that from the thermophilic eubacterium *Thermus thermophilus*, have been found to possess a Mn^{2+}-dependent reverse transcriptase activity at elevated temperature; because this type of enzyme exhibits reverse transcriptase and standard DNA polymerase activity, it is well suited to the one-tube RT-PCR format.

Ribonuclease (RNase) A family of resilient enzymes that rapidly degrade RNA molecules. Control of RNase activity is a key consideration in all manipulations directly or indirectly involving RNA.

Ribonuclease A (RNase A) A single-strand-specific endoribonuclease with activity directed toward the 3′ side of pyrimidine nucleotides (Py/pN), rendering a 3′ pyrimidine phosphate.

Ribonuclease H (RNase H) An enzyme with activity directed against the RNA component of an RNA:DNA hybrid.

Ribonuclease T1 (RNase T1) A single-strand-specific endoribonuclease that cleaves the phosphodiester bond at the 3′ end of guanine nucleotides (GpN) and, like RNase A, produces oligonucleotides with terminal 3′ phosphates.

Ribonucleic acid (RNA) A polymer of ribonucleoside monophosphates, synthesized by an RNA polymerase. RNA is the product of transcription.

Ribosomal RNA (rRNA) The predominant class of RNA in the cell. The highly abundant nature of rRNA makes it a useful indicator of the integrity, quality, and probable utility of a sample in downstream applications. The low complexity of this RNA species also makes it useful as a molecular weight marker for RNA electrophoresis.

Riboswitch An mRNA genetic control element that experiences a conformational change upon metabolite (ligand) binding. In so doing, a riboswitch is able to regulate the synthesis of the encoded protein.

Ribotyping An electrophoresis and/or DNA sequencing technique used to identify cells or microorganisms based on organism-specific variations in the base sequence of the ribosomal RNA genes.

Ribozyme An RNA molecule with the capacity to act as an enzyme; ribozymes are catalytic RNAs. Ribozymes may exhibit self-cleavage or, by folding into a characteristic three-dimensional shape, may induce the cleavage of a complementary mRNA molecule to which the ribozyme base-pairs.

RISC (RNA-induced silencing complex) A multiprotein endoribonuclease complex that mediates RNA interference by the cleavage of specific mRNA molecules. The RISC also carries an siRNA or miRNA which guides it to the complementary target mRNA.

RLM-RACE (RNA ligase-mediated RACE) A form of RACE in which an RNA oligonucleotide is ligated to full-length mRNA molecules and then reverse transcribed into cDNA. This approach has the advantage of producing only clones that represent the precise 5′ end of transcripts without the loss of any base sequence information.

RNA condensate See "Condensate."

RNAe (RNA enhancement) A method of posttranscriptional gene expression enhancement that results in an increase in protein synthesis. The method involves modifying naturally occurring long noncoding RNAs such that they overlap with the 5′ terminus of the cognate mRNA; the method enhances recruitment of ribosomes to the mRNA. Contrast with RNAi, below. Yao et al. (2015). *Nucleic Acids Res.* doi 10.1093/nar/gkv125.

RNA G-quadruplex A four-stranded structure that may be observed in transcripts exhibiting guanine-rich regions. RNA G-quadruplexes are inhibitory toward translation; these tenacious structures are unfolded by helicases and then maintained in the unfolded state by RNA binding proteins. Guo and Bartell (2016). *Science* 353, 6306. doi: 10.1126/science.aaf5371.

RNAi (RNA interference) Posttranscriptional gene silencing; occurs by coordinated destruction of specific mRNA molecules, thereby inhibiting translation of the encoded protein. As the gene itself is not targeted or modified, the RNAi effect is reversible, thereby facilitating functional studies of one gene at a time, especially related to developmental and disease pathways.

RNA polymerase Any of several enzymes capable of the synthesis of RNA polynucleotides by the process of transcription using DNA as a template.

RNA-seq (RNA sequencing) A transcriptomics tool based on next generation sequencing methods. RNA-seq is used to produce a global profile of transcriptional activity (qualitative and quantitative) in a biological source by sequencing all of the RNA molecules present under a particular set of conditions. RNA-seq is an often-sought-after cost-effective alternative to microarrays for assaying gene expression.

RNA World Hypothesis The belief that RNA as a molecule with replicative and catalytic capabilities preceded the arrival of proteins and DNA in the primeval world.

RT-PCR (reverse transcription PCR) A method for the analysis of gene expression by producing cDNA form an mRNA with reverse transcriptase and then amplifying it to a detectable level by PCR.

Running buffer The electrolyte solution used to conduct electric current during electrophoresis. RNA and DNA gels are submerged in running buffer to also dissipate heat. A running buffer is useful only to the extent that it resists pH changes and ion depletion during electrophoresis.

S1 nuclease An enzyme that degrades single-stranded nucleic acid molecules or any portion of a molecule that remains in single-stranded form post inter- or intramolecular base-pairing. The enzyme is zinc-dependent, shows optimal activity at an acidic pH, and has no known sequence affinities.

SAGE (serial analysis of gene expression) A gene expression profiling technique and bioinformatics tool for assessing the transcriptional activity of many genes under a defined set of experimental conditions. SAGE involves sequencing short cDNA molecules (tags) that have been randomly ligated together.

Saline sodium citrate (SSC) A salt solution frequently used for blotting of nucleic acids. It is also an essential component of various hybridization buffers and posthybridization washes. $20 \times$ SSC = 3 M NaCl; 0.3 M Na_3-citrate; adjust pH to 7.0.

Saline sodium phosphate-EDTA (SSPE) A salt solution frequently used for blotting of nucleic acids. It is also an essential component of various hybridization buffers and posthybridization filter washes. The phosphate in this buffer mimics the phosphodiester backbone of nucleic acids, thereby providing enhanced blocking, lower background, and higher signal-to-noise ratio on membranes during Northern and Southern analysis. 20 × SSPE = 3 M NaCl, 0.2 M $NaH_2PO_4 \cdot H_2O$, 0.02 M Na_2-EDTA; adjust pH to 7.4; autoclave.

scaRNA (Small Cajal body RNA) A nucleolar RNA associated with Cajal bodies found in the nucleus. These transcripts are part of the snoRNA family and are associated with maturation of snRNA.

Scorpions Fluorescent probes used in certain real-time PCR applications. Scorpions consist of an oligonucleotide primer linked directly to a fluorescent probe.

Senolytic A compound that promotes or directly causes the destruction of senescent cells. Examples include the drug Dasatinib and the plant flavonoid Quercetin. Senolytics have potential for treating aging-associated conditions by selectively clearing cells exhibiting the senescent phenotype. *See* Kirkland and Tchkonia (2020). doi: 10.1111/joim.13141.

Sense RNA (also + RNA) Any RNA molecule that has the same nucleotide sequence as naturally occurring mRNA. Although *antisense* RNA is capable of hybridization with mRNA on a filter or in situ, sense RNA is only useful as a probe in Southern analysis (in which both DNA strands are present); otherwise, sense RNA is generally used as a negative control to assess the magnitude of background (nonspecific) hybridization.

Sequencing by Synthesis (SBS) A contemporary approach to DNA sequencing that involves measuring a unique fluorescence signal or a change in ionic concentration that accompanies the insertion of each nucleotide during synthesis of DNA molecule that is complementary to a user-provided template.

Sevag Historical term for a mixture of chloroform:isoamyl alcohol (24:1). *See* Sevag, Lackman, and Smolens (1938). *J. Biol. Chem.* 124(2), 425–436.

shRNA (short hairpin RNA) A short RNA molecule with a characteristic hairpin shape that has a role in regulating gene expression via the RNA interference pathway.

siRNA (short interfering RNA) A short double-stranded RNA molecule with characteristic dinucleotide overhang at the 3′ end of each strand. These molecules have a role in regulating gene expression via the RNA interference pathway.

Single nucleotide polymorphism (SNP) A single base change that is mostly commonly associated with a variant phenotype and, in some cases, with the onset or progression of a disease.

Slot blot An array-type, membrane-based technique for the quantitation of specific RNA or DNA sequences in a sample. The sample is usually applied directly onto a filter by vacuum filtration through a manifold in either a slot or dot configuration (*see also* dot blot). Slot blots lack the qualitative component associated with electrophoresis-based methods and are generally considered low sensitivity assays.

Sodium dodecyl sulfate (SDS) An ionic detergent commonly used to disrupt biological membranes and to inhibit RNase.

Southern analysis (*also* Southern blotting) A technique for transferring electrophoretically separated DNA from an agarose gel matrix onto a filter paper for subsequent immobilization and hybridization. The information gained from Southern analysis is used to assess gene structure, organization, or copy number.

Spatial transcriptomics An umbrella term that refers to any of number of methods used to determine the histological distribution of specific cell types based on patterns of gene expression. Variants of some techniques allow subcellular localization of specific RNA species as well.

Specific activity The amount of radioactivity per unit mass of a radioactive material. It is most commonly expressed in curies per millimole of material (Ci/mmol).

SSC *See* saline sodium citrate.

SSH (suppression subtractive hybridization) A method for identification of differentially expressed genes. Briefly, the mechanics of SSH result in a hybridization-based physical subtraction of sequences shared by two or more cDNA populations. The sequences which remain after the subtraction are PCR-amplified and represent those genes that were up- or downregulated under a particular set of conditions. SSH is a powerful method for global analysis of gene expression, particularly when microarrays or RNA-seq are not available. Since no prior cloning is necessary, SSH is a true method for new gene discovery.

SSPE *See* saline sodium phosphate-EDTA.

Strand displacement amplification An isothermal alternative to PCR that is used for the exponential amplification of DNA.

Steady-state RNA The final accumulation of RNA in the cell or cytoplasm under a certain set of circumstances. For example, measurement of the steady-state abundance of a particular species of mRNA does not necessarily relate to the *rate* of transcription, nor about the age or turnover rate of specific RNAs. *Compare with* nuclear run-on assay.

Stringency A measure of the likelihood that double-stranded nucleic acid molecules will dissociate into their constituent single strands. In practical terms, stringency refers to the ability of single-stranded nucleic acid molecules to discriminate between molecules which they have a high degree of complementarity versus those that have a lower degree of complementarity. Stable hybridization is favored under high-stringency conditions only between nucleic acid molecules with a high degree of complementarity. As stringency is lowered, a proportional increase in nonspecific hybridization is favored.

Structural gene A gene that encodes for an mRNA or noncoding RNA, as opposed to regulatory elements such as operators or other promoter-type elements (and their respective products) that control the expression of a structural gene.

Sumoylation Addition of a SUMO (small ubiquitin-like modifier) protein to another protein. Sumoylation is a reversible posttranslational modification. Unlike ubiquitination, which marks proteins for destruction, sumoylation regulates diverse cellular functions, including apoptosis, cell cycle control, stress-related cellular responses, intracellular trafficking, and transcription regulation of gene expression.

SYBR Green, SYBR Gold Dyes for staining nucleic acids. Commonly provided as a 10,000 × stock solution in DMSO, SYBR dyes are diluted to a working concentration of 1 × in Tris buffer, such as 1 × Tris-acetate-EDTA (TAE) buffer or 1 × Tris-borate-EDTA (TBE) buffer. SYBR dyes have greatly reduced background fluorescence, higher sensitivity, and reduced mutagenicity when compared with ethidium bromide. SYBR Green I and SYBR Gold are used to stain DNA while SYBR Green II is optimized to stain RNA.

SYBR Green assay A platform for real-time PCR in which the dye SYBR Green I binds to double-stranded PCR products. The ensuing increase in fluorescence is in direct proportion to product accumulation in the reaction tube.

Systems biology The systematic study, integration, and mathematical modeling of the relationships between individual members of various categories of biomolecules (e.g., RNA, DNA, protein) and networks (e.g., signal transduction), and how their dynamic modulation influences both the cell and the organism.

T_a Annealing temperature. T_a, a hybridization temperature slightly lower than the T_m, promotes complete base pairing between complementary nucleic acid molecules. This temperature reduction strategy is used because when $T_a = T_m$, only 50% of all perfectly matched double-stranded molecules will form.

TAE buffer *See* Tris-acetate-EDTA buffer.

TAP (tobacco acid pyrophosphatase) An enzyme that removes the 5′ cap structure from eukaryotic mRNA. This enzyme is widely used in RLM-RACE.

TATA box (also Hogness box; Goldberg-Hogness box) An AT-rich regulatory element that is commonly associated with the promoter for eukaryotic RNA polymerase II. The TATA(A/T) box is usually found within 25 nucleotides upstream (−25 nt) from the actual transcription start site.

***Taq* DNA polymerase** Thermostable, highly processive DNA polymerase from the organism *Thermus aquaticus. Taq* polymerase is one of several enzymes used to support PCR.

TaqMan assay A real-time PCR platform in which a fluorescent probe is included in a reaction along with the primers. The accumulation of double-stranded PCR product leads to proportionate increase in fluorescence (due to probe degradation) which is measured at the end of each amplification cycle by a real-time thermal cycler.

Target Single-stranded DNA or RNA sequences that are complementary to a nucleic acid probe or primers. Target sequences may be immobilized on a solid support, may be available for hybridization in solution, or may function as template molecules for amplification by PCR.

TBE Buffer *See* Tris-Borate-EDTA buffer.

Technical replicates Two or more experimental samples, representing the same biological source, which are assayed at exactly the same, under the same set of conditions, as a means of assessing inter-sample variability and data interpretation. For example, RT-PCR reactions should be run in duplicate or in triplicate.

TEMED (*N,N,N′,N′*-Tetramethylethylenediamine) A quaternary amine used as a catalyst in the polymerization of acrylamide in the preparation of polyacrylamide gels.

Template A macromolecular informational blueprint for the synthesis of another macromolecule. All polymerization reactions, including replication, transcription, and PCR require templates, which dictate the precise order of nucleotides in the nascent strand. Primer extension-type reactions cannot proceed in the absence of template material.

Tetracycline An antibiotic that is used to select resistant bacterial cells in molecular cloning. Tetracycline interferes with translation by binding to the 30S ribosome subunit, thereby blocking access to the aminoacyl site.

Thermal cycler A temperature-programmable instrument in which PCR is performed.

Thymidine A nucleoside containing the base thymine.

Thymine A pyrimidine that usually base-pairs to adenine.

TIFF Tagged image file format; a type of software file used to store images.

T_m Melting temperature. That temperature at which 50% of all possible double-stranded molecules are dissociated into their constituent single strands. To maximize duplex (double-stranded molecule) formation, hybridization is performed slightly below the T_m of the duplex (*see* T_a). The lower the temperature, however, the greater the likelihood that imperfect duplexes (those that contain mismatches) will form.

TOPOcloning An innovative method of ligating DNA using the enzyme topoisomerase. The method is much faster than traditional ligations involving T4 DNA ligase, and is compatible with blunt end, sticky end, and TA cloning.

Touchdown PCR An excellent method for preventing nonspecific product formation. In touchdown PCR the initial annealing temperature (T_a) is *higher* than the T_m and is subsequently lowered, usually 0.5 degrees, in each passing cycle. Eventually the annealing temperature will drop below the T_m, thereby allowing amplification of only those template molecules which are exactly complementary (perfectly matched) to the primers.

Trailer sequence (also 3′ UTR) The untranslated sequence located just 3′ to the coding region of mRNA. In poly(A)$^+$ mRNA, the trailer sequence is located between the coding region and the beginning of the poly(A) tail.

Transcription The process by which RNA molecules are synthesized from a DNA template by RNA polymerase.

Transcription elongation The second of three stages of RNA synthesis, characterized by the covalent addition of nucleotides to the 3′ end of the nascent transcript.

Transcription factor Any of a number of molecules that binds to a DNA promoter region, recruits, or interacts with RNA polymerase, and has a regulatory role in the initiation of transcription.

Transcription initiation The first of three stages of RNA synthesis. Initiation begins with the binding of RNA polymerase and associated transcription factors to double-stranded DNA, followed by the local unwinding of the gene template.

Transcription start site (TSS) The location of the first transcribed nucleotide near the 5′ end of a gene.

Transcription termination The third and last stage of RNA synthesis. Transcription termination involves recognition of the point beyond which no additional nucleotides are to be added to the nascent transcript and ensuing release of the newly synthesized RNA and transcription apparatus from the DNA (gene) template.

Transcriptional noise Gene expression heterogeneity under conditions where uniform levels of expression are otherwise expected. Depending on the genetic locus, irregular gene expression (e.g., bursts of transcriptional activity) may also be reflected in non-uniform protein levels. Intercellular variability of this nature may be responsible for the emergence of subpopulations of cells. This phenomenon is known to occur in eukaryotic and prokaryotic cells.

Transcriptome The complete set of RNA molecules, coding and noncoding, produced by a cell under a particular set of conditions. This includes rRNA, tRNA, mRNA, miRNA, ncRNA, lncRNA, and so forth.

Transcriptomics The all-inclusive analysis of an organism's transcript inventory and how the population of transcribed RNA molecules changes as conditions change inside the cell.

Transfection The introduction of nucleic acid molecules into a eukaryotic cell, most often by chemical means or viral delivery, causing phenotype and genotype conversion of the recipient cell. Transfection is classified as either transient, meaning that there is eventually reversion of the cell to the pre-transfection phenotype, or stable transfection, in which the transfected DNA integrates into the chromosome of the host. Stable transfection of eukaryotic cells is technically challenging and difficult to achieve.

Transfer RNA (tRNA) A moderately abundant class of RNA molecules that shuttle amino acids to the aminoacyl site of the ribosome during protein synthesis. The total mass of tRNA in the cell is occasionally assayed to demonstrate that a particular experimental manipulation has not resulted in a change in overall transcription in the cell. While tRNA concentration may be suitable as a gel loading control, it is not well-suited as a transcription control because tRNAs are produced in the cell by RNA polymerase III while mRNAs are produced by RNA polymerase II.

Translation The process by which peptides are synthesized from amino acid monomers using the instructions encoded in an mRNA template. Translation occurs as mRNA is deciphered by the ribosomes.

Translation elongation The second of three stages of protein synthesis. Elongation involves the formation of all of the peptide bonds required to assemble the primary structure of a protein.

Translation initiation The first of three stages of protein synthesis. Initiation involves the association between mRNA and the ribosome, all of the biochemistry preceding this event, and the formation of the first peptide bond between the amino acids that will be located at the amino terminus of the nascent polypeptide.

Translation termination The third and last stage of protein synthesis, involving the release of both the completed polypeptide and the mRNA that directed its synthesis.

Translational Research A process or series of processes that are geared toward the enhancement of health outcomes. It involves moving basic, often multidisciplinary research and development outcomes to clinical trials that often address a pressing health issue. Translational research is sometimes referred to as a "bench to bedside" approach.

Triploid Having three complete sets of chromosomes (three of each chromosome). See also *diploid* and *haploid*.

Tris-acetate-EDTA buffer (TAE buffer) Common electrolyte reagent for the electrophoresis of DNA. 50 × TAE (per liter) = 242 g Tris base; 100 mL 0.5 M Na_2-EDTA, pH 8.0; 57.1 mL glacial acetic acid; autoclave or filter-sterilize. Working concentration is 1 × TAE.

Tris-borate-EDTA buffer (TBE buffer) Common electrolyte reagent for the electrophoresis of DNA, especially low molecular weight molecules. 20 × TBE (per liter) = 121 g Tris base; 61.7 g sodium borate; 7.44 g Na_2-EDTA; filter sterilize. Working concentration is 1 × TBE.

Ubiquitylation (*also* ubiquitination) The process of tagging a protein for destruction by the addition of ubiquitin, a highly conserved 76-amino acid protein.

UniProt An online repository of raw and protein sequence data located at www.uniprot.org.

Upstream Sequences in the 5′ direction (away from the direction of expression) from some reference point. For example, the 5′ cap in eukaryotic mRNA is located *upstream* from the initiation codon.

Uracil A pyrimidine base that usually base-pairs to adenine.

Uracil-N-glycosylase (UNG) An enzyme used to destroy dUTP-labeled PCR products from previous amplifications in order to eliminate its carryover contamination in future reactions.

Urea (NH_2CONH_2) Denaturant used as an alternative to formamide for disrupting RNA secondary- and higher-order folding, usually in conjunction with electrophoresis.

Uridine A nucleoside containing the base uracil.

Vanadyl ribonucleoside complexes (VDR; *also*, VRC) An inhibitor of RNase that may be added to gentle RNA lysis buffers to control RNase activity. VDR functions as an RNA analog. The use of VDR has fallen sharply out of favor with most RNA aficionados because even trace amounts of carryover VDR have been shown to be highly inhibitory toward reverse transcriptase, thereby making it unsuitable for RT-PCR applications. It is still used in sucrose-gradient centrifugation, other cell fractionation techniques, and some RNA in situ hybridization applications.

Vector A nucleic acid molecule such as a plasmid, bacteriophage, or phagemid into which another nucleic acid molecule (the so-called insert or foreign DNA) has been ligated. Vectors contain sequences that, in a suitable host, permit propagation of the vector and the DNA that it carries.

WebLogo A free web-based tool for creating a graphical display of nucleic acid or amino acid sequence alignments. Access at weblogo.berkeley.edu.

Well A depression in a gel into which the sample is pipetted just prior to the onset of electrophoresis. Wells are formed by the insertion of a comb into the gel matrix before it has polymerized (polyacrylamide) or solidified (agarose).

Western analysis (also Western blotting) A technique for transferring electrophoretically separated proteins from a polyacrylamide gel onto a filter paper for subsequent characterization by antibody detection. The information gained from Western analysis is used to assess qualitatively and quantitatively the prevalence of specific proteins.

Whole genome amplification (WGA) Amplification of the entirety of a genome when very limited amounts of starting material are available. Methods include fairly low-temperature annealing of 15-mer random primers or semi-degenerate oligonucleotides to ensure complete genome coverage. There are many forms of this methodology which render a highly uniform, high fidelity whole genome amplification, and some of which eliminate the sequence bias associated with PCR-based methods. These methods are popular when working with archived, formalin-fixed tissue samples, single cells, and many other applications.

Xenobiotic A chemical that is foreign to a cell or organism.

Xylene cyanol Commonly used tracking (loading) dye used to monitor the progress of nucleic acid electrophoresis.

Zoonotic virus A virus that is able to jump between species, e.g., between animals and humans.

Zwitterion An ion that carries both a positive and negative charge simultaneously. Zwitterions have a net neutral charge, though this is pH dependent.

Index

Note: Page numbers followed by "*f*" and "*t*" refer to figures and tables, respectively.

B

C

G

I

J

K

L

M

N

O

Q

R

S

T

U

V

9780323902212